River Networks as Ecological Corridors

River networks are critically important ecosystems. This interdisciplinary book provides an integrated ecohydrological framework blending laboratory, field, and theoretical evidence that changes our understanding of river networks as ecological corridors. It describes how the physical structure of the river environment impacts biodiversity, species invasions, population dynamics, and the spread of waterborne disease. State-of-the-art research on the ecological roles of the structure of river networks is summarized, including important studies on the spread and control of waterborne diseases, biodiversity loss due to water resource management, and invasions by non-native species. Practical implications of this research are illustrated with numerous examples throughout. This is an invaluable go-to reference for graduate students and researchers interested in river ecology and hydrology, and the links between the two. Describing new related research on spatially-explicit modeling of the spread of waterborne disease, this book will also be of great interest to epidemiologists and public health managers.

Andrea Rinaldo is Professor and Director of the Laboratory of Ecohydrology at the École Polytechnique Fédérale de Lausanne (EPFL) and Professor in the Civil, Environmental and Architectural Engineering department at the University of Padua. His research focuses on water controls on biota, for which he has received international recognition including membership of the US National Academies of Sciences and Engineering, the American Academy of Arts and Sciences, the Royal Swedish Academy of Sciences, and Accademia Nazionale dei Lincei in Italy. He is the co-author of *Fractal River Basins: Chance and Self-Organization* (with Ignacio Rodríguez-Iturbe, Cambridge University Press, 1997).

Marino Gatto is Professor of Ecology at Politecnico di Milano. His research focuses on ecological modelling, fish population dynamics and management, and parasite ecology. He was President of the Italian Society of Ecology from 2003 to 2006, and is a member of Istituto Lombardo Accademia di Scienze e Lettere and Istituto Veneto di Scienze, Lettere ed Arti.

Ignacio Rodriguez-Iturbe is Distinguished University Professor at Texas A&M University and J. S. McDonnell Distinguished University Professor (Emeritus) at Princeton University. He is a member of the US National Academies of Sciences and Engineering, the American Academy of Arts and Sciences, and the Vatican Academy of Sciences. He has been awarded the Stockholm Water Prize and the Bowie, Horton and Macelwane Medals from the American Geophysical Union. He is the co-author of *Fractal River Basins: Chance and Self-Organization* (with Andrea Rinaldo, Cambridge University Press, 1997) and *Ecohydrology of Water-Controlled Ecosystems: Soil Moisture and Plant Dynamics* (with Amilcare Porporato, Cambridge University Press, 2005).

River Networks as Ecological Corridors

Species, Populations, Pathogens

By

Andrea Rinaldo
École Polytechnique Fédérale de Lausanne

Marino Gatto
Politecnico di Milano

Ignacio Rodriguez-Iturbe
Texas A & M University

CAMBRIDGE
UNIVERSITY PRESS

CAMBRIDGE
UNIVERSITY PRESS

University Printing House, Cambridge CB2 8BS, United Kingdom

One Liberty Plaza, 20th Floor, New York, NY 10006, USA

477 Williamstown Road, Port Melbourne, VIC 3207, Australia

314–321, 3rd Floor, Plot 3, Splendor Forum, Jasola District Centre, New Delhi – 110025, India

79 Anson Road, #06–04/06, Singapore 079906

Cambridge University Press is part of the University of Cambridge.

It furthers the University's mission by disseminating knowledge in the pursuit of education, learning, and research at the highest international levels of excellence.

www.cambridge.org
Information on this title: www.cambridge.org/9781108477826
DOI: 10.1017/9781108775014

First published 2020

Printed in the United Kingdom by TJ Books Limited, Padstow Cornwall

A catalogue record for this publication is available from the British Library

Library of Congress Cataloguing in Publication data
Names: Rinaldo, Andrea, 1954– author. | Gatto, Marino, author. |
    Rodríguez-Iturbe, Ignacio, author.
Title: River networks as ecological corridors : species, populations,
    pathogens / Andrea Rinaldo, Marino Gatto, Ignacio Rodriguez-Iturbe.
Description: 1. | New York, NY : Cambridge University Press, 2020. |
    Includes bibliographical references and index.
Identifiers: LCCN 2019056059 (print) | LCCN 2019056060 (ebook) |
    ISBN 9781108477826 (hardback) | ISBN 9781108775014 (epub)
Subjects: LCSH: Ecohydrology. | Biodiversity. | Corridors (Ecology) |
Watersheds. | Water–Microbiology.
Classification: LCC QH541.15.E19 R56 2020 (print) | LCC QH541.15.E19
    (ebook) | DDC 577.6–dc23
LC record available at https://lccn.loc.gov/2019056059
LC ebook record available at https://lccn.loc.gov/2019056060

ISBN 978-1-108-47782-6 Hardback

Non quia difficilia sunt non audemus,
sed quia non audemus difficilia sunt

Not because things are difficult that we do not dare,
but because we do not dare they are difficult

Seneca, Letter to Lucilius (194,26)

Ἄριστον μὲν ὕδωρ
Greatest indeed is water

Pindar, Olymp. 1,1

It's all in the water
Mary Knapp Parlange

Contents

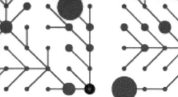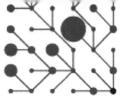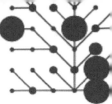

List of Boxes

Preface

This book draws together several lines of argument to suggest that an integrated ecohydrological framework that blends laboratory, field, and theoretical evidence focused on hydrologic controls on biota has contributed substantially to our understanding of the function of river networks as ecological corridors. This function is relevant to a number of key ecological processes. Jointly with other determinants that this book will discuss, these processes control the spatial ecology of species and biodiversity in the river basin, the population dynamics and biological invasions along waterways, and the spread of waterborne disease. As revealing examples, here we describe metapopulation persistence in fluvial ecosystems, metacommunity predictions of fish diversity patterns in large river basins, geomorphic controls imposed by the fluvial landscape on elevational gradients of species' richness, zebra mussel invasions of an iconic river network, and the spread of proliferative kidney disease in salmonid fish. Our main tenet is that ecological processes in the fluvial landscape are constrained by hydrology and by the matrix for ecological interactions (notably, the directional dispersal embedded in fluvial and host/pathogen mobility networks). Accounting for these drivers requires spatial descriptions that have produced a remarkably broad range of results that are worth being recapped in a book illustrating the coherent framework that produced them. In brief, this book investigates how the physical structure of the environment affects biodiversity, species invasions, survival and extinction, and waterborne disease spread, from the (somewhat narrow, yet, in our view, significant) perspective of ecosystems produced by fluvial processes and forms whose origins and features have long been a focus of study.

As we began exploring somewhat systematically the possible ramifications of these interests, either theoretical or grounded in relevant field or laboratory evidence, we realized that drawing together much scattered work would require a synoptic account of what ties the various topics together. This proved to be a trying task, however, especially given the pace with which novel and exciting contributions were (and are) emerging in supposedly disparate fields. It was thus necessary to conclude the writing of this book even in the face of continual changes in the field. However, this book is timely. We believe that it meets its initial aim, described in a nutshell in the introductory paragraph. More importantly, perhaps, our survey of the current cultural landscape of the discipline highlights key research questions that remain to be addressed, of which there are many. Although the material presented here is prone to aging quickly, owing to the broad interest that the topics are currently attracting, we think that showing the basic unity of the topics, at least from a conceptual viewpoint, is worth the effort, as this unity is largely overlooked in the literature to date. From the synoptic view given here, our line of thought should appear in its full coherence, if not in its perceived generality – or so, of course, we hope.

Over the course of several years, we have contributed to defining the field of ecohydrology which describes hydrologic controls on biological systems. Early examples were soil moisture dynamics controlling (when limiting, of course) plant growth in a semiarid landscape and flow distributions acting as the master variable of stream ecology. We sought to investigate how the very texture and fabric of river networks define the corridors for species biodiversity, population migration, and the spread of disease. We hoped to define general rules, because we have learned much in recent years about recurrent hydrologic properties that hold regardless of climate, vegetation, or exposed lithology – the universality of the geometry of nature in shaping the scale-free structure of the river basin. Thus we began constructing models of ecosystems in which these universal characters would feature explicitly. In early models, which we describe in the beginning of this book, each node in the model represented a metacommunity or a small ecosystem. Dissecting its behavior served us well, no less than simulating the ever-changing fish populations in the Mississippi–Missouri river network, finding that variable river runoff influences fish dispersal, habitat size, and metacommunity dynamics, ultimately affecting biodiversity. Regardless of whether the model was realistic, the patterns derived from a minimalist model – fed by a proper geometrical substrate for interactions – intrigued us. So, together with several collaborators over the years, such as groups developed at EPFL, Padova, Princeton, and Milano, at times collectively, we turned our attention to population migrations, using as a pivotal example the

American westward migration of the nineteenth century. The finding that a diffusion model of migration, in which migrants diffuse isotropically through space, could not adequately explain the rates of westward expansion gave us pause. In times in which freedom of migration was at stake, our interest grew by the day. A simple assumption that colonizers moved strictly along river networks showed a key improvement in modeling the speed of the migration wave. Remarkably, laboratory experimentation supported the theoretical and empirical result: directional dispersal has a life of its own and in turn defines key biodiversity properties.

Attention to spatial models of a different type of population migration (a waterborne disease epidemic) was the next logical step. Together with our collaborators, we published in 2008 the first spatially explicit model of cholera epidemics. In a January 2011 paper, our team used the model to predict the spread of cholera in Haiti following the devastating 2010 earthquake. The model proved prescient. From there, we focused progressively on better and more robust tools to assimilate data in real time to predict how the intertwined processes of pathogen transport via waterways and via human mobility affect the epidemiological connection of nodes – the human settlements where disease can develop and diffuse – and thus drive the dynamics of an infection.

This book was born out of a review paper that we wrote for the fiftieth anniversary issue of *Advances in Water Resources*. The material has been substantially expanded, however, especially from a methodological viewpoint. Methodological detours have largely been inserted into the main text as boxes, which, we hope, will help to maintain the flow of the presentation.

The motivations for writing this book extend further. The boy from Bangladesh shown in the cover photograph drank the water of the mighty Meghna River to convince one of us (AR) that no harm would result. This happened a few hundred meters away from the largest hospital in the world devoted to the treatment of diarrheal diseases, in a region with one of the highest endemicities of cholera (and where the pathogen originally evolved to irradiate globally). As it turns out, predicting such a simple outcome (will the boy get cholera?) is plagued by so much uncertainty – given the unknown concentration of the possible leakages or spillovers from the hospital's sewers, the rates of decay of bacteria in open waters, the hydrodynamic dispersion that commands the point concentration of pathogens in time (relevant because the threshold for infection is dose dependent), to name a few factors – that the resulting weakness of our predictions

is a permanent liability affecting our capability to put a price tag on ecosystem services (in this case, safe drinking water). This is a key point of our reflection.

In fact, an emerging literature, largely referred to as environmental economics, integrates development and environmental thinking by focusing on economic evaluation and on the development of the proper notion of sustainable development. An idea recently put forth, by Partha Dasgupta in particular, is that it is now clear that by "economic growth," we should mean growth in wealth – the social worth of an economy's entire set of capital assets, including its natural capital and hence the ecosystem services embedded in it – not only growth in gross domestic product (GDP) or the many ad hoc indicators of human development that have been proposed in recent years. Dasgupta writes that the very concept of wealth invites us to extend the notion of capital assets and the idea of investment well beyond conventional usage. As such, by "sustainable development," we should mean development in which wealth (per capita) adjusted for distribution does not decline in time. This has radical implications for the way in which national accounts are prepared and interpreted. ("These are still early days in the measurement of the wealth of nations, but both theory and the few empirical studies we now have at our disposal should substantially alter the way we interpret the progress and regress of nations" – excerpted from Dasgupta's "*The Nature of Economic Development and the Economic Development of Nature*" [1].)

As a matter of fact, we all seem still to be – so as to say – under the spell of Kuznets's paradigm, that is, the one suggesting that the growth in standard economic indicators would mirror a systematic reduction of social inequalities. More precisely, he said that inequality rises first, to be followed later with more capital accumulation by inequality reductions. To be fair, Kuznets was very cautious in defining the range of validity of his theory to hold only under special circumstances, and yet the main point that came across is that a stronger economy would lead to a better society – or so, of course, we thought, in the spirit of the liberalism inspired by the towering figure of the distinguished expatriate economist. Enter Thomas Piketty and his analysis of *Capital in the XXI Century*, following economists like Sen, Barbier, Fisher, and Turner, and the fake Kuznetsian world collapses, as it is clear (empirically and factually) that growth in wealth indicators does not correlate with reduction in inequalities; on the contrary, inequalities tend to be exacerbated. Wealth inevitably follows a power law distribution in a steady state of the dynamics of any economy implying that often (the percentage would depend on the scaling exponent, in fact) 80 percent of the wealth is in the hands of the richest 20 percent of the population. Piketty convincingly argues that the data upon which Kuznets's tenet was established are an artifact of the conditions faced by rapidly recovering societies post–World War II, and a reanalysis based on current data – including the original data – shows strikingly different results.

Why does all this matter for our book? Economic indicators that omit the depletion and degradation of natural resources and ecosystems are at best misleading – we must account for the depreciation of natural capital in appraising the wealth and poverty of nations. Natural capital must include the value of net losses to natural resources inflicted by the societies shaped by development thinking, that is, stacked against nature *sensu* Dasgupta, as opposed to those shaped by environmental thinking. Natural capital assets include minerals, fossil fuels, forests and similar sources of material and energy inputs, and – critically – those of all ecosystem services often impaired by purely economic development. If one views ecosystem services as incomes, that is, flows as opposed to reservoirs, one may see whole ecosystems as a capital asset. In any case, if we use up more natural capital to produce economic output today, then we have less for production tomorrow. At the same time, we are also squandering valuable natural capital – ecosystems provide important goods and services to the economy, such as recreation, flood protection, nutrient uptake, erosion control, water purification, carbon sequestration, and clear water to protect against waterborne disease. By converting and degrading ecosystems, we are depreciating important ecological capital endowments.

Economic indicators change dramatically when the depletion and degradation of natural resources and ecosystem services are accounted for. Unfortunately, natural capital tends to be undervalued because it is unique, hard to put a price tag on, poorly understood, and difficult to measure. Consider one example touched upon in Chapter 4 of this book: the case of schistosomiasis spread and water resources development. Schistosomiasis is a chronic parasitic waterborne disease that affects an estimated 779 million people at risk of contracting the disease, of whom 106 million (or 13.6 percent) live either in contexts where irrigation schemes have been constructed or in close proximity to large or small dam reservoirs. Compelling evidence has been gathered by studies that examined the relation between water resources development projects and the spread of schistosomiasis, primarily in sub-Saharan African

settings, where a stunning 90 percent of the world's burden is concentrated. This is clearly related to the expansion of the range of the obligatory intermediate host for the parasite (some species of freshwater snails) and by the reduced mean distance of human settlements from the nearest water body – infection occurs through direct contact with water through skin penetration of the parasite. As an illustration, one of our students doing fieldwork on the disease in Burkina Faso became infected because once, and only once, he had removed his protective gloves to pick up a pair of scissors that had fallen into water at the test site. It is beyond doubt that the development and management of water resources are important risk factors for schistosomiasis, and hence strategies to mitigate negative effects should become integral to planning, implementing, and operating future water projects.

But how? While the impact of improved agriculture is perceived immediately and directly by, say, the GDP of the region implementing the water resources use scheme, the social and economic costs of the increased burden of the disease are hard to quantify, let alone predict. A debilitating disease, schistosomiasis is a poverty-reinforcing neglected disease owing to its low mortality and impact on the poorest alone. While contemporary economics is considering the proper manner in which to account for the social and economic costs of future loss of the workforce, immaterial factors weigh in, and an assessment seems unlikely anytime soon. Pricing is key in some circles. A price tag means discounting the environment, and pricing is necessary inasmuch as declaring that some good cannot be priced is tantamount to assuming that its worth is next to nothing. Others contend that price is not value, and accepting that it is impossible to price all ecosystem services may be way toward giving them nonmonetary value. Material and immaterial values should then be considered in their own units (say, the number of species rescued from extinction or the number of cases avoided in epidemics). Not surprisingly, therefore, pricing biodiversity and ecosystem services has been termed the "never-ending story."

Key to all the above is our capability to assess and reliably predict the spread of the disease under different scenarios – economic and water development; human mobility and awareness of the mechanisms of infection (hinging on proper educational systems); improved or worsening water, sanitation, and hygiene conditions. In this sense, this book, focused as it is on the prediction power of spatially explicit epidemiology and centered on the robustness of the structure of waterways acting as a substrate for ecological interactions, provides, we hope, significant account of the progress recently made. Such progress, it is the authors' belief, is ready to be used for the betterment of society at large and to help in assessing the wealth or poverty of nations.

As an economy's GDP could be made to grow, and its related societal indicators made to improve for a time, by mining natural capital (say, by decimating forests, damaging soil, destroying key ecosystem services by depleting renewable resources or reducing biodiversity), there is no excuse for not using what we have learned to assess the true costs and benefits of development thinking and to rethink distributive justice, where a large share of the basis for environmental thinking could be made quantitative. This is the case, in our view, for river networks as ecological corridors. Will future large-scale water resources management plans include a reduction in the loss of biodiversity across scales in river basins? Could the structure of river networks be a template for the large-scale spread of waterborne disease? Are we capable of making a compelling economic argument for preventing water development schemes in light of the social and economic costs of the predicted increased burdens of disease they would bring? Do biological invasions, including historic population migrations that shaped human community compositions as we see them now, depend on the topology of river networks as the substrate for their dispersal? The answers to these questions are largely positive in our view, as the reader will gather from her reading, and at the heart of the future development of ecohydrology, a recent field of growing importance. Social discounting applied to public policies concerning the preservation of natural capital needs quantitative assessment and thus an "engineering" capable of producing reliable scenarios. Evaluations of the effects of learning-impairing disabilities brought by neglected waterborne diseases are not ethical primitives, nor do they have observable market value like the return rates on an investment. They need reliable projections, evaluations of management alternatives, and proper cost–benefit analyses. This is only possible if we are capable of evaluating material/immaterial and present/future commodities. Our ignorance of the true economic value of natural capital is often an unsurmountable barrier to proper policy analysis. Contributing to lifting this veil of ignorance seems like a worthwhile endeavor because no economy can survive without natural capital. In the case of the ecosystem services provided by the river basin, we show that the time is ripe for retooling our decision-making basis.

Large-scale water management plans currently do not consider protection of biodiversity. Why? Because bioversity protection has been perceived as mission impossible. Instead, we show that explicitly accounting for the constraining role of river networks acting as the substrate for the relevant processes and interactions makes a substantial step toward much improved predictability of quantitative evaluations. This is central to contemporary thought centered on how a society develops a collective consciousness of the meaning of the fair distribution of water, and thus of material and immaterial resources and ecosystem services, across different individuals and events of various natures across Time (capitalized, as Proust would have it).

A word of caution is in order at this point. We ask the reader to be forgiving, for, in this book, we have often resorted to using boxes, at times quite long but clearly highlighted. This stems from the fact that technical detours can distract from the main flow of the scientific discourse, which we believe shows a notable coherence. Some of these boxes are indeed highly technical and could be skipped, unless the reader is interested in replicating results or desires an in-depth understanding of a topic. We wondered whether it would be worthwhile to write much of the material at a somewhat lower technical level, as not many upper undergraduate courses exist on these topics. We have decided against doing so, but we also agreed to add road maps, especially in the openings of boxes aimed at more technical material, warning the reader about the difficulty of their contents. A few boxes are technical but kept at a fairly elementary level and, we hope, are self-contained: we placed them on purpose to encourage less specialized audiences to access most of the material contained in this book. Typically, they explain some of the basic background methodology not just for advanced undergraduates but also ideally for graduate students and researchers who may need a brush-up or are entering the field from different disciplines. They would be most welcome, in fact: we love contamination of different genres, as will become apparent to the reader.

In terms of audience, we believe that our target is mainly graduate students and researchers in earth and environmental sciences, in particular, in ecology and hydrology (and their meeting point, ecohydrology, i.e., water controls on living communities). All in all, we expect this book to appear on recommended reading lists for advanced undergraduate courses at a few universities that might have professors interested in these topics and in the peculiar synthesis of disciplines hitherto unconnected provided by the areas where this book has developed. We also wish to warn the reader that the diversity of phenomena described in this book, many of them pertaining to different disciplines with their own typical symbolic notations, made it impossible even to attempt a normalization of notation throughout the book. Thus we apologize that in different contexts, we allow ourselves to use the same symbol for different meanings.

As an example, we beg the reader to tolerate that the symbol ρ is used in Chapter 3 for denoting the density of an invading species, whereas it denotes the rate of loss of acquired immunity in Chapter 4. These are standard notations in the as yet unconnected fields of biological invasion in rivers and epidemiology of water-related disease.

Acknowledgments

We owe recognition to several institutions for their support and for the opportunities provided over many years. AR is grateful to the academic environment of EPFL for making possible the setup of a large research group in the last twelve years – a dream come true – and for providing a most exciting academic place where one feels the worth of seeking academic quality. Equally, the intellectual environment of the University of Padua, AR's alma mater and still his part-time employer, has been key to breeding the kind of intellectual curiosity that is the hallmark of this book. MG is deeply indebted to his mother institution, Politecnico di Milano, for providing ever-continuing support for his activity as an ecologist at one of the best technical universities in the world. Without the open-mindedness that characterizes the academic environment of Politecnico, pursuing this kind of interdisciplinary research would have been impossible. IRI is grateful to the superb academic environments of Princeton University and Texas A&M University, which made possible his pursuit of the topics described in this book. AR and IRI are grateful to the Hagler Institute of Advanced Studies at Texas A&M University for its support of AR's Faculty Fellowship during which the last part of this book was completed. We are also pleased to acknowledge the fundamental role in the conception of this book played by the funds granted to AR by the European Research Council (ERC) through an Advanced Grant (named, evocatively, like this book, hence the acronym RINEC). RINEC has been the incubator of our joint research on the subjects of this book, in particular because it fostered the exchange and free circulation of graduate students and postdocs among our labs. The establishment of a wet lab at EPFL, where many of the experiments mentioned here were carried out, was also possible because of the continuous support of the Swiss National Fund, which we gratefully acknowledge.

We feel we have to thank all collaborators and groups around the world who made this journey so enjoyable, counting indeed very few exceptions throughout the years. Without their contribution, none of this would have been possible. We insist, as we had done before, that the best reward in science is in its making, a global enterprise that brings people together and whose best part is in the mentoring and the willingness to share, such as we have experienced. Special gratitude goes to our former or current collaborators (most have been students and postdocs at one or more of our institutions), many of whom are now colleagues and who provided material to be assembled into this book: Paolo Benettin, Enrico Bertuzzo, Francesco Carrara, Luca Carraro, Renato Casagrandi, Serena Ceola, Jean-Marc Fröhlich, Jonathan Giezendanner, Andrea Giometto, Joseph Chadi Benoit Lemaitre, Theophile Mande, Lorenzo Mari, Rachata Muneepeerakul, Damiano Pasetto, Javier Perez-Saez, Anna Rothenbühler, and Samir Suweis. All photographs appearing in this book were taken by AR, except for the beautiful picture at the head of the references section of the woman crossing on foot the spillway of a small dam in Burkina Faso – epitomizing the difficult assessment of a concept of sustainable development that includes all ecosystem services in the balance. The photograph was taken by Jean-Marc Frölich of the ECHO Lab at EPFL, whom we gratefully thank.

This book is an outgrowth of several collaborations with colleagues with whom it has been a pleasure and a privilege to work: Florian Altermatt, Andrew Azman, Tom Battin, Melanie Blokesch, Giulio De Leo, Simon Levin, Amos Maritan, Marc Parlange, Suresh C. Rao, Riccardo Rigon, and Sanna Sokolow, to name only the most immediate ones. Many others, too many to mention here, featured in a disparate number of collaborations, and we thank them collectively, with gratitude. We also owe a great deal to a few colleagues who read drafts of the various chapters and commented on the structure, organization, and clarity of the material presented in this book. Their help has been fundamental, and thus our unmitigated gratitude goes to Tom Battin, Enrico Bertuzzo, Andrea Giometto, James W. Jawitz, and Lorenzo Mari.

Finally, our deepest gratitude goes to Elena, Maria Caterina, and Mercedes for the continuous support throughout the years that made it all possible.

Obviously, we leave it to the reader to judge the strength and coherence of our arguments, which seem quite compelling to us. Regardless of the outcome, putting this book together has been a very rewarding journey, both personally and intellectually.

1 | Introduction

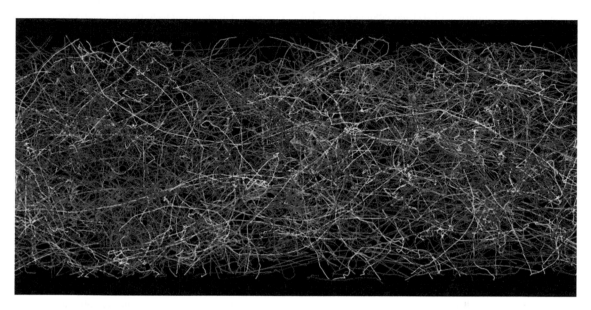

This introductory chapter outlines the *leit-motiv* of the book – dendritic substrates for ecological interactions, chief and foremost river networks in our case, bear important consequences for a number of processes, from patterns of biodiversity to controls on the spread of waterborne disease. In this chapter we discuss important methodological aspects of spatially explicit ecology that we use throughout this book. The image that we have chosen for the heading of this introductory chapter refers to accurate measurements of the behavior of the alga *Euglena gracilis* when exposed to controlled light fields. The superposed trajectories of individuals' movements, tracked in the Laboratory of Ecohydrology at EPFL, recall a Jackson Pollock painting. Laboratory studies of mesoscopic-scale movement and reproduction support theoretical work on directional dispersal in networked environments and give important bearings for the tenet of this book, as described in this chapter. Image courtesy of Andrea Giometto

1.1 The Context

Although natural ecosystems are characterized by striking diversity in form and function, they often exhibit deep structural similarities, at times emerging across scales of space, time, and organizational complexity [2]. One angle through which such features could be considered is via the necessary linkages among macroecological "laws" [3, 4], often expressing the scale invariance of ecological patterns of abundance or trait diversity subsumed by algebraic relations (popularly dubbed power laws), intended both as functional relationships among ecologically relevant quantities and probability distributions that characterize their occurrence [5]. Clearly, not

all ecological patterns exhibit scale-invariant properties; many well-defined characteristic scales exist in a broad spectrum of ecosystem dynamics. Yet many do, and there scaling theory offers a powerful tool to make way for coherent, unified descriptions capturing the essence of a process. In this chapter we introduce our main theme (highlighting the role of river networks viewed as ecological corridors that shape species and population distributions) in the context of spatially explicit ecological modeling.

One example, discussed in this chapter, concerns species' numbers and their abundance and size emerging in relation to broad ecosystem features like the topology of the substrate for ecological interactions [6–10].

We shall observe how emerging features, such as the distribution of species' persistence times at observation sites [9], are controlled more by the nature of the landscape where interactions occur than by many detailed features specific to the underlying ecosystem. A large body of empirical [9, 11] and laboratory [12–16] evidence is quoted to support such a view. We argue that dispersal constrained by specific habitat structures is a major determinant of the observed diversity patterns at both species and genetic levels [6, 8, 14, 17–20]. This result, a rather far-reaching one, is well captured by spatially explicit ecological approaches that we introduce in this chapter (Sections 1.2 and 1.3).

We contend that, within an ecohydrological framework, river network structures and their embedded hydrologic dynamics play an important role [21]. First, they provide supporting landscapes for ecological processes, many of which are essential to human life and societies. Historically, human settlements followed the river networks for the necessary water resources [22]; river networks are home to (and provide hierarchical habitat features for) freshwater fish [6–8] and stream ecology in general [23, 24] as well as pathways to life-threatening waterborne human diseases and zoonoses, that is, for human and animal hosts alike [25]. River networks may be also seen as meta-ecosystems that affect the metabolism of terrestrial organic carbon in freshwater ecosystems, an important part of the global carbon cycle [23, 26], and the amount of nutrients removed from streams and reservoirs affected by network structure and stream ecology therein [24, 27].

A broad research field exists where signatures of the hydrologic, ecologic, and geomorphologic dynamics of river basins coexist. This field has proved its importance by furthering our understanding of spatially explicit epidemiology and ecology (Chapters 2 and 3). Our ultimate goal is a comprehensive theory of how dendritic structures, their associated features, and interactions with external forcings (chiefly, hydrological stochasticity) shape emergent properties of various ecosystems. Such theory would help us address a wide variety of important questions: from conservation plans for freshwater ecosystems to optimal control for containing waterborne disease epidemics to proper inclusion of riparian systems into large-scale resource management [21, 28]. Understanding and control of biological invasions is also part of this scheme. While providing what we believe is a useful review, the novelty of this book lies in envisioning a research area where hydrology, ecology, and geomorphology intersect. We feel that important

advances will be made in this area in the near future. This book is by no means intended to provide closure on the role of river networks as ecological corridors; rather, it is a blueprint for future developments. Throughout its material, in fact, we suggest specific areas or open problems that appear to us to be particularly promising.

Incorporating ecological dynamics into riverine systems is not an easy task, given the variety of the taxa involved, their trophic positions, and the interactions between the different organisms ranging from competition to predation to parasitism. Very frequently, if the aim is to investigate population dynamics, the analysis is restricted to one or a few species or functional groups. This is what has been done, for example, when exploring zebra mussel invasions [29] or cholera dynamics [30]. If instead the aim is to investigate general patterns of biodiversity, one considers specific taxa or groups usually sharing the same trophic level, for example, fish or phytoplankton or riparian vegetation. In such a case, the main operating ecological interaction is interspecific competition, either indirect (e.g., exploitation of common food resources or nutrients) or direct (e.g., via interference). Available data usually comprise lists of presence/absence of species, possibly complemented by their relative abundances, the latter being averaged over time or simply measured in a given year. If the identity of the particular species is neglected, it is possible to derive species-abundance distributions, namely, the number of species that have a certain abundance or a certain abundance rank. Static models of species-abundance relations have long been proposed to achieve that goal (see, e.g., [31] for an excellent review). Dynamic models in which the observed relation is obtained as the long-term equilibrium of a model containing the basic time-dependent processes that shape community biodiversity are more recent. The processes shaping the maintenance of biodiversity are four fold [32]: selection, namely, the differences in the species fitnesses and therefore in their competitive ability, which operates in both ecological and evolutionary time; drift, namely, the inherent stochasticity that brings species to extinction and operates on an ecological timescale only when the size of the community is rather small; speciation, which counters drift and selection over evolutionary timescales; and dispersal, which counters local species extinction via the movement of organisms across space and acts on ecological timescales. Caswell's seminal paper on the related dynamic models [33] borrowed concepts of neutral molecular evolution and applied them to the ecological context. The organic development of a neutral theory of biodiversity was presented in a unified way only later [34]. The main tenet therein assumes that selection (i.e., differences in competitive ability, stated otherwise) is not operating, while drift is countered by speciation or dispersal. Concerning this last point, it is important to remark that almost all neutral theories are spatially implicit in that they consider either an isolated community whose survival is thus guaranteed by speciation, or a local nonisolated community whose survival is guaranteed by immigration from a "background" meta-community. A coherent theory that considers all four processes in a space-explicit framework distinguishing between ecological and evolutionary timescales is still lacking (but see [35] for a notable attempt). This book aims at partially filling this gap by presenting a series of models that are always space explicit and suited to specifically describing the peculiar structure (and thus connectivity) of river basins. We proceed step by step, first including the dendritic substrate of river basins into the neutral paradigm of biodiversity, then breaking perfect neutrality by adding either space-dependent carrying capacities of local communities or elevational niche apportionment. Species invasion and disease spread are subsequently investigated by paying greater attention to realistic details, though with a species-specific focus and within fluvial ecological substrates.

To set the context, we start with an example of the simplest dynamic model of biodiversity, the neutral one [34], which assumes that all species are competitively equivalent at a per capita level. It should be noted that some unrealistic assumptions of the neutral theory have attracted much criticism [36–38], for example, in terms of timescales, testability, and robustness; also, the neutral theory overlooks much species-specific ecological information, which is required when studying the dynamics of the system or of a set of particular species and the interactions among them [39] (Box 1.1). However, the neutral model has the advantage of letting us introduce the biodiversity-shaping processes one by one; in fact, the neutral theory switches off all the differences between species and all the interactions with the exception of strong competition for space (both intra- and interspecies), as we shall recall below. Being focused on competition for space, it is thus particularly suited to testing the fundamental differences between the spatial structure of river basins and 2D isotropic landscapes. Our first approach thus focuses on the quantitative assessment of the role of directionality and network structure on ecological organization, in particular on patterns of diversity distribution. We show, in particular, how the implemen-

tation of the neutral theory behaves in 2D lattices or 2D space-filling trees imposing directional dispersal [21, 40]. The investigation of the differences between the two substrates (the common name for the ecosystem landscape where interactions occur) proved important to later developments, chiefly laboratory ones (Section 1.2).

1.1.1 Neutral Theory of Biodiversity in a Nutshell

The neutral theory of biodiversity (NTB) was originally proposed [33] in complete analogy with the neutral theory of molecular evolution [41, 42], which assumes that gene mutations are selectively neutral, namely, that new genes are demographically equivalent to old genes, as they do not confer any advantage in terms of decreased mortality and/or increased fertility. The main advocate of NTB was Hubbell [34, 43], who greatly developed these ideas starting from his work on tropical forests, which typically display very high biodiversity. In NTB, genes are replaced by species, which all have the same demographic fitness. Mutations are replaced by the occurrence of new species. It is worth noting that in genetics, neutrality is rooted in specific biochemical mechanisms, for example, that different sequences of three nucleotides (codons) may code for the same amino acid. In ecology, instead, we do know that all species are different and have differential ecological functions and abilities. Moreover, in genetics, neutral theory is not advocated as the theory that can explain the whole of genetic diversity but as a theory that can explain the evolution of specific genes. Even the neutralists do not deny the importance of Darwinian selection in the origin of adaptations, although they think that most of the molecular diversity can be explained by random genetic drift, that is, the neutral model. In any case, neutral models in ecology may be seen as a limit approximation. The theory might hold when dealing with biodiversity within communities characterized by species with similar traits, for instance, those belonging to the same functional group or, more generally, the same guild. In these cases, in fact, we may conceive that the differences in demographic rates are not very large.

Neutral models and the pertaining theory have been fully developed by statisticians and population geneticists [41, 44–46] and blended into a coherent theory of biodiversity by Hubbell [34]. Thus, all the basic results of NTB can be found in the population genetics literature: just replace genotypes with species and mutants with new species. Neutral models are not space explicit and are traditionally phrased according to two possible paradigms: (1) biodiversity is studied at the regional scale, and the arrival of new species is due to immigration from outside the region; (2) biodiversity is analyzed at the continental level, and a new species can only arise via speciation. Clearly the two paradigms imply not only a different spatial scale but also a different timescale, because speciation is much rarer than the arrival of a new species from the surrounding regions, especially if the region is not very large. The extension of NTB to a space-explicit paradigm in riverine networks is actually one of the goals of the present book. In this section, we summarize the results of the space-implicit approach.

In NTB, abundances of all species fluctuate at random according to a birth–death stochastic process. The process may obey different rules (see below), but the most popular is the one in which the total number of individuals (no matter of what species) is constant across generations and equal to N. We might think of this situation as the one arising when considering territorial organisms, each occupying a fixed portion of the landscape, which is made up of N territories. At each time step a randomly chosen individual dies and is replaced by another individual, which may be the progeny of individuals belonging to a species already present in the community or an individual of a new species. It is often stated that NTB implies no interaction between species. This is not completely true: competition (of the so-called contest type) is actually quite strong, because each individual excludes any other individual from its own territory. However, neutrality is due to the assumption that intraspecific competition and interspecific competition have the same strength: no species has any advantage over another species in the process of replacing a dead individual.

The assumption of a constant number of individuals allows the derivation of the dynamics of a simple biodiversity index in the following manner (reported in [36]). Let f_t be the probability at generation t that two individuals of the community belong to the same species. We assume that the N individuals, before dying, produce progeny. The frequency of the progeny of each parent is $1/N$ because we assume neutrality, namely, that the fertility of each individual is the same independently of the species. Then each territory will be occupied at generation $t + 1$ by one individual chosen at random among all the progenies. However, with probability γ, this individual may be replaced by an individual of a new species (this might occur because of either mutation or immigration). Now, pick two individuals at random at generation $t + 1$. If they are the progeny of the same parent in the previous generation (which occurs with

probability $1/N$), then the probability that they belong to the same species is 1. If they are the progeny of different parents (which occurs with probability $1 - 1/N$), then the probability that they are of the same species is f_t. Also, with probability $(1 - \gamma)^2$, neither of the two individuals picked at random belongs to a new species. Therefore, one finally obtains

$$f_{t+1} = (1 - \gamma)^2 \left(\frac{1}{N} + \left(1 - \frac{1}{N} \right) f_t \right). \tag{1.1}$$

In the long run, f_t will approach the following equilibrium:

$$\bar{f} = \frac{(1 - \gamma)^2}{N - (N - 1)(1 - \gamma)^2},$$

which, by assuming that γ is very small, N is very large, and $N\gamma$ is finite, is well approximated by

$$\bar{f} \approx \frac{1}{1 + 2N\gamma}. \tag{1.2}$$

The quantity $2N\gamma$ is termed the fundamental biodiversity number θ. In fact, it is related to one of the most used biodiversity indices, Simpson's diversity index H. This is defined as the probability that two individuals of the same community belong to different species [47]. Thus, at equilibrium, Simpson's index of a neutral community is given by

$$H = 1 - \bar{f} = \frac{2N\gamma}{1 + 2N\gamma} = \frac{\theta}{1 + \theta}.$$

Most of the discussion around NTB is, however, focused not on a single biodiversity index but rather on the whole distribution of abundances defined as the distribution of the number of species $n_j(t)$ in the ecological community containing exactly j individuals at time t. As the abundances vary according to a birth–death stochastic process, $n_j(t)$ varies stochastically, of course. Under appropriate conditions, the values of $n_j(t)$ converge (in distribution) to a steady state for $t \to \infty$. This is termed the expected distribution of species abundance, say, ϕ_j, often used to fit empirical data. One can then compute the average of n_j, $E[n_j]$. Of course, data come from a sample of the community that does not presumably include all the N individuals of the community and all the S species. The assumption is made that the community has reached equilibrium, and the sample is so large as to justify the fact that the sampled n_j is close to the theoretical expected value of n_j, namely, ϕ_j (but see [46] for how to deal with small samples).

It is often stated that NTB implies that ϕ_j is the logarithmic series proposed by [48], so that the logarithmic series would be a sort of fingerprint of neutrality. However, this claim is not true in general. Three remarks are worthwhile before briefly going into the details: (1) depending on the rules of the stochastic process that governs the NTB model, different functions for ϕ_j can be obtained, one of which is the logseries; (2) numerous different mechanisms, other than NTB, can lead to the logseries distribution, and therefore, as clearly stated by [31], "an empirical species-abundance distribution cannot by itself give evidence on how to choose among them;" (3) many empirical species-abundance distributions have been examined since the 1940s, and, depending on the dataset, sometimes the best fit was the lognormal distribution, sometimes the geometric series, sometimes the logseries, and so on [47]. Therefore, researchers should refrain from using the logseries as a yardstick for neutral theory.

Let us now discuss how ϕ_j can be obtained. A key paper [45] set the problem within a wide context. The paper considers a model with Malthusian demography and a model in which the total size of the community is constant and equal to N. Here, we illustrate and discuss both.

Box 1.1 Deriving Species Abundance Distributions from Malthusian Models

First, consider the case of a Malthusian demography with constant birthrate β, constant death rate μ, and constant rate of demographic increase $r = \beta - \mu$. Neutrality implies that birth and death rates are equal across species. Each new species arises according to a Poisson process, with a constant arrival rate ν. The times between arrivals are independent, exponentially distributed with mean $1/\nu$. The only possible transitions of a certain species abundance j ($j = 1, 2, ..$) in an element of time dt are from j to $j-1$, j or $j+1$, and the transition rules are as follows: from j to $j-1$ with probability $j\mu dt$, from j to $j+1$ with probability $j\beta dt$, from j to j with probability $1 - j(\beta + \mu)dt$. If $j = 0$ (which is true for a candidate new species), the only possible transitions in time dt are from 0 to 0 or 1, the transition probabilities being as follows: from 0 to 1 with probability $j\nu dt$, from 0 to 0 with probability $1 - j\nu dt$. It should be clear that in Karlin and McGregor's approach the number of species $S(t)$ is not fixed *a priori* because there is a continuous turnover of species due to migration or speciation. This should be contrasted with, for example, the somewhat simpler approach of [49], in which the number of species S is fixed *a priori*. Each of the S species can become extinct and then start again owing to migration or speciation.

It is useful to recapitulate some properties of the simple birth–death Malthusian process [44]. Let $P_j(t)$ be the probability that a population started with one individual at time 0 contains j individuals at time t. Then $P_j(t)$ is a geometric series with a modified zero term:

$$P_0(t) = \frac{\mu(\exp(rt)-1)}{\beta\exp(rt)-\mu}$$
$$P_j(t) = (1 - P_0(t))(1 - u(t))u(t)^{j-1}, \quad \text{with } u(t) = \frac{\beta(\exp(rt)-1)}{\beta\exp(rt)-\mu}. \tag{1.3}$$

Note that $P_0(t)$ is nothing but the probability of extinction at time t. By letting t go to ∞, one obtains that eventual extinction is certain if $\beta \leq \mu$, while eventual extinction occurs with probability μ/β if $\beta > \mu$. It is easy to prove that the expected value of the abundance j varies as $\exp(rt)$. For this reason, the case $\beta > \mu$ is discarded because the expected value of the abundance of each species increases exponentially with time, which is clearly quite unrealistic. As for the average time to extinction for $\beta \leq \mu$, it is easy to prove (by integrating over time $1 - P_0(t)$, which is the probability that the time to extinction is less than t) that it is given by

$$t_{ext} = \frac{1}{\beta}(\ln\mu - \ln(\mu - \beta)). \tag{1.4}$$

Therefore, the time to extinction is infinite if $\beta = \mu, r = 0$.

Box 1.1 *Continued*

As the expected value of the abundance of each species is constant if $\beta = \mu, r = 0$, this first case seemed a good starting point for [45] and [46] as well as for [33]. Actually, it is possible to prove [45, 46] that in this case the number of species $n_j(t)$ containing exactly j individuals at time t has a Poisson distribution with expected value given by

$$E\left[n_j(t)\right] = \frac{v}{j\beta}\left(\frac{\beta t}{1+\beta t}\right)^j. \tag{1.5}$$

This is exactly the expected distribution ϕ_j of species abundance at time t and is a logseries of the kind advocated by [48]: $\alpha \lambda^j / j$, provided that one sets $\alpha = v/\beta$ and $\lambda = \frac{\beta t}{1+\beta t}$. Problems arise, however: (1) the mean of the total size N of the community (sum of the abundances of all the species) can be shown to increase linearly with time ($E[N(t)] = vt$); (2) the mean of the total number of species S increases logarithmically with time ($E[S(t)] = \frac{v}{\beta}ln(1+\beta t)$). This seems to be a sort of paradox, given that the probability of extinction is 1 for all species, but it can be explained as follows: the expected number of species grows to infinity because the average lifetime of each species is infinite even if each species becomes ultimately extinct. Therefore, the case $r = 0$ of the Malthusian model describes an ever-increasing community in both the total number of individuals and the total number of species, which is somewhat unrealistic. Also, if we let $t \to \infty$ in expression (1.5), we get for the species-abundance distribution

$$\phi_j = \frac{v}{j\beta} = \alpha/j,$$

that is, a hyperbolic distribution with just one parameter, which would be unable to fit most observed species-abundance distributions. The way out, advocated by [46] and [33], is to calculate the species-abundance distribution conditional on a fixed size N of the community; [46] has proved that it is given by

$$E\left[n_j | N\right] = \frac{v}{\beta j}\left(\begin{array}{c} v/\beta + N - j - 1 \\ N - j \end{array}\right)\bigg/\left(\begin{array}{c} v/\beta + N - 1 \\ N \end{array}\right). \tag{1.6}$$

Actually, it is this functional form and not the logseries that has been used by [33] to compare the patterns of species abundance generated by NTB against those generated by other traditional models, such as the broken-stick and the lognormal models.

The second case considered for the Malthusian model is $\beta < \mu$. If it is so, extinction is still certain, but the average extinction time t_{ext} is finite and given by Equation (1.4):

$$t_{ext} = -\frac{1}{\beta}\ln\left(1 - \frac{\beta}{\mu}\right),$$

in which $\beta/\mu < 1$. One should note that t_{ext} is also the average lifetime of each species. Thus, it is easy to understand [45] that the average number of species $E[S(t)]$ will converge to an equilibrium $E[S]$, which is simply given by the product of the arrival rate v times the average lifetime of each species t_{ext}. Therefore, in this case, the mean number of species of the community is finite and given by

$$E[S] = -\frac{v}{\beta}\ln\left(1 - \frac{\beta}{\mu}\right).$$

Box 1.1 *Continued*

We can now somehow suspect that the logseries is involved because the right-hand side of this equation is nothing but the sum of the following logarithmic series:

$$\sum_{j=1}^{\infty} \frac{\nu}{\beta j} \left(\frac{\beta}{\mu}\right)^j .$$

In fact, [45] have proved that the probability distribution of $n_j(t)$ converges for $t \to \infty$ to a Poisson distribution whose expected value is

$$E\left[n_j\right] = \nu \int_0^{\infty} P_j(t)dt.$$

After some boring calculations, based on the previous formulas for $P_j(t)$ (Equation (1.3)), we obtain

$$E\left[n_j\right] = \phi_j = \frac{\nu}{\beta j}\left(\frac{\beta}{\mu}\right)^j . \tag{1.7}$$

Thus the logseries is the expected value of the species-abundance distribution. Also, one can easily obtain the average size of the community,

$$E[N] = \sum_{j=1}^{\infty} j \frac{\nu}{\beta j} \left(\frac{\beta}{\mu}\right)^j = \frac{\nu}{\beta} \frac{\frac{\nu}{\beta}}{1 - \frac{\nu}{\beta}},$$

which closes our derivations.

Karlin and McGregor's approach [45] allows the calculation of the expected number of species (which are not always the same but have a continuous turnover) on the basis of the three fundamental rates: birth β, death μ, arrival or speciation ν (Equation (1.7)). Actually, the species-abundance distribution, the number of species, and the total number of individuals in the community depend on just two parameters: the ratios ν/β and β/μ. The logseries has also been obtained by a slightly different model [49]. They have assumed that there exists a given number S of potential species in the community. Each of these species may become extinct according to the above described birth–death process. Once extinct, it may be replaced by another species at a rate of occurrence ν_0. Therefore, the rate of arrival (*sensu* [45]) of a new species (no matter which) in the whole community is $\nu_0 S P_0$, where P_0 is the probability of extinction at equilibrium (which may be equated to the fraction of the S species that are extinct at equilibrium). As a matter of fact, the logseries obtained by [49] coincides with that obtained by [45], provided one sets $\nu = \nu_0 S P_0$.

The problem with this model (in both versions, by [45, 49]) is that it relies on β/μ being a number smaller than 1. However, β/μ is nothing but the average size of the progeny produced by one parent in the whole lifetime, namely, the fitness of each species (remember that because of neutrality, all the species have the same fitness). This immediately points out the weakness of the Malthusian model with $\beta < \mu$. Although the results are very elegant and lead to the logseries distribution, the theory relies on assuming that the ecological community consists of species that are all unfit. Such a community would be easily invaded by a new species with a fitness even slightly larger than unity at low density. As a matter of fact, a more realistic NTB model would require consideration of a community of non-Malthusian species exhibiting some sort of density dependence. Fitness can be assumed to be larger than unity for low abundance, declining with increasing abundance and smaller than unity above a carrying capacity. This would guarantee that all the species are equally fit, they do not increase or decrease disproportionately, and their time to extinction is finite (see, e.g., [50]), not infinite as in the Malthusian models with $\beta \geq \mu$. Obviously, the resulting expected value of the species-abundance distribution is no longer a logseries if one assumes density dependence. As far as we know, there is no simple function describing the distribution even for prototypical models of density dependence, such as the logistic model. However, this does not imply that the NTB assumption cannot be used. Very simply, expected distributions can be derived by extensive simulation of density-dependent models and compared with data.

The model that is most used in NTB, however, is the one in which the total community size is constant. This approach was pioneered by [45]. They basically assume a mechanism similar to the one that was illustrated with reference to Equation (1.1). However, they consider time units that are so small that at most one event can take place: one of the N individuals dies at random and is replaced by one individual of the same species or by one individual of another species with probability proportional to the relative abundance of each species. Then, with probability ρ, the replacing individual may mutate into another species. In practice, if $n_j(t) = N_j$, $j = 1, 2, \ldots r$, $\sum_{j=1}^{r} N_j = N$ (r being the number of possible species) and the transition is from species k to species i, then $n_k(t+1) = N_k - 1, n_i(t+1) = N_i + 1$. The species i may be one of the species already present in the community or a new species. Obviously, k may coincide with i, and in this case, nothing changes after one time unit. The number of species r is considered to be very large because it includes not only the species that are actually present in the community but also those that might arise because of speciation or immigration (of course, these latter species are characterized by $n_j(t) = 0$ and $n_j(t+1) = 1$ if t is the time of speciation immigration). Karlin and McGregor [45] have found that for $t \to \infty$ and $r \to \infty$ the expected species abundance distribution $\phi_j = E\left[n_j\right], j \geq 1$ is given by

$$\phi_j = \frac{1}{j} \frac{N\rho}{1-\rho} \left(\begin{array}{c} \frac{N}{1-\rho} - j - 1 \\ N - j \end{array} \right) \Bigg/ \left(\begin{array}{c} \frac{N}{1-\rho} - 1 \\ N \end{array} \right). \qquad \textbf{(1.8)}$$

It is interesting to observe that Equation (1.8) formally coincides with Equation (1.6) of Box 1.1 if one sets $\nu/\beta = \rho N/(1-\rho)$. Note that Equation (1.6) was obtained from a time-continuous stochastic process in which ν is the instantaneous rate of new species arrival in the whole community. Since $1/\beta = 1/\mu$ is the average generation time of each species, the quantity ν/β is the average number of new species arriving in the community per generation. Instead, Equation (1.8) derives from a time-discrete stochastic model in which ρ is the speciation probability per individual in a time unit. In any case, expressions (1.6) and (1.8) are equivalent in terms of data

fitting because they have the same form as a function of the abundance j.

Using Equation (1.8) [45], one obtains the probability that two individuals are of the same species. It turns out to be given by $1/(1 + N\rho - \rho)$, which for small speciation probability ρ, large N, and finite $N\rho$ is very well approximated by

$$\bar{f} = \frac{1}{1 + N\rho}.$$

This expression is the same as Equation (1.2), provided one sets $2\gamma = \rho$. The factor 2 is simply due to the fact that Equation (1.1) was obtained by assuming that two individuals may mutate at the same time, while [45] assume that at most one individual can mutate in each time unit.

1.2 Neutral Individual-Based Models on Networks (and Beyond)

One example, and an early suggestion that was instrumental in directing our thinking, stems from an application of the neutral model of biodiversity [34]. It deals with the quantitative assessment of the role of directionality and network structure on ecological organization. A word of caution is in order, as the exercise that we present here might be somewhat misleading. In fact, many factors other than network configuration and transport anisotropy are operating in nature, playing different but obviously relevant roles. However, inclusion of all factors, no matter how detailed and realistic, hardly seemed a good starting point for the pursuit of any generalizable signatures like the one we are taking on at this point [21]. Here, in fact, we first show results from a baseline, rather abstract theoretical model that focuses on the fundamental differences between the topology of river basins and 2D landscapes. The neutral theory offers the elements for a basic dynamics capable of maintaining biodiversity. Despite its bold (and in many cases unrealistic) assumptions, the neutral model has produced many important results, even after having been tested extensively against empirical data – many features shown by real systems do not require a more complicated model [51]. Patterns predicted by the neutral theory can also arise from nonneutral interspecific dynamics in the presence of some stochasticity and high species richness, thus widening the range of applicability of the theory. Despite its success, it is crucial to recognize that the neutral theory overlooks much ecological information, for example, species-specific information that is required when studying the dynamics of the system or of particular species and the interactions among them [39]. In any event, employing the neutral theory is generally justified as long as steady state biodiversity patterns are addressed (but see [39]). Here, we show how the implementation of the neutral theory behaves in 2D lattices or 2D space-filling trees imposing directional dispersal [21, 40]. Two different frameworks, namely, an individual-based model and a metacommunity model, are introduced to that end. Contact models are introduced much later, in Section 3.2.3.

Box 1.2 Species Diversity in Neutral Metacommunities

The main tenet of the unified neutral theory of biodiversity [34] assumes that selection, in this context the difference in competitive ability of species, is turned off in the making of species diversity, and – a bold statement indeed – that all species' vital rates are equivalent at a *per capita* level. The main ecosystem-forming processes are therefore simply drift, countered by speciation or dispersal. It should be noted here that originally all neutral theories were meant to be spatially implicit, in that they considered isolated communities whose survival depended on speciation, or a local connected community, whose survival was guaranteed by immigration from some background metacommunity providing immigration rates and their composition.

The neutral theory of biodiversity [34], with its minimal set of assumptions and parameters, has been the subject of a lively debate that peaked about 10 years ago, proving both influential [49, 51–53] and controversial [54–57] as an explanation of biodiversity patterns. Here, we are interested in testing and exploiting the theory (as in [8]) across ecosystems, not simply in two-dimensional landscapes or in mean-field contexts, to which other spatial aspects contribute only weakly [34, 49, 52, 57, 58]. This book, in fact, focuses on the search for implications of hydrologic controls placed by river networks functioning as ecological corridors, a highly constraining *milieu* where landscape effects matter decisively.

The overall context that we need to explore here shares the concerns of the early biogeographers: what conditions are to be met for a species to occupy a site and maintain a population there? In this context, three factors matter: dispersal ability, habitat suitability, and susceptibility to biotic filtering [59]. In words, a species must be capable to reach the site by accessing the region and disperse therein; the climatic drivers and, more generally, the abiotic environmental conditions must be ecophysiologically suitable for the species; and the biotic environment, the whole of the relevant biological interactions, must meet a minimum of species' needs. Dispersal capacity from areas where the species is endemic (or simply exists) is key. Its nature includes the biogeographic natural history of the species embedding all factors limiting its spread from the places where they first originated. This, naturally, includes barriers to migration, the roles of biotic and abiotic dispersal vectors, and the suitability of the landing site from all biological viewpoints [59] – in brief, all it takes in terms of the environmental conditions that a species needs to settle, grow, and maintain a viable population. If landscape effects are key, lesser importance lies in biotic interactions with other organisms, either favorable (like mutualism and commensalism) or unfavorable (like predation and competition), in shaping local communities. Biotic interactions may or may not include environmental constraints on communities, such as the concept that whole communities (and ecosystems thereof) may experience species composition limited by environmental carrying capacities or defining roles of ecosystem engineer species that manipulate their environment favoring other species [59]. Pinpointing the relative importance of the various effects is case specific and primarily requires examining ecological patterns along geographic and environmental gradients (Sections 2.2, 2.4, and 2.5). Moreover, the

Box 1.2 *Continued*

examination of specific population dynamics (Chapter 3) will explore whether individual populations of a species may persist in suboptimal conditions and, in such a case, what factors affect the intrinsic vital rates (say, growth and death rates), determining ultimately their steady state abundances. Fluctuations induced by population dynamics that prompt environmentally suitable sites to become unoccupied may also be a factor, ofter blurring a clear-cut interpretation of species–environment relations [59]. Despite stochastic fluctuations, however, species distributions in space are most often expected to respond to major features, perhaps not simply local environmental determinism [59], that control their potential and realized ecological niches (Box 2.9). Cases where this does not happen would be such that fluctuations in demography and the strength of biotic interactions are so large that species–environment relations would be clouded. While this is obviously not excluded in real ecosystems, the general aim of this book prevents inclusive efforts in that direction.

The threefold influence of dispersal, niche, and biotic interactions shapes species distributions in space (say, within specific dendritic and dendrite-derived landscapes in our case). It may be deconstructed in many cases of interest, viewing their components as separate entities treated like specific boundary conditions [59], and yet suitable conditions for a species lie at the intersection of the ensemble of factors that determine the individual suitabilities. Also, one must consider that obviously a species may not colonize a site for reasons other the the ones accounted for above. In particular, human disturbances, so as to say, may prevent the establishment of virtually any species, inasmuch as – accepting, for example, the idea that the size of the largest species surviving in an ecosystem is related to the ecosystem's size [60, 61] – habitat fragmentation has a long-term impact that may have long-established endemic species go locally extinct. However, once locally extinct, one may assess whether the species might be capable of recolonizing the same site once again, possibly because of dispersal [34] (see also Section 1.3).

How and where species have emerged from evolutionary processes may explain patterns of biodiversity at any scale. Speciation causes, whether allopatric (where geographic barriers split the range of ancestor species disrupting gene flow between the separated populations and ultimately leading to distinct species or subspecies) or sympatric (where divergence is due to ecological specialization), are known to be numerous (see, e.g., [62]). For the limited purposes of this book, it is sufficient to acknowledge that geographic or ecological speciation processes have occurred, and continue to occur at a slow pace (Section 1.3), because our focus is firmly placed on how speciation may shape future patterns of species distribution in complex landscapes shaped (or constituted) by fluvial processes. Rather than questioning whether (and how) species resulting from specific speciation processes would result in more/less specialized features, we shall sample a large number of neutral traits on noninteracting species and observe landscape effects under the null model provided by the neutral theory of biodiversity – with a few nonneutral ingredients at times selectively added to zoom in on the network perspective we pursue. It is also a matter of scale, of course. At continental scales, biogeographical history and dispersal limitation predominate, and environmental suitability plays secondary roles in explaining the geography of a focus species [59]. This perspective will have to be extended when studying biological invasions in networked environments (Chapter 3), because the effects of niche changes (Box 2.9) between native and invaded ranges may pitch in loudly when considering whether biological invasions are at all possible.

Box 1.2 *Continued*

The idea that dispersal plays a decisive role in shaping the distribution of species, and thus the diversity and composition of communities, will therefore guide our study of the effects of the dendritic and connected nature of fluvial ecological corridors. This is reminiscent of the very onset of the field of ecology, moving from early biogeographers' observations that significant similarities exist between flora and fauna found in separate continents [59]. In fact, long-distance dispersal (say, through past continental bridges) may explain some patterns as well as physical displacements not unlike those of spores of invaders displaced by long distances in the ballast water of tugged boats (Chapter 3) or pathogens brought into disease-free regions by asymptomatic, infected traveling individuals. To what extent would our ecologically narrow focus of dispersal across scales in specific dendritic substrates explain observed patterns of species distributions? Not only is this hard to assess but perhaps it is also immaterial to our scope. We take the extreme view of Hubbell [34] in assuming equivalent environmental niches across species and using dispersal alone to account for variations in species' compositions – thus implying that communities are predominantly shaped by dispersal together with speciation and extinction – as a null hypothesis, one that fosters distinctive effects of the substrate type, to test the effects of the environment in shaping species distributions.

One example seems appropriate at this point. In fact, the resemblance of ecological neutral theory to the more mature neutral theory of population genetics [42] prompted a flow of concepts and quantitative tools to be adapted by ecologists following early seminal work (see, e.g., [34, 39, 55, 58]). A proper network perspective was brought in only later [8, 35], clearly inspired by population genetics and statistical physics, in particular to derive novel tools for assessing species diversity in networks of communities (Section 1.1). In particular, neutral ecological dynamics in a network of communities were found to correspond to migration matrix models in population genetics [35]. In such a representation, a network of n local populations is dynamically described by a stochastic migration matrix, say, $\mathbf{M}$, whose elements m_{ij} quantify the fraction of individuals in a given subpopulation i that originated from a parent in subpopulation j in the previous generation. By continuity, one has $\sum_{j=1}^{n} m_{ij} = 1$, in analogy to connectivity matrices used in describing spanning trees (Box 1.3). Thus, by varying the elements of the matrix quantifiying the relative weight of each edge, and the size of the local populations, the underlying spatial structure of the metacommunity is studied in a true network perspective. Economo and Keitts [35], for instance, have studied directed networks (structures whose matrices yield $m_{ij} \neq m_{ji}$ according to the authors, although here we shall use a different meaning of the term) modified in that descendants of individuals from each node must be able eventually to reach every other node ($m_{ii} \neq 1$). In their approach, the speciation rate, say, ν, takes the place of a mutation reflecting the per generation probability of change in state for a single individual. An elaborate mathematical construction could therefore be used for ecological purposes [35]. The relevance for our scope lies in the substrates studied by the authors, topologically rather different, although unrelated to river networks (chain, island, and star graphs or randomly assembled networks). Significantly, the related results highlight the importance of the spatial structure of connectivities, coupled to the biological parameters of the neutral model, in determining species diversity of a local community, and among spatially separated communities at the scale of the entire metacommunity. Moreover, quite significantly for our purposes, topological differences in metacommunity structure were found to matter, as they strongly reflect different spatial arrangements and connectivities of the focal habitats – here subsumed by nodes of the metacommunity on whose

Box 1.2 *Continued*

physical and biological meanings in a fluvial context will require ecohydrological specifics. Whereas some communities were arranged in long chains, others were hierarchically clustered (purportedly reflecting patchy habitat distributions) or else characterized by drift. Overall, their theoretical results support unequivocally the results described in Sections 1.2, 1.3, and 1.4.

It must be stated up front, in concluding this introductory box on neutral approaches, that our aim by no means intends to diminish the importance of other effects that neutral approaches typically ignore. Yet neutral pattern does not imply neutral process [51], and the flexibility of the theoretical tools and the complexity (and sensitivity to the substrate form) of the distributions observed under the neutral perspective are argued to serve well the main the tenet of this book. This also holds, in particular, in the light of important validation from replicated laboratory experiments on removal and forced dispersal in natural and artificial communities (Section 1.6), common garden experiments, or controlled ecotones [59]. For instance, high-elevation plants in mountain environments have been moved to low elevations to test the hypothesis that they can grow and reproduce well outside their endemic domain, but may be still wiped out, systematically outcompeted by other plants more suited to the new conditions; or else, it may be vital to determine to what extent phenotypic differences between local populations along landscape gradients have a genetic basis [59], a circumstance we altogether neglect. The general idea, commonplace in statistical physics, is that one seeks the minimum ingredients in a model ecosystem that reproduce acceptably the features desired, in our case, large-scale patterns of biodiversity and abundance of species strongly affected by the topology of the substrate for ecological interactions, that is, the river network. This inspired our entire treatment of the subject.

In our neutral individual-based model, a 2D lattice is termed *savanna* to echo its ecohydrological background; that is, the ecological substrate is represented simply by a square lattice in which each site, or pixel, may be occupied only by one individual of a given species. As river landscape, we use within the same domain a space-filling tree – an optimal channel network (OCN) (Box 1.3) – built in a lattice of the same size as the savanna. In a space-filling network, all sites are channelized (Figure 1.1b), with certain implications for the size of the network relative to the drainage density [63] immaterial in this context. The dynamics at each time step are defined as follows. Species are arbitrarily distributed (in position and number) as an initial condition – the choice only affects convergence time to a stationary state. A randomly selected individual (i.e., at a randomly picked pixel) dies. With probability ν, termed the diversification rate, this site is occupied by a new species; with probability $1 - \nu$, the site is colonized with equal probability by an offspring of one of the neighbors. The two landscapes differ only in the definition of neighbors. For the savanna, in fact, the offspring that colonizes

the empty site is chosen among the individuals that occupy the four nearest neighbors (boundary effects at the edges of the lattice are removed by implementing the genetic algorithm proposed by [9, 40]). For the network landscape, the neighborhood of a pixel is constituted by the nearest pixels connected to it by the network connectivity or by all neighboring pixels with larger probabilities being assigned to those connected through the network. Notice that all pixels, except for the outlet, have only one downstream neighbor; source pixels have no upstream neighbors, whereas all the others have one or more upstream neighbors. The process is iterated until it reaches a stationary state.

The two upper insets of Figure 1.1 illustrate the resulting spatial biodiversity configuration in the two landscapes: pixels labeled by the same color represent individuals belonging to the same species. The lower panel shows the typical associated rank-abundance curves. The results are remarkably different. It can be noticed, in particular, that the configuration of the space and the directionality of the dispersal imposed by the network landscape determine a higher species

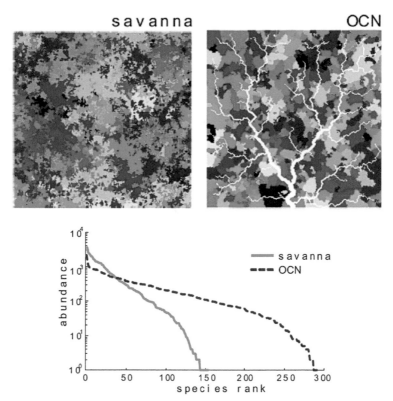

Figure 1.1 Comparison between neutral biodiversity patterns obtained by the neutral model described in the text within space-filling networks in a square domain: a savanna (a two-dimensional lattice) and a fluvial network where directional dispersal to nearest neighbors is regulated by an OCN connectivity matrix [63]. We refer here to 2D landscapes as savannas only for easier mental reference to the real world, as the neutral model used to produce the plot does not include all ecological features of real savannas. These results form the basis of our theoretical motivation, upon which additional realistic complications will be built. Species spatial patterns (upper insets) and their species rank-abundance curves are shown. The simulations are run on a 250×250 lattice with $\nu = 10^{-4}$. Figure after [21, 40]

richness. Moreover, the spatial configuration of the patches of the same species in the network landscape have sharp boundaries that resemble the boundaries of subbasins. Differences only arise because of the different connectivity imposed by the two landscapes.

Adding another factor typical of the dispersal in networked landscapes, such as biased transport (e.g., offsprings colonizing preferably downstream), would only enhance the observed differences. This has important consequences that we explore in this chapter.

Box 1.3 Optimal Channel Networks (OCNs) and Their Landscapes

Here, we introduce the models of river networks known as optimal channel networks (OCNs) [63–66], a framework for the study of river network morphology that is known to produce replicas of spanning trees (and their landscapes) filling an assigned space (see the example in Figure 1.2 and those in Appendix 6.2). The given domain is a lattice of sites to which an elevation and a connectivity are assigned.

OCNs are statistically indistinguishable from real river networks when subjected to truly distinctive tests [67] (see Figure 1.3 for examples of extraction of the river network proper from digital terrain maps [63]) as shown by a large body of literature summarized in Appendix 6.2.

Box 1.3 *Continued*

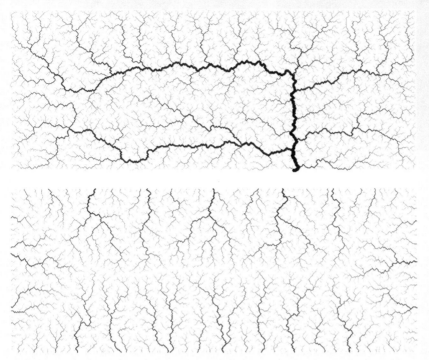

Figure 1.2 Two examples of OCNs of a rectangular 500 × 200 lattice. The two realizations differ only for the boundary conditions. In the top plot, in fact, at the boundary, no-flow conditions are enforced, except for a single outlet placed in anonsymmetric position to further elongate the main stream [64]; the lower plot is a multiple-outlet arrangement where catchments of different sizes compete for drainage for which exact solutions exist [63]. Image courtesy of Enrico Bertuzzo

The reader is referred to Chapter 6 (Appendices) for a proper introduction to the rationale and the theoretical background that support the OCN concept as an exact solution to a general landscape evolution problem. Here, we shall limit ourselves to a recollection of the main features of the OCN framework, in particular, its capability to produce statistically identical, realistic replicas of the substrates for ecological interactions implied by fluvial ecosystems. As shown in some detail in Appendix 6.2, the capability of OCNs to reproduce consistently all metric properties (say, the treelike arrangement or the detailed distributions of contributing area and upstream length at any point in a lattice) relevant to the ecological studies pursued is remarkable [67]. We emphasize that devising truly distinctive network statistics may be subtle. For instance, matching topological attributes, epitomized by Horton's numbers or Tokunaga matrices [63], has been termed inevitable for spanning trees [68] and therefore not distinctive of rather different fluvial forms. This consistency led in the past to curious claims of similarity of leaf vein architectures to river networks, where just by eyesight one could have concluded that they were completely different. Only when a comprehensive set of joint topological and metric statistics is reproduced (including all metrics relevant to the ecological studies pursued here) can finely distinctive comparisons of different network structures be enforced, matching field evidence on a grand scale [63]. This is a key feature of OCNs [64, 67, 69].

The OCN model is based on the fact that landscape evolution favors configurations of the system (that is, the connectivity of a given set of sites, each characterized by an elevation that determines topographic gradients and drainage directions) that minimize a function describing total energy dissipation. The derivation of an explicit form for such a function was of paramount importance. The latter uses, locally, landscape-forming flowrates Q_i and the drop in topographic elevation,

Box 1.3 *Continued*

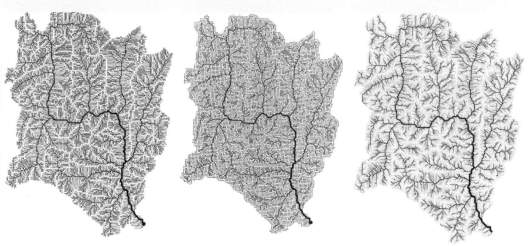

Figure 1.3 Extraction of fluvial forms from a digital terrain map (DTM) (after [63]). (left) Each site i of a lattice is provided by a DTM elevation, say, z_i, that approximates the topographic surface $z(x,y)$ in a process in which discrete horizontal tiles of height z_i over a datum cover the surface. One may therefore compute for each site i reasonable approximations Δz_i for the topographic gradient, a vector $\vec{\nabla}z(x,y)$, and for the topographic curvature, a scalar $\nabla^2 z(x,y)$. Drainage directions join neighboring sites via the steepest descent direction oriented along the gradient whose magnitude is the slope $|\vec{\nabla}z(x,y)| \approx \Delta z_i$, which is unique for each site. This image shows the set of all drainage directions in a 40 km^2 catchment in northern Italy covered by pixels of size 30×30 m^2. (center) Yet not all the pixels covering a topographic surface are actually channeled sites, as common sense suggests. Here we show, following [63], the network of channeled sites, where the discriminant for deciding whether a DTM site is channeled is the sign of the topographic curvature (i.e., $\nabla^2 z_i \geq 0$); that is, the terrain surrounding the site is concave up, indicating what geomorphologists call colluvium, or areas of deposit of eroded material. Here, dots indicate the unchanneled sites. (right) However, topographic concavity proves a necessary but not sufficient condition for channel occurrence, whereas it has been shown that a proper threshold for the onset of channelization – owing to a number of possible diverse mechanisms – is a combination of positive curvature and exceedence of a slope-dependent cumulated support area, that is, $\sqrt{A_i}\Delta z_i \geq \omega$ [70, 71], where ω is a suitable numerical constant that embeds all units used in the process. This image shows the application to the catchment at hand, where the value of ω has been calibrated on the "blue lines" of proper topographic maps and ground truthing [63]. Note that dashed lines show the drainage directions in unchanneled areas. The distinction is essential, not only to highlight the proper fluvial domain, but also in constructing OCN landscapes [64, 74, 75], because within fluvial domains a slope–area law $\Delta z_i \propto A_i^{-1/2}$ applies almost without exception across orders of magnitude in catchment size [63] (see also Appendix 6.2).

that is, potential energy or the leading form of energy in a gravity-dominated system. Thus energy dissipation $Q_i\Delta z_i$ becomes a nonlocal interaction because for landscape-forming discharges, one has $Q_i \sim A_i$ (where A_i is total contributing area at i; see Appendix 6.2.2), and the drop in elevation may be described by a slope–area relation $\Delta z_i \sim A_i^{-1/2}$ corresponding to universal features of the fluvial regimes distinct from unchanneled landscapes. The concavity of the function and the universality of the scaling exponent for the slope–area relation are well justified theoretically and empirically (see Appendix 6.2 and [63]). Thus, spanning, loopless network configurations characterized by minimum energy dissipation are obtained by selecting the configuration, say, s, that minimizes the functional $H(s)$, defined as

Box 1.3 *Continued*

$$H(s) = \sum_{i=1}^{N} A_i^{1/2}, \tag{1.9}$$

where i spans a lattice of N sites. Given that the master variable is A_i, total contributing area at any site i, it should be remarked that the entire topographic structure of a landscape contributes to determining drainage directions (the local topographic gradient, assuming that gravity is the leading geomorphic agent). Technically (Appendix 6.2), total contributing area at site i is given by a discrete integral equation, $A_i = \sum_j W_{j,i} A_j + 1$, where $W_{j,i} = W_{j,i}(s)$ is the generic element of a connectivity matrix (equal to 1 if $j \rightarrow i$ through a drainage direction, and 0 otherwise). In this case, $W_{j,i}$ is the element of an adjacency matrix derived from the topographic field where a site is connected directly only to its nearest neighbors by a drainage direction. Empirically, this result is also verified by a large body of field observations in fluvial domains (see, e.g., [70, 71] in the context of channel initiation; Figure 1.3).

Boundary conditions are required for the evolving optimal trees, as outlet(s) (single or multiple outlets along drainage lines) as well as no-flux or periodic boundary conditions [63, 72] must be imposed (Figure 1.2). The OCN model has been thoroughly explored, as it has produced various interesting results – and wonderful replicas of dendritic forms (see, e.g., Figures 1.2 and 6.14 of the appendix) whose statistical structure is indistinguishable from empirical ones regardless of vegetation cover, exposed lithology, geology, and climate.

OCNs entail the useful concept of replicas of statistically identical ecological substrates. In fact, the random search procedure needed to access the rich structure of the fitness landscape of trees (Section 6.2), each being a local minimum of the functional (1.9), produces for any search a different outcome. All realizations, however, are characterized by identical statistics of metric and topological properties, the latter much less distinctive than the former. This means that running 100 OCNs may be seen as the collection of 100 independent realizations of landscape evolution within the same domain. To construct the landscape associated with an OCN, we shall follow the procedure described in [73] and the theoretical insight developed later to justify it [72, 74, 75]. This is described in greater detail in Appendix 6.2.

Although the results from this simple neutral individual-based model are not to be taken as fully representative of ecological reality, they rather forcefully demonstrate that differences may arise in key biodiversity features simply because of the presence of the drainage network acting as ecological substrate. Thus, the potentially fundamental roles played by the river network warranted more refined modeling schemes to investigate other important issues. In that light, another set of results from a more structured neutral model [8, 40] proves revealing. In this model, the landscape is organized into local communities (LC), each of which contains a certain number of sites; every site contains only one individual. In this way, a genuine metacommunity model is defined.

At each time step, an individual, randomly selected from all individuals in the system, dies. With probability ν, the diversification rate, this will be occupied by a new species; with probability $1 - \nu$, the empty site will be colonized by an offspring of a species already existing in the metacommunity. In the latter case, the probability P_{ij} that the empty site in LC i will be colonized by a species from LC j is determined as follows [40]:

$$P_{ij} = (1 - \nu) \frac{K_{ij} H_j}{\sum_{j=1}^{N} K_{ij} H_j}, \tag{1.10}$$

where K_{ij} is the dispersal kernel, the fraction of offspring produced at LC j that arrives at LC i after dispersal; H_j is the habitat size of LC j, that is, the number of sites in

LC j; and N is the total number of LCs. All individuals in LC j have the same probability of colonizing the empty unit in LC i where the death took place. Note that the standard neutral theory is in a way improved by assuming that different LCs have different habitat sizes. Instead, neutrality is extended to dispersal; that is, the dispersal kernel of every species is assumed to be the same. In the context of metacommunity models, the dispersal kernel K_{ij} contains information on the landscape spatial structure and how individuals move about. Therefore, the key difference between lattice-like substrates (e.g., savannas in Figure 1.1) and river network metacommunities lies in their respective dispersal kernels. The dispersal kernels are typically assumed to take the form of an exponential decay [8] (Box 2.2). Note that, unlike in individual-based models, an offspring can now travel farther than its immediate neighbors in each time step.

In a metacommunity model, biodiversity patterns are measured by α-, β-, and γ-diversities [76, 77]. Here, α-diversity is a local description of biodiversity and γ- diversity a global one, both being inventory measures because they refer to the number of species; β-diversity is a differentiation diversity measuring the rate of change in, or the turnover of, the species, measuring how species compositions in local communities differ from one another. In the following, γ-diversity is defined as the total number of species in the entire metacommunity; α- diversity is a number of species in a randomly chosen LC – it is also useful to consider its mean value averaged across all LCs, denoted by $\langle \alpha \rangle$. The between-community diversity, or β-diversity, is a conceptual quantity that can be defined in many ways, all of which share the same general idea: the higher the β-diversity is, the more the local communities differ. Here, it is defined as $\gamma / \langle \alpha \rangle$.

The main results of the metacommunity model (not shown here; see supplementary material in [40]), namely, the rank-abundance curves and exceedance probability plots of abundance, prove in general qualitative agreement with the individual-based model described above (Figure 1.1). This result supports the importance of the topological structure of the network of possible interactions on the biodiversity composition and the configurations of local communities. Network structure and dispersal anisotropy decisively affect any biodiversity measures. In this case, the dispersal rate is defined as the fraction of propagules that is dispersed away from their birth local community, and the directionality is defined as the natural logarithm of the ratio between the fractions

of propagules at the nearest neighbors in the preferred and opposite directions of dispersal [40]. All three diversity measures – in both types of landscapes shown in Figure 1.1 – appear to be quite sensitive to dispersal anisotropy. River networks thus result in metacommunities with higher β-diversity, that is, more localized and heterogeneous ecosystems. This is due to a containment effect: in river basins, cross-subbasin dispersal is hindered by topographical divides, resulting in subbasins being more dissimilar from one another, echoing important field and theoretical evidence [6, 78].

1.3 Species' Persistence Times and Their Landscape

Another macroecological pattern proves relevant to our general tenet. Specifically, we study the distributions of local species persistence times, defined as the timespans between colonization and local extinction in a given geographic region. Empirical distributions pertaining to different taxa, in this case, breeding birds and herbaceous plants, have been analyzed in the above framework [9]. The framework critically accounts for the finiteness of the observational period of any field observations, a feature that is often overlooked and that may remarkably hinder the true features of persistence times. Their distributions, in fact, exhibit power law scaling limited by a cutoff determined by the rate of emergence of new species [9]. Note that, although generalizations are possible, the study of persistence (or lifetime) was conducted on trophically equivalent co-occurring species – in a broad sense, species that share resources and predators may be partitioned into equivalence classes where they play the same structural roles (e.g., [79]).

Theoretical investigations on how the scaling features depend on the topological structure of the spatial interaction networks prove worthwhile [9]. The lifetime τ of a species within a given geographic region is defined as the time incurred between its emergence and its local extinction. At a local scale, lifetimes are largely controlled by ecological processes operating at short timescales (e.g., population dynamics, immigration, contractions/expansions of species geographic ranges) as local extinctions are dynamically balanced by colonizations [11, 80, 81]. At a global scale, originations and extinctions are controlled by mechanisms acting on macroevolutionary timescales [60, 82]. From a theoretical viewpoint, the simplest baseline demographic model is a random walk without drift, according to which the abundance of a species in a geographic region has the same probability of increasing or decreasing by one individual at every time step. According to this scheme, local extinction is equivalent to a random walker's first passage to zero, and thus the resulting lifetime distribution has a power law decay with exponent 3/2 [5, 83].

A more realistic description can be achieved by accounting for basic ecological processes like birth, death, migration, and speciation, possibly in neutral mean field schemes (that is, dispersing individuals may end up anywhere with the same probability) [8, 34, 49]. This is done as follows. Consider a community of N individuals belonging to different species. At every time step a randomly selected individual dies and the space or resources are freed up for colonization. With probability ν the site is taken by an individual of a species not currently present in the system; ν is equivalent to a per-birth diversification rate and accounts for both speciation and immigration from surrounding communities. With residual probability $1 - \nu$ the dead individual is replaced by one offspring of an individual randomly sampled within the community [39, 84]. As such, the probability of colonization by a species depends solely on its relative abundance in the community. The asymptotic behavior $(t \gg 1)$ of the resulting lifetime distributions (i.e., $p_T(t)$) exhibits a power law scaling limited by an exponential cutoff:

$$p_T(t) \propto t^{-\alpha} e^{-\nu t}, \qquad (1.11)$$

with exponent $\alpha = 2$ [85]. In Equation (1.11), time is expressed in generation time units [34]; that is, it has been rescaled in such a way that the death rate is equal to 1. Notably, in the mean field scheme, the probability distribution depends solely on the diversification rate, which accounts for speciation and migration processes and imposes a characteristic timescale $1/\nu$ for local extinctions. While speciation rates are not expected to vary with the spatial scale of analysis, immigration rates decrease as the spatial scale increases. In fact, the possible sources of migration (chiefly dependent on the geometrical properties of the boundary and the nature of dispersal processes) are argued to scale sublinearly with the community size [9], which in turn is linearly proportional to geographic area [80, 86]. As continental scales are approached, migration processes (almost) vanish, and the diversification rate ultimately reflects only the speciation rate.

1.3.1 Network Topology and Persistence Times

We now provide evidence of different, nontrivial exponents observable as a function of the topology of the substrate. Instead of a mean field model, we use a space-explicit scheme in which dispersal limitation and the actual network of spatial connections are taken into account. The neutral game described above has been implemented in regular one-, two-, and three-dimensional lattices in which every site represents an individual, highlighted by a specific color that labels the species [9]. Key to our reasoning, we explore the patterns emerging from the application of the model to dendritic structures mimicking riverine ecosystems where dispersal processes and ecological organization are constrained by the network structure. To this end, we again use optimal channel networks (OCNs) (Box 1.3) as space-filling (within arbitrary domains) mathematical constructs that yield aggregation patterns and landscape forms statistically indistinguishable from real-life river networks [67]. To account for limited dispersal effects, only the offspring of the nearest neighbors of the dying individual are allowed to possibly colonize the empty site. In the networked landscape the neighborhood of a site is defined by the closest upstream and downstream sites. Limited dispersal, in fact, promotes the clumping in space of species, which enhances their coexistence and survival probability [39, 87].

Figure 1.4 shows the results of the neutral exercise described above – containing a remarkable message. In fact, in all the considered landscapes, lifetime (equivalently termed, at times, persistence time) distributions follow a finite-size power law behavior characterized by highly nontrivial scaling exponents smaller than that of the mean field. Power laws are inevitably limited by an upper exponential cutoff. The computational results reproduce perfectly the mean-field limit where all sites in the domain may contribute to the replacement with equal probability (yielding the exactly solved exponent $\alpha = 2$ [85]). Persistence time distributions deducted from the theoretical models change when dispersal kernels more general than dispersal from nearest neighbors are considered. As expected, as long as the mean dispersal distance remains small with respect to the system size, the distribution exhibits a longer transient regime but eventually ends up scaling like the one predicted by the nearest-neighbors dispersal [9]. Relaxing the neutral assumption [34] by implementing an individual-based competition/survival trade-off model [39] has also been tested. Specifically, species with higher mortality rates

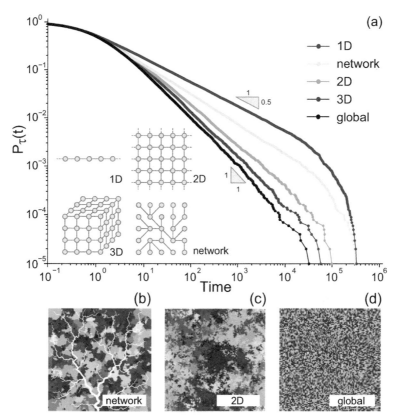

Figure 1.4 (a) Persistence exceedance probabilities $P_\tau(t)$ (the probability that a species' persistence τ is $\geq t$) for the neutral individual-based model [34, 39, 84] with nearest-neighbor dispersal implemented on the different topologies shown in the inset [9]. Note that in the power law regime, if $p_\tau(t)$ scales as $t^{-\alpha}$, $P_\tau(t) \propto t^{-\alpha+1}$. It is clear that the topology of the substrate affects macroecological patterns. In fact, the scaling exponent α is equal to 1.5 ± 0.01 for the one-dimensional lattice (red), 1.62 ± 0.01 for the networked landscape (yellow), and 1.82 ± 0.01 and 1.92 ± 0.01 respectively for the 2D (green) and 3D (blue) lattices. Errors are estimated through the standard bootstrap method. The lifetime distribution for the mean-field model (global dispersal) reproduces the exact value $\alpha = 2$ (black curve) [9]. For all simulations, $\nu = 10^{-5}$, and time is expressed in generation time units [34]. The panels in the lower part sketch a color-coded spatial arrangement of species in a networked landscape (b), in a two-dimensional lattice with nearest-neighbor dispersal (c), and with global dispersal (d). Figure after [9]

are assumed to hold less competitive ability in colonizing empty sites [86, 88]. It is important to note that the trade-off model also exhibits power law lifetime distribution with exponents indeed close to those shown by the neutral model. No major change arises.

We conclude that such theoretical results are thus robust with respect to both change in the dispersal range and relaxations of the neutrality assumptions. In particular, our results are not seen as a test for the neutrality hypothesis [34] for empirical distributions but rather as tools to reveal emerging universal and macroscopic patterns regardless of the detailed features of the particular model. Incidentally, a meaningful assessment of species'

local extinction rates is deemed valuable from a conservation perspective. Species lifetime distributions are in fact a robust tool to quantify the persistence of the species assembly currently observed within a given geographic area and, to some extent, predict the expected amount of future local extinctions. Mathematically, in fact, τ is defined as the time to local extinction of a species randomly sampled from the system regardless of its current abundance. Although these patterns cannot provide information about the behavior of a specific species or of a particular patch inside the ecosystem considered (e.g., a biodiversity hotspot), they can effectively describe the overall dynamical evolution of the ecosystem diversity.

Box 1.4 Derivation of the Probability Distribution of Persistence Times

Here we provide the exact derivation of the probability distribution of the variables τ' (persistence times that start and end within the observational window) and τ'' (the same variable, except that it comprises τ' together with all the portions of persistence times that start or end outside the observational window) (see Figure 1.5, after [9]). When dealing with observational data, the effect of the finiteness of the observed time window on the measured species lifetimes must be properly taken into account. In this theoretical framework, the probability vdt of observing a diversification event in a time step dt is assumed to be a constant; thus species emergence in the system due to migration or speciation is seen as a uniform point Poisson process with

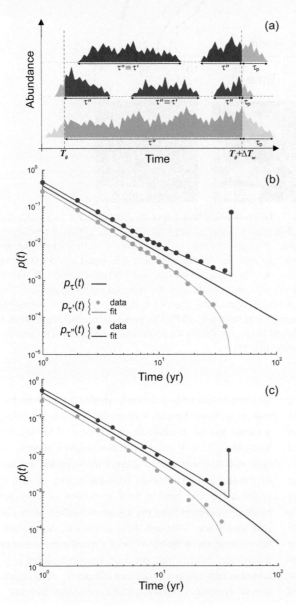

Figure 1.5 Empirical persistence time distributions. (a) A schematic of the variables that can be measured from empirical data over a time window ΔT_w: τ', persistence time that starts and ends inside the observational window, and τ'', which comprises τ' and all the portions of persistence times seen inside the time window that starts and/or ends outside. Residual persistence times τ_p are also shown. (b) Breeding birds and (c) herbaceous plants probability density function $p(t)$ of τ' (green), τ'' (blue), and τ (red). Filled circles and solid lines show observational distributions and fits, respectively. The best fit is achieved with $p_\tau(t) \propto t^{-\alpha}$, with $\alpha = 1.83 \pm 0.02$ and $\alpha = 1.78 \pm 0.08$ for breeding birds and herbaceous plants, respectively. Note that previous estimates [90] for (b) are revisited here in light of the new tools and of a longer dataset. The spatial scale of analysis is $A = 10,000$ km^2 and $\Delta T_w = 41$ years for (b) and $A = 1$ m^2 and $\Delta T_w = 38$ years for (c). The finiteness of the time window imposes a cutoff to $p_{\tau'}(t)$ and an atom of probability in $t = \Delta T_w$ to $p_{\tau''}(t)$, which corresponds to the fraction of species always present during the observational time. Variables $p_\tau(t)$ and $p_{\tau'}(t)$ have been shifted in the log-log plot for clarity. Figure after [9]

Box 1.4 *Continued*

rate $\lambda = \nu N$ (where N is total number of individuals in the system and λ has the dimensions of the inverse of a generation time). We term t_0 the emergence time of a species in the system and T_0 and $T_f = T_0 + \Delta T_w$ the beginning and the end of the observational time window, respectively. A species emerging at time t_0 will be continuously present in a geographic region for its lifetime τ until its local extinction at time $t_0 + \tau$.

We first analyze the distribution of τ'', the most complex case. The variable τ'' can be expressed as a function of the random variables τ and t_0, which are probabilistically characterized. We can distinguish four different cases (Figure 1.5a):

1. The species emerges and goes locally extinct within the time window.
2. The species emerges during the observations, and it is still present at the end of the time window.
3. The species emerges before the beginning of the observations and goes locally extinct within the time window.
4. The species is always present for the duration of the observations.

Or, mathematically,

$$\tau'' = \begin{cases} \tau, & \text{if } T_0 \leq t_0 \leq T_f \text{ and } t_0 + \tau \leq T_f, \\ T_f - t_0, & \text{if } T_0 \leq t_0 \leq T_f \text{ and } t_0 + \tau > T_f, \\ t_0 + \tau - T_0, & \text{if } 0 < t_0 < T_0 \text{ and } T_0 \leq t_0 + \tau \leq T_f, \\ T_f - T_0, & \text{if } 0 < t_0 < T_0 \text{ and } t_0 + \tau > T. \end{cases} \qquad (1.12)$$

The probability of observing τ'' conditional on a lifetime of duration τ has been derived exactly in the form [9, 89]:

$$p_{\tau''}(t|\tau) = \frac{1}{\mathcal{N}} \left(\delta(t - \tau) \int_{T_0}^{T_f - \tau} \Theta(T_f - T_0 - \tau) dt_0 \right.$$

$$+ \Theta(T_f - T_0 - t)\Theta(\tau - t)$$

$$+ \Theta(T_f - T_0 - t)\Theta(T_0 + t - \tau)\Theta(\tau - t)$$

$$\left. + \delta(t - (T_f - T_0)) \min[T_0, T_0 - (T_f - \tau)]\Theta(\tau - (T_f - T_0)) \right), \qquad (1.13)$$

where $\Theta(x)$ is the Heaviside function and $\mathcal{N}$ is the normalization constant made explicit in [9]. Marginalizing with respect to τ, we obtain the probability distribution of τ'':

$$p_{\tau''}(t) = \int_0^\infty p_{\tau''}(t|\tau) p_\tau(\tau) d\tau. \qquad (1.14)$$

Equation (1.13) combined with Equation (1.14) yields

$$p_{\tau''}(t) = \frac{1}{\mathcal{N}} \left((T_f - T_0 - t) p_\tau(t) \Theta(T_f - T_0 - t) \right.$$

$$+ \Theta(T_f - T_0 - t) \int_{t>0}^\infty p_\tau(\tau) d\tau$$

$$+ \Theta(T_f - T_0 - t) \int_{t>0}^{T_0 + t} p_\tau(\tau) d\tau$$

$$\left. + \delta(t - (T_f - T_0)) \int_{T_f - T_0}^\infty \min[T_0, T_0 - (T_f - \tau)] p_\tau(\tau) d\tau \right). \qquad (1.15)$$

Box 1.4 *Continued*

The last term of Equation (1.15) gives an atom probability in $t = \Delta T_w = T_f - T_0$ corresponding to the fraction of species that are always present during the duration of the observational window.

When comparing analytical and observational distributions, we assume that the system is at stationarity and unaffected by initial conditions, that is, that T_0 is far from the beginning of the process. Mathematically, this is obtained taking the limit $T_0, T_f \to +\infty$ with $T_f - T_0 = \Delta T_w$ in Equation (1.15), which finally takes the form

$$p_{\tau''}(t) = \frac{1}{\mathscr{N}} \Big((\Delta T_w - t) p_\tau(t) \Theta(\Delta T_w - t)$$

$$+ \Theta(\Delta T_w - t) \int_{t>0}^{\infty} p_\tau(\tau) d\tau$$

$$+ \Theta(\Delta T_w - t) \int_{t>0}^{\infty} p_\tau(\tau) d\tau$$

$$+ \delta(t - \Delta T_w) \int_{\Delta T_w}^{\infty} (\tau - \Delta T_w) p_\tau(\tau) d\tau \Big). \tag{1.16}$$

The variable τ' comprises only the first of the four cases listed in Equation (1.12). Thus the probability distribution $p_{\tau'}(t)$ follows directly from the first term of Equation (1.16):

$$p_{\tau'}(t) = \frac{1}{\mathscr{N}'} (\Delta T_w - t) p_\tau(t) \Theta(\Delta T_w - t), \tag{1.17}$$

where the proper normalization constant $\mathscr{N}'$ is derived in detail in the supplementary information of [9] and in [89].

1.3.2 Observational Distributions of Empirical Persistence Times

We empirically characterize species persistence time distributions by analyzing two long-term datasets covering very different spatial scales: (1) a 41-year survey of North American breeding birds [91] and (2) a 38-year inventory of herbaceous plants from Kansas prairies [92].

The North American Breeding Bird Survey consists of a record of annual abundance of more than 700 species over the 1966 to present period along more than 5,000 observational routes. The spatial location of the routes analyzed is shown in Figure 1.6. We consider only routes with a latitude less than 50° because density of routes with a long surveyed period drastically decreases above the fiftieth parallel. Noting that in many regions the survey started only in 1968, we discard the first two years of observations to have simultaneous records for all the regions in the system. The spatial extent of the observational routes allows us to analyze species persistence at different spatial scales, say, A. We consider 20 different scales of analysis with linearly increasing values of the

square root of the sampled area starting from $A = 10,000$ km^2 to $A = 3.8 \cdot 10^6$ km^2. We also analyze the whole system, which corresponds to an area of $A = 7.8 \cdot 10^6$ km^2.

For every scale of analysis, we consider several overlapping square cells of area A inside the system (for details on the possible combinations of overlaps, see [9]). A three-dimensional presence–absence matrix P can be constructed: each element P_{stc} of the matrix is equal to 1 if species s is observed during year t in at least one of the observational routes composing cell c. On the contrary, $P_{stc} = 0$ if species s is not observed in any of the observational routes composing cell c during year t. For every scale of analysis we discard the cells that (1) do not have a continuous record for the whole period (41 years) or (2) have more than 5 percent of their area falling outside the system. For every cell and every species we measure persistence from presence–absence time series, defined as the length of a contiguous sequence of presences (1). For every scale of analysis we consider all the measured persistences regardless of the species they belong to and the cell where they were measured.

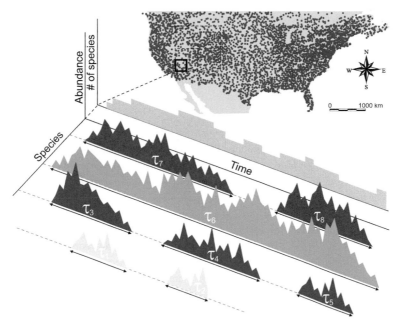

Figure 1.6 Species persistence times. Here, τ within a geographic region is defined as the time incurred between a species' emergence and its local extinction. Recurrent colonizations of a species define different lifetimes. The number of species in the ecosystem as a function of time (gray shaded area) crucially depends on species emergences and lifetimes. Here, two long-term datasets are analyzed, specifically about North American breeding birds [91] and herbaceous plants from Kansas prairies [92]. The inset shows the observational sites of the Breeding Bird Survey. By aggregating local information for a given geographic area, one may reconstruct species presence–absence time series that allow the estimation of empirical persistence time distributions. Figure after [9]

The herbaceous plant dataset [92] comprises a series of 51 quadrats of 1 m^2 from mixed Kansas grass prairies where all individual plants were mapped every year from 1932 to 1972. To meet the data quality standard required for our analysis as discussed above for the breeding bird data, we discard 10 quadrats and the first three years of observations. Owing to the limited number of observational plots in the herbaceous plant dataset, we limit our analysis to quadrat spatial scale $A = 1$ m^2. Analogously to the previous case, we reconstruct the matrix P from presence–absence data for every species, year, and quadrat.

When dealing with empirical survey data, the effect of the finiteness of the observational window on the measured species lifetimes must be properly taken into account (Figure 1.5a, after [9]). Given the probability density function of persistence times, the distribution of two additional variables must be measured from empirical data: (1) the lifetime τ' of species that emerge and go locally extinct within the observed time window ΔT_w and (2) the variable τ'' that comprises the lifetimes τ' and all the portions of species lifetimes that are partially seen

inside the observational time window but start and/or end outside (Figure 1.5a). The finiteness of the time window imposes a cutoff to $p_{\tau'}(t)$. On the contrary, $p_{\tau''}(t)$ has an atom of probability in $t = \Delta T_w$ corresponding to the fraction of species always present during the observational time. By matching analytical and observational distributions for $p_{\tau'}(t)$ and $p_{\tau''}(t)$, it is possible to infer the lifetime distribution $p_\tau(t)$ [9].

Remarkably, the lifetimes of breeding birds at different spatial scales of analysis and of herbaceous plants prove to be best fitted by a power law distribution with an exponent $\alpha = 1.83 \pm 0.02$ and $\alpha = 1.78 \pm 0.08$, respectively (Figure 1.5b,c). Both are significantly different from the existing baseline models. The scaling exponent and the diversification rate for the herbaceous plant lifetime distributions have been determined with a simultaneous nonlinear fit of observational and analytical $p_{\tau'}(t)$ and $p_{\tau''}(t)$. Confidence intervals are equal to the standard error of the fit. For breeding birds, we repeat the nonlinear fit for different spatial scales of analysis. The reported scaling exponent and the confidence interval have been obtained by averaging results across spatial scales. Details on the statistical procedures are in the original reference [9].

1.3.3 Scaling of Persistence Times and Species–Area Relations

The scaling exponent of the lifetime distribution is linked to ecosystem diversity. In fact, in the above framework, species emerge as a point Poisson process with rate $\lambda = \nu N$ and last for a lifetime τ. The mean number of species S in the system at a given time is therefore $S = \lambda \langle \tau \rangle$ [93], where $\langle \tau \rangle$ is the mean lifetime. Therefore, the smaller exponents found, say, for networked environments with respect to two-dimensional ones imply longer mean lifetimes and, in turn, higher diversity. This echoes recent results suggesting a higher diversity of freshwater versus marine ray-finned fishes [94, 95].

The spatial extent of the breeding birds dataset and the tools developed for the data analysis allow us to study how the lifetime distribution depends on the spatial scale of analysis (Figure 1.7a). As expected, while the scaling exponent remains the same, the diversification rate ν decreases with the geographic area A and is found to closely follow a scaling relation of the type $\nu \propto A^{-\beta}$, with $\beta = 0.84 \pm 0.01$ (Figure 1.7b), for a wide range of areas. Interestingly, this scaling form of the cutoff timescale

$1/\nu$ can be related to the species–area relationship, which characterizes the increase in the observed number of species with increasing sample area. Assuming that the number of individuals scales isometrically with the sampled geographic area [80, 86], that is, $N \propto A$, and given that $\langle \tau \rangle = \int t p_\tau(t)dt \propto \nu^{\alpha-2}$ (see [9]), one gets

$$S = \lambda \langle \tau \rangle \propto A^{1-\beta(\alpha-1)} = A^z. \qquad (1.18)$$

The observational values $\beta = 0.84 \pm 0.01$ and $\alpha = 1.83 \pm 0.02$ give an exponent $z = 0.30 \pm 0.02$, which is close to the species–area relation measured directly on the data for the same range of areas ($z = 0.31 \pm 0.02$; Figure 1.7c). Conversely, one could have used the observed species–area exponent to infer the scaling properties of the diversification rate.

From a conservation perspective, a meaningful assessment of species' local extinction rates is of great value. In particular, the distribution of species residual persistence times τ_p (say, that shown in Figure 1.5a) is a powerful tool to quantify the persistence of the species assembly currently observed within a given geographic area. Moreover, to some extent, it may be used to predict the expected amount of future local extinctions regardless of its current abundance. Therefore, the developed theoretical and operational tools derived in [9] allow one to infer persistence-time distributions, in particular for relatively short observational windows. The empirical verification

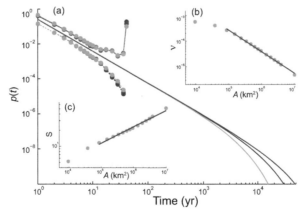

Figure 1.7 (a) Observational distributions $p_{\tau'}(t)$ and $p_{\tau''}(t)$ (interpolated solid circles) for the breeding bird dataset and corresponding fitted lifetime distributions $p_\tau(t) \propto t^{-\alpha}e^{-\nu t}$ (solid lines) for different scales of analysis: area $A = 8.5 \cdot 10^4$ km^2 (green), $A = 3.4 \cdot 10^5$ km^2 (blue), $A = 9.5 \cdot 10^5$ km^2 (red). Variable $\nu(A)$ provides the cutoff for the distribution, whose scaling exponent is unaffected by geographic area. Note that the position of the cutoff of $p_\tau(t)$ is inferred from the estimate of the atom of probability of $p_{\tau''}(t)$, which is more sensitive to the scale of analysis. (b) Scaling of the diversification rate ν with the geographic area $\nu \propto A^{-\beta}$, $\beta = 0.84 \pm 0.01$. (c) Empirical species–area relationship (SAR). The plot shows the mean number of species S found in moving squares of size A. We find $S \propto A^z$, $z = 0.31 \pm 0.02$. Slope and confidence interval have been obtained averaging 41 SARs, 1 per year of observation. Figure after [9]

serves well the major point, as one may derive key parameters simply from the observation of the atom of probability, that is, the number of species that are always present within a given geographic domain. The empirical verification that coarse graining highlights a clear scaling behavior for the decrease of the diversification ratio v with the domain A (over which local extinctions are evaluated) has broad implications for biogeography. In fact, such a decrease connects with species–area relations and thus to biodiversity [9].

Although these patterns cannot provide information about the behavior of a specific species or of a particular patch inside the ecosystem considered, they can effectively describe the overall dynamical evolution of the ecosystem diversity. The key connection with the central theme of this book is contained in Figure 1.4. Therein, the biogeographical characters of species persistence times are firmly linked to the structure of the spatial interaction networks defining the substrate for ecological interactions. In particular, the distribution of persistence times for systems operating on a river network is shown to hold a power law character across several orders of magnitude, inasmuch as the 2D lattice open to isotropic interactions – only with a much smaller scaling exponent (the slope of the curve in the log–log plot). This implies that long persistences are much favored. This has implications that add a new ingredient to a rich literature bearing major implications for the inventory of life on Earth.

1.4 Testing Directional Connectivity in the Laboratory

The abstract examples discussed in the previous sections of this chapter strongly suggest that directional dispersal has a major impact on the resulting biodiversity distributions. The examples shown there imply that the topology of the substrate for ecological interactions has a defining role for the distribution of species richness in space and time, regardless of the ecosystem's specific features and environmental drivers. Clearly, this suggestion has a fundamental importance in the way we look at river networks as ecological corridors. For instance, β-diversities (Section 1.2) computed separately for headwaters and confluences test the differences in species composition within the river network structure. Headwaters exhibit not only a higher variability in α-diversity, the local species richness (Section 1.2), but also higher β-diversities compared to confluences (as for field evidence, see [6, 96]). Therefore, differences in the loss of spatial correlation relative to lattice landscapes appear

even higher when only headwaters are considered in the comparison. Incidentally, these results reveal the crucial importance of headwaters as a source of biodiversity for the whole landscape. In natural systems, however, other local environmental factors may play a role in structuring ecosystems [19], for example, population turnover [97].

The neutral metacommunity approach [8] sheds light on the single effect of directional dispersal on biodiversity. Note that the patterns found in river network geometries are predicted to be even stronger in the presence of a downstream dispersal, which is typical for many passively transported riparian and aquatic species in river basins [7, 17]. Thus, the types of dispersal and disturbances used in these systems – abstract as they are here by design – are not specific to riverine environments but rather apply to a variety of heterogeneous and fragmented environments. The overarching theoretical result suggests that species constrained to disperse within dendritic corridors face increased spatial persistence and lower extinction risks. On the other hand, heterogeneous habitats sustain higher levels of biodiversity among local communities that can be altered by modifying the connectivity of the system, with broad implications for community ecology and conservation biology.

The above suggestions are far-reaching. In fact, species or populations whose ecological dynamics are constrained by directional dispersal would be inherently more predictable as the effects of other, uncontrollable heterogeneities would be less dominant. If true, spatially explicit metacommunity frameworks would be inherently more reliable precisely owing to such effects. However, field validations cannot prove or disprove such an ansatz, for a number of objective reasons (chiefly, the practical impossibility of replicating all ecological conditions in diverse topological substrates for the ecological interactions). Resorting to laboratory experimentation with well-established ecological microcosm landscapes [12, 13, 98, 99] was a logical step to explore the extent of the validity of the theoretical predictions.

1.4.1 Design of the Experiments

Biological communities of experiments can be arranged to occur in spatially structured habitats where connectivity directly affects dispersal and metacommunity processes. As the theoretical work suggests that dispersal constrained by the connectivity of specific habitat structures, such as dendrites like river networks, can explain observable features of biodiversity, experiments were

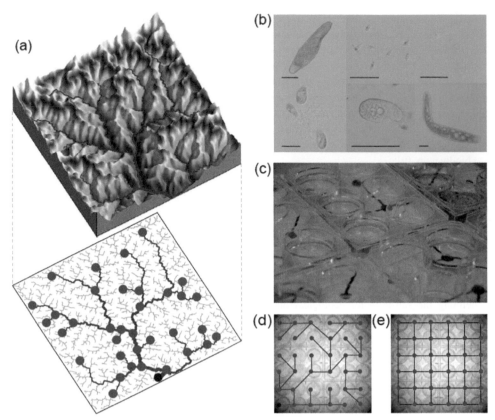

Figure 1.8 Design of the connectivity experiments carried out with well-established ecological microcosm landscapes. (a) The river network (RN) landscape (lower: red points label the position of local communities (LCs), and the black point is the outlet) derives from a coarse-grained optimal channel network (OCN) that reflects the structure of a river basin (upper). (b–e) The microcosm experiment involves 21 protozoan and rotifer species [14, 100]. (e) A subset of the species employed is shown to scale (for details, see SI Materials and Methods in [14, 100]) (scale bar = 100 μm). (c) Communities were grown in 36-well plates, where the dispersal protocol has been carried out rather accurately and with an appropriate number of replicas [12, 13, 99]. (d, e): Dispersal to neighboring communities followed the respective network structure: blue lines portray RN arrangements (d); black lines are for a 2D lattice with four nearest neighbors (e). Figure after [14]

designed to test experimentally the theoretical suggestions. Such experiments, specifically aimed at experimentally testing whether connectivity per se is capable of shaping diversity patterns in microcosm metacommunities at different levels [14, 100] (Figure 1.8), are described in this section.

For simplified landscapes, often described geometrically by linear or lattice structures (Figure 1.8), a variety of local environmental factors exist that create and maintain diversity among habitats [34, 49, 94]. Many highly diverse landscapes, however, exhibit hierarchical spatial structures that are shaped by geomorphological processes.

They are neither linear nor lattice-like, and therefore environmental substrates for ecological interactions shaped as trees may be appropriate to describe biodiversity of species living within fluvial dendritic ecosystems.

Local dispersal in isotropic lattice landscapes homogenizes local species richness and leads to pronounced spatial persistence (Figure 1.9). On the contrary, dispersal along dendritic landscapes leads to higher variability in local diversity and among-community composition, thereby confirming by replicated laboratory experimentation the theoretical prediction based on abstract models [14]. Although headwaters exhibit

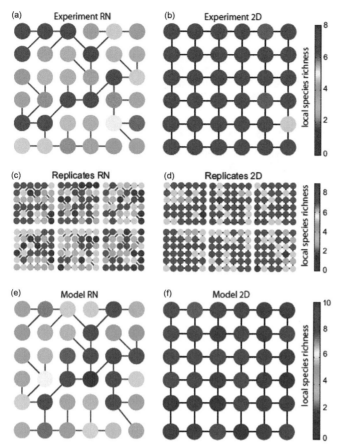

Figure 1.9 Experimental and theoretical local species richness in river network (RN) and lattice (2D) landscapes. (a, b) Mean local species richness (α-diversity, color coded; every dot represents a LC) for the microcosm experiment averaged over the six replicates. (c, d) Species richness for each of these replicates individually. (e, f) The stochastic model predicts similar mean α-diversity patterns (note different scales). The effect of the hierarchical dendritic architecture proves statistically significant – in fact, decisive [100].

relatively lower species richness, they are crucial for the maintenance of regional biodiversity. This result echoes prior theoretical evidence [6, 96]. Such results establish that spatially constrained dendritic connectivity is a key factor for community composition and population persistence [14]. Further laboratory work experimentally disentangled the effect of local habitat capacity (i.e., the patch size) and dendritic connectivity on biodiversity in aquatic microcosm metacommunities by suitably arranging patch sizes within river-like networks [100]. Overall, more connected communities that occupy a central position in the network exhibited higher species richness, irrespective of patch size arrangement. High regional evenness in community composition was found only in landscapes preserving geomorphological scaling properties of patch sizes (i.e., a patch volume proportional to the number of contributing nodes; see [100]. In these landscapes, some of the rarer species sustained regionally more abundant populations compared to landscapes with homogeneous patch

size or landscapes with spatially uncorrelated patch size [100].

Carrara et al. [100] experimentally singled out the interaction of dendritic connectivity and local habitat capacity (by suitably modulating patch size, i.e., proportionally to the contributing patches upstream of a node) on the diversity of microorganisms in dendritic metacommunities, which were mimicking network structure and patch connectivity of natural river networks. Specifically, the individual effects of connectivity and habitat capacity on protozoan and rotifer diversity were singled out by using three different configurations of patch sizes (Riverine, Random, and Homogeneous), connected following a river network topological and aggregation template (Figure 1.10) derived from OCNs (Box 1.3). An appropriate coarse graining procedure was enforced to reduce a complex construct to an equivalent 6 × 6-patch network, the size that made replicas feasible, yet preserving the characteristics of the original three-dimensional landscape [100]. In Riverine landscapes, local habitat

(a) "Riverine" landscapes

(b) "Random" landscapes

(c) "Homogeneous" landscapes

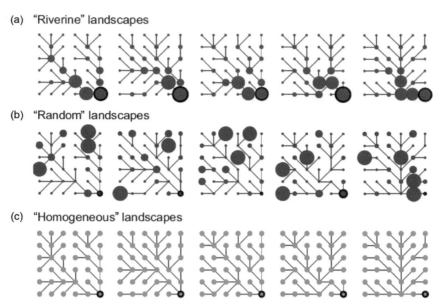

Figure 1.10 Spatial configuration of dendritic networks and corresponding patch sizes in the microcosm experiment. (a) Riverine landscapes (blue) preserved the observed scaling properties of real river basins. (b) Random landscapes (red) had the exact values of volumes as in the Riverine landscapes, randomly distributed across the networks. (c) In Homogeneous landscapes (green) the total volume of the whole metacommunity was equally distributed to each of 36 local communities. Patch size (size of the circle) is scaled to the actual medium volume. Five unique river-like (dendritic) networks were set up (columns: dispersal to neighboring communities followed the respective network structure, with a downstream bias in directionality toward the outlet community [black circled dot]). Figure after [100]

capacity correlates with position along the network and distance to the outlet (Figure 1.10a). Larger downstream communities thus consistently receive more immigrants from upstream communities. In the Homogeneous and Random landscapes (Figure 1.10b,c, respectively), local habitat capacity (i.e., the patch size) does not preserve any geomorphological hierarchy embedded in the scaling of total contributing area at any point in the network scaling as observed in natural river systems [63, 101]. We relate the habitat capacity of a metacommunity (MC) to physical properties that affect persistence of species. The largest organism to be sustained by an ecosystem is known to depend on habitat size (see, e.g., [102]) or habitat capacity [8]. Therefore, habitat capacity conceptualizes our ability to rank different landscapes in terms of their capacity to support viable populations. We implemented three different treatments of patch size configurations in each of the five OCN landscapes (Figure 1.10). (1) the Riverine landscape, in which the medium volume V_i (i.e., the patch size at the ith site) of the local community i (LC) preserves the scaling law observed in real river systems, $V_i \propto A_i^{1/2}$, akin to proper basin scaling properties [63, 103] – area A_i is the analog to the drainage area of the LC i, defined as the sum over all the volumes

V_j draining in that particular point i, that is, $\sum_{j \in \gamma(i)} V_j$ (where $\gamma(i)$ is every treelike path rooted in i), that is, $\forall j \to i$; (2) the Random landscape, in which the exact values of patch volumes V_i of the Riverine landscape were randomly scattered across the network (Figure 1.10b); and (3) the Homogeneous landscape, where the total volume of the whole MC is equal to the other two treatments, but where each LC had a constant average value $\bar{V} = 3.6$ mL. It should be noted that values $\bar{v}$ had to be binned, based on their original drainage areas, in four size categories (2, 3.5, 6, and 12 mL, respectively [100]). To test species coexistence in isolation, 72 communities were set up as subject to isolation treatment, in which the patch sizes were equal to the first two replicates of the Riverine configuration of the main experiment, but without dispersal. Environmental conditions, dictated in the well plates by the ratio of surface area to volume, were also changing between communities with different patch sizes, favoring different sets of species at a time (see [100], figures B1 and B2). The key question is, therefore, how does the system react across spatiotemporal scales without any disturbance/dispersal events? The species' ability to coexist in an isolation treatment has been tested under the same environmental conditions

[14, 100]. It was hypothesized therein that under stress (space saturation and/or reduced availability of bacteria), larger protozoans, such as *Blepharisma* and *Spirostomum* spp., could prey on smaller protozoans, such as *Chilomonas, Tetrahymena*, and *Colpidium* spp. (see table S1 of [14] for specifications of experimental species' traits). The latter appeared to be clearly inferior competitors. Note that predation could happen even at low protist densities and high bacterial densities.

The main results [100], rather relevant to the tenet of the present work, are briefly summarized here. Locally (i.e., at the patch scale), variation in patch size configuration in dendritic networks alters significantly community composition. Theory wants that species that are better competitors in a particular environmental condition eventually spread along the system, impeding other species' growth and thus exposing them to higher extinction levels [104]. Accordingly, in the experiment of Carrara et al. [100], a predominance was observed of the best competitor of the species pool in Homogeneous landscapes, whereas populations of larger species were reduced compared to heterogeneous configurations. It was found that habitat heterogeneity in Riverine and Random landscapes (Figure 1.10a,b) promoted both local species evenness and persistence of β-diversity compared to Homogeneous landscapes. This gave a causal, experimental proof of principles on how homogenization of habitat size along river networks can affect diversity. It has been noted [100] that at regional spatial scales the consequences of the spatial configuration of patch sizes on community composition in terms of degree of dominance and species turnover are significant – but far from obvious. Only in Riverine landscapes did LC

evenness increase consistently with increasing patch size and decreasing distance to the outlet [100]. This highlights the structuring power of hierarchy – a tenet of this book. Spatial environmental autocorrelation results in higher levels of regional evenness in Riverine ecosystems by increasing the population size of some of the rarer species and at the same time decreasing the total biomass of the more abundant species. Possibly, rare species with lower growth rates may be able to track their specific niche requirements more efficiently in Riverine landscapes, as observed elsewhere as well [104]. Higher population densities compared to the Random landscapes are assembled where the spatial autocorrelation was disentangled. Thus, alterations of river-like landscapes have strong effects on metacommunity dynamics and impact important regional diversity and evenness properties. Empirical and theoretical studies in community ecology and conservation biology must therefore consider this factor when species abundances are sought (or, more generally, when the ecology of communities hosted in stream ecosystems is studied).

The bulk of the experimental replicas suggested that altering the natural link between dendritic connectivity and patch size strongly affects community composition and population persistence at multiple scales, precisely as predicted by the neutral model and by every (neutral and nonneutral) metacommunity framework applied to the same topologically diverse matrices. We note that the same applies to broader classes of networked environments [96], including those artificially created by human or host mobility networks [105] that are so relevant to the spread of disease epidemics – to which we return later in this book.

Box 1.5 Aquatic Experimental Metacommunities

In the key experiments described above [14, 100], each local community (LC) within a metacommunity (MC) was initialized with nine protozoan species, one rotifer species, and a set of common freshwater bacteria as a food resource. The nine protozoan species were *Blepharisma* sp., *Chilomonas* sp., *Colpidium* sp., *Euglena gracilis, Euplotes aediculatus, Paramecium aurelia, Paramecium bursaria, Spirostomum* sp. and *Tetrahymena* sp., and the rotifer was *Cephalodella* sp.) (Carolina Biological Supply), whereas all other species were originally isolated from a natural pond and used for other studies [14]. All species are bacterivores, whereas some of them (*E. gracilis, E. aediculatus, P. bursaria*) can also photosynthesize. Others (*Blepharisma* sp., *Euplotes aediculatus, Spirostomum* sp.) not only feed on bacteria but also predate on smaller flagellates. Twenty-four hours before inoculation with protozoans and rotifer, three species of bacteria (*Bacillus cereus, Bacillus subtilis, Serratia marcescens*) were added to each community. LCs were located in 10-mL multiwell

Box 1.5 *Continued*

culture plates containing a solution of sterilized local spring water, 1.6 g L^{-1} of soil, and 0.45 g L^{-1} of Protozoan Pellets (Carolina Biological Supply). Protozoan Pellets and soil provide nutrients for bacteria, which are consumed by protozoans. Carrara et al. [14, 100] conducted the experiments in a climatized room at $21\,^\circ$C under constant fluorescent light in the Laboratory of Ecohydrology at EPFL. On day 0, 100 individuals of each species were added, except for *E. gracilis* (500 individuals) and *Spirostomum* (40 individuals), which naturally occur, respectively, at higher and lower densities. The species' intrinsic growth rates and carrying capacities from a demographic logistic model (Chapter 2) were derived in pure cultures in controlled, identical conditions [14].

1.4.2 Experimental Studies on the Role of Directional Dispersal and Habitat Size

We conclude this section on one of the *leit-motive* of this book by noticing that the experiments described in this section nicely complement the theoretical, experimental, and comparative work described in Sections 1.1 and 1.2. In brief, dispersal constrained by specific habitat structures is a major determinant of observable diversity patterns. Two major aspects have been addressed by theoretical and experimental studies. First, landscape connectivity was generally considered independently of local environmental factors, such as habitat quality, patch size, environmental disturbances, and intra- and interspecific competition. While there are indeed cases for which this simplification is appropriate, such as forests [34], island archipelagos [80], or natural ponds [99, 106, 107], it does not represent many natural landscapes, such as fluvial ecosystems [63], where local properties of the habitat and connectivity are intrinsically linked [96, 108]. It is customary for theoretical studies to adopt constant dispersal rates, symmetric kernels (but see [40]), and generally simplified landscape attributes (see, e.g., [10, 99, 109–112]). Many studies in stream ecology, in the wake of the highly influential river continuum concept [113], have considered linear conceptual models to analyze drainage basins. Such simplified linear environmental matrices, however, do not fully capture biodiversity patterns within dendritic ecosystems, as we have shown in this chapter.

In riverine ecosystems, landscape-forming discharges are related to total contributing drainage area, a byproduct of spatial aggregation, depth, and width of the active river cross section [63, 96, 101]. Thus the river network not only provides suitable ecological corridors for individuals to disperse but also regulates the availability of microhabitats that species may eventually exploit [104, 114]. Habitat capacity (i.e., river width/depth, reflecting patch size in experimental mesocosms arranged to reproduce riverine topologies and connectivity) scales with contributing area, dispersal is often biased downstream, and the distribution and intensity of disturbances are intrinsically linked with the position along the network through abrupt changes at confluences (see, e.g., [96, 97]). Spatial correlations therefore emerge between local properties and regional network descriptors in dendritic environments, where the hierarchical spatial organization of environmental heterogeneity is a fundamental driver of local species richness and community composition. Riverine ecosystems, which are among the most threatened ecosystems on Earth, are thus a prominent natural system for which disentangling the effect of local environmental conditions from that of connectivity of the landscape on overall diversity is key.

We measured species' persistence, density, and diversity patterns in terms of α-, β-, and γ-diversities, that is, related to local species richness, among-community dissimilarity, and regional species richness. We also considered community evenness in the above landscape configurations. Aquatic microcosms were employed, as in several other studies (see, e.g., [14–16, 114]), to offer a useful bridge between theoretical modes and comparative field studies, to test for general macroecological principles like the ones pursued here. Findings from such laboratory experiments, even if not directly comparable to natural systems, cast light on important underlying mechanisms that steer metacommunity dynamics in river systems, as we have noted consistently throughout this chapter.

It is also revealing that an experimental proof is provided to the fact that β-diversities increase in Riverine landscapes with increasing topological distance along the network. Such a pattern is observed in comparative studies on riverine diversity [8, 19], whereas no spatial correlation of LC similarity was found in Random and Homogeneous landscapes, suggesting that a combination of patch size and network position is needed to reproduce this pattern. Note that dispersal limitation alone cannot reproduce such an effect on community differentiation [19].

Key to our line of reasoning is that randomizing the patch size, that is, altering the hierarchical organization of the river network ecosystem epitomized by the patch size, is tantamount to altering the hierarchical riverine structure. This is an analog of fragmentation of landscapes, also in view of the experimental effects observed on α-diversity ([100] and figures 3a,b,d therein). At the same time, one may note that such an alteration opens up more diversified spatiotemporal niches, ultimately producing more distinct species compositions. Interestingly, β-diversity proved dependent on centrality in the Homogeneous landscapes with fixed habitat capacity. This strongly indicated that dendritic connectivity per se shapes both α- and β-diversity, as suggested by Carrara et al. [14]. Community composition turnover along centrality gradients was maximized for Riverine landscapes, and hierarchical patch size distribution enhanced the turnover provided by dendritic connectivity itself [100]. The experiments described here disentangled the effects of the intrinsic link of network position and habitat patch size on diversity and community structure (technically, its evenness). Evidence is thus provided of the interaction between species traits and population responses to spatiotemporal gradients of local environmental conditions in spatially structured habitats. For example, fast population growth, allowing rapid population responses to a more unpredictable environment, might favor species with higher intrinsic growth rates, and vice versa [100].

A note of caution relates to the fact that the above experiments, as in similar model systems (e.g., [114]), are conceptualized versions of natural ecosystems and do not allow direct extrapolation of our results to natural rivers. However, they enhance our understanding of complex systems in nature, where multiple processes are interacting on different scales and verify the theoretical suggestions derived by rather abstract – and therefore general – models. Dispersal rate, dispersal mode, and the strength of directionality are important factors in determining community patterns in theoretical models and natural communities. Different dispersal abilities, controlled by body size and dispersal mode, determine a change in the response of community similarity to environmental variation and geographic distance. This suggests that the relative importance of the two structuring forces may depend on the group of organisms and the spatial scale. However, with a major shakeup to the generality of our tenet, the river network itself provides at the same time the primary habitat for the species and suitable ecological corridors for individuals to disperse, resulting in a close match between the physical and the ecological scales. This correspondence is recognized as important for the ecosystems' resilience at different levels of ecological complexity. For a variety of species living in natural fluvial systems, out-of-network movements are likely to occur, leading to intercatchment dispersal [18, 19, 115]. In macroinvertebrates, active dispersers with a terrestrial stage should track environmental heterogeneity better than passive dispersers with only an aquatic stage [115], thus supporting the generality of our tenet. Moreover, the strength of directionality in river systems might be much stronger for passive dispersers, such as bacteria and protists, compared to macroinvertebrates, amphibians, or fish. We show in Chapter 2, for example, that a neutral metacommunity model endowed with a symmetric dispersal kernel suitably described fish biodiversity patterns in the entire Mississippi–Missouri river system [8]. Thus, when comparing or extrapolating the above results to natural systems, one needs to carefully assume that taxon-specific aspects of dispersal are fulfilled and patterns and processes may not scale directly across all species and landscapes sizes. However, the hints that emerge from the context shown here are rather compelling, in the authors' view. As the experiment described above adopted a diffusive downstream-biased dispersal between isolated habitat patches and at discrete time intervals [100], different dispersal strategies could not naturally arise. Competition–colonization trade-offs have been documented in protist studies that were adopting similar species ([116]; but see [114]). Such mechanisms of species coexistence, together with storage effects, may interact with the spatial structure to shape diversity and ecosystem productivity, as tested for example in bacterial metacommunities [117]. As the system was continuously perturbed away from stationarity by dispersal and emigration mortality, transient states were being observed. The duration of transient dynamics in the experimental mesocosm networks may depend nontrivially on the different patch sizes, which are sustaining different population sizes. System relaxation time to equilibrium could be investigated by implementing a metacommunity model with salient features of our experiment, but this has not yet been done.

A final note concerns the importance of how ecological selection is modulated across space. In fact, while uniform ecological selection across the network leads to higher diversity in downstream confluences, this pattern can be inverted by perturbations when population turnover (i.e., selective local mortality) is higher upstream than downstream [97]. Higher turnover in small headwater patches was shown there (theoretically and experimentally) to be capable of slowing down ecological selection, increasing local diversity in comparison to large downstream confluences. Such results provide a significant step forward in our understanding of the distribution of diversity in river-like landscapes and reinforce the importance of experimentally validated theoretical suggestions.

Because dispersal is constrained by the network pathway, the river network may become a trap for species when the dendritic system is exposed to habitat fragmentation and patch size alterations. Protecting highly connected communities could help to avoid extinctions of species with low reproductive rates. These species are prone to suffer more from environmental disturbances and require larger and well-connected habitats to persist. By preserving the natural hierarchy of spatiotemporal heterogeneity along river networks, fast-growing species and weak competitors alike are better able to persist. The results exposed in this experimental section not only causally demonstrate general ecological principles but also give insights for developing theoretical metacommunity models in dendritic environments and for future empirical studies focusing on riverine ecosystems. The rest of this book is centered on further illustrating the broad implications of such results.

1.5 Invasion Wave Fronts along Fractal Networks and Population Dynamics

We address here another key proof of the defining role of river network structure on the functioning of ecological corridors. Following a pathbreaking study of a quantitative model of migration [118], we consider the US colonization in the nineteenth century in light of the fundamental Fisher–Kolmogorov studies of reactive–diffusive transport [119, 120] extended to fractal river networks. Of interest for the general scope of this book, we consider mild generalizations of the original approach to include an embedded flow direction that biases transport (Chapter 3). We therefore explore here the properties of (possibly biased) reaction–dispersal models, in which reaction rates are described by a logistic equation as

commonplace in population dynamics. The relevance of the topic is somewhat obvious for the context of this introductory chapter, centered as it is on the role of hydrologic controls in biological invasion processes occurring within river basins (of species, humans included; of populations of invasive species; or else of propagules or infective agents, whereas the specifics of each depend on the choice of reaction and transport attributes).

A merit of this approach is the possibility to derive exact solutions for the relevant state variables, which are obtained along with general numerical solutions (a safe engineering tool to control the role of specific assumptions), which are applied to fractal constructs like the Peano basin, OCNs, and real rivers extracted from digital terrain maps (see [63], Section 1.2 and Appendix 6.2). In the next section, we show that the geometrical constraints imposed by the fractal networks imply strong corrections on the speed of traveling fronts. Such corrections may be enhanced, or else smoothed, by hydrologic bias. Moreover, we show from applications to real river networks that the chief morphological parameters affecting the front speed are those characterizing the node-to-node distances measured along the network structure. The spatial density and number of reactive sites thus prove to be vital hydrologic controls on biological invasions. We therefore argue that existing solutions, currently tied to the validity of the logistic growth model, might be relevant to the general study of species spreading along ecological corridors defined by the river network structure.

1.5.1 River Networks and Ecological Corridors: Migration Fronts, Hydrochory, Transport on Fractals

In this section, we explore how the topological structure of a river network – the substrate for ecological interactions in the experiments and theoretical studies addressed in the previous sections of this introductory chapter – affects biological invasions. While we look in detail to a specific biological process (the spreading of the Zebra mussel across the Mississippi–Missouri river system) in Chapter 3, specifically devoted to population dynamics interacting with a river network structure, here we present theoretical results that suggest a definite role of the (fractal) nature of river networks themselves. This result is particularly suggestive. In fact, it shows another way in which river networks have affected metahistory by decisively influencing the spread of humans. Therein the ecological corridors are meant to be the pathways the backbone (the mainstream) of rivers, and often settled near them to exploit water resources. It was thus quite interestingly argued in a quantitative manner that landscape heterogeneities must have played an essential role in the process of migration [22, 122]. The example chosen is one of the best-known modern range expansions: the colonization across the western United States during the nineteenth century [118]. A brief context: by 1790, the North American population of European origin was concentrated in the Atlantic region, but over the following decades, intense internal migrations led to displacement of the established population progressively westward. According to US Congress data and to several atlases, the average expansion rate for this transition between 1790 and 1910 was approximately 13.5 ± 0.8 km/yr [123]. A defining feature of the US transition westward was that settlers did not occupy all of the territory, as homogeneous models would inevitably assume, but rather followed the course of the greatest rivers and lakes [118, 123, 124] and settled near them to make use of their resources, energy, and transportation ease. Therefore, landscape heterogeneities, subsumed by the universal characters shown by fluvial networks regardless of scale, geology, exposed lithology, vegetation, or climate [63], have played an essential role in the process of migration. Such a condition is akin to the general case of dispersion of biological species along

rivers and streams [120], suggesting the ecological interest of the study for transport processes along directions defined spatially and topologically by river reaches as well as other natural or artificial hydrologic corridors.

One interesting byproduct of the analysis of migration fronts is the important role attributed to the structure of the network acting as the substrate for wave propagation. This indeed calls for specific structural models to be invoked. One must observe that mathematical models of natural forms as fractals involve nontrivial assumptions [63], in particular concerning the independence of results from the seeding point chosen for spreading material and species along the network where reaction and diffusion occur. This is seen as a corollary of the type of planar self-similarity shown by trees. The type of self-similarity observed for trees needs also proper finite-size corrections because upper and lower cutoffs in the aggregation structure reflect, respectively, the drainage density defining where channels begin and the loss of statistical significance of areas close to the overall basin size [64].

We first explore the dynamics of traveling waves that pertain to specific models of isotropic diffusion coupled to population dynamics (the Kolmogorov–Fisher model; Box 1.6). Variants of the basic theme, pertaining to extensions capable of handling fractal networks and biased transport, are given later (Box 4.1).

Box 1.6 Diffusion–Reaction Models and the Range Expansion Speed

The simplest way to describe the movement of organisms in space is to assume that it is random. If in addition one assumes that the population size N is large enough and makes some specific further hypotheses (which are better defined in Chapter 3), one can get the so-called equation of transport and diffusion, which is the simplest model for the dispersal of a population of organisms in space. We shall assume initially that space (indicated with x) is one-dimensional. A simple paradigm of movement is that individuals move right with a certain probability p and left with probability $1 - p$. Introduce the density (or concentration) of organisms in each position x at time t, namely, $\rho(x,t)\Delta x =$ number of organisms that at time t are located between $x - \Delta x/2$ and $x + \Delta x/2$. We can then write a balance equation in which individuals in a certain position at a subsequent time are the sum of a fraction p of individuals coming from the left plus a fraction $1 - p$ of individuals coming from the right. Under some specific assumptions that we better explain in Chapter 3, by letting Δx tend to zero in a suitable way, the balance equation results in the following partial differential equation:

$$\frac{\partial \rho}{\partial t} = -v\frac{\partial \rho}{\partial x} + D\frac{\partial^2 \rho}{\partial x^2}, \tag{1.19}$$

which is the advection (or transport or drift) and diffusion equation. If v is zero, one gets the equation of pure diffusion.

Box 1.6 *Continued*

The parameter D [L^2/T] is termed the diffusion coefficient. Note also that the transport term is proportional to $\frac{\partial \rho}{\partial x}$, that is, the concentration gradient. The transport term in Equation (1.19) has just a translation effect on the solution of the same equation. In other words, the solutions of Equation (1.19) can be obtained from solutions of the pure diffusion equation

$$\frac{\partial \rho}{\partial t} = D \frac{\partial^2 \rho}{\partial x^2} \tag{1.20}$$

provided that one replaces the space coordinate x with a moving reference frame coordinate, say, $x + vt$. For this reason, we will from now on focus exclusively on Equation (1.20). A solution of Equation (1.20) is completely determined if one specifies suitable initial conditions and suitable conditions at the space boundary. Suppose that a large enough number N of animals is released at a point in space (which we conventionally indicate with $x = 0$) and that the organisms can disperse without any barrier to their diffusion. One can easily verify by direct substitution into Equation (1.20) that the solution of the problem subject to an initial Dirac delta is nothing but a Gaussian distribution with respect to space, characterized by a time-increasing variance; more precisely, $\rho(x,t) = N p(x,t)$, where $p(x,t)$ represents the fraction of organisms per unit length that at time t are at distance x from the release site and is given by

$$p(x,t) = \frac{1}{\sqrt{4\pi Dt}} \exp\left(-\frac{1}{2}\frac{x^2}{2Dt}\right). \tag{1.21}$$

As time varies, the mean value – which is also the mode and the median – of $p(x,t)$ is always null because, without the transport term, organisms spread evenly to the right and the left. Instead, the variance grows linearly with time, and hence the concentration profile of individuals becomes flatter and flatter (see Figure 1.11). In particular, the standard deviation is given by $\sigma_x = \sqrt{2Dt}$ and thus increases with the square root of time. For $t = 0$ the variance vanishes, so the N organisms are all concentrated in the origin. Therefore the solution meets the initial conditions.

The diffusion equation is easily generalized to more than one spatial dimension. In particular, when organisms disperse in a plane, described by the spatial coordinates x and y, and the diffusion coefficient is the same in all directions (*isotropic diffusion*), the diffusion equation is

$$\frac{\partial \rho}{\partial t} = D \left(\frac{\partial^2 \rho}{\partial x^2} + \frac{\partial^2 \rho}{\partial y^2} \right). \tag{1.22}$$

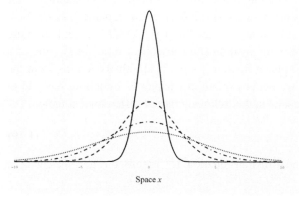

Figure 1.11 Time behavior of the solution to the diffusion problem when an initial number of organisms is released at $x = 0$. The graph shows various snapshots of the solution at different instants.

Space x

Box 1.6 *Continued*

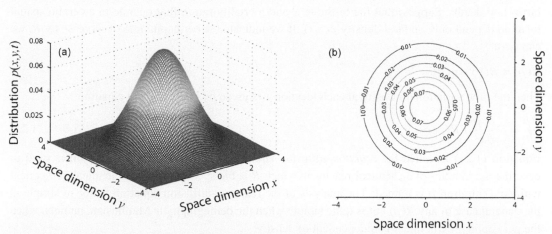

Figure 1.12 Solution to the diffusion problem in the plane when an initial number of organisms is released at the point of coordinates $x = 0$, $y = 0$. (a) Shape of $\rho(x, y, t)$ with $D = 2$ and $t = 0.5$. (b) Corresponding contour lines.

The solution to the problem of the release of N organisms in a point of a plane is just a bivariate normal distribution. More precisely, if we introduce the fraction of individuals per unit area $p(x, y, t)$, the solution of Equation (1.22) is given by $\rho(x, y, t) = Np(x, y, t)$ with

$$p(x, y, t) = \frac{1}{4\pi Dt} \exp\left(-\frac{1}{2}\frac{x^2 + y^2}{2Dt}\right).$$

At each time instant t, the fraction $p(x, y, t)$ has precisely the shape of a symmetrical bell (see Figure 1.12a). Contour lines are therefore circles (Figure 1.12b).

It is possible to prove (see [31]) that the fraction f_R of the population that lies outside the contour line of radius R at time t is given by

$$f_R = \exp\left(-\frac{R^2}{4Dt}\right).$$

Therefore, the radius of the contour line that contains, for example, 99 percent of the population at a given instant t satisfies the equation

$$0.01 = \exp\left(-\frac{R^2}{4Dt}\right)$$

and thus is given by

$$R = 2\sqrt{\ln 100}\sqrt{Dt} = 4.292\sqrt{Dt}.$$

It increases with the root of time, while the area that contains a certain fraction of the population (πR^2) grows linearly with time.

From the ecological viewpoint, things get much more interesting when, in addition to movement, we introduce the population demography. In other words, we consider that population density at

Box 1.6 *Continued*

each spatial location changes over time not only because of movement but also because of occurring births and deaths. Suppose that the birthrate β and mortality rate μ that operate in a certain spatial location depend only on local density $\rho(x,t)$. If we indicate the per capita rate of increase by R, we can write

$R(\rho) = \beta(\rho) - \mu(\rho)$.

Then the time variation of density at each location x is given by the following equation:

$$\frac{\partial \rho}{\partial t} = D\frac{\partial^2 \rho}{\partial x^2} + R(\rho)\rho. \tag{1.23}$$

Equation (1.23) is called the *reaction–diffusion equation* because this same equation is used to describe the kinetics of a chemical reactor in which R is the speed at which a chemical component with concentration ρ is formed. The analysis of the reaction–diffusion equation is not so simple in the general case of any $R(\rho)$ but is quite simple when the demography is Malthusian, namely, when the per capita growth rate R is independent of density:

$R(\rho) = r$ constant.

In fact, Equation (1.23) reduces to

$$\frac{\partial \rho}{\partial t} = D\frac{\partial^2 \rho}{\partial x^2} + r\rho$$

and can be easily solved via a change of variables. By introducing the new variable

$z(x,t) = \exp(-rt)\rho(x,t)$,

it is easy to derive that

$$\frac{\partial z}{\partial t} = -r \exp(-rt)\,\rho(x,t) + \exp(-rt)\frac{\partial \rho}{\partial t} = -rz + \exp(-rt)\left[D\frac{\partial^2 \rho}{\partial x^2} + r\rho\right] = D\frac{\partial^2 z}{\partial x^2}.$$

Therefore, z satisfies a pure diffusion equation whose solutions we have already learned how to derive.

Suppose that we introduce a new species or reintroduce one that was once present in the environment by releasing a number N_0 of individuals at time $t = 0$ at location $x = 0$. These organisms will disperse without barriers in space, will reproduce, and will die. How does the organisms' density vary in locations far away from the location of release? If $r > 0$, it is interesting to wonder whether the density at every location is going to increase or whether dispersal prevents the local growth of the population. From Equation (1.21) we know that

$$z(x,t) = N_0\frac{1}{\sqrt{4\pi Dt}}\exp\left(-\frac{1}{2}\frac{x^2}{2Dt}\right),$$

and thus we easily derive that

$$\rho(x,t) = N_0\frac{1}{\sqrt{4\pi Dt}}\exp\left(rt - \frac{1}{2}\frac{x^2}{2Dt}\right). \tag{1.24}$$

Figure 1.13 shows the evolution of density in successive time instants. At the release point $x = 0$, the density initially decreases because diffusion is prevailing, but in the long term, population growth

Box 1.6 *Continued*

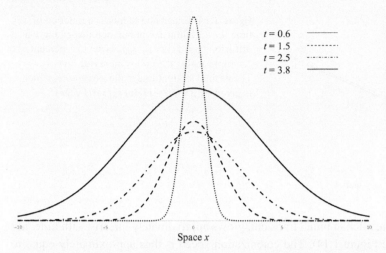

$t = 0.6$ ....
$t = 1.5$ ----------
$t = 2.5$ --·--·--
$t = 3.8$ _____

Space x

Figure 1.13 Time evolution of the solution to the problem of Malthusian growth and diffusion in an unbounded habitat. The figure shows various snapshots of the solution in successive instants.

prevails because the term $\exp(rt)$ has the most influence on the population evolution for large t. In spatial sites that were initially unpopulated, density is always growing: the locations are reached via diffusion, and then population keeps growing because of demography. It is obvious that the total number of organisms grows exponentially because

$$N(t) = \int_{-\infty}^{+\infty} \rho(x,t)dx = \exp(rt)\int_{-\infty}^{+\infty} z(x,t)dx = N_0 \exp(rt).$$

If we consider the problem from a more realistic viewpoint, that is, the organisms are released and can move in a two-dimensional environment, the solution is given simply by

$$\rho(x,y,t) = N_0 \frac{1}{4\pi Dt} \exp\left(rt - \frac{1}{2}\frac{x^2+y^2}{2Dt}\right).$$

From this formula we can gather how the population colonizes new space. It is possible to prove [31] that the number N_{out} of individuals in the population that at time t are outside a contour line of radius R_t is given by

$$N_{out} = N_0 \exp\left(rt - \frac{R_t^2}{4Dt}\right).$$

It is reasonable to assume that the presence of individuals is no longer detectable when N_{out} is below a certain threshold fraction f_{min} (e.g., 1 percent) of the number N_0 of initially released organisms. Therefore the radius within which the entire population is in practice contained satisfies the equation

$$f_{min} = \exp\left(rt - \frac{1}{2}\frac{R_t^2}{2Dt}\right)$$

and thus

$$R_t^2 = 4Drt^2 - 4Dt\ln f_{min}. \tag{1.25}$$

With the passing of time the first term of Equation (1.25) becomes much larger than the second; hence we can conclude that, if we include Malthusian growth in addition to diffusion, the radius

Box 1.6 *Continued*

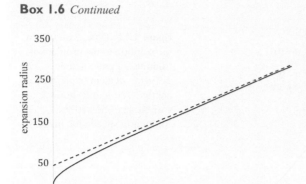

Figure 1.14 Expansion radius as a function of time for a Malthusian population dispersing via diffusion (solid curve); f_{min} is set to 1 percent. Asymptotically, the radius increases linearly. The dashed straight line is the asymptote, which is given by $2\sqrt{Dr}t - [D\ln(f_{min})]/(\sqrt{Dr})$.

marking population expansion, after an initial transient, grows approximately linearly with time, not with the root of the time (see Figure 1.14). The colonization speed is thus approximately equal to $2\sqrt{Dr}$. Equivalently, we can say that the area that contains virtually all the population increases with the square of time.

The Malthusian demographic model, however, is not very realistic. In fact, if the intrinsic rate of population increase r is positive and we wait for a sufficiently long time, the organisms' density at any point can become exponentially large. Instead, we know that intraspecific competition acts to limit the growth rate for high densities and even makes it vanish at the carrying capacity.

The mathematical treatment of the reaction–diffusion equation in the case of non-Malthusian demographics is a bit complicated and is not illustrated here (see Chapter 3 for details). We just report the main result, which, at any rate, is quite intuitive in light of what we learned for the Malthusian case. The fundamental assumption that we make is that the growth rate $R(\rho)$ is a decreasing function of density ρ, thus excluding the case of the Allee effect, which implies more complicated phenomena. For example, demographics may be logistic,

$$R(\rho) = r\left(1 - \frac{\rho}{K}\right),$$

with K being the carrying capacity. This is actually the problem originally studied by Kolmogorov [119] and Fisher [125], who derived the speed of traveling waves (solitons in the language of fluid dynamics, indicating waveforms propagating undeformed) in the logistic reaction–diffusion equation.

Let us analyze again what happens when a number of organisms is released in an environment without barriers. For the Malthusian demographics described in the previous section, the density at any point tends to increase exponentially with time. If there is logistic dependence, instead, density in the long run tends to the carrying capacity K at any point in space. Therefore, if organisms are released in a given location (with coordinate $x = 0$), the solution behavior is the one shown in Figure 1.15. Initially, diffusion prevails, but then the demographic growth leads to saturation in the zone close to the release site, and from that moment on, there occurs substantial propagation of two wave fronts: one to the right and one to the left. One can show that the front speed is $2\sqrt{DR(0)}$ (see Chapter 3 for details). Note, however, that this formula is the exact analogue of the asymptotic expansion speed for a Malthusian population: in fact, that radius was shown to grow as $2\sqrt{Dr}t$. We

Box 1.6 *Continued*

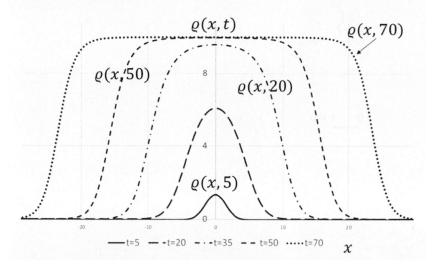

Figure 1.15 Behavior of the solution to the reaction–diffusion equation with density-dependent demography in an unbounded one-dimensional environment. The figure displays snapshots of $\rho(x,t)$ in successive time instants. The carrying capacity K at any site is 10.

should not wonder that the velocity of the front waves is influenced only by the value of $R(0)$, the demographic growth rate at very low density, because the density in close proximity to the wave front in the direction of propagation is indeed close to zero.

The 2006 model of Campos et al. [118] is our starting point. For general numerical calculations we shall adopt here (1) the topology and geometry of real rivers (e.g., Figure 1.16a) and (2) those of OCNs (Figure 1.16b; see Box 1.3). As specified therein, OCNs hold fractal characteristics that are obtained through a specific selection process from which one obtains a rich structure of optimal scaling forms that are known to closely conform to the scaling of real networks. Moreover, one may generate several statistically identical replicas of networks grown in the same domain. To derive exact results, we resort (as is usually the case in this context [118, 126, 127]) to Peano's directed network (Figure 1.16c), which is a deterministic fractal [121, 128] whose main topological and scaling features have been determined analytically.

Our starting point is an analysis concerning a reaction random walk (RRW) process through a Peano construct and OCNs. It is based on the following model. A particle, at an arbitrary node of the network, jumps, after a waiting time τ, to one of its z nearest-neighboring nodes with probability $1/z$. During the waiting time τ, the particles "react" by being assigned a change in density following the logistic equation. The determination of the wave front speed that this process develops along a network path

[129–131] is the starting point for our extensions. The network constructs for which our analsysis is carried out are shown in Figures 1.16 and 1.17.

Figure 1.18 illustrates the main result of [118]. It shows that the isotropic 2D diffusion–reaction front (Fisher–Kolmogorov model; see Boxes 1.6 and 3.5) propagates much faster than the wave forced to choose a treelike pathway whatever the value of the reaction rate. This proves that geometrical constraints imposed by a fractal network imply strong corrections on the speed of the fronts. It should be noted that it is not surprising that Peano and OC networks lead to similar results, because the speed of the front depends on topological features that are indeed quite similar for all the (rather different otherwise) networks shown in Figure 1.16 (see, e.g., [63, 64, 67, 69]). In fact, the wave speed is affected mostly by the main features of structure encountered by the front while propagating along the backbone of the network, chiefly the bifurcations. Hence topology, rather than the fine structure of the subpaths, dominates the process. Remarkably, it has been shown (see, e.g., [67]) that distinctive comparative features of different networks cannot be of topological nature but rather are of a metric character. As noted elsewhere in this

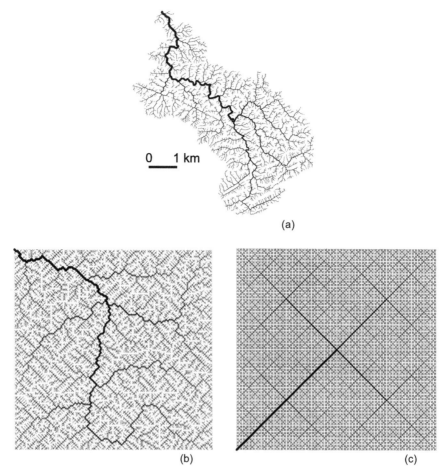

(a)

(b)   (c)

Figure 1.16 Examples of networks on which transport is considered: (a) a real river network, the Dry Tug Fork (CA), suitably extracted from digital terrain maps; (b) a single-outlet optimal channel network (OCN); (c) Peano's network. Figure after [7]

book (Appendix 6.2), the statistical inevitability of the recurrence of Horton numbers of bifurcation and length irons out important network features [68]. The theoretical *framework used for describing the above dynamical processes is based on continuous-time random walk (CTRW) processes [132] (briefly described in Box 1.7).*

Box 1.7 Continuous-Time Random Walk Model

In a dispersive process, focus is on particle density $\rho(x,t)$. The evolution of the density field $\rho(x,t)$ of being at position x at time t (assuming that $\rho(x,t) = 0 \; \forall t < 0$) is described by the relationship (see, e.g., [118])

$$\rho(x,t) = \int_0^t dt' \varphi(t') \int_{\mathbb{R}} dx' \Phi(x') \rho(x-x',t-t') + \int_0^t dt' \phi(t') f(\rho(x,t-t')), \qquad (1.26)$$

where $\Phi(x)$ is the length distribution for the jumps of a particle moving along the backbone of the network, to be specified in different manners, and $\phi(t)$ is the probability distribution of waiting (at least) a time t between two consecutive jumps, that is,

Box 1.7 *Continued*

$$\phi(t) = \int_t^\infty dt' \varphi(t'), \tag{1.27}$$

where we use an arbitrary imposed waiting time distribution $\varphi(t)$ at any node before jumping to an adjacent node either along the backbone or sideways to the branching structures.

The first term on the right-hand side of Equation (1.26) accounts for all dispersal processes, while the second term takes into account the reactive character of the process: the density of particles at position x grows according to the rate of increase $f(\rho(x,t))$. In the following we assume, as in the original context [118], a logistic reaction term of the type (see Box 1.6)

$$f(\rho) = a\rho(1-\rho) = R(\rho)\rho, \tag{1.28}$$

where $a = R(0)$ is the intrinsic population growth rate, assumed constant (for details about how Equation (1.26) could be derived directly from master equations, see, e.g., [129, 130]). The process described by Equation (1.26) yields a traveling wave (Figure 3.10) that connects the unstable state ($\rho = 0$) to the stable state ($\rho = 1$) and moves with constant celerity. If the initial condition $\rho(x,0)$ has compact support, for example,

$$\rho(x,0) = \begin{cases} 1 & x \le x_0 \\ 0 & x > x_0, \end{cases} \tag{1.29}$$

the front selects its minimal propagation speed v. Such speed (to be compared with that of Box 1.6) could be derived as

$$v = \min_s \frac{s}{p(s)}, \tag{1.30}$$

where $p(s)$ is the solution of the Hamilton–Jacobi equation

$$\frac{1}{\hat{\varphi}(s)} = \bar{\Phi}(p) + \frac{a}{s}\left(\frac{1}{\hat{\varphi}(s)} - 1\right). \tag{1.31}$$

In Equation (1.31), the function $\hat{\varphi}(s)$ is the Laplace transform of the waiting-time distribution $\varphi(t)$, while the function $\bar{\Phi}(p)$ is defined by the transformation

$$\bar{\Phi}(p) = \int_{-\infty}^{+\infty} e^{px} \Phi(x) dx; \tag{1.32}$$

thus, knowledge of the jump distribution $\Phi(x)$ and the waiting-time distribution $\varphi(t)$ allows the direct derivation of the speed v of the traveling wave.

Campos et al. [118] have considered a case of particular significance: a random walk through a Peano network (Figure 1.16c) for which exact solutions can be obtained. Such is the case in which a random walker at a certain site of the structure can jump, after a residence time τ, to each of its z_0 first neighbors with even probability $1/z_0$. A large body of literature exists on random walks on comblike structures [133] that address averaged properties relevant to macroscopic features. This is the case of traveling waves describing the movement of the walkers through any backbone identified by the seed position at $t = 0$, where each secondary branch emerging from a site of the backbone is assumed to introduce a waiting-time distribution of jumps from this site to the adjacent sites in the backbone. In the context of the study of propagation of fronts, waiting-time distributions are introduced by the branches, and their distribution is defined by the structure of the side structure encountered along the journey along a backbone.

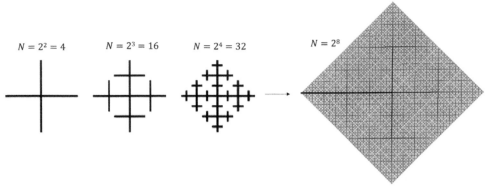

$N = 2^2 = 4$ $N = 2^3 = 16$ $N = 2^4 = 32$ $N = 2^8$

Figure 1.17 The iterative procedure needed to produce Peano's construct allows the computation of the topological properties defining the slowdown of the Fisher–Kolmogorov invasion speed. Here (from left to right) we show the initiation sensu Mandelbrot of the construction (N is the number of links, here 4) after the dissection of a unit segment that is the maximum length from source to outlet kept constant in all iterations. At the next stage of iteration, the new configuration is obtained by cutting each segment in half and by adding orthogonal protrusions of equal half-length, as shown here. We have enlarged the construct at the eighth stage of iteration to appreciate details of this deterministic fractal. Figure after [63, 128]

In this case, the branches in the Peano basin are such that when a walker reaches the backbone site corresponding to a first-order branch, it can jump to another site i of the backbone with probability $p_i = 2 \cdot 1/4 = 1/2$ or get into a secondary branch j with probability $p_j = 1/2$. Then, the waiting time is τ with probability $1/2$, while the waiting time is 3τ with probability $1/4$, and so on – only waiting times $\tau, 3\tau, 5\tau \ldots$ are allowed [118]. The rules of the random walk on such a fractal are summarized as follows [134]: when the particle gets trapped in a structure when it moves away from the backbone where traveling waves are monitored, the probability of the walker spending away from the backbone a time t is a convolution exactly solvable in this case (it is obtained from the distribution of total "away time," the sum of independent waiting times spent within the nodes invested by the invasion front), and the total probability distribution arises from the sum over all the possible waiting times t from 0 to ∞. The final result [118] is obtained analytically under the simplifying assumption that the probabilities of getting side-tracked into any lateral branch as a consequence of a bifurcation faced by propagating along a backbone are kept constant and that $\tau \ll t$, that is,

$$v = 2\sqrt{a \frac{\Delta x^2}{\tau \beta}}, \qquad (1.33)$$

where $a = R(0)$ (Box 1.6) and for $\Delta x \ll x$. One immediately notes the (dimensional and practical) analogies of $\Delta x^2/\tau$ with D (whose dimensions are L^2/T). Here Δx is the distance between first neighbors in the lattice underlying any network (i.e., the mean nodal distance) at the chosen level of resolution (hence the resulting maximum Strahler's order [63, 126]; see Figure 1.17), and x is the distance from the seed to the node where one expects the asymptotic behavior – say, after crossing more than five bifurcations. The parameter $\beta \geq 1$ determines the slowdown produced by a dendritic structure. It is prompted by the regular side trapping incurred at each bifurcation faced along the journey of a random walker – that is, the Fickian diffusion of matter or organisms – spreading along the backbone of the network. The parameter β can be computed from geomorphological attributes, and qualifies the departure of the case at hand from the speed of a wave traveling along a linear substrate (Box 1.8).

Box 1.8 Slowing a Traveling Wave by Branching Substrates

The slowdown effect may be computed exactly with reference to a deterministic fractal, the Peano network [63, 121, 128]. Figure 1.17 illustrates how a recursive construction underlies such networks, which are amenable to a number of exact calculations [63, 118, 126, 130, 135]. The branches' order

Box 1.8 *Continued*

in the Peano basin is a direct consequence of the total number of links indicated in Figure 1.17 [126]. Recapping: when a walker reaches the backbone site corresponding to a first-order branch, it can jump to another site of the backbone with probability $2 \times 1/4 = 1/2$ or get trapped into a secondary branch with probability $1/2$. Then, the waiting time is τ with probability $1/2$. Logically, the waiting time at the node becomes 3τ with probability $1/4$ (the myopic ant model, that is, a walker that has arrived at a certain site i of the backbone of the network structure may jump, after a residence time τ, to each of its nearest neighbors z_i with even probability $p_i = 1/z_i$). All possible cases are thus enumerated. In general [118], one obtains, for first-order branches,

$$\varphi(t) = \sum_{i=1}^{\infty} p_i \, \delta(t - [2i - 1]\tau),$$

where $\delta(t)$ is as usual the Dirac delta distribution. The analytical solution for branches of any order, from first-order ones (the leaves) to any order N, has been obtained [118] for a slightly simplified scheme. Summarizing in words a somewhat involved mathematical treatment [134], one observes the following for one-dimensional random walks:

- When the particle goes into a further structure by moving away from the backbone by random jumps, the probability that the walk takes a time (or a number of steps) t is a convolution of distributions along the individual reaches, as the random times spent in each geomorphic state are assumed to be statistically independent. Incidentally, in the Laplace approach, convolutions are treated as a product of transforms (Box 3.17), thereby greatly simplifying the mathematical manipulations. Also, Laplace transforms are straightforward moment-generating functions (Box 3.18). Ultimately, the overall probability distribution of travel times along a given direction, the backbone dictating the main avenue for biological invasions, arises from the sum over all the possible times t.
- When the particle reaches a crossing and must choose between two possible ways, the total probability is the sum of both probabilities. The sum of the probabilities of accessing confluent nodes (incoming or outgoing in a directed graph) is equal to 1, as no other path is available. Several exact manipulations are shown in [118].

Therefore, the waiting-time distribution for any different branch can be found [118]. The related jump-distance distribution has also been obtained for an isotropic random walk across the backbone in terms of the (constant) distance Δx between first neighbors in the lattice implied by the Peano construction (i.e., the spacing $1/N^2$ between any two adjacent nodes at the $N + 1$ level of iterative construction whose maximum distance from source to outlet is kept as a unit Figure 1.17 and [126]). For the asymptotic regime where $\tau \ll t$ and $\Delta x \ll x$, (to first order) one obtains Equation (1.33) with

$$\beta = \sum_{k=1}^{N} k p_k \; \geq 1, \qquad\qquad\qquad (1.34)$$

measuring essentially the mean probability faced by all bifurcating nodes along the backbone. The Fisher–Kolmogorov speed $v = 2\sqrt{aD} = 2\sqrt{R(0)D}$ is recovered when no secondary branches exist ($\beta = 1$ and $D = \Delta x^2/\tau$).

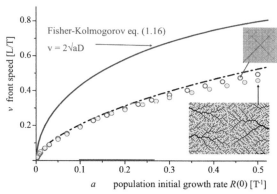

Figure 1.18 Front speed as a function of the growth rate $a = R(0)$ of the logistic equation (redrawn from [118]). The solid line is the exact solution of the continuous isotropic Fisher–Kolmogorov model of celerity in a linear landscape $v = 2\sqrt{a\Delta x^2/\tau}$ [119, 125] (see Box 1.6). The dashed line and the dots represent exact and numerical values for propagation along the backbone of Peano and OCNs ($v = 2\sqrt{a\Delta x^2/(\tau\beta)}$, with $\beta > 1$ [118]). Note that the dashed line illustrates an exact solution for an approximate model (the uniform comblike structure with equal link length) [118], whereas empty blue dots indicate the corresponding numerical simulation for the actual Peano topology. Numerical simulations of 100 ensemble-averaged OCNs are also shown (yellow dots) [7]

Thus, we conclude that the speed of the front moving across a fractal support is slower than that for a linear substrate. The topological complexity, epitomized by the sum of the transition probabilities p_i along the propagation backbone, defines the chance of getting slowed down. This result is shown in Figure 1.18, together with the results of several computational averages where OCNs were employed [7, 118] (details on the numerics are given in [7]). The most striking result is that the propagation of the invasion along *any* network structure seems to be slowing the traveling front comparably with respect to the benchmark Fisher–Kolmogorov prediction. This is partly due to the statistical inevitability of Horton's laws for treelike structures [68], because Peano's construct, real networks, and any OCN (like the shapes shown in Figure 1.16) have rather similar bifurcation numbers, even though all nodes in Peano show a four-way bifurcation that is unrealistic for real river networks. Moreover, radically different metrics for Peano, on one side, and real networks and OCNs characterize, say, the distribution of area and upstream length at any site or the elongation of the main stream subsumed by Hack's coefficient h [67] (see also Appendix 6.2). Clearly, the common topological structure of the substrate for the wave propagation –

the backbone of the trees – matters decisively for the similarity of the celerities of propagation of invasion waves in such different networks (Figure 1.18).

1.5.2 Modeling Human Range Expansions: Western Colonization in the Nineteenth-Century United States

Campos et al. closed their paper with a remarkable example [118]. It showed that topological constraints imply a major correction of the speed of a real migration front for which the model of Equation (1.33) likely applies. In fact, for some human prehistoric migrations, the wave-of-advance model was proposed as a sound scheme [22, 122], where population fronts spread into new regions and population density saturates to carrying capacity (constant or site dependent) behind the front. The erratic and anisotropic nature of the expansion along the direction of some chosen watercourse suggests that random walk models prove useful for the modelization of human migrations.

According to the available data and to revised atlases, the average expansion rate for the transition of the North American colonizers between 1790 and 1910 was 13.5 ± 0.8 km/yr [123]. The parameters in the approach [118] were estimated as follows. The time between jumps τ for biological migrations was speculated to be equivalent to the time between successive generations. For humans, the value $\tau = 25$ yr is usually taken [122]. The growth parameter $a = R(0)$ was computed directly by Lotka [136], who fitted the population versus time plot for the United States in the nineteenth century to a logistic curve, obtaining $a = 0.031 \pm 0.001$ per year. Regarding the distribution of jump lengths, yet another ingredient required by the solution to Equation (1.26), one observes that settlers did not always cover the same distance, so the distribution $\Phi(x)$ should include the possibility of different jump lengths, in contrast with the approximation implied by the solution equation (1.33). This may be accommodated by fitting the observed data to a continuous distribution. Otherwise, one may determine the average distance covered by settlers and use this value as Δx in Equation (1.33). Campos et al. [118] explored both possibilities by observing that in the case at hand, the jump distances covered by settlers were estimated from individual records obtained from the Migrations Project database (available at www.migrations.org). From there, 400 individual records were collected from the database and measured the distance covered by colonizers from their birthplaces until the place they were 25 years after-

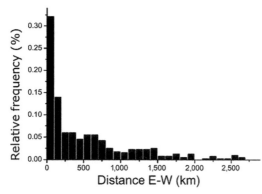

Observed Speed	13.5±0.8 km/yr
Fisher's prediction (N=0)	40.3±2.9 km/yr
Continuous Φ(x) (N=2)	19.4±3.2 km/yr
Continuous Φ(x) (N=5)	18.9±3.1 km/yr
Averaged Δx (N=2)	16.5±2.7 km/yr
Averaged Δx (N=5)	14.7±2.4 km/yr
Simulations on Peano	14.5±0.1 km/yr
Simulations on OCNs	14.4±0.1 km/yr

Figure 1.19 (left) Plot of the distribution of distances covered by migrants in the east–west direction during the nineteenth century, according to the 400 individual records taken from the Library of the US Congress archive. (right) Observed front speed and predictions obtained from theory and simulations on fractal basins for the case of US colonization. Figure after [118]

ward, that is, after the chosen time τ. Note that only the distances in the east–west direction were considered for the westbound US colonization example, in accordance with the 1D scheme underlying the present discussion. The jump length distribution obtained in this manner [118] is represented in Figure 1.19.

The last piece of information needed to run the example concerns an estimate of the fraction f of the population who remained at their birthplaces after 25 years, without migrating. Ferrie's work [137] based on the censuses of the nineteenth century allowed [118] an estimate of the needed value at $f = 0.3 \pm 0.05$. Wrapping up, the best fit for the data corresponded to an exponential decay distribution of the form

$$\Phi(x) = A \, e^{-x/\mathscr{L}}, \qquad (1.35)$$

where A is a normalization factor and $\mathscr{L}$ is a deterrence distance estimated at $\mathscr{L} = 640 \pm 23$ km. The resulting average distance was estimated at $\Delta x = 810 \pm 93$ km. By introducing the above empirical results (distributions and parameters) into the mathematical solvers of Equations (1.26) [118], one obtains the results for the speed shown in the left panel of Figure 1.19. The results for different values of the speed of propagation v provide a clear interpretation framework. The settlers, according to historical reports, moved mainly by following the major river valleys. Therefore, one could argue that details (like the level of detail in the description of the structure of the Peano network, that is, tertiary, quaternary, and higher-order channels) are not decisive in determining the overall dynamics of the migration process. Therefore even a low-order description like that akin to exact solution provides a good approximation [118]. In any

case, the main result highlighted by the US colonization example is that the geometrical constraints of the fractal networks involve strong corrections to the speed of the fronts, regardless of details about the network topology. The Fisher–Kolmogorov prediction significantly overestimates the observed speed of the migration front, while the results – whether found from simulations or through the theoretical predictions in Equation (1.33) – agree reasonably with the observations. This led Campos et al. [118] to conclude that colonization of the United States during the nineteenth century had been strongly affected by the landscape constraints, implying that heterogeneities faced along the migration directions reduced the propagation rate significantly. Limitations of the model abound, of course. For instance, the model assumes for simplicity that relevant heterogeneities are given only by river streams. However, other constraining factors come to mind, be they the effects of mountain reliefs, the crossing of deserts, or the nature of the terrain to be crossed by the caravans. These obvious limitations notwithstanding, the main features of the migrating fronts were captured beautifully, suggesting that perhaps the underlying ideas could still be extended to considering the whole territory as a fractal landscape where the settlers move, thus renewing interest in models based on transport through fractals like that presented above [118].

The main message of Figure 1.18 and the example of the colonization of the West in the nineteenth-century United States, resonate with the theoretical and empirical findings about the key role of directional dispersal on biodiversity patterns (Sections 1.2 and 1.3). These results show from yet another angle that the directionality imposed on any ecological interaction by specific network topologies – here simply the treelike, hierarchical and recurrent, characters of any river network – impacts significantly the resulting ecology. It is a remarkable result that grants fluvial ecological corridors the peculiar characteristics that we exploit in this book from a variety of angles.

Other issues of great cultural interest move from the premises of the above results. For example, in their seminal work, Ammerman and Cavalli-Sforza [22] analyzed the spread of farming technology through Europe in terms of a diffusive displacement of population termed demic flow. If considered as cultural diffusion, where technology passes on without significant movements of populations, demic flows raise the possibility that cultural, genetic, and linguistic traits with no intrinsic advantage may "hitchhike," that is, spread with the advancing farmers [138]. One wonders whether similar concepts and tools may be related to species diversification (Chapter 2), biological invasions of foreign species in large river systems (Chapter 3), or waterborne disease spread (Chapter 4). As posited above, historically, humans have chosen to live in proximity to rivers for water supply, energy, and navigation purposes, leading humans to follow the courses of rivers during migrations when establishing settlements [7]. However, humans progressively decreased their reliance on direct proximity to rivers by developing effective alternatives for water supply and shifting from waterborne transport modes to land and air transport [139]. It was shown convincingly that humans moved closer to major rivers only in preindustrial periods and have moved farther from major rivers after 1870 owing to the reliance on rivers only for trade and transport [139]. We shall not pursue these interesting topics any further. Suffice here to mention that the colonization example [118] allowed us to confirm empirically the speculation derived from abstract models that directional dispersal matters even for invasion dynamics, the subject of Chapter 3.

2 | Species

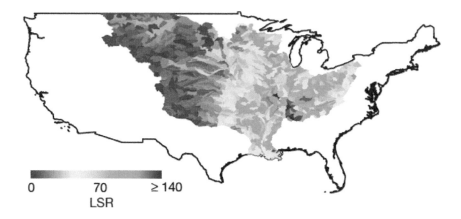

0   70   ≥ 140
LSR

In what follows we shall pursue not a review of the many facets of the subject, broad as they obviously are, but rather a specific choice of topics relevant to the general concepts outlined in Chapter 1 that find concrete applications here. The image chosen to subsume the contents of this chapter is a snapshot of the observational map of the Local (freshwater fish) Species Diversity index (whose color code scale is shown) within the Mississippi–Missouri river system (MMRS). The biogeographical data on fish shown here and used in the analyses were obtained from the NatureServe database of US freshwater fish distributions, which summarizes museum records, published literature, and expert opinion about fish species distribution in the United States and is tabulated at a predefined US Geological Survey (USGS) scale (from "Distribution of Native U.S. Fishes by Watershed"). The MMRS boasts one of the largest and most diverse river basin collections worldwide with resspect to heterogeneity of drained land cover and use across strong gradients of climate, geology, geomorphology, and hydrology, resulting in radically different stream ecosystems. One of the key exercises exemplifying the theses of the book is carried out with reference to the predictive power embedded in the proper geomorphological description of the fluvial system. The chapter's opening image, part of a figure in [8] representing local fish species richness in the domain of the MMRS, sets the tone for the rest of the chapter.

2.1 Fish Diversity, Hydrologic Controls, and Riverine Habitat Suitability

The first example of practical application of the ideas outlined in Chapter 1 deals with the prediction of fish diversity patterns in the large Mississippi–Missouri river system. Here a neutral metacommunity model of the type introduced in Box 1.2 was made more realistic by better describing the structure of the local community (LC) [8]. In fact, each LC is endowed with a carrying capacity that depends on its location (specifically, total contributing catchment area as a proxy for fluvial habitat size [63]) and

the relevant habitat. Each site within the LC hosts, not a single individual, but a small subpopulation of a certain species. The physical context is described in Figure 2.1.

The main tenet in [8] is the overarching theme of this work: river networks act as ecological corridors featuring connected and hierarchical dendritic landscapes (both for animals and plants). Their form and function, affected by directional dispersal whose relevance has been suggested in Chapter 1, thus present challenges and opportunities for testing biogeographical theories and macroecological laws. The directionality implied by the dendritic

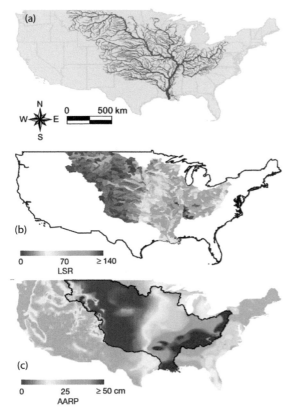

Figure 2.1 General characterization of the Mississippi–Missouri river system. (a) MMRS general layout, inclusive of topological and geometrical characterization and its localization within the conterminous United States [29]. (b) Local species richness (LSR), or α-diversity (Chapter 1), of freshwater fish in each direct tributary area (DTA) at the USGS HUC8 scale [8]. The biogeographical data on fish used in the analysis were obtained from the NatureServe [140] database of US freshwater fish species distribution. They are tabulated at the USGS HUC8 scale [141] (after [8]). (c) Annual average runoff production (AARP) (cm), the portion of precipitation drained by the river network at each site. The map in (c) is estimated from the streamflow data of small tributaries collected from approximately 12,000 gauging stations averaged over the period 1951–1980 [8].

ecological matrix will prove defining. In the specific case of riverine fish biodiversity, it is clear that patterns of local and basin-scale differences must relate to energy availability, habitat heterogeneity, scale-dependent environmental conditions, and river discharge [6]. However, any theoretical approach aimed at a comprehensive set of systemwide diversity patterns that strives to include all of the above details is hard to conceive, let alone assemble. Here each direct tributary area (DTA) is a

LC within a metacommunity. Each has a different fish habitat capacity, H, defined as the number of fish units sustainable by resources in that particular DTA; a fish unit can be thought of as a subpopulation of fish of the same species. To keep assumptions at a minimum, H is assumed to be proportional to the product of the DTA watershed area (following [8]) and the annual average runoff production (AARP), as an indicator of the quantity of resources available for fish [142–144]. The model uses the topological rather than Euclidean distances between DTAs (Box 2.2) representative of how far fish can travel across the network.

The procedure to assign different species to different DTAs is described in detail in [8] and is recapped in Box 2.2. Two matrices summarizing the data used in the analysis are necessary to reproduce all the results and are provided in the supplementary information of the original reference [8]. A first matrix reports the presence–absence data [140] of each of the 433 fish species in each of the 5,824 DTAs included in the study. The sum along each row of the matrix (composing 433 elements) gives the local species richness (LSR) of the corresponding DTA, and the sum along each column (5,824 elements) gives the occupancy of the corresponding species. The second matrix reports the topological distance between each pair of the 5,824 DTAs. Such matrices must be combined to produce the LSR profile as a function of topological distance from the outlet (Figure 2.1).

2.1.1 A Hierarchical Metacommunity Model of the MMRS Network

Model simulations of the MMRS proceed via a spatially explicit approach akin to the neutral approach described in Chapter 1. Every DTA is assumed always to be saturated at its capacity; that is, no available resources are left unexploited. Each DTA hosts a certain number of fish units that depends on the location. At each time step, a fish unit, randomly selected from all fish units in the system, dies, and the resources that previously sustained the unit are freed and available for sustaining a new fish unit. Let i be the DTA to which the unit belongs. With probability v, the diversification rate as in Section 1.2, the new unit will be occupied by a new species (the diversification is a rate per death). It is due to speciation, to external introduction of nonnative species, or to immigration (and reimmigration) of a new species from outside the MMRS. With probability $1 - v$, the new unit will belong to a species already existing in the system. In the latter case, with probability P_{ij}, the empty unit in DTA i will be colonized by a species from DTA j [8]. In this context, H_k is the habitat capacity of the DTA k, and N is the total number of DTAs (here $N = 5,824$, while the total number of fish units is $436,731$ [8]). All the fish units in DTA j have the same probability of colonizing the empty unit in DTA i where the "death" took place. Death must be interpreted as the local extinction of the species subpopulation hosted by the fish unit. The model results that we outline in the following pages are the average patterns after the system reaches a statistical steady state.

Specific dispersal kernels are assumed to determine how the fish units move within the river network [8]. K_{ij} in this context is the probability that a fish unit produced at DTA j arrives at DTA i after dispersal (Box 2.1). Here normalization constants are determined numerically such that no fish can travel out of the network (Box 2.1). The model is of a structured metacommunity type. The neutral theory of biodiversity [34] (whose strenghts and limitations have been discussed in Section 1.2) is implemented in the MMRS, using its network (Figure 2.1a) as the structure of the metacommunity. The model captures basic ecological processes: extinction, dispersal, colonization, and diversification. In all exercises shown here, the simulations are run until the system reaches a stationary state; the biodiversity patterns of interest are then determined and compared with the empirical patterns.

A necessary note pertains to the neutral assumption in this context, which is indeed strong because fish species obviously differ in their dispersal abilities. However, the functional equivalence between species is a key way in which the neutral theory of biodiversity departs from classical niche ecological models. It was assumed [8] that species equivalence holds to study how good a fit the neutral metacommunity model can produce to data in the absence of large-scale, detailed, species-specific information. As such, the model is engineered to single out the individual role of the dendritic structure of the topological substrate for ecological interactions (Section 1.2 and Boxes 2.1 and 2.2). An important innovation of the metacommunity model with respect to the zero-order model outlined in Chapter 1 concerns the imposition of hierarchical local habitat capacity at any site j, say, H_j. In fact, habitat capacity of DTA i, H_i, is determined by a properly normalized relation $H_i \propto AARP_i/WA_i$, where WA_i denotes total contributing (watershed) area at site i and $AARP_i$ is the average annual runoff production (Figure 2.1). Total contributing area at any point of a river network can be directly computed from suitably treated digital elevation maps [63] (Figure 1.3). The results are displayed in Figure 2.2. The statistically significant, major impact of modulated habitat capacity is clearly suggested by the comparison of the results from the modified neutral metacommunity model and those from zero-order modeling (Figure 2.2c).

Box 2.1 Model Simulations for MMRS Fish Diversity

Here, we shall describe the model implemented in the original work. Several variants may be set up, leading to analogous conclusions [8]. Here, every DTA is assumed to be saturated at its capacity; that is, no available resources are left unexploited by the fish community assemblage. The model dynamics proceed as follows (see for comparative purposes the rather similar description in Chapter 1). At each time step, a fish unit, randomly selected from all fish units in the system, dies, and the resources that previously sustained the unit are freed and available for sustaining a new fish unit. With probability v, termed the diversification rate, the new unit will represent a new species. As a note, the diversification is a rate per birth and is due to speciation, to external introduction of nonnative species, or to immigration (and reimmigration) of a new species from outside the MMRS. It may vary with ecosystem size [9]. With probability $1 - v$, the new unit will belong to a species already existing in the system (in this example, the MMRS). In the latter case, the probability P_{ij} that the empty unit in DTA i is colonized by a species from DTA j is determined by

$$P_{ij} = (1 - v) \frac{K_{ij} H_j}{\sum_{m=1}^{N} K_{im} H_m},$$

where K_{ij} is a dispersal kernel measuring the range of a species' colonization, H_k is the fluvial habitat capacity of DTA k typically established on the basis of scaling geomorphic relations [63], and N is

Box 2.1 *Continued*

the total number of DTAs ($N = 5824$ in the above exercise). We shall review the rationale for the choice of each of the parameters determining P_{ij} [8]. The procedure to assign different species to different DTAs from the available is also described therein.

Dispersal kernel. A dispersal kernel determines how fish units move within the river network. A wide ecological literature exists on its theoretical representation, a review of which is beyond the scope of this box. Suffice here to recall that a combination of several theoretical dispersal kernels has been used to achieve a good representation of dispersal kernels for seed dispersal and plant colonization [145]. Here, K_{ij} is assumed to be combination of back-to-back exponential and Cauchy distributions:

$$K_{ij} = C \left\{ a^{L_{ij}} + \frac{b^2}{b^2 + L_{ij}^2} \right\},$$

where K_{ij}, as implied above, is the probability that a fish unit born out of DTA j arrives at DTA i after the dispersal process; C is a normalization constant; and L_{ij} is the effective site-to-site network (at times termed chemical) distance, defined by the shortest path from i to j. Specifically, we distinguish here between downstream and upstream paths, for the selective effect that the streamflow downstream bias may exert on certain species, that is,

$$L_{ij} = ND_{ij} + w_u NU_{ij},$$

where ND_{ij} and NU_{ij} are the numbers (easily converted to lengths) of respectively downstream and upstream steps making up the shortest path from j to i; w_u is a weight factor that modifies the upstream distance subsuming the downstream bias effect of the streamflow ($w_u > 1$ implies downstream biased dispersal, thereby characterizing dispersal directionality, and vice versa); and $a < 1$ and b are constants that characterize the strength of the exponential and the Cauchy decays, respectively. Here, C is determined numerically so that, for every DTA j, $\sum_i K_{ij} = 1$, that is, no fish can obviously travel out of the network. Note that at the upstream ends of any river system, the source areas, this obviously applies. At the downstream end of the MMRS system, namely, at the outlet to the Gulf of Mexico, one must consider the effects of a marine body that acts as a barrier to freshwater fish. In the above exercise, the dispersal kernel of every species is assumed to be the same – a rather strong neutral assumption [34] – although fish species obviously differ in their dispersal abilities. This functional equivalence between species is, however, a key ingredient of the neutral theory of biodiversity, which departs from classical ecological models as noted in Chapter 1. Given our aim of understanding the peculiar role of the connected ecological corridors provided by the river network, it makes for a perfect starting point, as it avoids clouding the network corridor effect with a complex search for functional parameters. We thus assume species equivalence to study just how good a fit the neutral metacommunity model can produce to our data in the absence of detailed, species-specific information [8].

Average annual runoff production (AARP). Runoff is the fraction of the precipitation falling onto the catchment surface that is drained by the river network at a given control section. Runoff depends on the combined effects of all hydrologic processes, including rates and spatial distribution of precipitation, and cumulative effects of evapotranspiration and infiltration. The map in Figure 2.1b is estimated from streamflow data within the catchment, properly handled [8]. Streamflow within the

Box 2.1 *Continued*

network of tributaries has been collected from about 12,000 gauging stations averaged over the period 1951–1985 [146].

Direct tributary area (DTA). The DTAs in the study described here (reproduced from [8]) correspond to the HUC8-scale subbasins designated by the USGS [141].

Habitat capacity (H). Habitat capacity of the arbitrary DTA i, H_i, is computed, rounded to the nearest integer, by

$$H_i = C_H N \frac{AARP_i/WA_i}{\sum_{j=1}^{N} AARP_j/WA_j},$$

where WA_k is the total cumulative watershed area at site k, N (equal to $5,824$ in the study described here [8]) is the total number of DTAs, and C_H is the estimate (due to rounding) of average habitat capacity in a DTA.

Topological distance. The topological distance (at times termed chemical distance) measures distance along the network structure. Moving along the network structure, a unit increment in topological distance occurs when crossing from one DTA to another. In the study described here [8], one unit of topological distance corresponds to a physical distance of about 100 km.

The reference data were analyzed to produce spatial biodiversity patterns (Figure 2.2). The map of LSR shown in Figure 2.1b suggests that the DTA endowed with the maximum α-diversity (156 species) is observed somewhat midway through the MMRS, roughly identified as at the borders of the states of Alabama, Mississippi, and Tennessee. The sharp decrease empirically observed in species richness occurring around the 100°W meridian is known to correspond to sharp gradients of annual precipitation and runoff production [8] (Figure 2.1c). Although these gradients may partly explain the semi-arid climate and low fish diversity in the western half of the MMRS, Muneepeerakul et al. [8] argued that the western DTAs are low in fish diversity both because their climate is dry and because they are upstream portions of the river network. Figure 2.2a illustrates the frequency distribution of LSR, best fit to empirical data [8], whose two peaks at low and high values reflect the difference between the western and eastern halves of the MMRS. Figure 2.2b,c show computed and measured LSR as a function of the topological distance from the network outlet. It is remarkable that the LSR profile shows a significant increase in the downstream direction, except at the

outlet where the freshwater fish habitat capacities are significantly reduced by salinity, co-occurrence/intrusion by some freshwater-tolerant estuarine or coastal fish species, human disturbance, and pollution. In fairness, the model shows also a mild excess of LSR at the upstream end, contrary to data. The overall downstream increase in richness results from the converging character of the river network and is steepened by the dry–wet climatic gradient [8]. Neutral metacommunity patterns are equally good for rank-abundance curves and suitable measures of β-diversity (Section 1.2), where particularly significant is the resulting long-distance similarity in species composition maintained by species with extremely large occupancies, such as *Ictalurus punctatus* (channel catfish), *Ameiurus melas* (black bullhead), and *A. natalis* (yellow bullhead).

The neutral metacommunity model reproduces surprisingly well the general spatial biodiversity patterns of the MMRS freshwater fish once hydrologic controls, like the effects of average annual runoff production on fish habitat capacities, are enforced. A wide spectrum of observed biodiversity patterns are reproduced, as suggested by Figures 2.1 and 2.2. As a specific example, in addition to the general trend and magnitude, the model

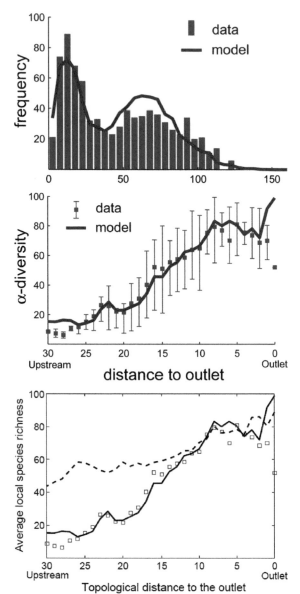

Figure 2.2 Patterns of LSR produced by the neutral metacommunity model in the MMRS system [8]. (top) Frequency distribution of LSR. (middle) LSR profile as a function of the instream distance measured in DTA units from from the outlet. The squares (average values) with error bars (ranging from the 25th to the 75th quantile) and bar plots represent empirical data, and the lines represent the average values of the model results. (bottom) As in the plot above, where for comparison a constant habitat capacity (dashed line) is employed. Figure after [8]

captures fine-structured fluctuations of the LSR profile (Figure 2.2). Simultaneous fits of diverse patterns (and others, such as the species–area relationship) make for a very demanding test for any model, especially one using few parameters, as in this case [8]. The model therefore provides rich ecological insight despite its simplicity.

One of the ecologically meaningful insights obtained from the theoretical exercise is that the parameters corresponding to the best fit imply that the spread of the average fish species is quite symmetrical, that is, it is not significantly biased either in the upstream or the downstream direction. The model results also suggest that, on average, most fish disperse locally (that is, to nearby DTAs), but a nonnegligible fraction travel very long distances. Given the broad range of environmental conditions covered by the continental-scale MMRS, the demonstration that a simple neutral metacommunity model coupled with an appropriate habitat capacity distribution and dispersal kernel can simultaneously reproduce several major observed biodiversity patterns has far-reaching implications. These results suggest that only parameters characterizing average fish behavior – as opposed to those characterizing biological properties of all different fish species in the system – in connection with habitat capacities and the connected structure, suffice for reasonably accurate predictions of large-scale biodiversity patterns. The neutral metacommunity model thus may represent a null model against which more biologically realistic models ought to be compared (*sensu* Akaike [147, 148]). Moreover, furthering our understanding of fluvial ecological corridors (related to general metapopulation persistence criteria for spatially explicit models [115, 149]) will allow us to improve in the alignment of ecological models and data.

Therefore, this first chosen example, which pertains to species' distributions at stationarity of a fluvial ecosystem, seems particularly insightful. It reveals that, through estimates of average dispersal behavior and habitat capacities objectively calculated from average runoff production (the minimum possible synthesis of the hydrologic regime), reliable predictions of large-scale spatial biodiversity patterns in a gigantic riverine system may follow. Indeed, it is remarkable that these results may follow from such a simple, almost unrealistic ecological insight. The success of selective relaxations of the neutral theory, here in a complex two-dimensional dendritic fluvial ecosystem, indeed suggests the forceful explanatory power of directional dispersal in metacommunity models. In fact, the highlighted dependence of habitat capacities from precipitation patterns and the connectivity provided by the MMRS suggest a direct linkage with (and insight from) large-scale forcings, such as the influence of global climate change, on biodiversity patterns.

2.1.2 Geomorphic Processes, the Frequency Concept, and Riverine Habitat Suitability

Despite several years of scrutiny since the onset of the related ecological framework developments, the dynamics of species constrained to disperse within river networks are often poorly captured by metapopulation or metacommunity models [6, 8, 150]. This is largely because often one tends to overlook spatial heterogeneities that are tied to the hierarchical structure of river networks, as this chapter strongly suggests. A framework for metapopulation dynamics (subject to local colonization–extinction dynamics and regional dispersal processes, as shown in Box 2.1) critically needs to prescribe – as clearly shown by Figure 2.2, for example – a size-structured riverine system. Only through such inclusion may one rightly compare theoretical predictions relative to fish species distribution data from the Mississippi–Missouri river system [8]. Ecological consequences abound, obviously. This scaling structure alone allows any reasonable model to predict a decline in species occupancy towards the upper reaches of the river and the suitability of an upstream-biased dispersal strategy for branching riverine systems, both of which are supported by empirical data [6, 150]. Metapopulation persistence is thus nontrivially related to the structure of riverine networks [115] (Section 2.2), and a relationship stems by necessity from the inclusion of the longitudinal variation of habitat heterogeneity within the network.

But how do we define a spatially explicit framework for hydrologic landscape-forming events responsible for shaping such longitudinal, macroscopical variation? Longitudinal gradients in width and depth of channel reaches obviously depend on the very fabric of the river network: the spatial aggregation structure, as the longitudinal addition of tributaries makes for predictable gains in runoff at a given instant in time [63]. Which are the events that trigger major and lasting alterations of the fluvial habitat? How can one single out a proper synthesis of the nexus between the frequency concept (i.e., the nexus of the frequency of exceedance of landscape-forming hydrologic events with the local shape of the fluvial domain [101]) and the geomorphic processes responsible for the shaping of the fluvial habitat at-

a-site? This issue was first considered in the seminal book on fluvial processes in geomorphology by Leopold, Wolman, and Miller [101]. Therein, it has been noted that the significance of different frequency distributions of factors affecting geomorphic processes (chiefly, a streamflow-determined shear stress threshold that defines the mobility of riverbed and bank material) is not easily evaluated because of the near-impossibility of collecting long-term, isochrone (local and catchment-wide) exhaustive empirical evidence. Everyone is familiar, however, with the major geomorphic alterations associated with hydrologic events "classified on the human scale as catastrophes" [101]. This is manifest by hurricane destruction of beaches, the demolition of tidal landforms by submergence, or the "gullying of hillsides by cloudburst rains" [101], to name a few diverse ones. Less easily recognized is the cumulative effect of floods or rains that occur once a year or perhaps less frequently, possibly once every three to five years.

It has been argued that the frequency distribution of shear stresses exceeding a threshold of mobility, determined by climate and meteorological events, follows a general lognormal distribution [101] and that the product of the weight of the material moved, multiplied by the frequency of the events moving it, will inevitably reach a maximum such that the largest quantity of material is transported by relatively frequent events rather than by extremes. The recurrence interval, or the inverse of the frequency, at which this maximum occurs is controlled by the maximum rates of change with the hydraulic stress fostered by hydrologic streamflows. It is the frequency interval of the occurrence of this maximum that determines the maximum amount of erosional work acting against the fluvial landscapes – thus the true and lasting landscape-forming events.

The process of channel formation and its relation to the fluvial habitat formation is determined by the time intervals during which the major portion of the erosional work is done. Therefore, hydrologic and geomorphic considerations can be tied together by considering a particular river section. Much empirical and theoretical evidence [101] indicates that the erosional and depositional work carried out by perennial streams, resulting in scour and fill and in the transport of debris of various origins underlying a regime shape, is mainly accomplished by flows near (or slightly above) the bankfull stage, that is, streamflows that occur less than 0.4 percent of the time, or roughly once a year. Such events are not to be considered catastrophic. Exceptions abound, but a sufficiently accurate general picture is consistently observed, to the point that one may infer a general rule that captures consistently a mean trend that is used in this book to characterize the hierarchical habitat size shown by riverine ecosystems.

As an example of an exception, in steep narrow valleys, the likelihood is relatively greater that events of major magnitude will be landscape forming in that the devastation of the channel and its valley are such that long intervening periods of more moderate flows have a less prominent effect [101]. Moderate flows may yield consistent repair effects, but only in a possibly unattainable long run (that is, larger than the return period – the inverse of the probability of exceedance – of a threshold catastrophic flood), tending to the norm dictated by the general empirical trends described later in this chapter. Landslides and debris flows are also major landscape-forming events, and they may blur further the general picture (topographic thresholds for the inception

of debris flow conditions stem from accurate digital terrain maps [70, 71]). Locally, the alterations induced by excess sediment production from hillsides may also trigger very long transients toward the predicted habitat sizes.

In conclusion, most fluvial processes are far more complex than the simple examples mentioned here may possibly describe. However, mean trends prove meaningful and time honored by decades of usage, leading the at-a-station relations that we address in Box 2.3 for the relation between total contributing area (a geomorphic feature objectively detectable from digital terrain maps [63]) and landscape-forming discharges [63, 151] and in Box 2.2 for the variation of hydraulic characteristics at a given cross section seen as a consequence of the fact that most rivers experience a wide range of flows from which landscape-forming events need to be sorted out.

Box 2.2 Stream Channel Geometry

In a fundamental study of river hydraulics, Leopold, Wolman, and Miller [101] suggested and demonstrated empirically how some characteristics of stream channels – depth, width, velocity – vary with discharge as simple power functions at a given river cross section. As we have seen in Section 1.2, this property has major implications for a quantitative evaluation of the hierarchical habitat sizes in fluvial ecosystems.

Power law relationships are also shown to exist among the previous variables and discharge when the channel characteristics are measured along the length of the river, under the condition that discharge at all points is equal in frequency of occurrence. An example highlighting the nature of these relationships is illustrated in Figure 2.3 (note that a straight-line relationship in a double logarithmic plot implies a power law type of functional relationship). The velocity, v, width, w, and depth, d, of flowrates Q are given by the relations

$$v = kQ^m, \qquad w = aQ^b, \qquad d = cQ^f, \tag{2.1}$$

where Q is the discharge; k, a, and c are proportionality constants; and m, b, and f are scaling exponents. Because discharge Q is approximately given by $Q \approx vwd$, Equation (2.1) implies for consistency that

$$m + b + f = 1.$$

When the analysis is performed for the variation of the hydraulic characteristics in a particular cross section of the river as a function of discharge (*at-a-station* type of analysis), Leopold et al. [101] found consistently average values of $b \approx 0.26$, $f \approx 0.40$, and $m \approx 0.34$. The example of relationships at a station shown in Figure 2.3 for the Powder River at Locate, Montana, has been reproduced consistently (see, e.g., [63]). The value $f = 0.4$ indicates that discharge increases much faster than depth of water in a stream. When the mean depth doubles, the discharge increases nearly

Box 2.2 *Continued*

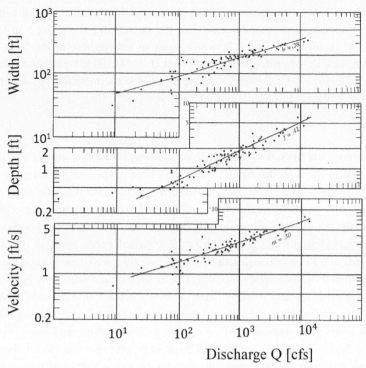

Figure 2.3 Relation of width, depth, and velocity to discharge Q, Powder River at Locate, Montana. Figure after [101]

6 times. The equation $d = cQ^f$ indicates that a rating curve – a graph of water stage versus discharge – should plot approximately as a straight line on log-paper, and indeed such plots are widely used in engineering practice.

In the analysis of hydraulic characteristics in different cross sections along the length of a river, the comparison is valid only under the condition of constant frequency of discharge at all cross sections. The analysis is termed in the downstream direction, and it is of special interest for interpreting the structure of the drainage network as a pattern that connects the elements of a basin. It has been further shown that similar power law equations are obtained whether one uses the mean discharge or bankfull discharge. The average values of the exponents obtained are $b = 0.5$, $f = 0.4$, and $m = 0.1$. Examples of the analysis in the downstream direction are shown in a number of references reviewed in [63, 101]. Proceeding downstream in a given river, discharge will increase because of the increasing area (except in very arid regions, where large losses in the downstream direction may exist). The interesting thing is that along a river channel, such a progressive increase is structurally related to the changes in width and water depth in the channel, regardless of where in the watershed or on what tributary the cross sections may be. The very low value of the exponent m indicates that the velocity tends to remain constant or increase slightly in the downstream direction as long as the discharge at all points is of similar frequency. This important finding has also been confirmed by a variety of field experiments (e.g., see, for a review, [63]).

Box 2.2 *Continued*

The above indicates that for mean annual discharge everywhere in the basin, the increase in depth compensates – or slightly overcompensates – for the decreasing river slope in the downstream direction with a net result that velocity is nearly constant everywhere in the basin. The nature of this compensation, in case of uniform flow, is reflected in Manning's equation relating velocity v to depth of flow (d), slope ($|\nabla z|$ assumed as the mean slope representative of a river reach), and channel roughness (embedded in Manning's coefficient n [$s/L^{1/3}$]):

$$v = \frac{1}{n}\, d^{2/3} \sqrt{|\nabla z|}, \tag{2.2}$$

where d approximates the hydraulic radius for large enough widths compared to depth. Equation (2.2) postulates that in the flow direction x, the component of the weight balances flow resistance and the effects of shear stresses. Thus the flow does not accelerate (i.e., $\partial v/\partial x = \partial v/\partial t = 0$), nor does it develop steady, gradually varied flow profiles ($\partial d/\partial x \simeq 0$). A note for ecologists: the fact that velocity remains relatively constant moving downstream notwithstanding major changes in the slope $|\nabla z|$ (at times referred to as S_0, which is the standard expression for channel slope under uniform flow conditions) is a consequence of decreasing flow resistance embedded in Manning's coefficient rather than a dynamic response of d. Under such conditions, usually applicable to large fluvial settings, one observes that velocity depends on the depth to the power 2/3 and on the slope to the power 1/2. Empirical observations of mean velocity at a series of gauging stations along the course of several rivers were obtained [101] from flood measurement data interpolated to represent floods of 50- and 5-year recurrence intervals. The main result (also reported within a context akin to that relevant to this book [63]) is that velocity remains essentially constant in the downstream direction at each flood frequency. This has noteworthy implications for the analytical tractability of a number of hydrological problems (see, e.g., Section 3.3 and Box 3.8), in particular when referring to landscape-forming discharges. The constant characteristic velocity v everywhere in the network is an important feature for hydrologic controls on biota. It is important to notice that a value of $m = 0$ in Equation (2.1) suggests values of 0.5 for the exponents b and f, which control the power law variation of width and depth with respect to discharge – thus of habitat size for stream ecosystems. For a more in-depth discussion of the contents of this box, the reader is referred to [63].

Box 2.3 Total Contributing Area At-a-Site as a Catchment Master Variable

Another regular factor among different river basins is the relation between discharge of a given frequency of occurrence and drainage area. Bankfull discharge has a recurrence interval averaging approximately 1.5 years [101], and its relation to total contributing area A is approximately given by an equation of the type

$$Q \propto A^{0.75}. \tag{2.3}$$

The mean annual discharge usually fills a river channel to about one-third of its bankfull depth, and it tends to have a similar frequency of occurrence among rivers of all kinds of different features. This flow is equaled or exceeded on average about 25 percent of the time. It represents roughly the

Box 2.3 *Continued*

discharge exceeded 1 day in every 4 over a large period of time. An important feature of the mean annual discharge is that even though it may not occur at all points in the drainage network on any given day of record, it is nevertheless closely approximated everywhere during many days of the year. For mean annual flows the exponent in the discharge versus area relationship is about unit, that is,

$$Q \propto A. \tag{2.4}$$

This is especially the case for river basins showing minor differences in mean annual rainfall for different locations inside the watershed. The lower exponent 0.75 at higher flows is a measure of storage in river valleys and the consequence of the fact that rains of high intensity rarely cover the entire basin but are instead irregularly spaced [101]. An example of application of Equation (2.4) is given in Figure 2.4 [151], where mean annual discharge is plotted versus area for all gauging stations on the Potomac river basin.

When studying the spatial distribution of streamflow-dependent quantities, it is usually the case that one does not have flow measurements throughout a major part of the links that make up the drainage network. Thus equations like (2.3) and (2.4) are of great utility because they allow the use of drainage area as a surrogate variable for discharge, whether bankfull or mean annual flow.

Microbial communities orchestrate most biogeochemical processes on Earth. In streams and rivers, surface-attached and matrix-enclosed biofilms dominate

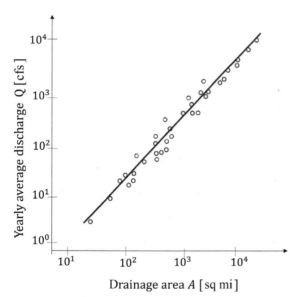

Drainage area A [sq mi]

Figure 2.4 Relation of discharge to drainage area for all gauging stations in the Potomac River. Line has slope 1. Figure after [151]

microbial life [152]. Despite their relevance for ecosystem processes (e.g., metabolism and nutrient cycling), it still remains unclear how features inherent to stream and river networks affect the fundamental organization of biofilm communities. The response of microbial community organization and persistence to human pressures that increasingly change the hydrological regime affects biodiversity dynamics in fluvial networks. One may therefore argue that streamflows are indeed the master variable of stream ecology, and their distribution (or their alterations thereof) carries critical ecological information [153]. Empirical, experimental, and theoretical studies in fluvial ecohydrology support such a statement. For example, hydrologic variability has been experimentally shown to affect invertebrate grazing on phototrophic biofilms [154]. As a result, stochastic flow regimes, characterized by suitable fluctuations and temporal persistence, offer increased windows of opportunity for grazing under favorable shear stress conditions. This was speculated to bear important implications for the development of comprehensive schemes for water resources management and for the understanding of trophic carbon transfer in stream food webs [154]. Such findings unravel small-scale trophic processes and how these may change as a streamflow regime is altered. Probabilistic approaches aiming at spatially explicit quantitative assessments of benthic invertebrate abundance as derived from near-bed flow variability also exist (e.g., [155]).

The increasing diversity of criteria of water resource management implies different alterations to natural streamflow regimes, which may, in turn, have severe effects on fluvial ecosystem structure and function [153, 156–158]. Indeed, effective management and restoration of fluvial ecosystems should include assessments of impacts on ecosystem processes. Alterations simply maintaining a minimum constant flowrate (even if equal to the mean of a fully stochastic distribution) as an environmentally conscious management strategy are inadequate to fully preserve ecosystem integrity.

Thus streamflows are key for the purposes of this book. Landscape and climate alterations foreshadow global-scale shifts of river flow regimes. A theory that identifies the range of impacts on streamflows resulting from inhomogeneous forcings across diverse regimes has been recently developed [159]. Therein, a measurable index was derived (Figure 2.5) that embeds climate and landscape attributes that discriminate erratic regimes with enhanced intraseasonal streamflow variability from persistent regimes endowed with regular flow patterns. Theoretical and empirical data (Figure 2.5) show that erratic hydrological regimes (high coefficient of variation of the streamflow Q, CV_Q) are resilient in that they hold a reduced sensitivity to climate fluctuations. The distinction between erratic and persistent regimes thus provides a robust framework for characterizing the hydrology of freshwater ecosystems and improving water management strategies in times of global change [159].

Box 2.4 Streamflow Distributions and River Flow Regimes

Analytical characterizations of flow regimes exist [160]. Daily streamflow dynamics are assumed to result from the superposition of a sequence of flow pulses triggered by precipitation, suitably censored by (catchment-scale) soil moisture dynamics [161]. In particular, the sequence of streamflow-producing rainfall events during a given season is approximated by a Poisson process [93] characterized by frequency of arrival λ_p and exponentially distributed rainfall depths with mean α. The reduced frequency of effective rainfall events λ with respect to the precipitation frequency λ_p [159] expresses the ability of the catchment to filter the rainfall forcing by exploiting certain inputs to fill the soil-water deficit created by plant transpiration. If subsurface environments were assumed to behave like a linear storage with rate constant k, each pulse h_i would determine a sudden increase of the streamflow followed by an exponential-like recession with rate k. Under these circumstances, the discharge at time t, $Q(t)$, is expressed by [160, 161]

$$Q(t) = A \sum_{t_i \le t} h_i \exp[-k(t-t_i)], \tag{2.5}$$

where A is the catchment area and the couples $(t_i; h_i)$ identify the arrival time and the depth of the ith pulse. The overall catchment-scale runoff Q is thus linked to the temporal evolution of the deep infiltration depths, $Y(t)$, originating from the relationship $Q(t) = kA \int_0^t Y(t-\tau) \exp[-k(t-\tau)]d\tau$ [160]. A Langevin equation for runoff could thus be obtained by taking the derivative of both sides with respect to t, yielding [160]

$$\frac{\partial Q(t)}{\partial t} = -kQ(t) + kA\,\xi_t, \tag{2.6}$$

where the first term on the right-hand side represents the deterministic exponential recession of $Q(t)$ due to the slow release of catchment water and the last term on the right-hand side represents the stochastic rate of runoff-forming events due to the inputs $\xi_t(\lambda) = \sum_j Y_j \delta(t-t_j)$. The intervals between subsequent deep percolation events, $t_{j+1} - t_j$, are exponentially distributed with parameter λ (exactly as the total effective rainfall). Accordingly, the master equation [161] for the probability distribution of discharge, $p(Q,t)$, can be written as [160, 161]

Box 2.4 *Continued*

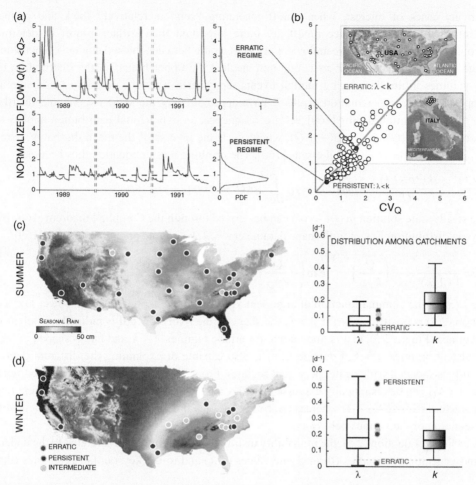

Figure 2.5 Analytical and empirical classification of river flow regimes as erratic or persistent. (a) Typical behavior of river flow dynamics in erratic and persistent regimes. Persistent regimes are characterized by enhanced frequencies of events that decrease the flow variability. (b) Ratio between the mean frequency of flow-producing rainfall events, λ [T^{-1}], and the inverse of their characteristic hydrograph recession time, k [T^{-1}], explaining most of the observed intraseasonal flow variability ($R^2 = 0.52$). This is supported by the scatterplot of the observed coefficient of variation CV_Q versus the corresponding estimate of the empirical values of $\sqrt{k/\lambda}$, representing the theoretical prediction of the analytical model (Equation (2.9)). Each circle identifies a given catchment during a season. Maps show the locations of the 44 study catchments used in [159]. (c, d) Spatial distribution of the flow regimes among the US study catchments during summer (June 1 to August 31) and winter (December 1 to February 28), supported by the corresponding box plot of the frequency distribution of λ and k. Based on the average value of $CV_Q = \sqrt{k/\lambda}$, catchment regimes are classified as persistent ($CV_Q < 0.9$), intermediate ($0.9 \le CV_Q \le 1.1$), or erratic ($CV_Q > 1.1$). Figure after [159]

$$\frac{\partial p(Q,t)}{\partial t} = \frac{\partial k Q p(Q,t)}{\partial Q} - \lambda p(Q,t) + \lambda \int_0^Q p(Q-z,t)b(z)dz, \tag{2.7}$$

where $b(Q)$ indicates the distribution of runoff increments $\Delta Q = kAY$ [160].

Box 2.4 *Continued*

In many cases of interest, when runoff-generating "soil" is relatively thick and permeable, deep infiltration and subsurface runoff are more important than surface runoff in determining the probabilistic structure of streamflows. Let r be the ratio between the soil storage capacity and the average daily rainfall. When $r \gg 1$, one may neglect the upper bound in the distribution of the water volumes infiltrating during rainfall events represented by soil saturation. Under the above assumption, the deep percolation depths Y have approximately the same exponential distribution of the effective rainfall depths [160]. As a consequence, an exponential distribution applies to the discharge jumps $b(Q) = \gamma_Q \exp(-\gamma_Q Q)$, where γ_Q is the inverse of the mean discharge increment due to incoming effective rainfall events [160]. The resulting master equation thus becomes

$$\frac{\partial p(Q,t)}{\partial t} = \frac{\partial k Q p(Q,t)}{\partial Q} - \lambda p(Q,t) + \lambda \gamma_Q \int_0^Q p(Q-z,t) \exp(-\gamma_Q z) dz, \tag{2.8}$$

whose steady state solution $p(Q, t \to \infty)$ can be derived through the Campbell theorem [160, 161] as a Gamma distribution $\Gamma(s,r)$ with shape parameter $s = \lambda/k$ and rate parameter $r = \alpha k$ [159]:

$$p(Q) = \frac{1}{\Gamma(\lambda/k)\gamma_Q k} \left(\frac{Q}{k\gamma_Q}\right)^{\frac{\lambda}{k}-1} \exp\left(-\frac{Q}{k\gamma_Q}\right), \tag{2.9}$$

where $\Gamma(x)$ is the Gamma function of argument x. Because the exponent of the power law term in Equation (2.9) is positive only for $\lambda > k$, the shape of the river flow pdf is radically different in the two regimes (Figure 2.5), that is, monotonic for erratic regimes $\lambda < k$ and hump shaped in the case of persistent regimes $\lambda > k$. Equation (2.9) is also capable of explaining the different degrees of variability associated with erratic/persistent regimes. From Equation (2.9), in fact, the coefficient of variation CV_Q can be analytically expressed as $CV_Q = \sqrt{k/\lambda}$ [159], implying that persistent regimes are characterized by $CV_Q < 1$, whereas erratic regimes featured $CV_Q > 1$. Figure 2.6 is a qualitative representation of the various behaviors.

According to Equation (2.8), the probability distribution of streamflows is related to the underlying soil and vegetation properties (through the values of λ) and to the key rainfall properties (through

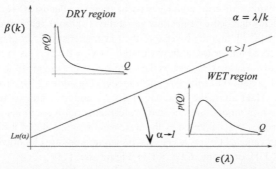

Figure 2.6 Qualitative representation of the dependence of the daily runoff pdf on the dimensionless parameters: $\alpha = \lambda/k$; $\beta(k)$ is a dimensionless catchment storage derived from the master equation (2.8), whose detailed specification is immaterial in this context. Suffice here to mention that β depends chiefly on the value of the recession constant k, and $\epsilon(\lambda)$ is a function of the frequency of runoff-producing rainfall events, a fraction of the Poissonian frequency of daily rainfall interarrival λ. The solid line represents the condition $\lambda = k$ (or $\alpha = 1$) that determines the transition from the dry to the wet regions in the (ϵ, β) plane. Figure after [160]

Box 2.4 *Continued*

both γ_Q and λ). It also depends on important geomorphic factors, such as the characteristic hydrograph recession time $(1/k)$ and the size of the river basin (A). When $\lambda/k > 1$ ("wet conditions"), the pdf of the runoff is bell shaped where low discharges, $Q \to 0$, are characterized by null probability (Figure 2.6), while for $\lambda/k < 1$ ("dry conditions"), $p(Q)$ is a mixed distribution with an atom of probability at $Q = 0$. The pdf otherwise monotonically decreases for $Q > 0$, approaching zero as $Q \to \infty$. The critical condition $\lambda = k$, which determines the shift between the "wet regime" and the "dry regime," can be expressed in terms of basic rainfall, soil, and vegetation properties by the use of Equation (2.9). Among the many consequences, one notes that drybeds are hotspots of carbon sequestration, and therefore whether or not $Q = 0$ occurs for spatially extended river domains has important consequences for biogeochemical cycles [152, 162].

Although significant assumptions restrict the generality of Equations (2.5) and (2.9), the analytical expression of the flow pdf in terms of three physically based measurable parameters embedding rainfall, soil, vegetation, and morphological attributes of the contributing catchment is very powerful. Most of these assumptions, however, can be suitably relaxed, allowing power law recessions, spatial/temporal variables k, and heterogeneous rainfall/landscape attributes to be tackled properly in the same framework [159].

An application relevant to the tenet of this section concerns the possibility of drawing spatially explicit maps of specific aquatic organisms' habitat suitability [155]. Hydrologic controls may be identified on basin-scale distributions, say, of benthic invertebrates. To synthesize hydrologic controls, one may use a habitat suitability curve [155] that describes the effects of any generic environmental variable on species behavior and distribution and is a fundamental tool to describe species habitat preferences. Independently validated relationships exist between bottom shear stress and suitability for benthic invertebrates that were combined with the shear stress probability distribution function derived by using the framework of Boxes 2.2–2.4. In fact, the average shear stress τ exerted by streamflow drained by a catchment of area A across a river cross section is approximately given by

$$\tau = \rho g \, d \, S_0,$$

where $\rho g = \gamma$ [N/m^3] is the specific weight of water; d is the average flow depth at a station, approximating hydraulic radius (Box 2.2); and $S_0 = \langle |\nabla z| \rangle$ is a suitably averaged local slope of the channel reach where approximate uniform flow conditions are assumed $(\partial d/\partial x \simeq 0)$. By employing uniform flow–like relations, and the scaling relations used for cross section features at-a-station (Box 2.2), one obtains relations of the type $\tau = C_\tau Q^e$, from which the analytical expression

of the site-specific probability distribution function of average bottom shear stress can be obtained as a derived distribution $(p(Q)dQ = p(\tau)d\tau)$. The final result is [155]

$$p(\tau) = \frac{\theta}{e\Gamma(\lambda/k)} \, (\theta\tau)^{\frac{\lambda}{ek}-1} \, \exp{(-\theta\tau)^{1/e}}, \qquad \textbf{(2.10)}$$

where $\theta = (\alpha A k)^{-e}/C_\tau$ is a parameter proportional to the mean streamflow increment due to incoming effective rainfall events. The proportionality constant C_τ may thus be estimated directly from hydrologic data by assuming that (1) flow conditions are uniform; (2) the river cross section is approximated by a rectangular shape of area wd [L^2], where w [L] and d [L] are river width and water depth, respectively; and (3) the river width is much larger than the water depth (i.e., $w \gg d$) [155]. Note that one has exactly

$$\langle Q \rangle = \lambda A \gamma_Q$$

and

$$\langle \tau \rangle = \left(\frac{1}{\theta}\right) \frac{\Gamma\left[e(\frac{\lambda}{ek})\right]}{\Gamma(\lambda/k)},$$

among other exact relations [155]. Figure 2.7 shows an application of the model Equation (2.10) to the Ybbs river network in Austria.

Network-wide hydraulic controls on river habitat suitability are thus within sight. In fact, the previously presented distribution $p(\tau)$ allows one to build a framework

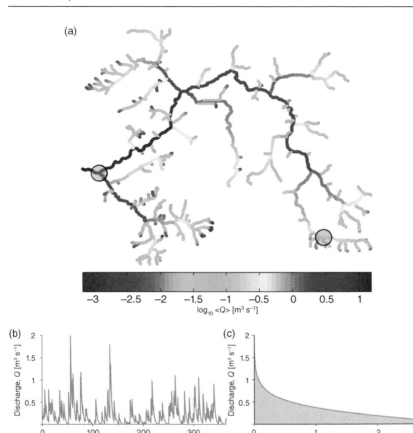

(a)

$\log_{10} <Q> \, [\text{m}^3 \, \text{s}^{-1}]$

Figure 2.7 Discharge variability along the Ybbs river network in Austria [155] as realizations of the modeling of a stochastic process (Equation (2.6)). (a) Average discharge $\langle Q \rangle$. (b, d) Time series. (c, e) Probability distribution functions for two locations, respectively named Goestling (light blue) and Ybbs South (orange). Figure after [155]

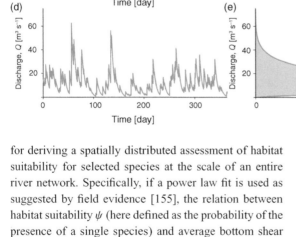

(b)

(c)

(d)

(e)

for deriving a spatially distributed assessment of habitat suitability for selected species at the scale of an entire river network. Specifically, if a power law fit is used as suggested by field evidence [155], the relation between habitat suitability ψ (here defined as the probability of the presence of a single species) and average bottom shear stress τ may be characterized as

$$\psi = c_\psi \, \tau(Q)^{e_\psi} .$$

As a consequence, the characterization of the pdf of random habitat suitabilities ψ to shear stress τ follows as [155]

$$p(\psi) = \frac{\theta_\psi}{e_\psi \Gamma(\lambda/k)} \left(\theta_\psi \tau \right)^{\frac{\lambda}{e_\psi k} - 1} \exp\left(-\theta_\psi \tau \right)^{1/e_\psi} ,$$

(2.11)

where $\theta_\psi = (\alpha A k)^{-e_\psi} / c_\psi$ is a function of the geomorphic and hydrologic properties of the river network and of the ecological traits of the considered species. Possible shapes of the analytical pdf of species habitat suitability (Equation (2.11)) may then be expressed in terms of different parameter combinations. Irrespective of whether a power law relation between habitat suitability and shear stress applies, the method provides a tool to sort out the spatial distribution of a species' relative occurrence probability, by using numerical simulation to transfer the time series of τ into a time series of ψ. In analogy with the performed analyses for discharge and bottom shear stress (Figures 2.5–2.8), the species habitat suitability temporal mean $\langle \psi \rangle$, variance σ_ψ^2, and coefficient of variation CV_ψ can be evaluated along any station of a river network

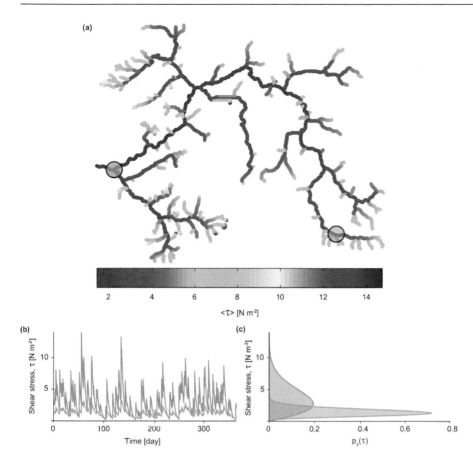

Figure 2.8 Shear stress variability along the Ybbs river network as realizations of the modeling of a stochastic process (Equation (2.10)). (a) $\langle \tau \rangle$. (b) Time series. (c) Probability distribution functions for the locations Goestling (light blue) and Ybbs South (orange). Figure after [155]

owing to the link $Q \propto A$ (Box 2.3), thus providing relevant information about benthic conditions in streams and rivers.

An example follows. From the spatial distribution of bottom shear stress along the Ybbs river network (Figure 2.8), a first-order ecological extension has been performed [155], aiming at the characterization of spatially explicit probability distribution functions of benthic invertebrate habitat suitability to shear stress. Invertebrate habitat suitability curves are derived for the Ybbs and expressed in terms of species density (i.e., normalized to the maximum observed areal abundance) as a function of shear stress (Figure 2.9). Suitability values lie between 0 and 1, where $\psi = 0$ represents the habitat state characterized by the absence of invertebrates and $\psi = 1$ corresponds to the maximum species density achievable. In accordance with the hydrological analysis, the summer generations of three mayfly species, namely, *B. muticus*, *Baetis rhodani*, and *Ecdyonurus venosus*, were considered. Recalling Equation (2.11), which defines an analytical relation between habitat suitability and bottom shear stress, proper parameters were estimated (see

Table 2 in [155]). In brief, *B. muticus* and *E. venosus* habitat suitability curves for the summer generation show a monotonically decreasing trend of normalized individual densities with increasing bottom shear stress, thus revealing in principle a preference for downstream reaches, characterized by relatively small shear stress values if compared to low-order branches (Figure 2.9). Conversely, *B. rhodani* presents a positive exponent of the power law relation, which results in a decreasing downstream suitability to shear stress. Average suitabilities ψ along the Ybbs river network, exclusively influenced by near-bed flow conditions, are estimated from Equation (2.11) for each of the considered species (Figure 2.9). When comparing the modeled habitat suitabilities for the three mayfly species, the analysis highlights higher ψ values for *B. rhodani* and *B. muticus*. Therefore, near-bed flow conditions along the Ybbs river network prove more favorable for these two mayfly species, as noted in the field.

The relevance of this approach, however imperfect in its assumptions that oversimplify important hydrological and ecological issues, lies – in our view – in the

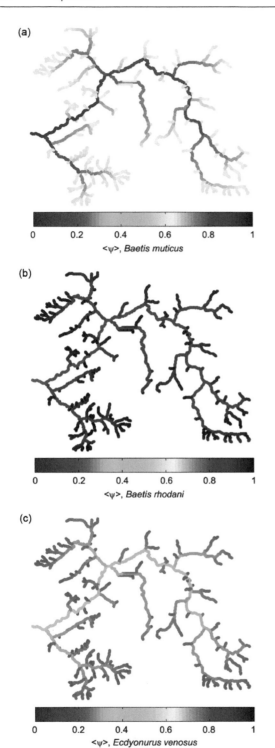

(a)

0   0.2   0.4   0.6   0.8   1
⟨ψ⟩, *Baetis muticus*

(b)

0   0.2   0.4   0.6   0.8   1
⟨ψ⟩, *Baetis rhodani*

(c)

0   0.2   0.4   0.6   0.8   1
⟨ψ⟩, *Ecdyonurus venosus*

Figure 2.9 Average habitat suitability based on shear stress along the Ybbs river network as derived from Equations (2.10) and (2.11) for the mayfly species (a) *Baetis muticus*, (b) *Baetis rhodani*, and (c) *Ecdyonurus venosus*. Figure after [155]

connection of the spatiotemporal patterns of hydrologic conditions along a river network to the habitat suitability, considering the whole range of discharges Q and shear stresses τ. The average $\langle\psi\rangle$ may thus be of use for conservation biology and water resources management. A notable feature is the shift in focus from the typical observation of benthic communities at the scale of a river reach to the level of entire stream networks. Conservation strategies based on metacommunity ecology in a dendritic landscape are therefore in sight [35].

Imprints of the river network structure on microbial co-occurrence networks [163] are mentioned in chapter 5. Their results suggest that hydrology and metacommunity dynamics, both changing predictably across fluvial networks, affect the fragmentation of the microbial co-occurrence networks throughout the fluvial network. Key to such a result is again the mapping of the parameter λ/k, where λ can be reasonably assumed to be constant and area-independent, whereas k was found to scale as a function of total contributing area A, that is, $k \propto A^\zeta$, where $\zeta < 1$ may assume a narrow range of values that depend chiefly on drainage density [63]. This approach has been used to study microbial co-occurrence networks in the Ybbs catchment [163] (see Chapter 5).

Given the extent of anthropogenic disturbance in most river networks, it is important to move beyond the scale of an individual reach for ecohydrological studies. Shifts in hydrological regime, for example, due to interbasin water transfer and damming [159] or to climate change, substantially affect biodiversity patterns [150, 155, 164]. Therefore, models of the type presented here offer a framework to assess such changes – for benthic invertebrate biodiversity, as shown in this section, but quite possibly for many other species as well.

2.2 Metapopulation Persistence and Species Spread in River Networks

As we hinted in the introduction to this book, we shall now broaden our aims by addressing general metapopulation persistence in river networks, following a rather general approach [115]. Indeed, river networks define ecological corridors characterized by unidirectional streamflow and a complex topological connectivity. In quite a few cases, rivers impose significant constraints to aquatic organisms that may strongly affect their movement, like downstream drift. Animals and plants manage to persist in riverine ecosystems, however, which in fact harbor high biological diversity. Here, following [115], we study metapopulation dynamics by analyzing stage-structured populations that exploit different dispersal pathways, both alongstream and overland. Key to the results contrived is the use of stability analysis (Appendix 6.1), from which a novel criterion for metapopulation persistence in arbitrarily complex landscapes described as spatial networks is discussed.

Central to the book's tenet is the demonstration of how dendritic geometry and overland dispersal can promote population persistence to explain the so-called drift paradox. This represents a long-standing issue in freshwater ecology [165, 166]. Strong unidirectional water flow, in fact, imposes a downstream drift to the movement of aquatic organisms, dependent on a number of specific factors. In the absence of mechanisms allowing for upstream colonization, the persistence of riverine populations would hardly be possible in the presence of a strong downstream bias in the movement of any species. This is the essence of the drift paradox, which proves particularly relevant to nonsessile organisms with relatively low self-propelled motion capacity, such as the larval stages typical of many freshwater species [115]. To explain the long-term persistence of such populations, several mechanisms have been proposed. As first empirically documented by [167] for Scandinavian freshwater ecosystems, many insect species compensate larval drift with upstream-directed flight of adults prior to oviposition, in what is termed Müller's colonization cycle [168]. As an alternative explanation, an excess production hypothesis has been put forth [169], by which drifting organisms would be those exceeding the balance of demography at the local scale. This, in turn, would implicitly assume that drift essentially represents an extramortality term in the mathematical description of the relevant ecology. Also, the broad range of heterogeneities found in the hydrodynamics of natural streams has been suggested to promote organism retention in hydrodynamic in-flow refugia [170–172].

It has been argued [115] that the passive movement of an aquatic organism in a river system is the end product of a large number of hydrodynamic effects resulting from the combination of advection, triggered by complex velocity profiles affecting the mean flow velocity and its fluctuations [173], and dispersion (say, the asymptotically diffusive effects of correlated displacements) determined by local natural streamflow heterogeneities [173]. The so-called active organismic movement, the net displacement occurring by swimming, crawling, or flying

(either directly or through some dispersal vector), and the geomorphological dispersion [174] resulting from the multiplicity of sources feeding the flow at a given riverine cross section further increase the overall macroscopic diffusive displacement of organisms. In many cases of ecological relevance, the latter may become the predominant factor. All these factors act within the distinctive landscape topology of fluvial environments, which is characterized by hierarchical branching geometries endowed with universal scaling features, as recalled in the introduction. Fluvial ecosystems are in fact perhaps the most representative examples of dendritic ecological networks [175]. As noted in Section 2.1, diffusive dispersal [165, 176] and river network topology [6, 164] have been recently proposed as key factors affecting the persistence of riverine populations. Topology is particularly important in constraining the dispersal of aquatic species lacking life stages that can disperse overland [6] (see also Chapter 3, in the context of biological invasions). This mode of dispersal has been both theoretically postulated [177] and experimentally observed [14] to facilitate the persistence of riverine populations. Dispersal can occur at different life stages, most frequently early in the life history of aquatic organisms. As an example, in a massive mark-recapture study [150] of two lungless salamander species in stream networks of Virginia, the newly metamorphosed (juvenile) salamanders had the highest probabilities of dispersing to other stream reaches, thus being primarily responsible for overland connections. While it is relatively common to find freshwater organisms that begin their life cycles as motile and reach maturity as sessile (e.g., mussels), there are notable exceptions – as in the case of parasites with complex life cycles that involve intermediate hosts with low motility (e.g., snails) and final hosts with high motility (e.g., fish; see [178]).

Despite their importance, diffusive dispersal, landscape geometry, stage-dependent movement, and exploitation of multiple dispersal pathways have been analyzed together only recently [115] to yield a comprehensive description of the conditions leading to the persistence and spread of riverine populations. In fact, classical approaches include only the analysis of reaction–advection–diffusion [165, 176], integro-differential [177], or integro-difference equations [166] in simple linear, one-dimensional landscapes. Dendritic geometries have been considered in simulation studies of individual-based models [6, 164] and in matrix population models applied to stage-structured populations in networks of habitat patches [179]. This

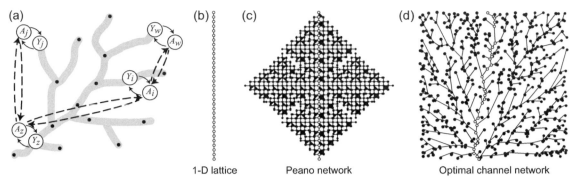

Figure 2.10 Schematic of the metapopulation model and of the river networks used in numerical analyses. (a) A hypothetical sketch of our multilayer network model (2.12); hydrological connections (in the example involving juveniles only) are in light gray, while overland connections (in the example of adults only) are represented as dashed arrows. (b–d) Different river network topologies; the backbone of each hydrological network has the same number of nodes ($n_b = 33$) and the same arbitrary length ($L_b = 33$) independently of topology and is indicated by the white-filled nodes. The southmost node is the network outlet. Figure after [115]

study is particularly interesting because it is devoted to the analysis of branching spatial structure and life history on the asymptotic growth rate of a riverine population – with clear implications for population persistence. Also, analytical results for the persistence of a population subject to advection and diffusion on a tree graph exist [180].

2.2.1 A Metapopulation Approach

Following [115], we consider a prototypical aquatic metapopulation living in a river network constituted by n nodes (Figure 2.10a). Each node is deemed representative of a river stretch – a distinct homogeneous enclave – where local ecological conditions are identical. In our abstract example, the species is assumed to hold two ecologically distinct developmental stages, the population of young (nonreproductive) individuals (Y) and that of adult (reproductive) individuals (A). Movement from node to node may occur through different pathways, either along the stream network or overland. Local demographic processes (birth, growth, and death) and dispersal dynamics in each node i of the river network are described by the following system of $2n$ ordinary differential equations:

$$\frac{dY_i}{dt} = -\mathcal{M}_Y(Y_i, A_i)Y_i - \gamma Y_i$$

$$+ \mathcal{N}(Y_i, A_i)A_i - \sum_{h=1}^{N_Y} l_h\left(Y_i - \sum_{j=1}^{n} P_{hji}Y_j\right)$$

$$\frac{dA_i}{dt} = -\mathcal{M}_A(Y_i, A_i)A_i + \gamma Y_i$$

$$- \sum_{k=1}^{N_A} m_k\left(A_i - \sum_{j=1}^{n} Q_{kji}A_j\right) \qquad (2.12)$$

where $\mathcal{M}_Y(Y_i, A_i)$ ($\mathcal{M}_A(Y_i, A_i)$) is the per capita mortality rate for juveniles (adults); γ is the rate at which young individuals become adult ($1/\gamma$ is thus the species' average duration of the juvenile phase); $\mathcal{N}(Y_i, A_i)$ is the natality rate of adults; l_h (m_k) is the rate at which young (adult) organisms undergo dispersal along the hth (kth) pathway ($h = 1 \dots N_Y$ [$k = 1 \dots N_A$] being the number of possible dispersal mechanisms for juveniles [adults]); and P_{hji} (Q_{kji}) is the fraction of young [adult] organisms moving from node j to node i through the hth (kth) dispersal mechanism available to juveniles (adults). We assume that the mortality (natality) rate is a monotonically increasing (decreasing) function of population density ($\partial\mathcal{M}_{Y,A}/\partial(Y_i, A_i) \geq 0$, $\partial\mathcal{N}/\partial(Y_i, A_i) \leq 0$ for any i), that is, that there is no Allee effect in which density dependence (i.e., a function of the state variable) affects the processes' rates. Note that the assumption of spatial homogeneity of the parameters can be relaxed and that the model can also be easily extended to describe populations with more complex age/stage structures (a list of mathematical symbols is provided in Table 2.1).

2.2.2 Connectivity Structures and Dispersal Mechanisms

Dispersal probabilities P_{hji} and Q_{kji} depend on the connectivity structure provided by the environmental

Table 2.1 Mathematical symbols used in the text and their definitions.

Symbol	Definition
	State variables
Y_i	Abundance of juveniles in node i
A_i	Abundance of adults in node i
	Network geometry
n	Number of nodes in the network
n_b	Number of nodes in the network backbone
L_b	Length of the network backbone
	Demographic parameters
$\mathcal{M}_Y$	Density-dependent mortality rate of juveniles
μ_Y	Mortality rate of adults at low population density
$\mathcal{M}_A$	Density-dependent mortality rate of adults
μ_A	Mortality rate of adults at low population density
$\mathcal{N}$	Density-dependent natality rate
ν	Natality rate at low population density
γ	Rate at which juveniles reach maturity
	Dispersal parameters
N_Y	Number of dispersal pathways available to juveniles
l_h	Dispersal rate of juveniles along the hth pathway
$\mathbf{P_h}$	Connectivity matrix for the hth juveniles' dispersal pathway
N_A	Number of dispersal pathways available to adult individuals
m_k	Dispersal rate of adults along the kth pathway
$\mathbf{Q_k}$	Connectivity matrix for the kth adults' dispersal pathway
	Persistence criterion
$\mathbf{X_0}$	Extinction equilibrium
$\mathbf{I_n}$	Identity matrix of size n
R_0	Reproduction number
$\mathbf{J}$	Jacobian matrix of size $2n$ associated with model (2.12) (Box 2.5)
$\mathbf{J^\star}$	Matrix of size n deducible from $\mathbf{J}$ (Box 2.5)
E_0	Dominant eigenvalue of matrix $\mathbf{J^\star}$ (Box 2.5)
	Figure 2.11
l_1	Juveniles' alongstream dispersal rate
$\mathbf{F}$	Hydrological connectivity matrix
b	Bias of alongstream dispersal (Box 2.8)
l_2	Juveniles' overland dispersal rate
$\mathbf{G}$	Connectivity matrix for overland dispersal (Box 2.7)
D	Average distance of the overland dispersal kernel (Box 2.7)
K	Total movement rate
ϕ	Fraction of the total movement rate allocated to overland dispersal

matrix and the dispersal mechanisms relevant to the metapopulation being investigated. As for connectivity, we consider three hypothetical network structures for theoretical analyses and two real river networks for paradigmatic case studies. The hypothetical networks considered here are a 1D lattice (Figure 2.10b); a deterministic fractal, namely, a Peano construct (Figure 2.10c); and an OCN (Figure 2.10d; see Box 1.3). While the lattice geometry clearly represents an oversimplification of real river networks, although often used to study population persistence in riverine ecosystems (see, e.g, [165, 177]), Peano's topological measures match those of real river networks and thus (see Section 1.5.1) closely represent the features relevant to biological invasions. OCNs represent a further step forward, however, in that their topological and metric properties are virtually indistinguishable from those of real river networks (Box 1.3).

As for dispersal pathways, the first and foremost mechanism to be considered in a riverine setting is alongstream aquatic dispersal, which may describe both hydrological drift and active movement along river corridors. Other mechanisms can be relevant to the dispersal of riverine populations as well. For instance, flying or human/animal-mediated transport processes [108] could be only partially constrained by river network geometry and flow direction (for empirical evidence concerning insect flight, see, e.g., [181, 182]), thus potentially providing aquatic organisms with suitable pathways for unbiased overland dispersal. This can be described by, for example, an exponential kernel [183], but other, possibly *ad hoc* mechanisms can obviously be introduced to describe dispersal in species-specific case studies.

Dispersal probabilities are subsumed into connection matrices, namely, $\mathbf{P_h} = \left[P_{hij}\right]$ and $\mathbf{Q_k} = \left[Q_{kij}\right]$. We assume that $\sum_{j=1}^{n} P_{hij} \leq 1$ and $\sum_{j=1}^{n} Q_{kij} \leq 1$ for any i, h, and k. Specifically, rowwise sums can be less than 1 in the presence of absorbing boundary conditions (Box 2.8) and/or costly dispersal [184], which both imply the nonconservation of the abundance of dispersing organisms. Finally, the union of the graphs associated with the matrices $\mathbf{P_h}$ and $\mathbf{Q_k}$ is assumed to be strongly connected, so that it is always possible for the individuals of the focal species to find a path between any two nodes of the river network via the available dispersal pathways.

2.2.3 Derivation of Species' Persistence Conditions

Irrespectively of parameter values, the state $\mathbf{X_0}$ characterized by $Y_i = 0$ and $A_i = 0$ for any i is a global extinction equilibrium for model (2.12). In the absence of an Allee effect, metapopulation persistence is related to the stability of this equilibrium. In fact, if X_0 is stable, the population cannot persist in any of the river network nodes. On the contrary, if X_0 is unstable, juvenile and adult abundances, even if initially small, are expected to grow – thus granting metapopulation persistence. The condition for the extinction equilibrium to switch from stable to unstable is that the Jacobian matrix J of system (2.12) linearized at X_0 has one zero eigenvalue. Population persistence can thus be assessed by analyzing how the eigenvalues of J vary with model parameters, connectivity structures, and dispersal mechanisms. Technicalities of the derivation of the relevant conditions are reported in Box 2.5, after [115].

Box 2.5 Derivation of the Persistence Criterion

As the model is a positive system (namely, its state variables can never become negative if the system is initialized at generic nonnegative conditions), the bifurcation of the trivial equilibrium X_0 from stable to unstable can only occur via an exchange of stability. This implies that global extinction ($Y_i = 0$ and $A_i = 0$ for any i) switches from stable equilibrium to being a saddle (i.e., an equilibrium with one unstable manifold) through a so-called transcritical bifurcation (Appendix 6.1). The condition for the bifurcation to occur is that the Jacobian matrix J of system (2.12) linearized at X_0, that is,

$$J = \begin{bmatrix} -\left(\mu_Y + \gamma + \sum_{h=1}^{N_Y} l_h\right)\mathbf{I_n} + \sum_{h=1}^{N_Y} l_h \mathbf{P_h^T} & \nu \mathbf{I_n} \\ \gamma \mathbf{I_n} & -\left(\mu_A + \sum_{k=1}^{N_A} m_k\right)\mathbf{I_n} + \sum_{k=1}^{N_A} m_k \mathbf{Q_k^T} \end{bmatrix},$$

has one zero eigenvalue. In the previous expression, $\mathbf{I_n}$ is the identity matrix of dimension n, while $\mu_{Y,A} = \mathcal{M}_{Y,A}(0,0)$ and $\nu = \mathcal{N}(0,0)$ are respectively mortality and natality rates at low population densities (the reader should refer to Table 2.1 for a description of the parameters and the other quantities involved). Note that J is a Metzler matrix (Appendix 6.1); that is, all its off-diagonal entries are nonnegative [185], therefore its dominant eigenvector is real and unique.

We start with the simple case of isolated populations ($l_h = m_k = 0$ for any h and k). In this case, population persistence is determined by the stability properties of the Jacobian matrix

$$J_0 = \begin{bmatrix} -(\mu_Y + \gamma)\mathbf{I_n} & \nu \mathbf{I_n} \\ \gamma \mathbf{I_n} & -\mu_A \mathbf{I_n} \end{bmatrix}.$$

Specifically, the extinction equilibrium X_0 is asymptotically stable if and only if the dominant eigenvalue of J_0 is strictly negative. The change of stability for X_0 is thus associated to the condition $\det(J_0) = 0$. Because of the block structure of J_0, it can be proved [186] that

$$\det(J_0) = \det(\mu_A(\mu_Y + \gamma)\mathbf{I_n} - \nu\gamma \mathbf{I_n}).$$

Therefore the condition $\det(J_0) = 0$ is obviously verified whenever

$$R_0 = \frac{\nu}{\mu_A} \frac{\gamma}{\mu_Y + \gamma} = 1,$$

and the population can persist if and only if $R_0 > 1$. Note that the condition for population persistence is identical at local and network scales because of the spatial homogeneity of ecological conditions and the lack of interconnections among nodes.

It is much more interesting, although obviously more complicated, to analyze system (1) when taking into account population dispersal ($l_h > 0$, $m_k > 0$ for some h and k). Let us first note that if the union of the graphs associated with matrices $\mathbf{P_h}$ and $\mathbf{Q_k}$ is strongly connected, then the graph

Box 2.5 *Continued*

associated with $\mathbf{J}$ is also strongly connected. Therefore we can apply the Perron–Frobenius theorem for irreducible matrices [187] and state that the dominant eigenvalue of $\mathbf{J}$ is the maximum simple real root of the characteristic polynomial. The condition for the transcritical bifurcation of the extinction equilibrium is that the dominant eigenvalue crosses the imaginary axis at zero, namely, the determinant of $\mathbf{J}$ is zero [115, 188]. When the global extinction equilibrium is stable (sufficient condition for global population extinction), all the eigenvalues have negative real parts, and $\det(\mathbf{J})$ is positive because $\mathbf{J}$ is a matrix of order $2n$. The global extinction equilibrium becomes unstable (necessary condition for population persistence) when $\det(\mathbf{J})$ switches from positive to negative or, equivalently, the dominant eigenvalue switches from negative to positive. For block matrices of the kind

$$\mathbf{M} = \begin{bmatrix} \mathbf{A} & \mathbf{B} \\ \mathbf{C} & \mathbf{D} \end{bmatrix},$$

in which all blocks are square and $\mathbf{CD} = \mathbf{DC}$, the following equality holds [186]:

$$\det(\mathbf{M}) = \det(\mathbf{AD} - \mathbf{BC}).$$

Therefore, writing $\mathbf{J}$ as

$$\mathbf{J} = \begin{bmatrix} \mathbf{J}_1 & \mathbf{J}_2 \\ \mathbf{J}_3 & \mathbf{J}_4 \end{bmatrix}$$

and noting that $\mathbf{J}_3 = \gamma \mathbf{I}_n$ is a scalar matrix (thus commuting with any other matrix), we get $\det(\mathbf{J}) = \det(\mathbf{J}_1 \mathbf{J}_4 - \mathbf{J}_2 \mathbf{J}_3)$.

From the previous equation, with lengthy yet straightforward algebraic manipulations, we can compute

$$\det(\mathbf{J}) = \det\left[\left(\mu_Y + \gamma + \sum_h^{N_Y} l_h\right)\left(\mu_A + \sum_k^{N_A} m_k\right)\mathbf{I}_n - \left(\mu_Y + \gamma + \sum_h^{N_Y} l_h\right)\left(\sum_k^{N_A} m_k \mathbf{Q}_\mathbf{k}^T\right)\right.$$
$$\left. - \left(\sum_h^{N_Y} l_h \mathbf{P}_\mathbf{h}^T\right)\left(\mu_A + \sum_k^{N_A} m_k\right) + \left(\sum_h^{N_Y} l_h \mathbf{P}_\mathbf{h}^T\right)\left(\sum_k^{N_A} m_k \mathbf{Q}_\mathbf{k}^T\right) - \nu\gamma \mathbf{I}_n\right]$$

$$= \det\left[\mu_A(\mu_Y + \gamma)\mathbf{I}_n + \mu_A\left(\sum_h l_h\right)\mathbf{I}_n + (\mu_Y + \gamma)\left(\sum_k m_k\right)\mathbf{I}_n \right.$$
$$+ \left(\sum_h l_h\right)\left(\sum_k m_k\right)\mathbf{I}_n - (\mu_Y + \gamma)\left(\sum_k m_k \mathbf{Q}_\mathbf{k}^T\right) - \left(\sum_h l_h\right)\left(\sum_k m_k \mathbf{Q}_\mathbf{k}^T\right)$$
$$\left. - \mu_A\left(\sum_h l_h \mathbf{P}_\mathbf{h}^T\right) - \left(\sum_k m_k\right)\left(\sum_h l_h \mathbf{P}_\mathbf{h}^T\right) + \left(\sum_h l_h \mathbf{P}_\mathbf{h}^T\right)\left(\sum_k m_k \mathbf{Q}_\mathbf{k}^T\right) - \nu\gamma \mathbf{I}_n\right]$$

$$= \mu_A^n(\mu_Y + \gamma)^n \det\left[\mathbf{I}_n + \frac{\sum_h l_h}{\mu_Y + \gamma}\mathbf{I}_n + \frac{\sum_k m_k}{\mu_A}\mathbf{I}_n + \frac{(\sum_h l_h)(\sum_k m_k)}{\mu_A(\mu_Y + \gamma)}\mathbf{I}_n \right.$$
$$- \frac{1}{\mu_A}\sum_k m_k \mathbf{Q}_\mathbf{k}^T - \frac{\sum_h l_h}{\mu_A(\mu_Y + \gamma)}\sum_k m_k \mathbf{Q}_\mathbf{k}^T - \frac{1}{\mu_Y + \gamma}\sum_h l_h \mathbf{P}_\mathbf{h}^T - \frac{\sum_k m_k}{\mu_A(\mu_Y + \gamma)}\sum_h l_h \mathbf{P}_\mathbf{h}^T$$
$$\left. + \frac{1}{\mu_A(\mu_Y + \gamma)}\left(\sum_h l_h \mathbf{P}_\mathbf{h}^T\right)\left(\sum_k m_k \mathbf{Q}_\mathbf{k}^T\right) - \frac{\nu\gamma}{\mu_A(\mu_Y + \gamma)}\mathbf{I}_n\right]$$

Box 2.5 *Continued*

$$= \mu_A^n \left(\mu_Y + \gamma\right)^n \det\left[\mathbf{I_n} - R_0\mathbf{I_n} - \frac{1}{\mu_Y + \gamma} \sum_h l_h \left(\mathbf{P_h}^T - \mathbf{I_n}\right) - \frac{1}{\mu_A} \sum_k m_k \left(\mathbf{Q_k}^T - \mathbf{I_n}\right) \right.$$

$$\left. + \frac{\sum_h l_h}{\mu_A \left(\mu_Y + \gamma\right)} \left(\sum_k m_k \mathbf{I_n} - \sum_k m_k \mathbf{Q_k}^T\right) - \frac{1}{\mu_A \left(\mu_Y + \gamma\right)} \left(\sum_h l_h \mathbf{P_h}^T\right)\left(\sum_k m_k \mathbf{I_n} - \sum_k m_k \mathbf{Q_k}^T\right) \right]$$

$$= \mu_A^n \left(\mu_Y + \gamma\right)^n \det\left[\mathbf{I_n} - R_0\mathbf{I_n} - \frac{1}{\mu_Y + \gamma} \sum_h l_h \left(\mathbf{P_h}^T - \mathbf{I_n}\right) - \frac{1}{\mu_A} \sum_k m_k \left(\mathbf{Q_k}^T - \mathbf{I_n}\right) \right.$$

$$\left. + \frac{1}{\mu_A \left(\mu_Y + \gamma\right)} \left(\sum_h l_h \mathbf{I_n} - \sum_h l_h \mathbf{P_h}^T\right)\left(\sum_k m_k \mathbf{I_n} - \sum_k m_k \mathbf{Q_k}^T\right) \right]$$

$$= \mu_A^n \left(\mu_Y + \gamma\right)^n \det\left[\mathbf{I_n} - R_0\mathbf{I_n} - \frac{1}{\mu_Y + \gamma} \sum_h l_h \left(\mathbf{P_h}^T - \mathbf{I_n}\right) - \frac{1}{\mu_A} \sum_k m_k \left(\mathbf{Q_k}^T - \mathbf{I_n}\right) \right.$$

$$\left. + \frac{1}{\mu_A \left(\mu_Y + \gamma\right)} \sum_h l_h \left(\mathbf{P_h}^T - \mathbf{I_n}\right) \sum_k m_k \left(\mathbf{Q_k}^T - \mathbf{I_n}\right) \right].$$

By introducing the matrix

$$\mathbf{J}^\star = R_0 \mathbf{I_n} + \frac{1}{\mu_Y + \gamma} \sum_{h=1}^{N_Y} l_h \left(\mathbf{P_h}^T - \mathbf{I_n}\right) + \frac{1}{\mu_A} \sum_{k=1}^{N_A} m_k \left(\mathbf{Q_k}^T - \mathbf{I_n}\right)$$

$$- \frac{1}{\mu_A(\mu_Y + \gamma)} \sum_{h=1}^{N_Y} l_h \left(\mathbf{P_h}^T - \mathbf{I_n}\right) \sum_{k=1}^{N_A} m_k \left(\mathbf{Q_k}^T - \mathbf{I_n}\right),$$

the bifurcation condition $\det(\mathbf{J}) = 0$ can finally be written as

$\det(\mathbf{I_n} - \mathbf{J}^\star) = 0.$

Let $\lambda_i^\star$ $(i = 1..n)$ be the eigenvalues of $\mathbf{J}^\star$. Then

$$\det(\mathbf{I_n} - \mathbf{J}^\star) = \prod_{i=1}^{n} (1 - \lambda_i^\star).$$

Therefore, $\det(\mathbf{J})$ switches from positive to negative when E_0, the dominant eigenvalue of $\mathbf{J}^\star$, switches from being smaller to being larger than one. It follows that the metapopulation can persist and spread (provided one sets a positive initial condition) if and only if E_0 is larger than 1. Note that assuming that the graphs associated with the matrices $\mathbf{P_h}$ and $\mathbf{Q_k}$ are strongly connected implies that if $E_0 > 1$, then persistence is granted in all the network nodes. To discuss differences in local persistence probabilities, a different approach would be needed, possibly based on interacting particle systems [189] in which local persistence/extinction can be precisely defined.

The dominant eigenvalue E_0 also sets a timescale for metapopulation dynamics and, in particular, for metapopulation extinction. From an engineering perspective, in fact, transient dynamics can be considered over a timespan that is approximately 5 times the time constant of the system [115], which, close to the persistence–extinction boundary, is given by $1/|\log(E_0)|$. On the extinction side of the bifurcation curve, where $E_0 < 1$, the timespan $-5/\log(E_0)$ thus represents the average time to metapopulation extinction. More in general, far from the persistence–extinction boundary, transient dynamics can be considered over approximately $5/|\lambda_{\max}(\mathbf{J})|$ time units.

2.2.4 Spatial Patterns of Species Spread and a Spatially Explicit Persistence Criterion

In our framework, the condition under which a species can invade a river network corresponds to that for metapopulation persistence and species persistence criteria. As such, if the global extinction equilibrium is unstable, the dominant eigenvector of matrix $\mathbf{J}$ pinpoints the direction in the state space along which the system trajectories, after a transient period due to initial conditions, will diverge from the equilibrium (Appendix 6.1). Specifically, the components of the leading eigenvector correspond to the evolving abundances of young or adult individuals in different locations of the river network. The analysis of the dominant eigenvector of the Jacobian of system (2.12) evaluated at $\mathbf{X_0}$ is thus key to understanding the early spatial patterns of species spread and can be useful – at least from a qualitative perspective – in studying the geography of aquatic invasions in riverine ecosystems.

As detailed in Box 2.5, the stability switch of the extinction equilibrium corresponds to the condition $\det\left(\mathbf{I_n} - \mathbf{J}^\star\right) = 0$, where $\mathbf{J}^\star$ is a matrix of size n, deducible from the $2n$-sized Jacobian $\mathbf{J}$ of (2.12), defined as

$$\mathbf{J}^\star = R_0 \mathbf{I_n} + \frac{1}{\mu_Y + \gamma} \sum_{h=1}^{N_Y} l_h \left(\mathbf{P_h}^T - \mathbf{I_n}\right)$$

$$+ \frac{1}{\mu_A} \sum_{k=1}^{N_A} m_k \left(\mathbf{Q_k}^T - \mathbf{I_n}\right)$$

$$- \frac{1}{\mu_A (\mu_Y + \gamma)} \sum_{h=1}^{N_Y} l_h \left(\mathbf{P_h}^T - \mathbf{I_n}\right) \sum_{k=1}^{N_A} m_k \left(\mathbf{Q_k}^T - \mathbf{I_n}\right).$$
$$(2.13)$$

In the previous expression, $R_0 = \nu\gamma/\mu_A/(\mu_Y + \gamma)$ is the quantity controlling population persistence in a nonspatial setting and can be interpreted as the average number of daughters successfully reaching maturity generated by one mother during her entire lifetime. The condition for an isolated population to persist is thus $R_0 > 1$. In the presence of dispersal, instead, metapopulation persistence is determined by the dominant eigenvalue $\lambda_{\max}(\mathbf{J}^\star)$. Specifically, the persistence–extinction boundary – that is, the curve or surface in the system parameter space that separates parameter combinations corresponding to metapopulation extinction from those corresponding to persistence [184] – is given by the condition

$$E_0 = \lambda_{\max}(\mathbf{J}^\star) > 1. \qquad (2.14)$$

In other words, the occasional introduction of some individuals in some network nodes results in a successful colonization if (and only if) $E_0 > 1$. In this case, the assumption of strong connectivity made above implies that persistence is granted in all the network nodes.

Criterion (2.14) shows that not only local demographic processes (first term on the right-hand side of Equation (2.13)) but also average net immigration from connected sites (second and third terms) is relevant to the persistence of riverine metapopulations. It also shows that the intertwining between different dispersal pathways may have nontrivial effects on metapopulation persistence or extinction (last term on the right-hand side of Equation (2.13)). As a matter of fact, the persistence condition is based on the dominant eigenvalue of $\mathbf{J}^\star$, which is not simply deducible from R_0 and the eigenvalues of matrices $\mathbf{P_h}$ and $\mathbf{Q_k}$. Note that, close to the persistence–extinction boundary, E_0 also sets a timescale for metapopulation dynamics and, in particular, for metapopulation extinction (see again Box 2.5). Criterion (2.14) can be extended to account for spatial heterogeneities in the model parameters whenever relevant for the underlying ecological processes.

2.2.5 The Role of Network Structure and Dispersal Pathways

As a basic test case to study persistence in a river network, we have analyzed a population in which adults are sessile and juveniles are subject to drift and alongstream dispersal ($l_1 > 0$, $l_h = 0$ for any $h > 1$; $m_k = 0$ for any k; $\mathbf{P_1} = \mathbf{F}$, with $\mathbf{F}$ being the hydrological connection matrix; see Figure 2.8b–d). Figure 2.11a (gray lines) shows that high values of alongstream dispersal and bias (defined as the difference between the probability of moving downstream and that of moving upstream) are always detrimental to species persistence and that network topology remarkably influences the fate of the metapopulation. Specifically, more complex networks (Peano, OCN) favor metapopulation persistence compared to simpler geometries (lattice) with the same backbone length. Quite interestingly, the largest relative differences emerging from the three contrasting topologies are found for high dispersal rates and low values of the transport bias. In these conditions, alongstream movement is significantly influenced by geomorphological dispersion, that is, by the intertwining of hydrodynamic dispersion within individual reaches and the morphology of the network structure [174].

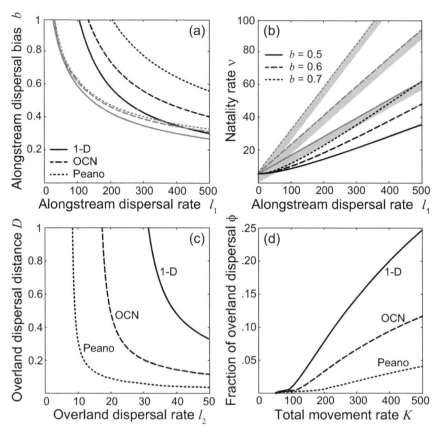

Figure 2.11 Persistence conditions for populations with sessile adults and juveniles dispersing via water and overland. Metapopulation persists for parameter combinations below (a) (above [b–d]) the persistence-extinction boundaries ($E_0 = 1$ contour lines, gray and black curves). All rates in year^{-1}. (a) Effect of aquatic dispersal parameters without (gray, $l_2 = 0$) (with [black, $l_2 = 5$]) overland dispersal. (b) Effect of transport and demographic parameters in an OCN without (gray) (with [black]) overland dispersal; gray-shaded areas indicate extinction debts longer than 10 years. (c) Effect of overland dispersal parameters ($l_1 = 400$, $b = 0.9$). (d) Effect of dispersal strategies ($l_1 = (1 - \phi)K$, $b = 0.8$, $l_2 = \phi K$, with $K = l_1 + l_2$). Other parameters: $\nu = 25$, $\gamma = 1$, $\mu_A = 1$, $\mu_Y = 5$, $\mathbf{P_1} = \mathbf{F}$, $\mathbf{P_2} = \mathbf{G}$, $D = 0.1$, $l_h = 0$ for any $h > 2$, $m_k = 0$ for any k. Figure after [115]

Changes in the flow regime can obviously affect the persistence of a metapopulation dispersing through water pathways. Figure 2.11b (gray lines) shows that the metapopulations that cannot compensate higher bias of aquatic dispersal with higher natality are doomed to extinction – that is, that downstream drift reduces metapopulation capacity *sensu* Hanski and Ovaskainen [190]. Close to the persistence–extinction boundary, the dynamics of the metapopulation are very slow, because $E_0 \approx 1$ (and $\lambda_{\max}(\mathbf{J}) \approx 0$). Therefore, extinctions may occur over long (yet still ecological) timescales, depending on the distance from the bifurcation curve characterized by $E_0 = 1$. This delay generates an extinction debt sensu [88]. As an example, the model predicts extinction for all the parameter settings

lying below the persistence–extinction boundaries in Figure 2.11b – yet in the lightly shaded regions, metapopulation extinction will take more than 10 years, approximately corresponding to 10 generation times for the population under study.

To analyze how different dispersal pathways can influence metapopulation persistence, we have studied populations in which juveniles disperse not only along the hydrological network but also overland ($l_1 > 0$, $l_2 > 0$, $l_h = 0$ for any $h > 2$; $m_k = 0$ for any k; $\mathbf{P_1} = \mathbf{F}$, $\mathbf{P_2} = \mathbf{G}$, with $\mathbf{G}$ being the connection matrix describing overland isotropic dispersal with characteristic dispersal length D; see Box 2.7). Figure 2.11a,b (black lines) show that overland dispersal can remarkably benefit riverine metapopulation persistence, in particular for high values of the bias of alongstream dispersal. Under these conditions, corresponding in fact to advection-dominated environments, overland dispersal can provide riverine populations with an effective means of upstream propagation, thus mitigating the downstream drift imposed on offspring and juveniles by passive hydrological transport. These results hold qualitatively for all the considered network topologies (not necessarily riverine; see Section 1.2 for some examples of 2D lattice geometries). However, it is apparent that topological complexity and the multiplicity of dispersal pathways operate synergistically (last term in Equation (2.13)), thus greatly favoring the persistence of metapopulations inhabiting complex river networks (Figure 2.11c). The effects of this synergism are very robust not only to changes of the demographic rates but also to variations in the exploitation of different dispersal pathways in relation to specific life histories (as detailed in [115]).

One wonders whether enhanced persistence due to the superimposition of different dispersal pathways is simply due to higher overall (i.e., alongstream + overland) dispersal. We have thus repeated some of the analyses above considering different dispersal strategies, defined as the combination of overland and aquatic dispersal operated by a population. Specifically, we assume that a fraction ϕ of the total movement rate $K = L_1 + L_2$ is allocated to overland movement, while the remaining fraction $1 - \phi$ is allocated to water-mediated dispersal. Figure 2.11d reports a systematic exploration of the parameter space (K, ϕ), each point of which represents a different dispersal strategy and shows that even relatively small fractions of total movement rate allocated to overland dispersal are sufficient to guarantee persistence. The exploitation of alternative dispersal pathways (specifically, of overland dispersal) can thus remarkably affect the fate of a population subject to downstream drift in a riverine ecosystem.

2.2.6 A Paradigmatic Example: The Persistence of an Amphibian Metapopulation in a River System

The framework presented above can be adapted to study the persistence of a real metapopulation in a river network. As a proof of concept, here we study the fate of a metapopulation of stream salamanders in the Shenandoah river network (Virginia, US; Figure 2.12a). Model (2.12) has been parameterized with demographic [191] and dispersal [150] data relative to the salamander species *Desmognathus fuscus* and *D. monticola* (technical details are in Box 2.6). The juveniles of these two amphibian species can move both along stream corridors and overland, while larvae and adults are almost sessile. Quite interestingly, juveniles' alongstream dispersal is known to be biased toward upstream sites in the river network. Despite the ongoing decline of amphibian abundances worldwide, populations of stream salamanders in eastern North America are reportedly stable – an observation that has been linked to their ability to exploit multiple dispersal pathways [150].

Box 2.6 Background on Amphibian Persistence

The analysis of the conditions that contribute to metapopulation persistence in a river network is an interesting topic per se [115]. However, our theoretical framework can be applied to analyzing the persistence of a real animal population living in a given river network. The ecological literature is rich in possible candidate species for the study of population persistence in river networks. To that end, stream insects would represent a natural choice. There are in fact several studies focusing on larval drift [169, 192–195] or adult (both along- and across-stream) flight [181, 182, 196–202]. These studies could be used to estimate aquatic dispersal bias and overland dispersal kernels for larvae and adults, respectively. However, observations encompassing the whole life cycle of a focal species would be needed, for example, to assess the relative importance of larval versus adult dispersal. Genetic relatedness is often used as a proxy for effective dispersal [203–210]. These studies too could be useful to parameterize stage-specific dispersal patterns (and possibly to evaluate the relative importance of different dispersal mechanisms) – yet this would require the development

Box 2.6 *Continued*

of a coupled genetic–ecological model for gene flow in a stream network [211, 212]. For these reasons we have decided to apply our framework to riverine organisms belonging to a different taxonomic group, namely, the stream salamanders *D. fuscus* and *D. monticola* that inhabit streams and rivers in the US state of Virginia. The life cycle of these species is well known [191]. Also, their ability to exploit alternative dispersal pathways has been observed and linked to metapopulation persistence [150]. As stated in the main text, in fact, the use of our modeling tools requires some basic knowledge of the demography of the focal population, its main dispersal pathways, and the underlying environmental matrix constituting its habitat. This example is also interesting because stream salamanders are actually not subject to downstream drift (quite the opposite, their movement patterns are strongly upstream biased), thus showing that our framework can be used to assess population persistence in a very wide range of environmental conditions.

Demographic parameters. As a first step in the process, we parameterize model equation (2.12) to reproduce the basic timescales of the life cycle of *D. fuscus* and *D. monticola*. All the relevant information can be retrieved from [191]. Female salamanders are characterized by a biennial reproductive cycle; an average clutch contains about 25 eggs, with a sex ratio close to $1:1$; eggs survive and become juvenile individuals with a 5/8 probability (the larval stage is short compared to the others and is thus not treated explicitly as an additional developmental stage). Using these data, we set the natality rate ν of adult individuals to $1/2 \cdot 25 \cdot 1/2 \cdot 5/8 = 3.9$ juveniles female^{-1} year^{-1}. The length of the juvenile stage can vary slightly between the two species as well as between male and female individuals, with an average duration of about 4 years. The parameter γ can be evaluated as the inverse of this duration and is thus set to $1/4$ years^{-1}. Average survival probability during the juvenile stage is about 20 percent. Therefore, the juvenile mortality rate can be estimated as $\mu_Y = -\gamma \ln(0.2) = 0.40$ years^{-1}. Adult survival 5 years after completion of the juvenile stage is on the order of 1 percent. The mortality rate of adult individuals is thus higher than that of juveniles and can be estimated as $\mu_A = -1/5 \ln(0.01) = 0.92$ years^{-1}.

Dispersal patterns. The second step concerns the characterization of dispersal rates and patterns. Following the mark-recapture study described in [150], we focus on juvenile dispersal (both along stream corridors and overland), while we treat adults (which basically do not disperse) and larvae (whose dispersal has been reported as not significant) as sessile. Campbell-Grant et al. [150] provide movement probabilities between a pair of locations (nodes 1 and 3 in figure S3a in [115]) along a stretch of the Shenandoah River (Virginia, US) as well as movement probabilities from these two sites to another location (node 2 in figure S3a in [115]) that is not directly connected through the hydrological network [150]. Assuming that there is no reproduction during the experiment, the simplest model to keep track of the movements of the organisms Y_i^m marked in a node m ($m = 1, 2$) is

$$\frac{dY_i^m}{dt} = -(\mu_Y + l_{1i} + l_{2i})Y_i^m + \sum_{j=1}^{n} l_{1j} P_{1ji} Y_j^m + \sum_{j=1}^{n} l_{2j} P_{2ji} Y_j^m,$$

where n is the number of nodes in the network ($n = 5$; note that we use a slightly more complex network structure than the one shown in [150] to better estimate the relative impact of along-stream vs. overland dispersal – see again figure S3a in [115]), whereas l_{ki} and $\mathbf{P_k}$ are, respectively, the

Box 2.6 *Continued*

(possibly) site-specific dispersal rates and the connectivity matrices for alongstream ($k = 1$) or overland ($k = 2$) movement. To describe connectivity, we assume that none of the considered nodes is terminal, that is, that nodes 1 and 2 have one upstream connection each and that node 5 has a downstream connection. The connectivity matrix for alongstream dispersal thus reads as

$$
\mathbf{P}_1 =
\begin{bmatrix}
0 & 0 & F_1^d & 0 & 0 \\
0 & 0 & 0 & F_2^d & 0 \\
F_3^u & 0 & 0 & 0 & F_3^d \\
0 & F_4^u & 0 & 0 & F_4^d \\
0 & 0 & F_5^u & F_5^u & 0
\end{bmatrix}
=
\begin{bmatrix}
0 & 0 & \frac{1+b}{2} & 0 & 0 \\
0 & 0 & 0 & \frac{1+b}{2} & 0 \\
\frac{1-b}{2} & 0 & 0 & 0 & \frac{1+b}{2} \\
0 & \frac{1-b}{2} & 0 & 0 & \frac{1+b}{2} \\
0 & 0 & \frac{1-b}{3} & \frac{1-b}{3} & 0
\end{bmatrix}.
$$

As for overland dispersal, we assume the following connectivity matrix:

$$
\mathbf{P}_2 =
\begin{bmatrix}
0 & 1/4 & 1/4 & 1/4 & 0 \\
1/4 & 0 & 1/4 & 1/4 & 0 \\
1/4 & 1/4 & 0 & 1/4 & 1/4 \\
1/4 & 1/4 & 0 & 0 & 1/4 \\
0 & 0 & 1/3 & 1/3 & 0
\end{bmatrix}.
$$

The set of 2×5 linear differential equations introduced above has then to be integrated from time $t = 0$ to $t = T$ [115] with $T = 1/12$ year, that is, the average time between mark and recapture in [150] with initial conditions $Y_1^1(0) = 1$ and $Y_2^2(0) = 1$ ($Y_i^m = 0$ otherwise). The values of the state variables at time $t = T$ give the distribution of the abundances of the marked salamanders in the five network nodes. These values can be directly compared to the connectivity values C_{ij} provided by [150] (see [115]; note that here we do not distinguish between the two species and between the two age transitions involving juveniles). For instance, the value of $Y_1^1(T)$ has to be compared to connectivity C_{11}, $Y_2^1(T)$ has to be compared to C_{12}, and so on.

We fit two versions of the aforementioned model, one ($\mathcal{M}_1$) with spatially homogeneous parameters ($l_{ki} = l_k$) and another ($\mathcal{M}_2$) with site-specific dispersal rates; [150] report in fact differences in the dispersal from upstream/downstream sites. We thus set $l_{ki} = l_u$ for upstream nodes and $l_{ki} = l_d$ for downstream nodes in $\mathcal{M}_2$. This spatially heterogeneous model does actually provide a slightly better fit than $\mathcal{M}_1$, but it is less parsimonious (three free parameters for $\mathcal{M}_1$, namely, b, l_1, and l_2; five free parameters for $\mathcal{M}_2$, namely, b, l_{1u}, l_{1d}, l_{2u}, and l_{2d}). To discount the effect of model complexity, we compare the performances of the two candidate models with the Akaike information criterion (AIC) with small sample corrections [213]. The AIC score of $\mathcal{M}_1$ is lower than that of $\mathcal{M}_2$ ($\Delta_{AIC} \approx 3$). We thus retain $\mathcal{M}_1$, for it offers a better trade-off between accuracy and complexity. The best-fit parameter values are $b = -0.49$, $l_1 = 1.58$, and $l_2 = 1.04$, which suggest that dispersal along stream corridors is strongly biased toward upstream locations and that both along- and across-stream connectivity pathways are important for the species. Figure S3b in [115] shows a comparison between among-site connectivity as estimated from data and as simulated by model $\mathcal{M}_1$. With the parameters that describe population demography and dispersal, and with the connectivity matrices $\mathbf{P}_1$ and $\mathbf{P}_2$ defined above, it is possible to assess population persistence in the small network shown in figure S3a in [115]. This calculation gives $E_0 < 1$, which means that the salamanders in the small five-node network are doomed to extinction if there is no immigration from other sites. However, this does not tell much about the fate of the metapopulation in its whole habitat, which is likely to be

Box 2.6 *Continued*

much larger than the area encompassing the five sites considered to assess connectivity – possibly extending to the entire system of streams connected to the Shenandoah river network.

Environmental matrix. The third step in the assessment of the persistence of the salamander metapopulation in the Shenandoah River thus concerns the description of the environmental matrix that underlies population dynamics. To that end, we assume that the habitat of the metapopulation coincides with the whole river network and proceed with its characterization. The extraction of channeled and unchanneled fluvial basin features and the delineation of catchment divides from remotely sensed and objectively manipulated geomorphic information is a well-established procedure in hydrology [63, 70, 214]. Drainage direction can be extracted from digital terrain models (DTMs) as the unique steepest-descent flowpath from each pixel to the outlet of the considered catchment. Flow accumulation (i.e., the drainage area of each pixel) can be readily computed from drainage directions. Several geomorphologic algorithms for river network delineation have been developed and widely validated [71, 215]. They all rely on information on drainage area and slope, which can also be straightforwardly derived from DTMs. Georeferenced information on river network structure is often available via open-access geographic information systems. In our case, in particular, the geometry of the Shenandoah River is derived from the US National Hydrography Dataset provided by the USGS (available online at http://nhd.usgs.gov/). The geometry of the Shenandoah river network used for the analysis of this case study is shown in Figure 2.12a.

Assessing the persistence of *D. fuscus* and *D. monticola* in the Shenandoah river network. Demographic traits, dispersal patterns, and network geometry play together to determine the fate of a population in a riverine system, as shown by our novel persistence criterion (Equation (2.14)). To apply the criterion to the populations of *D. fuscus* and *D. monticola* inhabiting the Shenandoah River, however, some further hypotheses have to be made. For instance, we have to assume that all the demographic rates estimated from [191] and the dispersal parameters b, l_1, and l_2 derived from [150] apply also to the population under study, that there is no spatial heterogeneity in the underlying ecological processes/parameters, and that the salamander metapopulation can actually inhabit the whole Shenandoah river basin. The biggest challenge is to scale metapopulation connectivity from a relatively small spatial scale up to a whole river network. In fact, because of the limited spatial extent of the original observations made by [150], it is possible to determine neither the average distance of the overland dispersal kernel nor the behavior of dispersing individuals close to the boundaries of the domain (the leaves and the outlet of the river network). As for alongstream connectivity, we use the geometry of the Shenandoah river network and assume $\mathbf{P}_1 = \mathbf{F}$ (Box 2.7). As for overland dispersal, we assume that organisms can move from a given node to any other node in its neighborhood with equal probability. We define the neighborhood of each node as the set of *its* nearest hydrological neighbors and *their* nearest hydrological neighbors (so that reaching some node in the neighborhood actually requires overland movement). We call $\mathbf{N}$ the resulting connectivity matrix for salamanders' overland dispersal (so that $\mathbf{P}_2 = \mathbf{N}$). As for boundary conditions, we leave them free to vary in a continuum ranging from completely reflecting (no cost of dispersal) to completely absorbing (dispersing individuals are lost to the metapopulation when they step out of the network) for both alongstream and overland dispersal. Specifically, we set (see Box 2.7)

$$\alpha_1 = \epsilon_s \frac{1 + 2b}{3} \quad \text{and} \quad \beta_1 = \frac{\epsilon_o}{n_1' + 1}$$

Box 2.6 *Continued*

for the outlet and

$$\alpha_i = 2\epsilon_s \frac{1-b}{3} \quad \text{and} \quad \beta_i = \frac{\epsilon_o}{n_i' + 2}$$

for the leaves, where ϵ_s [ϵ_o] is the fraction of dispersing individuals that are absorbed at each leaf (at the outlet) of the river network and n_i' is the number of nodes in the neighborhood of node i. All the above quantities are obtained assuming that each leaf is connected to two fictitious upstream sink nodes and the outlet to a fictitious downstream sink node (see Box 2.7). The persistence–extinction boundary ($E_0 = 1$) as a function of the parameters quantifying boundary effects (ϵ_s and ϵ_o) is shown in Figure 2.12b. For specific combinations of ϵ_s and ϵ_o, it is also possible to test the sensitivity of E_0 to changes of the parameter values. To that end, for each model parameter (say, θ), we compute $\theta' = \theta_0(1 + \delta)$, where θ_0 is the reference value of the parameter and δ sets the scale of parameter variability, and we repeat the computation of E_0 (after [115]; Figures 2.12c and 2.13).

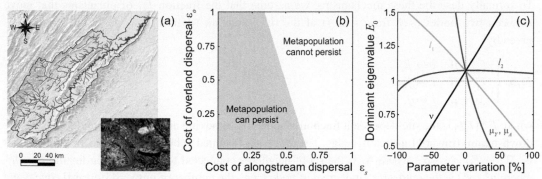

Figure 2.12 Persistence of stream salamanders in the Shenandoah river network (Virginia, US). (a) River network geometry; inset: *Desmognathus monticola* (from USGS). (b) Effect of dispersal cost on population persistence. (c) Sensitivity of E_0 to variations of the model parameters. Parameters: $\nu = 3.9$ juveniles adult^{-1} year^{-1}, $\gamma = 0.25$, $\mu_Y = 0.40$, $\mu_A = 0.92$, $l_1 = 1.58$, $\mathbf{P_1} = \mathbf{F}$, $b = -0.49$, $l_2 = 1.04$, $\mathbf{P_2} = \mathbf{N}$, $l_h = 0$ for $h > 2$, $m_k = 0$ for any k. All rates in year^{-1}. Figure after [115]

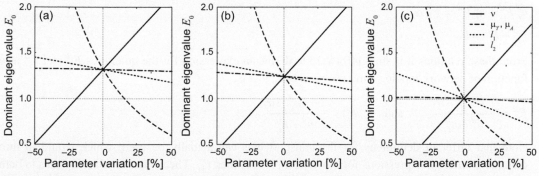

Figure 2.13 Sensitivity of stream salamanders'' E_0 to variation of the model parameters . (a–c) As in Figure 2.12c in the main text, with (a) $\epsilon_s = 0.25$ and $\epsilon_o = 0.25$, (b) $\epsilon_s = 0.25$ and $\epsilon_o = 0.5$, and (c) $\epsilon_s = 0.5$ and $\epsilon_o = 0.5$. Figure after [115]

Our analysis shows that if the cost of dispersal ϵ (here defined as the fraction of individuals that disperse outside their suitable habitat) is negligible, both alongstream (low ϵ_s) and overland (low ϵ_o), then the salamander metapopulation is predicted to persist ($E_0 \gg 1$). However, for increasing values of the cost of dispersal (possibly due to the alteration of the habitat template), the metapopulation can cross the persistence–extinction boundary and can thus be doomed to extinction (Figure 2.12b). It is also possible to test the sensitivity of E_0 to changes of the model parameters. Besides expected positive (negative) effects of increased natality ν (mortality μ_Y and μ_A) on E_0, increasing levels of overland dispersal l_2 can promote metapopulation persistence (as suggested by [150]), provided that the cost of overland movement is lower than that of alongstream dispersal. E_0 can actually peak for intermediate values of the overland dispersal rate (as in Figure 2.12c), a result that mirrors the intermediate dispersal principle of metapopulation ecology [184, 189].

Box 2.7 Alongstream and Overland Dispersal

The technical material here reproduces [115]. The most important dispersal mechanisms in river networks are flow-mediated drift and alongstream movement, both following hydrological pathways. To formally describe these mechanisms, we assume that the fraction F_{ij} of organisms that move between two nodes (say, from i to j) of the river network (described as an oriented graph) is given by

$$F_{ij} = \begin{cases} F_i^d & \text{if } i \to j, \\ F_i^u & \text{if } i \leftarrow j, \\ 0 & \text{if } i \nleftrightarrow j, \end{cases}$$

where F_i^d [F_i^u] is the site-dependent fraction of individuals moving along an outward (hydrologically downstream) (inward [upstream]) edge. The process is assumed to be possibly nonconservative, that is, $\sum_{j \in N_i} F_{ij} = 1 - \alpha_i$, where N_i is the set of neighbors connected to node i (of cardinality $n_i^d + n_i^u$, where n_i^d [n_i^u] is the outdegree [indegree] of node i, i.e., the number of outward [inward] edges) and $\alpha_i \geq 0$ is the node-dependent fraction of dispersing individuals that are lost to the metapopulation because of boundary conditions (BCs) and/or costly dispersal [184]. To close the specification of F_{ij}, we define the bias of aquatic dispersal as the difference between downstream (F_i^d) and upstream (F_i^u) movement probabilities [216] (pathwise partitioning). Therefore, in each node of the network, the following equalities must hold:

$$n_i^d F_i^d + n_i^u F_i^u = 1 - \alpha_i$$

$$F_i^d - F_i^u = b.$$

From these relations it is straightforward to derive an expression for the quantities F_i^d and F_i^u as a function of network topology and transport bias, that is,

$$F_i^d = \frac{1 + b n_i^u - \alpha_i}{n_i^d + n_i^u} \quad \text{and} \quad F_i^u = \frac{1 - b n_i^d - \alpha_i}{n_i^d + n_i^u}.$$

Note that different definitions for the transport bias can actually be used, which would in turn command different expressions for movement probabilities F_{ij}. The same formalism applies to both organisms subject to drift (high b) and organisms following stream corridors during active dispersal (low b; note that bias can be negative in the limit of preferential upstream dispersal).

Hydrological dispersal probabilities F_{ij} can account for proper BCs for the leaves and the outlet of the river network. Unless explicitly stated, all the numerical results described in this work refer

Box 2.7 *Continued*

to completely reflecting BCs for the leaves and completely absorbing BCs for the outlet of the river network. Therefore, α_i is larger than zero only at the network outlet (conventionally labeled as node 1). Technically, to compute its numerical value, we assume that $\alpha_1 = F_1^d$ with $n_1^d = 1$, that is, that there exists a fictitious sink node downstream of the outlet. From these hypotheses it follows that

$$\alpha_i = \begin{cases} \frac{1+bn_i^u}{1+n_i^u} & \text{if} \quad i = 1 \\ 0 & \text{otherwise.} \end{cases}$$

Other mechanisms can be relevant to the dispersal of riverine populations, for instance, human- or animal-mediated transport processes. Such mechanisms are only marginally (if at all) constrained by river network topology and flow direction, thus providing aquatic organisms with suitable pathways for unbiased overland dispersal. We assume that the fraction G_{ij} of organisms that move overland between two nodes of the river network (say, from i to j) is given by an isotropic exponential kernel [183, 217], that is,

$$G_{ij} = \frac{\exp\left(-\frac{d_{ij}}{D}\right)}{\sum_{k \neq i}^{n} \exp\left(-\frac{d_{ik}}{D}\right)},$$

where d_{ij} is the pairwise distance between any two nodes i and j and D is the shape parameter of the exponential kernel. All distances are normalized by the length of the backbone of the river network (see Figure 2.10 in the main text). Note that here isotropic overland dispersal is assumed to be conservative, that is, $\sum_j G_{ij} = 1$, unless explicitly stated. However, in the case of costly dispersal [184] or absorbing BCs, one can assume $\sum_j G_{ij} = 1 - \beta_i$, with $\beta_i \geq 0$ being the fraction of individuals that leave from node i and die during dispersal. Obviously, other, possibly ad hoc mechanisms can be introduced to describe case-specific dispersal pathways.

2.2.7 Spatial Patterns of Species Spread

The previous section, adapted from [115], shows how to determine conditions for population persistence in a river network, given its particular structure, partitioning a catchment into channeled and unchanneled portions, connected, however, to the main root, and the dynamics relevant to species' movements. We shall examine a few implications for species spread, here from a proper metapopulation geography viewpoint. We shall postpone until Chapter 3 such a study in the context of population dynamics and biological invasions.

At this point, we can safely state that metapopulation persistence within a river network is determined by many interactions between network geometry/ connectivity and the dominant dispersal mechanisms therein. The chief mechanism affecting persistence is alongstream movement driven by riverflow advection, with all its complexities [173], which plays such a

prominent role in the fluvial settings that are the subject of this book. The foremost result contrived theoretically in the context of metapopulation dynamics confirms previous results: dendritic geometries enhance metapopulation persistence in a river network. This result echoes recent theoretical and experimental findings that have linked dendritic topologies to long (and anomalously long) species persistence times and high local biodiversities, for example, when compared to unconstrained landscapes or simply regular lattices in any number of embedding dimensions [9, 13, 14, 89]. The reference work [115] has also formally confirmed a relevant ecological factor: in fact, overland dispersal can favor metapopulation persistence, especially for (but not limited to) species subject to hydrological drift. Also, the movement of organisms from their current range to a new area of suitable habitat (extrarange dispersal [108]) influences both metapopulation persistence and (as we shall see in Chapter 3) spatiotemporal invasion patterns, as indeed observed in the zebra mussel invasion of the Mississippi–Missouri river system [29]. We can thus conclude that diffusive dispersal, landscape geometry, and exploitation of multiple dispersal pathways may offer a multifaceted solution of the drift paradox for riverine populations. More generally, the model employed here, general by construction, and the overarching assumptions show that the above ingredients are key to understanding metapopulation persistence in real landscapes.

Although derived in the context of river systems, the persistence criterion proposed in this work can be adapted to populations living in different ecosystems, possibly characterized by high levels of spatial complexity. As an example, an interesting application would be the analysis of 2D lattice geometries, which would allow us to address the study of persistence conditions for terrestrial metapopulations. Preliminary explorations [115] confirm that, also in 2D lattices, metapopulation persistence is deeply related to the connectivity of the underlying environmental matrix, as well as to the dispersal mechanisms relevant to the metapopulation. The flexibility of our tools is essentially granted by the multilayer network framework [29] which generalizes previous network-based approaches in metapopulation ecology [183] and allows a hierarchical description of the interactions between ecological and spatial dynamics at different levels of organizational complexity. Overall, this subsumes the main tenet of this book.

The mathematical framework described in this section can be applied to real case studies whenever there exists sufficient information on the focal species. Basically, one needs to be capable of formalizing a demographic model for species' local-scale dynamics, and identifying the underlying environmental matrix constituting its habitat. In the context of this book, this means defining proper river stretches in a fluvial system, as opposed, say, to patches in a fragmented forest. This in turn implies a fundamental understanding of the natural gradients affecting river network drivers and controls (that is, gradients of streamflow, total contributing area or landscape-forming discharges, or habitat [63]). In addition, one should be capable of sorting out clearly the main dispersal pathways. In the absence of detailed information about the dynamics of the focal population (quite often the case for endangered species), scaling relations could assist in the definition of its demographic parameters [218].

Any generalist metapopulation model can therefore guide the analysis of persistence conditions for metapopulations living in realistic ecosystems, possibly subject to habitat alterations. Human activities inevitably represent the main driver for such alterations. Damming, for instance, is usually cited as a primary threat to the integrity of riverine habitats [219]. From an ecological perspective, the main effect of damming, in addition to changes of water quality and the ecosystem's species assemblage composition [155, 157], is a reduction of alongstream dispersal and migration, especially in the upstream direction, via a substantial modification of the natural streamflow distributions. This would in turn entail a profoundly affected hydrological dispersal. Our analysis has shown that increasing bias could reduce metapopulation capacity [190], leading to extinction of species that rely on aquatic dispersal where higher drift cannot be compensated by higher natality. In contrast, species that can disperse overland at some specific life stage are predicted to be more resilient to environmental changes, such as alterations of the flow regime or habitat fragmentation. Extinction debts [88] and average times to metapopulation extinction can also be quantified through the analysis of persistence–extinction boundaries [88, 190, 220, 221].

The presented framework could obviously be made even more realistic in many respects. In its present form, for instance, it does not account for the possible temporal variability of the environmental conditions, which, however, have already been proposed – along with spatial heterogeneity – as an important factor for population persistence in advective environments [165, 222, 223]. Incorporating spatial heterogeneity in the model parameters is relatively straightforward and does not imply major changes to the derivation of persistence conditions – although the algebra required proves rather involved (see, for example, Chapter 4 in the context of waterborne disease spread). On the contrary, adding seasonal variability would demand a considerably more elaborated mathematical treatment, possibly relying on Floquet [224] or Lyapunov [225] exponents. Another aspect that will certainly deserve future attention is demographic stochasticity, which has already been shown to play an important role in metapopulations dynamics close to the extinction threshold [184, 189].

Despite its limitations, the theoretical framework used to derive persistence conditions (i.e., the stability analysis of an ordinary differential equation network model) can be applied to study other ecological problems. We envisage that similar persistence criteria could in fact be usefully applied to designing natural reserves aimed at preserving ecologically important species, as already proposed for marine protected areas [226, 227] and fragmented landscapes [183]. In these cases, too, metapopulation persistence can be established by properly accounting for the relevant spatial interactions and studying the conditions under which the extinction equilibrium changes its stability properties. Eigenvector analysis could then assist in designing spatially calibrated conservation efforts. In an even broader perspective, extending the framework presented here to interacting functional groups would allow the study of the persistence of aquatic metacommunities (rather than metapopulations). With functional diversity being closely related to ecosystem functioning and

services [228], achieving a better understanding of how we can preserve it through suitably targeted actions would certainly represent a major accomplishment for current conservation ecology – at least, as addressed in this book, in the fluvial basin.

Finally, it is of interest for the purposes of this book to analyze the accuracy of the dominant eigenvector of matrix **J** as a descriptor of the spatial dynamics of early population spread. To convince the reader, it is useful to perform some numerical experiments (reported in [115]). Specifically, here we show the invasion of an OCN (Figure 2.10d) by the population of a species

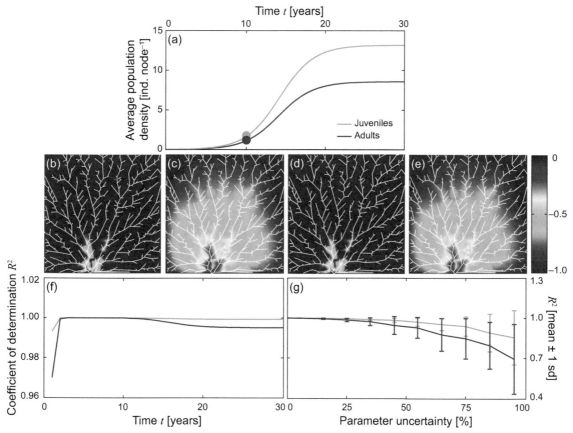

Figure 2.14 Spatiotemporal patterns of population spread in a river network. Juveniles are subject to downstream drift, while adults disperse overland. (a) Temporal evolution of the average density of juveniles and adults in the river network. (b) Normalized spatial distribution of juvenile density 10 years after the introduction of the species (dots in panel [a]. (c) As in (b) for adults' density. (d) Normalized components (juveniles) of the dominant eigenvector of matrix **J**. (e) As in (d) for the adults' components. (f) Comparison between eigenvector components and simulated spatial patterns of population density t years after species introduction. (g) Comparison between eigenvector components and simulated spatial patterns of population density 10 years after species introduction with uncertain model parameters (see supplementary material in [115] for details). Linear spatial interpolation has been performed in panels (b–e). Parameter values: $a_0 = 0.01$, $\nu = 25$, $\gamma = 1$, $\mu_A = 1$, $\mu_Y = 5$, $\eta = 0.05$, $l_1 = 500$, $\mathbf{P_1} = \mathbf{F}$, $b = 0.5$, $m_1 = 10$, $\mathbf{Q_1} = \mathbf{G}$, $D = 0.1$, $l_h = m_k = 0$ for any $h, k > 1$. All rates in year^{-1}. Figure after [115]

with juveniles and adults dispersing via water and over-land, respectively. The exercise, reproduced from [115], assumes that density dependence acts on adult mortality and that the species is initially absent from the river system where the invasion starts after the introduction of a small number of adult individuals at the network outlet. Such positions are imposed by setting $\mathcal{M}_Y = \mu_Y$, $\mathcal{N} = \nu$, $\mathcal{M}_A = \mu_A + \eta A_i$, $\mathbf{P}_1 = \mathbf{F}$, $\mathbf{Q}_1 = \mathbf{G}$, $l_h = m_k = 0$ for any $h, k > 1$. Also, the initial condition posits that $Y_i(0) = 0$ for all i, $A_1(0) = a_0$ and $A_i(0) = 0$ for all $i \neq 1$. The simulated temporal evolution of the average densities of juveniles and adults is shown in Figure 2.14a, while the spatial distributions of young and mature individuals in the initial phase of the colonization (10 years after injection) are reported in Figure 2.14b,c. Note that the spatial distributions of juveniles' and adults' densities are very different from each other, owing to differences in the dispersal patterns specific to each developmental stage. Nevertheless, the dominant eigenvector of matrix $\mathbf{J}$ is indeed found to be a good proxy for the spatial patterns of population density for both adults (Figure 2.14d) and juveniles (Figure 2.14e), with a coefficient of determination R^2 close to 1 [115].

It is also interesting to test whether the dominant eigenvector is a robust descriptor of the spatial dynamics. Figure 2.14f shows that, at least within the conditions of the present example, the values of R^2 for the plot of simulated versus "predicted" spatial patterns remain high ($R^2 > 0.97$) throughout the considered temporal window. One may conclude, therefore, that the agreement between the spatial patterns obtained through either the simulation of the model or the analysis of the dominant eigenvector of matrix $\mathbf{J}$ is not crucially dependent on the time chosen for comparison.

One may also wonder whether the predictions drawn from the dominant eigenvector are robust to some uncertainty in the underlying parameters. Random perturbations of the model parameters can be introduced as $\theta' = \theta_0(1 + \xi\delta)$, where θ_0 is the reference value of a generic parameter θ, ξ is a random variable drawn from a uniform distribution $\mathcal{U}(-1, 1)$, and δ sets the scale of parameter uncertainty. For a given value of δ, we generate independently a stochastic value θ' for each of the model parameters, compute the dominant eigenvector of matrix $\mathbf{J}$, estimate the coefficient of determination against the simulation obtained with the reference parameter values at time $t = 10$ years after the introduction of the species, and repeat this procedure 100 times. The results are shown in Figure 2.14g: the coefficient of determination remains as high as 0.9 for parameter uncertainty up to 60 percent (adults) or 80 percent (juveniles). Although the quantitative details may change for different parameter settings or different ecological dynamics, this sensitivity analysis suggests that the dominant eigenvector of matrix $\mathbf{J}$ can be a robust descriptor of spatial dynamics also in the presence of uncertainty.

We may thus safely conclude that eigenvector analysis can help identify the locations that are potentially more prone to the early phase of population spread. However, it can say little about the likelihood of an invasion, which could be triggered by an external event – such as the introduction of an alien species in a previously pristine ecosystem (Chapter 3).

Box 2.8 The Geography of Population Spread

In the above spatially explicit framework, the condition under which a population can persist in a river network corresponds to that for species spread (after [115]). This is specifically due to the assumption of strong connectivity. The geography of population spread, that is, the spatial localization of the sites that are colonized by a population in the early phases following its introduction in a river network, is thus determined by the dominant eigenvector of the Jacobian matrix $\mathbf{J}$, that is, the matrix that can be obtained by linearizing the system of equations close to the extinction equilibrium. In fact, if the global extinction equilibrium is unstable, the dominant eigenvector of matrix $\mathbf{J}$ pinpoints the directions in the state space along which the system trajectories, after a transient period related to initial conditions, will diverge from the equilibrium (Appendix 6.1). The dominant eigenvector is characterized by strictly positive components, according to the Perron–Frobenius theorem (Appendix 6.1) applied to Metzler matrices [115, 187], each corresponding in this case to the abundances of young or adult individuals in different locations of the river network.

Box 2.8 *Continued*

The dominant eigenvector of $\mathbf{J}$ can be computed by solving

$$\mathbf{J}\begin{bmatrix} \mathbf{y} \\ \mathbf{a} \end{bmatrix} = \lambda' \begin{bmatrix} \mathbf{y} \\ \mathbf{a} \end{bmatrix},$$

where λ' is the dominant eigenvalue of $\mathbf{J}$, and $\mathbf{y}$ and $\mathbf{a}$ are the components of the dominant eigenvector corresponding, respectively, to the abundances of juvenile and adult individuals. Writing again $\mathbf{J}$ as

$$\mathbf{J} = \begin{bmatrix} \mathbf{J}_1 & \mathbf{J}_2 \\ \mathbf{J}_3 & \mathbf{J}_4 \end{bmatrix}$$

as in Box 2.5, and recalling that close to the transcritical bifurcation through which the extinction equilibrium loses stability, the dominant eigenvalue of $\mathbf{J}$ is equal to 0, we get

$$\mathbf{J}_1\mathbf{y} + \mathbf{J}_2\mathbf{a} = 0$$
$$\mathbf{J}_3\mathbf{y} + \mathbf{J}_4\mathbf{a} = 0.$$

Assuming that $\mathbf{J}_1$ is invertible, from the first equation we have

$$\mathbf{y} = -\mathbf{J}_1^{-1}\mathbf{J}_2\mathbf{a}.$$

Plugging the former expression into the latter of the two equations above, we get

$$\left(\mathbf{J}_4 - \mathbf{J}_3\mathbf{J}_1^{-1}\mathbf{J}_2\right)\mathbf{a} = 0.$$

As $\mathbf{J}_2$ and $\mathbf{J}_3$ are scalar matrices (thus commuting with any other matrix), the previous expression can be worked out as

$$\left(\mathbf{J}_4 - \mathbf{J}_3\mathbf{J}_1^{-1}\mathbf{J}_2\right)\mathbf{a} = \left(\mathbf{J}_3^{-1}\mathbf{J}_4 - \mathbf{J}_1^{-1}\mathbf{J}_2\right)\mathbf{a} = \left(\mathbf{J}_1\mathbf{J}_3^{-1}\mathbf{J}_4 - \mathbf{J}_2\right)\mathbf{a}$$
$$= \left(\mathbf{J}_1\mathbf{J}_4\mathbf{J}_3^{-1} - \mathbf{J}_2\right)\mathbf{a} = (\mathbf{J}_1\mathbf{J}_4 - \mathbf{J}_2\mathbf{J}_3)\mathbf{a} = 0.$$

From the results presented in Box 2.5 we gather that

$$\mathbf{J}_1\mathbf{J}_4 - \mathbf{J}_2\mathbf{J}_3 = \mu_A(\mu_Y + \gamma)\left(\mathbf{I_n} - \mathbf{J}^\star\right),$$

with

$$\mathbf{J}^\star = R_0\mathbf{I_n} + \frac{1}{\mu_Y + \gamma}\sum_{h=1}^{N_Y} l_h\left(\mathbf{P_h}^T - \mathbf{I_n}\right) + \frac{1}{\mu_A}\sum_{k=1}^{N_A} m_k\left(\mathbf{Q_k}^T - \mathbf{I_n}\right)$$
$$- \frac{1}{\mu_A(\mu_Y + \gamma)}\sum_{h=1}^{N_Y} l_h\left(\mathbf{P_h}^T - \mathbf{I_n}\right)\sum_{k=1}^{N_A} m_k\left(\mathbf{Q_k}^T - \mathbf{I_n}\right).$$

Therefore we have

$$\mu_A(\mu_Y + \gamma)\left(\mathbf{I_n} - \mathbf{J}^\star\right)\mathbf{a} = 0.$$

If we remember that close to the transcritical bifurcation of the extinction equilibrium, the dominant eigenvalue E_0 of matrix $\mathbf{J}^\star$ is equal to 1, we can write

$$\mathbf{J}^\star\mathbf{a} = E_0\mathbf{a} = \mathbf{a}.$$

Box 2.8 *Continued*

We can thus conclude that close to the bifurcation through which the extinction equilibrium loses stability, the dominant eigenvector of matrix $\mathbf{J}^\star$ corresponds to the adults' components of the dominant eigenvector of $\mathbf{J}$ (the juveniles' components can be obtained by multiplying the adults' components by $-\mathbf{J}_1^{-1}\mathbf{J}_2$). This simple relationship between the dominant eigenvectors of $\mathbf{J}$ and $\mathbf{J}^\star$ holds only close to the transcritical bifurcation of the extinction equilibrium. In general, for parameter combinations for which the dominant eigenvalue of $\mathbf{J}^\star$ is larger than 1, the study of the geography of population spread requires computation of the eigenvalues and eigenvectors of matrix $\mathbf{J}$.

2.3 Elevational Gradients of Biodiversity in Fluvial Landscapes

Fluvial landscapes are the by product of steady state geomorphologic dynamics that balance fluvial erosion and tectonic uplift [63]. The landscapes that we study here are assumed to be dominated by the effects of fluvial erosion across scales, and hillslope effects and processes are altogether neglected. This is a somewhat restrictive assumption, even though river basins (i.e., landscapes carved by both fluvial and hillslope erosion) can be generated and studied within the same framework [63, 71, 74]. Suffice here to say that the mechanism for generating landscape replicas hinges on planar aggregation patterns coupled to a local rule relating the aggregated area (and hence landscape-forming discharges [63]) to drops in elevation. This avoids clouding the central idea – how realistic ecological connectivities and species suitabilities are shaped in the fluvial landscape – with unnecessary details at scales smaller than the mean distance one has to walk along the local steepest descent directions, on average, before reaching a channeled site, that is, the inverse of the drainage density [L^{-1}].

2.3.1 Fluvial versus Idealized Landscapes

We begin with an example, which we deem paradigmatic, dealing with geomorphic controls on elevational gradients of species' richness [229]. In this case, the lead role is taken by the structure of ecological interactions allowed by the available ecological substrate, the mountain landscape, which controls the gradients of biodiversity drivers. Key is the fluvial landscape as stipulated above, suitably constructed in three dimensions. The fluvial landscapes we adopt are either real topographies taken from digital elevation models [63] or replicated synthetic OCNs obtained within a given domain via the

transformation of its planar aggregation structure into a topographic landscape through the use of a slope–area relationship relating total contributing area at a point to the local topographic gradient [63]. OCNs [63–66] (Box 1.3) prove once more extremely versatile to that end.

The rationale for our choice of examples is simple. How biodiversity changes with elevation has long attracted the interest of researchers because it provides clues to how biota respond to geophysical drivers. Experimental evidence reveals that biodiversity in ecosystems significantly affected by the elevational gradients often peaks at intermediate elevations [230–237]. A factor that had been overlooked for a long time was that mountainous landscapes hold fractal properties [63, 121] with elevational bands forming habitat patches that are characterized by different areal extent and connectivity, well-known drivers of biodiversity. Specifically, the frequency distribution of elevation in real-life landscapes is distinctly hump shaped, with the majority of land situated at mid-elevations [63, 121] (Figure 2.15).

Elevational patterns showing a peak in elevational distributions at mid-elevation are ubiquitous in landscapes shaped by fluvial erosion when a sufficiently large region rather than a single slope or mountain is considered. It should be noted that this pattern is altered only if large areas outside runoff-producing zones (e.g., large plains) are included in the domain [63]. Mountains are no cones (nor are clouds spheres, nor coastlines simple broken lines – as in the celebrated prose of Benoit Mandelbrot [121]), and therefore simple 1D slopes are a highly misleading representation of nature's topographies, especially if connectivity depends on altitude. Mountain landscapes are in fact rather complex self-affine fractal structures [63]. Within an ecological context, this fact has seriously misled researchers interested in elevational trends and in natural

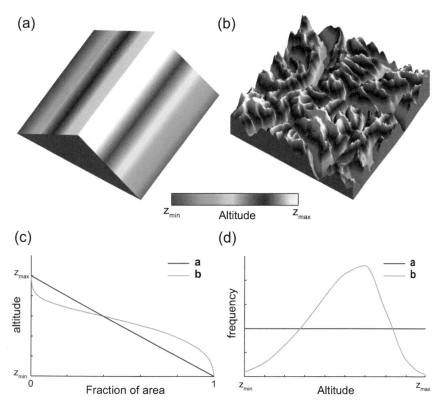

(a)

(b)

z_{min} Altitude z_{max}

(c)

altitude

z_{max}

z_{min}

0   Fraction of area   1

—— a
—— b

(d)

frequency

z_{min} Altitude z_{max}

—— a
—— b

Figure 2.15 Comparison between (a) an oversimplified, 1D topographic gradient elevation field and (b) a real-life elevation field (a fluvial landscape in the Swiss Alps, 50×50 km^2). (c) Hypsometric curve and (d) frequency distributions of elevation of the two landscapes. It is clear that 1D gradient experiments are unrealistic regardless of details on how the replicated real-life topographies are arranged. Figure after [229]

experiments conducted along gradients of elevation, because a mountain–linear slope analogy would suggest a monotonically decreasing distribution of elevation. This is definitely not the case for fluvial landscapes, and consequences are noteworthy. In fact, the area of available habitat within a given elevational band may have a direct effect on the global diversity of the regional community it hosts (measured by its γ-diversity [237–240]; see Chapter 1), as predicted by the species–area relationship [241]. The area of available habitat may also have an indirect effect on the LSR because local communities can be assembled from a more diverse regional pool of species that are fit to live at similar elevation [242].

Box 2.9 Fundamental and Realized Ecological Niche

We discussed earlier the concept of habitat suitability without a clear distinction among the various filters operating at a landscape scale. Abiotic filters, and, more generally, biotic and abiotic resources (e.g., food availability to animals), define the environmental niche, the set of climatic, physical, and ecological features that define the general conditions that a species needs to grow and maintain a viable population *sensu* Hutchinson [243]. Biotic factors (at equal or higher food-chain level than the focal species) comprise the set of interactions of the focal species with other organisms. Interactions may be positive/favorable (such as mutualism and commensalism) or negative/unfavorable (competition and predation), and overall a sensible approach assumes that the ecological processes shape community composition limited by some form of carrying capacity (see, e.g., [59]). By no means intending to review a broad and articulate subject, here we simply extend the previous concept to any "other" biotic effects, even including ecosystem engineers (i.e., species that modify the local physical environment by engineering it to favor other species). This is the case,

Box 2.9 *Continued*

for instance, of halophyte zonation shaping the topography of saltmarshes by organic soil production and the trapping of inorganic sediments in tidal environments [244] or trees providing canopy cover conditions for shade-tolerant species [59, 245].

As we have relentlessly repeated in this book, we limit our aims at sorting clearly and quantitatively landscape effects induced by the substrates for ecological interactions created by fluvial processes – in a broad sense, as this context not only includes the river network proper but also topographies including altitude generated by large-scale geomorphic processes inclusive of fluvial erosion and uplift in relative balance. To decouple the various effects, we need to make assumptions to avoid clouding the central aim with unnecessary detail. As such, models are engineered to single out the individual role of the dendritic structure of the topological substrate for ecological interactions or of landscapes carrying the imprinting of such processes. To do so, the main overarching assumption is that the strength of biotic interactions must not be incapable of hiding species–environment relationships.

The concept of realized niche has been widely used in ecology, and an account in this context seems appropriate. One wonders why a specific species is not observed everywhere, and conversely, as Humboldt and Darwin famously noted, why the same species could occur in sites marked by very different environmental conditions. The combination of such conditions is precisely what we define as habitat (see, for a comprehensive review, [59, 246]). The physiology of organisms is the determinant of the above conditions. The specialized physiological adaptation that species have locally undergone if biotic and abiotic conditions lasted enough to let evolution pitch in significantly results in possible maladaptation (and thus potentially local extinction) if such conditions change more or less abruptly. Physiological specialization usually results in a spectrum of possible responses for different species across environmental gradients, in most cases associated with a position along such gradients where the species performs at its very best. For instance, plant species across altitude gradients show a range – macroscopically the tree line – signaling the extent of their upper geographic range (typically dictated by environmental conditions like air or soil temperature in the growing season). Usually, however, the lower boundary of the species presence tends to be controlled by biotic factors [59]. It is normal, in the general case, to assume that the optimal position across the gradient (the peak of the fitness function corresponding to the physiological optimum) is associated with a progressive decrease in performance away from it. How gradual this is, of course, depends on the species, and the physiological response curves may exhibit unimodal or most often sigmoid responses when plotted against the driver environmental gradient – say, latitudinal or altitudinal (see Figure 2.16 for an example, where neutrality, *sensu* Hubbell [34], is implied by the simple shifting of the fitness curves defining the range and the performance of different species – physiological responses were assumed to be the same, thus as customary, drastically simplifying the complexity of ecosystems for a purpose). The width of the curve defines the fitness function, or the physiological response curves of a species along the actual environmental gradient, thus quantifying the physiological tolerance of the species, which in turn defines the species' range. In reality, the same species may show different optima and tolerances along different gradients, and the nature of the tolerance may vary significantly for different species depending on the gradient. Broad tolerance may be exhibited along a specific gradient, defining a generalist species in that respect. The transition from optimal to unsuitable conditions, reflected in negative population growth rates

Box 2.9 *Continued*

or the imbalance between birth and death rates of individual organisms, may be smooth or abrupt. It has been observed that the latter occur when metabolic pathways like photosynthesis experience threshold effects, leading to abrupt change. For iconic conifer trees along altitude gradients, for instance, the cambium activity that supports root growth stops suddenly below a threshold value of soil temperature during the growing season [59, 232]. On the contrary, gradients in water availabiliy tend to foster smoother transitions in lowering individual organisms' fitness [28].

The concept of the environmental niche of a species derives from the synoptic view of the combined effects of the relevant environmental controls on the species' physiology [247]. When considered together, the physiological response to several (say, n) environmental variables defines a multidimensional "volume" called the fundamental niche [243], that is, an n-dimensional hypervolume in a space whose independent variable, the coordinates, are the environmental variables having a direct influence on species viability (for a pictorial representation of the fundamental niche in a 3D space where the control variables are water, light, and temperature [exemplified therein by a minimalist sphere], see figure 3.5 in [59]). Notice that the fundamental niche concept applies to animal species [243] as well to plants and other organisms, such as fungi, protista, and bacteria [59]. For plants, the axes of the fundamental niche are typically resource variables (light, heat, water, and nutrient availability) and regulators (such as soil type, temperature ranges in given stages of growth, and extreme climatic conditions [246]). For animals, thermal limits, water and food availability, and habitat tend to be the main controls [59, 246]. Georeferencing environmental gradients mediated by the fundamental niche (see Figure 2.16 for a visual projection of altitude gradients) defines the species potential geographic range, the locations where the species could colonize an empty site via

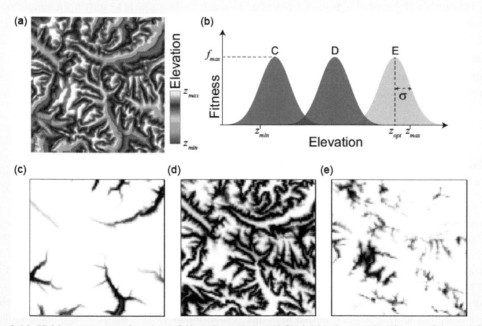

Figure 2.16 Habitat maps as a function of elevation. (a) A real fluvial landscape. (b) Fitness of three different species as a function of elevation. (c–e) Fitness maps of the three species shown in (b) Darker pixels indicate higher fitness. Figure after [229]

Box 2.9 *Continued*

its dispersal ability and the extant constraint, and maintain therein a viable population in the absence of biotic interactions.

The fundamental niche is an abstract concept in practice. On one hand, in fact, measuring the fundamental niche in the field is impossible [59], and *ex situ* information and limited measures only may be used to outline its boundaries. On the other, the ensemble of all processes constraining the actual presence of a species at a site, including the actual migration limitations in rugged terrains and in an ever-changing natural world, defines the realized niche, the one that can be observed (and mapped) in the field, based only in part on the true physiological response and rather subsuming all inferences of the ecosystem dynamics only partially observable due to practical limitations. This is reinforced by the fact that accurate surveys of all variables bearing direct physiological effects on a given species (the fundamental niche) would be unrealistic in large-scale settings, in part because of the need for prior laboratory evidence of actual metabolic rates, or individual fitness under analogous gradients, and other factors, for large assemblages of populations. The realized niche is therefore the observable one, the only viable concept for inferences, modeling, and predictions. Surrogate environmental variables (proxies) correlated to physiologically meaningful controls are thus often a practical and sensible choice [59].

Clearly, the chosen minimalist modeling approaches used in this book – though spatially explicit – do not cover properly biotic interactions, however placed across trophic levels, in an approach where one is left to assume that they are not dominant anyway. Elsewhere (Section 1.1) we have noted that, in reality, many factors other than network configuration and transport anisotropy are in place and play different roles. However, inclusion of all factors, no matter how realistic, is hardly a starting point for pursuit of generalizable signatures. Our research philosophy is to start from simple settings from which fundamental insights can be gathered and to later add complications, whose effects would then be distinguishable. This applies directly to the issue of characterizing habitats and niches in fluvial landscapes. In fact, in the sequel, we will employ elevation as the main niche dimension, assuming that it is the main driver of observed biodiversity. We are conscious that this is of course a simplifying assumption, although it has been used widely in the ecological literature (e.g., [248]).

2.3.2 Of Altitude-Specific Environmental Drivers

Because there exist two classes of environmental drivers, those that are altitude specific (such as atmospheric pressure and temperature) and those that are not (such as moisture, clear-sky turbidity and cloudiness, sunshine exposure and aspect, wind strength, season length, geology, and human land use) [232, 233], empirical results may hardly sort out unambiguously general rules, if such rules exist at all. Bertuzzo et al. [229] moved from this premise by exploiting universal self-affine features of elevation fields of fluvial landscapes, possibly obtained by planar aggregation structures where sensible local slope–area laws, reminiscent of the fluvial landscape,

are enforced [63]. Incidentally, the models discussed in this section may apply equally well by using real topographies where the slope–area laws break down at the drainage density threshold. The nature of fluvial landscapes and their physical evolution, in fact, provide universal invariance of patterns of connected areas at the same elevation [63].

The examples of biodiversity affected by the fluvial landscape structure that we address here thus describe how altitude-driven area connectivity fosters elevational gradients of species richness, in particular, the origins of empirically observed mid-altitude peaks. The analysis, as we said in the introduction, moves one further step away from neutrality by adding niche trade-offs (for a somewhat similar attempt, see [249]). In fact, the model employs an altitude-dependent adaptive fitness of otherwise equal vital rates of species. Also, as briefly discussed in what follows, connectivity is based on the altitude–area relations expected in general according to fluvial patterns. To investigate the role of the mountainous landscapes in shaping altitude gradients of species richness, a zero-sum metacommunity model is adopted [34, 250], namely, the system is always saturated. In this framework, the system comprises N local communities, which are characterized by their position in space and by their mean altitude. Only communities organized in an equally spaced two-dimensional lattice will be considered; however, the model could be readily adapted to account for other connectivity structures like those investigated in the previous section. Each LC assembles n individuals. Because of the zero-sum assumption, at any time, the system is populated by $N \cdot n$ individuals belonging to S different species. Each species is characterized by a specific altitude niche function that expresses, in this context, how the ability of a species to exploit resources varies with altitude. This relationship is modeled as a Gaussian function:

$$f_i(z) = f_{\max_i} \exp\left[-\frac{(z - z_{opt_i})^2}{2\sigma_i^2} \right], \qquad (2.15)$$

where $f_i(z)$ reflects the competitive ability of the individuals of species i at altitude z and z_{opt_i} is the optimal altitude of species i, that is, where $f_i(z)$ equals its maximum $f_{\max_i}$. The parameter σ_i controls the dispersion of the Gaussian function, that is, the niche width. In this example, the analysis is limited for simplicity to a neutral case in which all the species have the same parameters $\sigma_i = \sigma$ and $f_{\max_i} = f_{\max}$. Figure 2.16 illustrates how the niches of different species are modeled. While species differ for their altitudinal niches, all the other ecological traits (namely, birth, death, and dispersal rates) are identical, as in classical ecological neutral dynamics [34, 49].

Ecological interactions among individuals are simulated as detailed in Box 2.10. At each time step, a randomly selected individual dies. Its empty site is occupied by an offspring either of one of the individuals of the same LC, or of one of the four nearest-neighbor communities, or else of an additional individual belonging to a species not currently present in the system. This additional offspring is added with probability ν at each time step. The offspring is selected randomly with a probability proportional to the niche function $f_i(z)$ of all the candidate colonizing individuals evaluated at the elevation z of the LC of the dead individual. Needless to say, introduction of new species is aimed at

modeling both speciation and immigration from external communities.

In addition to the two landscapes illustrated in Figure 2.15, the model has been run in two other landscape structures (Figure 2.17b,c). Care is exerted in using landscapes holding the same frequency distribution of site elevation (i.e., the same hypsographic curve) of the mountainous landscape shown in Figure 2.15. The two alternative landscapes studied [229], in fact, are such that their hypsographic curves match those of mountainous landscapes of various origins. By analyzing sequentially the related effects, individual roles can be disentangled. All the landscapes are constructed by gridding different elevation maps in a regular 100×100 lattice ($N = 10^4$). The structure in Figure 2.15, replicated in the upper part of Figure 2.17, has been obtained by using a real-life elevation map where each pixel embeds the mean altitude of a 500×500 m region. The LC size n is set to 100. The system is initially populated by a single species and is simulated until a statistically steady state is reached (10^5 generations, where a generation is $N \cdot n$ time steps). Periodic boundary conditions are prescribed for the various landscapes. Notice that model results do not depend on the actual altitude range $[z_{\min}, z_{\max}]$ but only on the ratio $\sigma/(z_{\max} - z_{\min})$ [229].

Typical results of the metacommunity model are shown in Figure 2.17. Note that the general shapes of the elevational gradients of biodiversity in different subdomains of the same size are qualitatively similar in that they all show a peak at intermediate altitudes (upper inset of Figure 2.17). Whether local or averaged over larger domains, all landscapes produce to different extents hump-shaped α-diversity curves, yet only the one corresponding to a real landscape produces realistic variability akin to that produced by field evidence. The values computed therein depend on specific choices of the niche width σ and on other parameters. It has been shown, however, that the trends outlined therein are valid irrespective of parameter values (Figure 2.18) [229].

The hump-shaped patterns of species richness along elevational gradients shown in the previous figures are the by product of three concurring factors:

- the finiteness of the elevational range available;
- the mid-peak of the frequency distribution of elevations;
- differential landscape elevational connectivity (LEC; see Box 2.11) featured by fluvial landscapes.

To single out unequivocally landscape effects, one needs to disentangle the role of each of the three factors. To

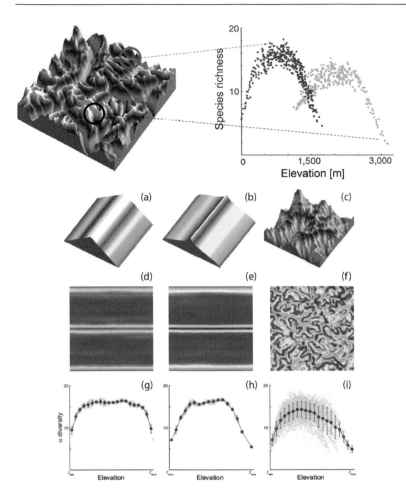

Figure 2.17 (top) Local Species Richness (LSR) in different subdomains of the same size of the general landscape of Figure 2.15b (Swiss Alps). The hump-shaped curve is evident, although the relative values of α-diversity are evidently site dependent. (a–c) Comparative landscape forms; note that (b) is constructed so as to yield the same hypsometric curve of (c), which is an OCN landscape (Appendix 6.2). (d–f) Maps of LSR color coded by the absolute values of α-diversity. (g–i) General plots relating LSR to elevation in the three domains, inclusive of the whole range of computed values, means, and variances. Figurer after [229]

that end, specific simulations via the zero-sum meta-community model have been run [229] over the three landscapes illustrated in Figure 2.17. For consistency, all elevation fields feature the same relief. Here, we summarize the results, detailed in the supporting information of [229]. In brief, the one-dimensional slope fosters relatively constant α-diversity along the elevational gradient, except for an expected decrease at the boundaries of the elevation range due to the finiteness of the elevation range. Indeed, while sites in the middle of the elevation range can potentially be colonized by species that live at (and have a preference for) higher and lower elevations, sites at the lowest (highest) extreme are only subject to the colonizing pressure from higher (lower) elevation. The pattern of α-diversity in the second landscape exhibits a more pronounced edge effect. This pattern is the result of the combination of the finiteness of the elevation range and the hump-shaped distribution of elevation. The pattern of α-diversity in a fluvial landscape presents a more pronounced peak for mid-elevation

sites and a marked variability of diversity for the same elevation. Note the relative small degree of smoothing of the elevational gradients after 500 realizations of the process [229].

Figure 2.18 explores the sensitivity of diversity patterns with respect to variations of the niche width σ and clarifies the role of the LEC. As the niche becomes wider (i.e., higher σ values are employed), the variability of α-diversity at the same elevation decreases. Such variability is attributable to the differential LEC values shown by different sites at the same elevation. As σ increases, species are less constrained by elevational barriers, and thus diversity is chiefly controlled by the abundance of sites with similar habitat in the system (i.e., the elevation frequency distribution). For large values of σ (Figure 2.18, bottom row), diversity patterns expectedly become insensitive to elevation and elevational gradients tend to flatten out. It is remarkable that LEC consistently captures the effect of niche width on elevational diversity.

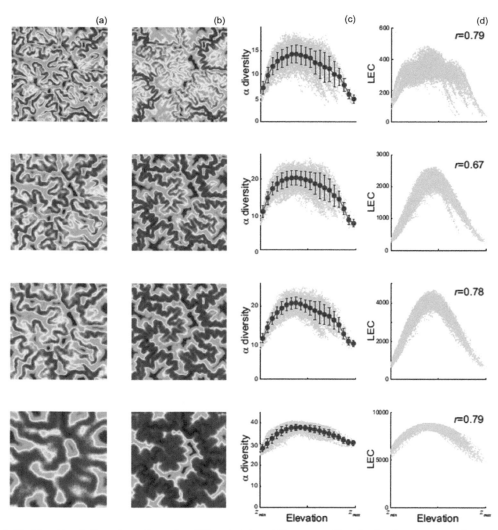

Figure 2.18 Elevational diversity patterns for different niche widths σ – α-diversity (a, c) and a network connectivity measure based on elevations. (LEC is an equivalent of the effective connectivity of any couple of sites $i \rightarrow j$ that accounts for the differences in elevation incurred in each intermediate planar step. In such a manner, LEC measures the likelihood for species to be able to settle in j crossing elevation-dependent unfavorable terrain. For a flat landscape, LEC reduces to the distance between the two sites measured along the planar path). (b, d) Spatial distribution (a, b) and elevational gradient (c, d). Text in panels (d) reports the Pearson correlation coefficient between local values of α-diversity and the LEC. Different rows show different values of niche width. From top to bottom, $\sigma/(z_{max} - z_{min}) = 0.1, 0.2, 0.3$, and 1. Simulations are performed over the same OCN landscape used in Figure 2.17c. Averages over 500 realizations of the metacommunity model are shown. Other parameters are $N = 104, n = 100, \nu = 1$. Figure after [229]

Finally, the sensitivity of diversity patterns with respect to the diversification rate ν is investigated in Figure 2.19, where results are compared to those obtained with a 10-fold reduction of ν. Therein, α-diversity decreases as expected [34]. However, the elevational gradients of diversity are similar over 1 order of magnitude variation of the diversification rate.

2.3.3 Fluvial Landforms and Biodiversity

Fluvial landforms show deep similarities across many orders of magnitude despite great diversity of their drivers and controls (e.g., relief, exposed lithology, geology, vegetation, or climate) [63]. Regardless of the self-affine nature of the elevation field as a whole, a marked heterogeneity of elevational distributions, and

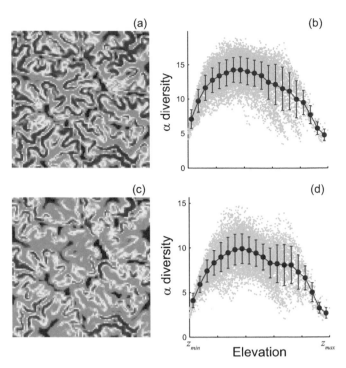

Figure 2.19 Elevational diversity patterns for different values of the diversification rate ν. (a, c) Spatial distributions and (b, d) elevational gradients of α-diversity for different values of the diversification rate: $\nu = 1$ (a, b) and $\nu = 0.1$ (c, d). Symbols as in Figure 2.17. Simulations are performed over the same landscape used in Figure 2.16. Averages over 500 realizations of the metacommunity model are shown. Other parameters are $N = 10^4$, $n = 100$, $\sigma/(z_{\max} - z_{\min}) = 0.1$. Figure after [229]

thus of ecological connectivity, characterizes the parts and the whole of real landscapes (Figures 2.17, 2.19, and 2.20). Nevertheless, the universality of the main attributes of the fluvial landscape naturally lends itself to the quest for general patterns of the ecological dynamics that such landscapes host [9, 19, 21]. Thus it is appealing that the α-diversity maps shown in Figure 2.17 reveal such clear spatial patterns, with valleys and mountaintops characterized by lower species richness. All geomorphic factors resulting in self-affine landscapes where peaks or troughs may occur at any elevation within the range simultaneously contribute to the formation of such patterns, regardless of other elevation-independent factors. A factor is also the finiteness of the landscape elevational range: sites at mid-elevation can potentially be colonized by species that live at (and are fit for) higher and lower elevations, whereas sites at the lowest (highest) extreme are only subject to the colonizing pressure from higher (lower) elevations. In addition to this boundary effect, the geomorphic structure of fluvial landscapes results in a mid-elevation peak in both area and connectivity across the landscape, both of which promote diversity. Specifically designed simulations have disentangled the role of each of these factors [229]. Results also show that without any of these effects, the model predicts

no gradients of diversity. Each geomorphic factor produces, independently or in combination with others, a hump-shaped pattern of species richness. Moreover, the differential elevational connectivity characteristic of fluvial landscapes results in a marked variability of diversity for the same elevation. The results reveal that similar mid-peak elevational gradients of diversity can be observed at different scales of observation even if the domains span different elevational ranges. This pattern is thus a direct consequence of the nature of the substrate for ecological interactions, as in all the examples pursued here – in particular, of the self-affine fractal structure of fluvial landscapes that reproduces statistically similar fluvial landforms across scales [63]. We thus show that deep similarities occur across many orders of magnitude despite great diversity of their drivers and controls (e.g., relief, exposed lithology, geology, vegetation, or climate). The universality of the main attributes of the fluvial landscape naturally lends itself to the quest for general patterns of the ecological dynamics that such landscapes host, which in turn is crucially affected by the nature of the ecological matrix where interactions occur. Results also show that, when the scale of observation is enlarged, different hump-shaped patterns are blended together, possibly producing a confounding effect [229], especially if the analysis is limited to the average diversity as

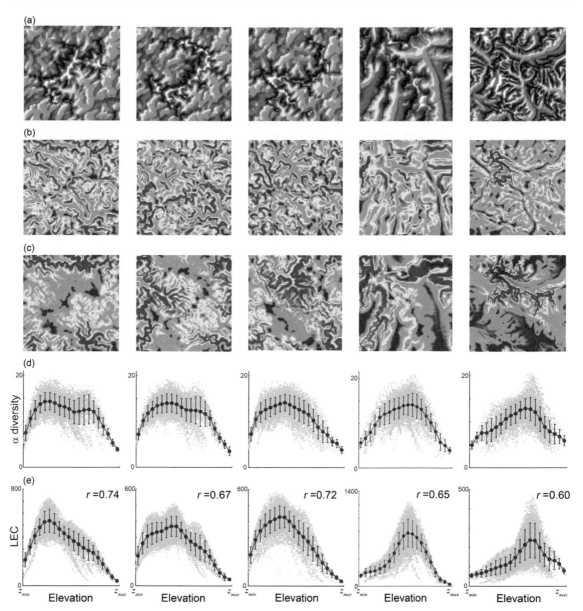

Figure 2.20 Diversity patterns for different fluvial landscapes. (a) Elevation fields. (b, c) Spatial distributions and (d, e) elevational gradients of (b, d) α-diversity and (c, e) LEC. The first three columns refer to landscapes derived from different realizations of the OCN over the same domain used in Figure 2.17. The last two columns represent two real fluvial landscapes derived from the DTM of the Swiss alpine region. Text in panels (e) reports the Pearson's correlation coefficient between local values of α-diversity and LEC. Averages over 500 realizations of the metacommunity model are shown. Parameters employed are $N = 10^4$, $n = 100$, $\sigma/(z_{max} - z_{min}) = 0.1$, $\nu = 1$. Figure after [229]

a function of elevation. This feature might help in understanding why elevational diversity is often found to be dependent on the scale of observation [236, 237, 239].

The above results, as in most of the examples put forward in this section, act as proof of concept. In fact,

we have presented results based on nearest-neighbor dispersal to highlight the role of elevational connectivity. Indeed, the effect of elevational isolation is expected to be reduced as dispersal limitation decreases because species can overcome elevational constraints with long-distance dispersal (e.g., dispersing from one mountaintop to another without going through unfavorable lower elevation habitats). Moreover, we have assumed that the landscape can be uniformly colonized, whereas real-life habitats are often composed by patches with different spatial connectivity and size [251]. Spatial and elevational connectivities can interact in complex ways to shape diversity patterns. While the modeling framework proposed can be easily generalized to accommodate both fragmented habitats and generic dispersal kernels, the coherent framework presented here has the potential to effectively describe how spatial and elevational connectivities shape diversity in complex 3D landscapes.

The above exploration of metacommunity patterns concludes our examples on how the ecological substrates affect in a decisive manner the ensuing biological diversity. It suggests that the specific spatial arrangement of sites at different elevation in fluvial landscapes suffices in inducing mid-peak elevational gradients of species richness without invoking specific assumptions, except that each species is fit for a specific elevation. In this framework, an elevation-dependent fitness applied to a real-life landscape translates into a fragmented habitat map. Such conceptualization lends itself to the application of classic concepts of metacommunity dynamics [251], according to which habitat size and connectivity are key drivers of biodiversity. We thus expect different metacommunity models to produce similar results.

Whether investigated by LEC measures (Box 2.11) without resorting to any particular biodiversity model or by full metacommunity approaches (as in Figure 2.17), geomorphic controls explain why linear landscapes or conelike structures fail to jointly reproduce the mean and the variance of elevational gradients of mean species richness, and the orders-of-magnitude smaller β-diversity, than real or realistic landscapes (Figure 2.20) – hence the kind of variability widely observed in nature. Needless to say, we do not dismiss as negligible other potential drivers of diversity, including those that often covary with elevation (e.g., habitat capacity, productivity, human disturbance). However, we argue that these drivers may act on top of the unavoidable effects provided by the geomorphic controls. A general consensus has thus been achieved that the analysis of elevational diversity should not seek a single overriding force but rather understand how different factors covariate to synergistically shape the observed patterns [233].

The above results strongly suggest, once more, that fluvial geomorphology embedded in the landscapes considered has an important role in driving emergent diversity elevational gradients of LSR – inasmuch as the topology of the planar substrates dominates the structuring of general biodiversity patterns.

Box 2.10 Zero-Sum Metacommunity Model

The model assumes that a system of N local communities is placed at the nodes of an equally spaced 2D lattice in which each cell is characterized by its elevation $z_i, i = 1 \ldots N$. Each LC assembles n individuals. The system is assumed to be at saturation (the zero-sum assumption [34]). Thus, at any time, the system is populated by $N \times n$ individuals belonging to S different species. Each species is characterized by a specific elevational niche that expresses, in this context, how its competitive ability varies with the elevation. This relationship is modeled by a Gaussian suitability function [252] (Equation (2.15)), that is, $f_i(z) = f_{\max_i} \exp\left[-(z - z_{opt_i})^2/(2\sigma_i^2)\right]$, where $f_i(z)$ reflects the competitive ability of the individuals of species i at elevation z, and z_{opt_i} is the niche position [252] of species i, that is, the elevation where $f_i(z)$ equals its maximum $f_{\max_i}$ (Figure 2.16b) and σ_i controls the dispersion of the Gaussian function. A neutral approach is adopted [34], save for the effects of elevation, in that the analysis is limited to the case where all species have the same parameters $\sigma_i = \sigma$ and $f_{\max_i} = f_{\max}$ as well as the same birth, death, and dispersal rates.

Ecological interactions among individuals are simulated as follows. At each time step, a randomly selected individual dies and is replaced by an offspring of one of the individuals living in either the same community or one of the four nearest-neighbor communities (von Neumann neighborhood) or by the offspring of an additional individual belonging to a species not currently present in the

Box 2.10 *Continued*

system. This additional competitor is added with probability v at each time step. To avoid edge effects, species with a niche position outside the elevation range of the system are also considered. Specifically, z_{opt_i} of newly introduced species is drawn from a uniform distribution spanning twice the relief. The introduction of new species is aimed at modeling both speciation and immigration from external communities [34, 39]. The system is initially populated by a single species, and it is simulated until a statistically steady state is reached (10^4 generations, where a generation is $N \cdot n$ time steps). Periodic boundary conditions are prescribed. Model results do not depend on the actual elevation range ($[z_{min}, z_{max}]$) but only on the ratio $\sigma/(z_{max} - z_{min})$. Moreover f_{max} does not affect the simulated dynamics and has thus been set to 1.

Box 2.11 Geomorphic Measures of Elevational Diversity: Landscape Elevational Connectivity

The LEC at site i, LEC_i, can be expressed as

$$LEC_i = \sum_{j=1}^{N} C_{ji},$$

where C_{ji} is a measure of the closeness of site j to i in terms of elevational connectivity. Such a measure should quantify how easily a species living in patch j can spread and colonize patch i. As the fitness is assumed to be elevation dependent, C_{ji} depends on how often a species, assumed to be adapted to the elevation of j, z_j, needs to travel outside its optimal fitness range in the path from j to i. C_{ji} thus depends on the elevation field but also on the width of the species niche, here described by the parameter σ. Indeed, for high σ, species are free to spread throughout the system regardless of elevational constraints. Conversely, for low values of σ, species are constrained to spread following paths with similar elevation without the ability to cross elevational barriers. The closeness C_{ji} can be expressed as

$$C_{ji} = \max_{p \in \{j \to i\}} C_{ji,p},$$

where $C_{ji,p}$ is a measure of the closeness of community j to i along path p (Figure 2.21). The overall closeness C_{ji} is defined as the maximum value of $C_{ji,p}$ along all the possible paths p from j to i (see a graphical representation in Figure 2.20). $C_{ji,p}$ is assumed to be proportional to the product of the probabilities of making each step of the path p. Let $p = [k_1, k_2, \dots, k_L]$, $k_1 = j$, and $k_L = i$ be the sites composing a path p form j to i. $C_{ji,p}$ can be expressed as

$$C_{ji,p} = \prod_{r=2}^{L} e^{-\frac{(z_{k_r} - z_j)^2}{2\sigma^2}}. \tag{2.16}$$

In Equation (2.16) the exponential form is proportional to the fitness that species adapted to site j have in site k_r.

In this application we focus only on isotropic nearest-neighbor dispersal, which implies that $d(k_{r-1}, k_r)$ is nonnull and constant for nearest-neighbor connections and can therefore be neglected

Box 2.11 *Continued*

(a)

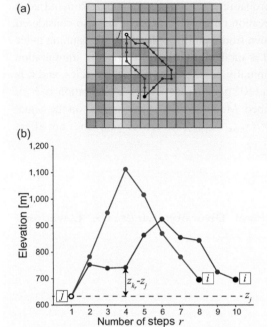

(b)

Figure 2.21 Sketch of the physical meaning and the computational procedure of the LEC (LEC_i) at site i. LEC proves a reliable proxy of the LSR computed by the full-fledged metacommunity model and is solely a geomorphic measure, in this case of elevational diversity. Here, we show examples of two possible paths connecting (a) an arbitrary site j to i with (b) their corresponding elevational profiles. We note that although the blue path is longer, the associated cost ($\sum_{r=2}^{L}(z_{k_r} - z_j)^2$) is smaller than that of the red path as it travels through sites with elevation more similar to that of the starting point (z_j). Figure after [229]

in Equation (2.16) (note that constants in the computation of $C_{ji,p}$ are immaterial, as we are not interested in the actual value of LEC_i but only in its distribution). However, the generality of the formulation in Equation (2.16) allows application of the method for arbitrary dispersal kernels. From an operational viewpoint, C_{ji} is computed by focusing on the quantity,

$$-\ln C_{ji} = \frac{1}{2\sigma^2} \min_{p \in \{j \to i\}} \sum_{r=2}^{L}(z_{k_r} - z_j)^2,$$

where the log-transformation maintains extremal properties. For each site j a graph is built with edges representing all possible nearest-neighbor connections among sites and weights equal to the square of the difference between the site elevation and z_j. The Dijkstra algorithm is then used to find the shortest path from j to all other sites (after [229]).

2.4 Metapopulation Capacity of Evolving Fluvial Landscapes

The form of fluvial landscapes is known to attain stationary network configurations that settle in dynamically accessible minima of total energy dissipation by landscape-forming discharges (Appendix 6.2). In Chapter 1 we have also advocated for a key role for the dendritic structure of river networks in controlling population dynamics of the species they host and large-scale biodiversity patterns. Here, we study the connections between the two problems by systematically investigating, following [253], the relation between energy dissipation, the physical driver for the evolution of river networks, and the ecological dynamics of their embedded biota. To that end, we use metapopulation capacities, λ_M, as a measure of species viability, which link the landscape structure with the population dynamics they host (Box 2.12).

Box 2.12 Metapopulation Dynamics in a River Network

To investigate the persistence and the probability of occupancy of species spreading along river networks, we make use of a well-established spatially explicit metapopulation model [183, 217, 251, 254, 255] that accounts for the fundamental ecological processes of colonization, extinction, and dispersal. The importance of the model owes much to the fundamental work of the late Ilkka Hanski, a towering figure in contemporary ecology.

Each pixel of the modeled landscape is assumed to be a patch that can be either occupied or not by the species considered. Patches thus have the same size Δx^2, the pixel size. We consider species that are constrained to disperse along the river network. Note that in the original formulation of the theory [217], dispersal had no preferential direction; that is, the probability of dispersing from patch to patch would depend on a suitable Euclidean distance of the patches (say, the distance between their centroids). In this specific case, however, we make two assumptions: first, we restrict our attention to species dispersing along the network structure – distances are thus measured in "chemical distance," that is, along the unique tree path that joins any two sites [63] (see also Appendix 6.2). Furthermore, for the scope of this introduction to metapopulation models, we shall not rescale distances by accounting for a bias imposed by the flow direction, as done in [8, 40], thereby focusing on metapopulations of active dispersers. The former assumption may be easily relaxed in the context of riverine ecological corridors as it would subsume dispersal beyond strictly fluvial pathways (Section 2.2.6). The latter assumption is meant as a proof of concept rather than a specific feature of a species (although it proved valid for large-scale patterns of fish biodiversity; see Section 2.1) and should not detract from the generality of our results as it could be relaxed to study the conditions of persistence under drift (see, e.g., [7, 115]).

In the spatial metapopulation scheme, the evolution of the probability $p_i(t)$ of the focus species being present in patch i at time t is a balance between colonization and extinction forces [183]:

$$\frac{dp_i(t)}{dt} = C_i(t)[1 - p_i(t)] - E_i(t)p_i(t), \tag{2.17}$$

where $E_i(t)$ is the extinction rate of the existing populations in patch i and $C_i(t)$ is the colonization rate of patch i when empty ($p_i = 0$). The effective colonization rate (the first term on the right-hand side of ((Equation 2.17)) accounts for the probability that the patch is empty ($1 - p_i(t)$).

We assume that all pixels are equivalent in terms of habitat suitability and habitat capacity and that they differ only in their degreess of connectivity. It follows that the extinction rate can be assumed as constant and uniform, $E_i(t) = e$, and that the colonization rate of patch i is a function of all possible contributors, that is,

$$C_i(t) = c \sum_{j \neq i} e^{-\alpha d_{ij}} p_j(t), \tag{2.18}$$

where d_{ij} is the distance (measured along the network) between sites i and j; $1/\alpha$ is the mean distance of dispersal, here assumed to be exponential [183]; and c is a constant. Depending on the focus species and its life cycle, colonization can be achieved through migration, movement, or dispersion of propagules. In the following, we will generally refer to dispersal processes and dispersal distance.

Under the above assumptions, Equation (2.17) reads

$$\frac{dp_i(t)}{dt} = c \sum_{j \neq i} e^{-\alpha d_{ij}} p_j[1 - p_i(t)] - ep_i(t). \tag{2.19}$$

Box 2.12 *Continued*

A key parameter is thus $\delta = e/c$, the ratio of extinction and colonization rate parameters. Equation (2.18) is simplified with respect to the original approach [217] in that "patches" are assumed to be of equal area and equal habitat suitability and to span an assigned domain (a condition posing serious constraints on the distribution of length between sites; see, e.g., Appendix 6.2.3), whereas the connectivity of the system affects directly the key metrics d_{ij}, that is, the distance of pixel i from any other site j.

The state $\mathbf{p_0}$ characterized by $p_i = 0$ for any i is a global extinction equilibrium for model (2.19). Metapopulation persistence is related to the stability of such equilibrium. If $\mathbf{p_0}$ is unstable, a small perturbation (e.g., the introduction of a few individuals) leads to a positive stable equilibrium $\mathbf{p}^*$ with $p_i > 0$ for any i. On the contrary, if $\mathbf{p_0}$ is stable, the species cannot persist, and any population is doomed to extinction. The condition for the extinction equilibrium to switch from stable to unstable is that the leading eigenvalue of the Jacobian matrix $\mathbf{J}$ of system (2.19), which is a positive system, linearized around $\mathbf{p_0}$ switches from negative to positive. If we define a matrix $\mathbf{M}$ consisting of elements $m_{ij} = \sum_{j \neq i} e^{-\alpha d_{ij}}$ for $i \neq j$ and $m_{ii} = 0$, the Jacobian reads $\mathbf{J} = c\mathbf{M} - e\mathbf{I}$ and the stability condition becomes [183]

$$\lambda_M > \delta, \tag{2.20}$$

where λ_M, termed metapopulation capacity, is the leading eigenvalue of the matrix $\mathbf{M}$. Because $\mathbf{M}$ is a nonnegative square and irreducible matrix, according to the Perron–Frobenius theorem (Appendix 6.1), it has a positive and simple maximum eigenvalue λ_M and a unique positive eigenvector associated to it.

The inequality (2.20) therefore provides the condition for long-term persistence of a species in a given landscape as a function of parameters proper to the species considered (δ and α). To compute λ_M for a given landscape, what matters is just the spatial scale of connectivity (set by the average dispersal distance $D = 1/\alpha$) and the spatial locations of the habitat patches, here identified as any node of the river network, identified from selected channeled pixels of the fluvial landscape [63] (Figure 1.3). For a given species, λ_M is a measure of the ecological suitability of the landscape, and it has powerful implications, as it allows comparisons of differently connected landscapes where the relative contribution of any site to all others is accounted through d_{ij}, that is, the distances available to ecological dispersal.

Hanski and Ovaskainen [183] have shown that a weighted average $\bar{p}^*$ of the equilibrium occupancy probability p_i^* values can be approximated by $\bar{p}^* = 1 - \delta/\lambda_M$. Therefore, when the conditions for persistence are satisfied, the higher the metapopulation capacity is, the higher is the expected occupancy of the population. Metapopulation capacity can be used specifically to rank different landscapes in terms of their capacity to support viable metapopulations. We shall employ this concept to analyze evolving fluvial landscapes and probe the parallel evolution of their capacity to support long-term persistence of arbitrary metapopulations.

Variations on the basic metapopulation scheme abound. The original formulation [217] identified patches differing only by their size, say, A_i where i indexes all suitable sites. The corresponding landscape matrix $\mathbf{M}$ consists of elements m_{ij}, defined as

$$m_{ij} = \frac{e^{-d_{ij}/D}}{2\pi D^2} \, A_i \, A_j \qquad \text{and} \qquad m_{ii} = 0. \tag{2.21}$$

Box 2.12 *Continued*

The maximum eigenvalue of the landscape matrix **M**, λ_M, gives the conditions for persistence of a species in a given landscape ($\lambda_M > e/c$) [183]. This may be (slightly) generalized. For instance, the fitness framework adopted in Section 2.5 (a stochastic patch-occupancy model [SPOM]; Box 2.4.4) corresponds to a modification of the concept of metapopulation capacity [183], as the landscape matrix **M** consists of elements m_{ij} that are slightly modified to account for fitnesses of the source or sink patches as

$$m_{ij} = \frac{e^{-d_{ij}/D}}{2\pi D^2} f_i f_j \qquad \text{and} \qquad m_{ii} = 0. \tag{2.22}$$

The maximum eigenvalue of the modified landscape matrix **M**, λ_M, gives, then, the conditions for persistence of a species in the given landscape, where patch abundance depends on a specific fitness (say, mean elevation) rather than sheer size. In the notable case of dendritic substrates, a first-order approximation may assume all fluvial sites equally fit for the focus species, and the patch area is constant, say, $A_i = A_j = A$, in Equation (2.21), thus yielding

$$m_{ij} = A^2 \frac{e^{-d_{ij}/D}}{2\pi D^2} \qquad \text{and} \qquad m_{ii} = 0. \tag{2.23}$$

where therefore the discriminating effects of dispersal would simply depend on the dendritic distances d_{ij} connecting any two sites i and j, which are unique within trees.

Because in any case **M** is a nonnegative square and irreducible matrix, according to the Perron–Frobenius theorem (Appendix 6.1), it always has a positive and simple maximum eigenvalue λ_M and a unique positive eigenvector associated to it. Therefore, to compute λ_M for a given landscape, what matters is often the spatial scale of connectivity defined by the average dispersal distance D and the spatial locations of the habitat patches. The latter may be labeled by their area or fitness or discriminated by their relative centrality (*sensu* Newman [256]) for the given landscape acting as a substrate for ecological interactions.

For a given species, λ_M is a measure of the ecological suitability of the landscape and has powerful implications, as it allows comparisons of differently connected landscapes where the relative contribution of any site to all others is accounted through d_{ij}, that is the distances available to ecological dispersal. Thus, metapopulation capacity can be used specifically to rank different landscapes in terms of their capacity to support viable metapopulations. We employ such a concept in Section 2.5 to analyze how different landscapes fare in their capacity to support long-term persistence of arbitrary metapopulations.

Finally, we note that whatever the metapopulation dynamics (i.e., regardless of the landscape attributes that determine the dominant ecological processes), it is possible to derive the contribution of the ith pixel to λ_M by computing the corresponding eigenvector:

$$\lambda_i \equiv x_i^2 \lambda_M,$$

where x_i is the eigenvector's ith entry. This permits us to understand the areas of suitability of a species defined by a combination of the parameters.

Specifically, Bertuzzo et al. [253] have studied how λ_M changes in response to evolving network configurations of spanning trees. Such sequences of configurations are theoretically known to relate network selection to general landscape evolution equations through imperfect searches for dynamically accessible states frustrated by the vagaries of nature (Appendix 6.2). In particular, relations were sought among the processes that shape metric and topological properties of river networks, prescribed by physical constraints in landscape evolution, and those affecting the landscape capacity to support metapopulations – with yet another view on biodiversity in fluvial ecosystems.

Fluvial landforms show empirically (and compellingly) profound similarities of the parts and the whole across several orders of magnitude regardless of major diversities in their drivers and controls, like geology, exposed lithology, vegetation, and climate [63]. Remarkably, one observes robust, approximate universality in the set of (mutually related) scaling exponents that describe mathematically the self-similar or self-affine metric or topological features of the fluvial landscape. River networks in runoff-generating areas are spanning trees: a unique route exists for landscape-forming discharges from every site to an outlet, and no loops are observed. OCNs are trees minimizing a functional describing total energy dissipated along drainage directions by landscape-forming discharges, which hierarchically accumulate toward the outlet of the basin (Box 1.3 and Appendix 6.2). As shown therein, OCNs are exact steady state solutions of the general landscape evolution equation under the small gradient approximation [74]. Any loopless network configuration that minimizes total energy dissipation corresponds exactly to stationary solutions of the general landscape evolution equation under reparametrization invariance in the small-gradient approximation [64, 69, 74, 257]. The large variety of dynamically accessible local optima and the universality of their scaling features akin to those observed in nature suggested several applications ranging from the design of experiments in the laboratory (e.g., [14, 100]) to a variety of explorations of network scaling [72].

Understanding the origins, and the needs for maintenance, of biodiversity in dendritic freshwater metacommunities is a primary goal of current ecological studies centered on population demography, population genetics, and community composition. In this context, river networks have been viewed as ecological corridors for species, populations, and pathogens of waterborne disease, the tenet of this book. Interestingly, the theoretical prediction for a key role of dendritic connectivity in shaping biodiversity patterns resists several generalizations, from individual-based to metacommunity models and for interactions/migrations ranging from nearest neighbors alone to long distances [21, 40]. Replicated experimental evidence that connectivity per se shapes diversity patterns

in protist microcosm metacommunities supports such a tenet [14, 100], as seen in Chapter 1. Empirical evidence to that end also exists [6, 96, 164]. Spatially constrained dendritic connectivity is thus accepted now as a key factor for community composition and population persistence in environmental matrices commonly found in many natural ecosystems, such as streams and watersheds. Such habitats are structured in linear, hierarchic arrangements where landscape structure and physical flows determine the directions and the range of organismic dispersal. Because dendritic ecosystems are typically embedded in landscapes structured by elevation fields, species' dispersal is often directionally biased (see, e.g., [99, 258–260]).

2.4.1 Optimality of Total Energy Dissipation and Species Viability

Figure 2.22 shows the results of a computational experiment carried out on a 128×128 lattice starting from an initial condition characterized by parallel-flow channeled hillslopes draining onto an orthogonal collecting channel (Figure 2.22a). The whole process of optimization is illustrated, including the high-T phase, when many changes were accepted even if the related energy was increased. Figure 2.22b shows an intermediate configuration along the minimum search process, characterized by a value of total energy dissipation comparable to the initial one. Figure 2.22c shows the OCN obtained after the algorithm has converged toward a local minimum of energy dissipation. Even with the naked eye, one recognizes the similarity of the parts and the whole that endows OCNs with features indistinguishable from natural forms distinctively marked by intertwined scaling exponents unlike chance-dominated trees [63, 67] (Appendix 6.2). Interestingly, along the process of minimizing total energy dissipation by shifting and sorting landscape matrices, the metapopulation capacity of the resulting landscapes increases. The evolution of λ_M mirrors, with changed sign, the convergence of total energy dissipation toward its local minimum. Figure 2.22 also shows the spatial distribution of the equilibrium occupancy probability p_i^* of the population of a species spreading in the corresponding networks. Note that the chosen intermediate state in the process of optimization, lowering only marginally the initial value of $H_\gamma(0)$ (Appendix 6.2 and Box 1.3), is characterized by a disordered aggregation process far from locally optimal either in terms of energy dissipation or in terms of metapopulation capacity. The occupancy probability

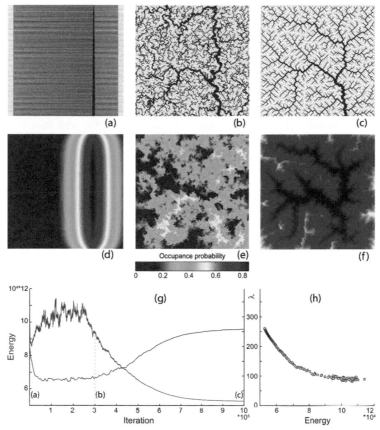

Figure 2.22 Linkage between the process of minimizing total energy dissipation, which leads to an optimal channel network (OCN), and the metapopulation capacity of the embedded fluvial landscape. (a) Initial network configuration, characterized by parallel flow directions collected by a central channel whose outlet is placed at the lower boundary. (b) Intermediate state with disordered structure and (c) final OCN. (d–f) Spatial distribution of the equilibrium probability of occupancy p_i^* (Box 2.12) for network configurations (a), (b), and (c), respectively. (g) Evolution along the iteration of the simulated annealing process of total energy dissipation $H_{1/2}(s)$ (red) and metapopulation capacity λ_M (blue). (h) One verifies empirically that energy minimization of a network configuration results in improved metapopulation capacities, emphasized by the one-to-one relation between λ_M and $H_{1/2}(s)$. Parameters used are OCNs, $L = 128$; simulated annealing, $T = H_{1/2}(0)/2 \cdot 10^5$; metapopulation model, $\delta = 50$, $D = 1/\alpha = 10$ pixels. Figure after [253]

of the OCN matrix is instead characterized by vastly improved values indicating higher chances of species persistence and occupancy. Another pattern is clearly distinguishable in Figure 2.22: occupancy probability increases moving from headwaters downstream, although with a final decrease toward the outlet of the catchment.

Figure 2.23 elucidates the mechanisms underlying the results presented above. It shows how the probability distribution of $R_i = \sum_{j \neq i} \exp(-\alpha d_{ij})$, that is, the sum along the rows of the landscape matrix $\mathbf{M}$, changes along with the optimization process leading to an OCN. Row sums R_i of $\mathbf{M}$ are convenient indicators because $\mathbf{M}$ is a nonnegative, irreducible square matrix, and a corollary

of the Perron–Frobenius theorem states that, under those conditions, one has

$$\min_i R_i \leq \lambda_M \leq \max_i R_i. \tag{2.24}$$

This corollary's implication is shown in Figure 2.22, where the red line represents the evolving value of λ_M. The physical meaning of R_i is the potential for a population occupying node i to disperse to any other site or, conversely ($\mathbf{M}$ being symmetrical), the potential of the population occupying any other site to reach node i. R_i can thus be thought of as a measure of the network closeness at i, and its distribution over all sites measures a collective closeness of the landscape. One sees that as energy is lowered, the amount of habitat that a population with a specific dispersal ability may reach from any given site increases.

Figure 2.24 illustrates how the previous result depends on the dispersal distance $D = 1/\alpha$. It shows the plot of the metapopulation capacity λ_M computed from 100 OCNs started from the same initial condition (Figure 2.22a) but choosing at random the outlet's position, for various values of the average dispersal distance $D = 1/\alpha$. The shaded area highlights the variability obtained for the various realizations.

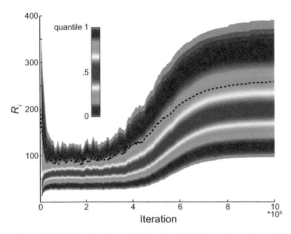

Figure 2.23 Evolution of the distribution of values of $R_i = \sum_j \exp(-\alpha d_{ij})$, $i = 1, N$, along with the optimization process shown in Figure 2.22. Grayscale code shows quantiles of the distribution. The red line represents the evolving value of λ_M, which is theoretically predicted to lie between the maximum and the minimum values of R_i. Figure after [229]

The green area in Figure 2.24 shows the metapopulation capacity λ_M for 100 random spanning trees, generated by performing $10N$ iterations of the search algorithm described in Appendix 6.2.1, where all changes – regardless of the energy attained by the evolving configuration – are systematically accepted provided that they maintained a proper tree structure. (See [63] for a description of the properties of these trees. Note that maintenance of the tree property of uniqueness of the path between any two sites is easy to check because otherwise the connectivity matrix could not be inverted.) Thus, in a sense, this experiment explores the entire range of options available to the evolving metapopulation capacity.

The convergence of λ_M for very large and very small dispersal distances regardless of the landscape matrix conforms with theoretical predictions. In fact, $1/\alpha \to \infty$ implies mean-field conditions with infinite dispersal. Thus, for any landscape, Equation (2.17) reduces to $dp_i(t)/dt = (N-1)cp_i(t)[1 - p_i(t)] - ep_i(t)$, $\forall i$, whose stability condition reads $(N-1) > e/c = \delta$, that is, $\lambda_M = N-1$. At the other extreme, $1/\alpha \to 0$ implies no dispersal at all, and Equation (2.17) reduces for any landscape to $dp_i(t)/dt = -ep_i(t)$, $\forall i$, according to which the extinction equilibrium is always stable and no population can persist, that is, $\lambda_M = 0$. For all intermediate cases, metapopulation capacities of OCNs are greater than those of random trees. Overall, it is clear that forms whose landscape matrix implies both chance and necessity at work produce higher viability for metapopulations to persist with respect to chance-dominated ones.

Figure 2.22 shows how minimizing energy increases metapopulation capacity. One thus wonders whether a local minimum of total energy dissipation, prescribed by

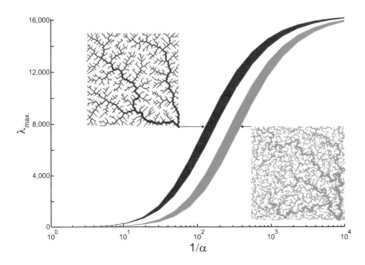

Figure 2.24 A comparative analysis of the metapopulation capacity λ_M for OCNs and random spanning trees as a function of the dispersal distance $1/\alpha$. Blue: OCN; green: random trees. Shaded areas represent the envelopes containing the relationships between λ_M and $1/\alpha$ for 100 different realizations (see text). Other parameters as in Figure 2.22. A species characterized by parameters δ and α can be thought of as a point in the illustrated space. Its population is predicted to persist if this point lies below the curve of a specific landscape ($\lambda_M > \delta$). Figure after [229]

the OCN search, corresponds to a local maximum of λ_M. This turns out not to be the case. Bertuzzo et al. [253] showed conclusively, in fact, that the search for network configurations that maximize λ_M (starting from an OCN as initial state) results in much higher metapopulation capacities. This implies different distributions of the occupancy probability with higher mean, that is, landscapes more supportive for metapopulations to persist. Crucially, it was shown therein [253] that the aggregation structures produced by maximization of metapopulation capacity are very different from the forms that we observe in nature for river networks, nor do they imply lower total energy dissipation. In any case, it must be remarked that Darwinian evolution (and the associated optimality criteria that have been the object of lively discussion in the past century) does not at all necessarily imply the maximization of metapopulation capacity, although this kind of maximization might vaguely recall the so-called K-selection [261], which implies maximization of species carrying capacity.

2.5 A Minimalist Model of Range Dynamics in Fluvial Landscapes

In keeping with our quest for the relations between the physical structure of the fluvial environments and the biodiversity they support, a long-standing question in ecology is whether a species distribution within a particular landscape is in balance with current climate-driven processes or else contains relict signatures of past climates and of species' adaptations. This echoes similar questions in geomorphology [262]. In a final development of this chapter, centered on biodiversity studies in fluvial ecosystems and landscapes, we report results of a study [263] addressing species range dynamics under warming temperatures. In line with the conceptual thread joining the themes of this book, the novelty and interest of the intensive computational exercise lie in the possibility of replicating statistically identical landforms offered by the OCN concept (Box 1.3 and Appendix 6.2). The full range of landscape effects is thus explored. The coherence with the other types of fluvial ecological corridors studied in this book derives once more from the landscapes generated by OCNs – a by product of the steady state balance of fluvial erosion and tectonic uplift (Box 1.3 and Appendix 6.2) at spatial scales where drainage density, and thus ecological and geomorphological processes in unchanneled areas, is not a defining factor [63]. Metapopulation dynamics (Box 2.12) are used where dispersal processes interact with the spatial structure of the landscape.

Predictive studies blending landscape and population ecology still face serious challenges when explicitly related to complex topographies (see, e.g., [229, 264]). Little dispute exists, however, on the major expected impacts on biodiversity in relation to climatic changes [265–270]. In this context, a metapopulation framework (Box 2.12) may address the features of species bound for extinction in response to climatic change. We seek to characterize the extinction dynamics [252] in the search for proof that geomorphic features of the environmental matrix matter selectively for species' survival. Specifically, the interplay of climate-induced upslope shifts of species habitats and the fluvial landscapes generates selective extinction and survival. We limit our analysis of the geomorphic effects to the sort of landscapes carved by fluvial erosion – a limited perspective, yet one addressing complex self-affine topographies carrying universal features regardless of climate, vegetation, and exposed lithology [63].

Geomorphic and altitude attributes may act as surrogates of habitat suitability at regional scales, owing to the interplay of environmental factors determining theoretical versus realized ecological niches [59, 88, 217, 233] (Box 2.9) governing species' fates under warming scenarios. As the rate of warming over the past 50 years (0.13 ± 0.03 °C per decade) is roughly twice that observed for the previous 50 years, extinction dynamics are likely to be challenged by evolving geophysical drivers almost everywhere [268]. This poses a special threat to biodiversity of mountain species owing to their high rate of local endemism [264, 271, 272]. Generally, extinction stems from species dynamics, when, for example, suitable habitat shrinks or is fragmented to the point that it cannot sustain a species' population [252]. Species that could potentially live in the new ecological landscapes created by climate change may not track the pace of displacements of their geographical range and go extinct [273, 274], and those persisting might still be unable to survive extinction debts [88].

Mechanistic models taking into account upward shifts of suitable species' habitat may predict possible effects of climate change, like climatically unsuitable occupied sites for high-mountain plants where suitable unoccupied sites increase [273] or upward shifts of native species and biological invasions [275]. While biodiversity drivers affecting species' niches are various and competing, air temperature emerges as a key player owing to its control over productivity of communities

[80, 86]. Mechanistic models, however, need to consider that some drivers change predictably with elevation, such as temperature, precipitation, and anthropogenic pressure [80, 236], while other relevant factors (such as moisture, clear-sky turbidity and cloudiness, sunshine exposure and aspect, wind strength, season length, and exposed lithology) are not elevation specific [233, 235]. Thus theoretical analyses are essential to understanding how biota respond to geophysical drivers and controls [276, 277].

A challenging problem in predicting biodiversity responses stems from landscape topography whose structure affects the drivers and whose connectivity alone calls for nontrivial diversity patterns, addressed in Chapter 1 [8, 9, 14, 229]. As we have shown in Section 2.4, deriving species distribution patterns directly from elevational gradients of potential drivers cannot sort out unambiguously expected changes. In fact, simplified landscapes, such as one-dimensional slopes, are in stark contrast with the geometry of nature [63, 121, 229, 264]. Mountains are no cones, as famously said by Benoit Mandelbrot: they are complex structures shaped by the interplay of fluvial erosion, unchanneled slope mass-wasting, and geologic uplift [63] that tend to display humps in their hypsometric curves (the frequency distributions of elevational area). Specifically, the relative proportion of landscape area almost inevitably peaks at mid-elevations whatever portion of a greater fluvial landscape one isolates. However, in a real landscape, one may find landscape peaks or valley troughs at the same elevation owing to the self-affine nature of mountain topographies [63], resulting in radically different levels of connectivity/isolation. Thus metapopulations, driven by connectivity and fitness-dependent species persistence in a landscape site, are dynamic: at rising air temperatures, assuming that fitness does not change (ecological and evolutionary timescales are kept separate in this exercise), species experience alterations of their suitable habitats depending on their initial range. As the proximity of areas with similar ecological characteristics depends on landscape morphology, crucial determinants of species persistence rely on metapopulation dynamics via extinction–colonization processes [217]. The choice of species traits that allow us to sort out landscape effects is discussed in Box 2.13.

Box 2.13 Fair Comparison Among Species: Trait Trade-Offs

It is clear that one cannot arbitrarily sample the parameter space of the metapopulation model. Although super-species exist that hold large niche width σ and high maximum fitness, and thus are capable of adapting to almost any change, allowing them into a comparative analysis of landscape effects would be misleading. This box formalizes the procedure that allows for a fair comparison among traits and species, aiming specifically at the capability of sorting out veritable landscape effects. This is done by sampling a restricted parameter space. The specific goal of the restriction is to foster comparative analyses of species with different attributes but comparable viability [263].

Consider a digital elevation model z_i (spanning all grid cells $i = 1, N$). Species suitability to a given elevation z_i is given as a Gaussian fitness function f_i (see Box 2.10) as in (Equation 2.15) $f_i = f(z_i) = f_{\max} \exp\left[-\frac{(z_i - z_{opt}^2)}{2\sigma^2}\right]$, where z_{opt} describes the elevation where the focus species shows its maximal fitness and σ is its niche width defining the elevation band in which the species can possibly thrive. Specifically, we assume a trade-off between niche width and fitness so that species have the same chance to survive in an idealized landscape with infinite elevation range and in the absence of dispersal limitations. As a result, generalist species (large σ) must have a lower maximum fitness than specialists (small σ). Fitness, in this case, is assumed to depend strictly on elevation (this condition can obviously be relaxed to make fitness landscapes more realistic and capable of contrasting real data) to narrow comparative landscape effects.

It has been suggested [263] that this may be achieved in a geomorphologically unbiased environment, that is, a landscape that does not favor some species under mean-field assumptions (infinite dispersal). This ensures that a hypersurface in parameter space is postulated that contains only species with comparable viability, such that differences in species fate computed in various landscapes

Box 2.13 *Continued*

are directly related to different geomorphologies. Thus, fair comparison among species is granted when, given a neutral landscape and a mean-field assumption (infinite dispersal), the metapopulation capacity λ_M (Box 2.12 after [217]) is approximately the same for all species considered. Only then does f_{max} ensure a fair comparison among species with different niche widths σ across a cut of the fitness landscape appropriate for similar competitors. The proof goes as follows [263].

We define a neutral landscape as an elevation field. The theoretical neutral landscape can be seen as a 1D landscape with a constant slope and an infinite length such that the metapopulation capacity is influenced by neither niche width nor optimal elevation. Within these assumptions, we center the landscape on the optimal elevation (i.e., $z(x = 0) = z_{opt}$) and consider a finite domain of size $[-L, L]$, with L large enough. The spatial discretization of the domain consists of $N + 1$ elements, $x_0 = -L, \ldots, x_{N/2} = 0, \ldots, x_N = L$, with the distance between two points defined as Δx (1 in [263]) and corresponding elevations $z_i(x_i) = x_i$ for $i = 0, \ldots, N$. In such a simple landscape, no species should obtain the advantage from the spatial structure of the elevation field. The metapopulation capacity, λ_M, defines the theoretical threshold of extinction to colonization ratio over which the population has no chance of surviving (i.e., a species persists in the domain if and only if $\lambda_M > e/c$) and is computed as the leading eigenvalue of the matrix $\mathbf{M}$, derived from the Jacobian of the system $\mathbf{J} = c\mathbf{M} - e\mathbf{I}$ [183]. The matrix $\mathbf{M}$ contains the information about the landscape and the quality of the patches. The elements of $\mathbf{M}$ are given by $m_{ij} = f_i f_j$ if $i \neq j$ and $m_{ij} = 0$ if $i = j$ (Box 2.12).

The values of the largest eigenvalue cannot be computed analytically; however, the Perron–Frobenius theorem (Appendix 6.1) provides an upper bound to the largest eigenvalue, that is, the maximum of the absolute values of the sum of a single row (or column) of the matrix $\mathbf{M}$. From its definition, the maximum of the row sum is obtained for the row corresponding to $z = z_{opt}$, that is, $i = N/2$:

$$\lambda_M(f) \leq \sum_{j=0}^{N} f(z_{opt}) f(z_i) \Delta x, \tag{2.25}$$

where f is the fitness function for a given species, characterized by a value of niche width σ.

Letting $L \to \infty$ and reducing the size of the single elements to zero, the series in Equation (2.25) converges to the following integral:

$$\int_{-\infty}^{\infty} f(z_{opt}) f(z(x)) \, dx$$

$$= \frac{1}{\sigma} \int_{-\infty}^{\infty} \exp\left(-\frac{(z(x) - z_{opt})^2}{2\sigma^2}\right) dx$$

$$= \sqrt{2\pi}, \quad \forall \sigma, \tag{2.26}$$

which, to constrain the largest eigenvalue of $\mathbf{M}$ to the same value for all niche widths, yields

$$f_{max} = \frac{1}{\sqrt{\sigma}}, \quad \forall \sigma. \tag{2.27}$$

Numerically, at decreasing values of Δx, the metapopulation capacity λ_M values associated with different species tend to the same constant (see Section 2.5.2 for numerical examples), while the numerical upper bound proves a good estimate of the theoretical value $\sqrt{2\pi}$. This proves that the

Box 2.13 *Continued*

ensemble of species characterized by the fitness in Equation (2.15) and trade-off as in Eqquation (2.27) has the same metapopulation capacity (i.e., same probability of surviving) in the given domain. Thus, comparing species having the proposed fitness function permits one to understand the real structural effects of the landscapes, given initially unbiased species in the sense of survival ability [263]. In fact, once the two parameters z_{opt} and σ are assigned, the landscape heterogeneity determines a spatially explicit fitness field for each species.

2.5.1 Landscape Effects on Metapopulation Extinction Dynamics

Each landscape z_i is tiled by pixels (i.e., unit patches) that may be occupied by a focus species. Each patch i is characterized by a certain quality, here a species-specific suitability to altitude that affects the fitness of the species $f(z_i)$, and in turn colonization and extinction rates, as the processes governing its occupancy of the landscape. Metapopulation theory (Box 2.3.4) is concerned with the persistence of the focus species in the landscape, assumed either to survive due to the balance between colonization and extinction processes or to go extinct. Each species is characterized by a set of parameters that defines its traits, a combination of niche width, σ, dispersal distance, D, optimal elevation z_{opt} (the Gaussian fitness function described by Equation (2.13)), and the colonization–extinction coefficients c and e. The core of the analysis is performed over synthetic landscapes to obtain suitable statistics. To check the effects of landscape connectivity, computational experiments on simplified topographies are also run [229, 264] (Figure 2.25). In fact, contrasting occupancy results in various landscapes, allowing us to systematically investigate how topography interplays with species' traits to concert their possible adaptation to altered drivers [263].

Tests of replicated landscape effects have been carried out on the domains shown in Figure 2.25a. Climate change is assumed to follow the worst-case scenario RCP 8.5 described in Box 2.14. The computational experiments consist of three phases (Figure 2.25c). Initially, a large initial pool of species – sampled by randomly selecting different combinations of parameters of the metapopulation model [252] subject to the constraint of equal viability (Box 2.13) – is selected and subjected to the present climate (termed "Initial Phase" in Figure 2.25c). The surviving species (termed regional, or native, species) are sorted out and used for the subsequent phase (termed "Climate Warming" in Figure in 2.25c). Then, climate change is enforced: species surviving the transient are those that are able to track change. Finally, these species are subjected for a long time to the new climate (termed "Post Climate Warming" in Figure in 2.25c): those that become extinct after the warming phase define the extinction debt. Results are shown in Figure 2.26.

As a general remark on the results, species endowed with large niche widths σ (relative to topographic relief) often successfully track climate change but do not survive afterward. In fact, under the imposed constraint on species trade-offs (Box 2.4.3), these species have a relatively low fitness everywhere, which only allows survival if enough surface is colonized. Strong colonization is thus required to grant species survival relative to large dispersal distances. Otherwise, colonization will not compensate the low fitness, and most such areas will slowly become unsuitable, leading to extinction debts [88]. This particularly affects strong dispersers, whereas weak dispersers are more subject to landscape fragmentation because they rely on close-by areas to persist. Species with a smaller niche and thus higher fitness are less affected by extinction debts but may go extinct before the end of the imposed temperature rise, that is, they might either track climate change, and thus thrive given the new conditions, or go extinct during the process, despite being suitable to the new conditions, because of their isolation. For such species, the local conditions are of utmost importance because the fitness in a single patch suffices to make them survive without help from surrounding occupied cells, which causes them to be particularly sensitive to changes in available and close-by suitable habitats.

An interesting finding [263] is that realistic landscape heterogeneities strongly impact species survival (Figure 2.26). In fact, the domain of parameters describing

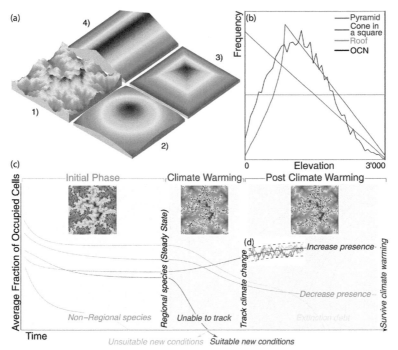

Figure 2.25 (a) Four synthetic landscapes z_i used in the experiment meant to single out geomorphic effects: (1) a virtual realistic landscape based on the OCN model (Box 1.2); (2) a cone-in-a-square; (3) a pyramid; and (4) a roof. (b) Hypsographic curves [264], that is, frequency distributions of the area available at the various elevations, here from $\min(z) = 0$ to $\max(z) = 3000$. (c) A sketch of the three phases of the experiment: an initial phase (green) to select regional species; a climate warming phase (red) characterized by an upward shift in species optimal elevations, which discriminates between species able and unable to track climate warming; and a postclimate warming phase (blue) exploring if the surviving species are suited to the new conditions or experience extinction debt. Species surviving the first phase are equivalently termed regional or native. The smooth lines depicted in the figure are meant as schematics of possible pathways of ensemble averages of several realizations, inset (d), for a given species. Insets (c) highlight snapshots of the spatial map of $f(z_i)$. Figure after [263]

traits of surviving species therein proves much smaller than that obtained for idealized landscapes, echoing the results in Section 2.4. The defining role of the achievable landscape connectivity is confirmed by noting that the OCN and the cone-in-a-square exhibit a rather similar hypsographic curve (Figure 2.25b) but rather different species fates (Figure 2.29). Local effects, such as mid-elevation plateaus or isolated peaks, influence the spatial projection of the niche, the geographical range, and the proximity of similar areas [278], thus locally reducing the fitness of the species. If a landscape has a fractal nature (more precisely, a self-affine topography) [63], highly fragmented connectivity is generated whereby species with larger niches but lower fitness struggle to subsist due to the weaker mutual rescue effect of occupied cells owing to meta-population processes of path colonization–extinction. Therefore, strong influences are waged by broad geomorphic features of a landscape not only for

biodiversity in equilibrium with the current climate [229, 264] but also for the impact of climate change on selective species survival, especially in realistic landscapes. The related extinction debt [88], which is hardly measurable as an ongoing process, is found to be a dominant feature in species extinction, making it difficult (if not impossible) to predict or measure changes before they actually happen. It is alarming that the effects of warming temperatures are not limited to the transient phase but rather are characterized by the disappearance of the occupation of large areas a long time after the enforcement of the model of climatic change. The timescales of the ecological response triggered by the imposed geographic range shift, in fact, prove much longer than the timescale of the warming period, as extinction debts may operate even centuries after the end of the shift.

The results shown in Figure 2.26 suggest that landscape features exert significant controls on the

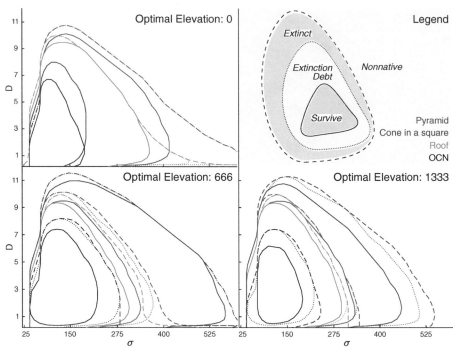

Figure 2.26 Parameter limits (niche width σ, dispersal D) of the initial pool of native (alternatively termed regional) species (dashed line), species present after climate change (dotted line), and species present at steady state after climate change (solid line). The colored domains in the legend (yellow, extinction debt; blue, inner circle where species survive; red, species going extinct during climate change; gray, nonnative species) are situated between the different line types. The figures show species with initial optimal elevations (z_{opt}) of 0, 666, and 1333 m. Figure after [263]

persistence of species with comparable traits. In all landscapes, in fact, and for all the optimal elevations characterizing the different species, only a subset of the trade-off species persists after the initial phase with constant climatic conditions (delimited by dashed curves in Figure 2.29). Such regional species are defined as shown in the scheme of Figure 2.25. The various landscape shapes exhibit differing presence patterns even if endowed with similar elevational distributions, such as the OCN and the cone-in-a-square landscapes (Figure 2.25b) [263]. In all considered landscapes, a subset of the parameters leads to the extinction of the species after climate change (see Figure 2.26). Notably, the extinction debt (measured by the area between the dotted and the solid curves in Figure 2.26) is a frequent cause of extinction for large niche widths and dispersal distances.

A landscape effect concerns the initial elevation of the maximum fitness of a species. In fact, part of the regional species having an initial optimal elevation facing a rising hypsographic curve (that is, an increase of area of suitable habitat resulting from the upslope shift; species in blue in Figure 2.29a,c) experience an increase in their occupancy, whereas species ending up around peaks after climate change are affected by a strong occupancy reduction (Figure 2.29a,c for $z_{opt} > (500/1,000)$ m). Thus, for a hump-shaped hypsographic curve (for which the majority of land is lying at intermediate elevation, that is, for OCNs, cones, and real landscapes), the fate of a species depends on the relative position of the peak elevation compared to its z_{opt}.

However qualitative, owing to the landscapes and the class of models used in this exercise, metapopulation range dynamics proved a useful tool for monitoring the time evolution of the ecological effects of climate change within a given domain. Generalizations would be straightforward. In fact, one may relax the current modeling constraints to allow species with small niche breadths to have large fitness, and vice versa. The invasion of native (say, from lower elevations unrepresented here) and nonnative species may also be allowed using the proposed methodology.

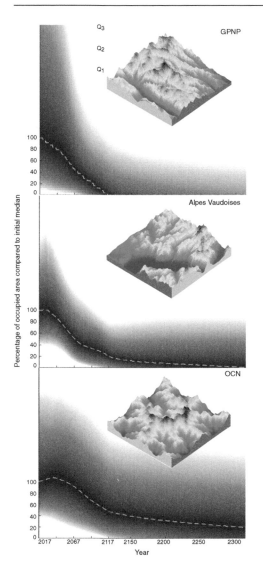

Figure 2.27 Results from an exercise computing landscape occupancy (during 2017–2117) and after climate change (after 2117) relative to the initial median occupancy of 4,000 equally viable species (sensu Box 2.13) for two real DTMs whose landscapes (GPNP and Alpes Vaudoises [263]) are used to test the tools rather than to contrast any field data, and one OCN realization for comparative purposes. The real landscapes (GPNP and Vaud) are extracted from the DEMs that were retrieved from the online Earth Explorer tool, courtesy of the NASA EOSDIS Land Processes Distributed Active Archive Center (LP DAAC), USGS/Earth Resources Observation and Science (EROS) Center, Sioux Falls, South Dakota (https://earthexplorer.usgs.gov/). The dashed line represents the median percentage of occupied area by regional (or native) species that decreases during and after climate change. Figure after [263]

Box 2.14 A Model of Climate Change

To simulate climate change impacts, the optimal elevation of each species is gradually shifted. The IPCC reports different possible long-term greenhouse gas concentration trajectories, termed representative concentration pathways [268]. The worst-case scenario, RCP 8.5, predicts an increase in temperature of approximately $\Delta T = 4°C$ over the next century (ΔT_c). We choose to uniformly change the optimal elevation of each species in the landscape: assuming a typical average global environmental lapse rate for air temperature of $\gamma_w = 1/150°C/m$ [279], the optimal elevation thus changes at a speed of $\Delta z/\Delta t = \Delta T/(\gamma_w \cdot \Delta T_c) = (4 \cdot 150)/100 = 6$ m/year, leading to a new optimal elevation $z_{opt}(t + \Delta t) = z_{opt}(t) + \left(\frac{\Delta T}{\gamma_w \Delta T_c}\right) \Delta t$ after each time step.

Another exercise has been run to explore the effects of warming temperatures on a batch of equally viable species in real and OCN landscape topographies z_i. Figure 2.27 shows the temporal changes in the occupied area compared to the initial median occupation for the same batch of equally viable species in three distinct landscapes. Because in this analysis species that can immigrate from outside the system (e.g., from lower elevations) are not considered, the focus of the analysis is on the fate of native species as defined in Figure 2.25. The exercise supports the view that, regardless of details in the fluvial landscape used as substrate for ecological interactions, a long period is required to stabilize the occupancy after the imposed trend of warming temper-

atures. Species occupancy continued to decrease after the climate stabilized, and the median occupancy tended to zero in two out of the three landscapes tested. The difference in outcome is due to geomorphic inferences of the different landscales, embedded in their elevational distributions of area. Hypsographic curves had already been used to asses the percentage of loss in occupancy after upward shifts of niche ranges in mountains [264]. However, the percentages of change in occupied area shown in Figure 2.27 are much larger than the percentage decrease predicted by the hypsographic curves (*sensu* Elsen [264]). Geomorphological attributes are thus decisive for survival of a pool of species of comparable viability.

Box 2.15 Spatially Explicit Stochastic Patch Occupancy Model

A spatially explicit stochastic patch occupancy model (SPOM) [252, 280] is described in this box (after [263]). The model is based on the balance between extinctions and colonizations and may be used to simulate the dynamics of a focus species in a given landscape.

The model proceeds as follows. At each step, the scheme, defining a Markov chain, allows unoccupied cells to be colonized by surrounding occupied cells and occupied cells to go extinct. The probabilities associated with the occurrence of these events are modeled with the following exponential distributions:

$$P(p_i(t) = 0 \rightarrow p_i(t + \Delta t) = 1) = 1 - \exp(-C_i \cdot \Delta t)$$

$$P(p_i(t) = 1 \rightarrow p_i(t + \Delta t) = 0) = 1 - \exp(-E_i \cdot \Delta t),$$

where E and C are the extinction and colonization rates, respectively; Δt is the simulation time step; and p is the binary state of occupancy of cell i, either 1 for occupied or 0 for empty.

As an example, here we briefly mention the case where the landscape matrix is affected by a heterogeneous distribution of the local fitnesses f_i, that is, the case where colonization and extinction mechanisms are directly related to a fitness field however assigned. Thus, the extinction rate is inversely proportional to the fitness of the species to the cell, that is, $E_i = e/f_i$, where e is the extinction constant. The colonization rate of an unoccupied cell is driven by the sum of the pressures from surrounding occupied cells, defined by an exponential kernel multiplied by the fitness associated to the source cells:

$$C_i = c \sum_{j \neq i} p_j \frac{e^{-d_{ij}/D}}{2\pi D^2} f_j, \tag{2.28}$$

with d_{ij} being the distance between cells i and j, D the average colonization distance (or dispersal), and c the colonization constant [263].

Obviously, parameter values matter a lot. For instance, in the exercises shown in Figure 2.26, the coefficients of extinction e and colonization c are kept constant (to $e = 0.02$ and $c = 15$ [263]), as these values only influence the scale of the two-dimensional distribution of parameters around the core of parameters of interest.

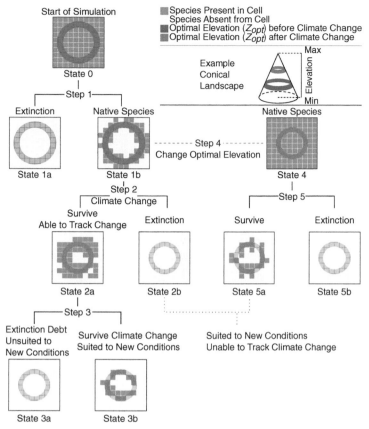

Figure 2.28 Overview of the different states and steps of the simulation. For simplicity, the landscape is displayed as a cone in this example: 100 random solutions of the SPOM model are generated for all different combinations of parameters and landscapes. In Step 1 of each run, SPOM reaches an equilibrium occupancy starting from full occupancy (State 0). If the species survives Step 1, it is considered a native species (State 1b) and climate change is applied (Step 2), leading to survival (State 2a) or extinction (State 2b). Extinction dept is evaluated by computing the equilibrium condition for the species surviving to climate change (Step 3). Additionally, a new branch of simulation is started for native species (Steps 4 and 5) by computing their equilibrium occupancy starting from full occupancy and considering the optimal elevation after climate change. This allows comparison of States 2b and 5a to find species unable to track climate change but that would have been able to survive the new conditions. Figure after [263]

2.5.2 Computational Experiments with SPOM

For each suitable species chosen, 100 OCN realizations were run to estimate probabilistically its fate by using the SPOM method (Box 2.15) [263]. To optimize the analysis and the interpretation of the results, a discrete pool of 4,000 species was assembled according to the trade-offs among traits (Box 2.13). In particular, species endowed with large (generalists) or small (specialists) niche widths were differentiated [281]. Specialists arguably depend more on the local characteristics of the terrain, whereas generalists mostly rely on the species' dispersal properties. Owing to its deliberate simplicity and minimum

parameter use, the metapopulation model allows one to study the changes in the spatial occupancy of a species due to climate change for several replicas of parameters and landscapes. Even with such simple structure, enough degrees of freedom exist to call for a demanding computational exercise to produce realistic patterns.

The pool of native species, that is, species that persisted in the landscape after an initial simulation under constant climatic conditions, was first selected (Figure 2.28). Owing to the stochastic nature of SPOM, the experiment is repeated 100 times for each set of species and landscapes. Native species are defined as the species that, at equilibrium (Step 1 in Figure 2.28), survive in the

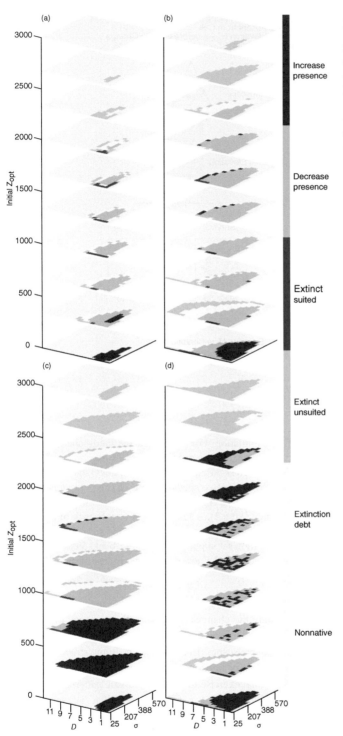

Figure 2.29 Fate of the considered species based on the most probable outcome after 100 random experimental runs. Each species is characterized by its niche width σ, dispersal distance D, and initial optimal elevation z_{opt}. Results are presented for the following landscapes: (a) OCN; (b) pyramid; (c) cone-in-a-square; and (d) rooflike landscape (see supporting information in [263] for the complete set of results)

landscape at least in 50 of the 100 random runs. Runs leading to species disappearing from the landscape are replaced by randomly selecting an equilibrium configuration from the remaining runs, such that all the native species have 100 starting configurations for the remainder of the experiment. We then apply climate change (Step 2) as an upward shift of the niche for approximately 100 years. Step 2 eliminates species unable to track climate change; climatic conditions are then frozen, and the experiment is run to steady state (Step 3) to identify the species unable to cope with the new temperatures, hence defining the extinction debt [88]. The last steps (Steps 4 and 5) consist in understanding whether the species that went extinct during the climate change phase would be able to survive given their new optimal elevation, had they been able to track climate change, or if they would have gone extinct anyway. For that, the native species are allowed once again to fully occupy the landscape, but their optimal elevations are set to be those after climate change. The simulation is then run until reaching a steady state (Step 5), allowing us to find the species suited to the new conditions. This method allows observations of species' transient states, and therefore to single out specific fates, such as the disappearance of species suitable for postclimate change conditions due to their inability to track favorable conditions or species surviving for some time after climate change only to go extinct in time. Note that the induced understanding of transient effects distinguishes metapopulation studies from habitat suitability and species distribution models (SDMs) [59].

The landscapes considered for the computational experiments, shown in Figure 2.25b, were chosen for their rather different elevational distributions of available area. Comparative studies (Figure 2.29) were run to generate different realizations in statistically identical replicas to properly study the statistics of the stochastic runs. Different landscape shapes were designed to probe distinctive geomorphic effects. In fact, the typical hump-shaped distribution of areas found in nature [229, 264] can also be constructed in unrealistic geometries (Figure 2.25a), thus allowing us to study the influences of the spatial complexity of the topography on its connectivity, proximity to similar area and fragmentation, and thus species distributions in response to climate change.

Given the parameter sets assumed to run the metapopulation model, three factors are found to govern the dynamics of the associated species at any site in a landscape: the initial occupancy, the distance to areas sharing similar fitness, and the available area around the optimal elevation. Their interplay governs the fate of the species in the landscape. Taken separately, we find that such effects may not suffice in predicting the fate of a species under imposed climate change. For example, the hypsographic curve alone does not suffice to subsume all processes governing species presence. Even in the simple, unrealistic geometric landscapes designed to distinguish geomorphic effects, the entanglement of those processes generates unforeseen outcomes. For instance, species with their niche partially outside the landscape elevation range before climate change may actually exploit increased area under their niche range projection even if the specific hypsographic curve is steadily decreasing (Figure 2.25). Occupancy thus reflects a complex balance between area availability, connectivity, and realized niche [59], in line with the main tenet of this book.

3 | Populations

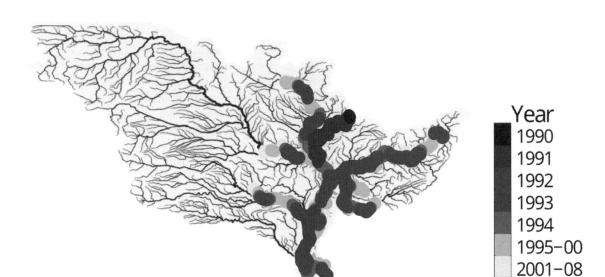

	Year
■	1990
■	1991
■	1992
■	1993
■	1994
■	1995–00
□	2001–08

This chapter's cover picture is a map of the progressive biological invasion of the Mississippi–Missouri river system by the Zebra mussel (*Dreissena polymorpha*). The MMRS once again is taken as the river network epitomizing the challenges posed by fluvial ecosystems to our ability to predict and control natural phenomena. This specific biological invasion, seeded by the accidental discharge of ballast water from European cargo ships in the Great Lakes region, posed (and poses) major ecological and economic threats, in particular owing to the huge population densities reached by local zebra mussel colonies and their unparalleled colonizing abilities along fluvial systems. This chapter first introduces mathematical tools needed to understand such phenomena, which we later use to develop a quantitative evaluation of the individual roles and the mutual interactions of drivers and controls of the Mississippi–Missouri invasion as iconic example. One interesting feature is the use, made in this example, of a multilayer network model accounting explicitly for zebra mussel demographic dynamics, hydrologic transport, and dispersal due to anthropic activities – echoing processes fundamental to the spread of waterborne or water-based disease (Chapter 6). In referring to one of the key exercises exemplifying the thesis of the book, the cover image stresses the ultimate goals of the study of river networks as ecological corridors.

3.1 Biological Invasions

Biological invasions, the spread in space and time of species originally alien to a given ecosystem defining patterns of progressive colonization, are increasingly frequent owing to the number of mechanisms available for seeds, spores, or organisms to survive long-distance displacements. Thus, should conditions be favorable, selective advantages evolutionarily developed in other contexts might be at hand, generating competitive advantages and the displacement or the local extinction of native species. Charles Elton [282] pioneered this field of

study while issuing an early warning that the accelerating rate of introduction of foreign species at the onset of the global village – and of the biological dislocations that typically follow – might have notable ecological and economic consequences. Here, we shall insist in particular on the fact that key to coping with invasive species is our ability to predict their rates of spread.

Biological dispersal is a key driver of many fundamental processes in nature [282–286]. Invasions control the distribution of species within an ecosystem and critically affect their coexistence. In fact, the spread of organisms along ecological substrates (or corridors, if effectively one-dimensional in space) governs not only the dynamics of invasive species but also the spread of pathogens and the shifts in species ranges due to climate or environmental change. Obviously, the subject comprises a major chapter of ecology, and we shall not deal with it here in a comprehensive manner. Our narrow perspective deals rather with directed networks of ecological corridors, that is, rivers and their ecosystems, where population invasions take place. As a result, we mostly deal with linear corridors where the longitudinal dimension of the domain is orders of magnitude larger than the other relevant dimensions, say, the depth and width characteristic of the channel cross sections. In what follows, therefore, we shall concentrate on a few cases of special ecological relevance: the determination of the celerity of population traveling waves in homogeneous or heterogeneous linear networked environments, population dynamics of invasive species in space and time in multiplex or multilayer networks, and experimental consequences on species distributions of the hierarchical structure of a river network.

The landscape structure of riverine habitats, the subject of this book, is inherently dendritic and hierarchical. As we have seen in Chapter 1, the dendritic connectivity alone imposes strong constraints on the dispersal of populations in the embedded ecosystems. In addition, important ecological consequences on spatial species distributions stem from the hierarchical structure implied by the dendrites, as the typical habitat – proportional to the local water volumes – increases predictably alongstream with total contributing area because landscape-forming flowrates are indeed proportional to it [8, 63] (Box 2.4). Moreover, flow conditions may induce a selective directional bias absent from other fluvial habitat types [287]. Therefore, geomorphological attributes define the characteristics of riverine habitats and thus influence the dynamics of populations within them [8, 156].

In the analysis of migration fronts, an important role must be attributed to the structure of the network acting as the substrate for traveling wave propagations. This calls for specific structural models to be used, possibly describing natural forms as fractals [63, 121]. In idealized cases, basic invariance properties may refer to the independence of the outcome of a biological invasion from the seeding point chosen for spreading material and species along the network where reaction and diffusion occur. This is seen as a corollary of the type of self-similarity shown by trees, although entailing somewhat complex issues in cases where loops are observed [69]. The type of self-similarity

observed for trees needs to be put in context. In fact, proper finite-size corrections arise because of upper and lower cutoffs that affect the statistics of the aggregation structure. These corrections reflect the impact of drainage density defining where channels begin (at small areas) and the loss of statistical significance of areas close to the overall basin size, respectively [63, 64, 74, 103, 257]. For general predictions of invasive properties, the fluvial substrates that we shall need to consider here are the topology and the geometry of real rivers extracted from digital terrain maps [63] and those of OCNs (Box 1.3), which may be used to generate statistically identical replicas indistinguishable from real rivers [67] (and thus generate proper statistics via different realizations obtained through a specific stochastic selection process from which one obtains a rich structure of scaling forms that are known to closely conform to real river networks; see Appendix 6.2).

A good starting point to introduce the specificities of this chapter is the analysis of a reaction random walk (RRW) process through dendrites, which was illustrated in Section 1.5 (adapted from [118]). It is based on the following model. A particle, at an arbitrary node of the network, jumps, after a waiting time τ, to one of its z nearest neighboring nodes, with probability $1/z$. During the waiting time τ, the particles react, that is, evolve the scalar property (mass, density) labeling the particle by following the logistic equation. The determination of the wave front speed that this process develops along a network path [130, 131] is the starting point for the ensuing considerations. Figure 3.1 illustrates the main result of the above premises. It shows that the isotropic diffusion–reaction front (Fisher's model [125]) propagates much faster than the wave forced to choose a treelike pathway. This proves that geometrical constraints imposed by a fractal network imply strong corrections on the speed of the fronts. It should not be surprising that Peano networks and OCNs lead to similar results, because the speed of the front depends on topological features that are indeed quite similar for all treelike networks (rather different otherwise) [63, 69]. In fact, it can be shown [118, 130] that the wave speed is affected mostly by the gross structure encountered by the front while propagating along the network, chiefly the bifurcations. Hence topology, rather than the fine structure of the subpaths, dominates the process. Suffice here to recall the main result described in Section 1.5, where we have shown an early proof that the branching structure faced by an invading species is distinctly retarded by the bifurcations. The basic model examined therein, realistic in a number of cases, such as the interpretation of patterns of human migrations, relies on a deterministic framework, the Fisher–Kolmogorov equation [119, 125], which we repeat here for completeness:

$$\frac{\partial \rho}{\partial t} = D \frac{\partial^2 \rho}{\partial x^2} + r\rho \left[1 - \frac{\rho}{K} \right], \tag{3.1}$$

where $\rho = \rho(x,t)$ is the density of organisms, r is the species' growth rate, D is a (constant) diffusion coefficient, and K is the carrying capacity of the population, also assumed to be constant (the latter assumption proves remarkably limited in the case of river networks, given their hierarchical structure and related habitat).

In fact, the role of the structure of river networks in modeling human-range expansions, that is, predicting how populations migrate when settling into new territories, has been studied through quantitative models of diffusion along fractal networks coupled with logistic reaction at their nodes [118]. An essential ingredient therein is the fact that settlers did not occupy all the territory isotropically but rather followed rivers and lakes and settled near them to exploit water resources. It was thus interestingly argued in a quantitative manner that landscape heterogeneities must have played an essential role in the process of human migration [22, 118, 130, 131]. It should be noted that the model proposed by Campos et al. [118] assumes simple diffusive transport to describe migration fluxes. This seems indeed reasonable in the case of human population migrations: the need for water resources should drive settlers regardless of the direction of the flow. Variations on the theme, for example, adding a bias to transport properties, would basically alter this interesting picture and will be investigated in Section 3.4. This was done on purpose: in fact, any other ecological agent (be it an aquatic organism or an infective agent of waterborne disease) would likely be affected by the flow direction while propagating within the network. Organisms can either move by their own energy (active dispersal) or be moved by water (passive dispersal). Most likely, movements along the flow direction would be favored, although movements against flow direction are completely admissible because of various ecological or physical mechanisms [8, 40]. All this is of great interest for the problem of hydrochory, that is, the transport of species along the ecological corridors that are shaped by the river network [7].

Recapping, Equation (3.1) is known to foster the asymptotic development of undeformed traveling waves of the density profile (see Box 1.5 and [120]). Mathematically, this implies that $\rho(x,t) = \rho(x - vt)$,

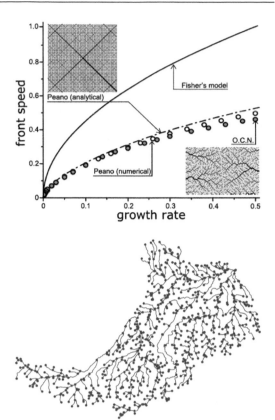

Figure 3.1 (left) A river network thought of as a directed graph, where nodes are sites of logistic population growth and edges are river reaches. (right) Invasion front speed as a function of the growth rate $[\mathrm{T}^{-1}]$ of the logistic equation. The solid line is the exact solution of the continuous isotropic Fisher–Kolmogorov model. The dashed line and the circles represent exact and numerical values for propagation along the backbone of Peano networks [63, 121, 126] and OCNs, respectively (after [7, 118], and Section 1.5.1).

where v is the speed of the advancing wave, a celerity in mechanics jargon. Fisher [125] proved that traveling wave solutions can only exist with speed $v \geq 2\sqrt{rD}$, and Kolmogorov [119] demonstrated that, with suitable initial conditions, the speed of the wave front is precisely the lower bound (Figure 3.1). The microscopic movement underlying the Fisher–Kolmogorov equation (3.1) is a Brownian motion [286]. We shall selectively relax in this chapter the assumptions underlying Equation (3.1).

Here, we shall also examine experimental evidence. In particular, we shall describe detailed experiments completely functional to the tenet of this book that were carried out in the ECHO Laboratory at the Ecole Polytechnique Fédérale de Lausanne. In fact, despite its

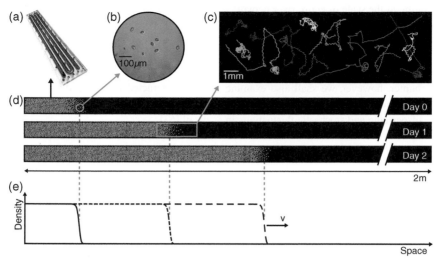

Figure 3.2 Schematic of the invasion experiments. (a) Linear landscape. (b) Individuals of the ciliate *Tetrahymena* sp. move and reproduce within the landscape. (c) Examples of reconstructed trajectories of individuals. (d) Individuals are introduced at one end of a linear landscape and are observed to reproduce and disperse within the landscape (not to scale). (e) Illustrative representation of density profiles along the landscape at subsequent times. A wave front is argued to propagate undeformed at a constant speed *v* according to the Fisher–Kolmogorov equation. Figure after [289]

relevance for understanding ecological processes, it has been argued that the subject suffers an acknowledged lack of experimentation, and current assessments point at inherent limitations to predictability even in the simplest ecological settings [288]. Giometto et al. [289] have instead shown, by combining replicated experimentation on the spread of the ciliate *Tetrahymena* sp. to a theoretical approach based on stochastic differential equations, that information on local unconstrained movement and reproduction of organisms (including demographic stochasticity) allows one to predict reliably both the propagation speed and the range of variability of invasion fronts over multiple generations. We shall briefly report their main findings in Section 3.4 because they are functional to our theses and validate our mathematical exercises. Figure 3.2 serves well in recapping how invasion experiments were carried out in the laboratory – in this case, employing protists as model organisms. Therein, individual organisms of *Tetrahymena* move and reproduce within the linear landscape constructed in the laboratory (shown in Figure 3.2a). Examples of measured trajectories of individual protists, shown in Figure 3.2c, support the view that a basic model of Fickian diffusion makes sense after an initial transient of the order of four to five widths of the channel [289]. In all the experiments carried

out at ECHO reported here, individuals are introduced at one end of the linear landscape and are allowed to disperse and reproduce within the linear channel (where the characteristic timescales of diffusive transport and reproduction have been carefully designed). Density profiles are measured as they unfold along the landscape at subsequent times. A traveling wave front is observed to develop and to propagate practically undeformed at constant, measurable speed; its similarities with (and departures from) the Fisher–Kolmogorov prediction (Section 1.5.1) will be central to the theses of this section.

The material in this chapter is organized moving from an introduction to the mathematical treatment of linear diffusion processes, in particular as relevant to the subject of this book. While far from being a complete treatment of the subject, this chapter collects results of perceived importance to the topics dealt with here. A notable example is the discussion of the hierarchical dispersion phenomena occurring in river networks, from molecular to turbulent diffusion to hydrodynamic dispersion affected by spatial gradients of velocity of the carrier flow enhancing dispersive processes [173]. Note that additional dispersive effects are induced at the scale of a river network by heterogeneity of the paths leading to the control section where phenomena are actually observed [174]. This leads to a proper geomorphologic dispersion operating at the scale of entire river networks and surely relevant to polydisperse sources of tracer or biomass or species. Traditional models of biological invasions assume that the environment (the substrate available to biological invasions) is a temporally constant entity. In this chapter, however, we shall also examine experimentally and theoretically the consequences for invasion speed of periodic and stochastic fluctuations in population growth rates and dispersal distributions. Many phenomena investigated here will be focused on one-dimensional spatial ecosystems, seen as edges of a directed graph whose connectivity is provided by the hierarchical structure of the river network. The latter is usually extracted automatically from digital terrain maps now ubiquitously available at unprecedented levels of detail [63].

Box 3.1 Modeling the Movement of Organisms via Diffusion

As we state in Box 1.6, the usual approach to describing the movement of organisms in space is to assume that it is random. Although we know that it is a gross approximation, it turns out that in many cases it leads to models that mimic reality reasonably well. With proper assumptions, which we detail below, one can derive the equation of transport and diffusion, namely, the simplest model for the dispersal of a population inside its habitat.

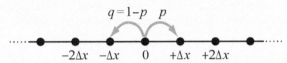

Figure 3.3 The scheme of random walk: an organism moves a step to the right with probability p or to the left with probability $1 - p$.

We start by dealing with one-dimensional space (indicated with x). First, suppose that both space and time are discrete (Figure 3.3). We denote by p the probability that an organism moves to the right with a step of length Δx in the time interval Δt and by $q = 1 - p$ the probability that an organism will move left. If we assume that, in every spatial position, there is a large number of individuals, we can also say that p is the fraction of organisms that moves to the right and q is the fraction of organisms that moves to the left (according to the law of large numbers). Similarly to Box 1.6, we introduce the density of organisms in each position x at time t, namely, $\rho(x,t)\Delta x$ = number of organisms that at time t are located between $x - \Delta x/2$ and $x + \Delta x/2$. The movement of organisms can be assumed to be equivalent to the motion of a tracer molecule or neutrally buoyant particle. Thus, if an ensemble of organisms begins its wandering at the origin at time zero, the density or concentration at location x at any later time $t > 0$ will be proportional to the likelihood of any one being in the neighborhood of that station. This usually goes under the name of Taylor's theorem [290]. Then we can state that

$$p\rho(x,t)\Delta x = \text{number of organisms that in the time interval } \Delta t \text{ move from location } x \text{ to}$$
$$\text{location } x + \Delta x$$
$$(1 - p)\rho(x,t)\Delta x = \text{number of organisms that in the time interval } \Delta t \text{ move from location } x \text{ to}$$
$$\text{location } x - \Delta x,$$

and we can then write the following balance equation:

$$\rho(x, t + \Delta t)\Delta x = p\rho(x - \Delta x, t)\Delta x + (1 - p)\rho(x + \Delta x, t)\Delta x.$$

Divide now by Δx and develop the left-hand-side and right-hand-side terms in Taylor series, thus getting

Box 3.1 *Continued*

$$\rho(x,t) + \frac{\partial \rho}{\partial t}\Delta t + \mathcal{O}(\Delta t^2) = p\left[\rho(x,t) - \frac{\partial \rho}{\partial x}\Delta x + \frac{1}{2}\frac{\partial^2 \rho}{\partial x^2}\Delta x^2 + \mathcal{O}(\Delta x^3)\right]$$

$$+ (1-p)\left[\rho(x,t) + \frac{\partial \rho}{\partial x}\Delta x + \frac{1}{2}\frac{\partial^2 \rho}{\partial x^2}\Delta x^2 + \mathcal{O}(\Delta x^3)\right],$$

where, with $\mathcal{O}(z)$, we indicate terms of order larger than or equal to z. If we simplify some of the terms left and right of the equal sign and divide by Δt, we obtain

$$\frac{\partial \rho}{\partial t} + \frac{\mathcal{O}(\Delta t^2)}{\Delta t} = (1-2p)\frac{\partial \rho}{\partial x}\frac{\Delta x}{\Delta t} + \frac{1}{2}\frac{\partial^2 \rho}{\partial x^2}\frac{\Delta x^2}{\Delta t} + \frac{\mathcal{O}(\Delta x^3)}{\Delta t}.$$

We must now let the space step Δx and the time step Δt tend to zero to obtain the appropriate equation in continuous space and time. However, some additional assumptions are needed to obtain the so-called diffusion approximation. More precisely, we have to assume that

$$\frac{\Delta x}{\Delta t} \to \infty \text{ absolute movement speed of a single organism,}$$

$$\frac{1}{2}\frac{\Delta x^2}{\Delta t} \to D \text{ finite positive constant,}$$

$$(2p-1)\frac{\Delta x}{\Delta t} = p\frac{\Delta x}{\Delta t} + (1-p)\frac{-\Delta x}{\Delta t} = \text{average population speed} \to v \text{ finite constant.}$$

In particular, note that the first and the third assumptions imply that $2p-1$ must necessarily tend to zero, or, stated otherwise, that the diffusive approximation is valid if the probabilities to move right or left differ only by a small amount, that is, differ very little from $1/2$.

Letting Δx and Δt tend to zero, one obtains

$$\frac{\partial \rho}{\partial t} = -v\frac{\partial \rho}{\partial x} + D\frac{\partial^2 \rho}{\partial x^2}, \tag{3.2}$$

which is the advection (or drift or transport) and diffusion equation where D is the diffusion coefficient and v the advection speed; $\frac{\partial \rho}{\partial x}$ is the population density gradient. If v is zero, one gets the equation of pure diffusion.

It must be remarked that the assumption that every organism is moving at very large speed (infinite in the limit) is crucial to obtaining Equation (3.2). In fact, if we had supposed that the speed of each organism were finite, then Δx would tend to zero like Δt, and then Δx^2 would tend to zero faster than Δt. As a result, the diffusive term would vanish, and only advection would be operating.

As the transport term in Equation (3.2) has just a translation effect on the solution of the same equation (one replaces the space coordinate x with the moving reference frame coordinate $x + vt$), the solutions of Equation (3.2) can be easily obtained from solutions of the pure diffusion equation

$$\frac{\partial \rho}{\partial t} = D\frac{\partial^2 \rho}{\partial x^2}, \tag{3.3}$$

on which we will focus from now on.

A solution of Equation (3.3) is completely determined by the specification of suitable initial conditions and suitable conditions at the space boundary. Providing initial conditions means to

Box 3.1 *Continued*

specify the organisms' density in each location x at the initial instant, conventionally denoted by 0, namely,

$\rho(x,0) = \rho_0(x) = $ a given function of space.

As regards the boundary conditions, there are several possible cases, depending on the spatial domain considered. In the so-called Cauchy problem, the domain is infinite $[-\infty, +\infty]$. From a practical standpoint, this condition is that of a population that can disperse without finding virtually any spatial barrier. Note that, in this case, there is no real boundary. As the only phenomenon that drives changes in the density of organisms at each point is just movement, the following important condition must be verified:

$$\text{Total number of organisms} = \int_{-\infty}^{+\infty} \rho(x,t)\,dx = \int_{-\infty}^{+\infty} \rho_0(x)\,dx = N\text{constant}.$$

Therefore the solution $\rho(x,t)$ must be a bounded function at each time instant such that

$$\lim_{x\to+\infty} \rho(x,t) = \lim_{x\to-\infty} \rho(x,t) = 0.$$

It is possible to prove that the solution of the Cauchy problem of Equation (3.3) is unique if we consider bounded functions only.

Alternatively, the domain can be finite. The simplest finite domain in a one-dimensional space is a segment of length L, namely, $x \in [0, L]$. One can imagine that the territory is an island or a wood that is very narrow, oblong, and of length L. Then, depending on the conditions that hold on the boundary of the domain, there can be different problems:

1. unsuitable habitat outside domain (absorbing boundary, Dirichlet problem),

$\rho(0,t) = \rho(L,t) = 0$

2. no flux through the boundary (reflecting boundary, Neumann problem),

$\dfrac{\partial \rho}{\partial x}(0,t) = \dfrac{\partial \rho}{\partial x}(L,t) = 0$

3. habitat is circle shaped like an atoll (periodic conditions, mixed Dirichlet–Neumann problem),

$\rho(0,t) = \rho(L,t)$

$\dfrac{\partial \rho}{\partial x}(0,t) = \dfrac{\partial \rho}{\partial x}(L,t).$

The diffusion equation is easily generalized to the case when we consider more than one spatial dimension. In particular, when organisms disperse in a plane, described by the spatial coordinates x and y, we can introduce the density of individuals per unit area $\rho(x,y,t)$, defined as

$\rho(x,y,t)dxdy = $ number of organisms that at time t are located in the square

of area $dxdy$ whose lower left corner has coordinates (x,y).

If the diffusion coefficient is the same in all directions (*isotropic diffusion*), the diffusion equation is

Box 3.1 *Continued*

$$\frac{\partial \rho}{\partial t} = D\left(\frac{\partial^2 \rho}{\partial x^2} + \frac{\partial^2 \rho}{\partial y^2}\right).$$
(3.4)

Obviously, the shape of the possible finite and connected domains in 2D can be much more varied than the simple segment in one dimension, although the three problems listed above (absorbing boundary, reflecting boundary, periodic conditions) still hold. In the problem with periodic conditions, the domain may be topologically equivalent to either a sphere or a torus. Simple analytical solutions can be obtained for simple domains like circles, squares, or spheres.

Box 3.2 Boundary Conditions and the Corresponding Solutions of the Diffusion Equation

In this box we derive relevant solutions of the diffusion equation (introduced in Box 3.3) under different boundary conditions. The first problem we consider is Cauchy's, namely, no barrier is posited to hinder the movement of N organisms initially distributed according to $\rho_0(x)$. In Box 1.6 the solution to initial conditions where the N organisms are all concentrated in the origin was shown to be, by direct substitution, a normal distribution with mean 0 and variance linearly increasing with time. Here we provide a general solution through a frequency domain approach. Introduce the angular frequency in space $\eta = 2\pi\nu$ (radians per unit space), where ν is the usual frequency (space^{-1}). Then a continuous function of space $f(x)$ can be decomposed in terms of its Fourier transform $F(f) = \phi(\eta)$ as

$$f(x) = \frac{1}{2\pi} \int_{-\infty}^{+\infty} \phi(\eta)\exp(j\eta x)d\eta$$

$$\phi(\eta) = \int_{-\infty}^{+\infty} f(x)\exp(-j\eta x)dx,$$
(3.5)

where j is the imaginary unit. In other words, the Fourier transform is nothing but the frequency spectrum of a function, which can thus be considered as the integral sum of its spectral components. It is easily derived that the Fourier transform is a linear operator, namely, that the Fourier transform of a weighted sum of functions is the weighted sum of the transforms. Also, the following properties hold for derivation and convolution:

$$F\left(\frac{df}{dx}\right) = \eta\,\phi(\eta),$$

$$F(f*g) = F(f)F(g) = F(h),$$

$$h(x) = f*g = \int_{-\infty}^{+\infty} f(z)\,g(x-z)\,dz,$$

together with the following useful transforms:

$$F(\delta(x)) = 1,$$

$$F(1) = 2\pi\delta(\eta),$$

$$F(\text{Norm}(x,\sigma)) = \exp(-\sigma^2\eta^2/2),$$

Box 3.2 *Continued*

where $\delta(x)$ is the Dirac delta and $\text{Norm}(x, \sigma)$ is the Gaussian distribution in space with mean 0 and variance σ^2. The spectrum of the Dirac delta is thus constant, that is, all the frequencies are present with equal weight in the pulse function. The transform of a normal distribution in x is a normal distribution in η, with the variance being the inverse of the variance of the function of x.

We can then apply the Fourier analysis to the solution $\rho(x, t)$ of the diffusion equation. Suffice to Fourier transform both sides of the equation. Let $\Phi(\eta, t)$ be the Fourier transform with respect to space of $\rho(x, t)$, and consider that the transform of the partial derivative of $\rho(x, t)$ with respect to time is nothing but the partial derivative of $\Phi(\eta, t)$ with respect to time. We then obtain the linear differential equation that governs the time evolution of the transform $\Phi(\eta, t)$:

$$\frac{\partial}{\partial t}\Phi(\eta, t) = -D\eta^2 \Phi(\eta, t)$$

$$\Phi(\eta, 0) = F(\rho_0(x)).$$

(3.6)

The solution of this equation is easy, namely,

$$\Phi(\eta, t) = \Phi(\eta, 0) \exp(-D\eta^2 t),$$

which shows that each spectral component $\Phi(\eta, t)$ is the product of a Gaussian function of the frequency η times the Fourier transform of the initial density distribution $\rho_0(x)$. Also, it is interesting to note that the amplitude of each spectral component decreases exponentially with time, but the components with higher frequency decrease at a rate that is much bigger (proportional to the square of the frequency). It is finally possible to obtain the solution of the diffusion equation by back-transforming and by remembering the convolution property:

$$\rho(x, t) = \int_{-\infty}^{+\infty} \rho_0(z) \text{Norm}\left(x - z, \sqrt{2Dt}\right) dz.$$

Therefore, the general solution of the diffusion equation is the convolution of the distribution at time 0 and the normal distribution with mean 0 and variance equal to $2Dt$. Note that this result contains as a special case the one in which the N organisms are concentrated in the origin. In fact, the initial distribution is $\rho_0(x) = N\delta(x)$. The Fourier transform of $\rho_0(x)$ is thus N, and the Fourier transform of $\rho(x, t)$ is $N\exp(-Dt\eta^2)$. So, by back-transforming, we obtain again the normal distribution with mean 0 and variance equal to $2Dt$ obviously multiplied by N.

The Fourier analysis is also key to solving the finite domain boundary problems (Dirichlet and Neumann). The basic idea is that $\rho(x, t)$ defined in the interval $[0, L]$ can be seen as part of a periodic function defined in $[-\infty, +\infty]$ and having a suitable period (see Figure 3.4). For the case of an absorbing barrier, this function has period $2L$ and vanishes at 0, L, and $2L$; for the case of a repelling barrier, the period is $2L$, and the gradient $\frac{\partial \rho}{\partial x}$ vanishes at 0, L, and $2L$. As for the circle-shaped habitat, the period is L, and no other constraint is required but periodicity of ρ and $\frac{\partial \rho}{\partial x}$. Before proceeding with the mathematical analysis of the three cases, we provide the intuitive results. With an absorbing barrier, the number of individuals in the domain will exponentially decrease to zero, because the individuals getting out of the boundary at 0 and L will never come back – one can think that they die as soon as they leave the favorable habitat. With a repelling barrier, the total number will always remain equal to N because the individuals cannot get out of the domain: the flux at 0 and

Box 3.2 *Continued*

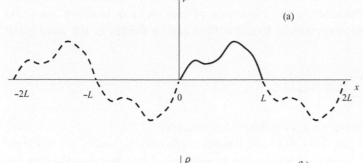

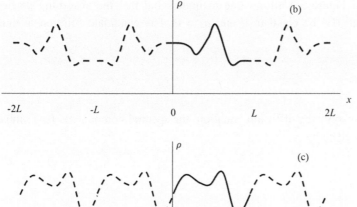

Figure 3.4 A function satisfying the boundary conditions of the diffusion equation in the interval $[0, L]$ can be considered part of a periodic function defined in $[-\infty, +\infty]$. (a) Absorbing barrier. (b) Repelling barrier. (c) Circle-shaped habitat.

L is zero. In the long run, the distribution of the organisms will become homogeneous. The same occurs with a circle-shaped habitat in which the origin 0 coincides with L and individuals go around the circle indefinitely.

As the problems with a finite domain have a characteristic dimension ($2L$ for the absorbing and repelling barriers and L for the circle-shaped habitat), the corresponding periodic functions have some specific spectral components. In other words, they can be seen as the sum of sinusoidal functions with frequencies that are multiples of $1/2L$ in the absorbing and repelling barrier cases and multiples of $1/L$ in the circle-shaped habitat problem. In other words, one can replace the Fourier transform given by Equation (3.5) with a Fourier series, which for a function $f(x)$ of period P is given by

$$f(x) = \sum_{k=-\infty}^{+\infty} c_k \exp(j\eta_k x), \qquad \eta_k = \frac{2\pi k}{P},$$

$$c_k = \frac{1}{P} \int_{-\frac{P}{2}}^{\frac{P}{2}} f(x) \exp(-j\eta_k x),$$

Box 3.2 *Continued*

with $2\pi/P$ being the fundamental frequency, of which all the other frequencies (harmonics) are integer multiples. Note that c_0 is nothing but the mean value of $f(x)$ and that c_{-k} is the complex conjugate of c_k. By remembering the expression of $\exp(j\eta_k x)$ in terms of sines and cosines, the decomposition of the real periodic function $f(x)$ can be written in the more usual way as

$$f(x) = MV + \sum_{k=1}^{\infty} a_k \cos(\eta_k x) + \sum_{k=1}^{\infty} b_k \sin(\eta_k x),$$

where MV is the mean value, $a_k = (c_k + c_{-k})/2$ and $b_k = (c_k - c_{-k})/2j$.

Obviously, when we substitute $\rho(x,t)$ for $f(x)$, the Fourier coefficients a_k and b_k are functions of time. A simple inspection of Figure 3.4 allows one to understand that the absorbing barrier boundary condition requires that MV be equal to 0 and $a_k = 0$. The candidate solution is thus given by

$$\rho(x,t) = \sum_{k=1}^{\infty} b_k(t) \sin\left(\frac{\pi k}{L} x\right).$$

Because this candidate must also solve the diffusion equation, the spectral components $b_k(t)$ must obey Equation (3.6) and are thus given by

$$b_k(t) = b_k(0) \exp\left[-D\left(\frac{\pi k}{L}\right)^2 t\right],$$

where $b_k(0)$ are the Fourier coefficients of $\rho_0(x)$. Therefore, the solution converges exponentially fast to 0, with the fundamental frequency being the mode that decreases at the lowest rate.

In the repelling barrier case, boundary conditions do not require that MV be zero. Instead, because the gradient $\partial \rho/\partial x$ must vanish at 0 and L, the coefficients $b_k(t)$ must vanish. The spectral components must again satisfy Equation (3.6) and thus converge to zero exponentially fast for any $k \geq 1$. So, the solution converges to the mean value of $\rho_0(x)$, which is obviously equal to N/L.

As for the circle-shaped habitat, the feasible frequencies are $\eta_k = \frac{2\pi k}{L}, k = 0, 1, 2 \ldots$, and boundary conditions do not require that either MV or $a_k(t)$ or $b_k(t)$ vanishes. However, the modes corresponding to the fundamental frequency and its harmonics ($k = 1, 2, \ldots$) decrease to zero exponentially fast (because of Equation (3.6)) so that the solution again converges to the mean value of $\rho_0(x)$, namely, N/L.

Results can be generalized to 2D space by using 2D Fourier transform and series. In the Cauchy problem, the solution is a 2D convolution with a bivariate normal distribution. An absorbing barrier problem is still characterized by an ever-decreasing population, while a repelling barrier implies population homogeneization. If the habitat is topologically equivalent to a sphere or a torus (mixed Dirichlet–Neumann problem), organisms will spread homogeneously on the surface.

3.1.1 Movement and Demographic Increase: Resulting Patterns

As we have briefly seen in Chapter 1 (Box 1.6), demography can be introduced into the diffusion model by resorting to a reaction–diffusion equation in which population density at each spatial location changes over time not only because of movement but also because of occurring births and deaths. We assume that the birthrate β and mortality rate μ that operate in a one-dimensional spatial location x depend only on local density $\rho(x,t)$. The per capita rate of increase R is given by

$$R(\rho) = \beta(\rho) - \mu(\rho).$$

The time variation of density at x obeys the equation

$$\frac{\partial \rho}{\partial t} = D\frac{\partial^2 \rho}{\partial x^2} + R(\rho)\rho. \tag{3.7}$$

Depending on the population being studied, the per capita rate of increase R can have different functional relationships linking the rate to the density. Among the most utilized in ecology, we find

$R(\rho) = r\,(1 - \rho/K)$ logistic, $r, K > 0$
(Fisher–Kolmogorov equation);

$R(\rho) = -r\ln(\rho/K)$ Gompertz, $r, K > 0$;

$R(\rho) = r(1 - \rho/K)^a$ generalized logistic, $r, K, a > 0$.

An analysis of the reaction–diffusion equation, when the demography is Malthusian, namely, when the per capita growth rate $R(\rho)$ is constant and independent of density, was already carried out in Box 1.6. There, we considered the Cauchy problem only. A full account is given now of the more interesting density-dependent case, including also other boundary conditions (see [291] for applications of reaction–diffusion equations to biology). We assume, however, that there is no inverse density dependence (the so-called Allee effect), in other words, that R is a decreasing function of density like in the three examples cited above. The Allee effect entails the occurrence of some interesting phenomena that are beyond the scope of this section. Also, we obviously assume that $R(0) > 0$ and that for sufficiently large ρ, the rate is negative, in other words, that R is zero at a certain density K, usually called the carrying capacity of the environment. At K, the competition between individuals of the same species becomes so strong that the birthrate equals the death rate.

The goal is to study the spatial patterns generated by Equation (3.7). To that end, we must find the stationary and stable solutions of the reaction–diffusion equation. If we impose stationarity (i.e., $\partial\rho/\partial t = 0$), we obtain the following second-order differential equation in space:

$$D\frac{\partial^2 \rho}{\partial x^2} + R(\rho)\rho = 0. \tag{3.8}$$

The solutions of Equation (3.8) can be found via nonlinear analysis (see Appendix 6.1 and Box 3.5) of the equivalent system:

$$V\frac{d\rho}{dx} = \sigma$$

$$\frac{d\sigma}{dx} = -\frac{R(\rho)}{\rho},$$

where σ is nothing but the spatial gradient of density. In Box 3.5 we derive the phase diagram of Equation (3.8). It is shown in Figure 3.5, which reports all the possible stationary patterns associated with the reaction–diffusion Equation (3.7). The arrows do not indicate increasing time, of course, but rather the spatial coordinate varying between $-\infty$ and $+\infty$. Note that the equilibrium $\rho = 0, \sigma = 0$ is a center, while the equilibrium $\rho = K$, $\sigma = 0$ is a saddle. This must be properly interpreted: the cycles surrounding the center are time-stationary solutions that are periodic in space; the stable manifold of the saddle is a stationary solution that tends to K for $x \to +\infty$.

Among these stationary solutions, the ones to be picked up are those that satisfy the boundary conditions. First, consider the Cauchy problem on the infinite domain $-\infty, +\infty$. The solution ρ must be nonnegative and bounded. It is thus easily seen that the only feasible

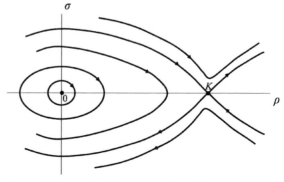

Figure 3.5 Portrait of the trajectories of Equation (3.8) in the state space $\rho - \sigma$; ρ is the density of organisms and σ the spatial gradient of density. Arrows indicate increasing spatial coordinate x.

solutions are $\rho(x,t) \equiv 0$ and $\rho(x,t) \equiv K$. All the other solutions turn out to be negative for some values of x and possibly unbounded. However, of these two stationary solutions, $\rho(x,t) \equiv 0$ is unstable and $\rho(x,t) \equiv K$ is stable (see Box 3.5). Therefore, for any initial condition $\rho_0(x)$ that is not identically equal to zero at any location, the population will converge to the carrying capacity K everywhere. The transient, however, can be very long and can give rise to interesting patterns, such as the traveling waves that will be analyzed in the next section.

Consider now the Dirichlet and Neumann problems. The Neumann problem (reflecting barrier) is easily solved. In fact, it requires that the gradient σ satisfy $\sigma(0) = 0$ and $\sigma(L) = 0$. This is possible only for $\rho(x,t) = 0$ and $\rho(x,t) = K$ in the interval $[0, L]$, because other possible solutions (the cycles shown in Figure 3.5) satisfying boundary conditions would be negative in part of the interval $[0, L]$. Again, it is possible to show (Box 3.5) that $\rho(x,t) = K$ is the only stable stationary solution.

The absorbing barrier case is more complicated and interesting. In this case, the boundary conditions require that $\rho(0,t) = \rho(L,t) = 0$. The feasible stationary solutions are $\rho(x,t) \equiv 0$ or a closed orbit surrounding the origin that has period $2L$ (see Figure 3.6), because half of that cycle, as indicated by the figure, meets the boundary conditions. The problem is whether one can find a cycle that has exactly period $2L$. To this end, note that the saddle equilibrium $\rho = K, \sigma = 0$ is characterized by one

branch of its stable manifold (see Appendix 6.1 for the definition) coinciding with a branch of its unstable manifold: this creates what is called a *homoclinic orbit*, which corresponds to a solution that tends to K for both $x \to -\infty$ and $x \to \infty$. It is the boundary of the region containing all the cycles surrounding the origin and is in a way a cycle of infinite period. Therefore, by continuity, the cycles surrounding the origin must have an increasing period length that tends to infinity as they get closer and closer to the homoclinic orbit. However, this does not imply that the small cycles close to the origin have a period length tending to zero. In fact, it is possible to compute the limit of the period by considering the eigenvalues of the linearized system around the center (see Box 3.5 for details), which turn out to be purely imaginary and equal to $\pm j\sqrt{R(0)/D}$. The modulus of the eigenvalues is the angular frequency associated with the infinitesimal cycle surrounding the origin, that is, $\eta = \sqrt{D/R(0)} = 2\pi/(2L)$. We thus finally obtain that the minimum period corresponds to $L = \pi\sqrt{D/R(0)}$. This result has one very important consequence: there exists a critical domain dimension $L_c = \pi\sqrt{D/R(0)}$ such that the only possible solution to the absorbing barrier problem is the extinction of the population ($\rho(x,t) \equiv 0$) if the domain length is too small, namely, if

$$L \leq L_c = \pi\sqrt{\frac{D}{R(0)}}.$$

In fact, in this case, there exists no cycle of the right period. The result can be understood in this way: there

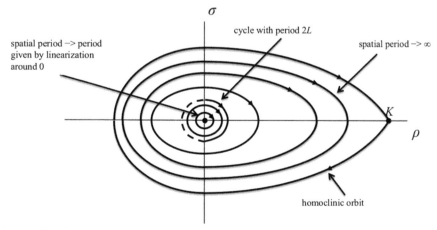

Figure 3.6 Closed orbits of Equation (3.8) surrounding the origin. Cycles have increasing spatial period moving from small cycles close to the origin toward cycles that are closer and closer to the homoclinic orbit of the saddle. The figure shows the case in which the habitat dimension L is larger than the critical value $L_c = \pi\sqrt{\frac{D}{R(0)}}$ and there is a nonnull solution to the absorbing barrier problem.

is a balance between the growth of organisms inside the domain ($R(0) > 0$) and the decrease of the population size due to the definitive exit of individuals through the boundary. If the domain length is too small, the number of organisms generated per unit time inside the whole length of the interval is smaller than the number getting out at the boundary. On the contrary, when $L > L_c$, extinction is unstable, and there exists a nonnull space-periodic pattern to which the population distribution converges. The larger L is, the closer is the cycle to the homoclinic orbit and to the saddle equilibrium $\rho = K, \sigma = 0$. Therefore, for large habitat size L, the solution is very close to the carrying capacity K everywhere, with the exception of the locations close to 0 and L, because the boundary conditions $\rho(0,t) = \rho(L,t) = 0$ must be satisfied.

Finally, the case of a circle-shaped habitat requires the periodicity conditions $\rho(0,t) = \rho(L,t)$ and $\sigma(0,t) = \sigma(L,t)$ for any t. Figure 3.6 shows that the two boundary conditions for ρ and σ cannot be simultaneously satisfied by any of the cycles surrounding the origin. The only possible solutions (see again Figure 3.6) are $\rho(x,t) \equiv 0$ and $\rho(x,t) \equiv K$. However, as we show in Box 3.5, only the solution $\rho(x,t) \equiv K$ is stable, while the other stationary solution is unstable. Obviously, in a circle-shaped habitat, the spatial distribution tends to homogenization.

The results just illustrated for one-dimensional space can be generalized to multidimensional space. In particular, in 2D, the Cauchy and Neumann (repelling barrier) problems still imply long-term homogenization toward the carrying capacity. The same result holds for sphere- or torus-like habitats (periodic Dirichlet–Neumann problem). For the Cauchy problem, however, similarly to the 1D case, the transient can be very long: given an initial distribution, for example, concentrated around the origin, the final homogeneous pattern will be reached via a traveling wave in two dimensions.

The absorbing barrier problem in 2D has a solution similar to the solution in 1D: there is a critical area below which the population is doomed to extinction, because the perimeter-to-area ratio is too large and the exit from the domain occurs through the perimeter while growth occurs inside the domain. However, in 2D, the shape of the connected and finite domain can vary considerably, and the critical area depends on shape too. The two simplest shapes are square and circular. In the first case (square with side L), the condition of absorbing barrier is

$$\rho(0,y,t) = \rho(L,y,t) = 0$$

$$\rho(x,0,t) = \rho(x,L,t) = 0,$$

while in the second case, the condition is that $\rho(x,y,t) = 0$ for all the points satisfying the relationships $x^2 + y^2 = R^2$ (circle of radius R).

For a square habitat, the critical size is

$$A_{cr} = 2\pi^2 \frac{D}{r}, \tag{3.9}$$

and for a circular habitat, it is

$$A_{cr} = 1.84 \ldots \pi^2 \frac{D}{r}. \tag{3.10}$$

It is no surprise that the critical area of a circular habitat is smaller than that of a square habitat, because the circle has the minimum perimeter-to-area ratio. Suppose you want to rescue a species threatened with extinction by placing a number of individuals within a reserve. Formulas 3.9 and 3.10 are fundamental to deciding the size and shape of the reserve. Of course, to determine the critical area, it is necessary to have at least rough estimates of the per capita growth rate and the diffusion coefficient. Figure 3.7 is an indirect confirmation of the theory of critical habitat size: William [292] analyzed several national parks in the west of United States and found that the species extinction probability significantly decreases with the park size.

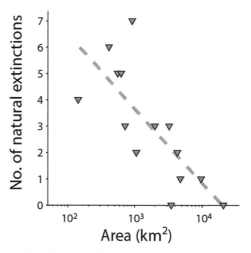

Figure 3.7 Number of originally present species that became extinct after setting up a park as a function of each park area. The analysis has been conducted for national parks in the western part of the United States. Redrawn after [292]

Box 3.3 Stationary Solutions of the Reaction–Diffusion Equation and Their Stability

The stationary solutions of the reaction–diffusion equation with density-dependent growth can be found via nonlinear analysis (see Appendix 6.1.6) of the equivalent system introduced in Section 3.1.1:

$$\frac{d\rho}{dx} = \sigma, \qquad \frac{d\sigma}{dx} = -\frac{R(\rho)}{\rho}, \tag{3.11}$$

where σ is the spatial gradient of density. Here we derive the phase diagram of this system. To this end, we first notice that, given the assumption that $R(\rho)$ is a decreasing function of ρ, the equation $R(\rho) = 0$ has just the solution $\rho = K$. Therefore, the system has two equilibria: (a) the extinction equilibrium $\rho = 0, \sigma = 0$, and (b) the carrying-capacity equilibrium $\rho = K, \sigma = 0$. We analyze the stability of both via linearization (see Appendix 6.1). The Jacobian of system (3.11) is

$$J = \begin{bmatrix} 0 & 1 \\ -\frac{R(u)+R'(u)u}{D} & 0 \end{bmatrix},$$

which, computed at the trivial equilibrium (a), yields

$$J_0 = \begin{bmatrix} 0 & 1 \\ -\frac{R(0)}{D} & 0 \end{bmatrix},$$

whose eigenvalues are

$$\lambda_{1,2} = \pm j\sqrt{\frac{R(0)}{D}}.$$

As the eigenvalues are purely imaginary, nothing can be stated about the nature of the equilibrium via linearization. Instead, at equilibrium (b), one has

$$J_K = \begin{bmatrix} 0 & 1 \\ -\frac{R'(K)K}{D} & 0 \end{bmatrix},$$

whose eigenvalues are

$$\lambda_{1,2} = \pm\sqrt{-\frac{R'(K)K}{D}}.$$

$R'(K)$ is negative, so both eigenvalues are real. Since one is negative and one is positive, this equilibrium is a saddle. The properties of the extinction equilibrium, though, can be ascertained in an alternative way. Note that

$$R(\rho)\rho\frac{d\rho}{dx} + D\sigma\frac{d\sigma}{dx} = 0,$$

which implies that

$$\int_{x_0}^{x} \left[R(\rho)\rho\frac{d\rho}{dx} + D\sigma\frac{d\sigma}{dx} \right] dx = \quad \text{constant.}$$

Box 3.3 *Continued*

If we integrate by changing variables and setting $\rho(x_0) = \rho_0$ and $\sigma(x_0) = \sigma_0$, we obtain

$$\int_{x_0}^{x} \left[R(\rho)\rho\frac{d\rho}{dx} + D\sigma\frac{d\sigma}{dx} \right] dx = \int_{\rho_0}^{\rho} R(\rho)\rho d\rho + D\int_{\sigma_0}^{\sigma} \sigma d\sigma$$

$$= \int_{\rho_0}^{\rho} R(\rho)\rho d\rho + \frac{D}{2}\left(\sigma^2 - \sigma_0^2\right) = \quad \text{constant.}$$

As ρ_0 and σ_0 are arbitrary, we can set them to 0 so as to have a function that vanishes at the extinction equilibrium. Because we rule out a possible Allee effect, the function $R(\rho)\rho$ is positive, unimodal, and concave in the interval $[0, K]$, which implies that its integral $V(\rho) = \int_0^{\rho} R(\rho)\rho d\rho$ is quadratic near $\rho = 0$. We conclude that orbits near the origin of the state plane $\rho - \sigma$ are the level curves of the function $V(\rho) + \frac{D}{2}\sigma^2$, which is quadratic close to the extinction equilibrium. Therefore, these orbits are closed curves, and the extinction equilibrium is a center surrounded by an infinity of concentric cycles.

We then obtain the phase diagram (see Figure 3.5) by sticking together the orbits surrounding the center and those surrounding the saddle. Actually, the trajectories of system (3.11) are nothing but the contour lines of the function $V(\rho) + \frac{D}{2}\sigma^2$. Remember that the arrows do not indicate increasing time, of course, but rather increasing spatial coordinate x, varying between $-\infty$ and $+\infty$. In fact, in Appendix 6.1, we use time t as the independent variable of the system of ordinary differential equations, but obviously time can be replaced by any other variable.

It is interesting to remark that, although linearization failed to provide the properties of the extinction equilibrium, it turned out that this equilibrium of the nonlinear system has the same stability properties of the linearized system, which is indeed a center! In fact, both eigenvalues are imaginary. It is thus possible to state a posteriori that very close to the origin, the linearized system provides a correct information about the dynamic properties of the nonlinear system. In particular, the modulus of the eigenvalues is the angular frequency associated with the infinitesimal cycle surrounding the origin, that is, $\eta = \sqrt{R(0)/D} = 2\pi/2L$. This result was fundamental to deriving the solution to the absorbing barrier problem presented in Section 3.1 and establishing the concept of critical domain size $L_c = \pi\sqrt{\frac{D}{R(0)}}$.

Studying the stability of stationary solutions, from now on indicated as $\tilde{\rho}(x)$, is not an easy task, because we must analyze the stability of solutions of partial differential equations, not of ODE like those considered in Appendix 6.1.6. We recall that the problem of stability concerns the fate of an initial condition $\rho_0(x) \neq \tilde{\rho}(x)$. The stationary solution $\tilde{\rho}(x)$ is asymptotically stable if, for sufficiently small $\epsilon > 0$, there exists $\delta_\epsilon > 0$ such that, for initially perturbed conditions satisfying $|\rho_0(x) - \tilde{\rho}(x)| < \delta_\epsilon$ for all x, (i) $|\rho(x,t) - \tilde{\rho}(x)| < \epsilon$ for all x and t, (ii) $\lim_{t\to\infty} |\rho(x,t) - \tilde{\rho}(x)| = 0$ for all x. However, if the considered stationary solution is homogeneous in space, that is, $\tilde{\rho}(x) = \tilde{\rho}$ constant, then the criterion for checking stability is rather simple and leads to the study of a linear reaction–diffusion equation. In fact, the following linearization criterion holds: suppose that we want to study the nonlinear reaction–diffusion equation

$$\frac{\partial\rho}{\partial t} = D\frac{\partial^2\rho}{\partial x^2} + F(\rho), \tag{3.12}$$

where $F(\rho)$ is a continuous function. Consider a value $\tilde{\rho}$ such that $F(\tilde{\rho}) = 0$. This is obviously a stationary and space-homogeneous solution of Equation (3.12) provided it satisfies the boundary

Box 3.3 *Continued*

conditions. Then the stability of $\bar{\rho}(x) = \bar{\rho}$ can be tested via linearization, namely, by studying the stability of the model (first-order reaction–diffusion or Malthusian reaction–diffusion)

$$\frac{\partial z}{\partial t} = D\frac{\partial^2 z}{\partial x^2} + rz, \tag{3.13}$$

where $z(x,t) = \rho(x,t) - \bar{\rho}$ and $r = \frac{dF}{d\rho}\big|_{\bar{\rho}}$. In Box 3.7 we study the dynamic properties of Equation (3.13) with different possible boundary conditions. We use those results to infer the stability properties of the nonlinear reaction–diffusion with $F(\rho) = R(\rho)\rho$. In particular, it turns out that the solutions of Equation (3.13) are simply the solutions of the pure diffusion equation multiplied by $\exp(rt)$, as we had already pointed out in Chapter 1. In fact, if we consider the Fourier analysis of a solution $z(x,t)$ (via Fourier transform for the infinite-domain Cauchy problem or via Fourier series for finite-domain problems), it is easy to see that each spectral component $\Phi(\eta,t)$ obeys the equation

$$\frac{\partial}{\partial t}\Phi(\eta,t) = (r - D\eta^2)\Phi(\eta,t),$$

where η is the angular frequency of the spectral component. Remember, however, that not all the spatial frequencies of the solutions to Equation (3.13) are to be considered, because boundary conditions must be satisfied. As regards this aspect, it must be remarked that if a function $\rho(x,t)$ satisfies the initial and the space-boundary conditions, so does the function $\exp(rt)\rho(x,t)$ (easy to check). Therefore, the feasible frequencies of the linear reaction–diffusion equation are the same as the feasible frequencies of the diffusion equation. For example, in the absorbing barrier problem, frequency $\eta = 0$ is not feasible. Therefore, the perturbation $z(x,t)$ increases exponentially if there exist feasible spectral components with frequency η such that $r - D\eta^2 > 0$ (instability criterion), while it decreases exponentially to 0 if $r - D\eta^2 < 0$ for all the feasible frequencies η (stability criterion).

First, consider the Cauchy problem on the infinite domain $-\infty, +\infty$. It has two space-homogeneous stationary solutions $\rho(x,t) \equiv 0$ and $\rho(x,t) \equiv K$. The Malthusian model resulting from linearization around $\rho(x,t) \equiv 0$ has $r = R(0) > 0$. As all the frequencies are feasible in the Cauchy problem, then sufficiently small frequencies, such that $\eta < \sqrt{\frac{R(0)}{D}}$, correspond to exponentially increasing spectral components. Therefore, the extinction equilibrium is unstable. On the contrary, the Malthusian parameter corresponding to $\rho(x,t) \equiv K$ is $r = K\frac{dR(K)}{d\rho} < 0$. Therefore, all the spectral components of the perturbation $z(x,t)$ are exponentially decreasing at a rate $r - D\eta^2 < 0$, and the space-homogeneous solution corresponding to carrying capacity everywhere is stable. Thus, for any initial condition $\rho_0(x)$ that is not identically equal to zero at any location, the population will converge to K.

Consider now the problems with a finite domain, the interval $[0, L]$. The feasible perturbations of a space-homogeneous solution $\bar{\rho}$ must satisfy the boundary conditions at 0 and L. We already know from Box 3.2 that this limits the feasible spatial frequencies η_k present in the Fourier series of $z(x,t)$.

The Neumann problem (reflecting barrier) is easily solved. In fact, the stationary solutions are $\rho(x,t) \equiv 0$ and $\rho(x,t) \equiv K$ in the interval $[0, L]$, both space-homogeneous. The values of the linearized Malthusian parameters are obviously those obtained above, but not all the frequencies are allowed in this problem. The feasible ones are $\eta_k = \frac{\pi k}{L}, k = 0, 1, \ldots$. Therefore, if $r > 0$, then $r - D\eta_k^2 > 0$, at least for $\eta_0 = 0$, so the extinction equilibrium is unstable. Instead, if $r < 0$, then

Box 3.3 *Continued*

$r - D\eta_k^2 < 0$ for any k, so the carrying-capacity solution $\rho(x,t) \equiv K$ is stable. Thus, for any initial condition $\rho_0(x)$ that is not identically equal to zero at any location, the population will converge to K.

The absorbing barrier problem has only one space-homogeneous solution, that is, $\rho(x,t) \equiv 0$. Its stability depends upon the length L of the habitat. In fact, the feasible frequencies are $\eta_k = \frac{\pi k}{L}$, $k = 1, 2, \ldots$. As r is positive, the maximum value of $r - D\eta_k^2$ is $r - D\eta_1^2 = r - D(\frac{\pi}{L})^2$. Therefore, if $r - D(\frac{\pi}{L})^2 < 0$, that is, if $L < L_c = \pi\sqrt{\frac{D}{R(0)}}$, then extinction is stable; otherwise, it is unstable. Thus, we have proved that the existence of a non-space-homogeneous positive solution (the half cycle of Figure 3.6) implies the instability of the extinction solution. This does not necessarily prove that the non-space-homogeneous solution is stable (an exchange of stability), but it is a good clue. The linearization criterion for space-homogeneous solutions cannot be applied, obviously, and the interested reader must consult, for example, [291] for the nontrivial proof that the non-space-homogeneous solution is indeed stable.

Finally, the case of a circle-shaped habitat possesses the two stationary solutions $\rho(x,t) \equiv 0$ and $\rho(x,t) \equiv K$. The values of the linearized Malthusian parameter are the usual ones, and the feasible frequencies are $\eta_k = 2\frac{\pi k}{L}, k = 0, 1, \ldots$. Therefore, if $r > 0$ then $r - D\eta_k^2 > 0$ at least for $\eta_0 = 0$, so the extinction equilibrium is unstable. Instead, if $r < 0$, then $r - D\eta_k^2 < 0$ for any k, so the carrying-capacity solution $\rho(x,t) \equiv K$ is stable. For any initial condition $\rho_0(x)$ that is not identically equal to zero at any location, the population will converge to K.

Box 3.4 The Linear Reaction–Diffusion Equation and the Sturm–Liouville Theory

As we stated in the previous box, the study of space-homogeneous solutions to the nonlinear reaction–diffusion equation

$$\frac{\partial \rho}{\partial t} = D\frac{\partial^2 \rho}{\partial x^2} + F(\rho) \tag{3.14}$$

is based on a linearization criterion that requires the analysis of the following linear reaction–diffusion equation:

$$\frac{\partial z}{\partial t} = D\frac{\partial^2 z}{\partial x^2} + rz, \tag{3.15}$$

where $z(x,t) = \rho(x,t) - \tilde{\rho}$ and $r = \frac{dF}{d\rho}\big|_{\tilde{\rho}}$. Here we study the dynamic properties of Equation (3.15) with different possible boundary conditions. We start by noting that the equation is actually a particular case of a more general partial differential equation, namely,

$$\frac{\partial \rho}{\partial t} + k(t)\rho = f(x)\frac{\partial^2 \rho}{\partial x^2} + g(x)\frac{\partial \rho}{\partial x} + h(x)\rho, \tag{3.16}$$

provided one sets $k(t) = -r$, $f(x) = D$, $g(x) = 0$, $h(x) = 0$. Actually Equation (3.16) includes an advection–reaction–diffusion equation in which the reaction rate (represented by $k(t)$ and $h(x)$) may be time and space dependent, while the diffusion coefficient (represented by $f(x)$) and the advection

Box 3.4 *Continued*

speed (represented by $g(x)$) are not constant but space dependent. Equation (3.16) has the usual initial conditions $\rho(x,0) = \rho_0(x)$ and boundary conditions of the kind

$$a\rho(0,t) + b\frac{\partial\rho(0,t)}{\partial x} = 0, \qquad a^2 + b^2 > 0,$$

$$c\rho(L,t) + d\frac{\partial\rho(L,t)}{\partial x} = 0, \qquad c^2 + d^2 > 0$$

or of the kind

$$\rho(0,t) = \rho(L,t)$$

$$\frac{\partial\rho(0,t)}{\partial x} = \frac{\partial\rho(L,t)}{\partial x}.$$

The first set of boundary conditions (sometimes termed Robin boundary conditions) is quite general and include as particular cases the absorbing ($b = 0, d = 0$) and repelling barrier ($a = 0, c = 0$) as well as more general conditions, like an absorbing barrier at 0 and a repelling barrier at L. The second set of boundary conditions refers to the circle-like habitat (periodic conditions).

We first note that linear combinations of solutions of Equation (3.16) with a given set of boundary conditions are also solutions of the equation and satisfy the same boundary conditions. It turns out that a general solution is the sum of all the elementary functions that are solutions of a special type, that is, separable solutions $\rho(x,t) = N(t)X(x)$. If we impose that they satisfy Equation (3.16) we obtain

$$\frac{\mathbf{L}X(x)}{X(x)} = \frac{\mathbf{M}N(t)}{N(t)}, \qquad\qquad (3.17)$$

where

$$\mathbf{L} = f(x)\frac{d^2}{dx^2} + g(x)\frac{d}{dx} + h(x)$$

$$\mathbf{M} = \frac{d}{dt} + k(t)$$

are two linear operators in the language of functional analysis. Since, by definition, $\mathbf{L}$ is independent of time and operates on $X(x)$, while $\mathbf{M}$ is independent of space and operates on $N(t)$, then both sides of Equation (3.17) must be equal to a constant λ, that is,

$$\mathbf{L}X(x) = \lambda X(x)$$

$$\mathbf{M}N(t) = \lambda N(t).$$

Therefore, λ must be an eigenvalue of both the operator $\mathbf{L}$ and the operator $\mathbf{M}$. The first of these equations defines a Sturm–Liouville problem for second-order ODEs, which we consider later. Suppose that the eigenvalues λ_k of $\mathbf{L}$ are found. Then the second equation is easily solved because it is given by

$$\frac{dN_k(t)}{dt} = (\lambda_k - k(t))\,N_k(t),$$

Box 3.4 *Continued*

whose solution is simply given by the exponential

$$N_k(t) = n_k \exp\left(\lambda_k t - \int_0^t k(\tau)d\tau\right),$$

where n_k is a suitable constant. Let $X_k(x)$ be the eigenfunction that corresponds to the eigenvalue λ_k (namely, the solution of the equation $\mathbf{L}X(x) = \lambda_k X(x)$). Then the elementary separable solution of Equation (3.16) is given by

$$\rho_k(x,t) = n_k X_k(x) \exp\left(\lambda_k t - \int_0^t k(\tau)d\tau\right).$$

An important remark is that if $X_k(x)$ satisfies the boundary conditions, so does $\rho_k(x,t)$. In fact,

$$a\rho(0,t) + b\frac{\partial\rho(0,t)}{\partial x} = n_k \exp\left(\lambda_k t - \int_0^t k(\tau)d\tau\right)\left(aX_k(0) + b\frac{dX_k(0)}{dx}\right) = 0$$

$$c\rho(L,t) + d\frac{\partial\rho(0,t)}{\partial x} = n_k \exp\left(\lambda_k t - \int_0^t k(\tau)d\tau\right)\left(cX_k(L) + d\frac{dX_k(L)}{dx}\right) = 0$$

for Robin's conditions and similarly for the circle-shaped habitat. Therefore, provided that the Sturm–Liouville problem is solved with all the $X_k(x)$ satisfying the boundary conditions, the general solution of Equation (3.16) is

$$\rho(x,t) = \sum_k n_k X_k(x) \exp\left(\lambda_k t - \int_0^t k(\tau)d\tau\right),$$

where the constants n_k are chosen so as to satisfy the initial condition, namely,

$$\rho(x,0) = \sum_k n_k X_k(x) = \rho_0(x).$$

That this is possible follows from the general properties of the solutions of the Sturm–Liouville problem [293]. Consider the problem

$$\mathbf{L}X(x) = \lambda X(x),$$

$$aX(0) + b\frac{dX(0)}{dx} = 0,$$

$$cX(L) + d\frac{dX(L)}{dx} = 0.$$

Then (a) all the eigenvalues of $\mathbf{L}$ in the problem with the given boundary conditions are real and their set is countably infinite; (b) the eigenvalues can be ordered as $\lambda_0 > \lambda_1 > \cdots . > \lambda_k > \cdots \to -\infty$ and the corresponding eigenfunctions (suitably normalized) $X_k(x)$ have exactly $k - 1$ zeros in the interval $[0, L]$; and (c) the eigenfunctions are orthonormal, that is,

Box 3.4 *Continued*

$$\int_0^L X_k(x)X_m(x)dx = 0, \qquad k \neq m,$$

$$\int_0^L X_k(x)^2 dx = 1,$$

and form a basis in the functional space of possible solutions, which means that any continuous function satisfying the boundary conditions can be expanded as a linear combination (possibly infinite) of the eigenfunctions. This guarantees that the constants n_k can be found because also the initial condition $\rho_0(x)$ must satisfy the boundary conditions and the eigenfunctions are a basis of the functional space.

We can apply all the above results to the simpler linear reaction–diffusion equation (3.15). In this case, $\mathbf{L} = D\frac{d^2}{dx^2}$, and the equation $\mathbf{L}X(x) = \lambda X(x)$ is actually the simple harmonic equation

$$\frac{d^2X}{dx^2} - \frac{\lambda}{D}X = 0,$$

which has the general solution

$$X(x) = \alpha\cos\left(\sqrt{-\frac{\lambda}{D}}x\right) + \beta\sin\left(\sqrt{-\frac{\lambda}{D}}x\right),$$

with α and β being suitable constants. To find eigenvalues and eigenfunctions, one must impose the boundary conditions, which will specifically determine both; however, we observe that the eigenfunctions are in any case sinusoidal functions with eigenvalues linked to the angular frequency η by the relationship $\lambda_k = -D\eta_k^2$. Therefore, we can state that the solution to the reaction–diffusion (3.15) is given by

$$z(x,t) = \sum_k n_k X_k(x)\exp\left[\left(-D\eta_k^2 + r\right)t\right].$$

The stability or instability of the space-homogeneous solution represented by $z(x,t) \equiv 0$ depends on the sign of r and the values of the feasible angular frequencies $\eta_k = \sqrt{-\frac{\lambda_k}{D}}$. The feasible angular frequencies are determined by the boundary conditions. If $r < 0$, the space-homogeneous solution is stable anyway. If $r > 0$, the stability depends on the smallest feasible frequency $\eta_0 = \sqrt{-\frac{\lambda_0}{D}}$. If $r - D\eta_0^2$ is negative, then all the $r - D\eta_k^2$ are negative, and $z(x,t)$ converges to 0 for $t \to \infty$ (stability of space-homogeneous solution). Otherwise, at least $r - D\eta_0^2$ is positive, and there is at least the initial condition $\rho_0(x) = \alpha\cos(\eta_0 x) + \beta\sin(\eta_0 x)$ (with α and β suitably chosen to satisfy boundary conditions) that will imply an exponentially increasing $z(x,t)$ (instability of space-homogeneous solution).

Periodic boundary conditions (circle-like habitat) differ slightly from Robin condition in that there can be eigenvalues with multiplicity 2, and their ordered sequence, instead of being mono-tonically decreasing, is monotonically nonincreasing. Table 3.1 recapitulates several eigenfunctions corresponding to different boundary conditions and possessing an analytical expression.

Table 3.1 Eigenfunctions corresponding to various typical boundary conditions for the linear reaction–diffusion equation

Boundary conditions	Eigenfunctions
$X(0) = X(L) = 0$; absorbing barrier	$X_k(x) = \sin\left(\frac{k\pi x}{L}\right)$ $\quad k = 1, 2, \dots$
$X'(0) = X'(L) = 0$; repelling barrier	$X_k(x) = \cos\left(\frac{k\pi x}{L}\right)$ $\quad k = 0, 1, 2, \dots$
$X(0) = X(L)\; X'(0) = X'(L)$; periodic boundary	$X_k(x) = \{1, \sin\left(\frac{k\pi x}{L}\right), \cos\left(\frac{m\pi x}{L}\right)\}$ $\quad k, m = 1, 2, \dots$
$X(0) = 0\; X'(L) = 0$; 0 is absorbing barrier, L repelling	$X_k(x) = \sin\left(\frac{(2k-1)\pi x}{2L}\right)$ $\quad k = 1, 2, \dots$
$X'(0) = h_1 X(0)\; X'(L) = -h_2 X(L)\; h_1, h_2 \geq 0$; Robin conditions	feasible angular frequencies = nonnegative roots of equation $\tan(\eta_k L) = \frac{\eta_k(h_1 + h_2)}{\eta_k^2 - h_1 h_2}$ $\quad k = 0, 1, \dots$
$\quad$ Case I $h_1, h_2 \neq 0$	$X_k(x) = \frac{\eta_k}{h_1}\cos(\eta_k x) + \sin(\eta_k x)$
$\quad$ Case II $h_1 \neq 0, h_2 = 0$	$X_k(x) = \frac{\cos(\eta_k(L-x))}{\sin(\eta_k L)}$ $\quad \eta_0 \neq 0$
$\quad$ Case III $h_1 \to \infty, h_2 \neq 0$	$X_k(x) = \sin(\eta_k x)$
	$\eta_0 = 0 \quad \eta_k \sim \left(\frac{2k+1}{2}\right)\frac{\pi}{L} \quad k \to \infty$

Note. We indicate the gradient $\frac{dX}{dx}$ by X'.

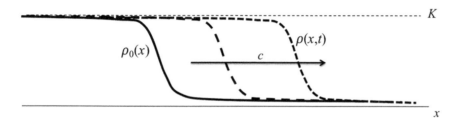

$\rho_0(x)$ $\quad c \quad \rho(x,t)$ $\quad K$

Figure 3.8 A traveling wave of the reaction–diffusion model Equation (3.7).

3.1.2 Traveling Waves and Invasion Fronts

To understand how organisms spread in the environment, it is good to start from the simple, though not unrealistic, situation in which the environment can be considered homogeneous and so large with respect to the initial distribution of the population that we can assume it is infinite. The paradigm that we will investigate is the reaction–diffusion equation of the previous section. In the next section, we will consider more sophisticated models that consider individuals as discrete entities (interacting particle systems, or IPS).

We consider again the infinite-domain Cauchy problem of equation

$$\frac{\partial \rho}{\partial t} = D\frac{\partial^2 \rho}{\partial x^2} + R(\rho)\rho \qquad (3.18)$$

with density-dependent demography ($R(\rho)$ is decreasing with ρ). We already know that any initial nonvanishing distribution will converge for $t \to \infty$ to the carrying capacity K, a space-homogeneous solution. However, as we have already pointed out, the transient toward the space-homogeneous pattern can be infinitely long and very interesting. In particular, the problem of invasion fronts, which was first proposed by Fisher [125] and Kolmogorov [119], investigates solutions with an initial distribution in which a portion of the habitat (e.g., the interval $[-\infty, x_0]$) is completely filled by organisms up to carrying capacity K. Specifically, we search for solutions of the type $\rho(x,t) = u(x - ct)$ with $c > 0$ and such that $\lim_{x \to -\infty} \rho(x,t) = K$ and $\lim_{x \to \infty} \rho(x,t) = 0$. Solutions are then steady (c is constant) progressing waves to the right, as shown by Figure 3.8. The speed c is sometimes termed celerity. Symmetrically, one may obtain waves progressing to the left, or, if organisms initially occupy a finite domain, an interval in 1D, two fronts, one progressing to the left and one to the right.

Finding the traveling waves implies use of so-called similarity solutions where one employs a moving reference frame $\xi = x - ct$. Hence $du/dt = -c\, du/d\xi$, $du/dx = du/d\xi$ and $d^2u/dx^2 = d^2u/d\xi^2$, and the reaction–diffusion equation is satisfied if

$$-c\frac{du}{d\xi} = R(u)u + D\frac{d^2u}{d\xi^2} \qquad (3.19)$$

with $\lim_{\xi\to-\infty} u(\xi) = K$ and $\lim_{\xi\to+\infty} u(\xi) = 0$. As usual, we transform a second-order ODE into a system of coupled first-order ODEs via the position $v = du/d\xi$ and $dv/d\xi = -cv/D - (R(u)u)/D$ and study their orbits. Equilibria of this system are (a) $v = 0, u = 0$, (b) $v = 0$, $u = K$. Their stability can be investigated via linearization. As for equilibrium (a), one can show (see Box 3.5) that it is stable, but it can be either a focus or a node, depending on the value of the celerity c. If $c \geq 2\sqrt{R(0)D} = c_{min}$, it is a stable node. Instead, equilibrium (b) is always a saddle (see Box 3.2.3).

We are searching for a solution that satisfies $\lim_{\xi\to-\infty} u(\xi) = K$ and $\lim_{\xi\to+\infty} u(\xi) = 0$ and is characterized by being nonnegative ($u(\xi) \geq 0$ for any positive and negative ξ). Therefore, it must be an orbit that connects equilibrium (b) with equilibrium (a) provided that (a) is not a focus, because otherwise the orbit would be spiralling around equilibrium (a), and $u(\xi)$ would be negative for some ξ. If $c \geq 2\sqrt{R(0)D}$, this orbit exists and is the heteroclinic connection (see Appendix 6.1.6) that connects the saddle (b) with the node (a), as shown in Figure 3.9.

Therefore, in principle, there exists an infinity of waves $u(\xi) = u_c(x - ct)$ corresponding to all the heteroclinic connections of ODE systems with $c \geq c_{min}$. One can wonder whether they are stable in the sense that, given initial conditions $\rho_0(x)$ with $\lim_{x\to-\infty} \rho_0(x) = K$ and $\lim_{x\to\infty} \rho_0(x) = 0$ and such that $\rho_0(x) \neq u_c(x)$, then $\lim_{t\to\infty} \rho(x,t) = u_c(x - ct)$ for all x. It turns out that for suitable (and finely tuned) initial conditions, all waves are stable. However, for "most" initial conditions (however imprecise this notation, it suffices for the purposes of this book), Kolmorogov showed that $\rho(x,t)$ converges to the traveling wave with speed c_{min}. The domains of attraction of waves with $c > c_{min}$ are small, and random perturbations will inevitably lead to a front progressing with the minimum speed $c_{min} = 2\sqrt{R(0)D}$. In a way, we recover the result that holds for Malthusian populations (see Box 1.6) provided we replace the Malthusian growth rate r with $R(0)$. As we had already observed, this can be explained by noting that the population density close to the wave front is very low: the organisms ready to invade a new habitat are indeed in almost Malthusian conditions.

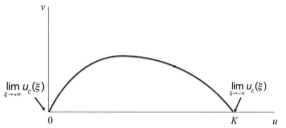

Figure 3.9 If $c \geq 2\sqrt{R(0)D}$, there exists a heteroclinic trajectory $u_c(\xi)$ connecting the carrying capacity equilibrium (a saddle) and the extinction equilibrium (a stable node). This trajectory represents a wave traveling with speed c.

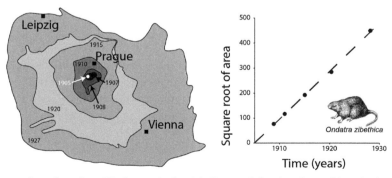

Figure 3.10 The expansion of muskrat (*Ondatra zibethica*) in Europe following the accidental release of five muskrats in the location shown as a white dot in panel (a), near Prague. Maps are redrawn after the original reported in [282]. (b) If we approximate the areas occupied by the muskrats with circles, their root increases linearly with time, as reported in the seminal paper [294]. By dividing the slope of the fitting line by $\sqrt{\pi}$, one obtains the average radius increase as kilometers per year. The estimated value is 11.8 km/year.

Table 3.2 Various expansion speeds for a few terrestrial and marine species

Terrestrial species	
Species name (common name)	**Observed speed(km/year)**
Impatiens glandulifera (Himalayan balsam)	9.4–32.9
Lymantria dispar (Asian gypsy moth)	9.6
Pieris rapae (small white butterfly)	14.7–170
Oulema melanopus (cereal leaf beetle)	26.5–89.5
Ondatra zibethica (muskrat)	0.9–25.4
Sciurus carolinensis (grey squirrel)	7.66
Streptopelia decaocto (collared dove)	43.7
Sturnus vulgaris (European starling)	200
Yersinia pestis (Bubonic plague bacterium)	400
Marine species	
Species name (common name)	**Observed speed(km/year)**
Botrylloides leachi (colonial tunicate)	16
Membranipora membranacea (lacy crust bryozoan)	20
Carcinus maenas (green crab)	55
Hemigraspus sanguineus (Asian shore crab)	12
Elminius modestus (acorn barnacle)	30
Littorina littorea (common periwinkle)	34
Mytilus galloprovincialis (Mediterranean mussel)	115
Perna perna (brown mussel)	95

Note. After [296].

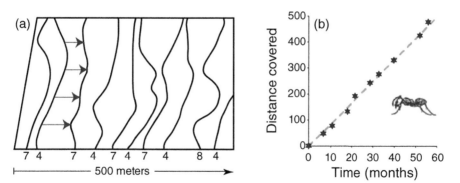

Figure 3.11 The expansion of Argentine ant *Iridomyrmex humilis* in a meadow close to San Diego (California), initially occupied by the Californian ant *Pogonomyrmex californicus*. The experiment took place between October 1963 and October 1968. (a) Solid lines represent the invasion wave fronts of the Argentine ant (left to right). Elapsed times (months) between one front and the next are indicated below each solid line. (b) Distance from release location averaged along the vertical transect as a function of time. Redrawn from [295]

When we consider more realistic two-dimensional environments, the theory, which we do not report here, provides exactly the same result. If diffusion is isotropic, environment is homogeneous, and the initial distribution of organisms is concentrated in a limited area close to a release point, there exists a stable traveling wave of circular shape whose radius increases with speed equal to $c_{min} = 2\sqrt{R(0)D}$.

There exists a large body of literature documenting the viability of the density-dependent reaction–diffusion model (in one or two dimensions) for describing the expansion front of biological invasions. Many studies refer to the introduction of alien species. The classical example is that of the muskrat (see Figure 3.10), a rodent native to North America that was imported to Bohemia at the beginning of the 1900s to exploit its fur. It looks like five animals managed to escape from a breeding farm near Prague in 1905. The muskrat population in the wild began then to increase colonizing space at a speed of about 12 km per year. Another relevant example, that of the Argentine ant *Iridomyrmex humilis* introduced in some areas of the western United States, is summarized in Figure 3.11. Note that the contour lines of the invasion fronts at different times support an approximate one-dimensional spatial model of propagation. Table 3.2 reports about a few measured values of expansion velocities empirically observed for both terrestrial and marine invasive species.

Box 3.5 Heteroclinic Connections and Traveling Waves

Finding a traveling wave of the density-dependent reaction–diffusion equation amounts to establishing whether there exists a feasible heteroclinic connection of the ODE system

$$du/d\xi = v \quad dv/d\xi = -cv/D - (R(u)u)/D. \tag{3.20}$$

We remind the reader that the heteroclinic connection must satisfy the following feasibility conditions: (i) $\lim_{\xi \to -\infty} u(\xi) = K$ and $\lim_{\xi \to +\infty} u(\xi) = 0$, (ii) $u(\xi) \geq 0$ for all ξ.

System (3.20) is characterized by the following equilibria: (a) $v = 0, u = 0$, (b) $v = 0, u = K$. Their stability can be investigated via linearization. To this end, we calculate the Jacobian of (3.20)

$$J = \begin{bmatrix} 0 & 1 \\ -\frac{R(u)+R'(u)u}{D} & -\frac{c}{D} \end{bmatrix},$$

and then compute the Jacobian at the two equilibria, obtaining

$$J_0 = \begin{bmatrix} 0 & 1 \\ -\frac{R(0)}{D} & -\frac{c}{D} \end{bmatrix}$$

and

$$J_K = \begin{bmatrix} 0 & 1 \\ -\frac{R'(K)K}{D} & -\frac{c}{D} \end{bmatrix}.$$

As $\det(J_K) = \frac{R'(K)K}{D} < 0$ because $R(u)$ is a decreasing function, equilibrium (b) corresponding to the carrying capacity is always a saddle. As for the extinction equilibrium (a), it turns out that $\text{tr}(J_0) = -\frac{c}{D} < 0$ and $\det(J_0) = \frac{R(0)}{D} > 0$. Therefore, equilibrium (a) is always stable. However, it might be a stable node or a stable focus. The phase diagram with the two possible cases is shown in Figure 3.12. Clearly, a feasible heteroclinic orbit is possible only if the origin is a stable node. This depends on the eigenvalues of J_0. They must be real, a fact that is simply ascertained by calculating the discriminant of the equation $\det(\lambda I - J_0) = \lambda^2 + (c/D)\lambda + R(0)/D = 0$, namely, $\Delta = (c/D)^2 - 4R(0)/D$. It must be nonnegative for the eigenvalues to be real, which implies

$$c \geq 2\sqrt{R(0)D} = c_{min}.$$

Box 3.5 *Continued*

This establishes the result that traveling waves do exist for the density-dependent reaction–diffusion equation and that their minimum speed is $c = 2\sqrt{R(0)D}$. Fisher [125] and Kolmogorov [119] have shown that this is also the speed of the "stable" traveling wave.

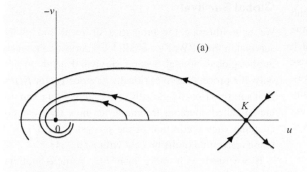

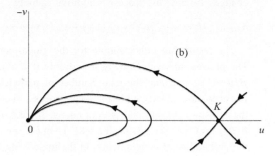

Figure 3.12 Phase portrait of system (3.20). (a) Origin is a stable focus if $c < 2\sqrt{R(0)D} = c_{min}$. (b) Origin is a stable node if $c \geq 2c_{min}$ and the trajectory joining the carrying capacity to the origin is the feasible heteroclinic orbit. Note that for readability, the v-axis is upside down.

3.2 Modeling Biological Invasions via Interacting Particle Systems (IPS)

The study of spatial patterns is central in ecology. The traditional approach is via continuous models in which both space and population size are treated as real variables. More recently, attention has shifted to a more realistic analysis in which both individuals and space are discrete, thus allowing for the inclusion of local and nonlocal stochastic effects. The field of mathematics that has provided the relevant results is interacting particle systems (IPS), which can be considered a development of the traditional theory of stochastic processes [297, 298]. Applications to ecology and, more generally to biology are very well described in [299] and [300], while a nice review is given by [301]. Here, we will introduce the fundamentals.

Consider a system of particles (cells, organisms, etc.) that reproduce, move, and die on a countable graph (e.g.,

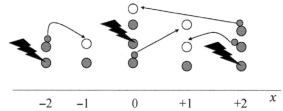

Figure 3.13 The scheme of an interacting particle system in a one-dimensional lattice. Large gray big cirlces represent mothers that might either die or reproduce (daughters are the small gray circles that then move and are hosted in a new site, where they are represented as white circles).

a lattice) whose vertices can be thought of as sites harboring local populations of particles. This is exemplified in Figure 3.13, which shows the simplest scheme of an interacting particle system. The concept can be extended to particles (organisms) of different types (species) that interact not only inside the same type but also across types or species. Here we will consider only the simple case of particles all belonging to the same species. There are two classes of processes that are in a way on the opposite sides of the spectrum: the branching random walk and the contact process. The continuous-time branching random walk (BRW) is a Malthusian birth–death process with random motion. There are no restrictions on the number of particles per site (a vertex of the graph), and birth and mortality rates are constant. Instead, in the contact process, when two or more particles hit each other in a certain site, they fuse (only one survives); thus there is at most one particle per site. In a way, the contact process incorporates a very strong density dependence with the *carrying capacity* in each site being equal to 1. These are the two processes that we analyze below.

3.2.1 Branching Random Walks

There are no restrictions on the number of particles per site. Here $p(x, y)$ are transition probabilities from site x to site y of a Markov chain (e.g., a random walk). A particle at x waits a random exponential time with rate β and then with probability $p(x, y)$ gives birth to a particle at y (such that $p(x, y) > 0$). A particle waits an exponential time with rate μ and then dies.

It is worthwhile to remark on the following:

- for every x, we have $\sum_y p(x, y) = 1$.
- the birth–death process is Malthusian. So, if there were no random motion, $\beta \leq \mu$ would imply certain extinction. Instead, $\beta > \mu$ would imply exponential increase of the average number of particles; remember from the theory of birth–death processes that the extinction probability starting from one particle is μ/β.
- Time can be rescaled, $t \to \mu t, \mu \to 1, \beta \to \beta/\mu$, so that the behavior of process depends only on β/μ.

A very important concept is the local and global survival of a spatial stochastic process. To this end, define the following:

$n_x(t, y)$ = number of particles in node y at time t for a process started with one particle in node x;
$N_x(t) = \sum_y n_x(t, y)$ = total number of particles.

We say that global survival occurs if there exists a positive probability that the total number of particles never goes to zero, namely,

$$\Pr\{N_x(t) \geq 1, \text{for all } t > 0\} > 0.$$

Instead, we say that there occurs local survival if there exists a positive probability that the original node is visited from time to time forever, namely,

$$\Pr\{\lim \sup_{t \to \infty} n_x(t, y) \geq 1\} > 0.$$

3.2.2 Properties of Local and Global Survival

We now illustrate the properties of local and global survival in the BRW. First of all, local survival obviously implies global survival. Second, survival depends on β/μ only. If a process survives (locally or globally) for $\beta/\mu = A$, it survives (locally/globally) for $\beta/\mu > A$ (one can use the so-called coupling technique to show this, namely, coupling each realization of the process with $\beta/\mu = A$ to the realization of the process with $\beta/\mu > A$).

If we start with more than one particle, nothing changes, because the process is additive, namely,

$$n_{x,z,w,\ldots}(t, y) = n_x(t, y) + n_z(t, y) + n_w(t, y) + \cdots$$

The results are rather simple for the global survival. The process survives globally if and only if $\beta > \mu$ (i.e., $\beta/\mu > 1$). Note that the total number of particles simply obeys the underlying birth–death process. Therefore, the transition to survival is continuous. In fact, define $h(\beta/\mu) = \Pr\{N_x(t) \geq 1, \text{for all } t > 0\}$. Then h is obviously the probability of nonextinction in the underlying birth–death process; thus $h = 0$ for $\beta/\mu \leq 1$, while, for $\beta > \mu$ ($\beta/\mu > 1$), we have

$$h = 1 - \frac{\mu}{\beta} = \frac{\frac{\beta}{\mu} - 1}{\frac{\beta}{\mu}}.$$

Local survival is more complicated. Consider separately random motion and birth–death, and let X_t = position of a particle that waits a mean 1 exponential time and then jumps from x to y with probability $p(x,y)$. Calculate $p_{xy}(t) = \Pr\{X_t = y | X_0 = x\}$ = probability of going from x to y exactly in time t. In particular, $p_{xx}(t)$ is the probability of being in x at time t starting from x. Obviously, it turns out $1 \geq p_{xx}(t) \geq \exp(-t)$.

The threshold for local survival is such that the transition is usually discontinuous. Consider the ratio

$$\ln [p_{xx}(t)] / \ln [\exp(-t)] = -\ln [p_{xx}(t)] / t$$

and calculate its asymptotic value

$$\lim_{t \to \infty} \ln [p_{xx}(t)] / t = -\gamma.$$

It can be shown [300] that local survival is guaranteed if $\beta/\mu \geq 1/(1 - \gamma) \geq 1$.

If $\gamma = 0$, the two thresholds (local vs. global) coincide. If $\gamma \neq 0$, then transition to local survival is discontinuous (see Figure 3.14). Define $g(\beta/\mu)$ = Probability of local survival = $\Pr\{\lim \sup_{t \to \infty} n_x(t, y) \geq 1\}$. Then we have

$g = 0$, for $\beta/\mu < 1/(1-\gamma)$,

$g = h$ = probability of global survival = $\dfrac{\frac{\beta}{\mu} - 1}{\frac{\beta}{\mu}}$, for $\beta/\mu \geq 1/(1-\gamma)$.

The calculation of the thresholds obviously requires the specification of the rules for the random walk. For example, if motion occurs on a finite graph according to an irreducible finite Markov chain, then there exists a stable stationary distribution π_x. In this case, $p_{xx}(t) \to \pi_x$ and $\lim_{t \to \infty} \ln[p_{xx}(t)]/t = 0$. Thus local and global survival coincide.

Among infinite graphs on which random walk can take place, the simplest is the one-dimensional lattice. Let p be the probability of moving right and $1 - p$ that of moving left. Then one can show [300] that $\gamma = 1 - 2\sqrt{p(1-p)}$.

Thus, if the random walk is symmetric ($p = 0.5$), then $\gamma = 0$ and local and global survival coincide, which is rather intuitive. If, alternatively, the random walk is asymmetric, local survival requires a birthrate higher than the mortality rate, precisely,

$$\beta/\mu \geq \frac{0.5}{\sqrt{p(1-p)}}.$$

In fact, the descendants can return to the location of their ancestor only if they are generated in sufficient numbers so that they can overcome the bias of the movement. This is particularly important in river networks where the bias is clearly given by the river flowing downstream. Suppose one wants to know whether a few specimens of a fish species that became extinct can be reintroduced upstream and lead to the permanent establishment of the population at least in the reintroduction site and in all the downstream reaches. The above threshold answers the question. Obviously, one is also interested in knowing the probability with which introduction is successful. This is given by the formula for the global survival, which coincides with local survival beyond the threshold, namely,

$$h = g = \frac{\frac{\beta}{\mu} - 1}{\frac{\beta}{\mu}}.$$

It is also interesting to compare the front speeds of the BRW with those of ADR (Box 3.6). Actually, the concept of *front* in IPSs must be specified because of the discrete nature of the model. The BRW theory provides results for the asymptotic speed of the leftmost and the

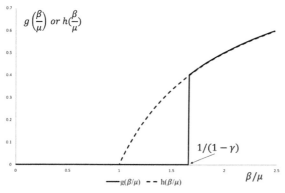

Figure 3.14 The thresholds of local and global survival.

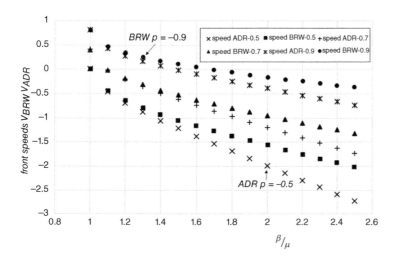

Figure 3.15 Comparison of front speeds v_{BRW}, v_{ADR} as a function of the ratio β/μ: advection–diffusion–reaction approximation versus branching random walk for different values of the probability, p, of moving right. Negative values of the speed indicate a receding front moving left, which implies local survival.

rightmost particles. The approach is not easy because it is based on the log-Laplace transform of a point process (see Box 3.7, based on the theory by [302]). Let $\psi(s)$ be the log-Laplace transform of the branching random walk. Then, the asymptotic speed of the leftmost particle (which defines the speed of the receding front) is

$$v_{BRW} = -\left(\frac{\beta}{\mu}+1\right)\inf_{s>0}\frac{\psi(s)}{s},$$

where

$$\psi(s) = \ln\left(\frac{\beta}{\mu}\right) - \ln\left(\frac{\beta}{\mu}+1\right)$$
$$+ \ln\left[1 + p\,\exp(-s) + (1-p)\,\exp(s)\right].$$

The function $\psi(s)/s$ is unimodal for $s > 0$ and $\beta/\mu > 1$ with a well-defined minimum that must be found numerically. The infimum for $\frac{\beta}{\mu} \to 1$ is $2p-1$, which is obtained for $s \to 0$. Thus, the front speed of the IPS coincides with the usual advection speed in this latter case. A comparison of the front speeds for the ADR equation and the BRW is shown in Figure 3.15. It is remarkable that the receding front speed of the BRW is not only always larger than the one estimated via ADR, but it is larger and larger for larger and larger ratios β/μ. Thus we can conclude that diffusion is not so good an approximation not only for large advection but also for large demographic growth rates.

Box 3.6 Comparing Branching Random Walks with Reaction–Diffusion Models

We can also compare the BRW with the more traditional model of advection–diffusion–reaction (ADR) shown in Equation (3.7) and repeated here for convenience:

$$\frac{\partial \rho}{\partial t} = -v\frac{\partial \rho}{\partial x} + D\frac{\partial^2 \rho}{\partial x^2} + (\beta-\mu)\rho, \tag{3.21}$$

where $\rho(x,t)$ is the concentration of particles at time t and space x per unit space. To mimic the IPS model, the usual approximations leading to the ADR equation must be slightly modified. In fact, the rules of the BRW are such that only the daughter particles move right with probability p and left with probability $1 - p$ by a step of length Δx. Instead, mothers either die or stay where they are while sending a daughter away. Therefore, the balance equation that generates ADR is as follows:

$$\rho(x,t+\Delta t) = \rho(x,t) - \mu\Delta t\rho(x,t) + \beta\Delta t(p\rho(x-\Delta x,t) + (1-p)\rho(x+\Delta x,t)).$$

By applying the usual approach (consider first-order terms with respect to time and up to second-order terms with respect to space), one obtains

$$\frac{\partial \rho}{\partial t} = (\beta-\mu)\rho + \beta\Delta x(2p-1)\frac{\partial \rho}{\partial x} + \frac{\beta\Delta x^2}{2}\frac{\partial^2 \rho}{\partial x^2}. \tag{3.22}$$

Thus the advection speed is given by $v_{adv} = \beta\Delta x(2p-1)$ and the diffusion coefficient by $D = \frac{\beta\Delta x^2}{2}$. Note that the average time between two birth events is $1/\beta$ and that, when a birth occurs, the new particle moves right or left by a step Δx. Therefore $\frac{\beta\Delta x^2}{2}$ is the variance of the step per unit time. Also, it is convenient to rescale space and time by setting $\Delta x = 1$ and $\mu = 1$ and replacing β by β/μ. Thus the advection speed is $v_{adv} = \frac{\beta}{\mu}(2p-1)$ and $D = \frac{\beta}{2\mu}$. Then the equation displays two traveling fronts, one with speed $v_{adv} + 2\sqrt{D(\beta/\mu-1)} = \frac{\beta}{\mu}(2p-1) + \sqrt{2(\beta/\mu-1)\beta/\mu}$ and the other one with speed $v_{adv} - 2\sqrt{D(\beta/\mu-1)} = \frac{\beta}{\mu}(2p-1) - \sqrt{2(\beta/\mu-1)\beta/\mu}$. Suppose that $p > 0.5$ so that the advection speed is positive. If the latter speed is positive, both fronts are moving right (or downstream) so that organisms released in a certain location will never return to that position, nor will their descendants. This condition is met if $v_{adv} > \sqrt{2(\beta/\mu-1)\beta/\mu}$ and is somehow equivalent to the failure of local

Box 3.6 *Continued*

survival in the branching random walk. We can compare the resulting thresholds for the ratio β/μ by using the ADR approximations for v_{adv} and D. The condition for local survival then becomes $\sqrt{2(\beta/\mu - 1)\beta/\mu} \geq (2p - 1)\beta/\mu$, that is,

$$\frac{\beta}{\mu} \geq \frac{1}{1 - 2(p - 0.5)^2}.$$

The thresholds obtained according to ADR and BRW are compared in Figure 3.16. It is remarkable that the thresholds are practically equal up to $p = 0.7$. They diverge for p getting closer and closer to 1. In particular, the BRW correctly states that the threshold tends to infinity for p tending to 1, because all the daughters move right and the mothers will die sooner or later. Instead, the ADR model displays a threshold of 2 at $p = 1$. This is due to the diffusion approximation, which implies solutions that are normally distributed in space and therefore are characterized by infinite tails.

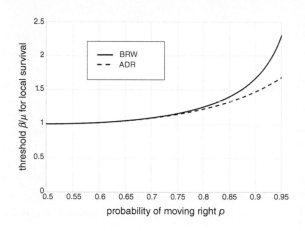

Figure 3.16 Comparison of the thresholds of local survival: advection–diffusion–reaction approximation versus branching random walk.

3.2.3 Contact Processes

The rules are the same as for the BRW (birth–death process with random walk), but births on occupied sites are suppressed. At most there is one particle per site. The theory has been mainly developed for d-dimensional lattices $\mathbb{Z}^d$ or for infinite homogeneous trees in which each site has d neighbors; we will consider lattices only.

The basic contact process (CP) can thus be described as follows. A particle at x waits an exponential time with rate β and then randomly gives birth to a particle on one of the $2d$ neighbors on the lattice ($p(x, y) = 1/2d$ if x and y are neighbors). If the neighbor is occupied, the birth is

suppressed. A particle waits an exponential time with rate μ and then dies. The contact process is thus coupled to a symmetric random walk.

For simplicity we will consider the one-dimensional infinite lattice. The contact process is actually a Markov chain with an uncountable infinity of states. In fact, the change of state between time t and time $t + h$ (with h suitably small) is, for instance, given by

$$\ldots 0\,0\,1\,0\,1\,0\,0\,0\,1\,0\,1\,1\,1\,0\,0\,0\,\mathbf{0}\,1\,0\,1\,1\,1\,0 \ldots \text{time } t$$

$$\ldots 0\,0\,1\,0\,1\,0\,0\,0\,1\,0\,1\,1\,1\,0\,0\,0\,\mathbf{1}\,1\,0\,1\,1\,1\,0 \ldots \text{time } t + h$$

in the case in which there has been one birth (evidenced in bold). Thus, in 1D lattices, each state is identified by a bi-infinite string of 0s and 1s (the configuration at each time). As known, real numbers of, for example, the [0,1] interval (an uncountable infinity) can be written as infinite strings of bits. The probability that there are two simultaneous events (birth or death) is zero; therefore, transition from one configuration to another implies change ($0 \to 1$ or $1 \to 0$) at one location only (this is called asynchronous updating). In other words, if one wants to simulate the contact process, one can proceed in this way.

Start with a given configuration of 0s and 1s at time t.

Pick a location x at random. Find out the number n_{occ} of occupied neighbors and the fraction of occupied $n_{occ}/2d$.

Extract a random death time D from the exponential distribution with rate μ and a random birth time B from the exponential distribution with rate $\beta n_{occ}/2d$.

If $D < B$, then

if x is occupied, change 1 to 0 at time $t + D$;

if x is empty, make no change at time $t + D$;

start again.

If $B < D$, then

if x is empty, change 0 to 1 at time $t + B$;

if x is occupied, make no change at time $t + B$;

start again.

The definition of local and global survival for the contact process is exactly the same as for the BRW; obviously $n_x(t, y)$, that is, the number of particles in node y at time t for a process started with one particle in node x can take on only values 0 or 1.

As we are considering the basic contact process on a lattice, which is coupled to a symmetric random walk, local and global survival coincide. This is not true for the contact process on trees (this is intuitive; for instance, in a binary tree, where $d = 3$, there is $1/3$ probability of moving toward the root and $2/3$ of moving toward the leaves). Survival depends on β/μ only. If a process survives (locally or globally) for $\beta/\mu = A$, it survives for $\beta/\mu > A$. Obviously, if $\beta/\mu \leq 1$, the process dies (couple the contact process with the corresponding BRW with random motion). However, contrary to the symmetric BRW in which $\beta/\mu > 1$ is necessary and sufficient for the global and local survival, the threshold for survival is much higher. There is no specific formula (even for the basic 1D CP), but some properties are known. First, the transition to survival is continuous, like in BRW, namely, $h(\beta/\mu) = \Pr\{N_x(t) \geq 1, \text{for all } t > 0\}$ is continuous such that $h(\beta/\mu)$ is zero for $\beta/\mu \leq$ threshold and strictly increasing for $\beta/\mu >$ threshold. In the 1D continuous-time contact process, simulations show that the condition

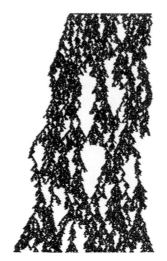

Figure 3.17 Simulations of the basic contact process: (left), above the threshold; (right) below the threshold. Figure modified after [299]

for survival is $\beta/\mu > 3.299$. Figure 3.17 shows two realizations of the contact process, one above and one below the threshold. That the threshold for survival is so much larger than the one in BRW should not be astonishing. In CP, density dependence is very strong because no more than one particle per site is allowed. In general, density dependence greatly increases the chances of extinction: remember, for instance, that in the simple birth–death process with density dependence, the probability of long-term survival is zero.

A major breakthrough was to understand that CP survival is basically a problem of percolation [300] in space-time. For instance, the 1D CP can be viewed as a problem of oriented percolation in a 2D lattice where the second spatial dimension is replaced by the temporal dimension: the contact process survives starting from one particle at x if there is a percolating cluster of site x.

A very important role in understanding the behavior of the contact process is played by the so-called largest stationary distribution. Suppose we start CP not with only one site being occupied but rather with all sites occupied, namely, from the state

$$\ldots .. 1\ 1\ 1\ 1\ 1\ 1\ 1\ 1\ 1\ 1\ 1\ 1\ 1\ 1\ 1\ 1\ 1\ 1\ 1\ 1 \ldots \ldots$$

Then the survival process is equivalent to percolation somewhere: it is well known from percolation theory that this is an on-off phenomenon, that is to say, below the threshold, percolation has probability zero, while above threshold, there exists one infinite percolation cluster with probability 1. Therefore, if $\beta/\mu \leq$ threshold, the

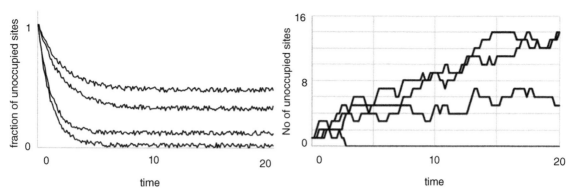

Figure 3.18 The duality property: the probability of extinction of the contact process (CP) starting from one single particle (right panel, which shows four realizations of the CP) can be found by simulating the CP starting from all sites occupied and considering the asymptotic fraction of occupied sites (the left panel reports simulation for different values of β/μ).

process dies; if $\beta/\mu >$ threshold, the contact process survives with probability 1. This should be contrasted with the result that the survival probability of a CP starting with one particle only is a continuous increasing function of β/μ.

To better understand the behavior of the contact process, consider that CP is a Markov chain with two fundamental stationary distributions. The first is related to the configuration

$$..... 0\ 0\ 0\ 0\ 0\ 0\ 0\ 0\ 0\ 0\ 0\ 0\ 0\ 0\ ......$$

being an absorbing state (a finite closed class).
Therefore $Pr\{.....00000000000000.....\} = 1$, $Pr\{$any other configuration$\} = 0$ is a stationary distribution corresponding to certain extinction. If $\beta/\mu \leq$ threshold, extinction is the unique stationary distribution. If, instead, $\beta/\mu >$ threshold and the starting configuration is "all 1s," the process converges to another stationary distribution, which is called largest stationary distribution. Owing to Markov chain properties, any other stationary distribution is a convex combination of the trivial distribution corresponding to extinction and the nontrivial largest distribution.

The nontrivial stationary distribution from all 1s is the largest stationary (LS) distribution in the sense that any other stationary distribution corresponds to a smaller number of 1s (a larger probability that a site will be empty). This stationary distribution (as any other distribution) is completely identified by the probabilities that there is 0 or 1 in any finite number of positions, namely, if $n_{LS}(x)$ is the number of particles at location x, by

$$Pr\{n_{LS}(x_1) = 0 \text{ or } 1, n_{LS}(x_2) = 0 \text{ or } 1, ...n_{LS}(x_k) = 0 \text{ or } 1\}.$$

The distribution is translation invariant. In particular, $Pr\{$site x is empty, i.e., $n_{LS}(x) = 0\} = p_E$ independent of x; $Pr\{$site x is occupied, i.e., $n_{LS}(x) = 0\} = 1 - p_E$.

A very important result is the duality property: the probability that the process survives starting from a single particle (namely, $h(\beta/\mu)$) is exactly the frequency of occupancy $1 - p_E$ in the largest stationary distribution. One can thus construct the function $h(\beta/\mu)$ by simulating the CP with all sites initially occupied and finding the asymptotic fraction of occupied sites (Figure 3.18).

What happens if the CP process starts from an arbitrary initial configuration ($\neq$ one site occupied or all sites occupied)? The main result is the complete convergence theorem: if the contact process does not die out, it will converge to the largest stationary distribution. The probability p_{ext} of dying out depends on initial conditions. We know from the duality property that $p_{ext} = p_E$ for one site initially occupied and $p_{ext} = 0$ for all sites initially occupied.

For intermediate initial conditions, the contact process will converge to the stationary distribution (convex combination)

$$p_{ext} \text{ Extinction} + (1 - p_{ext}) \text{ Largest Stationary}.$$

The probability of extinction p_{ext} must of course be calculated via simulation.

It is important to remark that stationary distributions display spatial correlations. The presence or absence of particles in nearby locations is correlated, so that the probability of site x being occupied is not independent of the probability that site $y \neq x$ be occupied. However, correlations decline exponentially fast with distance.

The basic contact process assumes that movement occurs only from a site to the nearest neighbors. It is interesting to explore what can happen if we relax this assumption. A first modification is the so-called propagule rain, in which all sites are neighbors: an organism born at site x is randomly sent to any other site. If this site is occupied, the newborn dies. In this way, spatial structure and correlation between adjacent states are destroyed! We can then define $p_x(t)$ = probability a site x is occupied independently of the state of the other sites. This allows us to derive a mean-field model for the contact process.

If we let L = total number of sites (which we will let tend to ∞ later), then

$$p_x(t+dt) = p_x(t)(1-\mu dt) + (1-p_x(t))\frac{\beta dt}{L-1}\sum_{y\neq x}p_y(t),$$

from which

$$\frac{dp_x}{dt} = -\mu p_x + \frac{\beta(1-p_x)}{L-1}\sum_{y\neq x}p_y.$$

If $L \to \infty$, $\sum_{y\neq x} p_y(t)/(L-1)$ is nothing but the average fraction u of occupied sites, so that

$$\frac{dp_x}{dt} = -\mu p_x + \beta u(1-p_x).$$

Finally, by summing up the left-hand sides of the above equations and averaging, we easily get the following mean-field equation:

$$\frac{du}{dt} = -\mu u + \beta u(1-u).$$

This is nothing but Levin's model [81], which was first introduced in the context of fragmented populations and inspired much of the subsequent work on metapopulations. It is a logistic-like model. If $\beta \leqslant \mu$, there is only one stable equilibrium: $u = 0$; if $\beta > \mu$, there is a nontrivial equilibrium: $u = 1 - \beta/\mu$. The mean-field model thus predicts a threshold of 1 for β/μ, which is too low compared to the bona fide contact model and coincides with the BRW threshold.

Durrett and Neuhauser [303] have explored the relationships between the contact process and reaction–diffusion. In particular, they have studied the case of fast stirring. Consider a normal contact process and introduce a space scale assuming that the lattice consists of sites a distance Δx apart. Assume that neighbor sites exchange

their contents at a rate proportional to Δx^{-2} (fast stirring), and let $\Delta x \to 0$. One obtains the Fisher–Kolmogorov equation:

$$\frac{\partial u}{\partial t} = D\frac{\partial^2 u}{\partial x^2} + \beta u(1-u) - \mu u, \qquad (3.23)$$

where $u(x,t)$ = concentration of occupied sites at location x at time t. Because of the fast stirring assumption, the diffusion coefficient is simply given by $D = 1/2$. Even in this case, the threshold of extinction is $\beta/\mu = 1$.

Therefore, in a way, neither the propagule rain nor the fast stirring approach implies any novel result because the threshold coincides with that of the BRW and ADR. In addition, the traveling wave speed in the fast stirring model coincides with the usual Fisher–Kolmogorov speed. In this regard, it is very interesting to explore the traveling speed of the bona fide contact process. Similarly to the BRW case, the theory provides results for the asymptotic speed of the leftmost and rightmost particles. The general results are quite weak.

Let us concentrate on the progressive wave, namely, the one corresponding to the rightmost particle. First of all, the β/μ threshold for survival (3.299 in the 1D case) is exactly the value at which the speed is zero and turns from being negative to being positive. More precisely, if we denote by r_t the position at time t of the rightmost particle in a realization of the CP started with all 1s in the interval $(-\infty, 0]$, then, with probability 1, $\lim_{t\to\infty} r_t/t = v$, with v being a velocity that is constant irrespective of the process realization. Second, the speed v is an increasing and continuous function of β/μ that changes sign exactly at the threshold. Third, if the CP is started with just one 1 in one site, there exists a traveling wave front for all those realizations that survive, and its speed is again v. Fourth, we can deduct from the previous results that, above the threshold, the process survives with probability 1, not only if it starts from 1s everywhere but also if it starts with 1s in the interval $(-\infty, 0]$; therefore the traveling wave converges in the long run to the largest stationary distribution.

The calculation of $v(\beta/\mu)$ is possible only via simulation. Ellner et al. [304] have estimated v in 1D and 2D. We have found that the function

$$v = \frac{1}{2}\sqrt{\frac{\beta}{\mu}\left(\frac{\beta}{\mu}-1\right)}$$

provides an excellent fit to their 1D simulation. Figure 3.19 reports a comparison of the speeds versus β/μ for the three models we have examined: the Fisher–Kolmogorov reaction–diffusion, the branching random

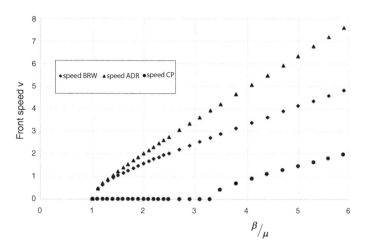

Figure 3.19 Front speeds as functions of the ratio β/μ for different models with underlying symmetric random walk: Fisher–Kolmogorov advection–diffusion–reaction (ADR) with zero advection speed, branching random walk (BRW), and basic contact process (CP).

walk, and the contact process. It is interesting to remark that density dependence does not influence the traveling wave speed in the diffusion approximation: in fact the speed does depend only on the diffusion coefficient and the Malthusian rate of population increase even in the case of logistic demography. Instead, the two IPSs, which differ because of density dependence (Malthusian demography in the BRW, local carrying capacity equal to 1 in the CP), display completely different features. The survival threshold is much higher in the contact process and the speed is considerably lower. This is basically due to the discrete nature of the models: they do incorporate demographic stochasticity (see also the ensuing Section 3.4), which of course very powerfully operates in the contact process where no more than one particle can be hosted in each site.

Box 3.7 Asymptotic Speed Calculation in Branching Random Walk

The asymptotic speed of the leftmost particle is based on the derivation by Shi [302] for the discrete-time branching random walk. The process is one in which at every generation a particle in position x generates, before dying, a random number of particles in different positions according to a given probability distribution. The continuous-time BRW can be coupled with Shi's process by assuming that (a) the time between one generation and the next is exponential with average $1/(\beta + \mu)$; (b) a particle alive at site x in a certain generation produces, after one generation, 0 particles (i.e., it dies) with probability $\mu/(\beta + \mu)$ and 2 particles (i.e., it survives and reproduces) with probability $\beta/(\beta + \mu)$; (c) if 2 particles are generated, one particle (the mother) is located at x while the second particle (the daughter) is located at $x + \Delta x$ with probability p and at $x - \Delta x$ with probability $1 - p$. In other words, if one starts with one particle in position x, then after one generation, with probability $\mu/(\beta + \mu)$ there is no particle, with probability $\beta p/(\beta + \mu)$ there is one particle in position x and one particle in position $x + \Delta x$, and with probability $\beta(1 - p)/(\beta + \mu)$ there is one particle in position x and one particle in position $x - \Delta x$.

Let $x_i(k)$ be the position of the ith particle at generation k. At generation 0, there is one particle in position 0. Then, according to [302], the asymptotic speed of the leftmost particle, namely, $\lim_{k \to \infty} \frac{1}{k} \inf_i x_i(k)$, is given with probability 1 by $-\inf_{s>0} \frac{\psi(s)}{s}$, where $\psi(s)$, is the log-Laplace transform of the process, namely,

$$\psi(s) = \ln E\left(\sum_i \exp(-s x_i(1)) \right).$$

Box 3.7 *Continued*

As the average time between one generation and the next is $1/(\beta+\mu)$, we can deduce that the average speed of the continuous-time BRW in the rescaled time is

$$v_{BRW} = -\left(\frac{\beta}{\mu}+1\right)\inf_{s>0}\frac{\psi(s)}{s}.$$

Given the rules stated above regarding the probability of the positions $x_i(1)$ of the particles after one generation, we obtain (after rescaling space so that $\Delta x = 1$)

$$\ln E\left(\sum_i \exp(-sx_i(1))\right) = \ln\left(\frac{\beta}{\mu}\right) - \ln\left(\frac{\beta}{\mu}+1\right) + \ln\left(1+p\exp(-s)+(1-p)\exp(s)\right).$$

Note that $\psi(0) = \ln\left(\frac{\beta}{\mu}\right) - \ln\left(\frac{\beta}{\mu}+1\right) + \ln 2$, which is the logarithm of the average number of particles generated by one particle in one generation. It is thus ≥ 0 for $\beta \geq \mu$.

3.3 Zebra Mussel Invasion of the Mississippi–Missouri River System

3.3.1 An Iconic Biological Invasion

A hydrologically noteworthy example of biological invasion within the same Mississippi–Missouri river system (MMRS) investigated in Section 2.1.1 [8] is that of the zebra mussel, *Dreissena polymorpha*, a freshwater bivalve native to Eurasia [29]. Owing to its adaptability to a wide range of environmental conditions, combined with dispersal abilities within fluvial systems that are unrivalled by other freshwater invertebrates, this invasive species managed to diffuse all over Europe and North America. After establishment, zebra mussel colonies can rapidly reach population densities on the order of tens (or even hundreds) of thousands of individuals per square meter and inflict huge ecological and economic damages. In fact, zebra mussel colonies may deeply alter invaded ecosystems by filtering large volumes of water, thus removing phytoplankton and boosting nutrients, and by severely impairing the functioning of waterworks [29]. The zebra mussel has thus become a prototypical example of invasive species, rightly included on the 100 World's Worst Invasive Alien Species list drawn up by the International Union for Conservation of Nature. One of the noteworthy features of zebra mussel invasions is the speed at which the species can spread over river networks. The example of the invasion of the MMRS is particularly revealing (see Figure 3.20, after [29]; see also, for appreciating the range of impacts, [305]).

The invasion, in fact, started from the Great Lakes region (Michigan), where some specimens were first sighted in the late 1980s after likely being introduced via ballast water sheddings by boats sailing from Europe. By 1990, *D. polymorpha* made its way into the Illinois River through the Chicago Sanitary and Ship Canal, which connects the Great Lakes watershed with the MMRS. By 1991, the zebra mussel had invaded the whole Illinois River and started diffusing into the Mississippi River. Afterward, the zebra mussel began its spread over the Mississippi River, reaching Louisiana along the backbone of the Mississippi River as soon as 1993. In the meantime, the species showed up also in selected reaches of the upper branch of the Mississippi River and in some of its most important tributaries (the Ohio, Tennessee, and Arkansas Rivers), as well as in other North American river systems (St. Lawrence, Hudson, and Susquehanna). The rate of spread of *D. polymorpha* has decreased remarkably after 1994, primarily because the species did not expand west of the one-hundredth meridian. However, the zebra mussel has been steadily infilling and colonizing new reaches of the MMRS during the last decade. Quite surprisingly, a few connected river systems (most notably, the Missouri River) had not been invaded until recently [306]. Nowadays, the zebra mussel occupies much of central and eastern North America.

D. polymorpha invasion patterns result from the interplay between local demographic processes occurring over long timescales and basin-scale transport phenomena taking place over much shorter timespans [29]. This is mainly due to the peculiarity of the species'

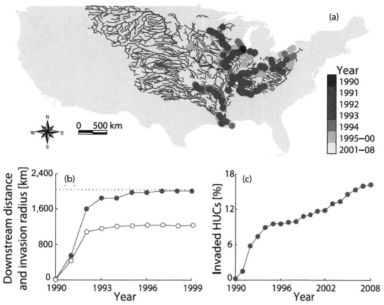

Figure 3.20 Synoptic view of the zebra mussel invasion pattern along the MMRS as recorded from field observations. (a) Spatiotemporal invasion pattern (first sightings) on the river network. (b) Progression of the invasion pattern (filled circles) and spatial extent of the spread (empty circles). Progression is evaluated as the distance traveled downstream by *D. polymorpha* along the backbone of the MMRS starting from the injection point (i.e., the distance traveled along the Illinois and Mississippi Rivers). Spatial extent is evaluated as the mean Euclidean distance between invaded sites on the river network and the injection point. The dotted line represents the length of the river network backbone. (c) Pervasiveness of the zebra mussel invasion, evaluated as the total fraction of invaded hydrologic units of the MMRS, identified by their code (HUC), as a function of time. Figure after [29]

life cycle, which can be roughly subdivided into two main periods: a short larval phase, lasting from a few days to a few weeks [307], and a relatively long adult stage, lasting up to three years in North America [308] (Box 3.8). Adults live anchored to a solid substratum, while larvae (also known as veligers) can be transported by the water flow, sometimes traveling for hundreds (or even thousands) of kilometers before settling [307]. Therefore, rivers represent the primary and natural pathway allowing species spread. However, anthropic activities can often result in (even anomalous) long-range dispersal from their species' current range. This, in turn, remarkably favors both the speed and the extent of the biological invasion [29]. Specifically, in the zebra mussel case, this has been known for a long time, as any human activity that involves the movement of a mass of water can be a potential vector for the spread of *D. polymorpha* [309]. Commercial navigation represents a major driver of mobility for the zebra mussel [29, 310]. For instance, large quantities of veligers are often shipped within the ballast water of commercial vessels. As ports are located even hundreds of kilometers apart from

each other, connections among them allow the species to disperse over very long distances and to colonize stretches of the river network that could not have been reached otherwise. Furthermore, empirical evidence suggests that recreational boating may be an important determinant of medium-range mussel redistribution [311–313]. A common mechanism associated with transient recreational boating is the transport of juveniles and adult mussels via macrophytes entangled on boat trailers [309]. This mechanism has been proposed as the most likely cause of *D. polymorpha* interbasin range expansion due to touristic boating Therefore, commercial navigation represents an efficient vector of long-distance dispersal, while touristic boating can provide a capillary mechanism for medium-range mussel relocation.

3.3.2 Fluvial Transport and the Demography of the Invader

Owing to the importance of the zebra mussel as an ecosystem invader, significant modeling effort has already been

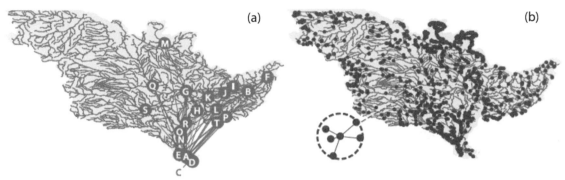

Figure 3.21 Drivers of the secondary dispersal of the zebra mussel (*D. polymorpha*). (a) The main fluvial ports (identified by letters within green circles) of the MMRS and the most important connections among them, which are, respectively, nodes and edges of the commercial network layer. (b) The main lakes, impoundments, and ponds of the MMRS. For exemplification, the inset shows the connections within the recreational network layer between one closed water body (marked in red) and its neighbors. Figure after [29]

devoted to the understanding of its demographic dynamics at a local scale [308] as well as to the description of the species spread along rivers [307, 314] and to the analysis of long-distance dispersal [108]. To single out the role of drivers and controls of the MMRS invasion, spatially explicit, time-hybrid, multilayer network models allow us to address the intertwining of hydrologic controls, acting through the ecological corridors defined by the river network, with long- and medium-range dispersal controlled by anthropogenic factors, which define secondary movement networks. Integrating multiple dispersal pathways is thus crucial to understanding zebra mussel invasion patterns and, in particular, the role played by human activities in promoting the spread. In particular, in this case, commercial navigation has been the most important determinant of the early invasion of the MMRS, and recreational boating can explain the long-term capillary penetration of the species into the water system. The spatially explicit ecohydrological model is described in detail elsewhere [29]. Suffice here to mention the basic unity of the approaches with the ones described in greater detail in Chapter 2. Of great relevance here is the multiplex network of ecological interactions adopted (later rediscussed in the context of pathogen/host mobility in Chapter 4). In fact, while veligers diffuse and settle along one-dimensional substrates provided by the fluvial ecological corridors, hence determining a traditional traveling wave invasion front dynamics (*sensu* [119, 125]), a much faster and heterogeneous spread can be envisioned and proved (Figure 3.21): the infection, in fact, can be propagated by veligers trapped in the ballast water of commercial or recreational boats, possibly moved by

land, surviving the trip and restarting from scratch their colonization process.

The scale of the problem and the complexity of the ecohydrological interactions addressed in this invasion are noteworthy [29]. As noted in Chapter 1, landscape heterogeneities, directional dispersal, and hydrologic controls shape ecological and epidemiological patterns. A novel factor we elaborate further in the context of the spread of waterborne disease in Chapter 6 concerns human-mediated dispersal processes. Such processes are proved to be of the greatest importance in many invasion problems, because they allow species to disperse beyond their ecological ranges, thus eventually shaping global biogeographical patterns (see, e.g., [108]). The importance of this approach – at least in the context of the logical development of this book – lies in the interplay of a selected ecological mechanism, in this case, density-dependent larval mortality affected by hydrologic transport, with human-mediated dispersal. Specific, measurable activities like port-to-port commercial navigation and recreational boating must be considered. Both are deemed of key importance in the spread of *D. polymorpha* in the MMRS.

The model described in this section succeeded in reproducing the zebra mussel invasion patterns observed in the MMRS at a regional spatial scale over a 20-year timespan (Figure 3.22). It offers a hindcasting exercise for the MMRS and a view to other types of problems (such as pandemics) of concern to river networks as ecological corridors. In fact, the proposed multilayer network approach could be applied to predict and control other potential invasions of the same or related alien species

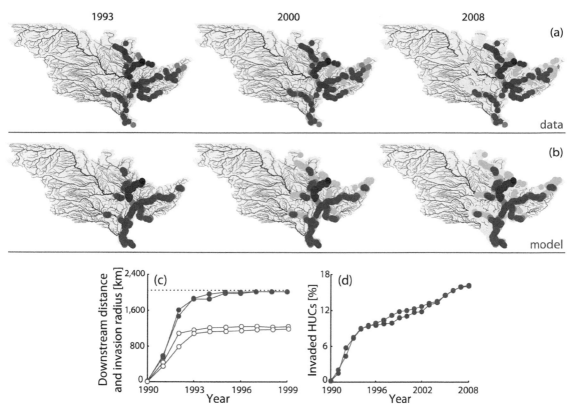

Figure 3.22 Zebra mussel invasion of the MMRS as simulated by the full multilayer network model (Boxes 3.9, 3.11, 3.10, and 3.11) (after [29]). Snapshots of the zebra mussel progression into the MMRS according to (a) data and (b) the model. Colors are shown as in the legend of Figure 3.20. (c, d) Progression of the invasion pattern (filled circles) and spatial extent of the spread (empty circles). Progression is evaluated as the distance traveled downstream by *D. polymorpha* along the backbone of the MMRS starting from the injection point (i.e., the distance traveled along the Illinois and Mississippi Rivers). Spatial extent is evaluated as the mean Euclidean distance between invaded sites on the river network and the injection point. The dotted line represents the length of the river network backbone. (c) Pervasiveness of the zebra mussel invasion, evaluated as the total fraction of invaded hydrologic unit codes (HUCs, i.e., the identification of a georeferenced reach) of the MMRS as a function of time. Red and blue lower insets refer to field data and simulation results, respectively. Parameter values are as in [29]

in other river systems, provided that a comprehensive analysis of extant ecological and hydrologic processes, and of relevant human-mediated transport mechanisms, is available.

The finding that a local precautionary control action could have remarkably delayed the zebra mussel invasion of the MMRS may thus assume a particular relevance. In the context of biological invasions, in fact, taking early action is difficult, yet extremely valuable. In the zebra mussel case, in particular, the containment and/or eradication of established colonies is very difficult and costly. Prevention should thus be favored over control, but this is possible only if quantitative tools to predict the development of the spread are available. To that

end, we point out the importance of a fully dynamical and spatially explicit approach to modeling biological invasions, for it allows combined predictions of both temporal and spatial patterns of species spread. In addition, we deem multilayer approaches an effective and general framework within which to model the spread of species that are characterized by well-identified, multiple dispersal pathways as usually found in biological invasions [108].

It is remarkable that only the inclusion of the three layers in the model allows a proper reconstruction of the zebra mussel invasion in terms of all the observed invasion indices (downstream distance, invasion radius, percentage of hydrologic units invaded), as shown in

Figure 3.22. More generally, as the coexistence of multiple propagation pathways is not unique to harmful invasive species, the idea of a multilayer network approach will be extended to other contexts, such as the spread of waterborne infectious diseases discussed in detail in the next chapter.

Biological invasions are inherently complex stochastic processes for which only a single realization is available for observation. As such, there exist obvious limits to our ability to predict them from models that necessarily need estimation of parameters from actual data (see [288] but also [289], as mentioned above). However, forecasting the main (or mean, in some ensemble-averaging sense) patterns of spread of alien species is becoming increasingly important, though more and more difficult, with anthropic activities rapidly coming abreast of (or even overwhelming) natural invasion pathways. For all these reasons, including important details about the role of ecological corridors and of related anthropogenic drivers is quite valuable. A thorough, quantitative understanding of the different processes that boost the spread of alien species is in fact our only hope to prevent and, to some extent, control biological invasions.

Box 3.8 Local-Scale Interannual Zebra Mussel Demography

Based on the life cycle of *D. polymorpha*, Mari et al. [29] proposed a new network model to describe the spread of the species in the MMRS. The primary, hydrologic network layer is an oriented graph made up of edges (channels) and nodes obtained via the spatial discretization of the river system (see below). Some of these nodes are special in that they represent closed water bodies connected to the water system. The nodes of the secondary layers are disjoint subsets of the nodes in the primary layer. In particular, the nodes of the second layer are the main fluvial ports of the MMRS (Figure 3.21a built around data about the most important fluvial ports of the MMRS in terms of total yearly traffic, according to the 2007 Report of the WCSC; see table 1 in [29]), while the nodes of the third layer represent the largest lakes, impoundments, and ponds connected to the water system (Figure 3.21b). The edges of the secondary layers are defined either by commercial navigation (second layer) or by recreational boating (third layer).

The local-scale dynamics of riverine zebra mussel populations can be described by adapting the nonlinear, discrete-time model with density dependence and age structure proposed by Casagrandi et al. [308], which has already been used to study mussel demographic dynamics. Let $n_k(x,t)$ (units as mussels/m^2) be the density of adult mussels in the kth age class at location x along the river network and time t (here, following [29], we define $1 \leq k \leq 3$ because the lifespan of *D. polymorpha* in North America does not typically exceed three years [315]). The interannual dynamics linking the abundances of mussels in two subsequent years is defined by the following set of equations [29]:

$$n_1(x,t+1) = \sigma_v v_s(x,t),$$
$$n_2(x,t+1) = \sigma_1 n_1'(x,t), \qquad\qquad (3.24)$$
$$n_3(x,t+1) = \sigma_2 n_2'(x,t),$$

where $v_s(x,t)$ [veligers m^{-2}] is the density of settling veligers, σ_v is the survival from the veliger stage to the first adult class (thus accounting for all the mechanisms that can affect mussel recruitment, such as the unavailability of suitable settling sites due to mechanical and/or chemical reasons), the σ_ks ($k = 1, 2$) represent the yearly individual survival probabilities from age k to age $k + 1$, and the n_k's are mussel densities evaluated after relocation due to recreational transport has possibly taken place (see the third network layer below).

Box 3.8 *Continued*

Equations (3.24) constitute the simplest ecologically well-founded model for local-scale dynamics of adult zebra mussels. However, they do not account for some other possibly relevant details of zebra mussel ecology at the local scale. In particular, local regulatory mechanisms such as competition for space or resources among adult mussels and density-dependent larval settlement [311] may be expected to play some role in determining size and structure of *D. polymorpha* colonies, which can reach very high population densities. We refrained from introducing such mechanisms into the models, however, to keep complexity at a minimum. In the same way, Allee effects, whose relevance for alien species dynamics has been the subject of theoretical studies [316, 317], and that have sometimes been envisaged as possible determinants of zebra mussels' invasion success, are not included in the present analysis. In fact, no mechanistic demographic models including Allee effects have been proposed so far for *D. polymorpha* populations [318, 319]. Because at large spatial and temporal scales, the abundances of generated, transported, and settling veligers is likely to be strongly related to local population densities and age structures, we relied on tested demographic models specifically developed for the zebra mussel [308, 311, 320, 321], none of which includes Allee effects.

Box 3.9 Hydrologic Transport and Settling of Veligers

The local abundance of settling veligers $v_s(x,t)$ is the result of within-year processes occurring after reproduction, namely, larval mortality, dispersal, and transport due to both river flow and human activities. In particular, the initial abundance of larvae $v_0(x,t)$ resulting from reproduction can be computed as

$$v_0(x,t) = \frac{1}{2} \sum_{k=1}^{3} f_k n_k(x,t),$$

where f_k (with $1 \le k \le 3$) is the average numbers of eggs released (and successfully fertilized) per adult female in class k, the sex ratio at birth being typically balanced [322]. Larval mortality and transport can be described by means of a partial differential equation (PDE) approach [314, 323]. From a technical viewpoint, coupling PDEs with Equations (3.24) makes the full model for the demography and spread of *D. polymorpha* a hybrid mathematical system [324].

Let $v(x,t,\tau)$ [veligers m^{-2}] be the abundance of veligers in the water column at spatial location x and year t, evaluated τ days after spawning. The abundance of settling larvae at the end of the dispersal period can be computed along each river stretch of the river network by integrating the following PDE between the time of spawning and that of settling:

$$\frac{\partial v(x,t,\tau)}{\partial \tau} = -\nabla \cdot \{u v(x,t,\tau) - \delta [\nabla \cdot v(x,t,\tau)]\}$$
$$- [\mu + \gamma N(x,t)] v(x,t,\tau) + \phi(x,t,\tau), \tag{3.25}$$

where

$$N(x,t) = \sum_{k=1}^{3} n_k(x,t)$$

Box 3.9 *Continued*

is the total density of settled adult mussels. In Equation (3.25), u [km day^{-1}] represents the mean water velocity, which is reasonably assumed to be spatially constant for both landscape-forming and average events, owing to the self-tuning geometry of open channel flows and to the different leading-flow resistance mechanisms from source to outlet [63]. As for δ [km^2 day^{-1}], it represents longitudinal dispersion, which in this framework also accounts for processes other than hydrodynamic dispersion (say, the macroscopic shear flow mixing processes typical of fluvial environments; e.g., [173]), such as local-scale, alongstream larval dispersal operated by aquatic animals or small sailing boats (for a related discussion on active and passive transport in fluvial environments, see, e.g., [325]). Note that for this reason, no direct measurement of hydrodynamic dispersion (e.g., with tracers) might be taken at face value for veliger dispersion.

Larval mortality may depend on both predation by filter-feeding adult mussels, which has been shown to be a key mechanism in the self-regulation of zebra mussel populations [308, 311, 321, 326], and other natural causes not related to the presence and abundance of the population itself [312, 323]. In Equation (3.25), μ [day^{-1}] corresponds to the baseline larval mortality rate, while γ [m^2 mussels^{-1} day^{-1}] represents the adult per capita filtering rate leading to density-dependent mortality. The source term $\phi(x,t,\tau)$ [veligers m^{-2} day^{-1}] is an external flux of veligers, accounting for larval input from upstream reservoirs. In particular, it represents veligers that enter the system from Lake Michigan through the Illinois River. Quite interestingly, in the absence of hydrological transport and input sources ($u = 0$, $\delta = 0$, and $\phi = 0$), models (3.24)–(3.25) would converge to the spatially implicit density-dependent model proposed in [308].

The integration of Equation (3.25) over the time horizon $[0, \tau_s]$, where τ_s [d] is the average timespan needed for a veliger to mature, gives the distribution $v_s(x,t) = v(x,t,\tau_s)$ of veligers that are ready to settle at the end of the larval phase in each year of the simulation. Solving Equation (3.25) over the MMRS requires ad hoc numerical techniques and suitably defined initial and boundary conditions. For each year t in the simulation horizon, we use the distribution of veligers released after reproduction as an initial condition for intra-annual larval transport, that is, $v(x,t,0) = v_0(x,t)$, while we adopt no-flux and absorbing boundary conditions at the headwaters and the outlet of the river network, respectively.

Box 3.10 Long-Distance Larval Dispersal Due to Inland Commercial Navigation

To give a straightforward yet quantitative description of larval movement due to trading navigation, we estimate the fluxes of waterborne commercial traffic among the main fluvial ports of the MMRS (Figure 3.21) with a production-constrained gravity model [105]. Notice that port-to-port vessel traffic is not the only mechanism related to commercial navigation that may be responsible for the spread of the zebra mussel. Barge traffic can in fact be a viable driver of medium-distance, alongstream mussel dispersal [310, 327]. Commercial barges are usually towed in small, variable groups and can use smaller ports than individual vessels. As a result, they are less amenable to a quantitative analysis of mussel transport at a regional spatial scale than traditional commercial traffic, for which federal data are available. We have thus chosen port-to-port traffic as a proxy for the complex set of mechanisms related to commercial navigation.

Box 3.10 *Continued*

Stemming from transportation theory, gravity models can be used to describe human movement when actual mobility data are not available [105]. Models of this family have already been applied to describe the spread of the zebra mussel among neighboring lakes [318, 328–331]. Here we assume that the flux of boats departing from a port (say, p) is proportional to the tonnage of goods passing through it, that is, its "mass" M_p. The fraction of boats that travel to a destination port q increases with M_q and decreases with the distance D_{pq} [km] between p and q evaluated along the river network, so that commercial traffic directed from port p to port q is proportional to

$$F_{pq} \sim M_p \frac{M_q/D_{pq}^{\xi}}{\sum_{z \neq p} M_z/D_{pz}^{\xi}},$$

where ξ is a positive parameter describing the decay of port-to-port boat fluxes with increasing distance among ports. By assuming that the flux of transported veligers is proportional to that of vessels, larval transport due to commercial navigation can be evaluated with the following model:

$$\frac{dv(i_p,t,\tau)}{d\tau} = -\alpha M_p v(i_p,t,\tau) + \sum_{q \neq p}^{n_p-1} \alpha M_q \frac{M_p/D_{qp}^{\xi}}{\sum_{z \neq q} M_z/D_{qz}^{\xi}}$$

$$\cdot v\left(i_q,t,\tau - \frac{D_{pq}}{u_b}\right) \exp\left(-\mu \frac{D_{pq}}{u_b}\right) \cdot H\left(\tau - \frac{D_{pq}}{u_b}\right), \qquad (3.26)$$

where i_p and i_q are the nodes of the second network layer corresponding to ports p and q, respectively; αM_p [day^{-1}] is the port-dependent rate at which veligers enter the ballast water of departing ships; n_p is the total number of ports; u_b [km day^{-1}] is the mean velocity of commercial boats; and $H(\cdot)$ is the Heaviside step function. We assume that veligers transported within boat ballast waters experience the same baseline mortality as those transported in the water flow, which translates into an exponential decay of larval abundance. We note that this assumption might lead to an overestimation of larval survival during transport, because we do not account for possible surplus mortality due to the noxious properties of ballast waters (but see below). We also note that in our model the effects of port-to-port larval shipping can be removed simply by setting $\alpha = 0$, which allows for the evaluation of different invasion scenarios.

Box 3.11 Medium-Distance Dispersal Due to Recreational Boating

Adult mussel transport due to recreational navigation can be evaluated by means of a gravity model accounting for the fluxes of boats that are trailered among neighboring lakes. Notice that boaters may occasionally move veligers as well [309, 332], yet the effects of incidental larval sheddings from recreational boats are far less noticeable, as veligers are expected to become diluted and suffer high mortality [333].

To estimate lake-to-lake boat fluxes, we extract the largest lakes, impoundments, and ponds connected to the MMRS from the GNIS (Geographic Names Information System) and the NHD (National Hydrography Dataset; see Figure 3.21b). Specifically, closed water bodies were selected

Box 3.11 *Continued*

with a minimum surface of 0.5 km^2. Then, the flux J_{lm} of boats that are trailered from one lake (say, l) to one of its neighbors (say, m) is proportional to

$$J_{lm} \sim B_l S_l \frac{S_m/d_{lm}^\chi}{\sum_{z \neq m} S_z/d_{lz}^\chi},$$

where S_l and S_m are the surfaces [km^2] of the two neighboring lakes, B_l is a measure of the number of recreational boats registered in the region [29], d_{lm} [km] is the Euclidean (i.e., aerial) distance between the two lakes, and χ is a shape parameter accounting for the decay of the number of travels with increasing distance among lakes. We assume that l and m are neighbors if they lie within a maximum Euclidean distance $d_{\max}$ [km]. If we further assume that the flux of transported mussels is proportional to that of boats, and follow the same line of reasoning used to obtain the gravity model of Equation (3.26), we obtain

$$\frac{dn_k(i_l,t,\tau)}{d\tau} = -\epsilon B_l S_l \, n_k(i_l,t,\tau) + \sum_{m \neq l}^{n_l-1} \epsilon B_m S_m \frac{S_l/d_{ml}^\chi}{\sum_{z \neq m} S_z/d_{mz}^\chi}$$
$$\cdot n_k(i_m,t,\tau-\tau_T)\sigma_r \cdot H(\tau-\tau_T) + \Phi(i_l,t,\tau), \qquad (3.27)$$

where i_l and i_m are the nodes of the third network layer corresponding to lakes l and m, respectively; $\epsilon B_l S_l$ [day^{-1}] is the lake-dependent rate at which mussels leave lake l; n_l is the number of neighbors for lake l; τ_T [d] represents the time of boat trailering from one lake to another; σ_r is the survival probability during transport between the two lakes; H is the Heaviside step function; and $\Phi(i_l,t,\tau)$ [mussels m^2 d^{-1}] is an input term due to an external flux of recreational boating. In particular, we consider an incoming flux of mussels from Lakes Michigan, Erie, and Ontario to nearby MMRS lakes (i.e., those located within a $d_{\max}$ radius from the Great Lakes). The effects of mussel relocation due to recreational activities can thus be estimated by integrating model (3.27) over the summer season, when most recreational trips occur. The solution of Equations (3.27) has then to be plugged into Equations (3.24). As in the commercial network layer, the effects of recreational boating on mussel transport can be ruled out by setting $\epsilon = 0$.

3.4 Demographic Stochasticity, Fluctuating Resource Supply, Substrate Heterogeneity

Mounting theoretical and experimental evidence [15, 226, 288, 289, 334] has suggested that demographic stochasticity, environmental heterogeneity [289] and biased movement of organisms [15] individually affect the dynamics of biological invasions and range expansions. This section addresses theoretical and experimental evidence on one-dimensional invasion corridors where different spatial arrangements of resources lead to different spread velocities – in particular, even if the mean resource density throughout the linear river-like landscape is kept constant. We organize the presentation of the relevant material by examining the effects on invasion speeds of

- demographic stochasticity (e.g., [289]);
- fluctuating environments (e.g., [335]);
- heterogeneity of resource distributions.

The third item is specifically focused on theoretical and experimental studies of the retarding role (i.e., slower invasion speeds) induced by either the geometry of branching heterogeneity [118, 131] or random space functions with assigned correlation structure describing the driving resource [15].

One wonders, what are the sources of uncertainty and variance in the spread rates of biological invasions? The search for processes that affect biological dispersal and sources of variability observed in ecological range expansions is fundamental for the study of invasive species dynamics, shifts in species ranges due to climate or environmental change, and, in general, the spatial distribution of species [282–285, 294, 296, 334, 336]. Dispersal is the key agent that brings favorable genotypes or highly competitive species into new ranges much faster than any other ecological or evolutionary process [125]. Understanding the potential and realized dispersal is thus key to ecology. When organisms' spread occurs on the timescale of multiple generations, it is the by product of processes that take place at finer spatial and temporal scales, the local movement and reproduction of individuals [334, 336]. The main difficulty in causally understanding dispersal is thus to upscale processes that happen at the short-term, individual level to long-term and broad-scale population patterns [2, 336]. Whether the variability observed in nature or in experimental ensembles might be accounted for by systematic differences between landscapes or by demographic stochasticity affecting basic vital rates of the organisms involved is an open research question [289, 334].

Modeling of biological dispersal established the theoretical framework of reaction–diffusion processes [119, 120, 125, 286, 294, 337], which now finds common application in dispersal ecology [334, 336, 338], control of the dynamics of invasive species [282, 294, 296], and related fields [118, 120]. As mentioned in Chapter 1, reaction–diffusion models have also been applied to modeling human colonization processes, such as the Neolithic transition in Europe [22] or the race to the West in the nineteenth-century continental United States [118]. The extensive use of these models and the good fit to observational data favored their common endorsement as a paradigm for biological dispersal [296]. However, certain assessments [288] suggested inherent limitations to the predictability of the phenomenon, owing to its intrinsic stochasticity. Therefore, single realizations of a dispersal event (as those addressed in comparative studies) might deviate significantly from the mean of the process, making replicated experimentation necessary to allow hypothesis testing, identification of causal relationships, and potential falsification of the models' assumptions.

3.4.1 Population Fluctuations and Different Kinds of Stochasticity

The numerical strength of populations often displays a pronounced variability in stark contrast with the outcomes of the deterministic models of population dynamics studied in the previous sections. The cause of this variability is in many cases unknown. It is nevertheless possible to distinguish between different types of stochasticity and provide a description for each type by using appropriate quantitative models. These models are useful in conservation ecology because they yield predictions of the probability of population trends and hence the assessment of its risk of extinction. Here, we shall provide only a basic treatment of the underlying ecology of fluctuations, considering the variations of the total population size only. In reality, average lifetimes matter: populations with longer average lifetimes, and thus larger numbers of age classes, are better able to absorb variations of the external conditions (e.g., food availability and the vagaries of weather) because often it is only the younger classes that actually suffer the most from this variability. This is often reflected in the indices of population variability. As the size N of a population is always a nonnegative variable, its variability over time can be effectively described by the coefficient of variation (CV), defined as the ratio of the standard deviation of the series of abundances σ_N and the average size of the population $\bar{N}$, that is, $CV = \sigma_N/\bar{N}$. It has been observed that, quite often, the CV is lower for long-lived species, although the differences between average life expectations do not always suffice to explain the different characteristics of population fluctuations.

The dynamics of each population are a mix of deterministic components (i.e., those that we can understand, predict, and measure) and stochastic components (i.e., those that we are not able to understand and/or predict and/or measure). Sometimes the deterministic components of the dynamics prevail. This is the case when, for instance, as a consequence of active policies of conservation, a population grows in a roughly logistic way, stabilizing at the carrying capacity around which it fluctuates with small stochastic components. The second case occurs when species undergo high adult mortality resulting in negative growth rates and an approximately exponential decrease. Also, deterministic factors can sometimes cause fluctuations and irregular oscillations that at first glance may seem completely stochastic. This is the phenomenon of deterministic chaos, which is linked to a strong density dependence (overcompensation). However, in populations with no age structure, seldom are intrinsic rates of demographic increase so large as to cause large chaotic fluctuations. In populations with long life expectations, one may expect chaotic fluctuations for lower intrinsic growth rates, which are, however, generally higher than the real ones. Therefore, when irregular fluctuations of plant and animal population numbers are observed, it is reasonable to assume that in most cases, these fluctuations are linked to truly stochastic factors rather than deterministic chaos.

As for the mechanisms generating these random fluctuations, we can distinguish two main types of stochasticity. The first is the so-called demographic stochasticity, which depends on random events operating at the level of a single individual mortality and reproduction. Like genetic drift, it acts with particular strength in small populations. The second type is environmental stochasticity, which depends on random events operating at the level of the environment in which populations live. This kind of stochasticity operates effectively and in a comparable way in both small and large populations.

To subsume the main features of demographic and environmental stochasticity, we describe here only simple cases, neglecting age and size structure. For the sake of simplicity, assume that, if the population reproduces sexually, the sex ratio is constant, so that it is sufficient to consider the dynamics of females. We will also assume that, even if the reproduction is annual, a continuous-time model can be used as a first approximation. So the baseline model for our consideration is as follows:

$$N(t+dt) = N(t) + r(t)dtN(t),$$

where $N(t)$ represents the number of adult females at time t (or the total population size for asexual species) and $r(t)$ is the rate of instantaneous demographic growth at time t. Note that the rate of population increase can vary with time, which explains the notation $r(t)$. The drivers of variations may be diverse: density dependence (i.e., the fact that the rate may vary with $N(t)$), demographic and/or environmental stochasticity, or interactions with other populations.

To discuss stochasticity from a quantitative viewpoint, one must preliminarily consider that the rate of demographic growth $r(t)$ is actually the average of the individual contributions of each individual to the reproductive output at time t and to the survival from time t to time $t + dt$. The contribution of the ith female (or asexual organism) to the change of population abundance between time t and time $t + dt$ is termed individual fitness. Rather than resorting to more sophisticated theoretical definitions, it is perhaps more appropriate to address the concept of individual fitness through a simplistic example. One of many unicellular organisms, say, the ith belonging to the considered population, in the time interval $[t, t + dt]$ may reproduce by fission, or may not reproduce but survive, or may not reproduce and die. The fitness is thus 2 or 1 or 0, respectively. If we denote by $w(i,t)$ the fitness of the ith organism at time t, we can then state that

$$N(t+dt) = \sum_{i=1}^{N(t)} w(i,t).$$

It is more convenient, however, in a continuous model like ours, to consider the contribution of each individual to the instantaneous rate of increase by setting $w(i,t) = 1 + r(i,t)dt$, where $r(i,t)dt$ can take on the value 1, 0, or -1. In this way, we can state that $r(t) = (1/N_t)\sum_{i=1}^{N(t)} r(i,t)$, thus establishing a strong link between the rate of population growth and individual fitnesses. It is also instructive to break down each stochastic variable $r(i,t)$ into the sum of two terms. The first term is the expected value of the contribution to the demographic rate of increase of the ith organism (say, $\bar{r}$), while the second term, δ, is a deviation from the mean. It is almost always reasonable to assume that, while the expected value of the fitness depends on season t, the deviation from the average value depends only on the characteristics of each individual. We thus write $r(i,t) = \bar{r}(t) + \delta_i$, both stochastic variables. By definition, $E[\delta_i] = 0$. Basically, the individual contribution to the rate of demographic increase is the sum of a random variable that reflects the effect of environmental stochasticity ($\bar{r}(t)$) and one that accounts instead for demographic stochasticity (δ_i). The contributions of the two different sources of stochasticity to the overall growth rate can be calculated as

$$r(t) = \frac{1}{N(t)} \sum_{i=1}^{N(t)} r(i,t) = \bar{r}(t) + \frac{\sum_{i=1}^{N(t)} \delta_i}{N(t)}.$$

In conclusion, the random variable $r(t)$, which is the average of individual contributions at time t, is the sum of two random variables, one depending on the year (environmental stochasticity) and the other depending on the variability between individuals (demographic stochasticity). Actually, $\bar{r}(t)$ is a stochastic process, while δ is simply characterized by its probability distribution. It is reasonable to assume that the two random variables are independent. The simplest statistical properties of $r(t)$ can be obtained by assuming that environmental stochasticity is a process without a trend, that is, stationary. In other words, we assume that the mean and variance of the process $\bar{r}(t)$ are independent of time, that is, $E[\bar{r}(t)] = \bar{\bar{r}} =$ constant and $Var[\bar{r}(t)] = \sigma_e^2 =$ constant. If we denote by σ_d^2 the demographic variance, that is, $Var[\delta]$ and consider the average and the variance of $r(t)$ conditional on the population size $N(t)$ at time t, then we directly obtain

$$E[r(t)] = \bar{\bar{r}}$$

$$Var[r(t)] = \sigma_e^2 + \frac{\sigma_d^2}{N(t)}.$$

Therefore, demographic stochasticity influences the variance of the growth rate through the factor $1/N(t)$ and therefore becomes immaterial for large populations. On the other hand, many data show that often the demographic variance is much larger than the environmental variance, so that for small populations, demographic stochasticity prevails. In this a case, a common approximation is the following stochastic logistic equation:

$$\frac{dN}{dt} = rN\left(1 - \frac{N}{K}\right) + \sigma\eta N,$$

where η is a standard white noise and the assumption is made that the intrinsic rate of increase r is affected by a stochasticity with standard deviation σ. If we suppose that this noisy term is the result of demographic stochasticity, then $\sigma = \sigma_d/\sqrt{N}$, and we finally get

$$\frac{dN}{dt} = rN\left(1 - \frac{N}{K}\right) + \sigma_d\eta\sqrt{N},$$

which we shall use in the following.

A further note concerns the fact that time-discrete models of invasion speeds in fluctuating environments have a long and distinguished history in ecology. In fact, our ability to predict rates of spread is often limited by the assumption that reproduction is continuous in time and the spreading environment is temporally constant. However, this often proves unrealistic [335]. Here, we briefly examine a time-discrete model of the consequences for invasion speed of periodic or stochastic fluctuations in population growth rates and in dispersal distributions, following [335, 339]. It was posited there that, in the face of environmental fluctuations, invasions may be modeled by discrete-time, continuous-space integro-differential equation models (Box 3.12), which have a long history of application in a number of disciplines, including epidemiology [335].

Box 3.12 Integrodifferential Models of Biological Invasions

The simplest nonlinear integrodifferential (IDE) model predicts the population density at site x, $n_{t+1}(x)$, at the $(t+1)$th generation, given the densities attained in the previous generations. Ordinarily, the basic IDE model takes the form

$$n_{t+1}(x) = \int_{-\infty}^{+\infty} k(y-x)\, F[n_t(y)]\, dy \qquad (3.28)$$

where the functions k and F can be understood as the descriptors of two temporally distinct processes: growth and dispersal [335]. In a first stage, growth occurs during a sedentary spell modeled by a nonlinear map like the compensatory model [335]:

$$F[n_t(x)] = \frac{\lambda n_t(x)}{1 + \lambda n_t(x)}, \quad \lambda > 0.$$

In the second stage, progeny disperse. The dispersal kernel $k(x)$ is, as usual, the probability density function for the distance that propagules attain by any mobility means (active or passive). The convolution operator in Equation (3.28) tallies the movement of progeny from all sites y of the domain to the current site x. As seen in Chapter 1, thin- or fat-tailed dispersal kernels (i.e., kernels with exponentially bounded tails or others characterized by algebraic decays) may be employed, possibly derived from mechanistic models of the dispersive process.

A typical solution of the IDE (3.28), fitted with an initial condition restricted to a finite portion of the 1D space, grows and spreads, eventually converging to a traveling wave endowed with constant speed c, valid for $\lambda > 1$, defined by [335]

$$c_{\min} = \min_{s \in S}\left\{\frac{1}{s}\log[\lambda\, m(s)]\right\}, \qquad (3.29)$$

where $\lambda = F'(0)$ is the population growth rate at low population density and $m(s) = \int_{-\infty}^{+\infty} k(x)\, \exp(sx)\, dx$ is the moment-generating function of the kernel $k(x)$. $S = [0, s_{\max}]$ is

Box 3.12 *Continued*

the set of all values of $s > 0$ for which the integral converges [335]. Specifically, the validity of Equation (3.29) is restricted to the case where F does not exhibit Allee effects (i.e., $0 \leq F[n] \leq \lambda n$ for $n \geq 0$). It is remarkable, in a development akin to the original Fisher–Kolmorogov approach, that Equation (3.28) gives the speed of the slowest nonnegative traveling wave solution

$$n_t(x) \propto e^{-s_{\min}(x - c_{\min}t)}$$

by linearizing it about $n = 0$, that is,

$$n_{t+1}(x) = \lambda \int_{-\infty}^{+\infty} k(y - x)\, n_t(y)\, dy. \tag{3.30}$$

The linear equation (3.30) has been extensively used to derive speeds of invasion in fluctuating environments [335]. Numerical simulations have shown that it provides a good approximation to the actual asymptotic speed. In simulations, the speed of invasion is measured by finding the location farthest from the origin with a population density larger than a critical density n_{cr} and determining how this location, say, x_t, changes with time [335]. In a constant environment the shape of the invasion wave does not change, so this operational definition applies as long as the critical density remains well below the carrying capacity. In fluctuating environments, the subject of this section, the shape of the wave at high population densities changes between generations, thereby requiring that $n_{cr} \ll 1$.

The model equation (3.28) can incorporate environmental fluctuations by making the population growth rate λ and the dispersal kernel $k(x)$ functions of time, thus obtaining [335]

$$n_{t+1}(x) = \lambda_t \int_{-\infty}^{+\infty} k_t(y - x)\, F[n_t(y)]\, dy, \tag{3.31}$$

with usual symbol notation. Periodic and stochastic environments have thus been studied.

To mimic the effects of a seasonal environment, a periodic variation in the intrinsic finite population growth rate λ_t, or in the parameters of a suitable dispersal kernel $k(x)$, or in both, Figure 3.23 shows a typical solution to model (3.31) within an environment characterized by a period of two years. A population initially concentrated at the origin evolves into a spreading wave that alternately advances and retreats at a fixed distance for each generation [335]. At every other generation, the solution resembles a traveling wave in a constant environment, having developed a constant shape moving with constant speed, in a traveling two-cycle wave. In this model, p-cycles develop whenever the environment fluctuates with period p [335].

Real environmental fluctuations ordinarily include both stochastic and periodic components. Environmental stochasticity can be incorporated into the framework outlined in Box 3.12 by randomly choosing the growth rates and dispersal kernels from a set of ecologically viable choices [335]. Suffice here to notice that the tractability of the problem stems, in such a case, from the fact that the kernels become independent and identically distributed (iid) random dispersal functions, and the growth rates become iid random variables independent of the kernels.

The framework outlined above has been used to tackle a number of ecological invasion problems [335].

3.4.2 Demographic Stochasticity and the Fisher–Kolmogorov Deterministic Model of Invasions

Giometto et al. [289] provided replicated and controlled experimental support to the theory of reaction–diffusion processes for modeling biological dispersal in a generalized context that reproduces observed fluctuations due to demographic stochasticity (Figure 3.24). In these

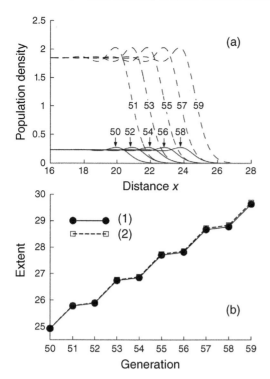

Figure 3.23 (a) Simulation of model equation (3.31) with a periodic shift of period 2 years and overcompensatory growth function $F = \lambda_t n_t \exp(-n_t)$ with a specific choice of kernel parameters (see [335]). (b) Predictions provided by the approximate analytical model equation (3.31) (2) (i.e., that the wave will advance even after a bad year) agree with numerical simulations of Equation (3.30) (1) (after [335]).

experiments, the freshwater ciliate *Tetrahymena* sp. was employed because of its short generation time and its history as a model system in ecology [12, 14]. The experimental setup consisted of linear landscapes (shown in the exemplifying Figure 3.2), filled with a nutrient medium, kept in constant environmental conditions and of suitable size to meet the assumptions about the relevant dispersal timescales (see also "Materials and Methods" in [289]). Replicated dispersal events were conducted by introducing an ensemble of individuals at one end of the landscape and measuring density profiles throughout the system at different times. Density profiles are shown in Figure 3.24 for six replicated dispersal events. Organisms introduced at one end of the landscape were observed to rapidly form an advancing front that propagated at a remarkably constant speed. The front position at each time was calculated as the first occurrence, scanning from the farthest end of the land-

scape, of a fixed threshold value of the measured density of organisms.

Traveling waves predicted by the deterministic Fisher–Kolmogorov equation imply a constant invasion speed. The mean front speed in the experimental replicas of [289] proves notably constant for different choices of the threshold density value chosen (Figure 3.24). What is remarkable is that demographic stochasticity shows up clearly, and the species' traits, r, K, and D, used to provide a baseline for the numerical solution of a stochastic version of the Fisher–Kolmogorov equation suited to reproduce demographic stochasticity (Equation (3.32)) were measured by independent, offline experiments on cultures reproducing temperature and the general conditions faced by the invading ciliates. In fact, in the offline local growth experiment, a low-density population of *Tetrahymena* sp. was introduced evenly across the landscape, and its density was measured locally at different times. Recorded density measurements were fitted to the logistic growth model (Section 3.1.1) to provide estimates for r and K [289]. In the offline local movement experiments, the time evolution of the mean square displacement msd of individuals' trajectories was computed to estimate the diffusion coefficient D in density-independent conditions, which proved akin to a diffusive model of linear growth in time of the msd [289] (see Section 3.5.1). Both the growth and the movement measurements were performed in the same linear landscape settings as in the dispersal experiment. Implicitly, one assumes therefore that demographic stochasticity in growth can be decoupled from diffusive effects shown by the ciliates' movements.

Giometto et al. [289] experimentally substantiated the Fisher–Kolmogorov prediction by including demographic stochasticity in the model to reproduce the observed variability in range expansions. Investigation of the movement behavior of *Tetrahymena* sp. shows that individuals' trajectories are consistent with a persistent random walk with an autocorrelation time $\tau = 3.9 \pm 0.4$ s. As the autocorrelation time for the study species is much smaller than the growth rate r ($\tau r \sim 10^{-4}$), an excellent approximation was provided [289]. The stochastic model equation reads

$$\frac{\partial \rho}{\partial t} = D \frac{\partial^2 \rho}{\partial x^2} + r\rho \left[1 - \frac{\rho}{K}\right] + \sigma \sqrt{\rho}\, \eta, \quad (3.32)$$

where $\rho(x,t)$ is the organisms' density [number L^{-1}]; $\eta = \eta(x,t)$ is a Gaussian, zero-mean white noise (i.e., with correlations $\langle \eta(x,t)\eta(x',t')\rangle = \delta(x - x')\delta(t - t')$, where $\delta(\cdot)$ is the Dirac's delta distribution); and $\sigma > 0$ is

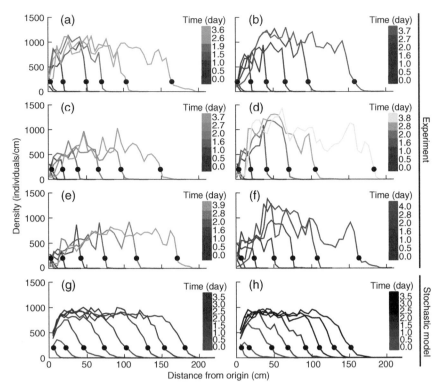

Figure 3.24 Density profiles of *Tetrahymena* sp. in the dispersal experiment and in the stochastic model. (a–f) Density profiles of six replicated experimentally measured dispersal events, at different times. Legends link each color to the corresponding measuring time. Black dots are the estimates of the front position at each time point. Organisms were introduced at the origin and subsequently colonized the whole landscape in 4 days (>20 generations). (g, h) Two dispersal events simulated according to the generalized model equation (1.32), with initial conditions as at the second experimental time point. Data are binned in 5-cm intervals, the typical length scale of the process. Figure after [289]

constant. Itô's stochastic calculus was adopted [289]. Note, in fact, that the choice of the Stratonovich framework would make no sense here, as the noise term would have a constant nonzero mean, which would allow an extinct population possibly to escape the zero-density absorbing state [289]. The square-root multiplicative noise term in Equation (3.32) is commonly interpreted as describing demographic stochasticity in a population and needs extra care in simulations. In particular, standard stochastic integration schemes fail to preserve the positivity of ρ. A split-step method to numerically integrate Equation (3.32) was employed [289].

These parameters inserted in the noisy Fisher–Kolmogorov equation (Equation (3.32)), which is solved numerically [289], accurately describe the dynamics at the front of the traveling wave in the dispersal events (Figures 3.24g,h). The comparison of the predicted front speed $v = 2\sqrt{rD}$ to the wave front speed measured in the dispersal experiment, v_O, yields a compelling agreement.

The observed speed in the dispersal experiment was $v_O = 52.0 \pm 1.8$ [cm/day] (mean±SE) [289], which we compare to the predicted one, $v = 51.9 \pm 1.1$ [cm/day] (mean ± SE). The two velocities are compatible within 1 standard error. A t-test between the replicated observed speeds and bootstrap estimates of $v = 2\sqrt{rD}$ gives a p-value of $p = 0.96$ ($t = 0.05$, $df = 9$). Thus, the null hypothesis that the mean difference is 0 was not rejected at the 5 percent level, and there is no indication that the two means are different. As the measurements of r and D were performed in independent experiments at scales that were orders of magnitude smaller than in the actual dispersal events, the agreement between the two estimates of the front velocity is deemed remarkable. A summary of the relevant results is shown in Figure 3.25 (after [289]). Therein, the range expansions measured in the dispersal experiments shown in Figure 3.24 are contrasted with the results of the stochastic model run with the parameters measured independently. The front

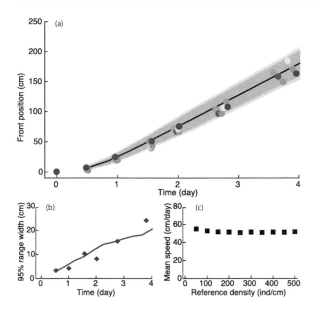

Figure 3.25 Range expansion in the dispersal experiment shown in Figure 3.24 and by the stochastic model equation (3.32). (a) Front position of the expanding population in six replicated dispersal events; colors identify replicas as in Figure 3.24. The dark and light gray shadings are, respectively, the 95 percent and 99 percent confidence intervals computed by numerically integrating the generalized model equation (3.32), with initial conditions as at the second experimental time point, in 1,020 iterations. The black curve is the mean front position in the stochastic integrations. (b) The increase in range variability between replicates in the dispersal experiment (blue diamonds) is well described by the stochastic model (red line). (c) Mean front speed for different choices of the reference density value at which we estimated the front position in the experiment; error bars are smaller than symbols. Figure after [289]

position of the expanding population is shown for the replicated dispersal events at 95 percent and 99 percent confidence intervals. Also shown are the increase in range variability between replicates in the dispersal experiment and the mean front speed for different choices of the reference density value at which we estimated the front position in the experiment. Emerging predictable features of replicated biological invasion fronts are thus shown.

The conclusions of this study are remarkable in that the Fisher–Kolmogorov equation correctly predicts the mean speed of the experimentally observed invading wave front, although its deterministic formulation prevents it from reproducing the variability inherent to biological dispersal. In particular, it cannot reproduce the fluctuations in range expansion between different replicas of the dispersal experiment. The generalization of the Fisher–Kolmogorov equation accounting for demographic stochasticity shows its ability to capture the observed variability as reported in Figure 3.25. The strength of demographic stochasticity is embedded in an additional species' trait σ $[T^{-1/2}]$. In the stochastic framework of Giometto et al. [289], the demographic parameters r, K, and σ were estimated from the local growth experiment, while the estimates of the diffusion coefficient D were obtained from the time evolution of the mean square displacement (*msd*) of individuals' trajectories. Local, independent estimates were thus used to numerically integrate the generalized model equation (3.32) with initial conditions as in the dispersal experiment, and it was found that the measured front positions

are in accordance with simulations. The estimates for the front speed and its variability in the experiment proved in good agreement with simulations – and thus demographic stochasticity can explain the observed variability in range expansion [289]. This result suggests that measuring and suitably interpreting local processes still allows us to accurately predict the bulk features of biological invasions. However, if, in addition, one has tools to characterize the stochasticity inherent in the relevant biological processes, the mean and the variability of range expansions can be defined. This is of obvious ecological interest, in particular for foreseeing worst-case scenarios for the spread of invasive species. Details on the movement behavior, biology, or any other ecological information are subsumed by three parameters describing the density-independent yet stochastic behavior of individuals riding the invasion wave.

Generalizations of the framework described here to organisms endowed with different biology (e.g., different ranges of growth rates and/or diffusion coefficients) are thus in sight [289, 340]. Other avenues of study exist in this field. However, we shall not review here a large existing body of literature (most often too mathematical for an attempt to reproduce the leading results in a book aimed at ecologists and hydrologists) that studied other effects, perhaps of lesser interest to fluvial ecological corridors. This is the case, for instance, of accelerating traveling fronts, that is, where the wave front remains substantially undeformed but the speed changes owing to alongstream alterations (see, e.g., [341–346]).

3.4.3 Slowing Invasion Speed by Heterogeneous Environments

Studies of species spread in heterogeneous linear landscapes have traditionally characterized invasion velocities as a function of some mean resource density throughout the invasion substrate (or, more generally, any property that affects reproductive rates r that feature in the homogeneous Fisher–Kolmogorov speed, $c \propto \sqrt{r}$). Higher-order moments of the spatial resource distribution are thus neglected. One wonders whether this is the case regardless of the degree of irregularity of the resource distribution. One also wonders whether other kinds of heterogeneity, epitomized by the presence of branched tributaries typical of any fluvial downstream journey, matter for the determination of the speed of traveling waves describing a biological invasion.

In this section, we specifically address the issue of determining the speed of the invasion fronts in spatially heterogeneous environments. We shall show through a number of relevant cases that heterogeneity affects decisively the front speeds, whether because of the effects of branching structures faced along the propagation backbone within a river network or because of resource autocorrelations – both are shown here to cause a reduction in the speed of traveling waves of species spread. Demographic stochasticity plays a role in the slowdown, as do other effects strengthened when individuals can actively move toward resources or enhance shear flow dispersion (i.e., when the process is asymptotically diffusive in the longitudinal dispersion and well mixed in the transverse directions and the dispersion coefficient depends on the strength of the small-scale mobility of individual organisms, which controls how fast they sample the heterogeneities of the advective velocity profiles; Section 3.5.3). Experimental work is shown to corroborate theoretically predicted reductions in propagation speeds. This was done, in particular, in microcosm laboratory experiments with the phototactic protist *Euglena gracilis*, where comparing invasions in linear landscapes endowed with different resource autocorrelation lengths was possible via the use of controlled light fields. The work presented here overviews various sources of heterogeneity; in particular, it identifies the key modulators and a few simple measures of landscape susceptibility to biological invasions. We thus support the view, based on a number of examples, that certain environmental attributes need to be considered for predicting invasion dynamics within naturally heterogeneous fluvial ecological corridors.

3.4.4 Biased Reaction–Diffusion Wave Fronts along Fractal Networks

The model of human migration described in Section 1.5 [118] assumes simple diffusive transport to purportedly describe migration fluxes. This indeed seems reasonable in the case of historic human population migrations – the need for water resources should have driven settlers during historic human migrations regardless of the direction of the flow. For species that live within fluvial ecological corridors, however, flow direction and its intensity, whether subsumed by the average flow velocity or by average shear stress (often assumed as the basis for suitability of a number of aquatic species), matter decisively. Bertuzzo et al. [7] investigated whether adding a bias to transport properties would have basically altered this framework.

This was done on purpose: in fact, any other ecological agent (be it an aquatic organism or an infective agent of waterborne disease) would likely be affected by the flow direction to propagate within the network. Organisms can either move by their own energy (active dispersal) or be moved by water (passive dispersal). Most likely, movements along the flow direction would be favored, although movements against the flow direction are completely admissible because of various ecological or physical mechanisms [40, 347]. This turns out to be of great interest for the problem of hydrochory, that is, the transport of species along the ecological corridors that are shaped by the river network itself.

3.4.5 Biased Random Walks on Fractals

Here, we illustrate, and adapt to the problem at hand, the discrete time-space biased random walk process through an oriented graph constituted by edges of equal length. This analysis will be useful in the next section, where we add the reaction term and study the properties of a reactive random walk (RRW) on oriented graphs (Section 1.5). Recapping, an oriented graph is a directed graph having no symmetric pair of directed edges. At every time step, τ, a walker can move with some probability from a node to one of the adjacent nodes, which are all the nodes that are connected to it through an inward or outward edge.

Consider the case of a graph in which every node has only one inward and one outward edge (i.e., a one-dimensional lattice). We define as P_{out} (P_{in}) the probability that a particle moves from a node to another along an outward (inward) edge. We analyze the case in which all particles move at every time step, hence $P_{out} + P_{in} = 1$. In greater generality, for a random walk process on a generic oriented graph in which every node can have an arbitrary number of inward and outward edges, we assume that a particle can move (following either an outward or inward edge) with a probability proportional to P_{out} and P_{in}, respectively. In this case, the probability P_{ij} for a particle to jump from node i to one of its neighbors j could be expressed as

$$P_{ij} = \begin{cases} \dfrac{P_{out}}{d_{out}(i)P_{out} + d_{in}(i)P_{in}} & \text{if } i \to j \\ \dfrac{P_{in}}{d_{out}(i)P_{out} + d_{in}(i)P_{in}} & \text{if } i \leftarrow j, \end{cases} \quad (3.33)$$

where $d_{out}(i)$ and $d_{in}(i)$ are, respectively, the outdegree and indegree of node i (i.e., the number of outward or inward edges of node, respectively). Obviously $\sum_{j=1}^{d(i)} P_{ij} = 1$, where $d(i) = d_{out}(i) + d_{in}(i)$ is the total degree of node i. We define $b = P_{out} - P_{in} = 2P_{out} - 1$ as the bias of the transport.

We particularize the results of Equation (3.33) to the case of the Peano network (whose construction and peculiar recurrent features were described in Chapter 1), because its deterministic spanning tree structure allows the derivation of exact results. Figure 3.26a shows the Peano basin at the third level of its construction process. Owing to the purported topological similarity between the Peano basin and a real river basin [7, 126, 127], we choose to assign to each edge of the graph a direction from the leaves to the root (namely, the outlet: point B in Figure 3.26a). In this way, we can also define cumulative flow through aggregated area, say, the number of upstream links connected to the current node. The nodes of this graph could be classified, on the basis of their total degree, in to first- and fourth-degree nodes. Every fourth-degree node has three inward edges and one outward edge. In the following, we define

$$P_+ = \frac{P_{out}}{P_{out} + 3P_{in}}, \quad P_- = \frac{P_{in}}{P_{out} + 3P_{in}}, \quad (3.34)$$

the probabilities that a particle, starting from a fourth-degree node, jumps to its downstream neighbor and the probability that it jumps to one of its three upstream nodes, respectively. The expressions for P_+ and P_- derive straightforwardly from Equation (3.33). Also, a particle starting from a first-degree node jumps to its downstream node with unit probability.

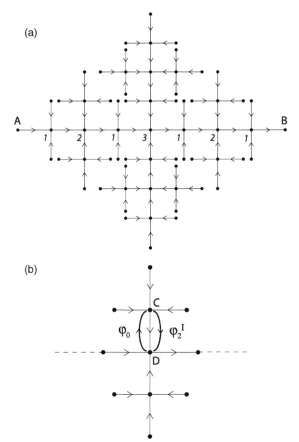

Figure 3.26 (a) Peano fractal basin [128] at the third level of its iterative construction process. The arrows show the edge direction. The line indicated as AB is the backbone of the network. The numbers indicate the order of the secondary branches emerging from the backbone. (b) A pair of symmetric second-order branches, typical of the bifurcation structure of the Peano construct. The function φ_2^I is the waiting time distribution for jumping from node C to node. The function φ_0 is the constant waiting time distribution for jumping backward from any node. Redrawn from [7]

3.4.6 Reaction Random Walks on Oriented Graphs

We focus on the analysis of a biased reaction random walk (RRW) process through a Peano network. At every time step τ, a particle moves from a node to one of its neighbors following the rules laid out in Equation (3.34). During the waiting time τ, the particle density at every node is assumed to grow following the logistic equation with intrinsic growth rate a and unit carrying capacity. If we observe the process only at the nodes of the backbone (line AB in Figure 3.26a), the secondary branches emerging from one of these nodes yield a waiting time distribution of jumping to adjacent nodes of the backbone that depends on the structure of the branches. The process along the backbone can be handled like a one-dimensional reactive continuous time random walk by a suitable discretization in time and space [118]. The Hamilton–Jacobi formalism [129, 130], described in the framework of the continuous-time random walk (CTRW) model (Box 1.7), allows the analytical computation of the speed of the traveling wave generated by a CTRW, with a density-dependent term of reaction. We first present here the general case in which the distribution of jump lengths, $\Phi(x)$, and the waiting time distribution, $\varphi(t)$, are continuous and independent; then, in the second part of this section, we particularize to the case at hand using distributions suitable for describing discrete space-time random walks on networks.

The method (Eqs. (1.30) and (1.31) in Section 1.5, Box 1.7) has been applied to the study of a biased RRW through a Peano drainage network [7]. To derive the speed of the front, v, we first derive the expression of the waiting time distributions due to the emerging branches of different order N. We remind the reader that the order N has the following properties [63]. The cardinality of the subtree rooted in the emerging branch is 4^{N-1}. A particle at a backbone node from which two first-order branches emerge can reach an adjacent backbone node after $2i + 1$ jumps, with $i = 0, 1, \ldots, \infty$, namely, after waiting a time $(2i + 1)\tau$, where i is the number of walks away from the backbone and back. The probability that a particle waits for a time t along one of infinite possible paths is a convolution of $(2i + 1)$ distributions:

$$\varphi_1^{(i)}(t) = (P_+ + P_-) \varphi_0 * (2P_- \varphi_0 * \varphi_0)$$
$$* (2P_- \varphi_0 * \varphi_0) * \cdots, \qquad (3.35)$$

with i convolution terms of the type $(2P_- \varphi_0 * \varphi_0)$. The function $\varphi_0(t) = \delta(t - \tau)$ is the waiting time distribution for the original random walk through the entire network. Note that $2P_-$ is the probability of stepping into one of the two emerging branches, whereas $(P_+ + P_-)$ is the probability of the last jump being to the left or to the right. Finally, the waiting time distribution for all possible paths is given by the sum of all the waiting time distributions of each single path weighted by the path probabilities. For a first-order branch, the Laplace transform of the waiting time distribution is

$$\hat{\varphi}_1(s) = \sum_{i=0}^{\infty} \hat{\varphi}_1^{(i)}(s) = (P_+ + P_-) \hat{\varphi}_0 \ (2P_- \hat{\varphi}_0^2)^i$$
$$= \frac{(P_+ + P_-)\hat{\varphi}_0}{1 - 2P_- \hat{\varphi}_0^2}, \qquad (3.36)$$

where $\hat{\varphi}_0(s) = \exp(-\tau s)$ is the Laplace transform of $\varphi_0(t)$.

Following rules already discussed in other contexts [134], the waiting time distribution for the backbone nodes from which a pair of branches of higher order emerges can be obtained exactly. We illustrate here the case of second-order branches. We first derive the waiting time distribution $\varphi_2^I(t)$ of jumping from the node adjacent to the backbone (point C in Figure 3.26b) to the backbone (point D) induced by the three first-order edges connected to it. Following the procedure and the rules illustrated before, the Laplace transform of this distribution becomes

$$\hat{\varphi}_2^I(s) = P_+ \hat{\varphi}_0 \sum_{i=0}^{\infty} (3P_- \hat{\varphi}_0^2)^i = \frac{P_+ \hat{\varphi}_0}{1 - 3P_- \hat{\varphi}_0^2}. \qquad (3.37)$$

Using the distribution in Equation (3.37), it is possible to derive the Laplace transform of the waiting time distribution induced by a pair of second-order branches as illustrated before:

$$\hat{\varphi}_2(s) = (P_+ + P_-) \hat{\varphi}_0 \sum_{i=0}^{\infty} (2P_- \hat{\varphi}_0 \hat{\varphi}_2^I)^i$$
$$= (P_+ + P_-) \hat{\varphi}_0 \frac{1 - 3P_- \hat{\varphi}_0^2}{1 - P_- \hat{\varphi}_0^2 (2P_+ + 3)}. \qquad (3.38)$$

With the above method, we can calculate the Laplace transform of the waiting time distributions up to fifth-order branches. The complexity of the expression $\hat{\varphi}_N(s)$ for $N > 5$ (N being the branch order) prevents us from using this method any further. Note that the Hamilton–Jacobi method assumes the waiting time distribution $\varphi(t)$ to be space invariant. Following previous suggestions [118, 131], we approximate the geometry of the system: specifically, we assume that all the branches encroaching on the backbone have the same waiting time distribution given by $\varphi(t) = \varphi_5(t)$. Later, the validity of this approximation will be tested numerically.

The length distribution for the jumps of a particle moving along the backbone of the network is given by

$$\Phi(x) = P_{out} \delta(x - \Delta x) + P_{in} \delta(x + \Delta x), \qquad (3.39)$$

where x increases toward the graph root and Δx is the constant length of the graph edges. Note that Equation

(3.39) allows us to apply the CTRW framework, which is usually employed to study continuous systems, to a discrete lattice process. Applying the usual transformation to Equation (3.39), one obtains

$$\bar{\Phi}(p) = P_{out}\, e^{p\,\Delta x} + P_{in}\, e^{-p\,\Delta x}. \tag{3.40}$$

Substituting Equation (3.40) into Equation (1.31) (Box 1.7) and solving it for the variable p, we can particularize the expression in Equation (1.30) to the case of the oriented Peano graph:

$$v = \min_{s} \frac{s}{\ln\left[\dfrac{\sqrt{c(s)+(c(s))^2-4P_{out}P_{in}}}{2P_{out}}\right]}, \tag{3.41}$$

where the function $c(s)$ has the following expression:

$$c(s) = \frac{1}{\hat{\varphi}(s)} - \frac{a}{s}\left(\frac{1}{\hat{\varphi}(s)} - 1\right). \tag{3.42}$$

The minimum in Equation (3.41) is then computed numerically.

Results for the Peano network are shown by solid lines in Figure 3.27 (redrawn from [7]), where the dimensionless speed of the front $v\tau/\Delta x$ is reported as a function of the dimensionless growth rate $a\tau$ for different values of the bias b. Note that for $b = 0$ ($P_{out} = P_{in} = 1/2$), we recover the case of the unbiased process already obtained [118]. For $b > 0$ (i.e., the walkers move preferably downstream), the speed of the front increases. However, one should mark the difference with the simple Fisher–Kolmogorov model with advection-diffusion-reaction [7]: the speed of the front is not given simply by the sum of the advection velocity in the backbone $u = (b\Delta x/\tau)$ plus the speed of the front of the unbiased process (i.e., with $b = 0$). In fact, there are two important new phenomena accounted for by the biased RRW model: (1) the front speed cannot physically exceed the particle velocity $\Delta x/\tau$ and (2)

the bias affects the transport not only on the backbone but also in the secondary branches, thus affecting the waiting time distribution $\varphi(t)$. The results shown in Figure 3.27 are obtained with $\varphi(t) = \varphi_5(t)$ by taking into account branches only up to the fifth order. However, as already noticed [118], the speed of the front converges rapidly as the order of the branches increases, and then it proves not particulary sensitive to the details of the self-similarity at all scales ($N \rightarrow \infty$) typical of fractal structures like the Peano network (which at times we term a "basin" to mark its directed graph character). Note, again from Figure 3.27, that the trivial result $v = \Delta x/\tau$ for $b = 1$ is recovered as the upper bound for the front speed.

Following [7], we now illustrate the results of extensive simulations of the biased RRW process on the structure of the Peano network [121, 128] and on real river networks, that is, the substrate for ecological interactions used earlier (Section 1.5). The state variable is the density of individuals $\rho(x,t)$ as in Section 1.5, one-dimensional in space along each network link. We start every simulation from an initial condition cast with $\rho = 1$ for the node (say, A) and its neighbor (see Figure 3.26a) and $\rho = 0$ for all the other nodes. At every time step τ, we first update the density ρ of all nodes through the master equation that obeys the rules of Equations (3.33) and (3.34), and then we let the node density grow for a time τ following the solution of the local logistic equation $d\rho/dt = a\rho(1 - \rho)$:

$$\rho(t + \tau) = \frac{\rho(t)}{\rho(t) + (1 - \rho(t))\, e^{-a\tau}}. \tag{3.43}$$

We observe the density only at the backbone nodes and start measuring the front speed only when it has reached a stable waveform that moves from A to B. In accord with theory [120], it was found that the front travels

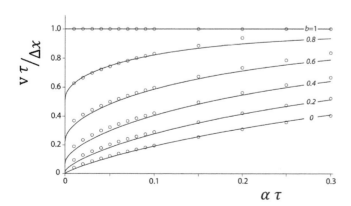

Figure 3.27 Dimensionless wave front speed ($v\tau/\Delta x$) obtained by the Hamilton–Jacobi method (lines) and by the analogous numerical simulations (circles) as a function of the dimensionless growth rate $a\tau$. Different lines from the bottom to the top refer to values of the bias b from 0 to 1 with step 0.2 (shown as insets). Redrawn from [7]

with constant speed along the backbone, maintaining its shape. Obviously, the front speed is not affected by the order of growth and movement. In our simulations, we recover the same results when the particles first grow and then move. Note that it is not strictly necessary to consider the process along the backbone. In fact, owing to the exact self-similarity of the Peano construct, one obtains the same results by assigning an initial condition (with compact support) at any site and following the process along the drainage path connecting that site to the outlet (point B in Figure 3.26a). In fact, by further simulation, we find that the front speed measured along any drainage path is the same, provided that the chosen path is long enough to develop a stable traveling wave. The relative independence of the characters of a traveling wave from the particular flowpath is somewhat reasonable, in retrospect, because the front speed along a path depends only on the sequence of the orders of secondary branches encountered along the path, which, for the Peano network, is the same by construction. Note that this is not an unreasonable assumption for real rivers [63], partly explaining the importance of the insight derived from the study of deterministic fractals.

Four sketches of the simulation of the colonization process are shown in Figure 3.28a for two values of the bias ($b = 0$ and $b = 0.4$, respectively) and for two different simulation times. Just from observing these sketches, one notes how the bias stretches the colonization cloud enhancing the front speed. Figure 3.28b shows the computed density at backbone nodes for three different time steps; a front can be recognized that travels with constant speed, maintaining its shape quite nicely. The continuous flux of particles from the secondary branches into the backbone may indeed lead to locally stable states with $\rho > 1$. Different values of the stable part of the front (peaks and troughs in Figure 3.28b) depend on different orders of the secondary branches flowing into the backbone.

Figure 3.27 also shows a comparison between numerical (circles) and analytical (solid line) results. The exercise goes beyond the trivial numerical control. In fact, while the reaction process is applied at every node in the simulations, the analytical Hamilton–Jacobi solution assumes that all the dispersing particles are concentrated in the backbone nodes, and thus the branches affect waiting time, not particle density. It is heartening that, despite this difference, the simulations exhibit good agreement with the analytical results for a wide range of values of the growth rate $a\tau$ [7].

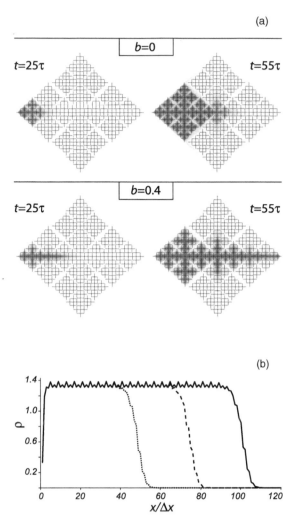

Figure 3.28 (a) Sketches of the numerical simulation of the colonization process through a fifth-order Peano basin. The gray color shading is obtained by space interpolation of the node density. The dimensionless growth rate employed is $a\tau = 0.5$; all other parameters are reported in the figure. (b) Traveling front of the density wave along the backbone. The different profiles are taken at the following time steps: 100 (dots), 150 (dash), and 200 (solid). The other parameters employed are $a\tau = 0.3$ and $b = 0.2$. Redrawn after [7]

We have also studied front speed propagation throughout the geometry of real river networks. Figure 3.29a shows the drainage network of the Tanaro river basin, a 8,000-km² catchment in northwestern Italy, extracted via suitable geomorphological criteria [63]. Initially, we only accounted for its topological structure, thereby neglecting its geometrical properties, such as edge lengths (i.e., we assume that all the edges have the same length Δx), to investigate how departures from the topological structure of Peano's network affect the ensuing reaction–diffusion processes. We note (Figure 3.29c) that at equal values of the dimensionless growth rate $a\tau$, the speed of the front developing through the topological structure of the real river (empty circles) is larger than that computed for Peano (solid circles). This is because at every junction along the drainage path in a real river, a traveling particle generally meets only one secondary branch instead of two, as in the case of the Peano basin. The effect is to increase the speed by an approximately constant amount.

We have also investigated how the distribution of edge lengths (i.e., the distance between subsequent reactive sites) affects the speed of the front. One should notice that in the above model, it is possible to consider networks constituted by edges of different lengths by subdividing each edge into subedges of equal length. Technically, we label as reactive only the nodes at the endpoints of the channel (see Figure 3.29a), so that at inner nodes, only transport occurs. This method introduces a new key parameter for the system: the mean ratio of the distance between two subsequent reactive nodes and the characteristic jump length of the traveling particle ($\langle L \rangle / \Delta x$). Note that this ratio is equal to 1 in all previous cases. Squares in Figure 3.29c show simulation results for a mean ratio of about 3. It turns out that the speed is smaller (and closer to the Peano case) for large values of the population growth rate. Diamonds (Figure 3.29c) show the results obtained by placing the same number of reactive nodes at random, not at network junctions (see Figure 3.29b). It was found [7] that the spatial distribution of the reactive nodes does not produce a meaningful variation of the front speed. Stars in Figure 3.29c show instead the front speed developing through the Tanaro river network when one-half of the reactive nodes are used. In this case, the ratio $\langle L \rangle / \Delta x$ is twice as big. Differences are indeed noteworthy. The front speed is greatly decreased, especially for large values of the population growth rate. We thus suggest that the number of reactive sites, and hence the distribution of node-to-node network distances, places a combined hydrologic–demographic control on the front progression. Other cases, not reported here for brevity, have been studied numerically. Of particular interest is the case in which the bias is assumed to depend on a state variable – notably, the bias increases with the number of nodes drained (i.e., total cumulative area). The bias thus increases downstream, as does the front speed. This determines a nonstationary behavior of the

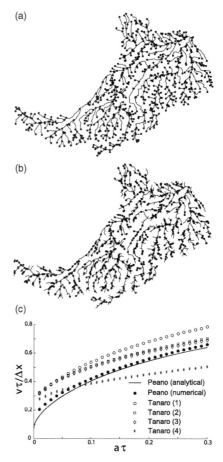

(a)

(b)

(c)

Figure 3.29 (a) Extraction of the River Tanaro network. Reactive nodes are imposed at the junctions. (b) The same network of case (a) with the same number of reactive nodes distributed randomly. (c) Dimensionless front speed ($v\tau/\Delta x$) as a function of the dimensionless growth rate $a\tau$ computed analytically for the Peano basin (solid line) and numerically for the Peano basin (full circles); the River Tanaro, taking into account only the topological structure of the network (Tanaro (1), empty circles); the River Tanaro, taking into account the distribution of the edge lengths with reactive nodes as displayed in (a) (Tanaro (2), squares), with reactive nodes as displayed in (b) (Tanaro (3), diamonds), and with one-half of the reactive nodes displayed in (b) (Tanaro (4), stars). All cases are computed with $b = 0.4$. Redrawn from [7]

system that would require extra machinery for a proper description.

In addition to Fisher–Kolmogorov's model and the Hamilton–Jacobi approach, a third analytical approximation is worth exploration. It is the approach that leads to the so-called reaction-telegraph equation (Box 3.13), which is a more realistic alternative to the diffusion approximation for describing the movement of organisms. In fact, biased transport could be modeled simply by adding to the left-hand side of the reaction–diffusion equation with logistic population growth a term taking into account the advective flux (i.e., $u\,\partial\rho(x,t)/\partial x$), where u is a reference constant advection velocity (if a more realistic scheme of advection field $u(x)$ is given, the same mathematical formalism may be retrieved; see Section 3.5.3). Then, by the change of variables $z = x - ut$, one can reobtain the mathematical form of the reaction–diffusion equation, whose solution for the front speed would then be $v = u + 2\sqrt{aD}$. However, the problem with the diffusion approximation is that starting from a discrete time-space RRW (i.e., the particle makes steps of length δx at every time step δt) and taking the limit $\delta x, \delta t \to 0$ while keeping $\delta x^2/\delta t = \text{constant} = 2D$, one obtains that the actual velocity of the particles during a jump ($\delta x/\delta t$) tends to infinity (see, e.g., [7]).

To avoid this unphysical assumption, it is possible to resort to a generalization of the one-dimensional reaction–telegraph model [348] in which we include biased transport. Consider a large number of particles moving, reproducing, and dying on a line. Particles make steps of length δx and duration δt moving at finite velocity $\gamma = \delta x/\delta t$. The walk is random, but correlated: a particle continues in its previous direction with probability p and reverses its direction with probability q.

The process is supposed to be Poisson, so that for small δt, we have $p = 1 - \lambda\delta t$ and $q = \lambda\delta t$, where λ is the rate of reversal. We assume that there is indeed a preferential direction, say, from left to right. Therefore, the rate of reversal λ_R of particles arriving from the right is bigger than the rate of reversal λ_L of particles arriving from the left. We also assume that new particles produced at a certain location (resulting from the difference between natality and mortality) move with equal probability ($1/2$) to the right or to the left. The equations of the model are then obtained by taking the limit $\delta x, \delta t \to 0$ while keeping $\delta x/\delta t = \text{constant} = \gamma$.

The biased reaction–telegraph model is more realistic with respect to Fisher's model because it assumes a correlated walk rather than a completely random one. Moreover, the model assumes that organisms move with constant finite velocity rather than with infinite velocity, as in Fisher's model. The key consequence of this basic conceptual difference is that the front speed admits an upper bound, as it cannot exceed the particles' velocity γ. This model is a better approximation of the discrete time-space RRW. Details on the reaction–telegraph model are provided in Box 3.13. A comparative analysis of the reaction–telegraph model with a RRW in a one-dimensional lattice in a range of parameters reasonable for the scale of the problems of this book shows that the two models lead to similar front speeds in a range of reaction rates most likely to cover practical interest [348].

Box 3.13 The Reaction–Telegraph Model

Consider a large number of particles moving, reproducing, and dying on a line. Particles make steps of length δx and duration δt moving at finite velocity $\gamma = \delta x/\delta t$. The walk is random but correlated: a particle continues in its previous direction with probability p and reverses its direction with probability q. This is a Poisson process [7], so that for small δt we have $p = 1 - \lambda\delta t$ and $q = \lambda\delta t$, where λ is the rate of reversal. We assume that there is indeed a preferential direction, say, from left to right. Therefore, the rate of reversal λ_R of particles arriving from the right is bigger than the rate of reversal λ_L of particles arriving from the left. We also assume that new particles produced at a certain location (resulting from the difference between natality and mortality) move with equal probability ($1/2$) to the right or to the left. The equations of the model are then obtained by taking the limit $\delta x, \delta t \to 0$ while keeping $\delta x/\delta t = \text{constant} = \gamma$. Technical details for the derivation of the speed of the front that this process yields are described in [7].

Let F be the rate of demographic growth. Let $\alpha(x,t)$ be the density at coordinate x and time t of particles that arrived from the left and $\beta(x,t)$ be the density of particles that arrived from the right. Then we can stipulate the following equations:

Box 3.13 *Continued*

$$\alpha(x,t+\delta t) = (1 - \lambda_L \delta t)\,\alpha(x-\delta x,t) + \lambda_R \delta t\,\beta(x-\delta x,t) + \frac{1}{2}\delta t\,F(x-\delta x,t) \tag{3.44}$$

$$\beta(x,t+\delta t) = (1 - \lambda_R \delta t)\,\beta(x+\delta x,t) + \lambda_L \delta t\,\alpha(x+\delta x,t) + \frac{1}{2}\delta t\,F(x+\delta x,t), \tag{3.45}$$

where δt, δx, λ_R, λ_L are the parameters introduced earlier. Expanding in Taylor series with respect to time and space and neglecting second- and higher-order terms, we obtain

$$\alpha + \delta t\,\alpha_t = (1 - \lambda_L \delta t)(\alpha - \delta x\,\alpha_x) + \lambda_R \delta t(\beta - \delta_x \beta_x) + \frac{1}{2}\delta t\,F - \frac{1}{2}\delta t\,\delta x\,F_x \tag{3.46}$$

$$\beta + \delta t\,\beta_t = (1 - \lambda_R \delta t)(\beta + \delta x\,\beta_x) + \lambda_L \delta t(\alpha + \delta_x \alpha_x) + \frac{1}{2}\delta t\,F + \frac{1}{2}\delta t\,\delta x\,F_x, \tag{3.47}$$

where the subscripts x and t indicate the partial derivatives with respect to space and time, respectively. Taking the limit as δt and δx go to zero with $\delta x/\delta t = \gamma$, we have

$$\alpha_t + \gamma\alpha_x = \lambda_R \beta - \lambda_L \alpha + \frac{1}{2}F \tag{3.48}$$

$$\beta_t - \gamma\beta_x = \lambda_L \alpha - \lambda_R \beta + \frac{1}{2}F. \tag{3.49}$$

It is convenient to introduce the total density of particles at location x: $S(x,t) = \alpha(x,t) + \beta(x,t)$ and the difference $R(x,t) = \alpha(x,t) - \beta(x,t)$. The rate F of demographic increase is supposed to be a unimodal, nonnegative function of the total density S, such that $F(0) = F(K) = 0$, where K is the "carrying capacity" or equilibrium population size. From Equations (3.48) and (3.49), we finally obtain

$$S_t + \gamma R_x = F(S) \tag{3.50}$$

$$R_t + \gamma S_x = \Delta\lambda\,S - 2\lambda R, \tag{3.51}$$

where $\Delta\lambda = \lambda_R - \lambda_L > 0$ is the directional bias and $\lambda = (\lambda_R + \lambda_L)/2$ is the average rate of reversal. Note that $\Delta\lambda/2\lambda \le 1$ in any case. We now search for a traveling wave moving from left to right with velocity c. To this end, we introduce the moving coordinate system $z = x - ct$, $c > 0$, into Equations (3.50) and (3.51), thus obtaining

$$-cS_z + \gamma R_z = F(S) \tag{3.52}$$

$$-cR_z + \gamma S_z = \Delta\lambda\,S - 2\lambda R, \tag{3.53}$$

and then

$$S_z = \frac{cF(S) + \gamma\Delta\lambda S - 2\gamma\lambda R}{\gamma^2 - c^2} \tag{3.54}$$

$$R_z = \frac{\gamma F(S) + c\Delta\lambda S - 2c\lambda R}{\gamma^2 - c^2}. \tag{3.55}$$

The singular points of Equations (3.54) and (3.55) are $(S,R) = (0,0)$ and $(K, \Delta\lambda K/2\lambda)$. The Jacobian is

$$J(S) = \frac{1}{\gamma^2 - c^2}\begin{bmatrix} cF'(S) + \gamma\Delta\lambda & -2\gamma\lambda \\ \gamma F'(S) + c\Delta\lambda & -2c\lambda \end{bmatrix}, \tag{3.56}$$

Box 3.13 *Continued*

where the prime indicates differentiation with respect to S. We must see whether it is possible to have a heteroclinic connection going from $(K, \Delta\lambda K/2\lambda)$ to $(0,0)$ and such that $S \geq 0$. This is equivalent to requiring that $(K, \Delta\lambda K/2\lambda)$ be unstable and that $(0,0)$ be a saddle or a stable node. Let us consider $(0,0)$ first. We must have either $det J(0) < 0$ (saddle) or $tr J(0) < 0$ and $0 < 4(det J(0)) < (tr J(0))^2$ (stable node). Introduce the intrinsic rate of demographic increase $a = F'(0)$, the dimensionless rate of increase $\rho = a/2\lambda$, and the dimensionless bias $b = \Delta\lambda/2\lambda$. Notice that both a and ρ are greater than zero and that $0 < b \leq 1$. It turns out that

$$det J(0) = \frac{2a\lambda}{\gamma^2 - c^2} \tag{3.57}$$

$$tr J(0) = \frac{ca + \gamma\Delta\lambda - 2c\lambda}{\gamma^2 - c^2}. \tag{3.58}$$

There exist two cases [7]:
1. $c > \gamma$; then $det J(0) < 0$ and $(0,0)$ is a saddle for any a or equivalently for any ρ.
2. $c \leq \gamma$; then we must verify the conditions for a stable node.
First condition: $tr J(0) < 0$ if and only if $c > \hat{c} = b\gamma/(1 - \rho)$. On the other hand, we must have $0 < c \leq \gamma$. Therefore, this imposes the further condition that $\rho \leq (1 - b)$.
Second condition: $4 det J(0) < (tr J(0))^2$ if and only if $G(c) = (1 + \rho)^2 c^2 + 2b(\rho - 1)\gamma c + (b^2 - 4\rho)\gamma^2 > 0$. The roots of $G(c)$ are

$$c_{\pm} = \frac{b(1 - \rho) \pm \sqrt{4\rho[(1 + \rho)^2 - b^2]}}{(1 + \rho)^2} \gamma. \tag{3.59}$$

As the dimensionless bias b is ≤ 1, then $b \leq 1 + \rho$, hence both roots are real. Therefore $G(c) > 0$ if and only if $c < c_-$ or $c > c_+$. Note that for $\rho = 0$, it turns out that $c_{\pm} = \hat{c} = b\gamma$, while for $\rho = 1 - b$, it turns out that $c_+ = \hat{c} = \gamma$. More generally, it is easy to verify that $c_{\pm} \leq \gamma$. Also, we have for $\rho < 1$

$$c_- = \frac{b(1 - \rho) - \sqrt{4\rho[(1 + \rho)^2 - b^2]}}{(1 + \rho)^2}\gamma \leq \frac{b(1 - \rho)}{(1 + \rho)^2}\gamma \leq \frac{b(1 - \rho)}{(1 - \rho)^2}\gamma = \hat{c}. \tag{3.60}$$

A cumbersome analysis [7] shows that $c_+ \geq \hat{c}$ for $\rho \leq 1 - b$. In fact, for $\rho < 1$, the inequality $c_+ \geq \hat{c}$ is equivalent in succession to

$$b(1 - \rho)^2 + \sqrt{4\rho[(1 + \rho)^2 - b^2]}(1 - \rho) \geq b(1 + \rho)^2$$

$$(1 - \rho)^2[(1 + \rho)^2 - b^2] - 4\rho b^2 = (1 + \rho)^2((1 - \rho)^2 - b^2) \geq 0.$$

The last inequality is true for $\rho \leq 1 - b$. Therefore, $c_- \leq \hat{c} \leq c_+$.

We can thus state, as in [7], that the singular point $(0,0)$ is a stable node provided $c \geq c_+$. The possible range of values for the dimensionless rate of demographic increase is $0 \leq \rho \leq 1 - b$.

We can conclude that the minimum value $c_{\min}$ of the velocity c for which the singular point $(0,0)$ is the limit of a heteroclinic orbit for $z \to +\infty$ is given by [7]

$$c_{\min} = \begin{cases} c_+ = \dfrac{b(1 - \rho) + \sqrt{4\rho[(1 + \rho)^2 - b^2]}}{(1 + \rho)^2}\gamma, & \text{for } \rho < 1 - b, \\ \gamma, & \text{for } \rho \geq 1 - b. \end{cases} \tag{3.61}$$

Box 3.13 *Continued*

Moreover, it is easy to verify that the singular point $(K, \Delta \lambda K / 2\lambda)$ is unstable. In fact,

$$det J(K) = \frac{2F'(K)\lambda}{\gamma^2 - c^2} \tag{3.62}$$

$$tr J(K) = \frac{cF'(K) + \gamma \Delta \lambda - 2c\lambda}{\gamma^2 - c^2}. \tag{3.63}$$

Remembering that $F'(K) < 0$ and $b = \Delta \lambda / 2\lambda$, we have that $det J(K) \leq 0$ for $c \leq \gamma$ and $tr J(K) > 0$ for $c > \gamma$ [7]. The final conclusion of Bertuzzo et al. [7] is that the minimum velocity for the existence of a heteroclinic connection between the two singular points is c_{min} given by Equation (3.61). Therefore, c_{min} is the traveling wave speed for a front moving from left to right. Note that the wave speed without demographic growth ($\rho = 0$) is $b\gamma$, which is the advection velocity. In fact, a particle in a certain position that arrived from the left has probability $\lambda_L / (\lambda_R + \lambda_L)$ of moving left and probability $1 - \lambda_L / (\lambda_R + \lambda_L)$ of moving right, while a particle that arrived from the right has probability $\lambda_R / (\lambda_R + \lambda_L)$ of moving right and probability $1 - \lambda_R / (\lambda_R + \lambda_L)$ of moving left. So the average velocity is

$$\frac{1}{2} \left[\gamma \left(1 - \frac{\lambda_L}{\lambda_R + \lambda_L} \right) - \gamma \frac{\lambda_L}{\lambda_R + \lambda_L} \right] + \frac{1}{2} \left[\gamma \frac{\lambda_R}{\lambda_R + \lambda_L} - \gamma \left(1 - \frac{\lambda_R}{\lambda_R + \lambda_L} \right) \right] = \frac{\Delta \lambda}{2\lambda} \gamma = b\gamma.$$

On the other hand, there is a threshold of the dimensionless growth rate $\rho = 1 - b$ beyond which the wave speed is equal to γ, the absolute velocity of a particle moving to a nearby location. Introducing the normalized wave speed $v = c_{min}/\gamma$, we have

$$v(b, \rho) = \begin{cases} \dfrac{b(1 - \rho) + \sqrt{4\rho[(1 + \rho)^2 - b^2]}}{(1 + \rho)^2}, & \text{for } \rho < 1 - b, \\ 1, & \text{for } \rho \geq 1 - b. \end{cases} \tag{3.64}$$

So the normalized wave speed depends on just two parameters: b, which is the normalized advection velocity, and ρ, which is the dimensionless rate of increase. Note that $1/a$ is the average time for the production of a new particle and $1/2\lambda$ is the average time between two direction inversions; ρ is just the ratio of the latter time to the first. Also, the threshold condition $\rho > 1 - b$ is equivalent to $1/a < 1/2\lambda_L$, namely, the average time for the production of a new individual must be smaller than the average time between two inversions in the direction opposite to the wave direction. Needless to say, the case $\lambda_L = \lambda_R = \lambda$ yields earlier results [348]. It can be proved that the velocity v_2 of the retrogressive traveling wave is simply $v_2(b, \rho) = -v(-b, \rho)$. Note that v_2 becomes negative for $\rho > b^2$. Also, it saturates to $-\gamma$ for $\rho = 1 + b$. Figure 3.32 shows the dimensionless front speed v/γ as a function of the growth rate a for a reaction telegraph model (solid lines) and for a RRW in a one-dimensional lattice (dashed lines). The latter is computed through the Hamilton–Jacobi formalism described in the context of the general model [118] by setting to zero the order of the branches (i.e., $\varphi(t) = \delta(t - \tau)$). Different couples of lines (solid and dashed) refer to different values of the bias b. The best fit between the two models is obtained taking $\tau = 1/\lambda$, where τ is the waiting time for the random walk and λ the average reversal rate for the telegraph model. This could be explained as follows: in the discrete random walk model, τ could be thought of as the time spent by a particle moving from left to right (from x to $x + \Delta x$) or likewise from right to left (from x to $x - \Delta x$), maintaining its direction. In the differential continuous biased telegraph model the mean

Box 3.13 *Continued*

times of preserved direction, from left to right- and vice versa, are $1/\lambda_R$ and $1/\lambda_L$, respectively. We can then take τ equal to the average of $1/\lambda_R$ and $1/\lambda_L$ weighted through the probability of going right ($\lambda_R/(\lambda_R + \lambda_L)$) and left ($\lambda_L/(\lambda_R + \lambda_L)$), respectively,

$$\tau = \frac{\lambda_R}{\lambda_R + \lambda_L}\frac{1}{\lambda_R} + \frac{\lambda_L}{\lambda_R + \lambda_L}\frac{1}{\lambda_L} = \frac{1}{\lambda}, \tag{3.65}$$

which is exactly the relation used. The two models lead to similar results for a large range of values of the growth rate a.

In a spatially unbounded domain, all the models analyzed (Fisher's, reaction telegraph, and biased RRW on networks) yield, if started from a compact support initial condition, two fronts that travel with velocity v_1 and v_2, respectively (see the sketch in the upper left inset of Figure 3.30). For unbiased processes the two fronts travel along opposite directions, whereas for a large enough bias they may travel along the same direction. The analysis presented up to this point refers to the computation of the speed v_1 of the first front that moves along the bias direction. For the telegraph and RRW models, the speed v_2 of the second front can be obtained as $v_2(b,a) = -v_1(-b,a)$. Note that for the derivation of v_2 for the backbone of a Peano network through the Hamilton–Jacobi

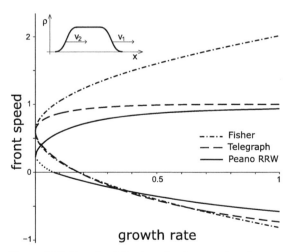

Figure 3.30 Speeds of the two fronts v_1 and v_2 for any value of the dimensionless logistic growth rate: Fisher–Kolmogorov model, dash-dotted line; telegraph model, dashed line; reaction random walk on Peano network, solid line. The bias in all cases is equal to 0.6. Dots of the Peano RRW are computed numerically. Redrawn from [7]

formalism, one has to consider an opposite bias only for the transport in the backbone, not in the branches (i.e., the Laplace transform of the waiting time distribution $\hat{\varphi}(s)$ remains the same for the derivation of both v_1 and v_2). For Fisher's model the speed of the second front is symmetrically given by $v_2 = u - 2\sqrt{aD}$ (where u is the advection velocity). Figure 3.30 shows the speeds v_1 and v_2 of the two fronts as a function of the logistic growth rate for (1) Fisher's model (dash-dotted line), (2) telegraph model (dashed line) (Box 3.13), and (3) RRW on Peano's network (solid line). The bias for the three cases is equal to 0.6. The comparison between the Fisher–Kolmogorov and RRW models is obtained by using the approximation $D = \Delta x^2/\tau$. The comparison between the telegraph and the RRW models follows the comparative rule discussed at the end of Box 3.13. For the range of parameters for which v_2 is positive for the RRW on the Peano network (and then $v_1(-b,a)$ is negative because $v_2(b,a) = -v_1(-b,a)$), we compute the speed numerically (dots in Figure 3.30) because the Hamilton–Jacobi formalism allows us to compute only positive speeds. It is noticeable that with biased transport, the telegraph and the network RRW models lead to an asymmetric behavior of the two fronts (i.e., $(v_1(a) + v_2(a))/2 \neq v_1(a = 0)$). This is to be contrasted with the classical Fisher model, in which the two fronts are perfectly symmetrical. However, the asymmetry is less remarkable in the Peano RRW model because the particles can disperse away from the backbone, thus slowing down the front progression.

A possibly important statistical measure related to the propagation of reactive particles throughout river networks has been identified [7]. Knowing the speeds v_1 and v_2 of the two fronts characteristic of a network, it is possible to compute the first colonization time, that is, the time that the front needs to propagate from one node to any other along the shortest path available in the river network. Every such path consists of a downstream and

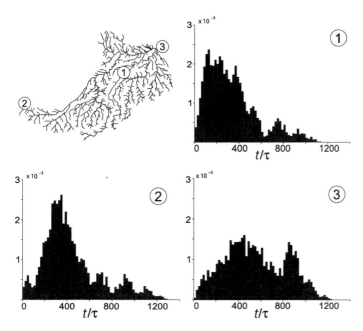

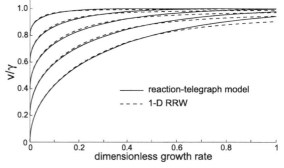

Figure 3.31 First colonization time distributions of all network sites starting from three different nodes. The position of the three nodes in the network is reported in the upper left inset. The speeds v_1 and v_2 employed correspond to values of $b = 0.4$ and $a\pi = 0.3$. Redrawn from [7]

an upstream part of length L_d and L_u, respectively. If we assume that the speeds do not depend on the particular path, that is, the bias is roughly the same, the travel time is simply given by $T = L_d/|v_1| + L_u/|v_2|$. Note that the front can reach all the nodes of the network only if v_2 is negative. Figure 3.31 shows the distribution of the times of first colonization from three different starting nodes to all the others for the Tanaro river network. It is interesting to note that this statistic is related to the dispersal kernel in [40]. Colonization times are shortest for a central node, say, node 1, and largest for the outlet, node 3. The initial condition seeded at any one node will therefore result in a specific distribution of invasions throughout the network, which is affected by geomorphology as well as by the nature of the reaction imposed by the logistic growth.

There exists a deep interplay of structural and dynamic controls that may operate at the network level – a tenet that again resonates with the main theme of this book. If the traveling particles in the above example were infective agents or organism propagules, thereby accounting for reactions different from the "simple" logistic growth considered here, our results would imply a definite hydrological role in the process of disease spread or ecological colonization. This seems an important quantitative insight into the understanding of the susceptibility of a landscape to biological invasions.

Summing up the issues outlined in this section (jointly with the main results shown in Section 1.5), we may draw a few general conclusions. First, we note that

Figure 3.32 Dimensionless front speed v/γ as a function of the dimensionless growth rate (a/λ or $a\tau$) for a reaction–telegraph model (solid lines) and for a RRW in a one-dimensional lattice (dashed lines). Different couples of lines (solid and dashed) from the bottom to the top refer to different values of the bias b from 0 to 0.8 with step 0.2.

adding a constant bias to a reactive transport model along one-dimensional corridors like a river network jointly acts with morphological effects in drastically modifying the speed of the migrating front of the traveling wave that describes the invasion process. Although our analytical and numerical results are derived by using the logistic growth as reactive component, we suspect that this might be a result of a general nature and of ecological interest even for other, more complex demographies. Second, exact solutions (although for simplified geometries/topologies) exist for the speed of the migrating front with biased transport for different invasion models. They provide an immediate order of magnitude of the phenomena of interest and a range of possible scenarios given reasonable values for the key parameters (the rate $R(0)$ of the density-dependent population growth and the hydrodynamic dispersion coefficient), even in one-dimensional comblike spatial settings. This is related to the result shown in Section 1.5: topological similarities among real rivers, optimal networks, and exact recursive constructs lead to similar behaviors in the spreading of populations, thus strengthening the predictive power of our analytical results.

We still need to address the issue of heterogeneity in the distribution of reactive sites, where the characteristic distance relative to a basic network length places further hydrologic controls on the invasion process. This is dealt with at a later stage in this chapter. Overall, however, a significant number of hydrologic controls are shown to affect migrating fronts of biological species that move along the ecological corridors defined by the river basin.

3.4.7 Heterogeneous Invasions: The Role of Spatial Resource Variability

The abundance of stream ecosystem populations often displays a pronounced variability that contrasts with the simple deterministic models of population dynamics. We have seen in previous sections how this can be ascribed to demographic stochasticity. As for environmental stochasticity, it can be described by a stochastic time process (Section 3.4.1), but in many cases, this variability is due to space-heterogeneous resource distribution.

Here, the focus is on biological invasions in landscapes characterized by temporally invariant but spatially heterogeneous resource distributions, which could, for example, reflect the spatial composition and the quality of soil, gap dynamics in forests, and subsequent light availability for understory plants and their associated herbivore fauna or habitat fragmentation due to human land use [349, 350]. The investigation of species spread in

spatially heterogeneous landscapes has mainly focused on the relationship between the mean invasion speed and the percentage of favorable habitat across the landscape [350, 351], showing that spread may not occur below minimal thresholds of suitable growth habitat. A limited number of empirical works have measured spread rates in heterogeneous and diverse habitats and compared realized spread distances in patchily distributed sites [352–354] or across landscapes with monotonic gradients [355]. Spatially heterogeneous landscapes, however, are characterized not only by the mean percentage of favorable habitat or the mean density of resources but crucially also by their spatial autocorrelation.

The centrality of spatial correlation has been proposed only recently [16]. Before a full-fledged theoretical and experimental study examining each contribution distinctly, one lacked insight into the role of spatial correlation of drivers and controls in slowing or accelerating the spread of species. One way of generating controlled resource distribution fields exploits heterogeneous light/resource distributions [15, 16]. For example, phototaxis, the process through which motile organisms direct their swimming toward or away from light, is implicated in key ecological phenomena, including algal blooms and diel vertical migrations. Such processes shape the distribution, diversity, and productivity of phytoplankton and thus energy transfer to higher trophic levels in aquatic ecosystems. Phototaxis also finds important applications in biofuel reactors and microbiopropellers. It has been argued [15] to serve as a benchmark for the study of biological invasions in heterogeneous environments owing to the ease of generating stochastic light fields. Here, following [15, 16], we illustrate accurate measurements of the behavior of the alga *Euglena gracilis* when exposed to controlled light fields. In a first set of experiments [15] and theoretical elaborations, the analysis of the phototactic accumulation dynamics of *E. gracilis* over a broad range of light intensities provided the proper backdrop for investigating questions related to the effects of heterogeneity of resource distributions. The theoretical computations [15] followed the classic Keller–Segel mathematical description of phototaxis (Box 3.14).

Box 3.14 The Keller–Segel Framework for Mathematical Taxis

The Keller-Segel mathematical framework for taxis provides an accurate description of both positive and negative phototaxis only when phototactic sensitivity is modeled by a generalized receptor law, a specific nonlinear response function to light intensity that drives algae toward beneficial light

Box 3.14 *Continued*

conditions and away from harmful ones. The proposed phototactic model captures the temporal dynamics of both cells' accumulation toward light sources and their dispersion upon light cessation. The model could thus be of use in integrating models of vertical phytoplankton migrations in marine and freshwater ecosystems and in the design of bioreactors.

The Keller–Segel framework [356, 357] consists of an advection–diffusion equation in a linear landscape (an ecological substrate rapidly well mixed in the other coordinate directions, as in shear flow dispersion, which we analyze in Section 3.5.3) for the cell density $\rho(x,t)$ [358] (neglecting cell division owing to the [relatively] short duration of the experiments with respect to the characteristic duplication time [15]):

$$\frac{\partial \rho}{\partial t} = \frac{\partial}{\partial x}\left[D\frac{\partial \rho}{\partial x} - \frac{d\phi}{dx}[I(x)]\rho(x,t)\right], \tag{3.66}$$

where $I(x)$ is the imposed field of light intensity, ϕ is a phototactic potential that is solely a function of the light field $I(x)$, and the term $d\phi/dx$ acts as an analog to a drift, or advection term. The phototactic velocity of displacement is therefore proportional to the gradients of the phototactic potential [15, 358]. The steady state accumulation of cells that satisfies Equation (3.66), computed over the spatial extent of a symmetric window $-L \le x \le L$ (in the experiments of Giometto et al. [15], $L = 6.25$ cm), yields the following distribution of normalized cell density $\bar{\rho} = \rho(x)/\rho(-L)$:

$$\bar{\rho} = \exp\left[\frac{\phi[I(x)]}{D}\right],$$

where at no loss of generality, $\phi = 0$ when $I = 0$ [15]. The temporal dynamics of density accumulation around a light source computed by solving the above framework to contrast experimental evidence are shown in Figure 3.33. The figure also shows an important proof of the diffusive behavior of the density profiles relaxing after switching off the light source (i.e., $I(x) \to 0$), which complies with the basic tenet of reactive–diffusive models.

Subsequently, in another set of experiments (Figure 3.34), the availability of light was manipulated as a limiting resource with a spatially heterogeneous distribution. Light is the most important resource for photosynthetically active organims, such as higher plants or algae, and these organisms are not only using light as their key resource for growth, but can also adjust their movement in response of light availability [15]. Such directional movement can be active movement (e.g., in flagellated algae) or indirected growth/budding of plant stolones [16] or growing yeast colonies [359]. For most plants, especially in forests, the availability of light is quite heterogeneous spatially, and mostly driven by gap dynamics [16]. Thus, gaps due to tree fall or anthropogenic activities create a patchy distribution of light as a key resource for understory plants and all of their associated interacting herbivores and pollinators. The experiments, planned within the general scope of the present book, were aimed at studying how the arrangement, specifically the autocorrelation of the light availability, can affect spread dynamics of organisms (Figure 3.35). While here, following [16], we focus on light as a resource, the theoretical work and the parallel experiments are more generic with respect to the specific resource creating the heterogeneity in habitat quality. The foremost result is that it is shown theoretically and experimentally that the invasion speed is affected not only by the mean amount of resources scattered along the landscape (or, analogously, by the percentage of suitable habitat) but also by their spatial autocorrelation structure. The foremost result is that environmental heterogeneity and demographic stochasticity jointly affect biological invasions. In particular, Giometto et al. [16] show that it is necessary to include demographic stochasticity in models of spread to properly understand biological invasions in spatially

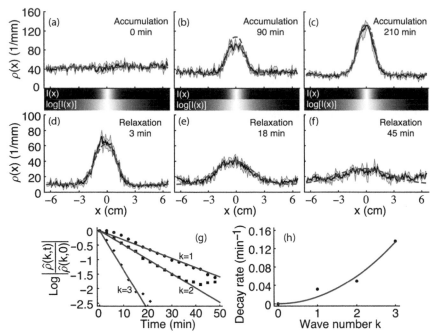

Figure 3.33 (a–c) Temporal dynamics of density accumulation around a light source placed at $x = 0$ cm and (d–h) relaxation of cell density peaks upon removal of light. (a–f) Experimental cell density profiles at different times. The shaded gray area is delimited by the maximum and minimum cell densities of three replicate experiments, and the black line denotes the mean. The red dashed line shows the theoretical prediction, Equation (3.66) (Box 3.14), using the experimentally controlled fields $\phi(I(x))$ and $I(x)$ [15], determined experimentally from the relaxation of density peaks (d–h). Density profiles are normalized to display the same mean abundance. The grayscale bars below (a–c) show the light intensity profile imposed during the accumulation; the gray level scales linearly (upper panels) or logarithmically (lower panels) with the intensity I, with the color white corresponding to the control level $I = 5.2$ W/m^2 and black to $I = 0.001$ W/m^2. The temporal decay of Fourier modes $\hat{\rho}(k,t)$ (the periodicity over a length L allows us to write a discrete Fourier transform (DFT), that is, $\rho(x,t) = \sum_{k=-N/2}^{N/2} \hat{\rho}(k,t) \exp(2\pi kx/L)$, where $N + 1$ modes are adopted in the DFT) (g) during the relaxation of the density peaks (d–h) is exponential only if the transport equation is genuinely diffusive (this in turn implies that $\log | \hat{\rho}(k,t)/\hat{\rho}(k,0) = -Dk^2t$; data in black and liner fit in red). A diffusive behavior in the absence of light fields is thus experimentally demonstrated [15].

heterogeneous environments (Box 3.15). This important result is shown theoretically in two models of invasion at different levels of biological detail (the speed of species spread decreases when the resource autocorrelation length increases) and is verified experimentally in a microcosm experiment with the flagellated protist *E. gracilis* by manipulating light intensity profiles along linear landscapes (light is an energy source for *E. gracilis*, as it has chloroplasts and can photosynthesize) (Figure 3.34).

Box 3.15 Modeling Spatial Resource Variability and Demographic Stochasticity

Species spread in heterogeneous linear landscapes is modeled via several methods. One is a stochastic generalization of the Fisher–Kolmogorov equation, including both demographic stochasticity and spatially heterogeneous resource distributions [289, 360, 361]:

$$\frac{\partial \rho}{\partial t} = D\frac{\partial^2 \rho}{\partial x^2} + r(I)\rho \left[1 - \frac{\rho}{K}\right] + \sigma\sqrt{\rho}\,\eta, \tag{3.67}$$

Box 3.15 *Continued*

where $\rho(x,t)$ is the density of individuals, D is the diffusion coefficient of the species driven by the active movement of individuals, r is the growth rate, K is the carrying capacity, σ is a parameter describing the amplitude of demographic stochasticity, and η is a Gaussian, zero-mean white noise (i.e., the noise has correlations $\langle \eta(x,t)\eta(x',t') \rangle = \delta(x-x')\delta(t-t')$, where δ is the Dirac's delta function). Itô's stochastic calculus is adopted, as appropriate for the demographic noise term [15, 16]. The growth rate $r(I) = r_0 I$ is assumed to be a function of the local amount of resources $I(x)$, which can assume two values: $I(x) = 1$ or $I(x) = 0$. Landscape heterogeneity is thus embedded in the resource profile $I(x)$. We studied the dimensionless form of Equation (3.67), which reads [15]

$$\frac{\partial \rho'}{\partial t'} = \frac{\partial^2 \rho'}{\partial x'^2} + \chi_I \rho' \left[1 - \rho'\right] + \sigma'' \sqrt{\rho'}\, \eta, \qquad (3.68)$$

where $t' = r_0 t$, $x' = \sqrt{\frac{D}{r_0}}x$, $\rho' = \rho/K$, $\sigma'' = \frac{\sigma}{\sqrt{K}(rD)^{1/4}}$, and $\chi_I(x')$ is the indicator function of the set of x' for which $I(x') = 1$. In the following we drop primes for convenience: one can recover the original dimensions by multiplying t by r_0 and x by $\sqrt{r_0/D}$ and by rescaling ρ and σ as indicated above. Numerical integration of stochastic partial differential equations with square root noise terms requires ad hoc numerical methods, as standard approaches like the first-order explicit Euler method inevitably produce unphysical negative values for the density ρ [360]. Therefore, Equation (3.68) was integrated with the split-step method [16].

Landscapes with various resource autocorrelation lengths were generated by imposing $I(x)$ to be composed of subsequent independent patches of suitable ($I(x) = 1$ and $r = r_0$) or unsuitable ($I(x) = 0$ and $r = 0$) habitats (Figure 3.35a). The length of each patch was drawn from an exponential distribution with rate μ. Therefore, each landscape was a stochastic realization of the so-called telegraph process with rate μ and autocorrelation length $c_L = 1/(2\mu)$. The mean extent of suitable and unsuitable patches in such landscapes is $1/\mu$. Because simulated landscapes were finite, only landscapes with mean resources equal to $\bar{I} = L^{-1}\int_0^L I(x)dx = 1/2$ were accepted. The autocorrelation length is confined in these experiments to a narrow window around $1/(2\mu)$. Examples of landscapes used in the simulations are shown in Figure 3.34 (red and blue paths).

Ninety-six landscapes were generated for each value of resource autocorrelation length c_L (Figure 3.34; see [16]). Equation (3.68) was integrated numerically for each landscape and for each value of $\sigma \in \{0.1, 0.2, 0.4, 0.6\}$, with initial density profiles localized at the origin. To avoid the extinction of the whole population, the left boundary was fixed at $\rho = 1$. For each numerical integration, the position of the front was tracked by fixing a threshold value of the density ($\bar{\rho} = 0.15$) and recording the farthest point from the origin where the cell density was higher than such value. The mean propagation speed for each value of the resource autocorrelation length was computed by fitting a straight line (least squares fit) to the mean front position versus time in the asymptotic propagation regime before any of the replicated invasions reached the end of the landscape.

The traditional way to compute the invasion speed in the deterministic Fisher–Kolmogorov equation is to analyze the traveling wave solutions of the equation (Section 3.1.2). Such a procedure cannot be adopted for Equation (3.68), owing to the stochastic noise term. A theoretical approximation to the mean front propagation speed, valid for large autocorrelation lengths c_L and σ, is obtained by characterizing the mean time taken to cross a patch of unfavorable habitat (where $I = r = 0$) of length z. This mean time has been shown [16] to depend on z and σ

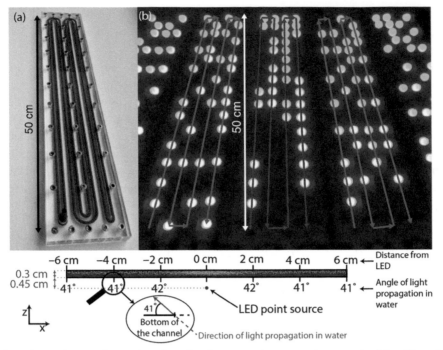

Figure 3.34 Experimental setup for testing the role of resource heterogeneity conceived by Giometto et al. [16]. (a) Linear landscapes used in the experiments [16] were channels drilled on a Plexiglass sheet. A gasket (orange rubber band) avoided water spillage. (b) Photograph of the LED strips used to control the distribution of resources for *E. gracilis*. The red and blue lines show the paths of landscapes with large and small resource autocorrelation length, respectively. (bottom) Sketch of the experimental setup. A LED point source (not to scale) was placed below the linear channels. Individuals of *E. gracilis* (green dots; not to scale) accumulated in the presence of light through phototaxis. Shown are distances from the LED and angles of light propagation in water, computed using Snell's law. The light direction component orthogonal to the channel was disregarded here, because the cells' movement dynamics in the vertical direction was dominated by gravitaxis, which resulted in the accumulation of cells at the top of the channel. Figure after [16]

as $\langle \tau \rangle (z, \sigma) = Cz^2 \exp[d \left(z\sigma^b \right)^a]$, where C, a, b, and d are constants, independent of z and σ. Additionally, the functional dependence of the variance of τ on z and σ yields an approximation to the variance of the total time taken by a front to colonize completely a landscape of finite length L. All details are in the original reference [16].

To test whether deterministic models actually predict a slowdown of the invading front for increasing resource autocorrelation length, numerical integration of the relevant equation (3.68) was conducted with $\sigma = 0$. Additionally, Equation (3.68) is numerically solved with $\sigma = 0$ and imposing a negative growth rate r in unfavorable patches where $I = 0$. It was found that the speed of invasion v in the model decreases with increasing resource autocorrelation length (Figure 3.35b). The mean front propagation speed, in heterogeneous landscapes where resource patch lengths are distributed exponentially with rate m, depends

on c_L and σ asymptotically (i.e., for large values of c_L and σ) as

$$v = \frac{L}{\frac{\mu L}{2} \int_0^L dz \langle \tau \rangle (z, \sigma) \mu e^{-\mu z}}$$
$$\simeq \frac{8c_L^2}{\int_0^\infty dz \, \langle \tau \rangle (z, \sigma) e^{-z/(2c_L)}}. \tag{3.69}$$

Figure 3.35 (after [16]) shows that Equation (3.69) correctly predicts the celerity of the invasion front for large values of c_L and σ. In heterogeneous landscapes with different spatial arrangements of favorable and unfavorable patches, if the percentage of space occupied by unfavorable patches is $f_0 \in (0, 1)$ and the distribution of the patches' lengths is $p_0(z)$, with mean $\int dz \, z p_0(z) = 1/\mu$, the asymptotic invasion velocity can be approximated as

$$v = \frac{1}{\mu f_0 \int_0^\infty dz \langle \tau \rangle (z, \sigma) p(z)}. \tag{3.70}$$

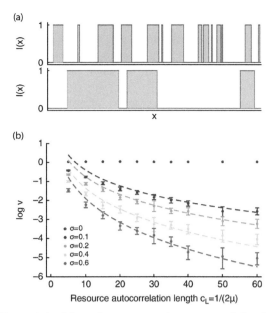

Figure 3.35 Mean front propagation computed for the dimensionless model (Box 3.15). (a) Examples of landscapes with different resource (i.e., light field $I(x)$) autocorrelation length c_L, generated via the telegraph process with rate m [16]. (b) The mean invasion speed computed in numerical integrations of the model equation (3.68) decreases with increasing resource autocorrelation length c_L (log-linear plot) for $\sigma > 0$ and is a decreasing function of the amplitude of demographic stochasticity σ (different colors according to legend). With $\sigma = 0$, the dynamics are fully deterministic and the mean front propagation speed does not decrease with c_L (gray dots). Error bars display the 95 percent confidence interval for $\log(v)$, computed with 2×10^3 bootstrap samples. Error bars for $\sigma = 0$ are smaller than symbols. Dashed lines show the mean front propagation speed computed according to the theoretical approximation (Equation (3.70)). Figure after [16]

It has been shown [16] that Equation (3.70) correctly predicts the speed of invasion in landscapes with percentages of unfavorable habitat different from $f_0 = 1/2$. Note that the speed of invasion according to Equation (3.70) is a function of the autocorrelation length if the landscapes consist of favorable and unfavorable patches generated as described in [16]. In general, however, the speed of invasion is not a one-to-one function of the resource autocorrelation length (or of other characteristic length scales of the landscape) but rather depends on the whole distribution of unfavorable patch lengths through Equation (3.70). The slowdown effect is due to the fact that, in the presence of demographic stochasticity,

long patches of unfavorable habitat act as obstacles for the spread of populations. The larger the extent of the unfavorable patch, the longer it takes for a population to cross it. The front propagation speed was also found to be a monotonically decreasing function of the amplitude of demographic stochasticity. Accordingly, integrating the model without demographic stochasticity ($\sigma = 0$ in Equation (3.68)) leads to no discernible slowdown of the front in strongly autocorrelated versus weakly autocorrelated landscapes, even when imposing negative values of the growth rate r in unfavorable patches where $I = 0$. These results demonstrate that the local extinctions caused by demographic stochasticity in unfavorable patches are responsible for the observed front slowdown.

Numerical integration of Equation (3.68) shows that the variability of the front position increases for larger values of c_L and σ. Such increased variability is caused by two factors: (1) two landscapes with identical resource autocorrelation lengths appear increasingly dissimilar for increasing values of the typical patch length $1/\mu$ and (2) the variance of the distribution of waiting times (i.e., the time to cross an unfavorable patch of length z) increases approximately quadratically with the mean time $\langle \tau \rangle(z, \sigma)$ [16]. These two observations can be used to approximate the fluctuations of the total time spent by the front to colonize a landscape of length L.

The theoretical and experimental investigation carried out by Giometto et al. [16] highlights a different type of slowdown of the speed of front propagation in biological invasions – that due to heterogeneous landscapes characterized by different resource autocorrelation lengths. The effect of resource autocorrelation length on the speed of invasion fronts is evident. In the experiments, the demographic and movement traits of the study species were fixed and were inherent properties of the species. The accompanying models allowed us to single out the individual role and the mutual interconnections of all processes included in the equations to the propagation dynamics in landscapes with different resource autocorrelation lengths. Such an investigation advances our current understanding of the spread of invading organisms in heterogeneous landscapes by addressing jointly the synergic effects on the dynamics of the spatial correlation of an environmental driver and of unavoidable demographic stochasticity. Arguably, no natural fluvial landscape is devoid of heterogeneous distributions of resources, nor of significant demographic stochasticity, and therefore incorporation of the two key elements is relevant to biological spread of populations in rivers seen as ecological corridors. A major result of the experimental validation

[16] is that demographic stochasticity is a key factor in the slowdown of front propagation in heterogeneous landscapes. In fact, the importance of including demographic stochasticity in theoretical models does not seem fully acknowledged [334] – to the point that fundamental limits to predictability have been invoked [288], a circumstance experimentally questioned later [289] – because of the many facets through which it affects species spread [15, 362].

Because the slowdown effect is only observed when demographic stochasticity is included in the models, from the bulk of the above investigations [16, 289, 334] one concludes that the stochastic birth–death dynamics are the main driver of the observed reduction in propagation speed, rather than the movement behavior of individuals in heterogeneous landscapes, which has received most attention in the literature so far [363, 364].

Previous studies have investigated the minimum percentage of suitable habitat that allows invasions to spread [349–351], suggesting that invasions cannot propagate in landscapes with mean resource density below a critical threshold. It now seems, complementarily, that the spatial arrangement of resources affects species spread even if the total amount of available resources is kept constant. Thus, it is not only the mean resource density that matters for the front propagation dynamics, because the autocorrelation structure of landscape heterogeneity alone also affects species spread.

It has been suggested [283] that the asymptotic speed of invasion according to the Fisher–Kolmogorov model in periodic landscapes (consisting of favorable and unfavorable patches of fixed size L_1 and L_2, respectively) may exhibit counterintuitive behaviors. It was shown therein that, for a fixed ratio L_2/L_1, the speed of invasion initially increases for small L_2 and then approaches an asymptotic fixed value. In contrast with the recent experimental and theoretical evidence [16], it has been predicted that the speed of invasion increases with increasing sizes of the unfavorable patches [283]. One should note, however, that the values of patch sizes L_2 for which the invasion speed increases before reaching the asymptote are in a regime where the diffusion process is sufficiently rapid compared to reproduction. The more recent results, instead, apply in the presence of patch sizes for which demographic effects are relevant, that is, in the presence of large unfavorable patch sizes. The above investigations extend previous work that addressed the effect of temporal environmental fluctuations on species spread [286, 339] by showing that the autocorrelation length of the resource distribution

should be added to the environmental factors that can slow species spread, along with temporal fluctuations of vital rates [335, 339], geometrical heterogeneities of the substrate [7, 118, 129, 131, 365], and demographic stochasticity [362].

Empirical real-world examples of invasions confirm the above general scheme. For example, Bergelson et al. [352] performed a field study with the invading weed *Senecio vulgaris* and found that the average spatial distance between two generations along linear transects increased when favorable patches were uniformly distributed in space (in the parlance of our work, the transect featured a small autocorrelation length), compared to transects with clumped patches (i.e., endowed with large autocorrelation length). Similar work exists [353] on the spread of the fungal pathogen *Rhizctonia solani*. Such work provides a complementary view to our investigation by evaluating the effect of the interdistance between favorable patches on the spread and identifying experimentally the existence of a percolation threshold at a critical level of interpatch distance.

These results have implications for species spread in natural environments, which are generally characterized by resources (seen as any field controlling vital rates, especially reproductive ones) being heterogeneously distributed. The perception of the kind of variability one might expect within an ensemble of realizations of front propagations may be gathered by the experimental data shown in Figure 3.36. One thus observes that the variance of individual invasion outcomes may be quite large, whereas the mean, as suggested here, retains reproducible characters.

Note that typical autocorrelation lengths of the resource distribution can be inferred from environmental data [366]. They can be used as a concise indicator for the propagation success of a species of interest, as argued above. Furthermore, the spatial availability of resources is often altered by human activity, reinforcing the fragmentation of landscapes. In fact, habitat fragmentation may decrease the autocorrelation length of the landscape significantly by introducing qualitatively different patches into the natural environment [12, 349, 350, 367]. The above results provide quantitative grounds to interpret field observations on the effect of environmental heterogeneity on species spread. For instance, Lubina and Levin [338] observed pauses in the spread of the California sea otter (*Enhydra lutris*) in the presence of habitat discontinuities. Such pauses and the corresponding piecewise-linear propagation of the front are also found in this context, which

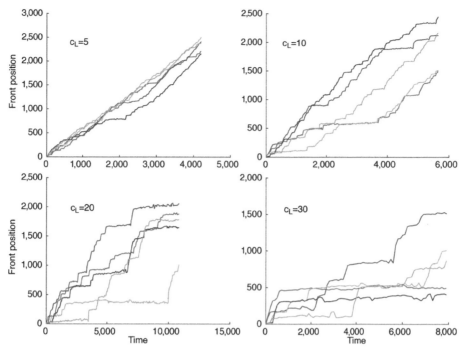

Figure 3.36 Examples of front propagation in numerical integrations of the model equation in landscapes with different resource autocorrelation lengths c_L and fixed amplitude of demographic stochasticity $\sigma = 0.2$. Figure after [16]

enables one to relate the mean spatial extent of habitat discontinuities to the average speed of invasion. An alternation between phases of halt and spread was also found in the range expansion of the cane toad (*Chaunus marinus*) in Australia [366]. Urban et al. [366] performed an in-depth analysis of the effect of environmental heterogeneity on the spread of the cane toad in the field and found a statistically significant effect of environmental heterogeneity and, most importantly, of the spatial autocorrelation of environmental variables on the realized patterns of invasion speed. They found this effect in nature in a realized (not replicable) invasion, and thus they could only correlate the realized spread dynamics and their reduction with the landscape autocorrelation. Here, in following [16], we have given a mathematical framework and an experimental proof showing that the slowdown effect caused by the spatial autocorrelation structure of the landscape is not an artifact of the mathematical model.

In conclusion, one needs to account for the intrinsic stochasticity of population dynamics to understand ecological processes occurring in spatially extended natural landscapes, which typically display various degrees of heterogeneity.

3.5 Mixing and Dispersion in River Networks

Fluid mechanics addresses a number of phenomena of great interest for the study of biological invasions and for their mathematical treatment [173]. Environmental transport is often labeled as mixing and dispersion, to highlight their range: diffusion in more than one dimension due to molecular or turbulent processes, and the one-dimensional asymptotic limit reached when one flow direction prevails on the others, which is the case when limiting our interests to fluvial domains. Suffice here to recall the one-dimensional character of open channel flow in rivers in intrinsic coordinates, where the longitudinal direction is aligned with the flow direction imposed by the boundaries. Under such premises, the longitudinal dimension x of the flow domain (x, y, z) exceeds by orders of magnitude the width and the depth of the average cross section – thus, after 4–5 widths downstream of a point injection of solute mass $M(x_0, y_0, z_0)$, complete mixing occurs along the width y, and after 6–10 depths the same occurs along the vertical direction z, leading to a concentration field $C(x, y, z, t) \sim \bar{C}(x, t)$. Note that we may or may not be foreseeing an active contribution of the dispersing organisms belonging to specific species, depending on their possible motility. In these cases, we speak of organisms' active dispersal, which may affect the dispersive character of fluvial transport. Fundamental mechanical processes underpin biological invasions, regardless of population dynamics or chemical and biological reactions, which we summarize in this section. Particular attention will be placed on dispersion in real rivers, that is, the phenomenon that describes the stretching and mixing of clouds of individual organisms caused by the combined action of shear and turbulence, jointly with macroscopic effects of the heterogeneous substrates open to invasion – the river network itself, whose recurrent geometrical and topological features have long been studied [63]. The self-organized structure of river networks bears consequences also on biological invasions, which we explore here, moving from basic processes and tools (largely excerpted from a classic textbook on this subject [173]).

3.5.1 Fickian Diffusion

Molecular diffusion of "particles" in a flowing environment is the starting point. Originally, it was straightforward to posit that the law regulating the diffusion of salt in its solvent, Fick's law, is identical to that of heat diffusion in a conducting body. Thus, if C is the concentration in units of mass per unit fluid volume, it may be defined at a point by the instantaneous limit

$$C = \lim_{\Delta V \to 0} \frac{\Delta M}{\Delta V},$$

where ΔM is the "tracer" mass in the reference elementary volume ΔV. In taking the limit, one cannot actually let $\Delta V \to 0$ but must rather confine the approximation of the continuum to a reference elementary size that is large compared to the particles composing ΔM. This reference elementary size, $\Delta V^{1/3}$, must be small enough so that continuous concentration gradients can be used in differential mass balance equations. $C = C(x, y, z, t)$ may be suitably averaged – in space or time – to suit specific needs.

Fick's law posits that the flux of transported mass q_x, that is, the mass crossing a unit area per unit time in the x-direction, is proportional to the gradient of concentration in that direction. For a one-dimensional diffusion process, Fick's law can be stated mathematically as

$$q_x = -D \frac{\partial C}{\partial x},$$

where D [L^2/T] is the proportionality coefficient and the minus sign indicates transport from high to low con-

centration. Obviously, in a three-dimensional coordinate system, the general statement of Fick's law entails the definition of a vector $\vec{q}$ of fluxes along coordinate directions $\vec{q} = (q_x, q_y, q_z)$ as $\vec{q} = -D\nabla C$ (if D is constant, thus portraying a homogeneous and isotropic process). If the mechanisms responsible for the erratic movement of particles or organisms were of molecular origin, we would speak of D as the molecular diffusion coefficient. Continuity implies equating the divergence of the fluxes ($\nabla \cdot \vec{q}$) in arbitrary elementary control volumes (dx, dy, dz) to the rate of change of mass per unit time. The result, should D be constant as reasonable for molecular actions, is the diffusion equation introduced in Box 2.1, that is,

$$\frac{\partial C}{\partial t} = D\nabla^2 C = D\left(\frac{\partial^2 C}{\partial x^2} + \frac{\partial^2 C}{\partial y^2} + \frac{\partial^2 C}{\partial z^2}\right),$$

which obviously reduces to its lower-dimensional versions (e.g., $\partial C(x,t)/\partial t = D\partial^2 C\partial x^2$) upon neglecting gradients or some spatial averaging.

A few solutions of the related boundary value problems, as relevant to the contents of this book, have been presented in Boxes 3.1, 3.2, and 3.4. In addition to those solutions, here we recall one additional result, needed to upscale dispersive processes, obtained by dimensional analysis [173]. The example is in one dimension, say, x. The solution describes the spreading by diffusion of an initial slug of mass (m) introduced at time zero at the origin of the x axis. Because the concentration $C(x,t)$ can only be a function of m, x, t, and D, and because the process is linear, C must be proportional to the mass introduced. In one dimension (units of concentration are mass per unit length), C must be proportional to m divided by a characteristic length. The units of D are [L^2/T], so a suitable length scale is $\sqrt{Dt}$. Thus dimensional analysis yields a relation of the type

$$C = \frac{m}{\sqrt{Dt}} F\left(\frac{x}{\sqrt{Dt}}\right),$$

and by defining $\eta = x/\sqrt{Dt}$, substitution into the 1D diffusion equation turns it into an ODE,

$$\frac{dF}{d\eta} + 2F\eta = 0,$$

which yields $F = C_0 e^{-\eta^2}$. The total mass contained in the system can be found by integrating the concentration along the whole x axis. By assuming that the total quantity of mass is constant at all times, we have $\int_{-\infty}^{\infty} C dx = m$ at any time, yielding $C_0 = 1$ and [173]

$$C(x,t) = \frac{m}{\sqrt{4\pi Dt}} e^{-x^2/4Dt},$$

which is the required fundamental solution. A few useful additional properties of the diffusion equation pertain to its spatial moments, that is,

$$M_q = \int_{-\infty}^{\infty} x^q \, C \, dx,$$

from which one immediately gathers $M_0 = M$, the mean $\mu = M_1/M_0$, and the variance $\sigma^2 = \int_{-\infty}^{\infty}(x - \mu)^2 C(x,t)dx = M_2/M_0 - \mu^2$. In the specific case of the general diffusion equation, one finds

$$M_0 = 1, \qquad \mu = 0, \qquad \sigma^2 = 2Dt,$$

the latter implying that under the diffusive approximation, the spreading grows linearly with time. One last remark pertains to the property, obvious for the Gaussian distribution, stating

$$\frac{d\sigma^2}{dt} = 2D, \tag{3.71}$$

which is also true for *any* concentration distribution, provided that it is dispersing in accordance with the diffusion equation in a one-dimensional system of infinite extent and that the concentration is always zero at $x = \pm\infty$.

Diffusion by continuous molecular movements postulates that the fluid is stationary and that mass transport is due to diffusion alone. If the fluid itself is moving with a given velocity field, say, $\mathbf{u} = \left[u_x, u_y, u_z\right]$, we may certainly assume that any particle within the fluid environment is displaced by taking on the local value of the velocity field. No other drift (say, a downward pull by gravity) affects such particles. We term the transport by the fluid advection, and assume that transport by the combined actions of advection and diffusion is an additive process, tantamount to assuming that diffusion takes place within the moving fluid exactly as it would have done if the fluid were stationary. We shall initially assume diffusion coefficients D constant in all directions.

Unsteady, spatially heterogeneous advection fields $\mathbf{u}(\mathbf{x},t)$ entail several important consequences. Stochastic models have been engineered to embed uncertainty in the definition of transport variables, whose aim is to relate measurable properties of heterogeneous quantities with predictions of C. The uncertainties intrinsic to the processes are surrogated by the introduction of probability statements: a particle moving within the control volume and driven by a hydrologic carrier flow has a trajectory which, at time $t \neq 0$, can be only partially known. Let $m(\mathbf{x}_0,t_0)$ be the initial mass of a water particle injected at time t_0 in the (arbitrary) initial position $\mathbf{X}(t_0) = \mathbf{x}_0$. Each trajectory is defined by its Lagrangian coordinate

$$\mathbf{X}(t) = \mathbf{X}(t;\mathbf{x}_0,t_0) = \mathbf{x}_0 + \int_0^t \mathbf{u}(\mathbf{X}(\tau),\tau)d\tau. \tag{3.72}$$

The notation emphasizes the Lagrangian character of the analysis where the local values of the concentration depend on the trajectory of the particle injected at a given point and time.

The spatial distribution of mass concentration in the (arbitrary) control volume V as a result of the injection of a *single* particle is given by $C(\mathbf{x},t;\mathbf{x}_0,t_0) = m\delta\left(\mathbf{x} - \mathbf{X}(t;\mathbf{x}_0,t_0)\right)$ [290], where $\delta(\cdot)$ is Dirac's delta distribution, $\int_V Cd\mathbf{x} = m$. The δ distribution (or delta function as sometimes referred to) is a generalized function that we will define here simply by two operational properties: $\int_{-\infty}^{\infty} d\mathbf{x} \, \delta(\mathbf{x}) = 1$ and $\int_{-\infty}^{\infty} d\mathbf{x} \, f(\mathbf{x})\delta(\mathbf{x} - \mathbf{x}_0) = f(\mathbf{x}_0)$. This result posits that, in the one-particle world, concentration (mass per unit transport volume) is nonzero only at the site where the particle is instantaneously residing (i.e., at its advective trajectory). Thus uncertainty in the dynamical specification of the particle is reflected in the transport process. The one-particle, one-realization picture may be generalized for applications relevant to a theory of river networks as ecological corridors, which is often characterized by spatially large injection areas and/or pronounced time variability. Nonpoint sources are characterized by the initial distribution $m_0(\mathbf{x}_0)d\mathbf{x}_0$, and an input mass flux may be defined by $\dot{m}(\mathbf{x}_0,t_0)d\mathbf{x}_0dt_0$. The resulting concentrations are elsewhere [63], and indices will be omitted implicitly, assuming that any boundary value problem of interest here may be referred to properly.

The rate of mass transport through a unit area in the y, z plane by the component of velocity in the x-direction is the quantity uC, because this is the rate at which fluid volume passes through the unit area ($u \times$ unit area = volume/unit time [173]), multiplied by the concentration of mass in that volume. The total rate of mass transport is the sum of the advective and diffusive fluxes. Thus one has

$$q = uC + \left(-D\frac{\partial C}{\partial x}\right),$$

where the first term on the right-hand side is the advective flux and the second the diffusive one. When this expression is substituted into the equation of mass conservation for an incompressible fluid ($\nabla \cdot \mathbf{u} \equiv 0$), we obtain the advection–diffusion equation

$$\frac{\partial C}{\partial t} + \mathbf{u} \cdot \nabla C = D\nabla^2 C.$$

The particular cases that we shall study here refer to the inception of hydrodynamic dispersion due to gradients of the advection field. Let x be the alongstream longitudinal direction, positive along the flow direction, and y a transverse coordinate direction. In practice, the two limiting solutions to the boundary value problems needed for the understanding of the concept of hydrodynamic dispersion are

$$\frac{\partial C}{\partial t} + u\frac{\partial C}{\partial x} = D\left(\frac{\partial^2 C}{\partial x^2}\right),\qquad(3.73)$$

where constant advection and diffusion are assumed; in the second term, the diffusive transport in the x-direction may be regarded as significantly smaller than the advective transport, so that

$$\frac{\partial C}{\partial t} + u\frac{\partial C}{\partial x} = D\left(\frac{\partial^2 C}{\partial y^2}\right).\qquad(3.74)$$

In the system described by the former, Equation (3.73), advective transport and diffusion operate in the same direction. For example, one could consider the problem of solute matter in a pipe filled with a fluid being displaced at a mean flow velocity u by another fluid with a tracer in concentration C_0. At time $t = 0$, a sharp front is given so that $C(x,0) = 0$ for $x > 0$ and $C(x,0) = C_0$, and otherwise for $x \leq 0$. By a coordinate change $x' = x - ut$, one obtains [173]

$$\frac{\partial C}{\partial t} = D\left(\frac{\partial^2 C}{\partial x'^2}\right),$$

which yields the same solutions to any BVP when viewed in a coordinate system moving at speed u:

$$C(x,t) = \frac{C_0}{2}\left[1 - \mathrm{erf}\left(\frac{x - ut}{\sqrt{4Dt}}\right)\right],$$

where $\mathrm{erf}(x) = \int_{-\infty}^{x} e^{-t^2}\, dt$ is the error function of argument x.

The second limit yet revealing example, Equation (3.74), represents instead the simplest case of transverse mixing of two streams of different uniform concentration flowing side by side. Since the input is constant, the solution must not depend on time, and the latter equation simplifies to

$$u\frac{\partial C}{\partial x} = D\left(\frac{\partial^2 C}{\partial y^2}\right),$$

with BCs $C(0,y) = 0$ for $y > 0$ and $C(0,y) = C_0$ for $y \leq 0$ and $C(x,\infty) \to 0$, $C(x,-\infty) \to C_0$. The solution derives directly from the previous one, that is,

$$C(x,t) = \frac{C_0}{2}\left[1 - \mathrm{erf}\left(\frac{y}{\sqrt{4Dt}}\right)\right].$$

The case of a maintained point discharge in a two- or three-dimensional flow is interesting because it is usually possible to simplify the problem by one dimension [173]. Suppose that a point source discharges mass at the rate m at the origin of a (x,y,z) coordinate system in a three-dimensional flow, and let the mean velocity be u in the x-direction. For simplicity, let us assume that the diffusion coefficient D is homogenous and isotropic in all directions, so that the diffusion equation becomes

$$\frac{\partial C}{\partial t} + u\frac{\partial C}{\partial x} = D\left(\frac{\partial^2 C}{\partial x^2} + \frac{\partial^2 C}{\partial y^2} + \frac{\partial^2 C}{\partial z^2}\right).$$

A general solution can be obtained by superposing point sources in space and time, using the fundamental solution given and the procedures described in [173]. In most practical cases, however, it is possible to reduce the three-dimensional problem to that of the spread of an instantaneous point source in two dimensions, for which we have already derived the relevant solution (Box 3.16).

Box 3.16 More on Superposition of Effects

If one considers the flow as consisting of a series of parallel slices of thickness dx bounded by infinite parallel $y - z$ planes, the slices are being advected past the source, and during the passage, each one receives a slug of mass of amount $\dot{m}\delta t$ in the time taken for the slice to pass the source ($\delta x/u$). The mass per unit area in the slice is

$$\frac{\dot{m}\delta x}{4\pi Dtu}\exp\left[-\frac{y^2 + z^2}{4Dt}\right],$$

and if we recognize that the location of the slice is given by $x = ut$ and that the three-dimensional concentration is the mass per unit area in the slice divided by the thickness of the slice, we have

$$C(x,y,z) = \frac{\dot{m}}{4\pi Dx}\exp\left[\frac{(y^2 + z^2)u}{\sqrt{4Dx}}\right].$$

Box 3.16 *Continued*

It should be noted, however, that the above solution is obtained by neglecting diffusion in the direction of flow. Diffusion in the flow direction produces spreading characterized by a length proportional to $\sqrt{2Dt}$ (the standard deviation of a diffusing cloud), and the distance from the source to the slice is $x = ut$. Therefore, the diffusion in the x-direction can be neglected if $ut \gg \sqrt{2Dt}$ or, alternatively, $t \gg 2D/u^2$. In practical problems of relevance to river networks, the value of t required to meet this condition is often very small, so the approximate solutions can be used without difficulty. The resulting approximate solution for spreading from a maintained point source in two dimensions can be shown [173] to be

$$C(x,y) = \frac{\dot{m}}{u\sqrt{4\pi Dx/u}} \exp\left[\frac{uy^2}{\sqrt{4Dx}}\right],$$

where $\dot{m}$ is now the strength of a line source in units of mass per unit length per unit time. This solution is useful in analyzing transverse mixing of a solute discharge from a point source into a river, to evaluate how many widths one must travel downstream to see complete transverse mixing. It should be used only when $t > 2D/u^2$, of course.

It is to be noted that the mathematical treatment of dispersal in this context has been referring only to passive particles. But what about motile species, of obvious interest to river networks as ecological corridors, or about species nonmotile per se but often traveling large distances upstream in a river attached to motile ones (such as *Vibrio cholera* pathogens native to coastal areas colonizing upstream reaches of rivers by traveling attached to the chitineous shells of motile zooplankton)? Clearly, the suite of nature's mechanisms is awesome, and each case shines separately. Suffice here to recall that active advection, propelled autonomously by the diffusing organism, may certainly be a factor clouding the bias induced by a gently flowing stream, in particular for freshwater fish. For an example related to the direction of movement (bias), studies carried out on salmonid fish report different results. Some, in fact, report that the predominant movement (active bias) is upstream [368, 369], while others report a predominantly downstream movement [370] or else no bias at all [371]. Another layer of complexity stems from the fact that the direction of movement could be influenced by factors like the strength of water flow drift or gradients of water temperature [372, 373]. Suffice here to mention that in each example proposed in this chapter or the next (for pathogen dispersal in waterways), the features of the organism at hand will be spelled out in detail,

based on trials for experimental work or on field observations.

In most cases pertaining to fish, for the definition of preferential movement directions, age classes matter for their relative motility. For instance, brown trout can move upstream and downstream, but typically we shall set no preference for young-of-the-year, while there is a positive bias for all other age classes (say, for mature individuals) with a positive bias (that is, a preference) for the upstream direction. In the case of adult fish, a positive bias may be identified in some cases in the downstream direction, because as adult fish grow, they tend to move downstream in the search for better habitats. Dispersal probability matrices can therefore be associated with specific species' ecologies. Obstacles can also be considered [374], as well as spawning and hatching seasons depending on environmental conditions, mainly thermal. Finally, it is relevant to some of the models described here that the attraction of fish schools to a certain destination river reach may be framed in the context of gravity models [105], in that spawning migration patterns may be defined according to a probability of fish movement from node i to node j of the network (then due back to node i), defined by the attractiveness factor of node j, which depends on its score for spawning suitability. A deterrence distance for spawning migration may also be assumed to be dependent on the alongstream distance between the two nodes.

3.5.2 Turbulent Diffusion

Here, following a classic treatment of the subject [173], we show under what conditions turbulent mixing can be described by a diffusion equation – with diffusion coefficients, however, greater than their molecular counterparts by orders of magnitude. Thus mixing in turbulent flow is way more efficient than that generated by still waters or laminar flows. Here, we treat the case of the spread of single clouds of particles, where the coefficient describing the rate of spread increases with the size of the cloud.

Velocities and pressures, ideally measured at a point in the fluid, are fluctuating, defining unsteady fields, and possess appreciable erratic components. As a conceptual example, Figure 3.37 illustrates a typical velocity signal, $u(t)$, observed at the center of a pipe in laminar and turbulent flow. In steady laminar flow the velocity is constant, and no fluctuations are perceived, whereas in steady turbulent flow we see random excursions above and below the constant mean. Turbulent flows occur when inertial forces exceed viscous, stabilizing ones and are epitomized by a dimensionless number, the Reynolds number, Re, exceeding a threshold – in the case of a pipe, $Re = \rho u d / \mu$ (where u is the mean longitudinal velocity shown in Figure 3.37, d is the diameter of the pipe where flow occurs, and $\mu / \rho \sim 10^{-5}$ [m²/s] is the kinematic viscosity of water at 4°) is Re $\gg 2,000$.

Obviously, a formal treatment of turbulence in fluid flows is beyond the scope of this book. However, its macroscopic effects define the range of diffusion coefficients that we may expect in practice for a neutrally buoyant organism drifted by river streamflow – hence a case relevant to the scope of this chapter. A consequence of the random motions induced by turbulent flow is that we can think of turbulent flow occurring in a range of scales of the induced motion of the neutrally displaced organism. To illustrate this concept, following [173], let us consider what happens if a straight plane of dye is painted on fluid particles across the cross section of pipe. In laminar flow, the plane is distorted by the velocity gradient with two possible scales. If the plane is painted near the inlet, the velocity is uniform near the center of the pipe and reduces to zero at the wall across a boundary layer. The only scale of distortion of the plane is the thickness of the boundary layer. If the initial plane is sketched further downstream, in a region where the pipe flow is fully developed, the dye is distorted into a parabolic surface extending over the diameter of the pipe. In this case the pipe size itself provides the scale of the extent of the distortion effect. In the case of turbulent flow, the dye plane can be distorted by the shape of the boundary layer near the inlet or by the fully developed velocity profile farther downstream, but it is also affected by the turbulent fluctuations of the advection field. Indeed, we may be able to observe radii of curvature of the dye plane varying all the way from the diameter of the pipe to very small scale. These strong local curvatures of the flow trajectories define the range of eddy sizes.

Consider now the spreading of a slug of tracer, or a group of marked particles, in turbulent conditions. The mass M of tracer is released at a fixed point within the turbulent flow field. The ensuing spread of the tracer by continuous movements is to be viewed by an observer moving with the mean (i.e., time averaged) velocity of the fluid. We thus assume that the turbulent characters are homogeneous and stationary – particles are blown apart in the same fashion everywhere at all times. The flow is assumed to be essentially two-dimensional, so that a sequence of photographs looking down on the tracer cloud as it spreads gives a proper representation of the phenomenon (the camera travels with the mean speed of the fluid, so no mean displacement is observed for the diffusing clouds). Figures 3.38a,b sketch two possible replicas (that is, identical experiments), where the photos recording the particles' position are recorded at equal times after release into the turbulent flow. Many replicas could be obtained similarly. The results of the two trials shown in Figures 3.38a,b show differences of two different kinds. Small-scale fluctuations cannot be identical for each cloud and distort the shape of the cloud to produce steep concentration gradients. Large-scale fluctuations, especially those substantially larger

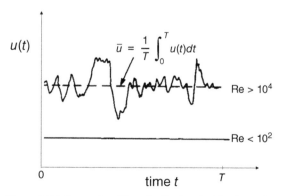

Figure 3.37 Example of longitudinal velocities in a pipe under laminar or turbulent conditions. Shown is a record of longitudinal velocity at the center of a pipe at large and small Reynolds numbers Re. Redrawn after [173]

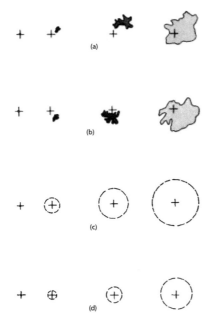

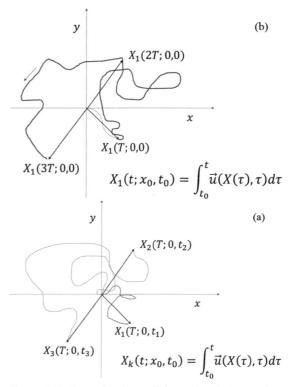

$$X_1(t; x_0, t_0) = \int_{t_0}^{t} \vec{u}(X(\tau), \tau) d\tau$$

$$X_k(t; x_0, t_0) = \int_{t_0}^{t} \vec{u}(X(\tau), \tau) d\tau$$

Figure 3.38 Diffusion in homogeneous, isotropic turbulence with zero mean velocity. (a) Spread of a single cloud at different times. (b) Spread of a replica of (a). (c) Spread of the ensemble mean of a number of replicas. (d) Spread of the ensemble mean of clouds after superposition of the individual centers of mass. Redrawn after [173]

Figure 3.39 Operational ergodicity and turbulent trajectories of marked neutrally buoyant particles. Note that $X = \mathbf{X}$ and $x = \mathbf{x}$ are vectors (here in two dimensions) labeled as scalars for simplicity and that the stationary homogeneous stochastic turbulent field $\mathbf{u}(\mathbf{x}, t)$ is unsteady and heterogeneous as usual in turbulent flow processes. (a) Trajectories $X_k(t; x_0, t_0)$ of three particles, single realizations of independent stochastic processes akin to turbulent ones, diffusing within a stationary homogeneous random velocity field. Each particle is released at the origin but at different initial times, each wandering for a time T. (b) Trajectory of a single particle wandering for a time $3T$, where three separate stop-image wanderings of duration T are tracked. Loosely defined, when a large enough collection of trajectories $X_k(T)$ in (a) and $X_1(kT)$ in (b) provide converging statistics, one speaks of operational ergodicity. Redrawn after [173]

than the cloud itself, displace the entire cloud. Each cloud of particles encounters a different set of large-scale motions, so the motion of the center of mass of each cloud is different [173].

Assume that we released a large number of clouds of particles, one after the other, and that observations describe the spread of each cloud over a sizable period of time. If the mean motion is subtracted, the average position of the center of mass will remain at the origin, although the center of mass of each cloud will in general be elsewhere from the origin, because of large-scale motions specific to each realization of the random process. Each separate cloud will eventually grow bigger than the largest eddies and average out their effects, while the center of mass of each cloud will tend to return to the origin when sufficiently long averaging of the random motions occurs. Two strategies exist to interpret the basic timescales incurred with the various phenomena. One is to average over all of the replicas to obtain an ensemble average. The other is to try to follow each replica separately for a very long time. As noted in [173], a major difficulty stems from the fact that the ensemble

average concentration at a point in space and time is likely to be an average of a large number of zeros (times when the tracer cloud does not cover the point) plus a few large values. This type of average may be meaningless to an organism subject to release of a pollutant, because the few high values of concentration may kill the organism (and the large numbers of zeros cannot bring it back to life [173]). On the other hand, not enough is known about turbulence to permit the computation of the spread of each realization or their peak concentration at any instant. In most cases, a statistical estimate of the size of an individual cloud suffices. The difference is illustrated by Figure 3.38, where, besides snapshots of two realizations of diffusing clouds (where darker colors imply larger concentrations), the growth of the ensemble average is shown (Figure 3.38c). Figure 3.38d shows instead an average obtained by superposing the centers of mass of each of the individual clouds and then averaging over the ensemble of releases. The mathematical implications of the operations involved are noteworthy, and it may be shown that the average extent of each individual cloud is smaller than the extent of the ensemble average (this is because the ensemble average includes the distribution of the centers of mass). Various physical settings may be sensitive to such a result, also appealing to common sense. The various kinds of averages, and most relevant concepts of statistical analysis of turbulence, are to be found elsewhere [173]. We shall be mostly concentrated here on the analysis of the spread of the ensemble average and of single clouds.

To fully understand the statistical implications of any theory of turbulence, one needs to recall the definition of the ensemble of trajectories defined by their random coordinates $\mathbf{X}(t)$ (Equation (3.72); see Box 3.17). The statistics of the particles' motion may be different if the average is made across the totality of experiments performed or else is an average over a limited period of time in each experiment. Figure 3.39 illustrates the first type of average; a number of particles are released at different times, and the displacement $\mathbf{X}_k$ of the kth particle at time T after its release is observed. Suffice here to note that $\mathbf{X}(t)$ is a random function whose statistical properties can be determined if a large number of experiments are carried out. An average carried out in this

manner on a suitably large number of replicas is called an ensemble average. Figure 3.39b illustrates the second type of average, illustrating the case where a single particle is released and followed through a large number of time increments each of duration T. The displacement $\mathbf{X}$ during each time increment is observed and recorded. This $\mathbf{X}$ is also a random variable, but its statistical properties in general differ from those obtained by an ensemble of experiments. If the time series average and the ensemble average coincide, in some limit, we term such a process ergodic.

In any one realization, instead of taking the initial position x_0 as fixed, we let $\mathbf{X}$ trace out an erratic line in space as t elapses in that realization. In this case, we can consider the point $\mathbf{X}$ as fixed and the point x_0 as a function of time. This view is that which an observer would see if she fixed her attention at a particular $\mathbf{X}$ in a particular experiment and at each time interpreted from where the particle originated, which is exactly at $\mathbf{X}$ at the time of observation. Notice that in hydrological applications, this perspective defines a Kolmogorov backward problem [375] (tantamount to asking "Given that a particle [or a given eDNA concentration; see Section 3.6] is observed here at time t, how many sites could be its origin at time $t = 0$?" – the polydisperse nature of many biological invasion processes suits this framework well). Thus, by fixing x_0, trajectories trace out a path of a particular particle, and by fixing the observation station $\mathbf{X}$, the inverse problem (possibly) reveals the initial coordinates of the particle presently at $\mathbf{X}$ at t. These viewpoints are central to turbulent diffusion because molecular diffusion merely smooths out sharp small-scale irregularities, and it is the turbulent motion of the fluid particles that really does the job of spreading the dispersing particles. Further relevant technicalities are shown in Boxes 3.17 and 3.18.

Box 3.17 Deterministic Boundary Value Problems and Probabilistic Dynamics

The trajectory $\mathbf{X}(t)$ of a marked particle advected by an arbitrary turbulent field $\mathbf{u}(\mathbf{x}, t)$ that tracks its position in time after labeling its initial position $\mathbf{x}_0$ at time t_0 has been defined in Equation (3.72). By sampling different realizations of the random process, for each release of a particle, there will be a different outcome reflecting the random nature of the trajectories in turbulent fluids. $\mathbf{X}(t)$ is seen as a stochastic process (i.e., a random function of time), and the position a particle takes at some time $t_0 + \tau$ may bear little correlation to the position at time t_0 if τ is large enough.

A key connection between deterministic boundary value problems and stochastic processes defining random trajectories of neutrally buoyant particles, carried by a flow field, is the physical

Box 3.17 *Continued*

interpretation of the probability density of the trajectory $\mathbf{X}(t)$ of a single realization of a single particle released at $\mathbf{x}_0$ at time t_0, that is, $p(\mathbf{X}(t)|\mathbf{x}_0, t_0)d\mathbf{X}$ is the probability that the trajectory is "around" any value $\mathbf{X}$ at time t. As shown before, with any particular choice of spatial coordinate, it is possible to evaluate the probability that the trajetory would happen to be there at any chosen time t, because one would need to solve the equation

$$\int_{-\infty}^{+\infty} \delta(\mathbf{x} - \mathbf{X}(t))p(\mathbf{X}(t)|\mathbf{x}_0, t_0)d\mathbf{X} = p(\mathbf{x}, t|\mathbf{x}_0, t_0) \propto \langle C(\mathbf{x}, t) \rangle,$$

from which a distinguished literature evolved [290].

The connection between the probability of the position in time of particles (many particles, many realizations – or any ergodic limit for one or both tending to infinity) displaced by the flow field and the deterministic equation of mass continuity is a fundamental physical interpretation. Here, we shall refer to the particle displacements $\mathbf{X}$ as stationary; that is, we assume that all the moments of the distribution of the displacement have become independent of the individual time origins and thus that they depend only on the time difference $t - t_0$. Processes are called stationary then. The analogy between the position of swarms of particles and the solution of diffusion equations for a mean concentration field centers the attention on the statistical mechanics of those particles. In addition, it brings attention on the covariances of the trajectories issuing from two different points at the same time. Particle displacement processes are called homogeneous if these same quantities depend only on the relative displacement $|X - \xi|$ and not on the initial position ξ. From the above positions, one may explain in particular the meaning of the various averages (see, e.g., Figure 3.38c). For instance, if one considers the average of an ensemble of clouds (equivalent to many particles, many realization experiments of diffusion of a slug of mass M), the expected position of the center of mass of the ensemble of several clouds, each with its centroid $\bar{X}$, is thus

$$\langle \bar{X} \rangle = \frac{1}{M} \int_{-\infty}^{\infty} \int_{-\infty}^{\infty} \int_{-\infty}^{\infty} dx\,dy\,dz \; x \; C(x, y, z, t),$$

whose variance is the sum of two effects (the individual cloud variance plus the variation embedded in the different realizations). We are interested in the diffusion of the ensemble mean concentration because ergodicity implies that the ensemble average, in nonanomalous diffusion problems that are not the subject of this book, is never far from any single realization, that is, $C \sim < C >$. Key, with respect to this problem, is recognizing that the problem of finding the ensemble mean size of the cloud is equivalent to finding the ensemble mean square displacement of the fluid particles passively advected by the fluid flow.

Box 3.18 Turbulent Diffusion and Correlations of the Flow Field

We shall briefly cover here the classic analysis of G. I. Taylor [290]. Let U be the velocity of the particle, immersed in a one-dimensional velocity profile aligned in the x-direction. The field may be assumed to have zero mean without loss of generality. One may thus assume a homogeneous fluid flow, that is, $U(t)$. Without loss of generality, we shall assume $\mathbf{x}_0 = x_0 = 0$ and $t_0 = 0$. Then the location of the particle, $X(t)$, at the arbitrary time t is

Box 3.18 *Continued*

$$X(t) = \int_0^t U \, dt.$$

The equal-time covariance $X^2(t)$ of two independent displacements thus reads

$$X^2(t) = \left(\int_0^t U \, d\tau_1 \right) \left(\int_0^t U \, d\tau_2 \right),$$

whose ensemble mean is

$$\langle X^2(t) \rangle = \int_0^t \int_0^t \langle U(\tau_1) U(\tau_2) \rangle \, d\tau_1 d\tau_2.$$

The ensemble average $\langle U(\tau_1) U(\tau_2) \rangle$ is physically interpreted as the covariance (i.e., the ensemble average over a large number of trials) of the product of the velocity of a single particle at time τ_1 multiplied by the velocity of the same particle at time τ_2. Because the turbulent field is assumed to be in stationary conditions, this can only be a function of the difference between $\tau_2 - \tau_1$. Therefore, we can define a correlation coefficient as [173]

$$R_U(\tau_2 - \tau_1) = \frac{\langle U(\tau_1) U(\tau_2) \rangle}{\langle U^2 \rangle}$$

where $\langle U^2 \rangle =< U(0)U(0) >$ is termed the (square of the) intensity of the turbulent field. R_U is an autocorrelation function, assumed to be stationary, which is Lagrangian in character, in the sense that the velocity field is sampled following the motion of an individual particle rather than that of a fixed point in space (the Eulerian perspective). So the evolving "moment of inertia" (in the language of the mechanics of many particle systems) of the diffusing cloud (whose rate of evolution defines the asymptotic diffusion coefficient) becomes

$$\langle X^2(t) \rangle = \langle U^2 \rangle \int_0^t \int_0^t R_U(\tau_2 - \tau_1) \, d\tau_1 d\tau_2.$$

The remainder of the analysis [173, 290] is exact. By changing the variables of integration ($s = \tau_2 - \tau_1$ and $\tau = (\tau_2 + \tau_1)/2$), one gets

$$\langle X^2(t) \rangle = \langle U^2 \rangle \int_0^t (t - s) \, R_U(s) \, ds,$$

noting that if a three-dimensional flow field would be operating, as noted by [173], one would get the moments in the other coordinate directions as well ($< Y^2 >, < Z^2 >$) and the cross-covariances for clouds lacking any symmetry. The beauty and the generality of Taylor's analysis stem from the possibility of exploring exactly the properties of two limit cases, those for $t \to 0$ and $t \to \infty$, respectfully, for which the correlation of Lagrangian velocities issued at the same spot is either perfect or null. For perfectly correlated velocities, one has $R_U \sim 1$, giving the limit growth for initial times, that is,

$$\langle X^2(t) \rangle = \langle U^2 \rangle t^2.$$

The asymptotic limit for $t \to \infty$ implies the limit of the correlation for velocities that become progressively negligible. Grossly, one has $R_U \to 0$, but a more relevant quantity is the correlation

Box 3.18 *Continued*

scale, or macroscale, Λ that physically describes the time required by a homogeneous turbulent field to lose memory of its initial position, that is,

$$\Lambda = \int_0^\infty R_U(s)\,ds,$$

which has to be finite. With the additional, reasonable condition that $\int_0^\infty s R_U(s)ds$ is also bounded, one obtains the limit

$$\langle X^2(t)\rangle \;\to\; 2\langle U^2\rangle \Lambda t,$$

from which the asymptotic diffusion coefficient (i.e., constant) is derived as

$$\frac{d\langle X^2(t)\rangle}{dt} = 2\langle U^2\rangle\Lambda,$$

which shows that after some time, long enough to assume that memory of the initial features of the flow field are forgotten by the particle advected passively by the turbulent flow ($t \gg \Lambda$), the variance of the ensemble-averaged concentration distribution for clouds dispersing in a stationary homogeneous field of turbulence grows linearly with time. The time required to reach such regime is proportional to the Lagrangian timescale Λ.

Thus a properly diffusive equation for turbulent flow fields – in the sense of linearly growing spatial variance in time, that is, the rate of size growth of the "cloud" – operates only asymptotically. Framing the transient thus becomes important. In fact, Fick's law leads to the derivation of the diffusion equation whose fundamental solution is the Gaussian distribution. A property of the diffusion equation is that the variance of a concentration distribution grows linearly with time (i.e., Equation (3.71)). After some start-up time, the random walk generates a Gaussian distribution of displacements; that is, the Gaussian distribution implies that the diffusion equation describes processes in which the variance of a spreading cloud of molecules undergoing a random walk grows linearly with time. Taylor's analysis shows that after some start-up time, the variance of a spreading cloud of particles in stationary homogeneous turbulent motion also grows linearly with time and therefore suggests that we can define a turbulent mixing coefficient, analogous to the molecular diffusion coefficient but larger by orders of magnitude, defined by the relation (analogous to Equation (3.71))

$$\epsilon_x = \frac{1}{2}\frac{d\langle X^2(t)\rangle}{dt} = \langle U^2\rangle\Lambda, \tag{3.75}$$

where obviously ϵ has units $[L^2/T]$. Thus, mathematical details aside (in particular related to whether the linear growth of the variance is a sufficient condition for a diffusion equation to hold), one can therefore confidently write that the spread of the ensemble mean concentration may be described by a diffusion equation whose simplest three-dimensional form is

$$\frac{\partial C}{\partial t} = \epsilon_x \frac{\partial^2 C}{\partial x^2} + \epsilon_y \frac{\partial^2 C}{\partial y^2} + \epsilon_z \frac{\partial^2 C}{\partial z^2}. \tag{3.76}$$

Equation (3.76) is written for zero mean flow velocity. If the flow field has a finite mean velocity, it is necessary to add the related terms (e.g., $u\partial C/\partial x$), as was done for Fickian diffusion. Relative diffusion of clouds will not be treated here for the nature of the fluvial fluid flows of relevance to biological invasions.

3.5.3 Shear Flow Dispersion

Taylor's framework [290] has been extended to a variety of environmental flows, and it can be used to give a reasonably accurate estimate of the rate of longitudinal dispersion in rivers, the objective of our analysis. Common to the upscaling experienced by large-scale flows is spreading in the direction of flow affected primarily by heterogeneities in the velocity profiles developing within the river cross sections. Flows exhibiting velocity gradients are often referred to as shear flows, because gradients and shear stresses are linked.

In Box 3.17 we have shown how the rate of spreading is determined by turbulent velocity fluctuations. Mixing by a turbulent homogeneous field is asymptotically described by a constant diffusion coefficient $\langle U^2 \rangle \Lambda$ [L^2/T], where Λ is the Lagrangian correlation scale in time and $\langle U^2 \rangle$ the intensity of the turbulent field, subsumed by the variance of the pulsating velocities. The message is clear: the mixing process is mathematically the same as operated by small-scale mechanisms driven by a molecular diffusion D, only much more effective, because the force of the mixing is measured by the diffusion coefficient $K \gg D$. A simple example of direct calculation of K is reported in Box 3.19.

Box 3.19 Longitudinal Dispersion in a Parabolic Shear Flow

A simple case to elucidate the basis of shear flow dispersion is the case of shear flow developing within a uniform laminar flow between two planes of infinite width. This two-dimensional flow field is guided by the two walls separated by a distance h in such a way that velocities are directed only in the x-direction. However, gradients are induced by shear stress forced by the forced null velocities at the solid wall, resulting in a longitudinal velocity profile $u(y)$. Whatever the form of $u(y)$, the mean velocity is calculated as $\bar{u} = \frac{1}{h} \int_0^h u(y) \, dy$, and the deviation of the velocity from the cross-sectional mean is defined as $u'(y) = u(y) - \bar{u}$. Let the flow transport be a solute whose steady state concentration field is $C(x, y)$. Let D denote the molecular diffusion coefficient, assumed as customary, as constant, and as isotropic. The mean transverse concentration at any cross section is therefore $\bar{C}(x) = \frac{1}{h} \int_0^h C(x, y) \, dy$, and in analogy, $C'(x, y) = C(x, y) - \bar{C}(x)$. Because the flow is maintained only in the x-direction, the diffusion equation becomes

$$\frac{\partial (C' + \bar{C})}{\partial t} + (u' + \bar{u}) \frac{\partial (C' + \bar{C})}{\partial x} = D \left[\frac{\partial^2 (C' + \bar{C})}{\partial x^2} + \frac{\partial^2 (C')}{\partial y^2} \right],$$

which can be simplified by a transformation to a coordinate system moving at the mean flow velocity, that is, $\xi = x - \bar{u}t$, yielding

$$\frac{\partial (C' + \bar{C})}{\partial t} + u' \frac{\partial (C' + \bar{C})}{\partial \xi} = D \left[\frac{\partial^2 (C' + \bar{C})}{\partial \xi^2} + \frac{\partial^2 (C')}{\partial y^2} \right],$$

which allows us to view the flow as an observer moving with the mean velocity. The rate of spreading along the flow direction x due to the velocity profile $u(y)$ greatly exceeds that due to molecular diffusion. In fact, assuming this to be proved, we must then neglect the related diffusive terms, that is,

$$\frac{\partial C'(y)}{\partial t} + \frac{\partial \bar{C}(x)}{\partial t} + u' \frac{\partial C'(y)}{\partial \xi} + u' \frac{\partial \bar{C}(x)}{\partial \xi} = D \left[\frac{\partial^2 C'(y)}{\partial y^2} \right].$$

Taylor [290] proposed to discard three out of four terms of the left–hand side (for a technical proof of the order of magnitude of the terms discarded, see [173], p. 84), yielding the tractable problem

$$u'(y) \frac{\partial \bar{C}(x)}{\partial \xi} = D \left[\frac{\partial^2 C'(y)}{\partial y^2} \right] \quad \text{with} \quad \frac{\partial C'}{\partial y} = 0 \quad \text{at } y = 0, h. \tag{3.77}$$

Box 3.19 *Continued*

Solution is obtained by direct integration,

$$C'(y) = \frac{1}{D}\frac{\partial \bar{C}(x)}{\partial x} \int_0^h dy \int_0^h u' dy + C'(0),$$

and the diffusive mass flux M is given by

$$M = \int_0^h u'C' dy = \frac{1}{D}\frac{\partial \bar{C}(x)}{\partial x} \int_0^h u' dy \int_0^h dy \int_0^h u' dy \propto h\frac{\partial \bar{C}(x)}{\partial x},$$

because the added term $\int_0^h u'C'(0)dy = 0$. The total mass transport in the streamwise direction is thus proportional to the concentration gradient in the streamwise direction, which is the result postulated for molecular diffusion. The main difference is that the coefficient of proportionality depends on the whole field of flow.

Several relevant analytical or numerical solutions exists, mediated by suitable interpretation of the actual gradients of velocity deviations from the mean advection $\bar{u}$. So, if one posits $M = -hK\partial \bar{C}(x)/\partial x$, the related dispersive equation reads

$$\frac{\partial \bar{C}}{\partial t} + \bar{u}\frac{\bar{C}}{\partial x} = K\frac{\partial^2 \bar{C}}{\partial x^2}, \tag{3.78}$$

from which all evaluations of total mass flux, including those of dispersive origins – whether driven by molecular processes or by macroscopic shear flows across scales – are derived. Equation (3.78) is particularly relevant to the analysis of dispersion in environmental flows, such as river, estuarine, and coastal waters [173]. Among the foremost results, suffice here to mention a few exact results (see [173] for a complete derivation) where small-scale molecular fluxes proportional to the local concentration gradients via a molecular diffusion D have the merit of forcing the dispersing particles to sample the heterogeneity of the velocity flow. As a result, after a long enough transient, is the complete mixing along the directions transverse to the main flow field, resulting in a concentration field constant in the transverse direction and a diffusing-like one in the direction of the shear advection flow.

In two parallel plates of infinite extent, separated by a distance h, where Newtonian shear and a constant velocity gradient are generated by a laminar flow to infinitely wide plates, the top plate is moving relative to the bottom one, and the space between the plates is filled with fluid. For simplicity, assume that the top plate is moving in one direction with velocity $U/2$ and that the bottom plate is moving in the other direction with velocity $U/2$. In laminar flow the velocity distribution between the plates is given by $u(y) = Uy/h$ from $-h/2$ to $h/2$. Integration leads exactly to $\bar{u} = 0$ and

$$K = \frac{U^2 h}{120D},$$

which shows quantitatively that the engine for sampling the macroscopic heterogeneity of the shear flow field is still the basic mechanism of molecular diffusion proportional to D via local concentration gradients à la Fick. Molecular diffusion is essential, however, for otherwise a single molecule would not be perturbed from a single trajectory in the x-direction, thus being unable to sample the shear flow. The asymptotic dispersive process, however, is only qualitatively of the same

Box 3.19 *Continued*

type, because it is orders of magnitude more efficient because the rate of change of the variance σ of the Gaussian concentration field solution to Equation (3.78) is

$$\frac{d\sigma^2(t)}{dt} = 2K + 2D, \qquad \text{where} \quad K \gg D,$$

as readily shown by using the concentration moment method [173].

A summary of results that are useful to our description of longitudinal dispersion in river stretches closes this section.

In general, after a tracer has become adequately mixed across the entire cross section of width W and depth d, the final stage of the macroscopic transport process is the reduction of alongstream gradients of concentration by longitudinal dispersion. Biological invasions along waterways, the key concern of this chapter, underpin cases where longitudinal dispersion is important. We therefore need to discuss how to estimate the longitudinal dispersion coefficient K in a natural stream for use in the one-dimensional dispersion equation (3.78) for $\bar{C}(x,t)$, the concentration fully mixed over the cross section. In turn, the drivers are the mean advective velocity $\bar{u}$, taking the role of the constant downstream drift, and K, the proper longitudinal dispersion coefficient as a by-product of turbulence and of the heterogeneities of the velocity field. This describes longitudinal dispersion when $x \gg L$, that is, after a length L from the injection required for complete mixing over the width W, $L > 0.1\bar{u}(2W)^2/\epsilon_t$, (where ϵ_t is the transverse turbulent diffusion coefficient, $\epsilon_t \simeq d\sqrt{gdS_0}$, and S_0 is the reach slope; see Section 2.1). Notice that we have assumed here $W \gg d$, so the limiting mixing length would be given by the transverse (as opposed to vertical) mixing, a condition normally met by real rivers [101].

Elder's analysis of dispersion due to a logarithmic velocity profile $u(y)$ in a prismatic open channel flow [173] led to the result

$$K = 5.93 \, d\sqrt{gdS_0}.$$

The expectations that Elder's relation might describe what happens in real rivers proved inadequate to describe field observations. Experiments were performed in various streams under a broad spectrum of hydrologic conditions, and the observed values of $K/(du\sqrt{gdS_0})$ were almost inevitably orders of magnitude larger than the the-

oretical value of 5.93, and actually in the range 140–500. Field results collected in particularly complex braided channel dispersion reached values as high as 7,500, whereas the lowest value recorded, observed for a very straight, artificial, and lined canal, led to $K \approx 9\,d\sqrt{gdS_0}$. Thus the empirical evidence suggests that Elder's result does not apply to real streams [173]. Approximate procedures often work reasonably well. One result worth mentioning for its ability to capture the order of magnitude of actual longitudinal dispersion is [173]

$$K = 0.011 \, \frac{\bar{u}^2 W^2}{d\sqrt{gdS_0}}$$

(with usual symbol notation), which is the reference value for the hydrodynamic dispersion in real rivers used in this book. In practice, K is ordinarily in the range $10 \div 100$ [m^2/s].

3.5.4 Geomorphological Dispersion

Here, we will place a special emphasis – for reasons that will become evident in the context of our efforts to interpret environmental DNA measurements at a station (Section 3.6) – in the dispersive effects of geomorphologic nature related to the heterogeneity of the paths from any network site to the common outlet. The closure of the network, the measurement site, acts in fact as a trapping state implicit in the hierarchical structure of any river network, which is seen as a directed graph [63]. We shall study the dispersive effects produced by the river network when measuring at the closure the effects of a polydisperse instantaneous injection of marked particles (representing matter or organisms transported by streamflow). Genetic material shed locally by sessile or moving sources – hence passive to transport but reactive owing to its decay, prompted by a number of mechanisms examined later – are particular cases of ecological and hydrological interest. These particles, however initially distributed along the various reaches, are hierarchically transported by the treelike river network and disperse under the action of hydrologic drift and shear flow dispersion. We shall assume here that a linear advective–diffusive regime is appropriate owing to the low concentrations of particles tracked. Under such an assumption, superposition of local concentrations stemming from different sources but arriving at the measurement station at the same time applies.

A dispersive effect of a completely different nature would be seen by an observer measuring at-a-point concentrations in times of initially polydisperse mass injections. It is related to the effects of the different lengths and connections of the drainage paths that may lead to the measurement section. This effect has been termed geomorphological dispersion [174] and can be characterized analytically.

Transport from nonpoint sources observed at a fixed position x_0 in space implies a distribution of travel times for particles (injected anywhere at $t = 0$) that are collected at x_0 at any subsequent time t. Travel times are the collection of the particles' times of residence in the network since their injection, calculated when the particles manifest at the measurement section, that is, the distribution of arrivals. Such distribution reflects the structure of the pathways leading to the trapping state. In fact, the heterogeneity of travel times to the measuring station subsumes that of the paths leading to it, together with the spatial distribution of the sources of the released mass. The net result is an asymptotically diffusive effect that is imprinted in the shape of the river network and in the main characters of its hydrodinamic dispersion, to which it is additive [174, 375]. Here, we shall confine our mathematical derivations to the case of conservative, passive matter for simplicity, as the focus is firmly on the impact of river network morphology on the dynamics of biological invasions. An outlook of how this property may inform further research topics relevant to eDNA transport in river networks are given inChapter 5.

Box 3.20 Formulation of Transport by Travel Time Distributions

Let m be the initial mass of a reactive particle ($m(\mathbf{X}, t)$) injected at time t_0 in the (arbitrary) initial position $\mathbf{X}(t_0) = \mathbf{x}_0$. Each trajectory is defined by its Lagrangian coordinate $\mathbf{X}(t)$. The multiparticle, multirealization concentration field is therefore $C(x, t) \propto p(x, t)$, that is, it is proportional to the probability that m is in $[x, x + dx]$ at time t. Note that indices related to boundary conditions, say, point versus nonpoint, steady or time varying, will be omitted, thereby implicitly assuming that the formulation adopted may be referred to all cases without affecting the formalism. Let $p(\mathbf{X})d\mathbf{X}$ be the probability that the particle is in $(\mathbf{X} - d\mathbf{X}, \mathbf{X} + d\mathbf{X})$ at time t (notice that the functional dependence $p(\mathbf{X})$ implies $p(\mathbf{x}, t)$ in terms of Cartesian coordinates, because of the evolution of the trajectory with time). The ensemble average (i.e., many realizations) concentration $\mathbf{C}(\mathbf{x}, t)$ over all possible paths is then given by the classic relation [290]

$$\bar{C}(\mathbf{x}, t) = \int_{-\infty}^{\infty} m\, \delta(\mathbf{x} - \mathbf{X})\, p(\mathbf{X})\, d\mathbf{X}. \tag{3.79}$$

A particular case of (3.79) may be thought of as significant for hydrologic transport processes, that is, that of the transport of conservative matter where no distinction is drawn between the carrier and the carried particles in terms of travel times. This will be the case if neither the advection field is affected by the concentration of solute nor the solute undergoes mass exchange phenomena, thereby somewhat restricting the margins of ecological analysis. Mathematically, the assumption of passive solute postulates that the mass m with which each particle is labeled is conserved throughout the transport process, that is, $m(\mathbf{x}, t) \approx m$. The passive behavior is further described by the lack of dependence of the velocity field characterizing $\mathbf{X}$ on the actual values of the concentration of mass in the field.

Mixing and dispersion mechanisms are due to the combined effects of the geometry of the transport volume, of molecular diffusion, and of the heterogeneous velocity fields. Integration of

Box 3.20 *Continued*

Equation (3.79) yields Taylor's [290] relation conceived for stationary turbulence, but of general validity:

$$\bar{C}(\mathbf{x},t) = m\, p(\mathbf{x},t), \tag{3.80}$$

which links the stochastic displacement mechanics of individual particles with the concentration of a swarm of them carried by the flow field. The distribution $p(\mathbf{x},t)$ is the displacement probability density function.

Decoupling the flow field $\mathbf{u}(\mathbf{x},t)$ from the properties of dispersing matter is allowed in the above framework of passive transport, save for density dependence, which may be important for fish under certain conditions [376], overlooked for now. The Lagrangian coordinate of a single particle is defined by the relationship $\mathbf{X}(t;\mathbf{x}_0) = <\mathbf{X}> + \mathbf{X}'(t;\mathbf{x}_0) + \mathbf{X}_B(t)$. Taking the velocity field as $\mathbf{u}(\mathbf{x},t) = \bar{u} + \mathbf{u}'(\mathbf{x},t)$, where $\mathbf{u}'$ stands for velocity fluctuations, we obtain $\langle\mathbf{X}\rangle = \bar{u}t$ and $\mathbf{X}'(t,\mathbf{x}_0) = \int_0^t \mathbf{u}'(\mathbf{X}_\tau,\tau)d\tau$. In general, $\mathbf{X}_B(t)$ is an isotropic Brownian motion component either of molecular origin or condensing information motility of the particles, seen as active dispersants. Mathematically, one has $\langle\mathbf{X}_B(t)\rangle = 0$ and $\langle\mathbf{X}_B(t)^2\rangle = 2Kt$, where K is a homogeneous diffusion coefficient for the Brownian motion component, in this case, hydrodynamic dispersion in Equation (3.78) (see Box 3.17).

One can establish the relation between the displacement pdf $p(\mathbf{x},t) = p(x,t)$ for the case at hand. Let $\mathcal{V}$ be the transport volume in which a control section is defined. We assume that the time t in which a particle crosses the control section is unique and, most importantly, that all particles injected in $\mathcal{V}$ ensuing from $\mathbf{x}_0 \in \mathcal{V}$ must transit the predefined absorbing barrier. Owing to the uncertainty characterizing $\mathbf{X}$, the arrival time at the absorbing barrier is a random variable T characterized by a probability density function $f(t)$ and a distribution function $P(T<t) = F(t;\mathbf{x}_0,t_0)$, that is, the probability that the particle originated at $\mathbf{x} \in \mathbf{x}_0$ at $t = t_0$ has already crossed the trapping state at time t. The link of the Eulerian and the Lagrangian approaches is defined by the relationship

$$P(T<t) = 1 - \int_{\mathcal{V}} p(\mathbf{x},t;\mathbf{x}_0,t_0)d\mathbf{x}, \tag{3.81}$$

which states the obvious: the complement of the probability that T is less than t (that is, the particle is still in the transport volume) is the sum of its probabilities of being everywhere in $\mathcal{V}$ at time t. Upon substitution, the fundamental relations for the probability of exceedence and the distribution $f(t)$ of travel times within a network of one-dimensional links are

$$P(T \geq t) = \frac{1}{m}\int_{\mathcal{V}} \bar{C}(x,t)dx = \frac{\langle M(t)\rangle}{m} \tag{3.82}$$

$$f(t) = \frac{dP(T<t)}{dt} = -\frac{1}{m}\frac{d\langle M(t)\rangle}{dt},$$

where $\langle M(t)\rangle$ is the (ensemble mean) mass still in the transport volume at time t normalized by the initial mass injected m.

Equation (3.83) connects the complete, deterministic boundary value problem solving the mass balance equation with the probabilistic formulation of transport by travel time distribution. In the simplest case, the only way for the injected mass to escape the control volume is by exiting through

Box 3.20 *Continued*

discharge at the outlet. In this case, from mass continuity and $t > 0$, one has $d\langle M\rangle/dt = -Q_M(t)$ (where $Q_M = Q(t)\mathscr{C}(t)$ is the mass flux exiting the control volume, that is, flow discharge Q [L^3/T] times the so-called flux concentration $\mathscr{C}$ [M/L^3]). Therefore one obtains the key probabilistic statement of transport at the scale of whole networks by

$$f(t) = \frac{1}{m}Q_M(t). \tag{3.83}$$

For instantaneous unit injections ($m = 1$), the probability density of travel times is therefore the mass flux at the control section.

Box 3.21 Travel Time Distributions in River Network Reaches

We shall now derive, following [174], the travel time distribution for the transport problem where the mean convection ($\bar{u}$) and the dispersion coefficient K are taken as their constant asymptotic values. Such a solution proves useful in ecohydrologic applications. In this case the probability $p(x,t)dx$ for a particle of being in $(x, x + dx)$ at t is computed by the general model of longitudinal dispersion described by Equation (3.78), where the constant mean convection is directed along the alongstream direction x. Equation (3.78) requires specification of two conditions for the spatial coordinate x. A reference solution (matters of boundary conditions particular to each problem will be discussed later in this box) involves a Dirac-delta input at $x = 0$ (say, the displacing particle is at $x = 0$ at $t = 0$ with probability 1) and $p(\pm\infty,t) = 0$. Also, $p(x,0) = 0$, except at the origin, where a unit pulse of flux at $x = 0$ is enforced as

$$\left|\bar{u}p(x,t) - K\frac{\partial p}{\partial x}\right|_{x=0} = \delta(t). \tag{3.84}$$

The above condition specifies that the total flux (convective and diffusive) at the input control surface is impulsive. The general solution in the above conditions is the standard Gaussian model:

$$p(x,t) = \frac{1}{\sqrt{2\pi Kt}}\exp\left(-\frac{(x-\bar{u}t)^2}{4Kt}\right). \tag{3.85}$$

One can thus write

$$f(t) = -\frac{d}{dt}\int_{\mathscr{V}}p(x,t)dx. \tag{3.86}$$

We now turn to the control volume $\mathscr{V}$. In there we consider $x \in (-\infty, L)$, L being the x-coordinate of the control section acting as an absorbing barrier. The fact that the displaced particle once its coordinate x exceeds $x = L$ cannot return into the transport volume is defining mathematically the absorbing barrier assumption. A reflecting barrier is assumed at $x = -\infty$ describing the fact that the particle cannot leave the transport volume except through the control section at $x = L$. Thus one has

$$f(t) = -\frac{d}{dt}\int_{-\infty}^{L}dx\,p(x,t).$$

Box 3.21 *Continued*

The general solution for the case of an absorbing barrier at $x = L$ and a reflecting barrier at $x = -\infty$ is thus obtained by integrating the marginal Gaussian distribution $p(x,t)$ in space from $x = -\infty$ to $x = L$ and taking its time derivative by Leibnitz's rule:

$$f(t) = \frac{1}{2}\left(\frac{L}{t} + \bar{u}\right)p(L,t),$$ (3.87)

with

$$p(L,t) = \frac{1}{\sqrt{2\pi Kt}}\exp\left(-\frac{(L-\bar{u}t)^2}{4Kt}\right).$$ (3.88)

A more general solution can be obtained by the above procedure whenever K depends on time t [174] (recall that we set $t_0 = 0$ for simplicity of notation without loss of generality). In this case, we define

$$X_{\|}(t) = 2\int_0^t K(\tau)d\tau.$$ (3.89)

It has been shown [174] that the travel time distribution $f(t)$ with the same set of boundary conditions becomes

$$f(t) = \left(\frac{L-\bar{u}t}{2X_{\|}(t)}\frac{dX_{\|}}{dt} + \bar{u}\right)p(L,t),$$ (3.90)

which reduces to Equation (3.87) if K becomes asymptotically constant. It can be shown, by numerical comparisons, that Equations (3.87) and (3.90) with a constant reference value K yield quite similar travel time distributions for a broad range of variation in $K(t)$ [174].

Travel time distributions at the outlet of a system whose input mass is disperse over the fluvial domain are given by the weighted average of the individual distributions from any source point to the outlet (the measuring station). The path probability, $p(\gamma)$, requires some attention, as the only obvious properties are that $p(\gamma) \geq 0$ and the completeness of the inspection of all possible paths Γ, that is, $\sum_{\gamma \in \Gamma} p(\gamma) = 1$. As a result, one may write, in general,

$$f(t) = \sum_{\gamma \in \Gamma} p(\gamma,t)\, f_{\gamma}(t),$$

where γ indexes all possible paths belonging to Γ. This implies consideration of any path connected by the directed graph that is implied by the gravity-driven drift imposed by hydrodynamic transport. The weight $p(\gamma,t)$ of each individual travel time distribution along the directed path γ is determined as follows. By necessity, the path probability will be proportional to the input mass

along that path, normalized by the total mass input. In this example, for simplicity, we shall refer to source areas and channel states where downstream routing occurs. We thus need to define the collection Γ of all individual paths γ triggered by a rate of injection of mass $\dot{M}_{\gamma}(t)$ dispersing in the river network up to the basin outlet. The time-dependent weighting function of the individual pathways is therefore

$$p(\gamma,t) = \frac{\dot{M}_{\gamma}(t)}{\sum_{\gamma \in \Gamma} \dot{M}_{\gamma}},$$ (3.91)

from which several cases of practical use are derived, including the simplest case of an instantaneous injection of a spatially uniform mass input proportional to the area A_k drained by the kth reach (leading to $p(\gamma,t) \sim p(\gamma) = A_k/A$, where $\gamma = A_k \to c_k \to \cdots \to x_{\Omega}$, and $A = \sum_k A_k$ is the total catchment area, as posited by the original approach centered on a geomorphological theory of the instantaneous unit hydrograph [377, 378]). The collection of connected transitions $\gamma = x_1, x_2, \cdots x_\Omega$ defines a feasible path (where we define as Ω the closure of the network identified by a gauging station) consisting of the set of all feasible routes to the outlet, that is, $x_1 \to x_2 \to \cdots \to x_\Omega$. The above rules specify the spatial distribution of pathways available for hydrologic transport of spatially distributed source mass through an arbitrary network of river reaches.

Our closure of the mathematical model hinges on the general determination of the travel time spent by a particle along any one of the above paths. Each is composed by the sum of the travel times within every transition downstream of it. The time T_x that a particle spends in state x is assumed to be a random variable described by probability density functions (pdfs) $f_x(t)$. Different states, say, x and y, yield travel times T_x and T_y that in general have different pdfs, $f_x \neq f_y$, and we assume that T_x and T_y are statistically independent for $x \neq y$.

Box 3.22 Distribution of the Sum of Independent Random Variables

Consider two random variables X_1, X_2 concentrated in $(0, \infty)$ whose values are mutually dependent. The marginal probabilities $F_1(x_1), F_2(x_2)$ (where $F_1(x_1) = \int_0^{x_1} f_1(x) dx$, where $f_1(x_1)$ is the probability density function (pdf) of X_1; analogously for X_2) thus do not contain information on the probabilistic linkage of X_1, X_2. One thus introduces the joint probability $F_{X_1 X_2}(x_1, x_2)$ to express the probability $P(A)$ that simultaneously the event $A: X_1 \leq x_1$ and $X_2 \leq x_2$ occurs. Formally, one has

$$F_{X_1 X_2}(x_1, x_2) = \int_0^{x_1} dx \int_0^{x_2} dy \, f_{X_1 X_2}(x, y),$$

where $f_{X_1 X_2}$ is the pdf of A. Notable extensions of standard probability imply that the expected value

$$< g(x_1, x_2) > = \int_0^{x_1} \int_0^{x_2} f_{X_1 X_2}(x, y) g(x, y) \, dx dy$$

applies to any function g, including those yielding the standard moment of order r, s, that is,

$$\mu_{rs} = < (x_1 - \mu_1)^r (x_2 - \mu_2)^s >$$

$$= \int_0^{x_1} \int_0^{x_2} f_{X_1 X_2}(x, y)(x - \mu_1)^r (y - \mu_2)^s \, dx dy$$

(where μ_1 is the mean of the marginal distribution of X_1, analogously for X_2). The common treatment of correlation through the study of the covariance $\mu_{11} = < (x_1 - \mu_1)(x_2 - \mu_2) >$ follows from the implication that if X_1, X_2 are independent, their covariance is null. Note that the reverse is not true, as can be shown by considering $X_1 = X$, with $X = N(0, \sigma)$, a normal zero-mean Gaussian variable, and $X_2 = X^2$: in fact, their covariance, $< X^3 >$, is always zero under the normal assumption, even though obviously the variables are not independent but rather deterministically linked.

It is reasonable to extend the above construction to the case of N random variables $X_1, X_2, \ldots X_N$, possibly with $N \to \infty$. The formalism $\forall X_i$ is the natural expansion of the joint pdf of two variables and will be omitted for simplicity at no loss of generality.

A useful result is derived under the banner of statistical independence, which has important hydrological implications. In fact, as per the postulates of axiomatic probability theory, the event $A: X_1 \leq x_1$ and $X_2 \leq x_2$ can be seen through Bayes's law characterizing $P(A) = P(X_1 \leq x_1, X_2 \leq x_2)$ as

$$P(X_1 \leq x_1, X_2 \leq x_2) = P(X_1 \leq x_1 \mid X_2 \leq x_2) P(X_2 \leq x_2)$$

or, alternatively, by specifying the conditional probability of the second event. The statement of statistical independence is tantamount to assuming that $P(X_1 \leq x_1 \mid X_2 \leq x_2) \approx P(X_1 \leq x_1)$,

Box 3.22 *Continued*

yielding the important result $P(X_1 \leq x_1, X_2 \leq x_2) = P(X_1 \leq x_1)P(X_2 \leq x_2)$ and hence, from (1), $f_{X_1 X_2}(x_1, x_2) = f_1(x_1)f_2(x_2)$.

Let us assume now that one is interested in the random variable $Y = X_1 + X_2$, where X_1, X_2 are random variables however distributed. For direct implications on the theory of the hydrological response, we assume that X_1, X_2 are concentrated in $(0, \infty)$, that is, are definite positive variables (say, describing travel times in different geomorphic "states"). We are then interested in the probability measures $F_Y(y) = P(Y \leq y) = \int_0^y f_Y(x)dx$ with obvious symbol notation. By noting that a value of y is a straight line of equation $y = x_1 + x_2$ in the (x_1, x_2) plane of the possible values of the independent variables, one extends the above equation as

$$F_Y(y) = P(Y \leq y) = \int_0^y dx_1 \int_0^{y-x_1} dx_2 f_{X_1 X_2}(x_1, x_2),$$

where the integration is extended over the entire domain bounded by the limit line $y = x_1 + x_2$ (note that in the case of random variables concentrated in $(-\infty, \infty)$, the extension is nontrivial, as the first integral extends from $(-\infty, \infty)$ and the second from $(-\infty, y - x_1)$). If one assumes that the variables are statistically independent, one has $f_{X_1 X_2}(x_1, x_2) = f_1(x_1)f_2(x_2)$, and the above results simplify to

$$F_Y(y) = P(Y \leq y) = \int_0^y dx_1 f_1(x_1) \int_0^{y-x_1} dx_2 f_{X_2}(x_2)$$

$$= \int_0^y dx_1 f_1(x_1) F_2(y - x_1).$$

By applying Leibniz's differentiation $(d/dx \int_{\alpha(x)}^{\beta(x)} f(x,y)dy = d\beta/dx f(x,\beta) - d\alpha/dx f(x,\alpha) - \int_{\alpha(x)}^{\beta(x)} \partial f(x,y)/\partial x dy$ with $d\beta/dx = 1$, $d\alpha/dx = 0$, and $\partial f(x,y)/\partial x = 0$), one has

$$f_Y(y) = \frac{dF_Y(y)}{dy} = \int_0^y dx \, f_1(x)f_2(y-x) = f_1 * f_2$$

(where $*$ denotes the convolution operator for definite positive random variables), which is an important result to be extensively applied in the theory (travel time to the outlet is the sum of the [random, independent] travel times in each successive geomorphic state crossed by the water particle in her route to the exit).

Suppose now that the (random) time T spent in an arbitrary control volume $\mathcal{V}$ is composed by three clearly identifiable random components, such as the time spent in hillslope and two serially connected channel states, say, T_1, T_2, T_3, linked by topographic steepest descent such that $T = T_1 + T_2 + T_3$. All T_i are assumed to be i.r.v., that is, independent, arbitrarily distributed random variables with pdfs $f_{T_i}(t)$. The pdf of their sum T is thus given by

$$f_T(t) = \int_0^t d\tau_1 \, f_{T_1}(\tau_1) \int_0^{t-\tau_1} d\tau_2 \, f_{T_2}(\tau_2) \, f_{T_3}(t - \tau_1 - \tau_2)$$

$$= f_1 * f_2 * f_3,$$

which is easily extended to the n-folded convolution of n i.r.v. $T_1, T_2, \ldots T_n$ as $f_{X_1} * f_{X_2} * \cdots * f_{X_n}(t)$. Note that it is $\int_0^\infty f_{T_1}(t)dt = \int_0^\infty f_{T_2}(t)dt = \cdots = \int_0^\infty f_T(t)dt = 1$; that is, the convolution operator is mass conserving.

A feasible path is a sequence of gravity-driven directed transitions. If such a path originates in the reach x_n, it is defined by the collection of unique transitions (states joined by the directed graph structure) $\gamma \in \Gamma$, where Γ is the set of all states joined by each path from the source to the common outlet Ω. Thus the specific path originated at x_n is replicated by generating a set Γ of transitions of the type $\langle x_n, x_{n+1}, \ldots x_\Omega \rangle$ implying the corresponding transitions $x_n \rightarrow x_{n+1} \rightarrow \cdots \rightarrow x_\Omega$. For each possible path γ, we define the travel time from source to outlet, T_γ, as $T_\gamma = T_{x_n} + T_{x_{n+1}} + cdots + T_{x_\Omega} = \sum_{x_\omega \in \gamma} T_{x_\omega}$. From the statistical independence of the random variables T_{x_i}, it follows (Box 3.22) that the derived distribution $f_\gamma(t)$ of the the sum of

the residence times T_{x_i} is the convolution of the connected individual travel time pdfs:

$$f_\gamma(t) = f_{x_n} * f_{x_{n+1}} * \cdots * f_{x_\Omega}(t).$$

Thus the pdf of the travel times $f(t)$ at the outlet of a system whose input mass is distributed over the entire domain is obtained by randomization over all possible paths Γ, yielding [174, 377, 378]

$$f(t) = \sum_{\gamma \in \Gamma} p(\gamma, t)\, f_{x_n} * f_{x_{n+1}} * \cdots * f_{x_\Omega}(t). \tag{3.92}$$

An example of complete enumeration of possible paths is given in Box 3.23.

Box 3.23 Example of Complete Enumeration of Fluvial Pathways and Their Travel Times

A complete example of determination of the complete pdf of travel times where the simplest path probability is chosen is reported in Figures 3.40 and 3.41. Therein the set of all possible transitions $\Gamma :< \gamma_1, \cdots, \gamma_5 >$ is

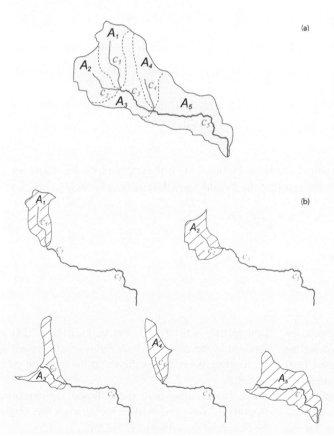

(a)

(b)

Figure 3.40 (a) Example of a hierarchical geomorphological structure of a river basin and notation for the derivation of the network travel time distribution. The set Γ of all possible paths to the outlet defined by the geomorphic structure is made up by 10 states, namely, five source areas (labeled by o_k, $k = 1, 5$) and five channels (e.g., transitions from source areas A_i to their outlet channels c_i and then to ensuing transitions $(c_i \rightarrow c_k \rightarrow \cdots \rightarrow c_5)$ toward the measurement site Ω, the endpoint of channel c_5). Here we assume that all source areas, A_1 to A_5, are potentially acting as generators of solutes to the mobile phase ultimately collected at the outlet. (b) The set of independent paths available for hydrologic runoff is highlighted. Figure after [379]

Box 3.23 *Continued*

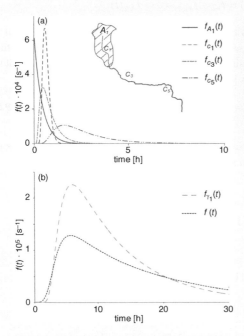

Figure 3.41 Individual travel time distributions within reaches may differ from one another, reflecting their lengths or their hydrodynamic conditions. (a) Individual travel time distributions along the path $A_1 \rightarrow \cdots \rightarrow c_5$. (b) Travel time distribution $f_1(t)$ obtained by convolution of the individual pdfs along the hierarchical path originated in A_1, together with the overall travel time distribution $f(t) = \sum_\gamma p(\gamma) f_\gamma(t)$. Figure after [379]

$\gamma_1 : A_1 \rightarrow c_1 \rightarrow c_3 \rightarrow c_5$

$\gamma_2 : A_2 \rightarrow c_2 \rightarrow c_3 \rightarrow c_5$

$\gamma_3 : A_3 \rightarrow c_3 \rightarrow c_5$

$\gamma_4 : A_4 \rightarrow c_4 \rightarrow c_5$

$\gamma_5 : A_5 \rightarrow c_5,$

where, evidently, $\Omega = 5$. Thus, in the example shown in Figure 3.40, one simply plots, for whatever choice of individual distributions $f_{x_k}(t)$, the result of the Γ-fold convolution, in this case,

$$f(t) = \frac{A_1}{A} f_{A_1} * f_{c_1} * f_{c_3} * f_{c_5}(t) + \frac{A_2}{A} f_{A_2} * f_{c_2} * f_{c_3} * f_{c_5}(t)$$

$$+ \frac{A_3}{A} f_{A_3} * f_{c_3} * f_{c_5}(t) + \frac{A_4}{A} f_{A_4} * f_{c_4} * f_{c_5}(t) + \frac{A_5}{A} f_{A_5} * f_{c_5}(t),$$

with usual symbol notation.

As discussed in Box 3.21, two boundary conditions are required for the boundary value problem involving the diffusion equation for the resident concentration $C(x,t)$ in each connected reach of the network. By assuming prevailing advection (epitomized by a catchment Peclet-like number introduced in [174], $Pe = \bar{u}L/K$, with usual symbol notation), the simplest possible form for the Laplace transform of $f(t)$, that is, $\hat{f}(s)$ (Box 3.24), is obtained in the case for which $Pe \rightarrow \infty$, corresponding to reflecting boundary conditions at the injection point (Box 3.5).

The linear character of the diffusion approximation supports the following relation for the mass flux Q_M at the closure of the network [377–379]:

$$Q_M(t) = \sum_{\gamma \in \Gamma} \int_0^t d\tau \, \dot{M}_\gamma(\tau) \, f_\gamma(t-\tau). \qquad (3.93)$$

Applications of the above framework in surface hydrology are numerous (see, for a review, [63]). Figures 3.42–3.44 show the results of the numerical integration of Equation (3.93) in the case of a spatially uniform mass injection, lasting for a concentration time $L/\bar{u}$, over a Peano network at the eleventh stage of construction (shown in the insets). Three different Peclet-like numbers $\bar{u}L/K$ have been tested, including the case of pure advection (i.e., $K = 0$ or $Pe \to \infty$). The smoothing role of the hydrodynamic dispersion that often irons out the fluctuations characteristic of the deterministically recurring motifs is apparent [63].

A few exact results may be derived for the general case. For simplicity of notation, we shall assume here the simplest case of the instantaneous unit pulse of mass uniformly distributed over the fluvial domain, for which the pdf of travel times reads

$$f(t) = \sum_{\gamma \in \Gamma} p(\gamma) f_{x_\omega} * f_{x_\omega + 1} * \cdots * f_{x_\Omega}(t), \qquad (3.94)$$

where each path γ corresponds to the transitions $x_\omega \to \cdots \to x_\Omega$ belonging to the set of all possible paths Γ that portray the geomorphology of the chosen river network.

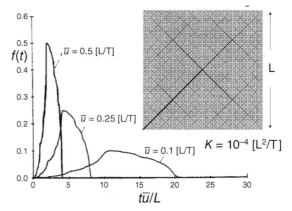

Figure 3.43 (a) Mass flux at the outlet of the Peano basin shown in the inset (the order of the basin used for the actual computations is equal to 8) in the case of pure convection ($K = 10^{-4}$ [L^2/T]). Mass pulses are uniformly distributed along all reaches and last the whole concentration time in the different cases. The length scale is normalized by the maximum path from source to outlet L. Time is normalized by the time to the control section from the most distant source at unit velocity (average arrival from a distance L is at $t = 1$ for $\bar{u} = 1$ [L/T]). Redrawn after [174]

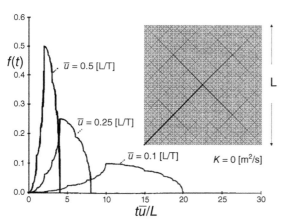

Figure 3.42 (a) Mass flux at the outlet of the Peano basin shown in the inset (the order of the basin used for the actual computations is equal to 8) in the case of pure convection ($K = 0$). Mass pulses are uniformly distributed along all reaches and last the whole concentration time in the different cases. The length scale is normalized by the maximum path from source to outlet L. Time is normalized by the time to the control section from the most distant source at unit velocity (average arrival from a distance L is at $t = 1$ for $\bar{u} = 1$ [L/T]). Redrawn after [174]

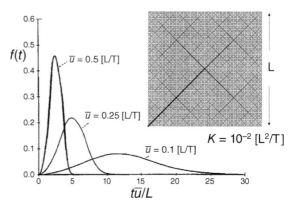

Figure 3.44 (a) Mass flux at the outlet of the Peano basin shown in the inset (the order of the basin used for the actual computations is equal to 8) in the case of pure convection ($K = 10^{-2}$ [L^2/T]). Mass pulses are uniformly distributed along all reaches and last the whole concentration time in the different cases. The length scale is normalized by the maximum path from source to outlet L. Time is normalized by the time to the control section from the most distant source at unit velocity (average arrival from a distance L is at $t = 1$ for $\bar{u} = 1$ [L/T]). Redrawn after [174]

Box 3.24 Properties of the Laplace Transform of the Relevant Transport Equation

To obtain manageable solutions, it is convenient to solve Equation (3.94) using Laplace transform techniques. As such,

$$\hat{C}(s,x) = \mathscr{L}(C) = \int_0^\infty C(x,t)e^{-st}\, dt, \tag{3.95}$$

and the transformed equation reads

$$s\hat{C}(s,x) + \bar{u}\frac{\partial \hat{C}(s,x)}{\partial x} - K\frac{\partial^2 \hat{C}(s,x)}{\partial x^2} = 0, \tag{3.96}$$

where obviously two boundary conditions are required. The above follows from the continuity of the function C and the operational property of the Laplace transform:

$$\int_0^\infty e^{-st}\frac{\partial C(x,t)}{\partial t}\, dt = s\hat{C}(s,x) - C(x,0), \tag{3.97}$$

after integration by parts and making use of $C(x,0) = 0$ for $x \neq 0$.

Let then $\hat{f}(s)$ be the Laplace transform of $f(t)$. From Equation (3.94), one obtains [174]

$$\hat{f}(s) = \sum_{\gamma \in \Gamma} p(\gamma) \prod_{x_\omega \in \gamma} \hat{f}_{x_\omega}(s), \tag{3.98}$$

where the sum over all the paths Γ from the inlet to the outlet is weighed by their path probability. The question of whether Equation (3.98), which gives an exact rule for computation of travel time distributions, is amenable to analytic solution rests on the definition of $f_{x_\omega}(t)$, which in turn is the travel time probability density that matter injected at the inlet of the state x_ω reaches the endpoint of the reach in a time t. Equation (3.98) is derived through the property that the convolution operator reduces to the product of Laplace transforms:

$$\mathscr{L}(f_1 * f_2 * \cdots f_n(t)) = \hat{f}_1(s)\hat{f}_2(s)\cdots\hat{f}_n(s). \tag{3.99}$$

This operational rule is needed in the following.

The Laplace transform, say, $\hat{f}_X(s)$, of the travel time distribution $f_X(t)$ follows from Equation (3.82) as

$$\hat{f}_X(s) = -s\int_0^L \hat{C}(s,x)\, dx. \tag{3.100}$$

The general solution for the Laplace transform of first-passage distributions $\hat{f}_X(s)$ in the reaches is a characteristic function of the first-passage distributions f_X of overall length L. It is obtained by solving Equation (3.96):

$$\hat{f}_X(s) = \bar{A}e^{L\theta_1(s)} + \bar{B}e^{L\theta_2(s)}, \tag{3.101}$$

where the arguments of the exponentials follow from the roots of the characteristic equation (3.96):

$$\theta_{1,2}(s) = \bar{u} \pm \frac{\sqrt{\bar{u}^2 + 4sK}}{2K}. \tag{3.102}$$

The general solution for $\bar{A}, \bar{B}$ yields a rather untransparent expression [174]. Therein, a discussion of the different solutions (3.101) as a function of the type of boundary conditions is provided.

Box 3.24 *Continued*

Regarding a choice of a functional form of (3.101), two points are relevant when referring to channel networks. On one hand, in fact, one wonders – following [174] – whether the imposition of the reflecting barrier at $-\infty$ rather than close to the injection point has any effect. This is important since the mathematics would be considerably simpler. Intuitively, one would tend to think that such an assumption may be valid especially for long river channels and large average flow velocities. In the dynamic conditions arising in real channels, the average basin Peclet number is usually large and thus the simplest solution applies. Straightforward numerical Laplace inversion shows that at $Pe = 10$ (here the spatial scale is defined by the distance L to the absorbing barrier located at the endpoint of the reach), setting a reflecting barrier at $x = 0$ or at $-\infty$ yields very similar pdfs (Figure 3.45). Therefore the solution with reflecting barrier at infinity, which is considerably simpler, is well approximated by the solution with a reflection at the injection point already at $Pe = 10$ (Figure 3.46).

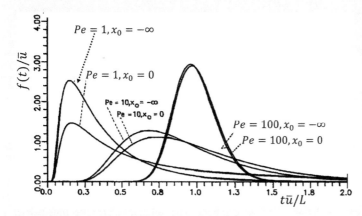

Figure 3.45 The effect of a reflecting barrier on travel time distributions within individual reaches at differing river Pe numbers. The solution with reflecting barrier at infinity, which is considerably more simple, proves to be already well approximated by the solution with a reflection at the injection point at $Pe = 10$, attained without difficulty in most fluvial settings. Figure after [174]

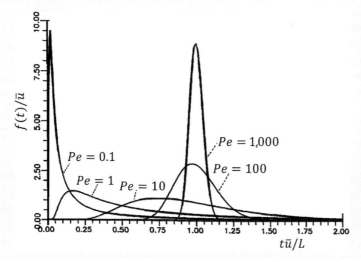

Figure 3.46 The form of the travel time distributions that solve the general transport equation of parameters $\bar{u}, K$ at various $Pe = \bar{u}L/K$ numbers. It should be noted that at increasing predominance of convective over diffusive forces ($Pe \to \infty$), the solution tends to a spike-like Dirac delta distribution characteristic of pure convection. Figure after [174]

Box 3.25 Laplace Transforms as Moment-Generating Functions

It is interesting to observe that Equation (3.94) is a moment-generating function for the distribution of the random variable T defined as the arrival at the outlet of a particle injected anywhere in the river network at time $t = 0$ via

$$< T^n > = \int_0^\infty t^n f(t)dt = (-1)^n \frac{d^n \hat{f}(s)}{ds^n}\big|_{s=0}, \qquad (3.103)$$

where $< T^n >$ is the nth moment of the arrival time distribution $f(t)$, whose physical meaning of $f(t)$ is the mass flux due to a unit instantaneous injection arbitrarily distributed in space according to a path probability $p(\gamma)$, where γ labels any transition from source to outlet in a hierarchical system [174].

The moment-generating (or characteristic) function is derived by expanding the exponential contained into the Laplace transform

$$\hat{f}(s) = \int_0^\infty e^{-st} f(t)\, dt = \int_0^\infty \left(1 + st + \frac{(st)^2}{2!} + \frac{(st)^3}{3!} + \cdots \right) f(t)\, dt. \qquad (3.104)$$

Assuming that the series within the integral is uniformly convergent, one can differentiate under the integral sign. Thus

$$\frac{d^n \hat{f}(s)}{ds^n} = \int_0^\infty \left(t^n + st^{n+1} + \frac{s^2 t^{n+2}}{2!} + \cdots \right) f(t)dt,$$

from which, by setting $s = 0$, Equation (3.103) is recovered.

The key result is obtained by solving exactly the network transport problem in the Laplace transform domain [174]. To that end, a choice of model equation for the transport dynamics in a single reach, say, j, of length L_j is needed, together with proper boundary and initial conditions. As shown in Box 3.24, following the proof in [174], for individual reaches j of length ℓ_j, a convenient general form of the L-transform for the travel time distribution is simply

$$\hat{f}_{\ell_j}(s) = \exp[-\ell_j \theta(s)],$$

where $\theta(s) = [-\bar{u} + \sqrt{\bar{u}^2 - 4sK}/(2K)$ solves the longitudinal dispersion problem equation (3.97) with suitable boundary conditions. To that end, Figure 3.45 illustrates how the solution of the boundary value problem is affected by the choice of boundary conditions. In particular, it shows the effects of a reflecting barrier on travel time distributions of individual reaches within individual reaches of length ℓ at differing river Peclet numbers $Pe = $

$\bar{u}\ell/K$ numbers. The solution with reflecting barrier at infinity, which is considerably simpler, proves to be well approximated by the solution with a reflection at the injection point at $Pe \geq 10$, attained without difficulty in most fluvial settings – for a back-of-the-envelope calculation, for a real-life setting, one may assume $\bar{u} \sim O(1)$ [m/s] and $\ell \sim O(10^3)$ [m], from which a range of longitudinal dispersion coefficients $K < O(10^2)$ is derived for a firm validity domain. The latter range is in concordance with most field and theoretical estimates of K (Section 3.5.3).

Substitution into Equation (3.98) yields the general L-transform of $f(t)$:

$$\hat{f}(s) = \sum_{\gamma \in \Gamma} p(\gamma) \exp\left[-\sum_{j \in \gamma} \ell_j \theta(s) \right], \qquad (3.105)$$

with usual symbol notation, and noting that Equation (3.105) is valid regardless of possible assumptions on the relevant features that might be specific to reach j of the functional dependence of the hydrodynamic parameters $\bar{u}$ and K in $\theta(s)$. In the simplest (and yet meaningful [174]) case where they do not depend on j being a constant, inverse transformation yields exactly

$$f(t) = \frac{1}{4\sqrt{\pi K t^3}} \sum_{\gamma \in \Gamma} p(\gamma) L(\gamma) \exp\left[-\frac{(L(\gamma) - \bar{u}t)^2}{4Kt}\right],$$

(3.106)

where $L(\gamma) = \sum_{j \in \gamma} \ell_j$ is the total distance from source to outlet for each path $\gamma \in \Gamma$.

For the derivation of the various moments $< T^N > = \int_0^\infty t^n f(t)\, dt$, moment-generating functions (Box 3.25) allow for a general solution. Here, we shall show only two of them, – the mean $E(T) = <T>$ and variance $Var(T) = <T^2> - <T>^2$ – and discuss why the network substrate for transport implies a geomorphological dispersion effect. They are as follows [174]:

$$E(T) = \sum_{\gamma \in \Gamma} p(\gamma) \sum_{j \in \gamma} \left(\frac{\ell_j}{\bar{u}_j}\right)$$

$$Var(T) = 2 \sum_{\gamma \in \Gamma} p(\gamma) \sum_{j \in \gamma} \left(\frac{\ell_j K}{\bar{u}_j^3}\right)$$

$$+ \sum_{\gamma \in \Gamma} p(\gamma) \left(\sum_{j \in \gamma} \frac{\ell_j}{\bar{u}_j}\right)^2 - \left(\sum_{\gamma \in \Gamma} p(\gamma) \sum_{j \in \gamma} \frac{\ell_j}{\bar{u}_j}\right)^2,$$

(3.107)

where the notation $\bar{u}_j$ is meant to emphasize the dependence of $\bar{u}$ on the features of the jth link.

In the important case where $\bar{u}$ and K are constant for every link j, we obtain

$$E(T) = \frac{1}{\bar{u}} \sum_{\gamma \in \Gamma} p(\gamma) L(\gamma)$$

(3.108)

$$Var(T) = 2 \frac{K}{\bar{u}^3} \sum_{\gamma \in \Gamma} p(\gamma) L(\gamma)$$

(3.109)

$$+ \frac{1}{\bar{u}^2} \left(\sum_{\gamma \in \Gamma} p(\gamma) L^2(\gamma) - \left(\sum_{\gamma \in \Gamma} p(\gamma) L(\gamma)\right)^2\right),$$

where $L(\gamma) = \sum_{j \in \gamma} \ell_j$. Higher-order moments may be calculated similarly [174].

These results lead to interesting speculations. In particular, the variance $Var(T)$ of the travel times to the control section is made up by two individual contributions: one involving hydrodynamic dispersion and another that is

dispersion free, computed by the last two terms, for example, in Equation (3.110). The first contribution is the weighted sum of the variances of every path to the outlet. In the case of constant hydrodynamic parameters, it reduces to the variance along the mean length from source to outlet $\bar{L} = \sum_{\gamma \in \Gamma} p(\gamma) L(\gamma)$, that is, $2\bar{L}K/\bar{u}^3$. The remaining terms produce a geomorphologic effect: the contribution to the variance of the travel time distribution due simply to constant convection along paths of different lengths. This effect is null only for explosion patterns rooted in a single outlet, for which only one type of route is possible. Moreover, Schwartz's inequality proves that this contribution is strictly positive [174]. Both effects are argued to be operating in the eDNA study presented in Section 3.6.

A geomorphological dispersion coefficient is thus defined via the relationship [174]

$$\frac{\bar{u}^3 Var(T)}{2\bar{L}} = K_G + K$$

because the two effects are additive, leading to the final result

$$K_G = \frac{\bar{u}}{\bar{L}}\left[\sum_{\gamma \in \Gamma} p(\gamma)\left(\sum_{j \in \gamma} \ell_j\right)^2 - \left(\sum_{\gamma \in \Gamma} p(\gamma)\sum_{j \in \gamma} \ell_j\right)^2\right].$$

(3.110)

In analogy with the theory of diffusion by continuous movements [290], as hydrodynamic dispersion soon clouds molecular diffusion, at the scale of a river network, geomorphological dispersion tends to overwhelm the hydrodynamic dispersion mechanisms operating at the scale of single river reaches. Thus, part of the variance of the arrivals at the control section is explained by the weighted sum of the variances accumulated along the individual routes to the network outlet. In very particular cases (e.g., radial equal streams converging at the end), this would constitute the total variance as the rates of arrival through different paths would coincide. A major contribution to spreading of the rates of arrival, however, derives from the heterogeneity of the flowpaths leading to the measurement section. Equation (3.110) is useful because one may predict when this contribution is bound to prevail. In particular, a well-studied geomorphological measure, the width function of the network (Box 3.26), becomes central to this task.

Box 3.26 Geomorphological Width Functions

A fundamental property of any river network [63] is that there exists a unique one-dimensional path connecting any pair of points within the dendritic structure. In particular, the flowpath from any point to the basin outlet is uniquely determined. When studying the structural characteristics of a dendritic river network, and its viability for transport or movement, the arrangement of the flowpaths to the outlet from any point of the domain drained by the network is important. This arrangement is characterized by the width function $W(x)$ of the network, first introduced by Shreve [380]. The width function defines the relative proportion of network sites, say, n_i/N, equally distant from the common outlet – n_i is the number of sites at distance x_i, and N is the total number of sites. Note that $\sum_{i=1}^{N} n_i = N$. The distance x_i for the ith class of nodes, however, must be measured along the tree structure to reconnect lengths and travel times rather than, say, radially (Figure 3.47). This definition thus denotes the length along the unique channel network pathways between two junctions defining a link or between a junction and a source. Details matter for the proper definitions when accurate digital terrain maps are used to extract the river networks, however, and these are reviewed elsewhere [63]. Figure 3.48 shows examples of width functions extracted from real basins of different size, climatic context, and geologic characteristics. The functions look erratic, but many elements of a common fundamental structure exist [63].

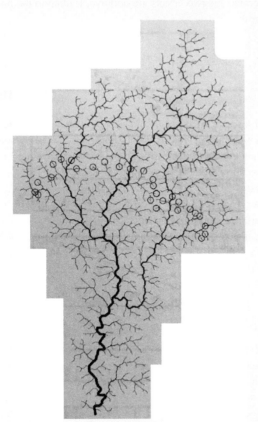

Figure 3.47 Definition of width function $W(x)$, that is, $W(x) = n(x)/\sum_{x=1,L} n(x)$ (where $n(x) = n(i\Delta x) = n_i$ is the number of sites placed at distance x from the outlet when pixels of size Δx^2 tile the landscape [Figure 1.3] and L indexes the maximum distance from source to outlet), the relative proportion of network sites placed at distance x from the outlet. $N = \sum_{x=1,L} n(x)$ is the total number of network nodes. Open circles indicate one snapshot of network sites equally distant from the outlet, distance being measured along the tree structure. Figure after [63]

Box 3.26 *Continued*

The connections of width functions with the hydrologic response of a catchment were first explored by Kirkby [381], followed by several applications relevant to basin-scale transport by travel time distributions (initially elucidated in [262, 377, 378, 382]). The basic result for the mass response of a basin (subsumed by the distribution of arrivals at the outlet, $f(t)$, following an instantaneous unit impulse uniformly distributed in space), under the assumption of constant along-network drift $\bar{u}$, is

$$f(t) = \bar{u}\frac{W(ut)}{L},$$

where t is time, in suitable units; $\bar{u}$ is an effective celerity of propagation of mass transport, related to a physical flow velocity [262], assumed constant and defining a dynamic scale for the process; L is a geometrical scale, here conveniently the maximum length from source to outlet; and $W(x)$ is the (dimensionless) width function of the basin. In laypeople's terms, the arrival of marked particles injected at time $t = 0$ follows gravity-driven pathways identifiable with the same steepest descent directions that are used to extract the river network from DTMs (Figure 1.3).

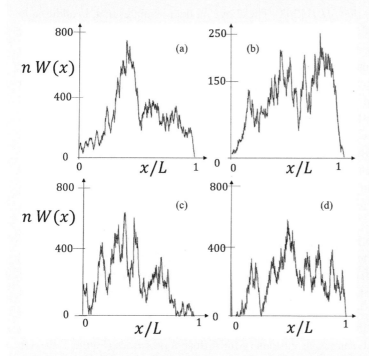

Figure 3.48 Examples of the distribution of the absolute number of network nodes $N\,W(x)$ plotted versus distance x measured in pixel units (i.e., $x = 1$ corresponds to 30 [m] in this case). Four real basins are chosen to give an overall impression of the sample size of data extracted by 30×30 m^2 digital terrain maps (DTMs): (a) Schoharie river basin $(2,408$ km$^2)$. (b) Nelk $(440$ km$^2)$. (c) St. Joe $(2,834$ km$^2)$. (d) Racoon $(448$ km$^2)$. The main geometric and geomorphic characters of these basins are described in [63]. Notable is the remarkable diversity of the broad features of the functions shown, reflecting the outer shape of the different catchments (due to geologic controls and competition for drainage), and yet the consistent statistical features of their fluctuations across gradients of climate, exposed lithology, and state of vegetation [63]

Arrivals at the outlet are thus labeled by the distance to travel at the onset of their journey along the network, and the proportion of arrivals equals the proportion of the nodes equally distant from the outlet if the initial distribution is uniform in space. However unrealistic it may seem, the fact that each water particle undergoes the same drift $\bar{u}$ along the local direction of the links of the network, this assumption is time honored in reproducing well key characters of the hydrologic response [63, 174, 262, 379, 381]. Because no distinction is drawn between channeled and unchanneled sites

Box 3.26 *Continued*

in the context of this book, the issue of defining celerities is greatly simplified. Generally, when strictly fluvial domains are investigated, $\bar{u}$ may be viewed as the celerity of flood propagation (the downstream speed at which a mobile observer has to move to observe $dQ(x,t) = 0$, where $Q(x,t)$ $[L^3/T]$ is the flow discharge), which may be assumed as roughly 3/2 the average flow velocity [174]. The latter is known to range around $O(1)$ [m/s]. Thus the fundamental geometric and dynamic scales are well identified. Width functions $W(x)$ of real basins reflect both common recurrent characters (independent of climate, geology, vegetation, and the shape of the boundaries) and some features peculiar to the shape of the basin [63, 262].

Randomization of the only parameter, $\bar{u}$, provides a generalization of the first-order approximation above (see [63, 262] for a complete derivation):

$$f(t) = \bar{u}\frac{<W(\bar{u}t)>}{L},$$

where $<W(\bar{u}t)> = \int_0^t W(x)dP(x,t)$, where P is the solution to Kolmogorov's backward equation: $dP(x,t) = (dx/\sqrt{4\pi Dt^3})\exp[-(x-\bar{u}t)^2/4Dt]$ [262]. The above model postulates that, in addition to a constant drift u, the dynamics of any water particle are affected by noise epitomized by a simple Brownian motion $X_B(t)$ biased by u in the direction of the network ($<X_B> = 0$, $<X_B^2> = 2Dt$). In the general case, one can derive analytically the moments of the distribution $f(t)$ and, in particular, the mean and variance of the arrival time distribution equation (3.110). The leading result is that the morphological contribution to the variance of the hydrologic response is null only for very particular (and unrealistic) networks, reflecting the heterogeneity of the geomorphic paths from injection to the outlet. In the inverse problem, typical of eDNA studies (Section 3.6), once a basic decay timescale T_D is identified – for instance, the exposure to UV activity or to turbulence that would render a genetic sample decayed beyond recognition – the possible source of genetic material may come from a plurality of sites, whose travel time to the measurement site does not exceed T_D. Thus width functions are key to estimating species distribution and abundance in river networks.

The morphology of river networks contains important informations about the features of hydrologic transport. High-frequency features of the geomorphologic width functions (Figure 3.48) show recursive characters notwithstanding the huge diversity of geologic, climatic, vegetational, and geomorphic features exhibited by real river networks reflecting the common patterns of self-organization criticality that river networks exhibit consistently [63, 383]. These high-frequency modes do not bear fundamental implications on transport, as they are rapidly smoothed out by the filtering effects of the fluvial dynamics, as customary for diffusive processes (Figure 5.49). On the contrary, the main characters of the hydrologic response are imprinted in the low-frequency modes of the width function. They reflect the gross availability of contributing areas at isochrone distances from the outlet – thus, apart from dynamic fluctuations, synchronous arrivals at the outlet of simultaneous injections. However, isochrone distances are greatly affected by different dynamic specifications reflecting heterogeneity and hydrodynamic dispersion (Section 3.5.3). Since deterministic constant drifts are a crude representation of the dynamics in river channels, stochastic averaging rooted in fluvial dynamics has been used to explain the physical mechanisms responsible for smoothing the fluctuations experienced by at-a-station measurements [262].

In geomorphologic representations extracted from DTMs, every pixel (of size Δx^2) may identify a source and prescribe a unique path to the outlet. Thus the path probability may be effectively substituted by the relative proportion of pixels at isochrone distances from the outlet; that is, the path length $L(\gamma)$ is replaced by the arbitrary distance x to the outlet measured along the network, and $p(\gamma)$ is replaced by the geomorphological width function $W(x)$ (Box 3.26). Both can be remotely extracted from DTMs and objectively manipulated. One may thus write [262]

$$E(T) = \sum_{x=1}^{N} W(x) \frac{x}{\bar{u}} \qquad (3.111)$$

$$Var(T) = 2 \sum_{x=1}^{N} W(x) \frac{xK}{\bar{u}^3} \qquad (3.112)$$

$$+ \sum_{x=1}^{N} W(x) \left(\frac{x}{\bar{u}}\right)^2 - \left(\sum_{x=1}^{N} W(x) \frac{x}{\bar{u}}\right)^2,$$

where N indexes the number of distances making up the maximum length from source to outlet, $N\Delta x$. We note that the above equations hold even in the case where velocity depends on position, that is, $u(x)$ [174]. We also note that the morphological contribution to the variance of the hydrologic response (the second and third terms in Equation (3.113)) is negligible only in very particular cases, the by product of unrealistic topographies.

3.6 Estimating Species Distribution and Abundance in River Networks Using Environmental DNA

Organisms leave traces of DNA in their environment (eDNA), such as cells in mucus or feces. When extracted from water or soil, eDNA can be used to track the presence of a target species or the composition of entire communities. In rivers, eDNA dynamics are modulated by transport and decay. Here, we show how the use of hydrologically based models opens new avenues for reconstructing the upstream distribution and abundance of target species throughout a river network from at-a-station eDNA measurements that identify a detectable concentration of genetic material attributable to that species. Key to predictive power of the models we shall discuss is the filtering effect operated by fluvial transport along individual pathways from source to gauging station. We validate our method by estimating the catchment-wide biomass distribution of a sessile

invertebrate and its parasite, causing disease in salmonids. This work, largely excerpted from [376], intends to show the potential of hydrologic arguments to unlock the power of eDNA sampling (and prospectively of metagenomics) for monitoring biodiversity across broad geographies in a way hitherto unfeasible with traditional survey approaches.

As stated above, all organisms leave traces of DNA in their environment, and eDNA measurements are already in use to track occurrence patterns of target species. They rely on use of polymerase chain reaction (PCR), which allows the exponential amplification of DNA sequences. Applications are especially promising in rivers, where eDNA can integrate information about populations upstream. However, they are currently limited to determining the presence or absence of target species (say, specific pathogens of waterborne disease as an indication of potential epidemics) with the caveat about absence that the source species may go undetected – for example, owing to concentrations dropping below detection limits or for the odds of sampling probabilities. The dispersion of eDNA in rivers is modulated by complex processes of transport and decay through the dendritic river network, and until very recently, we lacked a method to extract quantitative information about the location and density of populations contributing to the eDNA signal. Here, following the work of Carraro et al. [376], we present a general framework to reconstruct the upstream distribution and abundance of a target species across a river network, based on observed eDNA concentrations and hydrogeomorphological features. The exercise shown here describes how the model, even in its current minimalist formulation, captures well the catchment-wide spatial biomass distribution of two target species: a sessile invertebrate (the bryozoan *Fredericella sultana*) and its parasite (the myxozoan *Tetracapsuloides bryosalmonae*). The general method presented in this section is designed to easily integrate general biological and hydrological data toward spatially explicit estimates of the distribution of sessile and mobile species in fluvial ecosystems based on eDNA sampling.

3.6.1 Of eDNA and Rivers

Environmental DNA (eDNA), present as loose fragments, as shed cells [384, 385], or in microscopic organisms [386, 387], can be extracted from matrices such as water or soil and used to track the presence of target species or the composition of entire communities [388, 389].

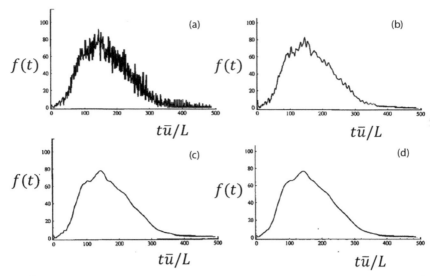

Figure 3.49 The effects of hydrodynamic dispersion on the response to an instantaneous unit pulse of mass uniformly distributed in every link of a real network whose width function $W(x)$ (Box 3.2.3) has been extracted from a digital terrain map (pixel size 30×30 m^2) as indicated by [262]. Here the dispersive process is modeled as in Equation (3.78). The hydrodynamic dispersion coefficient is set to (a) $K = 0$, (b) $K = 0.05$ [L^2/T], (c) $K = 0.1$ [L^2/T], and (d) $K = 0.2$ [L^2/T]. Pixel units (one unit is 30 m) are used in this case. Figure after [262]

Approaches using eDNA for qualitative species detection have proved their value in management and conservation, improving the measurement of biodiversity in a replicable and consistent manner [390] and facilitating the detection of rare, invasive, or parasitic species [391–396].

Environmental DNA in river water carries a record of the species present upstream, but the interpretation of this signal is a complex issue [397]. Once released to the environment, eDNA undergoes selective decay. Nucleic acids incur progressive damage (e.g., due to biological activity or pH [398, 399]) during hydrological advection, retention, and resuspension [400, 401]. These processes result in alterations that affect eDNA detection in environmental samples. The magnitude of the decay is highly dependent on the nature of the flow regime and the substrate type [402]. Furthermore, eDNA has polydisperse properties due to its origin in diverse organic sources (e.g., spores, cells, tissues, feces), which complicates the evaluation of decay rates [403]. The eDNA sampled at any point within a dendritic network of sources is the outcome of diffuse eDNA release from points upstream, modified by decay processes during transport that are governed by network connectivity, in which each path to the observation point may be described by different hydromorphological conditions. As a result, while it is straightforward to link a positive PCR test with the presence of the target species at some (unknown)

distance upstream, quantification of species densities and the location of populations is currently impossible because, besides a number of potentially confounding factors affecting eDNA shedding (e.g., animal behavior, movement, physiology and size [404, 405]), it requires consideration of the effects of the dynamics of eDNA transport along river branches and the deconvolution of the hierarchical aggregation of the various network branches. Here, we describe a recently established and generally applicable framework to interpret quantitative eDNA point measurements in rivers and relate them to the spatial distribution of the DNA sources, jointly with estimates of the density distribution of the target species throughout the river basin.

The framework proposed by Carraro et al. [376] stems from fundamental mass balance relationships. It is intended for use in river networks discretized into "nodes," that is, river stretches of suitable length within which hydrological conditions, as well as the target species density and hence its eDNA production, can be considered homogeneous (see Section 4.5). Within such nodes, the basis for the spatially explicit model contrasting measured eDNA concentrations is given by

$$\widehat{C_j} = \frac{1}{Q_j} \sum_{i \in \gamma(j)} A_{S,i} \, \exp\left(-\frac{L_{ij}}{v_{ij}\tau}\right) p_i, \qquad (3.113)$$

where $\widehat{C_j}$ [NL^{-3}] is the eDNA concentration at node j predicted by the model; $\overline{Q_j}$ [L^3 T^{-1}] is a characteristic (e.g., median) water discharge at node j; $\gamma(j)$ indexes the set of nodes upstream of j connected by the river network; $A_{S,i}$ [L^2] is the source area (or the spatial extent of the habitat) of the target species pertaining to node j; L_{ij} [L] is the alongstream length of the path connecting i to j; τ [T] is the inverse of the decay rate (i.e., a characteristic decay time) subsuming the damage rates to genetic material experienced during hydrologic transport [401, 402], possibly constant for all eDNA fragments irrespective of the different hydrological and environmental conditions along all paths to the sampling site [406]; $\overline{v_{ij}}$ [LT^{-1}] is the average flow velocity along the path connecting i to j; and p_i [NL^{-2} T^{-1}] is the eDNA production at i per unit habitat area and unit time. We assume that p_i is proportional to the target species density at node i [395, 406]. In the case of sessile species dwelling in riverine habitats, $A_{S,i}$ can be considered equal to the riverbed area of the river stretch i; for terrestrial species, such as certain diatoms that colonize unchanneled areas and have been used to track surface runoff [407], $A_{S,i} = A_{L,i}$, where $A_{L,i}$ is the directly contributing area [63] to node i (i.e., the total contributing area at site j, A_j, is the sum of all directly contributing areas in the path upstream of it, $A_j = \sum_{i \in \gamma(j)} A_{L,i}$).

Equation (3.113) is based on the hypothesis that eDNA undergoes first-order exponential decay along the downstream path from the source i to the measurement node j [400]. As an alternative characterization of decay, we may introduce the parameterization $\lambda = \overline{v_{ij}}\tau$, where λ is a decay length that is assumed to be constant irrespective of hydrological regimes and heterogeneities in morphological conditions across the watershed [376]. This alternative formulation, allowing one to avoid the effective calculation of $\overline{v_{ij}}$, is justified by the observation that water velocities in catchments for a large range of flow regimes generally show modest longitudinal gradients, as high velocities potentially prompted by steeper slopes in the upper reaches are limited by increased flow resistance [408]. In previous studies, estimated decay lengths range from the order of magnitude of a few meters for experimental flumes [401] to that of kilometers observed in real catchments [402]. Suffice here to note that all estimates of decay times (converted to travel lengths as noted above) from field measurements [376, 402] and mesocosm experiments [400, 406] point to values much greater than the mean distance between significant confluences in real catchments, at least in runoff-producing areas [63]. This implies that a plurality of sources could be contributing

to detectable eDNA concentrations at each measurement site, thus prompting the need to resort to approaches, such as (3.113), that take into account the structure of the network.

While not all of the above assumptions are equally valid in the general case, this framework represents a flexible general theory of spatially explicit eDNA source tracking. Some of these assumptions may be easily relaxed. For instance, the purely convective treatment of the decay of genetic material outlined above may be the subject of more refined formulations of the transport problem. In particular, travel times from source to measurement site can be made explicitly dependent on the hydrodynamic and geomorphological dispersion induced by the hierarchical nature of the network [408]. One obvious extension would characterize the transport of eDNA from polydisperse sources as a shear flow dispersion with decay along the river network structure. Assuming that turbulence and shear flows would rapidly mix the emitted eDNA concentration across width and depth (say, after distances of the order of 10 widths), the eDNA concentration at some downstream site x would be represented by $\bar{C}(x,t)$ (Section 3.5.3) – which (properly additive from various sources) would be the one that sampling would measure at site (x,t). The architecture of the dispersion in a linear reach would be the same, save for an extra (linear) term that modified the Laplace transform formalism in a straightforward manner, that is, the basic equation (3.78) would be modified as

$$\frac{\partial \bar{C}}{\partial t} + \bar{u}\frac{\partial \bar{C}}{\partial x} = K\frac{\partial^2 \bar{C}}{\partial x^2} - \chi\bar{C} \qquad (3.114)$$

(where χ [T^{-1}] is a [local] linear decay rate) for any reach crossed by the eDNA. The manipulation of the polydisperse source would follow directly the geomorphological dispersion formalism (Section 3.5.4). Matters of boundary conditions would become important, however. In particular, the higher level of detail required by the transport mechanism outlined in Equation (3.114) postulates the need to specify the time distribution of the initial emission of eDNA concentration at the source in x_0, that is, $C(x_0,t) = f(t)$, a largely unknown one, much less keen to be simply assumed to be constant and proportional to the local species abundance as implied by Equation (3.113). Whether this approach will prove superior to the one outlined in [376] thus remains to be seen.

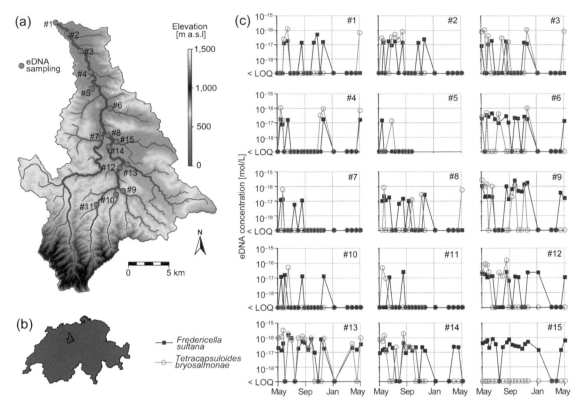

Figure 3.50 Environmental DNA sampling of *F. sultana* and *T. bryosalmonae*. (a), Map and color-coded digital elevation map of the study region showing the extracted river network and locations of the eDNA sampling sites. (b) Location of the study region within Switzerland. (c) Measured eDNA concentrations of *F. sultana* and *T. bryosalmonae* at the 15 sampling sites during the period May 2014–2015 (LOQ = limit of quantification). Figure after [376]

3.6.2 eDNA-Derived Spatial Distribution of the Source Biomass

To derive estimates of biomass distribution, the general source area model described above was coupled with a species distribution model [59]. This combined approach provides a versatile method to capture the influence of the catchment-wide ecological, hydro morphological, or geological drivers promoting eDNA production (and reflecting species density) within the defined river stretches. Local eDNA production can be expressed by means of the exponential link

$$p_i = p_0 \exp[\boldsymbol{\beta}^T \mathbf{X}(i)], \tag{3.115}$$

where p_0 is a baseline eDNA production value constant in space, $\mathbf{X}(i)$ is a vector of covariates evaluated at site i, and $\boldsymbol{\beta}$ is a vector of parameters requiring calibration that identifies the effect of such covariates on eDNA production and thereby on the distribution of the tar-

get species. Covariates included in the vector $\mathbf{X}(i)$ will depend on the particular system and data available. Their extent can be either local (e.g., only pertaining to the area directly contributing to a given stretch) or nonlocal, that is, related to the whole catchment area upstream of the stretch.

The above framework has been tested by contrasting joint field measurements of eDNA concentrations of the myxozoan parasite *T. bryosalmonae* and its primary host, the freshwater bryozoan *F. sultana*, across various locations within the Wigger watershed of Switzerland (Figure 3.50) [376]. *T. bryosalmonae* is the causative agent of proliferative kidney disease (PKD), a high-mortality disease affecting salmonid fish populations (Section 4.5). PKD is recognized as one of the leading causes of declines in brown trout populations in Europe. It affects diverse salmonid populations in North America and is a major aquaculture pathogen [409, 410]. The water samples used for eDNA detection of both *F. sultana*

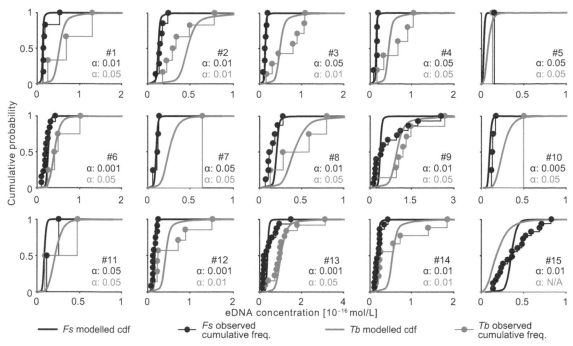

Figure 3.51 Comparison between the observed cumulative frequencies (dots) of measured eDNA concentrations at the 15 sampling sites for *F. sultana* (*Fs*; purple) and *T. bryosalmonae* (*Tb*; green) and the cumulative distribution function obtained by the model (solid lines); α values (color coded to match the solid lines) indicate the confidence level at which, according to a two-sample Kolmogorov–Smirnov test, the null hypothesis that the two samples (modeled and observed) come from the same distribution cannot be rejected. Higher values of α indicate a better fit. Tested values for α were 0.05, 0.01, 0.005, and 0.001. N/A = not applicable (i.e., at site 15, no positive values of eDNA concentration for *Tb* were detected). Note that some of the cases where the two distributions are not different at $\alpha = 0.05$ are characterized by a limited number of positive eDNA detections. Figure after [376]

and *T. bryosalmonae* were collected at roughly monthly intervals at 15 sites over 12 months (one 500-mL sample per sampling occasion and site) (Figure 3.50c). *T. bryosalmonae* eDNA is likely to be largely derived from spores shed into the environment. Parasite spores, released into water by infected bryozoans, infect brown trout through skin and gills and proliferate in the kidney. To complete the life cycle, spores infective to bryozoans are excreted in the urine of infected fish. These two types of spores are genetically indistinguishable yet differentiated in terms of function, which poses further challenges for modeling. The *T. bryosalmonae* eDNA concentration may thus be a product of the genomic contents of the two types of spores originating from very different transport sources, that is, from an immobile source (bryozoans) coupled with a mobile source (fish). In this particular case, a comparative analysis of field-measured eDNA for both *F. sultana* (sessile source of eDNA) and *T. bryosalmonae* (eDNA that could

jointly originate from sessile and mobile hosts) proves particularly instructive as a demonstration of the potential of the framework proposed in [376].

The model has been implemented and calibrated as described in Box 3.27. Note that the possibility that samples with low eDNA concentration may be interpreted as zeros (see Figure 3.50c) has been accounted for by introducing the nondetection probability ϕ_j, a monotonically decreasing function of $\widehat{C_j}$. The covariates included in vector $\mathbf{X}(i)$ were local elevation, contributing area, and the fractions of contributing area covered by moraine, peat, or superficial water (e.g., lakes, ponds, or wetlands) upstream of site i. These five covariates were chosen as representative of morphological and geological features of the catchment [376]. Overall, the model proved accurate at reproducing observed eDNA concentrations (Figures 3.51 and 3.52a) [376]. The median eDNA concentration of the different sampling sites (Figure 3.52a) is well reproduced by the model

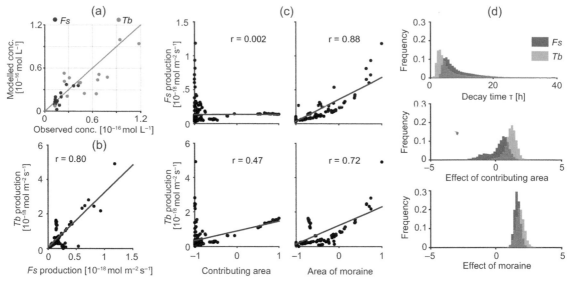

Figure 3.52 Correlation between eDNA concentrations, predicted species density, and covariates. (a) Correlation between observed and modeled eDNA concentrations for *F. sultana* (*Fs*; purple) and *T. bryosalmonae* (*Tb*; green). Each point corresponds to a sampling site. *X*- and *y*-values, respectively, are taken from the medians of the observed and modeled cumulative distribution functions displayed in Figure 3.51. (b) Correlation between medians of predicted *F. sultana* and *T. bryosalmonae* parasite *T. bryosalmonae* eDNA production. Each point represents one river stretch. The trend line is displayed in red. Pearson's correlation coefficients *r* are also reported. (c) Correlations between medians of predicted eDNA production for the two target species and the covariates representing the contributing catchment area and fraction of the upstream catchment covered by moraine. Correlations with other covariates are reported in [376]; maps of covariate values are presented in figure S3. (d) Posterior distributions for the decay time τ and values of the β coefficients associated with the covariates reported on the axis labels. The posterior distributions for the other covariates are reported in [376]

(Pearson's correlation coefficient $r = 0.84$ for *F. sultana*, $r = 0.75$ for *T. bryosalmonae*). According to two-sample Kolmogorov–Smirnov tests [411] (Figure 3.52a), the null hypothesis that observed positive eDNA concentrations drawn from the predicted distributions cannot be rejected at 5 percent confidence level for 33 percent (*F. sultana*) and 71 percent (*T. bryosalmonae*) of the sampling sites. At the 1 percent confidence level, these percentages rise to 73 percent and 100 percent, respectively.

Box 3.27 Habitat Suitability Models, Bayesian Calibration, and Model Settings

The application of a habitat suitability model [59] proved useful in the Wigger case study (see details reported in [376]). The chosen covariates have been checked for multicollinearity [412], and all variance inflation factors for the five retained covariates proved below the rule-of-thumb threshold of 10 [376]. The three geological covariates were obtained from the vectorized geological map of Switzerland provided by the Swiss Federal Office of Topography (Swisstopo). Covariates were normalized (i.e., linearly transformed into vectors within the range $[-1; 1]$). Figure 3.53 shows the values of the covariates plotted onto the maps of the river network, extracted by standard software whose details are reported in [376].

The choice of parameters has been carried out by Bayesian calibration, as briefly described in what follows. Details are in [376]. The average velocity $\overline{v_{ij}}$ along the path between nodes i and j

Box 3.27 *Continued*

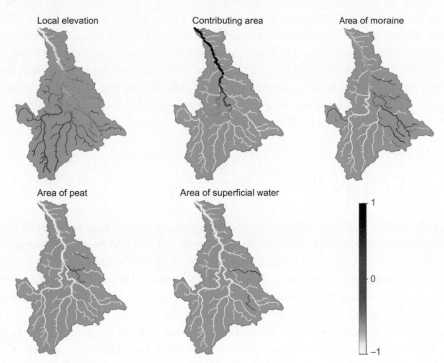

Local elevation   Contributing area   Area of moraine

Area of peat   Area of superficial water

Figure 3.53 Maps of normalized covariate values used in the model of the river Wigger (CH) whose hydro-geomorphological attributes are described in [376]

was calculated as $\overline{v_{ij}} = \sum_{k \in P_{i \to j}} l_k / \sum_{k \in P_{i \to j}} (l_k/v_k)$, where $P_{i \to j}$ indexes the path connecting i to j, while l_k and v_k are, respectively, the length and the average water velocity of stretch k. Velocities v_k were calculated by assuming nearly uniform flow conditions for each stretch and at all times via Manning's equation. The following further assumptions were made: water discharges measured by the Swiss Federal Office for the Environment in Zofingen (corresponding to site 3 in Figure 3.50a) were used to calculate discharges across all network stretches based on the assumed proportionality between discharge and contributing area; river cross sections were assumed to be rectangular with width much larger than depth; river widths w_k were estimated for all stretches based on aerial images; and Manning's roughness coefficient was taken as $n = 0.033$ m$^{-1/3}$ s and deemed representative of the flow resistance in the whole river network. Source areas $A_{S,i}$ were assumed equal to river bed surfaces $l_i w_i$.

As sampled eDNA concentrations for both *F. sultana* and *T. bryosalmonae* did not show a clear temporal pattern (Figure 3.50c), measured values were considered as realizations of a random variable whose distribution is constant in time. Measured eDNA concentrations $C_{j,t}$ above LOQ observed at site j and time t were assumed to be log-normally distributed, that is, $\ln\left(C_{j,t}\right) = \ln\left(\widehat{C_j}\right) + \epsilon_{j,t}$, where $\epsilon_{j,t} \sim N(0, \sigma^2)$, or alternatively, $C_{j,t} = \widehat{C_j} \exp\left(\epsilon_{j,t}\right)$. Samples with low eDNA concentration may fall below the LOQ or induce sampling errors and be interpreted as indicating

Box 3.27 *Continued*

absence, owing to the small sampling volumes and the lack of replicates for a sample taken at a given site and time. To account for the number of samples where the target eDNA goes undetected, the minimalist assumption was made that the probability φ_j of not detecting eDNA from a sample collected at site j is a monotonically decreasing function of the eDNA concentration $\widehat{C_j}$ predicted by the model, $\varphi_j = \exp\left(-\widehat{C_j}/C^*\right)$, where C^* is a concentration scale (species specific but constant in space) that requires calibration. The likelihood of nondetection at j is therefore equal to φ_j, while the likelihood of a positive observation reads $(1 - \varphi_j)\phi\left(\ln\left(C_{j,t}/\widehat{C_j}\right)/\sigma\right)$, where $\phi(\cdot)$ is the standard normal probability density function. Thus, the overall likelihood reads as

$$L(\mathbf{C}|\boldsymbol{\beta}, p_0, \tau, \sigma, C^*) = \prod_{j=1}^{M}\left[\varphi_j^{N_j}\left(1 - \varphi_j\right)^{D_j}\prod_{t=1}^{D_j}\phi\left(\frac{\ln\left(\frac{C_{j,t}}{\widehat{C_j}}\right)}{\sigma}\right)\right],$$

where $\mathbf{C}$ indicates the full set of eDNA concentrations observed at any time and site; M the number of sampling sites; and N_j and D_j, respectively, the number of null and positive observations at site j, and where $t = 1, \ldots, D_j$ spans all positive observations at site j. The sampling of the likelihood L was performed by means of a Metropolis-within-Gibbs algorithm [413]. For all free parameters, prior distributions were chosen as flat. Further details are reported in the SI appendix of [376]. A simplified version of the likelihood has been employed instead in [376].

To complete our examination of the model taken as a template of the general case as proposed in [376], Figure 3.52 shows the distribution of nondetection probabilities $\varphi_j = \exp(-\widehat{C_j}/C^*)$ for

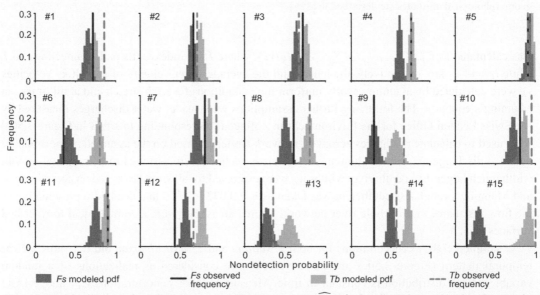

Figure 3.54 Distribution of nondetection probabilities $\varphi_j = \exp(-\widehat{C_j}/C^*)$ for all 15 sites and both target species. Observed nondetection frequencies (i.e., the number of null measured eDNA concentration values over the total number of samples taken at that site) are also reported. Figure after [376]

Box 3.27 *Continued*

all 15 sites and for both target species. Observed nondetection frequencies (the number of null measured eDNA concentration values over the total number of samples taken at that site) are also reported therein. Additionally, Figure 3.54 shows the distributions of nondetection probabilities, and Figure 3.55 illustrates the actual shapes of the posterior distributions of the model parameters, to complement the results highlighted in Figure 3.52.

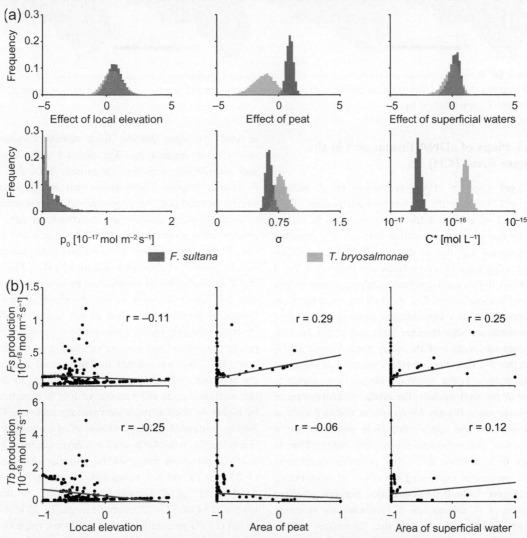

Figure 3.55 Posterior distributions of model parameters and correlations of predicted eDNA production with covariate values. (a) Posterior distributions for parameters not shown in Figure 3.52d. (b) Correlations between medians of predicted eDNA production for the two target species and the covariates not shown in Figure 3.52c. Each point represents one river stretch. The trend line is shown in red, with Pearson's correlation coefficient reported in the top right corner. Figure after [376]

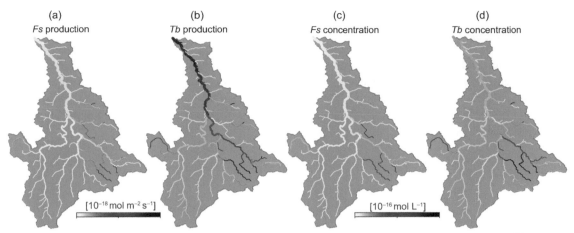

Figure 3.56 Predicted species distributions. Maps of (a, b) predicted production p and (c, d) concentration $\widehat{C}$ of eDNA for *F. sultana* (*Fs*) and *T. bryosalmonae* (*Tb*). Estimates were obtained as the medians of the distributions predicted by the model. Figure after [376]

3.6.3 Maps of eDNA Production in the Wigger River (CH)

Predicted maps of eDNA production for *F. sultana* (Figure 3.56a) identify the southeastern portion of the watershed as a hotspot for bryozoans. This is mainly due to the positive correlation between the presence of moraines and the production of bryozoan eDNA (Figures 3.52c,d). This correlation was uncovered, but it is not yet fully clear what the underlying causes of this positive association are. The upstream bryozoan reservoir explains the eDNA concentration patterns observed at the downstream sites (Figures 3.50c and 3.52c), because the estimated values of the decay time (Figure 3.52d) allow for the detection of eDNA material at distances comparable with the maximum length from source to outlet of the river system. The predicted distribution of *T. bryosalmonae* (Figure 3.56b) mirrors that of *F. sultana* in headwaters and upper sites, while higher values of production are estimated toward the outlet. This is shown by a positive shift in the posterior distribution of the parameter expressing the effect of contributing area (Figure 3.52c,d). The correlation between predicted densities of *F. sultana* and *T. bryosalmonae* is strong (Figure 3.52b) and suggests that bryozoans release disproportionately more spores compared to fish hosts, which release spores in their urine. Thus, we suggest that a full description of the spatial distribution of PKD-infected fish [376] might be unnecessary to understand the bulk of the distribution of the parasite sources when the eDNA signal is dominated by locally abundant colonies of infected bryozoans, as in the case

at hand. We argue that the much stronger correlation observed between predicted densities of *T. bryosalmonae* and contributing area in comparison with that for *F. sultana* (Figure 3.52c) posits that the density of overtly infected (i.e., spore producing) bryozoans tends to increase along downstream directions. In fact, the positive correlation between PKD prevalence and total contributing area in river networks has been shown to be a by product of network connectivity [376]. The fact that *T. bryosalmonae* is mostly shed by bryozoans rather than fish seems plausible in this specific host–parasite system, as parasite maturation in fish kidney tubuli is observed relatively rarely, compared to the prolific spore production within large parasite sacs inside the bryozoan host [414]. We note further that the spatial match of the bryozoan and fish populations is unlikely to drive this relationship, as the biomass of fish is expected to be higher in deeper, more downstream sections [415]. Finally, estimated median values of the decay time (Figure 3.52d) were 4.0 hours for *T. bryosalmonae* (with a 25–75 percentile range of the posterior distribution of 2.7–7.0 h) and 6.9 hours (25–75 percentile range: 5.0–11.1 h) for *F. sultana*, corresponding to decay lengths of 14 km (25–75 percentile range: 10–25 km) and 25 km (25–75 percentile range: 18–40 km), respectively (obtained by assuming an average flow velocity of 1 ms^{-1}), in agreement with previous findings [402].

The framework presented here proved capable [376] of interpreting both eDNA data incidentally shed from benthic populations (*F. sultana*, with likely sources of eDNA from faecal pellets and sloughed cells) and eDNA from spores released into the water (*T.*

bryosalmonae). Although different forms of eDNA may be differently impacted by environmental factors, such as temperature and pH, the choice of formulation involving a single parameter expressing the decay time for both species appeared satisfactory for capturing the integrated eDNA transport dynamics at the catchment scale. Further considerations with regard to this aspect are presented in the SI Appendix of [376]. Another strength of this approach is the possibility of applying adequate parameterizations for eDNA production in an explicit manner to accommodate the nature of the link between the target species density and its biological and environmental filters along hydrologic pathways, such as the environmental conditions or the density of species with which it interacts. Such parameterization provides a simple and versatile means to assimilate field data and to integrate population or species distribution models. Accurate field validations of the current assumptions are needed to generalize this framework, and this could be achieved by relatively simple experimental designs.

For instance, the displacement and decay of genetic material from nonnative known biomasses placed in well-differentiated positions (say, within a catchment where hydrologic and geomorphologic drivers are known) could be a key factor. Subsequent sampling at downstream sites, where eDNA would be contributed by sources at known distances, could then be used to assess the strengths and weaknesses of each assumption underlying the proposed approach. Tracking the source area and the local biomass density of target species via downstream eDNA measurement is possible, provided that a suitable spatially explicit framework is used to interpret the field data. Key is accounting for the filtering produced by the progressive damage occurring during hydrological transport and harnessing it to recover spatial information on species distributions. The integration of quantitative eDNA measurements, hydrogeomorphological scaling, and ecological models presented here has opened a novel direction in ecohydrological studies by unlocking the great potential of remote monitoring using eDNA.

Box 3.28 Methods for eDNA Data Collection, Interpretation, and Use

Characterization of the study area was carried out by obtaining the morphology of the watershed from a 25-m resolution digital terrain model, from which the river network was extracted by means of the Taudem method [215]. Flow directions were determined by following steepest descent paths. Pixels were considered to belong to the channeled portion of the landscape if their drained area was greater than or equal to 0.5 km^2. To subdivide the catchment into units where reasonably constant local morphological conditions apply, stream reaches longer than 5 km were split into equally long stretches. In total, the river network was divided into 166 stretches hierarchically arranged according to the network connectivity, each of them associated with a cluster of pixels directly draining into the stretch. Further details are in [376, 416].

eDNA data collection was carried out by collecting stream water samples in 15 locations (Figure 3.50) along the river network of the Wigger watershed, Switzerland. For each site, a total of twenty-one 500-mL samples were taken at approximately biweekly (or monthly during December, January, and February) intervals (except site 5, a connected pond that was artificially drained after twelve samples were taken). Presterilized (10 percent bleach followed by UV-B treatment) plastic bottles were used to collect water from the river by submerging the bottle with a gloved hand. The samples were transported to the laboratory on ice and filtered using a gentle vacuum on the same day into 5-cm-diameter, 0.45-μm-pore-size individually packaged sterile membrane filters (Merck Millipore). A vacuum pump with a borosilicate glass filtration setup was used and sterilized between each sample in 10 percent bleach followed by three clean water rinses. Negative controls were created by filtering MilliQ water through a sterile filter at the start and end of each filtration session, as well as once during the filtration (after sample 7). Filter papers were placed in 2-mL bead beating tubes (obtained from the kit described below) and frozen at −80°C until extraction. Prior to extraction, filter papers were cut with sterilized scissors to break them up. eDNA was extracted from all filter

Box 3.28 *Continued*

papers, including controls, using a PowerSoil DNA kit (MO BIO Laboratories) in a dedicated clean laboratory (free of PCR products). The kit includes a bead beating step and a separate inhibitor removal step. The eDNA was eluted in 60 μL of Solution C6 and subsequently preserved at -20°C. Samples were only removed from the freezer for analysis and remained at room temperature for a maximum of 2 hours. Details on *F. sultana* and *T. bryosalmonae* eDNA measurement and further characterizations of the study area are reported in [376].

Other relevant details on *F. sultana* and *T. bryosalmonae* eDNA measurements; PCR amplification and target quantification; and techniques of eDNA sampling in rivers are reported in [376].

4 | Waterborne Disease

This chapter completes our line of thought in that the analysis of the implications of the dendritic support for ecological interactions is completed here with the study of the spread of infections of waterborne disease. The reader will note a significant acceleration in the formal complications because, in addition to the directional dispersal of pathogens induced by waterways (whether natural or artificial), a *multiplex* network approach emerges as a key ingredient, that is, the superposition of diverse networks of interactions. In fact, human mobility networks are a fundamental driver of disease spread, in particular owing to the huge pathogen loads displaced by symptomatic and asymptomatic infected individuals. The connections with metapopulation and metacommunity ecology will be highlighted, as well as the deep relation with the core business of this book, placed in context within the introductory chapter. The water-related (WR) diseases that we shall consider are epidemic cholera, endemic schistosomiasis, and proliferative kidney disease in fish. They are representative of limit cases, or classes of problems, and encompass by design a broad spectrum of pathogens, hosts, and life cycles. As we shall discuss in our outlook, Chapter 5, several applications are either available or in the making, because the techniques to render a disease transmission model spatially explicit, by using river networks as the substrate of ecological interactions, is (now) quite straightforward. This journey started by looking at river networks as ecological corridors for epidemic cholera spread, whose first spatially explicit transmission model was developed by our group back in 2008. This photograph, taken in Carrefour, Haiti, at the height of the 2010 cholera epidemics, renders the context.

4.1 Introduction to the Ecology of Waterborne Disease

In the introduction of their remarkable book on the infectious diseases of humans, Anderson and May [417] recall that policy and public health management are often based on theory – not always the best theory. And surely enough science renews itself continuously to provide better theories, better predictions, and better assessments of disease dynamics and their controls. The epidemiological literature abounds with accounts of infectious

diseases invading human communities that in many cases produced (and produce) devastating effects on human populations worldwide. In this chapter, we strive to bring the reader to what is currently considered the best theory in a number of areas that involve fluvial ecological corridors at some point of the life cyle of pathogens and/or hosts. "Best," of course, does not mean perfect or even satisfactory in some cases.

Regarding humans, despite many remarkable medical advances, infectious diseases are still a leading cause of death, representing more than 25 percent of the annual toll [418–421]. At some point, some of these diseases were believed to be definitively defeated. However, they often resurface, possibly in drug-resistant variants that pose significant epidemiological challenges. Also, new diseases emerge. In the past 50 years alone, we count AIDS/HIV, Ebola hemorrhagic fever, the acute respiratory syndrome SARS, and the coronavirus disease COVID-19, to name the most visible diseases. A global map of emerging or reemerging ones has been compiled [422] from a database of infectious diseases that emerged between 1940 and 2004, recording 335 events. The number of deaths linked to them has been increasing since 1940, reaching a peak between 1980 and 1990 in conjunction with the AIDS pandemic. As noted by [422], the majority (more than 60 percent) of infectious events are zoonoses (i.e., diseases related to pathogens of non-human animal origin), and most of these zoonoses (more than 70 percent) are due to pathogens whose reservoir is wildlife. This shows, as a matter of fact, that there is a *continuum* that links wildlife, domestic animals, and humans [423] with an exchange of pathogens between compartments. Therefore, it is now clearer and clearer that parasite ecology is not just a chapter of ecology that is basic to understanding ecosystems functioning but is also a fundamental pillar for designing policies of public health protection. The traditional approach to the treatment of infectious diseases based on pharmacological treatment must necessarily be complemented by both an understanding of the natural environment of which man is part (just a part, of course) and policies of protection and management of ecosystems – relying on theory, including the compartments of parasites and pathogens.

Anderson and May introduced the fundamental distinction between the dynamics of microparasites (mainly bacteria and viruses) and macroparasites (basically helminth worms and arthropods) [417]. The dynamics of diseases caused by microparasites is a major chapter of epidemiology and ecology. The essential feature of microparasites is that they reproduce within the host and have a life cycle shorter than the host average lifetime. Therefore, we can capture the dynamics of many diseases through a compartmental model in which hosts are divided into several categories with respect to their ability to be infected and to infect. Depending on the transmission mode, the compartments can be of various kinds. For directly transmitted diseases (human-to-human, animal-to-animal), it is sufficient to account for a single host species divided into several compartments. In waterborne diseases, the microparasite can be hosted in both humans (or animals) and water, which is thus contaminated and becomes a source of infection for the hosts. In vector-borne diseases (such as malaria), the pathogen needs two species for completing its life cycle: some stages are hosted by a vector species, and the infection of the focal host is never direct but is due to the encounter with a vector; noninfected vectors become carriers of the pathogen via the encounter with an infected host.

Directly transmitted diseases are most relevant to our tenet. This kind of diseases caused by microparasites can be divided into two major classes: there are diseases that, once contracted, leave recovered individuals in a semipermanent or permanent immunization state. There are instead diseases that can be contracted several times by the same individual. Examples of the first class are measles and chickenpox, examples of the second are influenza and common cold. Actually, this subdivision is not so sharp in reality: for the same disease, the degree of immunization can vary from individual to individual and from species to species. In diseases without any immunization, individuals are usually divided into susceptible (indicated with S) and infected (and thus infectious, indicated with I). The recovered from the disease become immediately susceptible again. Models describing diseases of this kind are therefore called SI. In diseases with partial or complete immunization, a third category must be introduced, that of recovered and immune (indicated by the letter R). In this case, they are called SIR models. After a certain time, the recovered may lose their immunization and become susceptible again. Sometimes, these models are thus called SIRS. In the following, we mainly illustrate SI models, which are simpler, though less realistic, and then illustrate some results concerning SIR models. In addition, there exist diseases with a long incubation period during which organisms are infected, but are still unable to transmit the disease. In these cases, it is necessary to introduce a further infection class: that of the exposed (i.e., infected but not yet infectious, indicated with E). It is then also possible to have models of SEI or SEIR type. For simplicity, here we will not discuss these two types of models. For those who want to better investigate the matter, [424] provide a broad overview of the different microparasite models.

The dynamics of diseases caused by macroparasites deserve some further reasoning. In diseases caused by macroparasites, the dynamics of parasites cannot be neglected because their average lifetime is comparable to that of the hosting organisms. Macroparasites grow inside the host but reproduce by releasing infective stages (typically eggs and larvae) into the surrounding external environment, for example, through host defecation. The disease transmission occurs then through ingestion by a new host of these infective stages. Many herbivores are for example infected by intestinal worms whose eggs or larvae are ingested during grazing. Here, we first consider macroparasites with relatively simple life cycles, for example, without intermediate hosts, and then we include more complex life cycles.

An important feature of macroparasites is that their distribution inside the host population can be highly heterogeneous: most of the animals can harbor few or no parasites, and a minority can instead accommodate a large number of parasites. Since the higher the parasite load is, the greater is the damage inflicted to the host, such heterogeneity must be properly taken into account. It should also be borne in mind that, although there are some exceptions, the death of a host causes the death of the hosted parasites.

In particular, waterborne (WB) and water-related or -based (WR) diseases are infections in which the causative agent (or its vector/host) spends at least part of its life cycle in water – and in this book we shall restrict our attention to specific ecosystems like waterways and river networks to highlight the role of the ecological corridors embodied by dendritic ecosystems (and by their peculiar influence on the ecology of pathogens and hosts). A wide range of micro- (viruses, bacteria, protozoa) and macroparasites (mostly flatworms and roundworms) is responsible for WB and WR infections, which are generally caused by exposure to (or ingestion of) water contaminated by pathogenic organisms. WB and WR diseases still represent a major threat to human health, especially in the developing world. As an example, diarrhea, commonly associated with WB pathogens, is responsible for the deaths of about 1.5 million people every year, thus representing one of the leading causes of death among infants and children in low-income countries [420]. Most of that burden is attributable to unsafe water supply, lack of sanitation, and poor hygienic conditions, which either directly or indirectly affect exposure and transmission rates [421].

Here we focus on WR diseases that are to some extent mediated by fluvial waters and the related ecohydrology and affected by the dispersal patterns implied by the structure of waterways like river networks or artificial drainage systems (Figure 4.1). We shall consider as substrate for disease spread a network constituted by a set of nodes (human settlements, each characterized by its known population H_i in demographic equilibrium), connected by edges. These edges (Figure 4.1) define the possible pathways affecting disease spread, namely, the connectivity matrix defining gravity-driven hydrologic transport where pathogens diffuse (basically the river network proper), and the mobility matrix that affects disease spread through the displacement of susceptibles and infected individuals. In fact, we shall see formally later in this chapter that one contracts WR diseases either via contact with water, or by ingestion of pathogens which is water-related anyhow (in addition to possible human-to-human transmission mechanisms). In microparasitic diseases, the key state variables of the description of their evolution in space and time are the number of infected and susceptible individuals in each node i, that is, $I_i(t), S_i(t)$, as well as the concentration of pathogens in accessible water, $B_i(t)$, and, if necessary, the number of recovered invididuals $R_i(t)$. The demography of compartments like $I_i(t)$ and $S_i(t)$ specifies the features of the dynamics of the disease. Figure 4.1 illustrates a possible arrangement of nodes and edges in a fluvial domain, here characterized by human settlements concentrated in network nodes regardless of their population size.

Perhaps the best known of WB diseases is cholera. Cholera is caused by the bacterium *Vibrio cholerae*, which colonizes the human intestine and was discovered in 1854 by Filippo Pacini [425] during an outbreak in Florence (Italy). Its WR spreading mechanisms were famously identified in the late nineteenth century by John Snow, a Victorian anesthesiologist based in London, using early geographic information systems, that is, mapping reported cases and drinking water sources. Indeed, the transmission of the disease is mediated by water. In fact, *V. cholerae* is a natural member of the coastal aquatic microbial community and can survive outside the human host in the aquatic environment. Therefore, the disease can spread from the coastal region, where it is autochthonous, inland through waterways and river networks. The infection is always caused by ingestion of water (or food) either contaminated by *V. cholerae* present in a natural reservoir (primary route) or contaminated by humans (secondary infection), and thus the role of the aquatic environment is crucial for the disease's transmission and spread. Because of poor water sanitation, cholera remains a global threat to public health, especially in developing countries [420, 421], as unfortunately shown by the catastrophic epidemic outbreak that recently struck Haiti (e.g., [426, 427]), where almost 10,000 people died between October 2010 and December

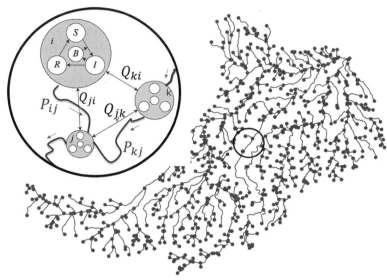

Figure 4.1 Sketch of network models for spatially explicit descriptions of WR disease spread in space and time. Nodes (red) are human communities, here placed for simplicity directly on a waterway (they may be otherwise characterized by their average distance from the nearest stream, should exposure to possible infection depend on that). (inset) Symbols within the enlarged node refer to any SIRB-like model accounting for the demography of susceptible (S), infected (I), and recovered (R) individuals; bacteria in water reservoirs (B); and riverine (P_{ij}) and human mobility (Q_{ij}) connectivity matrices.

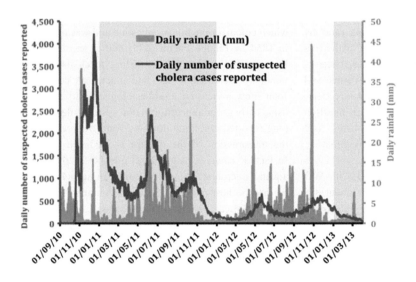

Figure 4.2 The 2010–2013 course of the cholera epidemic in Haiti (total number of weekly reported cases country-wide) plotted synoptically with weekly rainfall intensities. Figure after [427]

2016. Figure 4.2 reports the initial phase of the epidemic. Precipitation is also shown, because the reappearance of the disease in Haiti is often associated with heavy rain events. Mechanistic explanations for this phenomenon exist [428, 429] and are discussed later within the context of the book's tenet.

By no means is our treatment of the subject complete. For example, vector-borne diseases are not dealt with here, despite their obvious conceptual and practical importance. They are transmitted by vectors between humans or from animals to humans. The vectors are living organisms, often bloodsucking insects, that ingest disease-producing micro-organisms while taking a blood meal from an infected host (human or animal) and later inject the micro-organisms into a new host during a subsequent blood meal. Mosquitoes are the best-known vectors, but ticks, flies, fleas, and snails are also responsible for many of these diseases. Every year, globally, there are more than 1 billion cases and more than 1 million deaths from vector-borne diseases, such as malaria, dengue, schistosomiasis, human African trypanosomiasis, leishmaniasis, Chagas disease, yellow fever, Japanese encephalitis, and onchocerciasis. Vector-borne diseases account for over 17 percent of all infectious diseases. Among such diseases, suffice here to mention that malaria is the best studied. It is caused by protozoans of the genus *Plasmodium*. It is transmitted by the biting of infected *Anopheles* mosquitoes, which breed in fresh (or less often brackish) water. Because of the importance of water in the development cycle of the parasite vector, malaria is also cataloged as a WR disease [420], yet it will be overlooked in this context owing to the lack of evidence of a specific role of the hydrological and thus ecological connectivity that is the focus of this book.

4.1.1 Epidemiological Models of Microparasitic Diseases: A First Assessment

Several modeling approaches exist to investigate how a disease spreads within a population. The main ingredients of any model of disease spread – whether epidemic or endemic, depending on the relative number of new cases, attack rates, and the speed of the related demographic changes – are more or less the same. Usually, one assumes that the demography is in equilibrium – that is, background natality and mortality are not dramatically affected in the timescale of the outbreak by the mortality due to the disease (if any). Thus, normally, the overall population of a community suffering disease spread within itself is considered constant. However, variations are possible (like using a logistic model), especially for nonhuman hosts. Susceptible individuals are members of the population (human or animal) that may acquire the disease because of lack of immunity, whether genetic or acquired (permanently or temporarily). Susceptibles become infected at a rate that is termed the force of infection, which very much depends on environmental conditions and the nature of the infection. Human-to-human transmission mechanisms are examined as much as those environmentally mediated by the ingestion of

water or food or by skin penetration by pathogens through water contact. Obviously, reproducing sensible transmission mechanisms via modeling becomes awkward when one or more intermediate hosts need to be taken into account. Most often, in such cases, the ecology of the host(s) must be considered, because at the very least, the number of hosts available to close the infection cycle is important. Surely, where and when such hosts are present and affect the spreading process is a complex ecohydrological problem in many cases of interest, a relatively new one where spatial effects – if anything, habitat suitability of species involved in the disease cycle – must be taken into account.

Here we first examine the basic modeling approaches for microparasites. Refinements, and the related significant acceleration in the technicalities required, follow.

SI Models

When surviving a disease grants permanent immunity (most often the case with direct transmission diseases, but never with waterborne ones), the dynamics of infection are fully characterized by the temporal changes of two variables, S and I, that is, the density (measured, for instance, as number of organisms per square kilometer) of the susceptibles and the infected individuals. We first examine SI models owing to their simplicity. A general formulation of a model with continuous-time reproduction and no immunization is as follows:

$$\frac{dS}{dt} = \nu_S S + \nu_I I - \mu S - iS + \gamma I$$

$$\frac{dI}{dt} = iS - (\mu + \alpha + \gamma) I, \tag{4.1}$$

where t represents time (measured in appropriate units, e.g., years); ν_S and ν_I are the birthrates of susceptibles and infecteds, respectively; μ is the mortality rate in the absence of disease; i is the rate at which the susceptibles are infected; α is the mortality rate caused by the disease (sometimes called virulence); and γ is the recovery rate. The birthrate of the infected ν_I is obviously lower than that of the susceptible (ν_S). An assumption that is often reasonable and that we will use here is that this rate is negligible ($\nu_I = 0$), namely, that organisms do not reproduce during the course of the disease.

It is important to introduce some terms commonly used in epidemiology. The flow of new infected iS is called the incidence rate of the disease (measured in the number of new infected per unit time). The ratio I/N, with $N = S + I$ = total density of organisms, is called prevalence of the disease and is nothing more than the fraction of the population that is sick and therefore infected and infectious.

Model (4.1) takes different forms and has different dynamic behaviors depending on how one specifies the various demographic and epidemiological parameters that characterize it. The first important specification concerns the population demography in the absence of infection. The two most common assumptions are that hosts follow either Malthusian dynamics or logistic dynamics. In the first case, the rates of birth (v_S) and death (μ) are constant; in the second case, the per capita growth rate of the population $v_S - \mu$ is a linear and decreasing function of density N. The second specification concerns the incidence rate of the disease. The parameter i represents the probability per unit time that a susceptible becomes infected; it is given by the product of three factors:

1. the probability $c(N)$ that a susceptible contacts another organism in a time unit;
2. the probability I/N (prevalence) that this other organism will be infected;
3. the probability of actually becoming infected and infectious, which from now on we suppose to be constant.

Depending on the assumptions for the contact rate $c(N)$, several models of disease transmission can be obtained. In particular, we can make two extreme hypotheses:

1. The number of contacts per unit time is proportional to the density N of organisms, for example, because the disease is airborne, such as is the case with influenza.
2. The number of contacts per unit time is constant, for example, because the disease is sexually transmitted and the number of sexual contacts depends on the behavior of each organism, not on population density.

In case 1, the rate of infection i is proportional to $N \times I/N$, thus ultimately to the density of infected I (law of mass action or density-dependent transmission). In case 2, the infection rate is proportional to I/N or the frequency of infection (frequency-dependent transmission). Actually, both assumptions are unrealistic: density-dependent transmission is unrealistic for very high values of the density N, because it is unthinkable that the number of contacts that each organism has per unit time does not saturate to a maximum value, while the frequency-dependent transmission is unrealistic for very low values of N, because it is unthinkable that, in such a case, the number of contacts per unit time remains constant (it must necessarily tend to zero for N tending to zero). A possible functional form of $c(N)$ reconciling assumptions 1 and 2 is the one that we also use in the case of a predator's functional response, namely, $c(N)$ increases with N but saturates to a maximum value. The consequences of the different assumptions are examined in Boxes 4.1 and 4.2. The main results are that Malthusian populations can be regulated by disease and that the establishment of the disease is controlled by a fundamental epidemiological parameter called the basic disease reproduction number R_0. More precisely, a Malthusian population that would grow exponentially, if infected by a disease that does not provide immunization, does instead go to a finite equilibrium abundance, in this way being regulated neither by limited resources nor by predators but by disease. In populations that are regulated by limited resources, instead, the establishment of the disease is regulated by an important parameter (termed R_0) related to the carrying capacity of the population. Specifically, R_0 is the average number of secondary infections caused by one infected individual introduced in a disease-free population at carrying capacity. If it is larger than 1, the disease can establish; otherwise, it cannot.

Box 4.1 SI Model with Malthusian Demography and Density-Dependent Transmission

Since the infection rate i is proportional to the density of the infected, the SI model assumes in this case the following form:

$$\dot{S} = rS - \beta IS + \gamma I$$
$$\dot{I} = \beta IS - (\mu + \alpha + \gamma) I, \tag{4.2}$$

where $r = v_S - \mu$ is the instantaneous rate of Malthusian population growth and β is the coefficient of disease transmission from infected to susceptible (measured in time^{-1} number of infected^{-1}). Suppose now that the population can grow exponentially in the absence of the disease ($r > 0$).

Box 4.1 *Continued*

We show that the microparasite is able to regulate the population, that is, to prevent the host population from exhibiting Malthusian growth. To this end, we note that model (4.2) has two equilibria states, obtained from setting to zero both time derivatives:

1. $\bar{S}_0 = 0$, $\bar{I}_0 = 0$;

2. $\bar{S} = \dfrac{\mu + \alpha + \gamma}{\beta}$, $\bar{I} = \dfrac{r(\mu + \alpha + \gamma)}{\beta(\mu + \alpha)}$.

Like in the Malthusian model without disease, the first equilibrium is unstable, but the real novelty is represented by the presence of a second equilibrium in which the total number of organisms is finite and given by

$$\bar{N} = \bar{S} + \bar{I} = \frac{(\mu + \alpha + \gamma)(\nu_S + \alpha)}{\beta(\mu + \alpha)}.$$

It can be proved (for example, by using the method of linearization) that this equilibrium is stable. Figure 4.3 shows the isoclines for model (4.2) and the evolution of some trajectories. The only trajectory that diverges exponentially is the one with initial conditions $I = 0$; all the others converge toward the stable equilibrium, thus demonstrating the effectiveness of the microparasite disease as demographic regulator. Note that the regulation effect is much more effective (lower $\bar{N}$) if the transmission coefficient β is greater, and it is much less effective (higher $\bar{N}$) if the recovery rate γ is larger, because this allows the replenishment of the susceptible class. However, high recovery rates guarantee the absence of oscillations in the dynamics toward equilibrium. Particularly interesting is the dependence of $\bar{N}$ from the disease virulence (mortality rate α). Indeed, the population size at the equilibrium appears to be either growing for all αs or decreasing for small αs and increasing for large αs, with a minimum value for intermediate virulence. Therefore, very virulent diseases do not

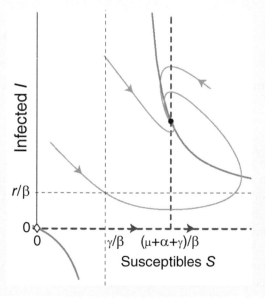

Figure 4.3 Isoclines and trajectories of the *SI* model (4.2) with Malthusian demography and density-dependent transmission (dashed isocline, $\dot{I} = 0$; bold isocline, $\dot{S} = 0$). For small recovery rates γ, the dynamics show damped oscillations, while for high γs, fluctuations are absent.

Box 4.1 *Continued*

regulate at all the population growth, as one might think. This becomes clearer by calculating the prevalence at equilibrium. It is given by

$$\frac{\bar{I}}{\bar{N}} = \frac{r}{\nu_S + \alpha}.$$

So, the larger is the disease virulence, the smaller is the fraction of infecteds in the total population. Basically, for very virulent diseases, the reservoir of the infected is very small because infected organisms die very quickly. Fortunately, terrible diseases, such as the hemorrhagic fever caused by the Ebola virus, cannot spread very effectively.

Box 4.2 SI Model with Logistic Demography and Density-Dependent Transmission

For the sake of simplicity, let us assume that the mortality rate μ is constant and that logistic demography is exclusively determined by the birthrate being dependent on density. The model *SI* then becomes

$$\dot{S} = rS\left(1 - \frac{S+I}{K}\right) - \beta IS + \gamma I$$

$$\dot{I} = \beta IS - (\mu + \alpha + \gamma)I,$$

(4.3)

where K is the population's carrying capacity in the absence of infection. It is easy to find out that, under certain conditions, model (4.3) has three equilibria: the first is the trivial equilibrium $\bar{X}_0 = [\bar{S}_0 = 0, \bar{I}_0 = 0]^T$ (where the superscript T indicates matrix transposition); the second is the equilibrium corresponding to the healthy population at its carrying capacity ($\bar{X}_K = [\bar{S}_K = K, \bar{I}_K = 0]^T$); the third equilibrium $\bar{X}_+$ corresponds to a partially infected population. In fact, it can be easily found that both time derivatives of model (4.3) vanish if

$$\bar{S}_+ = \frac{\mu + \alpha + \gamma}{\beta}$$

$$\bar{I}_+ = \frac{rK(\mu + \alpha + \gamma)}{\beta K(\mu + \alpha) + r(\mu + \alpha + \gamma)}\left(1 - \frac{\mu + \alpha + \gamma}{\beta K}\right).$$

(4.4)

For this equilibrium to make sense in biological terms, it is required that $\bar{I}_+$ not be negative, that is, that the following inequality holds:

$$R_0 = \frac{\beta K}{\mu + \alpha + \gamma} \geq 1.$$

(4.5)

The quantity R_0 is termed the basic reproduction number of the disease and is one of the key parameters of epidemiology. In fact, since $1/(\mu + \alpha + \gamma)$ is the mean residence time in the infected class, R_0 can be interpreted as the average number of secondary infections produced by one infected individual introduced in a healthy population consisting of K individuals.

If R_0 is less than 1, the equilibrium with a partially infected population cannot exist. It can be shown that for $R_0 \leq 1$, the equilibrium $\bar{X}_K$ is stable whatever the initial conditions, and so the disease spontaneously fades out. On the contrary, if R_0 is greater than unity, the equilibrium $\bar{X}_+$ (see Eq. (4.4)) is the only stable equilibrium, and therefore the microparasitic disease is endemic.

Box 4.2 *Continued*

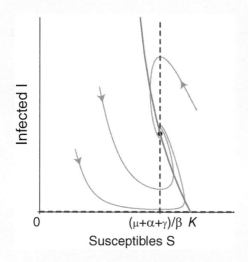

Figure 4.4 Isoclines and trajectories of the *SI* model (4.3), with a logistic growth and density-dependent transmission (dashed isocline, $\dot{I} = 0$; bold isocline, $\dot{S} = 0$).

Figure 4.4 shows the isoclines of model (4.3) and the evolution of the trajectories in the latter case. It is interesting to notice that the larger the carrying capacity of the host population or the larger the transmission rate β is, the larger is R_0. On the contrary, the larger the recovery rate γ or the greater the virulence α is, the smaller is R_0. Once again, we remark that very lethal diseases are not as dangerous as one may think because they cannot spread very easily or they cannot even spread ($R_0 < 1$).

It is also interesting to calculate the prevalence at the equilibrium. It is given by

$$\frac{\bar{I}}{\bar{N}} = \frac{r}{\mu + \alpha + r} \frac{\beta K - (\mu + \alpha + \gamma)}{\beta K}.$$

The percentage of infected is thus decreasing with the virulence α of the disease.

Logistic demography and saturating transmission are the next logical development. If the contact rate is constant, then transmission is termed frequency dependent. As mentioned above, the hypothesis that the number of contacts remains constant even for N tending to 0 is unrealistic. A hypothesis that couples density-dependent transmission for small N and frequency-dependent transmission for intermediate or large N is one in which we assume that the contact rate $c(N)$ is increasing yet saturating with N. We can use an expression similar to that of the predator's functional response of the second type, which is widely used in population ecology, namely, $c(N) \propto N/(\delta + N)$, where δ is the half-saturation density. It follows that the infection rate i is given by

$$i = \beta \frac{N}{\delta + N} \frac{I}{N} = \beta \frac{I}{\delta + N},$$

where β is an appropriate coefficient of proportionality (measured as time^{-1}) that depends also on the probability of actually contracting the disease after having contacted an infected individual. The resulting SI model with logistic demography is

Box 4.2 *Continued*

$$\dot{S} = rS\left(1 - \frac{S+I}{K}\right) - \beta\frac{IS}{\delta + S + I} + \gamma I$$

$$\dot{I} = \beta\frac{IS}{\delta + S + I} - (\mu + \alpha + \gamma)I. \tag{4.6}$$

The analysis of the model is easier after a change of variables. Instead of using S and I, it is convenient to introduce the total density of hosts $N = S + I$ and the prevalence $x = I/(S+I)$. Exploiting the relationships

$$\dot{N} = \dot{S} + \dot{I}$$

$$\dot{x} = \frac{1}{N}\dot{I} - \frac{I}{N^2}\dot{N},$$

one gets

$$\dot{N} = \left[r(1-x)\left(1 - \frac{N}{K}\right) - (\mu + \alpha)x\right]N$$

$$\dot{x} = \left[\left(\beta\frac{N}{\delta + N} - r\left(1 - \frac{N}{K}\right) - (\mu + \alpha)\right)(1-x) - \gamma\right]x. \tag{4.7}$$

Like in the previous density-dependent case, this model has a trivial equilibrium

$$\bar{X}_0 = \left[\bar{N}_0 = 0, \bar{x}_0 = 0\right]^T$$

and an equilibrium corresponding to the population being disease-free at its carrying capacity

$$\bar{X}_K = \left[\bar{N} = K, \bar{x} = 0\right]^T.$$

The nontrivial isoclines (see Figure 4.5) have the following expressions:

$$\dot{N} = 0 \quad \Rightarrow \quad x = \frac{r\left(1 - \frac{N}{K}\right)}{\mu + \alpha + r\left(1 - \frac{N}{K}\right)}$$

$$\dot{x} = 0 \quad \Rightarrow \quad x = 1 - \frac{\gamma}{\beta\frac{N}{\delta + N} - r\left(1 - \frac{N}{K}\right) - (\mu + \alpha)}.$$

They are displayed in Figure 4.5. It follows that there is a third equilibrium at which the population is partially infected only if the nontrivial isoclines intersect at values of N in the range $(0, K)$ and at values of prevalence x between 0 and 1. It is easy to verify that the condition required for this to occur is

$$\beta\frac{K}{\delta + K} - (\mu + \alpha) > \gamma.$$

In this case, the third equilibrium is stable (see the trajectories in Figure 4.5) and the trivial equilibria are unstable. If the above condition is not verified, the third equilibrium does not exist and the stable equilibrium is, of course, $\bar{X}_K$.

Box 4.2 *Continued*

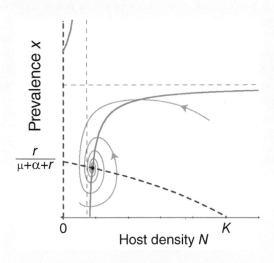

Figure 4.5 Isoclines and trajectories of the SI model (4.7) with logistic demographics and saturating transmission. Unlike Figure 4.4, the system dynamics are not represented in the SI state space (density of infected vs. density of susceptible) but in the state space $N - x$ (disease prevalence vs. total population density) (dashed isocline, $\dot{N} = 0$; bold isocline, $\dot{x} = 0$).

Even in this case, we can define the basic reproduction number of the disease R_0, that is, the average number of secondary infections produced by one infected individual introduced into a healthy population (which therefore consists of K individuals). As the number of new infections produced in a unit time is $iK = \beta \frac{1}{\delta + K} K$ and the mean residence time of the infected individual in the infected class is $1/(\mu + \alpha + \gamma)$, it follows that

$$R_0 = \frac{\beta K}{(\delta + K)(\mu + \alpha + \gamma)}.$$

It is then easy to verify that the condition for the existence of the third equilibrium is still nothing but the condition $R_0 > 1$. If the reproduction number of the disease is smaller than 1, then the disease cannot spread in the population. Note that for very small values of the half-saturation density δ, that is, when the disease is practically frequency dependent, one gets

$$R_0 \cong \frac{\beta}{\mu + \alpha + \gamma}.$$

Therefore, we can conclude that the spread of a frequency-dependent disease (e.g., AIDS) within a population is not facilitated by the size K of the population, in contrast to what has been found above for density-dependent transmission, where R_0 is directly proportional to the demographic carrying capacity of the population (see Eq. (4.5)).

SIR Models

The dynamics of infection are fully characterized by the behavior over time of the variables S, I, and R, that is, the density of susceptible, infected, and recovered (and permanently or temporarily immune). A general formulation of a SIR model follows:

$$\dot{S} = \nu_S(S + R) + \nu_I I - \mu S - iS + \gamma R,$$
$$\dot{I} = iS - (\mu + \alpha + \rho)I, \tag{4.8}$$
$$\dot{R} = \rho I - (\mu + \gamma)R,$$

where ν_S and ν_I are the birthrate of susceptibles and infecteds, respectively; μ is the mortality rate in the absence of the disease; i is the rate at which susceptibles become infected; α is the mortality rate induced by the disease (virulence); ρ is the recovery rate resulting in immunization; and γ is the rate of immunity loss. It is assumed that the recovered have the same birth and death rates of the susceptible and that their progeny does not inherit immunity and is therefore susceptible. Even in the case of SIR models, we make the assumption that the birthrate of the infected v_I is negligible, that is, that individuals do not reproduce during the disease course.

The main difference with a disease without immunization is that a Malthusian population cannot always be regulated by the disease. The demographic regulation is possible only if the recovery rate is below a certain threshold (for details, see Box 4.3).

The only case analyzed here is the one with Malthusian dynamics and density-dependent transmission. Our purpose is in fact to investigate if a disease that provides complete or partial immunity is able to control an otherwise exponentially growing population.

Box 4.3 SIR Model with Malthusian Demography

The SIR model is the following:

$$\dot{S} = rS + v_S R - \beta SI + \gamma R,$$
$$\dot{I} = \beta SI - (\mu + \alpha + \rho)I, \tag{4.9}$$
$$\dot{R} = \rho I - (\mu + \gamma)R,$$

where $r = v_S - \mu > 0$ is the instantaneous Malthusian rate of population growth and β is the coefficient of disease transmission from infected to susceptible. Like in the corresponding SI model, there are two possible equilibria: the trivial equilibrium $S = 0$, $I = 0$, $R = 0$, which is always unstable because the population dynamics of the disease-free population are Malthusian, and a second equilibrium given by

$$\bar{X}_+ = \begin{bmatrix} \bar{S} = \dfrac{\mu + \alpha + \rho}{\beta} \\ \bar{I} = \dfrac{r(\mu + \alpha + \rho)(\mu + \gamma)}{\beta\left[(\mu + \alpha + \rho)(\mu + \gamma) - (v_S + \gamma)\rho\right]} \\ \bar{R} = \dfrac{r\rho(\mu + \alpha + \rho)}{\beta\left[(\mu + \alpha + \rho)(\mu + \gamma) - (v_S + \gamma)\rho\right]} \end{bmatrix}.$$

The constraint that $\bar{I}$ and $\bar{R}$ be positive, however, requires that the denominator of their expression be positive, namely, that the recovery rate be such as to satisfy the inequality

$$\rho < \frac{(\mu + \alpha)(\mu + \gamma)}{r}. \tag{4.10}$$

It can be shown that if the recovery rate is below the threshold given by (4.10), the population dynamics converge toward the equilibrium $\bar{X}_+$; otherwise, the only equilibrium is the trivial one, which is unstable. We can therefore conclude that, in contrast to diseases that do not provide immunity, those that confer partial or complete immunity are able to control the population growth only if the recovery rate is not too high.

4.1.2 Models of Macroparasitic Diseases

In diseases caused by macroparasites, the dynamics of parasites cannot be neglected because their average lifetime is comparable to that of the hosting organisms. Macroparasites grow inside the host but reproduce by releasing infective stages (typically eggs and larvae) into the surrounding external environment, for example, through host defecation. The disease transmission occurs then through ingestion by a new host of these infective stages. For example, many herbivores are infected by intestinal worms whose eggs or larvae are ingested during grazing. Macroparasites may have a relatively simple life cycle, but in a number of cases of interest to WB disease, intermediate hosts play a major role in the spread of disease.

An important feature of macroparasites is that their distribution inside the host population can be highly heterogeneous: most of the animals can harbor few or no parasites, and a minority can instead accommodate a large number of parasites. The higher the parasite load is, the greater is the damage inflicted to the host. This must be properly taken into account in evaluating the impact and spread of WB parasitic disease. It should also be borne in mind that, although there are some exceptions, the death of a host causes the death of the hosted parasites.

The classical model of macroparasite diseases is the one in [430]. Let us denote by H the number or density of the hosts, by P the total number or density of adult parasites inside hosts, and by L the number or density of the younger stages of the parasite (e.g., larvae) living outside the hosts. The average parasite load is then P/H. However, the distribution of loads can be heterogeneous, so let us denote by p_i the proportion of hosts harboring i parasites ($i = 0, 1, 2, \ldots$). Note that this implies $\sum_{i=0}^{\infty} i p_i H = P$. By assuming that

- new adult parasites are recruited by casual encounters between the free-living infective stages and the hosts;
- the mortality of every host linearly increases with the number i of its parasites;
- the death of a host causes the death of all adult parasites hosted in it,

one can write the following coupled equations for the host–parasite dynamics:

$$
\begin{cases}
\dot{H} = (\nu - \mu) H - \sum_{i=0}^{\infty} \alpha i p_i H \\
\dot{P} = \beta L H - m P - \sum_{i=0}^{\infty} (\mu + \alpha i) i p_i H,
\end{cases}
\tag{4.11}
$$

where ν and μ are the birth and death rates of hosts in the absence of the disease, β is the per unit time probability that a host encounters a larval parasite, m is the mortality rate of the adult parasites inside the host, and α is the additional mortality rate due to the presence of each parasite inside the host. The two mortality terms of Equations (4.11) that are linked to the parasite load must be properly explained. As for the equation of hosts, it must be noticed that the proportion of hosts containing i parasites (p_i) suffers an additional mortality that amounts to $\alpha \cdot i$, and then the total number of additional deaths per unit time in the host population is in fact $\sum_{i=0}^{\infty} \alpha i p_i H$.

As for the equation of parasites, it should be noted that adult parasites do not only die because of the intrinsic mortality m; as we specified above, they also die together with a dying host that harbors them. Since a host containing i parasites suffers a total mortality $\mu + \alpha \cdot i$ that leads to the death of all the i hosted parasites, it follows that the total number of deaths per unit time among adult parasites equals $\sum_{i=0}^{\infty} (\mu + \alpha i) i p_i H$.

The ultimate form taken by model (4.11) is specified in different ways, depending on the assumptions made for the various demographic and epidemiological parameters that characterize it. As in the microparasitary models, the first important specification concerns the population demography in the absence of infection. The two most common assumptions are that hosts follow either a Malthusian or a logistic demography. The second specification relates to the density L of the free-living parasite stages, which live outside the hosts (e.g., in the grass). Obviously, L must be an increasing function of the number of adult parasites that produce these stages (for example, as eggs, which are then defecated by hosts) by reproducing inside the host. On the other hand, L must be a decreasing function of the number of hosts, because the greater H is, the greater is the ingestion of the parasite free-living stages, with a consequent depressing effect on their density in the surrounding environment. A functional form that is widely used (and that can be justified by a simple model of the L dynamics, which is not reported here) is the following:

$$
L = \frac{\theta P}{H + H_0},
$$

where θ is the parasite fertility and H_0 is a positive parameter specifying for which values of H the density of the parasite free-living stages is significantly reduced, as a result of ingestion by hosts.

The third and most important specification is that concerning the nature of the distribution p_i of parasite loads inside the host population. A simple statistical distribution that is well suited to a large amount of data is the negative binomial distribution [31]. It is characterized by two parameters: the mean M and a clumping parameter k. Figure 4.6 shows several shapes of the negative binomial distribution for different values of k. The variance σ^2 of the negative binomial distribution is linked to the mean by the following relationship:

$$
\sigma^2 = M + \frac{M^2}{k}.
$$

When the clumping parameter k is large, the variance is almost equal to the mean, as in the Poisson distribution:

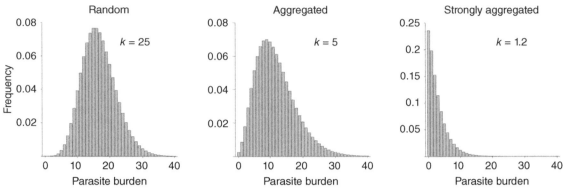

Figure 4.6 Shape of negative binomial distributions for different values of the clumping parameter k.

parasite loads are more or less evenly distributed around the mean. But in most cases, k is small (less than 5) and the distribution is clumped, namely, a few hosts contain a number of parasites that is much higher than the average parasite load $M = P/H$, and many hosts instead contain quite a small number of parasites. Assuming that the parasite distribution is a negative binomial, we can specify the following terms in Equations (4.11):

$$\sum_{i=0}^{\infty} i p_i = M = P/H$$

$$\sum_{i=0}^{\infty} i^2 p_i = M^2 + \sigma^2 = M^2 + M + M^2/k$$

$$= \frac{P}{H} + \frac{k+1}{k}\frac{P^2}{H^2}.$$

Assuming a logistic demography, we can finally formulate the following model for host-macroparasite dynamics:

$$\dot{H} = rH\left(1 - \frac{H}{K}\right) - \alpha P$$

$$\dot{P} = \frac{\lambda P H}{H_0 + H} - (m + \mu + \alpha)P - \alpha\frac{k+1}{k}\frac{P^2}{H}, \tag{4.12}$$

where r and K are, respectively, the intrinsic rate of increase and the carrying capacity of the host population, and $\lambda = \beta\theta$.

The fundamental result of the macroparasitic model (see Box 4.4) is that the establishment of the disease is still related to the basic reproduction number R_0, which is given by $R_0 = \lambda K/(H_0 + K)(m + \mu + \alpha)$. As noted above, if $R_0 > 1$, the disease can establish.

Box 4.4 Modeling of Macroparasitic Disease

Like in microparasite models, we now try to determine under which conditions the host population is permanently infested by macroparasites. To this end, we determine the possible equilibria of model (4.12). In addition to the trivial equilibrium ($\bar{H}_0 = 0$, $\bar{P}_0 = 0$), they turn out to be

- $\bar{H}_K = K$, $\bar{P}_K = 0$;
- $\bar{H}_+ = H^*$ and

$$\bar{P}_+ = \frac{r}{\alpha}H^*\left(1 - \frac{H^*}{K}\right), \tag{4.13}$$

where H^* is the unique positive solution of the equation

$$\frac{\lambda H}{H_0 + H} - (m + \mu + \alpha) - \frac{k+1}{k}r\left(1 - \frac{H}{K}\right) = 0. \tag{4.14}$$

It is easy (yet boring) to reduce expression (4.14) to a quadratic equation and determine that there is a positive solution H^* only if $(\lambda/[K + H_0]) - (m + \mu + \alpha) > 0$. This inequality can be interpreted in light

Box 4.4 *Continued*

of an epidemiological parameter that we already met when we dealt with microparasite diseases: the basic reproduction number of the disease, R_0. In fact, since the average lifetime of an adult parasite is $1/(m + \mu + \alpha)$, the expected number of secondary infections (i.e., of new adult parasites) caused by one parasite introduced into a disease-free host population at its carrying capacity turns out to be

$$R_0 = \frac{\lambda K}{H_0 + K} \frac{1}{m + \mu + \alpha}.$$

If $R_0 < 1$, the only nontrivial equilibrium is $(\bar{H}_K, \bar{P}_K)$. It is stable, and so the disease cannot develop. If $R_0 > 1$, the equilibrium $(\bar{H}_+, \bar{P}_+)$ exists and is stable, as shown in Figure 4.7, which displays the isoclines of model (4.12) and the phase portrait with some trajectories. In this case, the disease can develop, and it permanently settles into the host population.

It is interesting to analyze how R_0 depends on the different parameters involved: it increases, of course, with the parasite fertility (parameter λ); it increases and saturates with the host population size K; and it decreases with the mortality of both hosts and parasites. In particular, R_0 decreases with the mortality α inflicted by macroparasites on their hosts, and this shows once again that very lethal diseases cannot easily settle in animal populations. It is to be remarked that R_0 does not depend on the clumping parameter k. Such a parameter is, however, important in determining the parasite average load: in fact, from Equation (4.14), we can deduce that H^* decreases with increasing k and from Equation (4.13) that the average parasite load P^*/H^* decreases with H^*; thus, ultimately, it increases with k. Therefore greater clumping (k small) corresponds to a lower average parasite load.

Until now, we have assumed that the influence of macroparasites on their hosts basically consists of increasing mortality proportionally to the parasite load. However, in many cases, for example, the red grouse population, *Lagopus lagopus scoticus*, hosting the nematode worm *Trichostrongylus tenuis* [431, 432], the macroparasite exerts its action by decreasing host fertility. We can easily analyze the consequences of this phenomenon by reformulating model (4.11) as follows:

$$\dot{H} = \left(\nu - \sum_{i=0}^{\infty} \varepsilon i p_i - \mu \right) H$$
$$\dot{P} = \beta L H - m P - \sum_{i=0}^{\infty} \mu i p_i H,$$

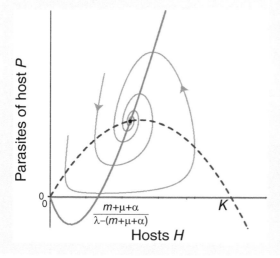

Figure 4.7 Isoclines and trajectories of the host-macroparasite model Equation (4.12) with logistic demography and influence of the macroparasite on the host survival.

Box 4.4 *Continued*

where we assume that the birthrate decreases proportionally to the parasite load i (with a proportionality constant equal to ε). By making the same assumptions about demography and density of juvenile stages of the parasites that led to formulating model (4.12), we get

$$\dot{H} = H\left(1 - \frac{H}{K}\right) - \varepsilon P$$

$$\dot{P} = \frac{\lambda PH}{H_0 + H} - (m + \mu)P.$$

(4.15)

The analysis of model (4.15) is simple and interesting. In addition to the trivial equilibrium $\bar{X}_0$ and the equilibrium corresponding to the healthy population $\bar{X}_K$, a third equilibrium $\bar{X}_+ = \left[\bar{H} = H^*, \bar{P} = P^*\right]^T$ can exist, at which

$$H^* = \frac{m + \mu}{\lambda - m - \mu} H_0 \qquad P^* = \frac{r}{\varepsilon} H^*\left(1 - \frac{H^*}{K}\right).$$

It is easy to see that this equilibrium is feasible ($H^* > 0$, $P^* > 0$) if the condition holds that the basic reproduction number of the disease R_0 is greater than unity, where

$$R_0 = \frac{\lambda K}{H_0 + K} \frac{1}{m + \mu}.$$

Even if feasible, however, this equilibrium is not always stable. In particular, it can be shown that it is unstable if $H^* < K/2$. The two situations ($H^* > K/2$ and $H^* < K/2$) are illustrated in Figure 4.8 through the graph of isoclines and a sketch of some trajectories. In the case in which $H^* < K/2$, there is no stable equilibrium, and the host and parasite populations converge toward a stable limit cycle (periodic solution) of appropriate period. Note that this situation can be established when the parasite fertility θ or the contact rate β is large (which implies a large λ) or when the half-saturation constant H_0 is small. Of course, if $R_0 < 1$, the equilibrium $\bar{X}_+$ is not feasible, and the disease cannot establish itself in the population, so that the only stable equilibrium is the one corresponding to the disease-free population $\bar{X}_K$. Possible self-sustained oscillations obtained with model (4.15) are corroborated by empirical evidence [431].

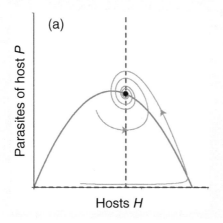

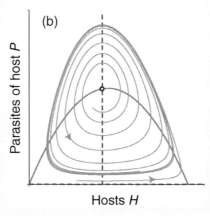

Figure 4.8 Isoclines and trajectories of the model Equation (4.15) in the case of (a) stable steady state and (b) periodic regime with a limit cycle.

Macroparasites are often large in size at their adult stage, in particular helminth worms such as nematodes (e.g., roundworms and protostrongylids) and cestodes (e.g., tapeworms). They may have a simple life cycle like the one we have just described. In other cases, macroparasites do need not only a definitive host for developing into an adult stage but also an intermediate host inside which they develop particular juvenile stages that are then ingested by the final host (see Box 4.16 for the outstanding case of schistosomiasis, which has a particular relevance to the tenet of this book). Even more complex life cycles are possible, yet they are not described herein.

4.1.3 Early Models of WB Disease Spread

Early attempts to model WB diseases, in particular cholera dynamics, date back to the work by [433]. More recently, [434] proposed a simple system of three ordinary differential equations where, in addition to the compartments of susceptible (S) and infected (I) that characterize traditional microparasitological models, one equation accounts for the population dynamics of free-living bacteria (B), that is, bacteria that can be found in a common water reservoir. The standard SIB model can be stated as

$$\frac{dS}{dt} = \mu(H - S) - \beta BS,$$

$$\frac{dI}{dt} = \beta BS - (\mu + \alpha + \rho)I, \qquad (4.16)$$

$$\frac{dB}{dt} = \theta I - \delta B,$$

where μ is the human natural mortality rate (which is assumed to match the birthrate, so that the human population is at demographic equilibrium in the absence of disease, with H being the population size), β is the rate of exposure to water contaminated by the pathogen, α is the additional mortality rate induced by cholera on infected hosts, ρ is the recovery rate from disease (recovered individuals are assumed to gain long-lasting immunity in this simplified model), θ is the contamination rate (i.e., the rate at which bacteria or viruses shed by infected people reach the water reservoir), and δ is the mortality rate of free-living pathogens. Note that, as a first approximation, we neglect the reproduction of pathogens in the water body. In fact, the only inflow of new pathogens into the water reservoir is the one due to the shedding of infected people.

In addition to the disease-free equilibrium ($S = H$, $I = 0$, $B = 0$), system (4.16) has an endemic nontrivial equilibrium with coordinates

$$\bar{S} = \frac{\delta\phi'}{\beta\theta}, \quad \bar{I} = \frac{\mu}{\phi'}\left(H - \frac{\delta\phi'}{\beta\theta}\right), \quad \bar{B} = \frac{\mu\theta}{\phi'\delta}\left(H - \frac{\delta\phi'}{\beta\theta}\right),$$

where $\phi' = \mu + \alpha + \rho$. Of course, this equilibrium is biologically meaningful only if $\bar{I}$ and $\bar{B}$ are positive, namely, only if

$$H > \frac{\delta\phi'}{\beta\theta}.$$

By introducing the basic reproduction number

$$R_0 = \frac{\beta\theta H}{\delta\phi'},$$

it is possible to show that if $R_0 < 1$, the disease-free equilibrium is stable, while the endemic equilibrium is not feasible; conversely, if $R_0 > 1$, the disease-free equilibrium is unstable, while the nontrivial equilibrium is feasible and stable. Note that $1/\phi'$ is the average residence time of hosts in the infected class, while $1/\delta$ is the average residence time of pathogens in water. Thus, if one infected individual is introduced in a community of size H, he or she will produce θ/ϕ' pathogens during the infectious period, and these pathogens will in turn infect $\beta H/\delta$ susceptibles during the pathogen lifetime in water. Therefore, R_0 can be again interpreted as the average number of secondary infections produced by one infected individual introduced in a healthy population consisting of H individuals. For example, typical reproduction numbers for the Haitian cholera epidemic ranged between 2.5 and 4 [435]; that is, they were quite large. Lower values of R_0 were recorded for other cholera epidemics, for example, the one that occurred in Zimbabwe during 2008–2009 [436].

Generalizations of the model Equation (4.16) are possible. In particular, demography can be made logistic, for instance, and the disease transmission, instead of being linearly dependent on pathogen concentration B, can be a saturating function of B. Disease establishment is still related to a modified expression of the basic reproduction number R_0. Even more important is that the model can be made space explicit by introducing a suitable network of nodes (i.e., communities of hosts) that are connected by both the fluvial and the mobility network.

4.1.4 Generalized Reproduction Numbers for Spatially Explicit Models of WB Disease Epidemics

Outbreaks of waterborne diseases pose challenging problems because infection patterns are influenced by spatial structure and temporal asynchrony. Spatial modeling is needed to tackle such patterns, with the support of widespread data mapping of the relevant hydrological forcings and data basic to epidemiology, such as transportation infrastructure, population distribution, and sanitation state. The precise conditions under which a waterborne disease epidemic can start in a spatially explicit setting are relatively complex to sort out. Here we show formally, following [437], that the requirement that all the local reproduction numbers R_0 be larger than unity is neither necessary nor sufficient for outbreaks to occur when local settlements are connected by networks of primary and secondary infection mechanisms.

The basis of our analysis is a spatially explicit nonlinear differential model (Box 4.5) that accounts for both the hydrological and the human mobility networks [428, 438]. The hydrological networks can be those of river basins, extracted from landscape topography [63], or those of human-made water distribution and sewage systems (or both). Depending on spatial resolution,

network nodes can be cities, towns, or villages. In the ith community of size H_i, with $i = 1, n$ (n is the number of nodes), the state variables at time t are the local abundances of susceptibles, $S_i(t)$, and infected/infective individuals, $I_i(t)$, and the concentration of pathogens, $B_i(t)$, in water reservoirs of volume W_i. Infectives release pathogens at a site-dependent rate p_i. Susceptibles are exposed to contaminated water at a rate β_i and become infected according to a saturating function of B_i ($K =$ half-saturation constant). Connections between communities are described by two matrices [428, 438], $\mathbf{P} = \left[P_{ij} \right]$ (hydrologic network) and $\mathbf{Q} = \left[Q_{ij} \right]$ (human mobility), with $i, j = 1, n$. Pathogens die at a rate μ_B and move in water from node i to node j with a probability P_{ij} at a rate l, depending on downstream advection and other water-mediated transport pathways (e.g., attachment to plankton). They are also spread by human mobility: individuals leave their home node i with probability m_S for susceptibles and m_I for infectives (usually $m_S \geq m_I$), reach their target j with probability Q_{ij}, and then come back to i. Consistency requires $\sum_{j=1}^{n} P_{ij} = 1$ and $\sum_{j=1}^{n} Q_{ij} = 1$ for any i. We assume that the union of $\mathbf{P}$- and $\mathbf{Q}$-associated graphs is strongly connected, namely, the infection can spread to any community along either network.

Box 4.5 SIB Dynamics on Networks

On timescales shorter than the typical duration of acquired immunity, the epidemiological dynamics of susceptibles S_i and infected/infectives I_i in the ith community, with $i = 1, n$, and transport of pathogens (as represented by their concentrations B_i) over the networks are described by the following set of ODEs (t is time):

$$\frac{dS_i}{dt} = \mu(H_i - S_i)$$

$$- \left[(1 - m_S)\, \beta_i f(B_i) + m_S \sum_{j=1}^{n} Q_{ij} \beta_j f\left(B_j\right) \right] S_i, \tag{4.17}$$

$$\frac{dI_i}{dt} = \left[(1 - m_S)\, \beta_i f(B_i) + m_S \sum_{j=1}^{n} Q_{ij} \beta_j f\left(B_j\right) \right] S_i$$

$$- (\gamma + \mu + \alpha)I_i, \tag{4.18}$$

$$\frac{dB_i}{dt} = -\mu_B B_i - l\left(B_i - \sum_{j=1}^{n} P_{ji} \frac{W_j}{W_i} B_j \right)$$

$$+ \frac{p_i}{W_i}\left[(1 - m_I)\, I_i + \sum_{j=1}^{n} m_I Q_{ji} I_j \right]. \tag{4.19}$$

Box 4.5 *Continued*

The evolution of susceptibles (first equation) is a balance between population demography and infections due to contact with pathogens. The host population, if uninfected, is at demographic equilibrium H_i (the size of the ith local community), with μ being the baseline mortality rate of humans. The parameter β_i is the site-dependent rate of exposure to contaminated water, and $f(B_i)$ (dose–response function) is the probability of becoming infected due to exposure to concentration B_i of pathogens. In accordance with much literature on cholera [434], we adopt the hyperbolic and saturating function $f(B_i) = B_i/(K + B_i)$ (K is the half-saturation constant). The dynamics of infectives (second equation) are a balance between newly infected individuals and losses due to recovery or natural/pathogen-induced mortality, with γ and α being recovery and disease-induced mortality rates, respectively. The dynamics of recovered individuals are neglected, because waterborne diseases confer at least temporary immunity, and its loss neither determines conditions for disease onset nor affects its evolution in the immediate development following the epidemic peak. The dynamics of free-living pathogens concentration (third equation) assumes that bacteria or protozoa are released in water (e.g., excreted) by infective individuals and immediately diluted in a well-mixed local water reservoir of volume W_i at a site-dependent rate p_i. Free-living pathogens are also assumed to die at a constant, site-independent rate μ_B. As regards the hydrological transport, the spread of pathogens over the river network is described as a biased random walk process on an oriented graph [325]. Here, we assume that pathogens can move at a rate l from node i to node j of the hydrological network with a probability P_{ij}. The rate depends on both downstream advection and other possible pathogen transport pathways along the hydrological network, for example, short-range distribution of water for consumption or irrigation or bacterial attachment to phyto- and zooplankton. Possible topological structures for the hydrological network range from simple one-dimensional lattices to realistic mathematical characterizations of existing river networks. The nodes of the human mobility network are assumed to correspond to those of the hydrological layer, whereas edges are defined by connections among communities. We also assume that susceptible and infective individuals can undertake short-term trips from the communities where they live toward other settlements. While traveling or commuting, susceptible individuals can be exposed to pathogens and return as infected carriers to the settlement where they usually live. Similarly, infected hosts can disseminate the disease away from their home community. In many cases, infected individuals are asymptomatic and thus are not barred, or are only partially barred, from their usual activities by the presence of the pathogen in their intestine. Human mobility patterns are defined according to a connection matrix in which individuals leave their original node (say, i) with an infection-dependent probability (respectively, m_S for susceptibles and m_I for infectives, usually with $m_S \geq m_I$), reach their target location (say, j) with a probability Q_{ij}, and then come back to node i. Topological and transition probability structures for human mobility network used in epidemiology can be based on suitable measures of node-to-node distance, like in gravity models [105], on the actual transportation network or on models based on conceptually different interactions, such as Erdős–Rényi random graphs, scale-free networks, or small world–like graphs [439].

In addition to the assumptions enumerated above, the above equation postulates specific timescales for the use of reported cases and for the overall duration of the epidemic compared to the mean duration of acquired immunity. In fact, should daily cases be reproduced, an additional compartment for exposed persons, say, $E_i(t)$, who are incubating the disease and thus provide a delay of a few days for the transfer from the compartment of susceptibles to that of infected proves useful

Box 4.5 *Continued*

[429, 440, 441]. Moreover, if long-term trends are studied, say, multiseason studies lasting for a few years, then by necessity, a further compartment, $R_i(t)$, of recovered individuals is reinserted into the relative compartment of susceptibles $S_i(t)$ at a rate $\rho R_i(t)$, where ρ^{-1} is of the order of three years for cholera but varies generally as a function of the specific disease and strength of the infection [428].

Figure 4.9 shows the epidemiological, demographic, and sanitation data needed for a comparative analysis of computed and reported cases for the 2000–2001 South Africa cholera epidemics that unfolded along the Thukela River (Kwa-Zulu Natal) and its aftermath, after proper calibration [437].

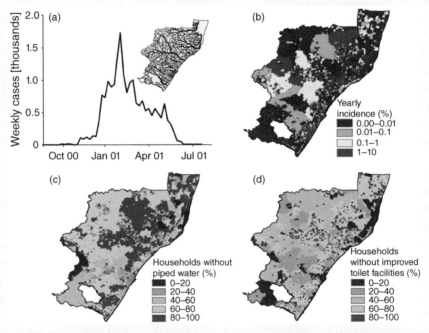

Figure 4.9 Epidemiological, demographic, and sanitation data for the Thukela-Kwa Zulu Natal cholera epidemic. (a) Total incidence data (weekly cases) recorded in the Thukela river network from October 2000 to July 2001. The inset shows the Thukela river network. (b) Yearly cholera incidence, evaluated as the number of reported cases in each health district from October 2000 to September 2001 divided by population size. (c) Spatial distribution of households without access to piped water. (d) Spatial distribution of households without access to improved toilet facilities [437]

Disease onset is determined by the instability of the disease-free equilibrium $\mathbf{X_0}$ ($S_i = H_i, I_i = 0$, $B_i = 0$ for all i). If the communities were isolated (no hydrological or mobility connections), the onset condition in each community [434] would require that the local basic reproduction number

$$R_{0i} = \frac{p_i H_i \beta_i}{W_i K \mu_B (\gamma + \mu + \alpha)} > 1,$$

where γ is the recovery rate and μ and α are, respectively, human baseline and disease-induced mortality rates. For the spatially connected system, instead, onset in the

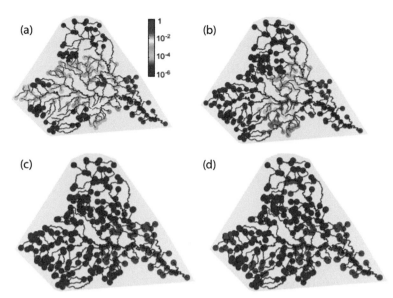

Figure 4.10 Data and model predictions of cholera epidemic along the Thukela river, South Africa, network [437]. (a) Normalized spatial distribution of recorded cases cumulated during the epidemic onset phase. (b) Spatial distribution of cases as predicted by the dominant eigenvector. (c) Spatial distribution of local basic reproduction numbers. Locations i in red (blue) are characterized by $R_{0_i} > 1$ ($R_{0_i} \leq 1$). (d) Cholera cases as in (b). Red (blue) dots indicate communities with more (less) than 10 reported cases during disease onset. It is clear that the condition $R_{0_i} > 1$ is neither necessary nor sufficient to establish an outbreak whenever mobility of pathogens and suscptibles/indected is significant [437]

metacommunity does not correspond to one or more local reproduction numbers > 1 (see Figure 4.10 for an example). The stability analysis of the disease-free equilibrium (Box 4.6) shows that disease can establish in the community when the dominant eigenvalue Λ_0 of a generalized reproduction matrix $\mathbf{G_0}$ is larger than 1. Define $\phi = \gamma + \mu + \alpha$ and introduce the diagonal matrices β, $\mathbf{H}$, $\mathbf{p}$, $\mathbf{W}$, $\mathbf{R_0}$ (whose diagonals consist of parameters β_i, H_i, p_i, W_i, R_{0_i}, with $i = 1, n$, respectively). $\mathbf{W}^{-1}$ is also diagonal with elements equal to $1/W_i$, thus $\mathbf{R_0} = \mathbf{pH}\beta\mathbf{W}^{-1}/(K\mu_B\phi)$. Define the matrix

$$\mathbf{T_0} = (1 - m_I)(1 - m_S)\mathbf{R_0} + m_I(1 - m_S)\mathbf{R_0^I}$$
$$+ (1 - m_I)m_S\mathbf{R_0^S} + m_S m_I \mathbf{R_0^{IS}},$$

which is a transmission matrix accounting for different probabilities of movement. $\mathbf{R_0^I}$, $\mathbf{R_0^S}$, and $\mathbf{R_0^{IS}}$ are matrices that correspond, respectively, to metacommunities with (1) infectives only, (2) susceptibles only, or (3) both infectives and susceptibles being mobile. More precisely, we have

$$\mathbf{R_0^I} = \frac{\mathbf{pQ}^T \mathbf{H}\beta\mathbf{W}^{-1}}{K\mu_B\phi}, \qquad \mathbf{R_0^S} = \frac{\mathbf{pHQ}\beta\mathbf{W}^{-1}}{K\mu_B\phi},$$

and $\quad \mathbf{R_0^{IS}} = \dfrac{\mathbf{pQ}^T \mathbf{HQ}\beta\mathbf{W}^{-1}}{K\mu_B\phi}.$

$\mathbf{T_0}$ thus depends on mobility $\mathbf{Q}$ through the terms $\mathbf{HQ}$ (movement to a community), $\mathbf{Q}^T\mathbf{H}$ (movement from a community), and $\mathbf{Q}^T\mathbf{HQ}$ (movement to and from).

The generalized reproduction matrix is

$$\mathbf{G_0} = \frac{l}{\mu_B + l}\mathbf{P}^T + \frac{\mu_B}{\mu_B + l}\mathbf{T_0}, \tag{4.20}$$

and the onset of the disease is triggered whenever $\lambda_{\max}(\mathbf{G_0})$ switches from being less to being larger than 1, that is,

$$\Lambda_0 = \lambda_{\max}(\mathbf{G_0}) > 1. \tag{4.21}$$

$\mathbf{G_0}$ depends (linearly) on the hydrological matrix $\mathbf{P}$ and (nonlinearly) on the human mobility matrix $\mathbf{Q}$. Therefore, the two networks interplay in a complex manner to determine disease outbreak and spread.

Box 4.6 Stability of Disease-Free Conditions: Generalized Reproduction Numbers

To analyze stability, we consider the Jacobian of the linearized system evaluated at the disease-free equilibrium $\mathbf{X_0}$ ($S_i = H_i, I_i = 0$, $B_i = 0$ for all i), which is given by [437]

$$
\mathbf{J} = \begin{bmatrix} j_{11} & 0 & j_{13} \\ 0 & j_{22} & j_{23} \\ 0 & j_{32} & j_{33} \end{bmatrix},
$$

where

$$j_{11} = -\mu \mathbf{U_n},$$
$$j_{13} = -m_S \mathbf{HQ\beta} - (1 - m_S)\mathbf{H\beta},$$
$$j_{22} = -\phi \mathbf{U_n},$$
$$j_{23} = m_S \mathbf{HQ\beta} + (1 - m_S)\mathbf{H\beta},$$
$$j_{32} = \frac{m_I}{K}\mathbf{pW}^{-1}\mathbf{Q}^T + \frac{1 - m_I}{K}\mathbf{pW}^{-1},$$
$$j_{33} = -(\mu_B + l)\mathbf{U_n} + l\mathbf{W}^{-1}\mathbf{P}^T\mathbf{W}.$$

Note that the variables for pathogen have been scaled as $B_i^\star = B_i/K$. Because of its block-triangular structure, the Jacobian has obviously n eigenvalues equal to $-\mu$; therefore, instability is determined by the eigenvalues of the block matrix

$$
\mathbf{J}^* = \begin{bmatrix} j_{22} & j_{23} \\ j_{32} & j_{33} \end{bmatrix}.
$$

$\mathbf{J}^*$ is a proper Metzler matrix [185], namely, its off-diagonal entries are all nonnegative and at least one diagonal entry is negative (Appendix 6.1). Thus its eigenvalue with maximal real part (dominant eigenvalue) is real. If the union of the graphs associated with matrices $\mathbf{P}$ and $\mathbf{Q}$ is strongly connected, then the graph associated with $\mathbf{J}^*$ is also strongly connected. Therefore one can apply the Perron–Frobenius theorem [187] for irreducible matrices and state that the dominant eigenvalue is a simple real root of the characteristic polynomial. The condition for the transcritical bifurcation of the disease-free equilibrium is that the dominant eigenvalue crosses the imaginary axis at zero, namely, the determinant of $\mathbf{J}^*$ is zero [188]. Actually, when the disease-free equilibrium is stable, all the eigenvalues have negative real parts, and $\det(\mathbf{J}^*)$ is positive because $\mathbf{J}^*$ is a matrix of order $2n$. So the disease-free equilibrium becomes unstable when $\det(\mathbf{J}^*)$ switches from positive to negative, or equivalently, the dominant eigenvalue becomes zero.

As $\mathbf{U_n}$ obviously commutes with any matrix, we have (see [186])

$$
\det(\mathbf{J}^*) = \det\Big[\phi(\mu_B + l)\mathbf{U_n} - \phi l\mathbf{W}^{-1}\mathbf{P}^T\mathbf{W}
$$

$$
- \frac{m_S m_I}{K}\mathbf{pW}^{-1}\mathbf{Q}^T\mathbf{HQ\beta}
$$

$$
- \frac{m_I(1 - m_S)}{K}\mathbf{pW}^{-1}\mathbf{Q}^T\mathbf{H\beta}
$$

$$
- \frac{(1 - m_I)m_S}{K}\mathbf{pW}^{-1}\mathbf{HQ\beta}
$$

$$
- \frac{(1 - m_I)(1 - m_S)}{K}\mathbf{pW}^{-1}\mathbf{H\beta}\Big].
$$

Box 4.6 *Continued*

Since $\mathbf{H}$, β, and $\mathbf{W}^{-1}$ are diagonal, thus commuting, matrices, we can state that $\mathbf{R}_0 = \dfrac{\mathbf{p}}{K\mu_B\phi}\mathbf{H}\beta\mathbf{W}^{-1} = \dfrac{\mathbf{p}}{K\mu_B\phi}\mathbf{W}^{-1}\mathbf{H}\beta$ and rework the determinant of $\mathbf{J}^*$ (see the supporting information to [437]). The condition $\det(\mathbf{J}^*) = 0$ is thus given by

$$\det\left\{\mathbf{U_n} - \frac{l}{\mu_B+l}\mathbf{P}^T - \frac{\mu_B}{\mu_B+l}\left[(1-m_I)(1-m_S)\mathbf{R}_0\right.\right.$$
$$+ \frac{m_S m_I}{K\mu_B\phi}\mathbf{p}\mathbf{Q}^T\mathbf{H}\mathbf{Q}\beta\mathbf{W}^{-1}$$
$$+ \frac{m_I(1-m_S)}{K\mu_B\phi}\mathbf{p}\mathbf{Q}^T\mathbf{H}\beta\mathbf{W}^{-1}$$
$$\left.\left.+ \frac{(1-m_I)m_S}{K\mu_B\phi}\mathbf{p}\mathbf{H}\mathbf{Q}\beta\mathbf{W}^{-1}\right]\right\} = 0.$$

In addition to the matrix $\mathbf{R}_0 = \dfrac{\mathbf{p}}{K\mu_B\phi}\mathbf{H}\beta\mathbf{W}^{-1}$, we can now introduce three other matrices of reproduction numbers, namely,

$$\mathbf{R}_0^{\mathbf{I}} = \frac{\mathbf{p}\mathbf{Q}^T\mathbf{H}\beta\mathbf{W}^{-1}}{K\mu_B\phi}, \qquad \mathbf{R}_0^{\mathbf{S}} = \frac{\mathbf{p}\mathbf{H}\mathbf{Q}\beta\mathbf{W}^{-1}}{K\mu_B\phi},$$

and $\qquad \mathbf{R}_0^{\mathbf{IS}} = \dfrac{\mathbf{p}\mathbf{Q}^T\mathbf{H}\mathbf{Q}\beta\mathbf{W}^{-1}}{K\mu_B\phi},$

corresponding to metacommunities with infectives only being mobile, susceptibles only being mobile, and both infectives and susceptibles being mobile, respectively. If we account for the different probabilities of movement in the metacommunity, we can define a transmission matrix averaged over nonmobile individuals, mobile infectives, and mobile susceptibles as

$$\mathbf{T}_0 = (1-m_I)(1-m_S)\mathbf{R}_0 + m_S m_I \mathbf{R}_0^{\mathbf{IS}}$$
$$+ m_I(1-m_S)\mathbf{R}_0^{\mathbf{I}} + (1-m_I)m_S\mathbf{R}_0^{\mathbf{S}}.$$

Therefore, the bifurcation of the disease-free equilibrium corresponds to the condition

$$\det\left(\mathbf{U_n} - \frac{l}{\mu_B+l}\mathbf{P}^T - \frac{\mu_B}{\mu_B+l}\mathbf{T}_0\right) = 0.$$

Equivalently, the dominant eigenvalue Λ_0 of the matrix

$$\mathbf{G}_0 = \frac{l}{\mu_B+l}\mathbf{P}^T + \frac{\mu_B}{\mu_B+l}\mathbf{T}_0, \tag{4.22}$$

which is a convex combination of $\mathbf{P}^T$ and $\mathbf{T}_0$, must equal unity. Actually, the disease-free equilibrium switches from being stable to being a saddle, thus triggering the start of the disease, whenever the dominant eigenvalue of $\mathbf{J}^*$ switches from positive to negative, and hence whenever Λ_0 switches from being less than 1 to being larger than 1.

Technically, our approach is similar to that of the next-generation matrix (NGM) [442–444], which has been used for compartmental models rather than spatially explicit ones. Both methods are based on dominant eigenvalue analysis [445]. More specifically, however, we use a bifurcation analysis [188] to determine how the eigenvalues of the Jacobian at $\mathbf{X}_0$ vary with the model

Box 4.6 *Continued*

parameters. As the system is positive and $\mathbf{X_0}$ is characterized by null values of I_i and B_i, the bifurcation can only occur via an exchange of stability, that is, the disease-free equilibrium switches from stable node to saddle through a transcritical bifurcation (the Jacobian has one zero eigenvalue).

Our main result is therefore that the onset of the disease can be triggered whenever $\lambda_{\max}(\mathbf{G_0})$ switches from being less to being larger than 1, namely,

$$\Lambda_0 = \lambda_{\max}(\mathbf{G_0}) > 1. \tag{4.23}$$

$\mathbf{G_0}$ is the sum of two matrices, one depending (linearly) on the hydrological matrix $\mathbf{P}$ and the other (nonlinearly) on the human mobility matrix $\mathbf{Q}$ [437]. Therefore, the two networks interplay in a complex manner to determine disease outbreak and spread. An example of application of the above techniques to the determination of local and generalized reproduction numbers is reported in Figure 4.10.

4.1.5 The Geography of Disease Spread

The geographical distribution of disease onset can be linked to the dominant eigenvector of $\mathbf{G_0}$ [437]. When Λ_0 is slightly larger than unity and no other eigenvalue has modulus > 1, $\mathbf{X_0}$ is a saddle and the dominant eigenvector corresponds to the unique unstable direction in the state space along which the system orbit will diverge from the equilibrium. Once the eigenvector is suitably projected onto the subspace of infectives and normalized, its n components are the relative proportions of the infectives in the n communities.

Box 4.7 Geography of Disease Onset

The geography of disease onset, that is, the spatial localization of the sites that are hit with more strength during the early phase of the epidemic, is determined by the dominant eigenvector of the Jacobian matrix $\mathbf{J}^*$ to which $\mathbf{G_0}$ is linked, as shown in Box 4.6. The eigenvector lies in the subspace (of dimension $2n$) $S_i - H_i = 0$ ($i = 1, n$) and has strictly positive components I_i and B_i according to the Perron–Frobenius theorem for nonnegative matrices [187]. The dominant eigenvector of $\mathbf{J}^*$ can be computed by solving

$$\mathbf{J}^* \begin{bmatrix} \mathbf{i} \\ \mathbf{b} \end{bmatrix} = \lambda \begin{bmatrix} \mathbf{i} \\ \mathbf{b} \end{bmatrix},$$

where λ is the dominant eigenvalue of $\mathbf{J}^*$ and $\mathbf{i}$ and $\mathbf{b}$ are the components of the dominant eigenvalue corresponding, respectively, to infectives and pathogens. Writing $\mathbf{J}^*$ as

$$\mathbf{J}^* = \begin{bmatrix} \mathbf{A} & \mathbf{B} \\ \mathbf{C} & \mathbf{D} \end{bmatrix},$$

we get

$\mathbf{A}i + \mathbf{B}b = \lambda i$

$\mathbf{C}i + \mathbf{D}b = \lambda b.$

Box 4.7 *Continued*

Because close to the transcritical bifurcation through which the disease-free equilibrium loses stability, the dominant eigenvalue λ of $\mathbf{J}^*$ is equal to 0, from the first equation, we have

$$\mathbf{i} = -\mathbf{A}^{-1}\mathbf{Bb},$$

therefore the second equation can be written as

$$-\mathbf{CA}^{-1}\mathbf{Bb} + \mathbf{Db} = 0.$$

Since $\mathbf{A}$ is a diagonal matrix with equal diagonal entries, with simple algebraic manipulations, we can write the previous equation as

$$(\mathbf{AD} - \mathbf{CB})\,\mathbf{b} = 0.$$

Thus the result is

$$\mathbf{G_0 Wb} = \mathbf{Wb}.$$

Therefore, we can conclude that, close to the transcritical bifurcation of the disease-free equilibrium, where the dominant eigenvalue Λ_0 of $\mathbf{G_0}$ is equal to 1, the dominant eigenvector $\mathbf{g_0}$ of matrix $\mathbf{G_0}$ corresponds to the pathogens' components of the dominant eigenvalue of $\mathbf{J}^*$ multiplied by the volumes of the corresponding water reservoirs ($\mathbf{g_0} = \mathbf{Wb}$). The infectives' components $\mathbf{i}$ of the Jacobian matrix can thus be computed as

$$\mathbf{i} = \frac{m_S \mathbf{HQ}\beta + (1 + m_S)\mathbf{H}\beta}{\phi} \mathbf{W}^{-1}\mathbf{g_0},$$

and they can be used to effectively portray the geography of disease onset. Notice, however, that this simple relationship between the dominant eigenvector of $\mathbf{G_0}$ and the infectives' components of the dominant eigenvector of $\mathbf{J}^*$ holds only close to the transcritical bifurcation of the disease-free equilibrium. In general, for parameter combinations for which the dominant eigenvalue of $\mathbf{G_0}$ is significantly larger than 1, the study of the geography of disease onset requires the computation of the eigenvalues and the eigenvectors of matrix $\mathbf{J}^*$.

Whenever the dominant eigenvalue is sufficiently larger than 1, there may be other eigenvalues of $\mathbf{G_0}$ with modulus > 1 and more than one unstable direction of $\mathbf{X_0}$. However, after a short-term transient due to initial conditions, the disease will mainly propagate to the nodes corresponding to the largest components of the dominant eigenvector. These communities are those where the number of infectives will be highest during the onset and will thus act as the main focus of the disease.

An example of comparative analysis of patterns of disease computed by the full-fledged integration of the equations in Box 4.5 applied to the Haiti cholera outbreak case study [437] is shown in Figure 4.11.

4.1.6 Disease Spread in Theoretical Networks

Box 4.8 shows tests of the onset of spatial effects from the numerical solution of a spatially explicit model of epidemic disease spread. The SIB model on networks adopted (see Eqs. (4.19) and Appendix 6.3 for their numerical solution) is particularized by setting the key mobility parameter $m_S = m$ and $m_I = 0$ (infected individuals do not engage in commuting) and by neglecting pathogen spread through waterways ($P_{ij} = 0$) leaving human mobility (specified by the matrix Q_{ij}) as the sole transport mechanism responsible for the spread of

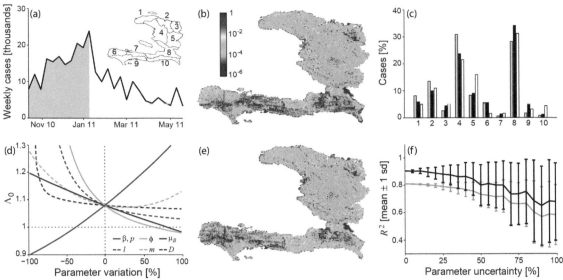

Figure 4.11 Data and model predictions of Haiti epidemic. (a) Total incidence data (weekly cases) from November 2010 to May 2011. (b) Fine-grained spatial distribution of recorded cases cumulated during the onset phase of the epidemic (defined as period from beginning to peak, gray in panel (a)). (c) Comparison of coarse-grained data (department level, labels as in inset of panel (a)) on spatial distributions of cumulated cases (gray bars, onset phase; black bars, whole period) versus predictions based on the dominant eigenvector (white bars); R_O^2 = coefficient of determination for the onset phase = 0.81, R_T^2 = coefficient of determination for the whole period = 0.90. (d) Sensitivity to parameter variations of the dominant eigenvalue of $\mathbf{G_0}$; the dotted horizontal line indicates the value below which the disease cannot start. (e) Fine-grained spatial distribution as predicted by dominant eigenvector (*SI text*); $R_O^2 = 0.92$, $R_T^2 = 0.95$. (f) Sensitivity to parameter variations of correlations between spatial distribution as predicted by dominant eigenvector and actual spatial distribution of cumulated cases (gray, onset phase; black, whole period). Parameter values and details are in the original work [437]

infections. Q_{ij} is determined by the attractivity of node j for the mobile fraction mH_i of the population of node i, that is, $Q_{ij} \propto H_i H_j \exp(-d_{ij}/D)$, where the proportionality constant is determined by normalization [446], d_{ij} is the distance computed along the network between nodes i and j, and D is a commuting deterrence distance [438, 446] (i.e., a gravity-like model [105]). The substrate for disease spread is Peano's construct [128] (Figure 1.17), chosen as a paradigmatic case of a theoretical network because of its exactly known topological

and metric properties [63, 121, 126, 135]. Other parameters' role is discussed in [446].

The onset of spatial effects is determined by the emerging differences between the solution as posited and a spatially implicit account (theoretically corresponding to the case $m = \infty$) where the domain is treated as a single SIB node with total population H equal to the sum of the human populations concentrated at its nodes, $H_{tot} = \sum_{i=1,N} H_i$ [446].

Box 4.8 Tests of Spatial Effects in Disease Spread

Figure 4.12 shows a test comparison aimed at defining conditions for which spatial effects need to be considered in SIB models on networks [446]. As noted in Box 1.8, the topology of the bifurcations along the backbone of a tree is a key determinant of the speed of biological invasions, and therefore the spread of epidemic cholera on a Peano network is significant also for generalizations of the results to real fluvial networks.

Box 4.8 *Continued*

The N nodes shown in the inset are placed at the intersections of the edges. At each node i, a local community is composed by fixed population size $H = S_i(t) + I_i(t) + R_i(t)$ (recovered are considered immune for the timescale of the epidemic), whose initial local reproduction number $R_0 = 3 > 1$. The numerical model was run in time with the same initial conditions $I_i(0) = 0$ for $i \neq k$, where k, the seeding node, was placed in the example shown here at one endpoint and assigned an inoculum $I_k(0) = I_0$. Note that a randomly chosen starting node with the same inoculum would not change the resulting picture [446]. With any reasonable choice of all other model parameters, the key parameter in this example proves m, the fraction of the nodal population H_i that engages in mobility from node i to node j, defined by the matrix Q_{ij}.

At low values of the transport rate m, the spatial variability is reflected in epidemic evolutions that significantly depart from well-mixed ones ($m \to \infty$). The secondary peaks of infected individuals that emerge at low transport rates depend on the number of susceptibles that are reached by the epidemic wave per unit time. The latter depends, in turn, on the number of nodes that are located at the same distance from the outlet (and thus reached simultaneously by the epidemic wave). The relative proportion of the number of nodes placed at a given distance x from the source is indeed a well-known geomorphological feature of the Peano network, that is, its width function $W(x)$ (Box 3.26), which maps exactly a binomial multiplicative process of rate $1/4$ [126].

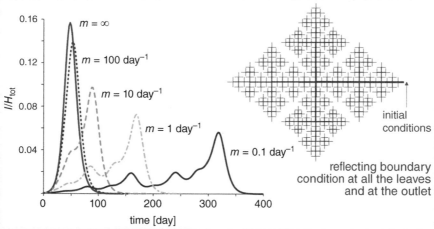

Figure 4.12 Example of spatial effects in SIB models on networks (Eqs. (4.19)). The plots show the temporal evolution of the total number of infected individuals $I = \sum_{i=1,N} I_i(t)$ normalized by the total population $H_{tot} = NH$ in a Peano network with $1,025$ nodes populated by a uniform number H of initially susceptible individuals. Different lines correspond to different values of the fraction of mobile individuals analog to a transport rate m. The $m = \infty$ (thick solid line) case corresponds to the spatially implicit model (Box 4.13). Effects of increased fractions m of mobile susceptibles in an idealized community of equal-sized human settlements placed at the nodes of a Peano network. The plot shows the instantaneous total number of infected individuals computed by adding the local values, that is, $I(t) = \sum_{i=1}^{N} I_i(t)$ for all N nodes of the Peano construct ($H_{tot} = NH$). It is evident that the local model ($m = \infty$) reproduces well the overall progression of the infection only for very large mobilities that effectively well mix the domain rendering spatial effects globally immaterial. Figure after [446]

Box 4.8 *Continued*

The main result in Figure 4.12 is that the same network may be seen by an epidemic as essentially a single human settlement (i.e., described by a spatially implicit scheme of SIB demography) if dispersive factors dominate, in this case, a strong mixing component due to high mobility m (high values of the fraction of susceptible population engaging in travels to neighboring nodes). Should this mixing effect be weaker, the shape of the network has a role, resulting in pronounced spatial effects, all other conditions being equal. This suggests the importance of multilayer networks in many practical cases that only spatially explicit approaches capture. Moreover, mixing effects are enhanced in the test case by the uniform nodal population assumed to be invaded by the wave of susceptibles turning infected. Needless to say, for real settings, one may choose to characterize human settlements by their actual population and spatial location.

Gatto et al. [437] have also studied the onset of instability of disease-free equilibria and the geography of disease spread in theoretical landscapes, in particular in the Peano network that pervaded this book for the reasons put forth in Chapter 1 (Figure 1.17). The study of a theoretical landscape as a substrate for epidemics requires assumptions about the spatial distribution of the population. The simplest choice is, of course, a uniform distribution ($H_i = h$ for all i). However, a uniform population distribution represents a rather crude simplification of the observed spatial arrangement of human communities. Empirical observations show systematically that a much more realistic model for the frequency of size of human settlements is provided by the so-called Zipf's law [5], according to which the size distribution H_i of human communities is a random variable whose probability distribution can be well represented by a power law distribution, namely, by $P(H_i \geq h) \propto h^{-2}$. Gatto et al. [437] have performed the analysis of the disease geography by sampling the size of each local community from a power law distribution. To allow fair comparison with the results of the case endowed with spatially uniform population size, the normalizing constraint imposes $\sum_{i=1}^{n} H_i = nh$. Mobility is in any case assumed to follow a gravity model.

The results shown in Figures 4.13c,d have been obtained with a Zipf-like population distribution. Note that, in this case, for each analyzed parameter setting, 500 independent realizations of sample size n have been extracted from the population distribution, and the n population sizes have been distributed randomly in the landscape. Therein colors code the fraction of different realizations for which onset conditions are met.

The theoretical framework allows for the description of the geography of disease onset in theoretical landscapes as well. In particular, the predictive ability of the dominant eigenvector of matrix $\mathbf{G_0}$ can be tested against numerical simulations of the epidemiological model implemented on a Peano-like network. To initialize the simulation, one assumes that the outbreak starts close to the outlet of the river network (bottom right corner of the embedding domain), as is often observed in real-world epidemics, with 0.1 percent of the local population representing the initial infective pool. Figure 4.14 shows the dominant eigenvector of $\mathbf{G_0}$ and the simulation of epidemic onset in a Peano network with either spatially homogeneous (Figures 4.14a–d) or Zipf-like (Figures 4.14e–h) population distributions. Disease emergence has been identified numerically from model simulations as the week after which daily incidence and its first and second time derivatives exceed 0.5 percent of the respective maximum values recorded in the simulation. Not only different population distributions but also different models of human mobility can be applied and analyzed in this framework of disease spread in theoretical or real metapopulations, as reported in [437].

4.2 Seasonal Environmental Forcings and Epidemicity of Spatially Explicit Waterborne Epidemics

The transmission of waterborne pathogens is a complex process that is heavily linked to the spatial characteristics of the underlying environmental matrix as well as to the

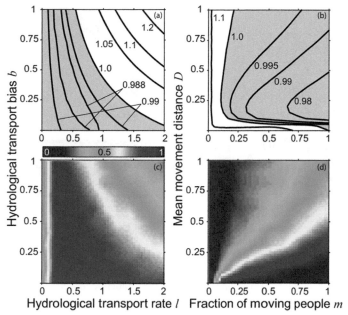

Figure 4.13 Onset conditions of a waterborne disease epidemic in a Peano network with gravity-like mobility. (a) Effects of hydrological transport parameters for different local reproductive numbers (labels of isolines) in a homogeneous population $H_i = h$. Epidemics can start for combinations lying on the left of the curves with $R_{0i} \geq 1$, whereas subthreshold epidemics (gray shading) can be triggered for combinations lying in between the relevant isolines. (b) Same as (a) for various human mobility parameters. Epidemics can start for combinations lying on the right of the isolines. (c) As in (a), with $R_{0i} = 0.95$ and Zipf-like population distribution. Different colors code the fraction of realizations (different population distributions) for which onset conditions are met. (d) As in c, with reference to various mobility parameters. Other parameter values are $\beta = 1$, $\alpha = \mu = 0$, $\mu_B = 0.23$, $m_S = m_I = 0.5$ (a and c), $D = 0.1$ (a,c), $l = 0.5$ (b,d), and $b = 0.2$ (b,d). Figure after [437]

temporal variability of the relevant hydroclimatological drivers. A time-varying, spatially explicit network model for the dynamics of waterborne diseases may be tackled by using Floquet theory [447] (see also Appendix 6.1.4), which underpins local stability of periodic dynamical systems.

The basic question is to identify conditions for pathogen invasion and for its establishment in systems characterized by fluctuating environmental forcings [447], thus extending to time-varying contexts the generalized reproduction numbers obtained for spatially explicit epidemiology of waterborne disease [437]. Temporal variability may have various diverse effects on the invasion threshold, as it can either favor pathogen invasion or make it less likely. Environmental fluctuations characterized by distinctive geographical signatures can produce diversified, highly nontrivial effects on pathogen invasion. In addition, as we shall briefly summarize here, the threshold conditions derived

up to now that guarantee the failure of long-term establishment of a WB disease (the so-called endemicity) do not guarantee that a transient (and possibly large) epidemic can actually hit a certain region even if the disease cannot become endemic there. Therefore, appropriate conditions must be derived that provide threshold conditions for epidemicity. These are based on the concept of reactivity in dynamical systems [448] that will also be illustrated in this section.

The first basic issue is to generalize to time-varying environments spatially explicit models of waterborne disease dynamics (Box 4.5) whose generalized reproduction numbers have been studied in Box 4.6. As noted there, the model describes local epidemiological, demographic, and ecological dynamics; pathogen transport along waterways; and the effects of short-term human mobility on disease propagation. Network nodes represent human communities (villages, towns, or cities) of assigned population, arranged in a given spatial

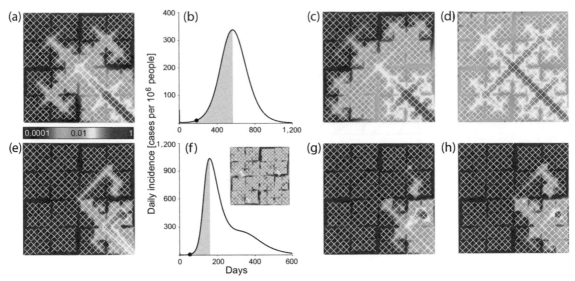

Figure 4.14 Dominant eigenvector of $\mathbf{G}_0$ and simulation of epidemic onset in a Peano network with (a–d) homogeneous or (e–h) Zipf-like population distributions. (a) Dominant eigenvector (associated to $\Lambda_0 = 1.001$). (b) Simulated temporal pattern of daily incidence. (c) Simulated spatial distribution of weekly cases during the emerging phase of the outbreak (black dot in (b). (d) Simulated spatial distribution of cumulative cases up to disease peak (gray shading in (b). (e–h) As in (a–d), with Zipf-like population distribution (f, inset). The dominant eigenvalue is $\Lambda_0 = 1.11$. The Peano construct is here used at the fifth stage of iteration (Figure 1.17). The network outlet is located in the lower right corner of the spatial domain. Parameter values include $\beta = 1$, $\alpha = \mu = 0$, $\mu_B = 0.23$, $l = 1$, $m_S = m_I = 0.125$, $D = 0.05$. Figure after [437]

setting and connected by hydrological pathways and by fluxes of human mobility. Specifically, here we shall show results from a model where we assume that human communities constitute the nodes of an OCN (Appendix 6.2). The OCN is embedded in a square of arbitrary side 1. Let, as usual, $S_i(t)$ and $I_i(t)$ be the local abundances of susceptible and infected individuals in each node i of the network at time t, and let $B_i(t)$ be the concentration of pathogens (e.g., bacteria, viruses, or protozoa) in local water reservoirs. Epidemiological dynamics and pathogen transport over the hydrological and human mobility networks among n nodes can be described by a set of $3 \times n$ coupled ordinary differential equations described in Box 4.9.

Box 4.9 Time-Varying SIB Model

We consider the following space-explicit version of the SIB model:

$$\frac{dS_i}{dt} = \mu(H_i - S_i) - \mathscr{F}_i S_i,$$

$$\frac{dI_i}{dt} = \mathscr{F}_i S_i - (\gamma + \mu + \alpha) I_i,$$

Box 4.9 *Continued*

$$\frac{dB_i}{dt} = -\left[\mu_{B_i}(t) + l_i(t)\right] B_i + \sum_{j=1}^{n} l_j(t) P_{ji}(t) \frac{W_j(t)}{W_i(t)} B_j + \frac{p_i(t)}{W_i(t)} \mathcal{G}_i,$$

$$\mathcal{F}_i = \left\{ [1 - m_i(t)] \beta_i(t) \frac{B_i}{K + B_i} + m_i(t) \sum_{j=1}^{n} Q_{ij}(t) \beta_j(t) \frac{B_j}{K + B_j} \right\},$$

$$\mathcal{G}_i = \left\{ [1 - m_i(t)] I_i + \sum_{j=1}^{n} m_j(t) Q_{ji}(t) I_j \right\}, \tag{4.24}$$

with usual symbol notation (Box 4.5) (save for the explicit dependence on time t of epidemiological and hydrological parameters). In addition, we have emphasized a possibly node-specific mortality rate for free-living pathogens in the water reservoir, μ_{B_i}, and a site-dependent pathogen transport rate $l_i(t)$, which assumes a key importance in the process. Model (4.24) represents a broad scheme of waterborne disease transmission across a broad range of spatial and temporal scales.

Some of the parameters of model equations (4.24) – namely, those related to human demography (μ) and the physiological response to the disease (α, γ, K) – are reasonably assumed to be constant over the spatial and temporal scales relevant to an outbreak (implicit in the choice of neglecting a compartment for recovered individuals). The size of local human communities (H_i) is assumed to possibly vary in space but not in time, at least over timescales of epidemiological interest. Some other parameters – namely, the exposure and contamination rates $\beta_i(t)$ and $p_i(t)$, the pathogen mortality rate $\mu_{B_i}(t)$, the hydrological transport rate $l_i(t)$, and the hydrological bias $b_i(t)$ implicit in the definition of the matrix P_{ij}, the fraction of traveling people $m_i(t)$ – can vary in both time and space. Human mobility is defined by the matrix Q_{ij}, here chosen according to a gravity model (see, e.g., Box 4.8) where $Q_{ij} \propto H_i H_j \exp\left(-d_{ij}/D_i(t)\right)$, with d_{ij} the along-stream distance between i and j and the average human mobility (or deterrence) distance $D_i(t)$. Also, the volume of the local water reservoir $W_i(t)$ can vary in time and/or space. The analysis referred to herein [447] restricted its analysis to periodical fluctuations linked to seasonal environmental drivers. The period of the seasonal environmental fluctuations is assumed to be one year without loss of generality. While sinusoidal oscillations may provide a simplistic representation of the natural frequencies of the relevant environmental drivers, they nevertheless capture the essence of the issue addressed here: whether a time-varying framework affects macroscopically the conditions for waterborne pathogen invasion and the resulting patterns of infection.

In this model, as usual, and in many similar ones, pathogens can invade the system if (and only if) the disease-free equilibrium $\mathbf{X_0}$ (the state of model (4.24) where $S_i = H_i$ and $I_i = B_i = 0$ for all $i = 1, \ldots, n$) is unstable under the assumption of periodic parameter fluctuations. To analyze the stability of $\mathbf{X_0}$, one must consider a linearized system [447] (Appendix 6.1.4). Specifically, $\mathbf{X_0}$ is unstable (thus allowing pathogen invasion) if and only if the maximum Floquet exponent $\xi_{\max}$ is larger than zero [447, 449]. The computation of the Floquet exponents is highlighted in Box 4.10.

Box 4.10 Computation of Floquet Exponents

The time-varying linearized system is

$$\frac{d\Delta S}{dt} = -\mu \Delta S - [\mathbf{U_n} - \mathbf{m(t)}]\mathbf{H}\beta\mathbf{(t)B}^* - \mathbf{m(t)HQ(t)}\beta\mathbf{(t)B}^*,$$

$$\frac{d\mathbf{I}}{dt} = [\mathbf{U_n} - \mathbf{m(t)}]\mathbf{H}\beta\mathbf{(t)B}^* + \mathbf{m(t)HQ(t)}\beta\mathbf{B}^* - (\gamma + \mu + \alpha)\mathbf{I}, \qquad (4.25)$$

$$\frac{d\mathbf{B}^*}{dt} = -\left[\mu_{\mathbf{B}}\mathbf{(t)} + \mathbf{l(t)}\right]\mathbf{B}^* + \mathbf{W}^{-1}\mathbf{(t)P^T(t)W(t)l(t)B}^*$$

$$+ \frac{1}{K}\mathbf{p(t)W}^{-1}\mathbf{(t)}\left[\mathbf{U_n} - \mathbf{m(t)} + \mathbf{Q^T(t)m(t)}\right]\mathbf{I},$$

where the superscript T indicates matrix transposition; $\mathbf{U_n}$ is the identity matrix of dimension n; $\Delta S = [S_1 - H_1, \ldots, S_n - H_n]^{\mathbf{T}}$; $\mathbf{I} = [I_1, \ldots, I_n]^{\mathbf{T}}$; $\mathbf{B}^* = [B_1/K, \ldots, B_n/K]^{\mathbf{T}}$; and $\mathbf{H}$, $\beta\mathbf{(t)}$, $\mathbf{p(t)}$, $\mu_{\mathbf{B}}\mathbf{(t)}$, $\mathbf{l(t)}$, $\mathbf{m(t)}$, $\mathbf{W(t)}$ are diagonal matrices whose nonzero elements are made up by the parameters H_i, $\beta_i(t)$, $p_i(t)$, $\mu_{Bi}(t)$, $l_i(t)$, $m_i(t)$, $W_i(t)$, with $i = 1, 2, \ldots, n$, respectively [447].

Because of the block-triangular structure of the Jacobian of system (4.25), in which all but the first n elements of the first n columns are identically equal to 0, we can evaluate the stability of the disease-free equilibrium $\mathbf{X}_0$ by applying Floquet theory to the subsystem that includes the linearized equations for the "infected" compartments of the model [447, 450], namely, infective abundances I_i and pathogen concentrations B_i^*. Specifically, under the assumption of periodic parameter fluctuations, $\mathbf{X}_0$ is unstable if and only if the maximum Floquet exponent $\xi_{\max}$ of the subsystem of linearized equations is larger than zero.

Using matrix notation, the dynamics of the subsystem close to $\mathbf{X}_0$ can be expressed as

$$\frac{d}{dt}\begin{bmatrix} \mathbf{I} \\ \mathbf{B}^* \end{bmatrix} = \mathbf{J_0(t)}\begin{bmatrix} \mathbf{I} \\ \mathbf{B}^* \end{bmatrix},$$

where

$$\mathbf{J_0(t)} = \begin{bmatrix} \mathscr{A} & \mathscr{B}\mathbf{(t)} \\ \mathscr{C}\mathbf{(t)} & \mathscr{D}\mathbf{(t)} \end{bmatrix}$$

is the Jacobian of Equation (4.24) evaluated at the disease-free equilibrium. The entries of $\mathbf{J_0(t)}$ are thus given by

$$\mathscr{A} = -(\gamma + \mu + \alpha)\mathbf{U_n},$$

$$\mathscr{B}\mathbf{(t)} = [\mathbf{U_n} - \mathbf{m(t)}]\mathbf{H}\beta\mathbf{(t)} + \mathbf{m(t)HQ(t)}\beta,$$

$$\mathscr{C}\mathbf{(t)} = \frac{1}{K}\mathbf{p(t)W}^{-1}\mathbf{(t)}\left[\mathbf{U_n} - \mathbf{m(t)} + \mathbf{Q^T(t)m(t)}\right],$$

$$\mathscr{D}\mathbf{(t)} = -\left[\mu_{\mathbf{B}}\mathbf{(t)} + \mathbf{l(t)}\right] + \mathbf{W}^{-1}\mathbf{(t)P^T(t)W(t)l(t)}.$$

Floquet exponents can be numerically evaluated by integrating the matrix differential equation [447]

$$\frac{d\mathbf{F(t)}}{dt} = \mathbf{J_0(t)F(t)}$$

over one period (from $t = 0$ to $t = T = 1$) with the identity matrix of size $2n$ as initial condition for matrix $\mathbf{F(t)}$ (i.e., $\mathbf{F(0)} = \mathbf{U_{2n}}$). Floquet multipliers ρ_i ($i = 1, \ldots, 2n$) can then be computed as the eigenvalues of the monodromy matrix $\mathbf{F(T)} = \mathbf{F(1)}$, and Floquet exponents ξ_i are defined as $\xi_i = \ln(\rho_i)/T$ (with $T = 1$, as the period of the environmental fluctuations is assumed to be one year).

Box 4.10 *Continued*

To illustrate how Floquet multipliers/exponents can be computed numerically in practice, it is perhaps worth lingering over the simplest case considered in this work, that is, the local epidemiological model (i.e., model (Equation (4.24)) with only one node). Assuming that the periodically varying parameter is the exposure rate β (with period $T = 1$ year), the dynamics of infected hosts and pathogen concentration close to the disease-free equilibrium are given by

$$\frac{d}{dt}\begin{bmatrix} I \\ B^* \end{bmatrix} = \begin{bmatrix} -(\gamma + \mu + \alpha) & \beta(t)H \\ \frac{p}{KW} & -\mu_B \end{bmatrix}\begin{bmatrix} I \\ B^* \end{bmatrix},$$

where

$$\beta(t) = \beta_0[1 \pm \epsilon \sin(2\pi t)].$$

To evaluate Floquet exponents, we have to solve the matrix differential equation

$$\frac{d\mathbf{F}(t)}{dt} = \begin{bmatrix} -(\gamma + \mu + \alpha) & \beta(t)H \\ \frac{p}{KW} & -\mu_B \end{bmatrix}\mathbf{F}(t)$$

from $t = 0$ to $t = T = 1$ with initial condition

$$\mathbf{F}(0) = \begin{bmatrix} F_{11}(0) & F_{12}(0) \\ F_{21}(0) & F_{22}(0) \end{bmatrix} = \begin{bmatrix} 1 & 0 \\ 0 & 1 \end{bmatrix}.$$

The monodromy matrix $\mathbf{F}(1)$ is thus simply given by the final values of each of the four F_{ij}s (with $i, j = 1, 2$). The maximum Floquet multiplier $\rho_{\max}$ is then computed as the dominant eigenvalue of $\mathbf{F}(1)$, while the maximum Floquet exponent $\xi_{\max}$ is defined as the natural logarithm of $\rho_{\max}$ (actually divided by $T = 1$). The implementation of the numerical algorithm for the local case is reported in the original paper [447].

4.2.1 Floquet Theory for River Network Models

Mari et al. [447] went on with the computation of Floquet multipliers/exponents that can be performed also for spatially explicit network models, namely, Equations (4.24) with a realistic network structure, in particular, an OCN.

Obviously, Floquet exponents vary as a function of the model parameters – and so obviously do pathogen invasion conditions. To proceed step by step, we illustrate the analysis of these conditions starting from the simple case of model (4.24) with no spatial structure and with periodic fluctuations of one of the model parameters expressed by a sinusoidal oscillation

$$\theta(t) = \theta_0[1 \pm \epsilon \sin(2\pi t)].$$

A reference local reproduction number needs to be determined. To obtain a spatially implicit version of model (4.24), it is sufficient to consider the dynamics in the ith node and disconnect it from its substrate for interactions by setting $l_i(t) = 0$ and $m_i(t) = 0$ for all times. This is tantamount to disregarding pathogen transport and human mobility. The stability of the disease-free equilibrium $\mathbf{X_0} = [\ H\ 0\ 0\ ]^T$ is thus determined by the maximum Floquet exponent of the linearized subsystem accounting for the local dynamics of infectives and pathogen concentrations, where local pathogen invasion conditions in periodic environments have been found [447] (see also Box 4.10). It has been shown that simplistic conclusions (say, stating that seasonality favors – or plays against – pathogen invasion) cannot be drawn because the effect of seasonality will depend on which of the drivers (of either biotic or abiotic nature) is made to fluctuate seasonally.

We remark that in a time-constant and spatially implicit environment, the condition for pathogen invasion is provided by $R_0 > 1$, where $R_0 = (\beta_0 p_0 H)/[\mu_{B_0}(\gamma + \alpha + \mu) KW_0]$. In a periodically fluctuating environment, one might still calculate an approximate reference value $\widetilde{R}_0$ by using the values of the model parameters averaged over time. However, the condition $\widetilde{R}_0 > 1$ will no longer correspond to pathogen invasion because it is the maximum Floquet exponent that must be evaluated instead.

Mari et al. [447] still term this approximate $\widetilde{R}_0$ as the basic reproduction number, because it allows a straightforward comparison of the invasion thresholds obtained in a periodic environment, and in a spatially explicit setting, against those relevant to the simple case of constant parameters and no spatial structure [447]. As usual, the assessment of the invasion thresholds can be usefully complemented by the analysis of the initial dynamics of infection conducted via simulation [447], introducing environmental fluctuations that increase infection risk at the beginning of the simulation timespan.

Figure 4.15a reports the invasibility curves computed for different parameter values and different assumptions regarding a single time-varying parameter, taken in the sinusoidal form shown above, in an OCN river network in which population distribution is assumed to be spatially homogeneous ($H_i = H$ for all i). If $\epsilon = 0$ (constant parameter values), pathogen invasion is possible if and only if $\widetilde{R}_0 > \lambda_0$, where λ_0 is the value of R_0 for which the dominant eigenvalue Λ_0 of the generalized reproduction matrix $\mathbf{G}_0$ of size $3 \times n$ [437, 451] is equal to 1. This matrix represents the time-invariant counterpart of the matrix $\mathbf{J}_0(t)$ defined above (Box 4.10). Such a matrix therefore accounts not only for epidemiological dynamics but also for spatial connectivity. Note that, for the parameter setting shown in Figure 4.15, λ_0 is larger than 1, implying that pathogen transport and human mobility impose a stricter invasion condition than the one derived in the simple case of no spatial connectivity and no temporal variability ($\widetilde{R}_0 = R_0 > 1$) [447]. This result indicates that the spatiotemporal dynamics of the epidemiological process in dendritic networks share some basic properties with contact processes (Section 3.2.3) or spatially explicit metapopulations [115] (Section 2.2). More importantly, Figure 4.15a shows that, if $\epsilon > 0$, fluctuations of the environmental conditions subsumed by fluctuations of the model parameters can exert a remarkable influence on pathogen invasion in a spatially explicit setting. Pathogen invasion can in fact be either favored (i.e., expected to occur for values of $\widetilde{R}_0 < \lambda_0$) or made more difficult, depending on the parameter assumed to vary in time (Figure 4.15a). Specifically, periodic fluctuations of the parameters that appear at the denominator of $\widetilde{R}_0$ [447] (namely, pathogen mortality and water volume) favor pathogen invasion, while fluctuations of the other parameters considered here play against it. Remarkably, for sufficiently large values of $\widetilde{R}_0$, fluctuations of pathogen

mortality or water volume can induce pathogen invasion not only for $\widetilde{R}_0 < \lambda_0$ but also for $\widetilde{R}_0 < 1$, that is, for values of the parameters that would not allow it in a local model (a case of subthreshold epidemic [447]).

These results hold true not only for a spatially homogeneous population distribution but also for more realistic spatial arrangements. For instance, the size of each local community can be set so as to follow the so-called Zipf's law [5], according to which the probability density function of local population sizes decays proportionally to H^{-2}. The remarkable feature of Zipf's distribution is that it is suggested to accurately describe the sizes of cities of any country in a purportedly universal fashion regardless of geographical, environmental, social, or economic conditions, once properly corrected for the finite size effect induced by total population [5] (see [446, 451] for relevant epidemiological applications). For the sake of simplicity, Mari et al. [447] did not introduce any correlation of population size and position in the river network, which would not be expected to alter the basic results presented here. Also, to ease comparison with the spatially homogeneous case, the average reservoir volumes were assumed to be proportional to the size of the local communities ($W_i = cH_i$), and the total population abundance was set to be the same in the two different spatial arrangements (i.e., $\sum_{i=1}^{n} H_i = nH$). In brief, Mari et al. [115] showed that a heterogeneous population distribution is expected to favor pathogen invasion, as already found in a time-invariant setting [437, 451].

Interesting results can be derived when epidemiological parameters are not only time varying but also space dependent. Specifically, we assume that the space/time-varying parameter (generically termed $\theta_i(t)$) is described by

$$\theta_i(t) = \bar{\theta}_i \left[1 \pm \epsilon \sin(2\pi t + \phi \pi \delta_i) \right],$$

where $\bar{\theta}_i$ is the average value of $\theta_i(t)$, $0 \leq \epsilon < 1$ and $\phi \geq 0$ quantify amplitude and lag of space-time oscillations, and δ_i is a parameter suitably accounting for the lag dependence on the position of node i within the spatial domain. A space-dependent temporal delay in environmental fluctuations ($\phi > 0$) can determine a heterogeneous spatial distribution for the time-varying parameter in large river basins. This heterogeneity may have important implications for pathogen invasion, as shown in Figure 4.15b, in which $0 \leq \delta_i \leq 1$ is supposed to increase from the bottom (i.e., where the network outlet lies; see the inset of Figure 4.15c) to the top of the spatial domain (north–south [N-S] direction), so that $\theta_i(t)$ is characterized by early peaks moving

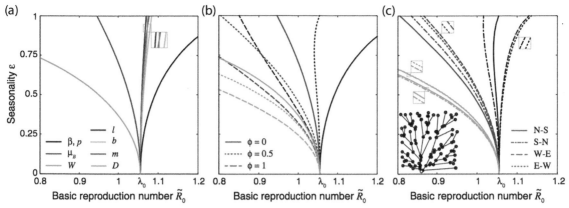

Figure 4.15 Analysis of pathogen invasion conditions for spatiotemporal periodic fluctuations of the parameters (Eq. (4.24)) in the full network model applied to a realistic river network. Pathogens can invade for parameter combinations on the right of the invasibility curves. (a) Effect of synchronous ($\phi = 0$) fluctuations of different parameters (color coded). (b) Effect of different lags ϕ for δ_i following a north–south gradient. (c) Effect of different gradient directions for the spatiotemporal fluctuations of the model parameters ($\phi = 0.5$). In panels (b) and (c), colors are the same as in panel (a). Population is assumed to be homogeneously distributed ($H_i = H$) over the river network (inset of panel (c)). The network has been obtained from extracting from an OCN the 100 nodes with largest cumulative drainage area, and the average volume of local water reservoirs is proportional to local population size ($W_i = cH_i$). Parameter values are fully specified in the original paper. Figure after [447]

northward from the outlet. Increasing spatiotemporal asynchrony of the seasonal fluctuations (represented by increasing values of ϕ) favors pathogen invasion, with subthreshold epidemics being possible for sufficiently large values of ϵ and ϕ. Different spatial patterns of parameter fluctuations (represented by different choices of geographic gradients δ_i) may also matter. Figure 4.15c shows in fact that invasion thresholds can be quite different for different directions of the spatial perturbations. A clear difference emerges for latitudinal versus longitudinal patterns, with the former leading to lower invasion thresholds for all tested time-varying parameters. As for latitudinal patterns, it does not seem possible to infer a general rule, with N-S variations leading to higher invasion thresholds than those in the S-N directions for some parameters (β, p), to lower thresholds for others (μ_B) – and being almost indistinguishable for other parameters (W). As for longitudinal patterns, which are orthogonal to the main direction of the water flow, the invasion thresholds obtained for west–east (W-E) versus E-W perturbations are remarkably similar. This finding is not surprising given the almost symmetric structure of the OCN used in the numerical experiments. Quantitative details aside [447], most of these findings are valid for different choices of the baseline epidemiological parameters.

4.2.2 Geography of Periodic Disease Spread

Finally, it has been noted [447] that defining the conditions for pathogen invasion represents an important step toward a better understanding of waterborne disease outbreaks. However, the identification and possible prediction of the spatial patterns of the initial epidemic spread are crucial in operational terms, for example, for early allocation of health care staff and medical supplies. In the simpler case of time-constant parameter values, the spatial signature of an epidemic outbreak has been linked to the dominant eigenvector of a generalized reproduction matrix accounting for epidemiological processes and the relevant pathogen relocation mechanisms [437, 451] (Section 4.1.5). In fact, when the disease-free equilibrium proves unstable, the dominant eigenvector of the generalized reproduction matrix underpins the direction in the state space along which the system orbits, after a transient, will diverge from the equilibrium. Floquet theory suggests that in the presence of time-varying parameters, the spatial patterns of pathogen invasion and epidemic outbreak will be given, respectively, by the infectives' and pathogens' components of the dominant eigenvector of the monodromy matrix $\mathbf{F}(1)$ defined in [447]. Note that, in analogy with the time-constant case, transient dynamics can (at least partially) cloud the

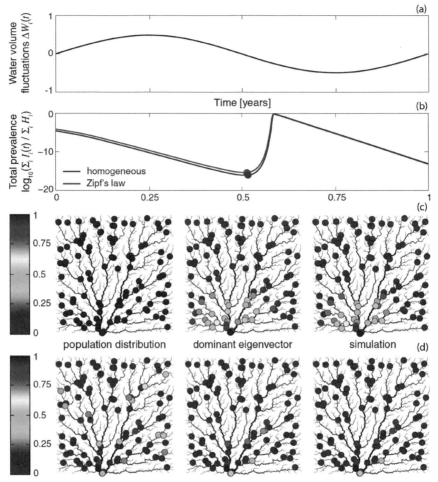

Figure 4.16 Spatial patterns of epidemic outbreak under the Floquet framework. (a) Fluctuations around the reference value of the local water reservoir volumes $W_i(t)$. (b) Temporal pattern of total disease prevalence (semilog scale); the filled dots represent epidemic emergence as defined in [447]. (c) Population density map (left, homogeneous population), components of the dominant eigenvector of the monodromy matrix $\mathbf{F}(1)$ (middle, $\xi_{max} = 2.04$), and spatial distribution of disease prevalence at epidemic emergence as simulated by model Equation (4.24) (right). (d) As in panel (c), for population density following Zipf's law ($\xi_{max} = 4.82$). The quantities shown in panels (c) and (d) have been normalized (so that the maximum has unit value). Parameters and computational simulation details are specified in [447]

predictions of Floquet theory, especially in the case of fast-developing epidemics that emerge soon after pathogen introduction.

The predictions of Floquet theory applied to the spatial patterns of epidemic spread can be easily verified by performing numerical simulations of model (4.24) and contrasting the infectives' components to the spatial distribution of disease prevalence at epidemic emergence, here defined as the point in time in which the sign of $d\left(\sum_{i=1}^{n} I_i\right)/dt$ switches from negative to positive in response to the periodic oscillation of the time-varying

parameter (e.g., the volume of local water reservoirs in Figure 4.16; see Figures 4.16a,b). The comparison between eigenvector components and simulation results can be performed not only for a homogeneous population distribution (Figure 4.16c) but also for other, more complex spatial patterns, even in the case of outbreaks without endemism. As an example, Figure 4.16d is obtained under the hypothesis that the size of local human communities follows Zipf's law. Other patterns of population distribution are analyzed in [447], leading to analogous conclusions.

The agreement between the components of the results is very satisfactory for every tested population distribution (coefficient of determination $R^2 \geq 0.98$ for all cases [447]). Similar correspondences were found also in the case of periodic fluctuations initially leading to an increase of infection risk (coefficient of determination $R^2 \geq 0.95$ for all cases). Note also that, despite differences in the underlying population distributions, the time series of total disease prevalence shown in Figure 4.16b are very similar. This is connected to the observation that epidemic emergence is triggered by increasing local pathogen concentrations resulting from a decrease in the volumes of the local water reservoirs. Once started, for $\epsilon \gg 0$, the course of the epidemic is very fast. As the attack ratio approaches very high values (>90 percent), the spatial distribution of human communities loses importance in comparison with less explosive epidemics.

The framework presented here [447] could be made more realistic in many respects. First, seasonal fluctuations are not simple sinusoidal signals. The erratic patterns often observed in time series of, say, temperature or rainfall intensity, which are known to influence pathogen demography, should mirror water availability and thus exposure/contamination rates. Overcoming this limitation would need a mathematical framework considerably more involved than Floquet theory, as it would require the (typically numerical) computation of the Lyapunov exponents associated with the disease-free equilibrium or the use of suitable approximations to evaluate the long-term growth rate of the pathogen population [447]. We also remark that the techniques adopted in this paper cannot be reliably used to derive long-term epidemiological patterns. This task would in fact require the determination of the attractor(s) of the model and the analysis of their dynamical characteristics (e.g., via bifurcation analysis; see Appendix 6.1). To that end, the model should be further extended to include partial/waning immunity because individuals recovered from waterborne diseases usually can become susceptible again after a certain time.

Despite the inevitable limitations, we believe that the generality of the mathematical approach used in [447] to derive invasion conditions for waterborne pathogens in a spatially explicit network model subject to seasonal fluctuations could possibly be applied to other diseases (not necessarily waterborne) as well as to other geographic settings (not necessarily river networks). We thus suggest that the above results, general as they are because they reduce exactly to particular cases already dealt with for spatially homogeneous and/or time-invariant conditions,

may define a framework (as opposed to a model) for the realistic description of waterborne pathogen invasion and the geography of epidemic spread addressed by spatially explicit, multilayered, and time-varying network models.

4.2.3 The Transient Spread of Epidemics

Determining the conditions that favor pathogen establishment in a host community is key to disease control and eradication. However, focusing on long-term dynamics alone, as we have examined in the previous sections for epidemic cholera, may lead to an underestimation of the threats imposed by outbreaks triggered by short-term transient phenomena. Achieving an effective epidemiological response thus requires one to look at different timescales, each of which may be endowed with specific management objectives [452]. Epidemicity thresholds have been shown to exist for prototypical examples of waterborne and water-related diseases, the broad and diverse family of infections transmitted either directly through water infested with pathogens or by vectors or hosts whose life cycles are associated with water bodies. Here, following Mari et al. [453], we shall typify this situation by capitalizing on our previous mathematical description of cholera spread. We shall confine our attention to methods that may be applied to diseases relevant to the tenet of this book, that is, the study of river networks as ecological corridors for pathogens.

Technically, while conditions for endemicity are determined via stability analysis (the exchange of stability of disease-free conditions, analyzed via the properties of a modified Jacobian of the relevant dynamics, in analogy with the procedure outlined in the previous sections [437]), epidemicity thresholds may be defined more generally. This may be done via a generalized reactivity analysis, a method that allows us to study the nature and the possible damping of short-term instability properties of ecological systems however perturbed [454]. Understanding the drivers of waterborne and water-related disease dynamics over timescales that may be relevant to epidemic and/or endemic transmission has been argued to represent a challenge of most waterborne epidemiology applications [452], because large portions of the developing world are still struggling with the burden imposed by these infections, as we have argued in several places in this chapter.

In our previous analysis of the models of the dynamics of infectious waterborne disease, emphasis has been placed on the definition of necessary conditions for pathogen invasion and long-term persistence, derived through linear stability analysis (Section 4.1.4). We have seen that necessary conditions for endemic establishment are usually stated in terms of the basic reproduction number, R_0, with $R_0 > 1$, being the required condition for sustained disease spread (see, e.g., [434, 455]), or its spatially explicit generalizations, namely, the maximum eigenvalue of the generalized reproduction matrix $\mathbf{G_0}$ being larger than unity [437]. This type of analysis yields asymptotic long-term results, while short-term dynamics are often disregarded – with consequences [456]. However, as noted in [454], should transient epidemiological phenomena fade out slowly, the resulting epidemic could stretch over timescales comparable with the decision-making horizon for health care emergency policies. Thus, while the study of the asymptotic properties of disease transmission models can crucially assist in the design of control strategies aimed to permanently reduce (or even eliminate) pathogen transmission, it ought to be usefully complemented by the analysis of transient phenomena of potential epidemiological interest.

The importance of transitory changes in disease prevalence has a long epidemiological history that we shall not review here. Suffice here to mention that the relevant type of transient dynamics, sometimes referred to as stuttering transmission, may typically occur when a pathogen is weakly passed on from infected to susceptible hosts, because of either the self-limiting nature of the process (like in the case of zoonoses with inefficient between-human transmission) or the implementation of control strategies [452]. In these cases, transmission cannot be sustained indefinitely over time. Therefore, temporary outbreaks can be thought of as the expression of an epidemiological system's short-term instability to perturbations. Transient epidemic waves are possible, in fact, if suitable perturbations to the attractor of an epidemiological system can temporarily grow before eventually vanishing. In this respect, a simple measure of a system's short-term instability to small perturbations was

originally proposed by Neubert and Caswell [454, 457]. Specifically, they introduced the concept of reactivity, defined as the maximum instantaneous rate at which perturbations to a stable steady state can be amplified. Revived by that seminal paper, the analysis of the transient behavior in biological systems has been recognized as key to long-term ecological and epidemiological understanding (see, e.g., [458–463]). Here, we shall simply review the concept of reactivity in epidemiological contexts [457, 463], a useful method to characterize the transient dynamics associated with a stable equilibrium of a nonlinear system of ODEs. This is a subject that is surely worth pursuing in further studies (Chapter 5).

Let the dynamics around a stable equilibrium of a nonlinear epidemiological model be characterized by the general linearized dynamic equation

$$\frac{d\mathbf{x}}{dt} = \mathbf{A}x,$$

where $\mathbf{A}$ is the $n \times n$ state matrix and $\mathbf{x}$ is the vector defined by the differences between the system state and the equilibrium. Nonzero perturbations $\mathbf{x}(0) = \mathbf{x}_0$ are in any case absorbed in the long term because the equilibrium is stable. However, they might prompt a reactivity of the system in the short term. The measure of reactivity is defined as the maximum initial amplification rate of small perturbations away from equilibrium, evaluated over all possible types of local perturbations. In such a general context, reactivity corresponds to the dominant eigenvalue of the Hermitian part of matrix $\mathbf{A}$, that is,

$$H(\mathbf{A}) = \frac{(\mathbf{A}+\mathbf{A}^T)}{2},$$

where T indicates matrix transposition. Through this definition, it is possible (and relatively straightforward) to evaluate a system's short-term response to small perturbations as a function of the model parameters. Via appropriate application of this paradigm to epidemiological WB models (Box 4.11), it is possible to implement the concept of epidemicity, which is again related to the dominant eigenvalue of a suitable matrix.

Box 4.11 Finding the Epidemicity Threshold

Neubert and Caswell's [448] measure of an ecological system reactivity relies on the Euclidean norm of the system state, so that all variables are given equal weight in the assessment of the reactivity properties of an equilibrium. The original definition of reactivity was in fact given by the following relation:

Box 4.11 *Continued*

$$\text{reactivity} = \max_{x_0 \neq 0} \left(\frac{1}{\| x \|} \frac{d \| x \|}{dt} \right) \Bigg|_{t=0},$$

where $\| \cdot \|$ indicates the Euclidean norm of the full state vector. Reactivity was thus originally defined as the maximum initial amplification rate of a small perturbation. As noted by Mari et al. [463], this may represent a possible setback in epidemiological applications. Therein, in fact, one would rather focus on the reactivity properties of key components of the state space pertaining to infection (such as infected and/or infectious hosts, environmental pathogen concentrations, and/or parasite loads), while neglecting the others (e.g., susceptible hosts or parasite predators). A weighted measure of reactivity was recently proposed [463] as an extension of the original idea. It is based on the dynamic analysis of a suitably defined system output, obtained as a linear transformation of the state variables tailored to epidemiological coupled ODEs. In epidemiological applications, in fact, the system output may correspond to one or more linear combinations of the state variables that pertain to the infection-related components of the specific model [463].

To this end, introduce a suitable (i.e., ecologically motivated) linear output transformation $\mathbf{y} = \mathbf{Cx}$ for the linearized (or linear) system $d\mathbf{x}/dt = \mathbf{Ax}$, where $\mathbf{C}$ is a real, full-rank $m \times n$ ($m \leq n$) matrix defining a set of independent linear combinations of the system's state variables. A stable equilibrium point is defined as g-reactive if there exist small perturbations that can lead to a transient growth in the Euclidean norm of the system output. To be epidemiologically meaningful, the output transformation should include all the infection-related variables of the system. Mari et al. [463] showed that the classification of a stable equilibrium point as g-reactive is based on the sign of the dominant eigenvalue of a matrix that can be easily obtained from the output and state matrices. Specifically, a stable steady state is g-reactive if

$$\lambda_{\max} \left[H(\mathbf{C}^T \mathbf{CA}) \right] > 0, \tag{4.26}$$

where $\lambda_{\max}(\mathbf{B})$ is, as usual, the maximum eigenvalue of the arbitrary matrix $\mathbf{B}$, and

$$H(\mathbf{C}^T \mathbf{CA}) = \frac{(\mathbf{C}^T \mathbf{CA} + \mathbf{A}^T \mathbf{C}^T \mathbf{C})}{2}$$

is the Hermitian part of the matrix $\mathbf{C}^T \mathbf{CA}$. The maximum eigenvalue is real because the matrix $\mathbf{C}^T \mathbf{CA}$ is real and symmetric. If the condition (4.26) is met, then some perturbations $\mathbf{x}_0$ exist that are temporarily amplified in the system output. These are perturbations for which the growth rate of the output norm is positive, that is,

$$\| \mathbf{CAx}_0 \| > 0.$$

These perturbations belong to the so-called g-reactivity basin of the equilibrium defined by the quadratic form

$$\mathbf{x}_0^T H(\mathbf{C}^T \mathbf{CA}) \mathbf{x}_0 > 0.$$

Trajectories that originate in a neighborhood of the stable (yet g-reactive) steady state within its g-reactivity basin will generate a transient epidemic wave, while perturbations that lie outside the basin will monotonically decay in the system output without producing an outbreak. Therefore,

Box 4.11 *Continued*

initial conditions (i.e., perturbation structure) may indeed play a crucial role in the development of short-term epidemics. In this framework, all models of epidemic spread, whether spatially implicit or explicit, may be reanalyzed, as done by Mari et al. [452, 463].

Cholera, for instance, has been analyzed in this framework. Here, we refer to a simplified model to highlight the features of the method more clearly, following [452]. The ruling mathematical problem elucidated here is given by Equations (4.16) [434], in which we assume conventionally $H = 1$. To study the g-reactivity of the system (4.16), we look at the properties of the steady states of the system using the output matrix

$$\mathbf{C_c} = \begin{bmatrix} 0 & a & 0 \\ 0 & 0 & b \end{bmatrix},$$

so that the output transformation is $\mathbf{y}_c = [aI \ bB]^T$ with a and b being two positive parameters, which may weight, for example, the socioeconomic costs associated with cholera and its control, either in terms of human infections (a) or environmental contamination (b). The susceptible component of the state space is neglected in the system output, and epidemicity is evaluated only by considering transient amplifications of the prevalence of infected hosts in the human community and of the concentration of pathogens in the water reservoir [452].

Model (4.16) has two steady state solutions. The first one is the disease-free equilibrium $S = 1$, $I = 0$, and $B = 0$. The second one is an endemic equilibrium with

$$S = \frac{\delta\phi'}{\beta\theta}, \quad I = \frac{\mu}{\phi'}\frac{(\beta\theta - \delta\phi')}{\beta\theta}, \quad B = \frac{\mu}{\phi'}\frac{(\beta\theta - \delta\phi')}{\beta\theta},$$

where $\phi' = \mu + \alpha + \phi$. Clearly, the endemic equilibrium is feasible if $\beta\theta - \delta\phi' > 0$. Linear stability analysis [434, 452] shows that this is the condition for the disease-free equilibrium to be unstable and, at the same time, for the endemic equilibrium to be stable. With this notation, the basic reproduction number reads $R_0 = \beta\theta/(\delta\phi')$, and an exchange of stability between the disease-free and the endemic equilibria occurs for $R_0 = 1$. For $R_0 > 1$ the former is unstable, while the latter is positive and stable. The stable disease-free equilibrium is instead g-reactive if

$$E_0 = \frac{(a^2\beta + b^2\theta)^2}{4a^2b^2\delta\phi'} > 1,$$

marking the condition for epidemicity, as proved in [452]. If either exposure or contamination is low ($\beta \to 0, \theta \to 0$, respectively), then disease spread will halt shortly ($R_0 \to 0$). On the contrary, transient epidemic waves could be triggered by small perturbations to the disease-free equilibrium even if either the exposure or the contamination rate were close to zero [452]. As an example, consider a scenario in which water contamination by infected humans could effectively be prevented ($\theta = 0$). In this case, one has $E_0 = (a^4\beta^2)/(4b^2\delta\phi')$, which can be larger than one implying that the disease-free equilibrium is g-reactive depending on parameter values, specifically if $\epsilon > 2(b/a)\sqrt{\delta\phi'}$. A suitable influx of bacteria into the accessible water reservoir could thus result in a sizable number of infections before the pathogen eventually decays. This is expected in any nonendemic environment where further contamination by infected humans is prevented. Conversely, a positive perturbation of the prevalence of infected hosts would rapidly fade out without any transient amplification.

Box 4.11 *Continued*

These simple observations further emphasize that the structure of perturbations (corresponding to different initial conditions) may play a defining role in the development of transient epidemic outbreaks.

Obviously, it can be shown [452] that if the disease-free equilibrium is unstable, namely, if the disease can become endemic ($R_0 > 1$), then necessarily, $E_0 > 1$, whatever the values of the parameters a and b; that is, the establishment of the endemism begins in any case with an epidemic.

Box 4.11 has shown an application of epidemicity to a simple model of cholera without spatial structure. However, more recently, Mari et al. [453] have applied g-reactivity tools to studying epidemicity thresholds in general space-explicit WB models. They have discovered that the start of an epidemic is linked to the dominant eigenvalue of an epidemicity matrix. This dominant eigenvalue generalizes the simple condition given in Box 4.11 for the space-implicit model. The generalized epidemicity matrix is linked to the various matrices that describe the network structure (including both hydrologic and human mobility connectivities) and the epidemiological parameters.

Suitably designed epidemicity indices may thus represent an important complement to the definition of threshold for invadibility and endemicity based on stability/endemicity analysis, that is, typically on the basis of the basic reproduction number R_0 of local infection models. As we have shown in Section 4.2, generalizations of R_0 are needed for epidemiological problems endowed with strong spatial [437, 451] and/or temporal [447] heterogeneity. In a similar way, the g-reactivity analysis shown here following [452] can be applied [453] to space/time-explicit settings to derive generalized epidemicity conditions accounting not only for local epidemiological processes but also for spatial coupling mechanisms that are relevant to pathogen dissemination, such as human mobility and hydrological connectivity. This generalization, together with the possibility of applying g-reactivity analysis to relatively large problems, will be essential to providing scalable, operational tools for the analysis of real-world applications.

4.3 Epidemic Cholera

Mathematical models may provide insight into the course of an ongoing epidemic in time for aiding real-time emergency management and optimally allocating health care resources. They may, if suitably reliable, also rank alternative interventions by their impact [428, 464]. A comparative analysis on the ex post reliability of predictions of the iconic 2010–2011 Haiti cholera outbreak will be described here. It illustrated, in fact, the impact of different approaches to modeling the spatial spread of *V. cholerae* and the various mechanisms of cholera transmission during an outbreak.

As a major cholera epidemic spread through Haiti (see, e.g., [427, 465]), leading to 170,000 reported cases and 3,600 deaths at the end of 2010 [427, 466], four independent modeling studies appeared almost simultaneously during the unfolding epidemic [30, 467–469], each predicting the subsequent course of the epidemic and/or the impact of potential management strategies. The spate of mathematical models confirmed earlier suggestions that their impact on public health practice was (and is) gaining momentum [440, 470]. Indeed, mathematical models of infectious diseases, once properly tested for reliability, can provide key insights into the course of an epidemic in time for action, thus averting deaths and reducing the number of infected patients through a sensible allocation of resources, possibly including vaccines [471, 472]. Because the Haiti cholera models were published early in the course of the epidemic, which as of October 26, 2011, had gone on to produce an estimated 485,092 cases, 259,549 hospitalizations, and 6,712 deaths [421, 427, 466], the subsequent course of the epidemic allows an assessment of the reliability of the early predictions and a related discussion of the lessons learned, with a view of where the field is headed.

Most significantly, before 2010, cholera had not been reported in Haiti for hundreds of years [473]. It was thus likely that the population had no significant prior exposure or acquired immunity to the disease, and therefore the entire population at the time of the outbreak could be assumed susceptible to the disease. Haiti also lacked preparedness for this epidemic and suitable health infrastructure through which to combat it [428, 465, 474, 475], not only because of the notable disease-free timespan that the island enjoyed but also because the social, economic, and infrastructural situation was disastrous in the aftermath of a deadly earthquake earlier in 2010 that killed more than 300,000 people [427, 465, 466]. Although there has been some debate as to the source, most experts agree that the first cases were autochthonous, brought into Haiti from a distant geographic source [465, 476], and that these cases seeded the subsequent epidemic, which originated within the Centre department and then spread to all of the Haitian departments, exhibiting complex spatial and temporal patterns [427] (Figure 4.17).

Once a cholera epidemic starts, infected patients excrete huge numbers of *V. cholerae* bacteria that spread in complex spatial patterns easily ironed out by local SIRB-like descriptions – either through water pathways (via active and passive dispersal [325]) or through human mobility networks involving both susceptibles and infected individuals [438]. Poor sanitation, which characterized Haiti after the disastrous 2010 earthquake, facilitates both types of spread and fosters the abundance of microorganisms in the water system. Some, such as *V. cholerae*, are extremely versatile and can quickly adapt to new environments. Being primarily an aquatic bacterium, *V. cholerae* can persist indefinitely in rivers, estuaries, and coastal regions without any need for human passage. The incidence of cholera in such ecosystems fluctuates as a function of climatic forces (in particular, ENSO) and changes as extensively described primarily for the region around the Bay of Bengal, where it originated in the first place [477]. Because of weak sanitary infrastructure and favorable environmental conditions, it was obvious from the onset that cholera would continue to be a threat in Haiti as well as in many other developing countries [426].

Premises, methods, and results of models of the 2010–2011 Haiti epidemic cholera thus matter, to derive lessons that directly affect the predictive value of model outputs on pathogen dispersal mechanisms, model-guided field validations, data requirements, and model identification [428]. On the basis of such an analysis, we intend to highlight shortcomings of past approaches and discuss mechanisms of disease transmission. It should be noted that in the reference work [428], best-performing models are identified via formally rigorous comparative criteria from available epidemiological and hydrological time series.

Of interest to the general tenet of this book, to explain resurgences of a cholera epidemic, one needs to include the waning acquired immunity and a mechanism explicitly accounting for rainfall as a driver of enhanced disease transmission. The latter may be due to either an overload of pathogens in the water reservoir, say, by washout of open-air defecation sites and spillovers/leakages from treatment centers [428], or enhanced exposure to the infection [429]. A formal comparative analysis measures the added information provided by each process modeled, discounting for the added parameters. The construction and the discussion of a generalized model for the Haitian epidemic cholera – iconic and unique in its kind for the reasons discussed below – and the related estimation of uncertainty are thus proposed here as a significant example. The models are applied to a year-long dataset of reported cases then available. Here we discuss the lessons learned from the Haiti modeling study, and existing open issues are placed in perspective. This section suggests that, despite differences in methods that can be tested through model-guided field validation, spatially explicit mathematical and numerical models of large-scale outbreaks emerge as an essential component of future cholera epidemic control.

4.3.1 A First Assessment of the Haiti Cholera Outbreak

All four models of the 2010–2011 Haiti cholera epidemic [30, 467–469] address the coupled dynamics of susceptibles, infected individuals, and bacterial concentrations in the water reservoir in a spatially explicit setting of local human communities. The entire Haitian population was assumed to be susceptible at the outset of the epidemic. Each of the models assumed that the rate at which susceptibles become infected is dependent on the *V. cholerae* concentration in available water and, in turn, that new free-living bacteria are produced by infected individuals through fecal contamination of water. The main differences among the models stemmed from assumptions about pathogen redistribution mechanisms among the different human communities. Box 4.12 illustrates the basic elements of the earliest model of space-time cholera evolution. Figure 4.18 shows the results of a three-month projection based on calibration of the parameters of the model in a preceding two-month timespan of spatially distributed reported infections [30].

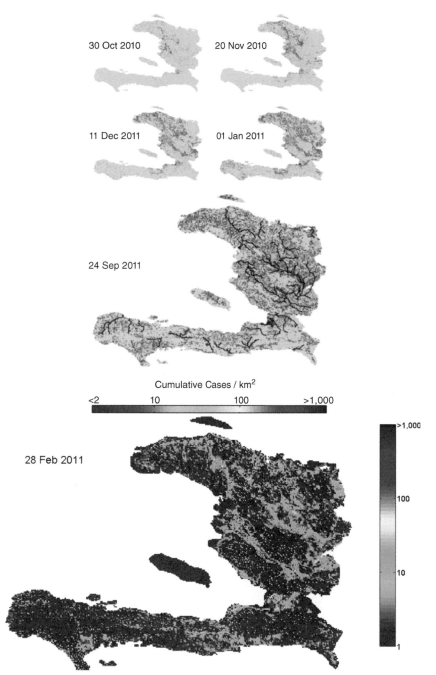

Figure 4.17 Map of the spatiotemporal evolution of cumulative reported cholera cases at the country scale [466]. Infections are projected at the (pixel) scale of the population database (see [428], figure 2C of the main text) with the assumption of spatially uniform incidence within departments. The early propagation of the disease from the Centre department mainly along the Artibonite River is clearly shown. The subsequent, fast outbreak in the most densely populated region (Port-au-Prince) is also evident. Note that the patterns of total cumulative reported cases mirror fluvial pathways (rivers are shown in black in the map of September 4, 2011), either because settlements follow colonization in need of water resources [7, 118] or because fluvial pathways define the main directions of pathogen spread (*sensu* [427]). Figure after [428]

Box 4.12 Early Spatially Explicit Model of Epidemic Cholera

The basic epidemiological model is a spatially distributed version of the SIB model (Section 4.1), which is simpler than the one developed later (Box 4.5). It is adapted to information on an outbreak that may be gathered remotely or based on local weekly reported cases during its onset [30, 325]. In this metacommunity framework, Haiti has been subdivided into $n \sim 500$ local communities (representing the fourth administrative level). For each community, the dynamics of susceptibles and infected to the disease neglect the compartment of recovered individuals owing to a loss of acquired immunity on a longer timescale than that of the predictions (a few months in this case). Recovered individuals are thus considered immune. A key parameter, the pool of susceptibles at time $t = 0$ (the onset of the outbreak) is assumed to coincide with the total population. In fact, Haiti has been known to be cholera-free for hundreds of years before 2010, and current inhabitants, never exposed to cholera, were lacking any immunity [476].

The key assumptions of the model are as follows: the rate at which susceptibles become infected depends on the concentration of pathogens in the available water and, in turn, new free-living bacteria are produced by infected individuals through fecal contamination. As for pathogen transport, the epidemic information of the Haitian outbreak available at the early stages [30] (reported cholera cases aggregated at the department level) does not allow the assessment of the relative role of transport mechanisms (hydrology vs. human mobility). Moreover, the complete absence of any historical reference for a cholera epidemic in that region [476] further complicated any modeling attempt but perhaps reflects the most common circumstances in which emergency management of cholera outbreaks occurs worldwide. For these reasons, Bertuzzo et al. [30] resorted to a simplified pathogen dispersion scheme inclusive of passive and active transport through waterways and the net effect of the mobility of asymptomatic infected individuals. According to this dispersive model, the flux of pathogens incoming into a destination community from a source community decays with the distance separating them and is proportional to the product of the pathogen concentration in the source community times the population size of the destination community. The last assumption borrows concepts from the so-called gravity models of transportation theory [105] arguing that the dispersion model is able to capture the features of the two main transport mechanisms analyzed: a short-range water contamination and a long-range anisotropic transport induced by human mobility.

Epidemiological dynamics and pathogen transport are therefore described by the following set of ordinary differential Equations [30]:

$$\frac{dS_i}{dt} = \mu(H_i - S_i) - \beta_i \left(\frac{B_i}{K + B_i} \right) S_i, \tag{4.27}$$

$$\frac{dI_i}{dt} = \beta_i \left(\frac{B_i}{K + B_i} \right) S_i - (\gamma + \mu + \alpha) I_i, \tag{4.28}$$

$$\frac{dB_i}{dt} = -\mu_B B_i + \frac{p}{W_i} I_i - \ell \left(B_i - \sum_{j=1}^{n} P_{ji} \frac{W_j B_j}{W_i} \right), \tag{4.29}$$

with the usual symbol notation of SIB models. The parameter β represents the exposure rate to contaminated water, and $B_i/(K + B_i)$ is the probability of becoming infected due to the exposure to a concentration B_i of vibrios, K being the half-saturation constant [434]. Infected individuals recover at a rate γ or die, with natural or cholera-induced mortality at rates μ and $\alpha \ll \mu \sim 0$ to maintain demographic equilibrium, respectively. Infected individuals contribute to the concentration

Box 4.12 *Continued*

of free-living vibrios at a rate p/W_i, where p is the rate at which bacteria excreted by one infected individual reach and contaminate the local water reservoir W_i [325]. Bacteria die at a constant rate μ_B and undergo dispersal at rate ℓ. The probability P_{ij} of pathogen transport from node i to node j of the network is defined as $P_{ij} \propto H_j e^{-d_{ij}/D}$, where the dispersion to node j depends on its population size, H_j, and decays with the distance d_{ij} between the two nodes i, j, exponentially cut off by a deterrence distance D (the proportionality constant is obtained by normalization [30]). The scheme can be solved numerically for arbitrary networks of human communities (Appendix 6.3). Initial conditions for the case of Haiti are $I_{i \neq k}(0) = 0$ for all nodes, except for the inoculum k (where a suitable number $I_k(0) > 0$ is assigned – typically, the reported cases from the earliest cholera outbreak report), $I_i(0) = B_i(0) = 0$ for all nodes.

We shall first examine the reliability of our own original scheme [30], later discussing the main differences from the other approaches [467–469] and the impact of these differences on predictions. Human mobility was described (Box 4.12) by a diffusive process calibrated on spatial data on reported infections, and no climatic drivers were assumed (for instance, rainfall played no role in the transmission process). Figure 4.19 compares the projected course of the epidemic as published in January 2011 with the actual reported case counts reported at the end of September 2011 [30]. Highlighted (dark gray) in the figure is the dataset used for calibration, which is limited to the end of December 2010 before the first decline of the incidence of the disease. The original prediction ran up to the end of May 2011 (solid line in Figure 4.19). The five-month forecast, judged in retrospect, was quite robust and could have been used to make practical decisions and act in time. To facilitate a further assessment of our model reliability, we have extended the original prediction to the end of September 2011 (Figure 4.19, dashed line). Whereas the order of magnitude of total cumulated infections is captured up to September 2011, important features are clearly missed, such as the June–July revamping of weekly incidence (which is likely correlated to seasonal rainfall; Figure 4.19 [top]).

Various approaches dealt with the Haiti cholera outbreak. They may be differentiated on the basis of their treatment of spatial transmission mechanisms. One study [469] treated each Haiti administrative department independently without explicitly considering the spread of cholera among them, whereas the other models [30, 467, 468] explicitly modeled inter- and intradepartmental pathogen redistribution. When each

department was treated as an independent, spatially implicit unit, thus embedding in the parameters of the local model the effects of, say, human mobility and pathogen redistribution through fluvial pathways [469], the result was an overparameterization of the model (see [428] for a technical assessment). Different spatial resolutions also characterize the metacommunity models (10 local communities in [467], 11 in [469], 560 in [30], and on the order of 20,000 in [468]).

For those studies that provided predictions of the subsequent course of the outbreak, the projections can be tested against current data, yielding a first assessment of validity and the limitations of different modeling assumptions. For example, Andrews and Basu [469] forecast a toll of 779,000 cases and 11,100 deaths from March 1 to November 30, 2011, that significantly overestimated the course of the epidemic; 324,405 new reported cases (accounted for as proposed in [469]) and 2,040 deaths were actually reported between March 1 and October 26 [466]. This is clearly a case of a major error incurred in the description of the spread of an epidemic when overparameterizing its model. The differences with observational data stem from the nature of the modeling assumptions that undermine the predictability of the approach. In fact, disregarding interdepartmental pathogen dispersal mechanisms implies the independent fitting of the model parameters to 11 separate departments. Therefore, model parameters are also charged with the effects of long-range transmission mechanisms, including human mobility, which is empirically known to be reaching well beyond departmental domains [478]. Regardless of the choice of calibration techniques, the large number of parameters (33 vs. 5 in [30]) increases dramatically the uncertainty of early projections.

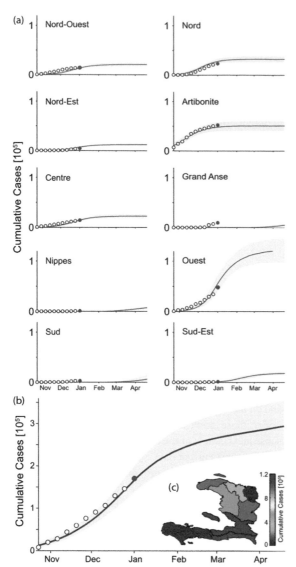

Figure 4.18 Predicted evolution of the Haiti epidemic up to May 2011, based on reported cases as of January 1, 2011, via numerical solution of Equations (4.27)–(4.29). Predicted (solid line) and reported (circles) cumulative cases for (a) the 10 administrative departments and (b) the whole country. Data reported in red circles were not used for model calibration as they became available upon submission. Gray shaded areas show the 5th–95th percentiles of the model prediction. (c) Spatial distribution of the predicted cumulative reported cases up to May 1, 2011. Figure after [30]

The role of asymptomatic carriers may be clarified by the detailed examination of one scheme [467] that employs transmission mechanisms similar to those that will be described in Box 4.13 (save for the inclusion of a human-to-human contagion) but neglects the role of inapparent infections. Asymptomatic carriers are thought to be a critical factor in cholera epidemics [478], particularly in Haiti, because of their number, unimpaired mobility, and thus major role in long- and short-range disease transmission. Inapparent infections are estimated at $75 \div 83$ percent of the total [478, 479]. Moreover, they lead to some degree of acquired immunity, resulting in temporary reductions in the number of susceptibles in a given region, thus potentially affecting significantly the uncertainty of modeling approaches. Note that the model by [467] with realistic values of the basic local reproduction number R_0 fitted nicely the initial phases of the epidemic – but would predict an excessive number of reported cases at later stages. To overcome this limitation, the authors proposed an ad hoc sixfold reduction of the effective reproduction number in the first three months of the epidemic, owing to disease-control interventions. If the compartment of susceptibles is not depleted otherwise, an equal decrease in transmission rates is implied, which seems unrealistic compared with the sanitation interventions analyzed. Adopting a model in which inapparent infections are accounted for avoids the need to force effective reproduction numbers to decrease in time because of unspecified disease-control measures [428]. An interesting mathematical cholera transmission model, individual based and of stochastic nature [468], focuses on the effects of vaccination strategies for epidemic cholera in Haiti. It addresses the same basic transmission processes as those in [30, 467] but also includes a one- to five-day latent period, a hyperinfective state of freshly shed bacteria, and a model of human mobility that incorporates remotely sensed population density data at 1 km^2 resolution and the localization of major rivers and highways. The study does not attempt to tune processes by matching observed space-time distributions of reported cholera cases [468] but is rather based on model parameter values and ranges from the literature. The main shortcomings of this approach include a limited capability of reproducing past observed infections and thus of reliably predicting future epidemic evolutions.

4.3.2 A Second Assessment of the Haiti Cholera Outbreak

The information accumulating on the Haitian outbreak at later stages allowed a thorough reanalysis of primary and ancillary transmission mechanisms, thereby benchmarking the various approaches [428]. Key to the result

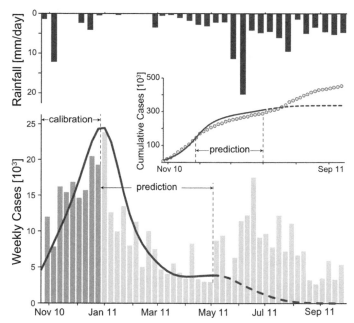

Figure 4.19 (top) Daily decadal rainfall intensity, averaged over the entire Haiti region. (bottom) Weekly reported cases (1) (gray bars) compared with the simulated incidence pattern (solid line) computed by the model in [30]. Data from each department were collected until September 30, 2011. The calibration dataset (dark gray) was limited to the total reported cases available until December 2010. The solid line shows the published early prediction [30] that was run until the end of May 2011. To facilitate the assessment, the original prediction has been extended to the end of September 2011 (dashed line). (middle) Simulated and reported weekly cumulated cases. Figure after [428]

of such reanalysis is a comparative study of a suite of models of different complexity that are described in Box 4.13. Because the analysis was extended to a one-year timespan, the loss of acquired immunity (i.e., a flux from the pool of recovered cases back to the pool of susceptible individuals) cannot be ignored (from SIB to SIRB models in the above jargon). A loss of immunity acquired by surviving a cholera bout whose extent depends on the severity of the infection is thus accounted for by all models compared there. This, in turn, prompted various revisions of the metacommunity models of transmission among human settlements exposed to the infection by including specific mechanisms of hydrological transport, as suggested by empirical observation of the downstream spreading of early infections along the Artibonite River [480]. Pathogen dispersal along waterways is thus described (1) by a careful extraction of the river networks (Figure 4.20) from digital terrain maps (DTMs), through suitable geomorphologic criteria, and (2) as a biased random walk process on an oriented graph [325, 428]. Pathogen redistribution is also enhanced by contamination of the water reservoir driven by heavy seasonal rains.

Inclusion of such overloads, which could occur not only in sites where open-air defecation sites are the norm but also in locations served by combined sewer networks where overflow may occur, was prompted by the clear empirical correlation, observed in Haiti in June and July, between weekly rainfall and enhanced infections (Figure 4.2). We considered two options: an increase of contamination rates depending on rainfall intensity and a mechanistic account of the washout of open-air defecation sites by surface runoff [428]. Human mobility patterns were explicitly modeled (Boxes 4.13 and 4.15, and references therein).

Observations of fast intercatchment transmission of the infection that would not be explained by water pathway pathogen dispersal and of actual individual displacements in times of cholera support this assumption. Mobility patterns are described by a layer of nodal connectivity (Figure 4.20d and Box 4.15). With suitable spatial resolution, human settlements may be placed at nodes of the hydrological network [428] (Figure 4.20), and edges are measured by the distances connecting them. We assume that susceptible and infected individuals engage

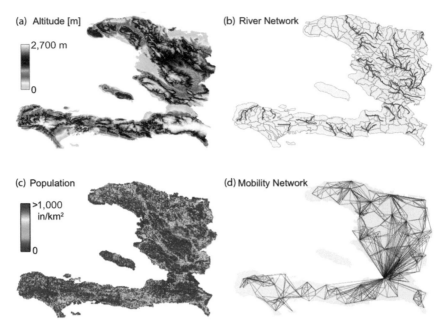

Figure 4.20 (a) Color-coded digital terrain elevation map (DTM) of Haiti. (b) The subdivision of Haitian territory in hydrological units (subbasins) extracted from a suitable DTM, as a result of the convergence of several geomorphological criteria [63]. (c) Spatial distribution of population density obtained from LandScan remote sensing, which is translated into a georeferenced spatial distribution of nodes i endowed with population H_i (Box 4.13). (d) A relevant subset of the network of human mobility, here portrayed synthetically by the four largest outbound connections for each node. Figure after [428]

in short-term trips from the communities where they live toward other settlements. While traveling or commuting, susceptible individuals can be exposed to pathogens and return as infected carriers to their settlement [438, 468]. Similarly, infected hosts can disseminate the disease away from their home community – in many cases, infected individuals are asymptomatic and thus are not barred from their usual activities. Connectivity structures and fluxes of human mobility have long been studied in epidemiology [481, 482], often on the basis of gravity-like models where the flux between two communities owing to human mobility is proportional to the product

of the respective populations and decays with the distance separating them [105, 435]. A choice of a model of this kind (Box 4.13) is indirectly supported by empirical studies [483] that tracked daily Haitian average movements through mobile phones to determine likely new areas for cholera outbreaks far from the site where the disease was first detected. The study suggested that outbound travels from the source area were frequent and that most of the country received persons from the affected area, whereas the vast majority of individuals leaving the source area traveled to just a few large recipients that included surrounding communal sections.

Box 4.13 Examples of Spatially Explicit SIRW Models

Rinaldo et al. [428] considered several variants of the original model Equations (4.27)–(4.29) (shown in Box 4.12) and ranked the performance of the various refinements via the Akaike information criterion (AIC) [147] by contrasting epidemiological data gathered on the Haitian epidemics. Our understanding of primary and ancillary transmission mechanisms much improved as a result.

Box 4.13 *Continued*

Modifications of the model include evidence not available at the time of the early predictions; in particular,

- although the evidence was clear of a major role of river networks in the early transmission of the disease [480], explicit inclusion of pathogen transport by the fluvial network was not contemplated at first – explicitly addressing hydrologic transport of cholera pathogens was therefore a compelling necessity [428];
- proxies provided by mobile phone data suggest their fundamental role in tracking human mobility and thus the sustained and long-range spread of the disease [483];
- with multiyear epidemiological data, loss of acquired immunity needed to be considered as a factor (i.e., a flux from the pool of recovered back to the compartment of susceptible individuals), with characteristic time estimated in $1 \div 3$ years, depending on the severity of the infection [484];
- there is a significant correlation between seasonal rainfall patterns and the resurgence of the epidemic [466] (Figure 4.2) – assumptions on how increased water contamination relates to rainfall-mediated pathogen overloads were thus needed.

From a technical perspective, to account for all the above mechanisms (see Figure 4.21), Equations (4.27)–(4.29) may be modified as follows:

$$\frac{dS_i}{dt} = \mu(H_i - S_i) - \mathscr{F}_i(t)S_i + \rho R_i,$$

$$\frac{dI_i}{dt} = \mathscr{F}_i(t)S_i - (\gamma + \mu + \alpha)I_i,$$

$$\frac{dR_i}{dt} = \gamma I_i - (\rho + \mu)R_i,$$

$$\frac{dB_i}{dt} = -\mu_B B_i - l\left(B_i - \sum_{j=1}^{n} P_{ji}\frac{W_j}{W_i}B_j\right)$$

$$+ \frac{p}{W_i}\left[1 + \phi J_i(t)\right]\mathscr{G}_i(t). \tag{4.30}$$

A few notes on the added ingredients follow.

Hydrological Transport of Vibrios

As in model (4.29), the evolution of bacterial concentration in the water reservoir is a balance between vibrio excretion/dilution, natural mortality, and spatial dissemination. However, the fourth equation of model (4.30) explicitly accounts for two different pathogen dispersal pathways, respectively, through either hydrological transport over river networks or human mobility (similarly to what we showed in Box 4.5). Pathogens have an intrinsic mobility rate l and move from node i to node j of the hydrological network with probability P_{ij}. For the present scale of analysis, we assume $P_{ij} = 1$ if j is a downstream nearest neighbor of node i, and zero otherwise, thus reducing the transport scheme to a deterministic advection-dominated process. Details on its application to the Haitian context are in the supporting information of [428].

Box 4.13 *Continued*

Human Mobility Patterns

We assume that all individuals (susceptibles, infectives, and recovered) can undertake short-term trips from the communities where they live toward other settlements. Note, however, that only the movement of susceptibles and infectives is important for disease dynamics. In fact, while traveling or commuting, susceptible individuals can be exposed to pathogens and return as infected carriers to the settlement where they usually live. Similarly, infected hosts can disseminate the disease away from their home community. It should be remembered that in many cases, infected individuals are asymptomatic and thus are not barred from their usual activities by the presence of the pathogen in their intestine. Human mobility patterns are defined according to a connection matrix in which individuals leave their original node (say, i) with a probability m, reach their target location (say, j) with a probability Q_{ij}, and then come back to node i. Human mobility is accounted for in the first, second, and fourth equations of model (4.30). In the revised model, in fact, a contact rate

$$\mathscr{F}_i(t) = \beta \left[(1-m)\frac{B_i}{K+B_i} + m \sum_{j=1}^{n} Q_{ij} \frac{B_j}{K+B_j} \right]$$

is defined to account for both local (first term) and mobility-related (second term) infections. $\mathscr{F}_i(t)$ is in fact the total contact rate of the disease, potentially depending on all bacteria present in the system, that is, bacteria that might be uptaken at node i by susceptibles who do not move from their home site or, at each of the other nodes j, by those susceptibles who do move away from it. In the same way, to account for both local and mobility-related pathogen shedding by infected individuals, a total infective pool

$$\mathscr{G}_i(t) = (1-m)I_i + m \sum_{j=1}^{n} Q_{ji} I_j$$

is considered, which describes infected individuals who are actually active at node i at time t.

Topological and transition probability structures Q_{ij} for human mobility networks used in epidemiology are quite varied [481, 482]. They can be based on "attractivity" like in gravity models [105] on the actual transportation network [485] or on theoretical network models. Most reported mobility networks, however, refer to specific settings, possibly empirically derived for developed countries, and should be thoughtfully adapted to different socioeconomic and environmental contexts. Gravity models, in particular, have long been applied in the epidemiological literature to describe the impact of human mobility on the emergence of a suite of diseases [105, 486], including influenza [487], HIV [488], and measles [489].

From a mathematical perspective, a gravity-like model for human mobility can be defined by specifying connection probability as

$$Q_{ij} = \frac{H_j e^{-d_{ij}/D}}{\sum_{k \neq i}^{n} H_k e^{-d_{ik}/D}},$$

where the attractivity factor of node j is subsumed by its size, while the deterrence factor is assumed to be dependent on distance and represented by an exponential kernel (with shape factor D). Other details, especially computational, are in [428].

Box 4.13 *Continued*

Loss of Acquired Immunity

The loss of acquired immunity is explicitly accounted for in the first (susceptible pool) and third (recovered individuals) equations of model (4.30). We assume that immunity from cholera is lost at a rate ρ, thus leading to a replenishment of the susceptible compartment. The parameter ρ is initially set at $1/\rho \approx 3$ [years] [484]. The sensitivity of model results with respect to variations of this parameter is in [428].

Rainfall as a Driver of Increased Water Contamination

To address the seasonal revamping of infections, we assume that the baseline contamination rate p can be increased by rainfall (whose intensity is denoted by $J_i(t)$ in model (4.30)) through a synthetic runoff coefficient ϕ. This is in turn due to extra loads of pathogens brought into the water reservoir by seasonal rainfall, which can wash out large loads of pathogens from open-air defecation sites into waterways through surface runoff. This infection mechanism is speculated to be properly described neither by the mass balance equation of the water reservoir (in times of heavy rainfall, one would reasonably expect the water reservoir to increase in volume, thus decreasing its concentration at constant pathogen loads) nor by empirically augmented exposure probabilities, but rather from added pathogen loads to the current reservoir volume. Within the Haitian context, the clear empirical correlation between weekly rainfall and new reported infection cases – started at the end of May 2011 [466] – leaves little doubt about the origins of the transmission (see, e.g., Figure 4.2). A comparative study on the worth of different modeling approaches is reported in Section 4.6.1. If forecasting of cholera infections needs to extend to multiple seasons, say, to study the probability of extinction of an epidemic, suitable projections of rainfall fields must be generated (see SI in [428] for details).

Hyperinfectivity

A somewhat debated topic in the literature concerns the description of hyperinfective bacterial stages [490]. There exists, in fact, experimental evidence suggesting that the passage of *V. cholerae* through the human intestine increases pathogen infectivity [490, 491]. Increased infectivity reportedly lasts from a few hours up to one day. At later times, bacteria enter a stage characterized by normal infectivity. A mathematical model accounting for the hyperinfective bacterial state has been proposed a few years ago [492], and since then, and despite being subject to some criticisms [493], hyperinfectivity has often been accounted for in cholera modeling exercises, including attempts to describe the Haiti epidemic [468, 469]. The class of model Equation (4.30) can be effortlessly extended so as to include a hyperinfective bacterial stage into the general architecture of cholera transmission (Figure 4.21). To that end, the mathematical description of pathogen dynamics in the water reservoir has to be breaken into two separate differential equations, respectively, for hyperinfective and regular bacteria, that is,

$$\frac{d\mathcal{B}_i}{dt} = -\xi \mathcal{B}_i - l\left(\mathcal{B}_i - \sum_{j=1}^{n} P_{ji} \frac{W_j}{W_i} \mathcal{B}_j\right) + \frac{p}{W_i} \mathcal{G}_i(t)$$

Box 4.13 *Continued*

$$\frac{dB_i}{dt} = \xi \mathscr{B}_i - \mu_B B_i - l\left(B_i - \sum_{j=1}^{n} P_{ji} \frac{W_j}{W_i} B_j\right),$$

where $\mathscr{B}_i$ is the concentration of hyperinfective pathogens in the water reservoir and ξ is the rate at which vibrios lose hyperinfectivity (here we assume $1/\xi = 1$ [day]). The total contact rate $\mathscr{F}_i(t)$ has to be modified accordingly [428].

A Mechanistic Description of Water Contamination

A driver of water contamination that certainly plays a key role in the spread of the disease is rainfall. By extending the basic model (Figure 4.21), we can specifically focus on the assumption that overall water contamination increases during rainfall events owing to extra loads of pathogens brought into the water reservoir by seasonal rainfall. In the fourth equation of model (4.30), it is in fact assumed

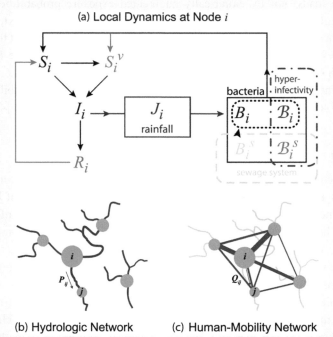

Figure 4.21 Scheme of the nodal SIRW model and of the spatially explicit multilayer network model [428]. (a) Block diagram of the ith site epidemiological model (Box 4.13). (b, c) Schematics of hydrologic (b) and human mobility (c) networks. Human communities of different sizes are assumed to be concentrated in nodes (green circles). Connections between nodes may be due either to hydrological pathways (i.e., river branches), which host active and passive fluvial transport of pathogens, or the displacements of susceptibles or asymptomatic infected individuals (c). Figure after [428]

Box 4.13 *Continued*

that a local rainfall event of intensity $J_i(t)$ directly enhances water contamination through a synthetic runoff coefficient ϕ. A proper review of the various possible mechanisms is provided also in Section 4.6.1.

Let $B_i(t)$ and $B_i^s(t)$ be bacterial concentrations, respectively, in the water reservoir and in the sewage system of node i at time t. The sewage system is intended in its broadest possible acception, including the lack of it thereof, that is, open-air defecation sites. We assume that (1) excreted vibrios actually reach the sewage system and that (2) the pathogens contained in the sewage system can be released to the water reservoir because of leakage/washing out (at a rate that can depend on rainfall intensity). With these hypotheses in mind, we can thus write the following submodel for the balance of *V. cholerae* concentrations:

$$\frac{dB_i^s}{dt} = -\mu_B^s B_i^s + \frac{\epsilon}{W_i^s}\mathscr{G}_i(t) - f(J_i(t))B_i^s$$

$$\frac{dB_i}{dt} = -\mu_B B_i + f(J_i(t))B_i^s - l\left(B_i - \sum_{j=1}^n P_{ji}\frac{W_j}{W_i}B_j\right),$$

where ϵ represents the vibrio excretion rate of infected individuals; μ_B^s is the bacterial mortality within the sewage system; W_i^s is the volume of water stored in the sewage system ($W_i^s = c^s H_i$) [325]; and $f(J_i(t))$ is a function describing the flux of pathogens from the sewage system to the water reservoir owing to a rainfall event at the ith node, characterized by intensity $J_i(t)$. Note that the vibrios contained in the sewage system are assumed not to undergo hydrological dispersal. For the sake of simplicity, we further assume a linear relationship between rainfall intensity and pathogen flux from the sewage system to the water reservoir, that is, $f(J_i(t)) = f_0 + f_1 J_i(t)$, where f_0 and f_1 are positive constants.

We single out here the best-ranked model according to the AIC (see SI of [428]):

$$\frac{dS_i}{dt} = \mu(H_i - S_i) - \mathscr{F}_i(t)S_i + \rho R_i, \tag{4.31}$$

$$\frac{dI_i}{dt} = \mathscr{F}_i(t)S_i - (\gamma + \mu + \alpha)I_i, \tag{4.32}$$

$$\frac{dR_i}{dt} = \gamma I_i - (\rho + \mu)R_i, \tag{4.33}$$

$$\frac{dB_i}{dt} = -\mu_B B_i - l\left(B_i - \sum_{j=1}^n P_{ji}\frac{W_j}{W_i}B_j\right)$$

$$+ \frac{p}{W_i}[1 + \phi J_i(t)]\mathscr{G}_i(t), \tag{4.34}$$

with the above symbol notation. While several model parameters can be estimated from the literature, a few parameters are to be obtained by a suitable calibration procedure contrasting model simulations with the reported cases – in this case, in each Haitian department, as recorded in the

Box 4.13 *Continued*

epidemiological dataset. These parameters are five, namely, the ratio $\theta = p/(Kc)$, the hydrological dispersal rate l, the human mobility rate m, the average deterrence distance D, and the contamination parameter ϕ [428]. The mathematical models are solved numerically on arbitrary spatial arrangements of human settlements and connectivities, usually by Runge–Kutta schemes [428, 494] (Appendix 6.3).

A final note deals with the modifications needed to possibly assess the impact of vaccination campaigns. The model equations need to be adapted [495] to consider vaccinated individuals via SIRBV schemes (Figure 4.22). We assume that vaccination targets individuals independently of their cholera infection history; that is, both susceptible (S) and already immune individuals (R) are eligible with the same probability. Vaccinated individuals already immune are assumed to remain completely immune, while vaccinated susceptibles benefit from a leaky immunity with vaccine efficacy $0 < \eta < 1$, thus reducing the susceptibility by η. Oral cholera vaccine (OCV) is assumed to provide immunity with a delay of one week after administration of a single dose. Owing to the short time horizon of the forecast, no assumption is made for the duration of the vaccine-induced immunity, and individuals remain immune for the duration of this study. At each node i, the number of vaccinated individuals at time t is subdivided into three separate categories in addition to S_i, I_i, and R_i, denoted as vaccinated susceptible individuals, V_i^S; vaccinated individuals that become infected, V_i^I; and vaccinated recovered, which have a complete immunity, V_i^R. The equations are modified accordingly [440, 495], as shown schematically in Figure 4.22, where the number of susceptible individuals is computed as $S_i = H_i - I_i - R_i - V_i^S - V_i^I - V_i^R$. The model assumes a linear ramp-up of vaccine uptake, meaning that the daily number of OCV doses distributed in each community, v_i, is computed by equally deploying the available doses during the days of the campaign. Normally, vaccines are evenly distributed among susceptible and recovered individuals. A constraint $v_i/(S_i + R_i) < 0.9$ is imposed to bound the rate of vaccination when only a few susceptible

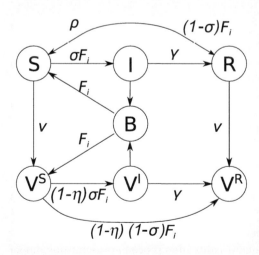

Figure 4.22 Scheme of the SIRBV model [495]. Schematic of the SIRB model at node i with the three additional compartments for vaccinated individuals: vaccinated susceptible individuals (V^S), vaccinated infected (V^I), and vaccinated recovered (V^R). Compartments I and V^I contribute to the local bacterial concentration (B). Compartment S of the ith node is exposed to a force of infection F_i and enter either compartment I at a rate σF_i or R with rate $(1-\sigma)F_i$, respectively. These probabilities are rescaled by a factor $1-\eta$ to account for vaccinated susceptibles. Figure after [495]

Box 4.13 *Continued*

and recovered individuals are remaining, thus avoiding possible convergence issues in the numerical solver. To model the leaky immunity of vaccinated susceptible individuals, the force of infection F_i is decreased by a factor $1 - \eta$, as shown in Figure 4.22. The system of ODEs is modified straightforwardly and is not reported here for brevity (see [495]).

It is of interest that the reference study [428] has also comparatively tested many disease transmission mechanisms. The revised models considered in particular enhanced community-wide transmission due to a hyperinfectious *V. cholerae* state (freshly shed cholera pathogens requiring much lower concentrations to cause infection) [468, 478, 490, 492]. Latent stages and human-to-human transmission have been ruled out by a formal model comparison ([428] and "SI Materials and Methods" therein). Other cofactors of disease transmission, judged of lesser importance, are discussed in the original reference [428].

Parameter calibration is performed via Markov chain Monte Carlo (MCMC) techniques [496, 497]. To compare the ability of different models (with different added complexity and parameters) to reproduce the spatiotemporal epidemic patterns observed in Haiti, performances of different candidate models have been formally ranked according to Akaike's information criterion (AIC) [147]. AIC is a model-selection procedure that explicitly takes into account the trade-off between model accuracy and complexity. The ranking is based on AIC scores that measure the goodness of fit, discounting for the different number of calibration parameters. Four candidate models have been specifically tested [428]: (1) a simple model with no pathogen hyperinfectivity [492] and a single water compartment (Box 4.13), (2) the same model with pathogen hyperinfectivity, (3) a model with two water compartments (water reservoir and sewage system) but no hyperinfectivity, and (4) the same as in (3) but including vibrio hyperinfectivity. AIC scores show that the optimal model is clearly identified. According to the results of model identification, bursts of infection can best be explained by accounting for larger concentrations in the water compartment due to massive pathogen loads brought by hydrologic washout. AIC not retaining the modeling of hyperinfective stages of the *V. cholerae* bacterium may be surprising given its

importance in other approaches. Our result would confirm earlier remarks [498] suggesting that the timescale of hyperinfectivity is so short that all that matters for modeling purposes would be the overall rate of transmission resulting from the many mechanisms that underlie it, especially given the complexity of the spatial linkages involved. This result cannot be generalized, however, as discussed in [428]. In any case, the exercise described herein evokes a general case, one in which research assumptions are tested by formal model comparison.

Figure 4.23 shows the results of the best-ranking model and its estimation uncertainty. Optimal parameters and their credible intervals are reported in [428]. The model can better reproduce the timing and the magnitude of the epidemic in the 10 Haitian departments adopted as a base discretization, including the seasonal June–July resurgence, in particular in the most populated and affected regions (Artibonite and Ouest). The capability to describe the spatial and temporal patterns of the reported infections grew considerably with the information gained after the early predictions, although the short-term prognostic value of the early model [30] remains noteworthy. Thus, different levels of model sophistication might serve well for evolving insight into the course of an ongoing epidemic. In particular, the model including rainfall drivers and waning immunity allows us to draw predictions for long-term cholera dynamics in Haiti. Figure 4.24 shows a multiseason projection up to January 2014, obtained by using suitable rainfall field predictions [428]. We have chosen to show the example with an average duration of the acquired immunity of three years (it must be noted, however, that in such cases, model-guided field studies on the rate of loss of acquired immunity become crucial). The related predictions are fairly consistent in suggesting significant fall bursts of infection, stemming from seasonal rainfall and from the timing of the replenishment of the pool of susceptibles from previous infections due to immunity waning.

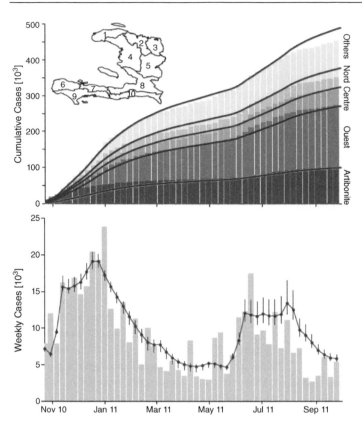

Figure 4.23 Simulated evolution of the Haiti epidemic cholera from October 2010 to September 2011 by the revised complete model that includes bursts of infections caused by pathogen loads brought into the water reservoir by hydrologic washout. The final choice follows from the ranking of the performances of different candidate models according to the Akaike information criterion. (upper) Weekly cumulated reported cases are visualized as the sum of the reported cases in each department (gray bars), fitted by the simulation of the revised model at the department level (blue solid line). The performance of the model at the department level is also shown (blue solid lines). (inset) Haitian departments are listed as follows: 1, Nord-Ouest; 2, Nord; 3, Nord-Est; 4, Artibonite; 5, Centre; 6, Grand-Anse; 7, Nippes; 8, Ouest; 9, Sud; and 10, Sud-Est. (lower) Evolution of reported new weekly cases (gray bars) along with the simulated incidence pattern of the revised model (solid line). Error bars highlight the range of uncertainty due to parameter estimation. Figure after [428]

The underestimation of the predicted infections in late fall 2011 (for which data were then – in 2012 – available) is likely explained by the fact that rainfall patterns projected from the end of September missed the extreme rainfall events that actually occurred in the first decade of October. Note finally that the above projection is a worst-case scenario that assumes no improvement in sanitation and ignores any decrease in exposure upon learning from past experience.

The encouraging outcomes of early predictions of the 2010–2011 Haiti cholera outbreak and the broadened capabilities of a generalized approach based also on later observations suggest a number of implications worth discussing in this context.

Despite their capabilities, several limitations and open issues remain toward a general predictive model of epidemic cholera. One important limitation for long-term predictions is our relatively poor knowledge of community-wide loss rates of acquired immunity (here characterized by a deterministic rate parameter; see Box 4.13). Susceptibles decrease because of infections and

mortality and increase through the loss of immunity of recovered cases. The dynamics of recovered individuals play a crucial role in the long run. Although at short timescales, it is reasonable to assume complete immunity of recovered patients, waning or boosting of acquired immunities at longer time scales may have a substantial impact on epidemic dynamics. Acquired immune responses may vary in relation to age group, *V. cholerae* strain (e.g., serogroup, biotype, and serotype) [499], the severity of the contracted infection [500], and the intertwined dynamics of endemic versus epidemic cholera [501].

On a population level, the buildup of durable immunity usually results in a lengthening of the interepidemic intervals. However, complex cyclic climatic forcings may control interarrivals of cholera outbreaks. Unfortunately, cholera elicits only a temporary immunity whose uncertainty affects the long-term replenishment of the pool of susceptibles. Thus, the fraction of the population susceptible to the disease at the onset of subsequent outbreaks will likely be predictable only with great uncertainty.

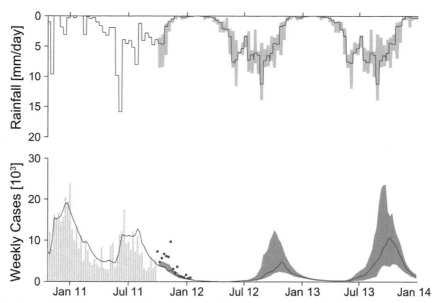

Figure 4.24 Multiseason evolution of Haiti epidemic cholera (from October 2010 to January2014) simulated by the best-performing model. Reported new weekly cases (gray bars) are shown along with the simulated incidence pattern (solid line). (upper) Rainfall is predicted starting from October 1, 2011. The range of uncertainty due to the uncertainty in rainfall forecast is highlighted by the shading. Red dots highlight cases reported after September not used for calibration at the time of the prediction. Note the agreement between the epidemic fading in the data and the model projection, with the exception of an unpredicted infections peak in late fall 2011. This exception is likely explained by the extreme rainfall events that occurred in the first decade of October that were missed by the rainfall patterns projected from the end of September. Figure after [428]

Worst case scenarios (i.e., assuming again the entire Haitian population being susceptible) seem too crude an approximation for predicting a meaningful deployment of suitable medical supplies and staff.

In a similar vein, the minimalist assumption of a constant ratio of asymptomatic to symptomatic infections during the course of the epidemic, to which modeling results prove quite sensitive, is clearly an approximation. Extended epidemiological evidence and relatedly improved modeling are needed. Also, individuals with blood group O are more prone to severe cholera symptoms than other blood group individuals, although the mechanism underlying this association is still under debate [502, 503]. Hence, the distribution of blood groups among the population might need to be considered where significant differences emerge. Besides blood group, moreover, the susceptibility to cholera depends on local intestinal immunity (from previous exposure or vaccination); bacterial load; and intrinsic host factors, such as stomach pH (gastric acid provides a barrier) [504], whose community-wide evaluations are

difficult. One further issue concerns the modeling of the rapid patient discharge from treatment centers. Infected patients are released when fewer than three watery stools over the past six hours are observed. This action is reasonable, given the need for space in emergency hospitals, but poses the problem that released patients are treated as recovered. However, they still excrete significant quantities of vibrios and thus contribute to enhanced spread of the disease, a process that is not accounted for in any of the current models. Although backtracking the mobility of early discharges seems possible on the basis of treatment center acceptance data, a large-scale application seems impractical. A related issue concerns the actual sanitation within treatment centers, for example, the isolation of their latrines from the local water cycle, especially during acute phases of the epidemic. From a modeling viewpoint, all this information is currently combined into calibration parameters at the loss of predictive power.

Another issue concerns the proper depletion of the pool of susceptibles due to intervention strategies. In this context, two main interventions are currently being discussed: vaccination and the extended use of antibiotics [505, 506]. Whereas the idea of mass vaccination in Haiti has split experts [507, 508], the impact of parameter uncertainty on the effects of vaccinations is certainly in need of an assessment. A more uniform opinion concerns the use of antibiotics. With respect to this type of intervention, the World Health Organization generally recommends the administration of antibiotics only to severe cases of cholera [420, 421, 509]. However, soon after the onset of the cholera epidemic in Haiti, many researchers demanded a widening of these regulations toward an extended use of antibiotics [506, 510]. In favor of a broader use of antibiotics for all hospitalized cases, regardless of the severity of the symptoms, is the shorter duration of the acute phase and the reduced shedding of *V. cholerae* bacteria into the environment [511]. The new policies therefore recommend also treating moderately ill patients with antibiotics [474]. Significant differences in the use of antibiotics existed among treatment centers. The most extreme examples were health partners who used antibiotics in a prophylactic manner to protect family/community members affiliated with cholera patients [427, 428]. The most common antibiotic used in Haiti is a single dose of doxycycline, which proved very effective for cholera treatment [504]. However, drastic increases of doxycycline/tetracycline-resistant isolates were reported, referring to a cholera outbreak in Zambia [512] where a clear correlation was observed between the development of resistant strains and large-scale use of antibiotics (either for treatment or for prophylaxis). In fact, many drug-resistant strains of *V. cholerae* have been described within recent years due to the spread of antibiotic resistance genes by horizontal gene transfer (HGT) (see, e.g., [513]). That antibiotics themselves, as well as signals within *V. cholerae*'s environmental niche, induce diverse mechanisms of HGT [514–518] strengthens the fear of the development of antibiotic resistance in Haiti. It is therefore crucial to monitor the resistance pattern of the still circulating *V. cholerae* strains. Regardless of deeper questions about possible development of specific resistance by the bacterial strains, we note that only very detailed information, possibly well beyond reach, about clinical practice in space and time would allow us to better frame the related parameters of the transmission model.

Many significant results applied to the specific understanding of the Haiti iconic epidemic prompted a substantial improvement in the way spatially explicit epidemiology makes its way into public health practice. One important result concerns real-time projections of cholera outbreaks through data assimilation and rainfall forecasting [440]. Real-time forecasting frameworks that readily integrate new information as soon as it is available and periodically issue an updated forecast have obvious relevance and follow a well-worn path followed by oceanographers and meteorologists. The framework presented two major innovations for cholera modeling: the use of a data assimilation technique, specifically an ensemble Kalman filter, to update both state variables and parameters based on the observations and the use of rainfall forecasts to force the model. The exercise of simulating the state of the system and the predictive capabilities of the novel tools, set at the initial phase of the 2010 Haitian cholera outbreak using only information that was available at that time, served as a key benchmark. Results suggest that the assimilation procedure with the sequential update of the parameters outperforms traditional calibration schemes based on MCMC constructs. Formal ranking of models contrasting Haiti epidemic data has also been carried out [435], suggesting that spatially connected models have generally better predictive ability than disconnected ones. It is a commonsense solution as well, as the parameters of the disconnected models must include information that the spatial schemes embed directly.

We may safely conclude, therefore, that in a forecasting mode, properly assimilated spatially explicit models usefully predict the spatial incidence of cholera at least one month ahead, thus providing significant lead time to set up the deployment of medical supplies and staff and to evaluate alternative strategies of emergency management. For instance, a by-product of the confidence in modeling predictions concerns the implications for the course of large-scale cholera epidemics of an extended use of rice-based, as opposed to glucose-based, oral rehydration therapy [519].

Other applications are of great potential interest. Big data and, in particular, mobile phone data are expected to revolutionize epidemiology, yet their full potential is still untapped. A significant step forward was taken by developing (cholera) epidemiological models that account for the spatiotemporal patterns of human mobility derived by directly tracking properly anonymized mobile phone users [483, 520]. Such data allow us to investigate, with an unprecedented level of detail, the effect that mass gatherings can have on the spread of waterborne diseases like cholera. Identifying and understanding transmission hotspots opens the way to the implementation of novel disease control strategies whose review, within general epidemiological models, is beyond the scope of the book. To this issue we have dedicated Section 4.3.4 and a few final considerations in our outlook (Chapter 5).

Finally, spatially explicit WB disease infection models have also radically changed the very concept of basic reproduction number R_0, as pointed out in the previous section. Indeed, challenging problems arise when observed infection patterns show spatial structure and/or temporal asynchrony. These features are ironed out by spatially implicit schemes that ignore spatial effects by assuming a uniform distribution of susceptible and infected individuals across the domain of interest. Because of the current ease of mapping hydrology, sanitation, and transportation infrastructures, in addition to the actual population distributions and proxies of their WASH conditions (a by-product of remotely acquired and objectively manipulated information), the conditions leading to waterborne disease outbreaks can be studied in a spatially explicit framework. However, the adoption of spatially explicit schemes is only recent and mainly with reference to cholera [437, 447, 451]. We expect, however, that similar approaches will follow soon for other diseases, because the underlying mathematical framework is general [464, 521].

4.3.3 On the Probability of Extinction of the Haiti Cholera Epidemic

As of 2018, Haiti was unlikely to be cholera-free [495]. Therefore, refined versions of the models described in the previous sections were used to forecast the future risk and to evaluate the relative worth of alternative control strategies, although the last reported cholera case was recorded in January 2019 (cfr https://news.un.org/en/story/2020/01/1056021). However, understanding disease transmission during the lull phases that characterize the dry season is of crucial importance to understanding extinction dynamics, and a deterministic approach may fall short of reliability expectations. In fact, the particular case definition used by the national cholera surveillance system, which differed from the standard adopted by WHO [522], could have actually led to an overestimation of the cholera burden during these phases. Haitian reports

also include children under the age of five, who are excluded by WHO standards because of the high prevalence of acute diarrhea caused by infections other than cholera in this age group [494]. Diarrhea in adults is also likely to be reported as cholera independently of its actual origins because laboratory confirmation could not be performed for every single patient due to the huge number of cases. Rebaudet et al. [473] estimated a background noise of diarrhea cases not related to cholera of about 250 per week. This noise source is not so relevant when analyzing outbreak peaks, but it may become important during lull phases of the epidemic, which are seen as true windows of opportunity toward cholera eradication [473, 494]. We thus need reliable tools to predict the evolution of disease transmission when a limited number of new infections brings massive uncertainty into deterministic predictions. Stochastic models are needed in this context.

Yet deterministic models are fundamental when the background signal of reported infections is large, typically at the onset of an outbreak within a context where most of the exposed individuals are susceptible and the population is large. For instance, years after its appearance in Haiti, as seen in Section 4.2, cholera caused directly more than 8,500 deaths and more than 800,000 infections. Revamped by hurricanes and tropical storms, it is feared that cholera has developed into an endemic condition. How this happened is not completely clear. In fact, no definitive evidence exists for a stable environmental reservoir of pathogenic *V. cholerae*, suggesting the possibility that the transmission cycle of the disease is largely maintained by bacteria freshly shed by infected individuals [446, 494]. Should this be the case, cholera could in principle be eradicated from Haiti. A framework for the estimation of the probability of extinction of the epidemic based on current information on epidemiological dynamics and health care practice is thus an interesting exercise for the scope of this book, because – as we shall see in this section – a spatially explicit stochastic framework needs be set up in parallel to (and analog of) the deterministic one described in Box 4.13. Cholera spreading will thus be modeled by an individual-based, spatially explicit stochastic scheme that accounts for the dynamics of susceptible, infected, and recovered individuals hosted in different local communities connected through hydrologic and human mobility networks.

As noted in our brief introduction to cholera, the bacterium colonizes the human intestine, but it can also survive outside the human host in the aquatic environment. Moreover, *V. cholerae* can be a natural member of the aquatic microbial community in certain regions of the world where cholera is endemic [477, 523, 524]. A few studies in Haiti have screened environmental samples for toxigenic *V. cholerae*. One such study, conducted in the final months of 2010 when the epidemic was peaking, found the strain in only five samples out of eighteen in running freshwater [525]. Other environmental studies were staged after the peak of the epidemic. The absence of clear field evidence of an established, thriving environmental community of toxigenic *V. cholerae* led some authors to conclude that the transmission cycle of the disease is maintained by bacteria freshly shed by infected individuals [473] and thus that the disease could be targeted for eradication.

In this context, a framework for estimating the probability that the variability of the dry season, together with the inherent demographic stochasticity of disease transmission, led to the extinction of the epidemic outbreak was put together [494]. The analysis postulates that an environmental reservoir of *V. cholerae* cannot self-sustain indefinitely without input from infected individuals.

Most of the models described previously for the Haiti outbreak approximate the number of infected and susceptible individuals with continuous variables, and for good reason: this is an assumption often made in mathematical epidemiology when the number of individuals in each class is large enough to neglect the demographic stochasticity emerging from the intrinsic discrete nature of individual processes. Continuous models are computationally efficient and are also suitable to be calibrated by contrasting simulations and epidemiological records. Prior to [494], a notable exception has been the agent-based model proposed by Chao et al. [468], which, however, was not calibrated against field and epidemiological data but generated scenarios based on parameter values taken from literature. Other possible approaches to modeling stochasticity in disease transmission are the use of Langevin-type differential equations [436, 526] and of spatiotemporal random fields accounting for multiple sources of uncertainty [527, 528].

The tools of the deterministic trade are those outlined in Sections 4.13 and 4.2.2 (see Box 4.13). This is a particularly important stage of the work for the possibility of calibrating during outbreaks (endowed with high figures of attack rates) by contrasting epidemiological, hydrological, and GIS data, the parameters that feature in both

the deterministic and the stochastic models. The parallel stochastic formulation, necessary to deal with the low number of infected individuals characteristic of endemic infection patterns, is discussed in Box 4.14. The deterministic and the stochastic models can be intertwined. In fact, the deterministic formulation presented in Box 4.13 is used to calibrate the model parameters. Specifically, use is made [494] of the data relative to the first phase of the epidemic, from November 2010 to the end of 2012, truly used as a training set. During this period, the number of infected individuals is large and can thus be reasonably approximated by a continuous variable (the suitability of this hypothesis has been thoroughly verified). Simulations of the deterministic formulation are $O(10^3)$ times faster than the stochastic ones, thus allowing the use of efficient iterative calibration schemes (see [494] for a complete description of the parameters identified). Note that by introducing a dimensionless bacterial concentration $B_i^* = B_i / K$, it is possible to group three model parameters (namely, the rate p at which bacteria excreted by one infected individual reach and contaminate the local water reservoir, the per capita volume of water reservoir c, and the half-saturation constant K) into a single dimensionless parameter $\theta = p/(cK)$, which is one of the calibrated parameters. The ratio θ epitomizes all the parameters related to contamination and sanitation – the higher the value of θ is, the worse become the sanitation conditions and the resulting contamination of the environment.

As initial conditions for model simulations, it is assumed that, as of October 2010 ($t = 0$), the values of $I_i(0)$ match for all nodes i the reported cases detailed in Piarroux et al. [480]. Also, (1) the initial bacterial concentration in the ith water reservoir, $B_i(0)$, is assumed to be in equilibrium with the local number of infected cases, that is, $B_i^*(0) = \theta I_i(0)/(H_i \mu_B)$, and (2) the whole population is assumed to be susceptible at the beginning of the epidemic, that is, $S_i(0) = H_i - I_i(0)$, because of the documented lack of any preexisting immunity [480, 505].

Suitable calibration approaches are usually based on MCMC sampling. Specifically, Bertuzzo et al. [494] used the DREAM algorithm (Differential Evolution Adaptive Metropolis, available online at http://jasper .eng.uci.edu/software.html) [497], a popular and efficient implementation of MCMC that runs multiple different chains simultaneously to ensure global exploration of the parameter space and adaptively tunes the scale and

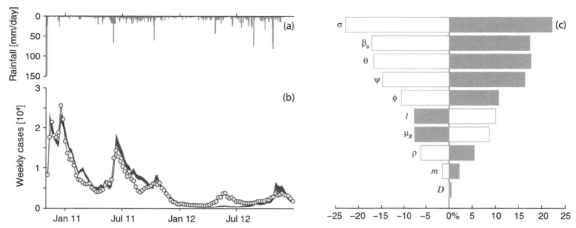

Figure 4.25 Calibration of the continuous deterministic model. (a) Time series of mean daily rainfall averaged over the Haitian territory. (b) New weekly cases as reported in the epidemiological records (gray circles) and simulated by the model (blue line). The blue shaded area shows the 5th–95th percentile bounds of the uncertainty related to parameter estimation. (c) Effects of parameter variations. Variations are shown in total cholera incidence during the calibration phase (October 2010–December 2012) produced by ±20 percent variations of the calibrated parameters. Shaded or open bars represent, respectively, positive or negative variations of the relevant parameters. Figure after [494]

orientation of the jumping distribution using differential evolution and a Metropolis-Hastings update step [494]. The DREAMZS variant of the DREAM algorithm [529] may also be conveniently employed. The algorithm is initialized with broad flat prior distributions [494] for parameter values and is allowed to run up to convergence ($O(10^5)$) iterations). The goodness of each single simulation is computed as the residual sum of squares between weekly reported cholera cases in each of the Haitian departments as recorded in the epidemiological dataset and simulated by the model [494]. The corresponding fit is illustrated in Figure 4.25, where results are aggregated at the country level (but notice that the calibration was performed by simultaneously fitting data at a higher spatial resolution – the departmental level [494], 1,130 data points, RSS = 2.41×10^8, Nash–Sutcliffe index = 0.79). The deterministic model shows a good agreement with the outbreak data, in particular by capturing the timing and the magnitude of the peaks and the response to heavy rainfall events that promote a seasonal recrudescence of the epidemic.

Calibration results indicate a likely lifetime of bacteria in the environment lasting about one month [494]. Similar values have previously been obtained through calibration for the same case study [467] based on clinical evidence. The fitted value of the symptomatic ratio is close to the qualitative estimate of the World Health Organization [530], according to which only 25 percent of infected individuals reportedly show symptoms – and among these, just around 20 percent develop profuse diarrhea requiring medical attention.

The estimation of immunity loss rate ρ is a weak point of all models so far because of the near-impossibility of large-scale field trials that would require blood samples taken. The calibrated value found by contrasting reported infections corresponds to an average immunity duration of less than one year, which is shorter than the multiyear estimates usually reported in the literature [484, 530]. Notice, however, that these estimates refer to the duration of the immunity conferred by symptomatic infections, while in our framework, the mean immunity $1/\rho$ refers to the average duration resulting from both symptomatic and asymptomatic infections. The latter are indeed expected to confer shorter protection, for example, a few months according to King et al. [478]. To disentangle the impact of each parameter on the dynamics of the epidemic outbreak, a sensitivity analysis of the model outcomes with respect to variations of the parameter values is appropriate. In particular, parameters were allowed to vary (20 percent variation with respect to the best-fit parameter set) one by one through repeated model runs. Variations of simulated total cholera incidence in the calibration phase are compared to the best-fit simulation (Figure 4.25c).

Box 4.14 On Stochastic Cholera Models

In the stochastic formulation of the spatially explicit cholera model, the population of each node is assumed to be made up by identical individuals classified according to their epidemiological status [494]. Accordingly, the numbers of susceptible, symptomatic, infected, and recovered individuals hosted in node i at time t are discrete stochastic variables denoted as S_i^r, I_i^r, and R_i^r, respectively. The superscript r is used to differentiate random variables from their deterministic analogs S_i, I_i, and R_i in Box 4.13. The concentration of *V. cholerae* at site i, $\mathscr{B}_i$, is modeled as a continuous stochastic variable instead because the number of bacteria is expected to be large enough to allow a continuous representation. Time is a continuous variable. The state of the system is described by the vector $(\mathbf{S}^r, \mathbf{I}^r, \mathbf{R}^r)$, where $\mathbf{S}^r = \left(S_1^r, S_2^r, \dots, S_n^r\right)^T$ and the other vectors are defined analogously. All events involving human individuals (births, deaths, and changes of epidemiological status) are treated as stochastic events that occur at rates that depend on the state of the system. Possible events are (1) birth; (2) death of a susceptible individual; (3) symptomatic infection; (4) asymptomatic infection; (5) death of a symptomatic infected individual for causes other than cholera; (6) cholera-induced death of a symptomatic infected individual; (7) recovery of a symptomatic infected individual; (8) death of a recovered individual; and (9) immunity loss of a recovered individual.

Transitions and rates of occurrence of all possible events in the arbitrary node i can be defined by the sequence of events, transitions, and rates, defined as follows:

$$
\begin{aligned}
\text{birth} \quad & (S_i^r, I_i^r, R_i^r) \rightarrow (S_i^r + 1, I_i^r, R_i^r) \quad \text{with rate } v_i^1 = \mu H_i, \\
\text{susceptible death} \quad & (S_i^r, I_i^r, R_i^r) \rightarrow (S_i^r - 1, I_i^r, R_i^r) \quad \text{with rate } v_i^2 = \mu S_i, \\
\text{symptomatic infection} \quad & (S_i^r, I_i^r, R_i^r) \rightarrow (S_i^r - 1, I_i^r, +1, R_i^r) \quad \text{with rate } v_i^3 = \sigma \mathscr{F}_i S_i, \\
\text{asymptomatic infection} \quad & (S_i^r, I_i^r, R_i^r) \rightarrow (S_i^r - 1, I_i^r, R_i^r + 1) \quad \text{with rate } v_i^4 = (1-\sigma)\mathscr{F}_i S_i, \\
\text{death of infected} \quad & (S_i^r, I_i^r, R_i^r) \rightarrow (S_i^r, I_i^r, -1, R_i^r) \quad \text{with rate } v_i^5 = \alpha I_i, \\
\text{cholera death} \quad & (S_i^r, I_i^r, R_i^r) \rightarrow (S_i^r, I_i^r, -1, R_i^r) \quad \text{with rate } v_i^6 = \mu I_i, \\
\text{recovery of infected} \quad & (S_i^r, I_i^r, R_i^r) \rightarrow (S_i^r, I_i^r, -1, R_i^r + 1) \quad \text{with rate } v_i^7 = \gamma I_i, \\
\text{death of recovered} \quad & (S_i^r, I_i^r, R_i^r) \rightarrow (S_i^r, I_i^r, R_i^r - 1) \quad \text{with rate } v_i^8 = \mu R_i, \\
\text{loss of immunity} \quad & (S_i^r, I_i^r, R_i^r) \rightarrow (S_i^r + 1, I_i^r, R_i^r - 1) \quad \text{with rate } v_i^9 = \rho R_i,
\end{aligned}
$$

$$(4.35)$$

with symbol notation as in Box 4.13.

The stochastic model outlined above defies analytical solutions in terms of the probability distributions of the state variables (or their moments thereof). To investigate the properties of the system, Bertuzzo et al. [494] resorted to a Monte Carlo approach to simulate many different trajectories (realizations) of the process with a stochastic simulator algorithm (SSA) [531]. The SSA assumes that event occurrence is a Poisson process whose rate v is defined by summing the rates of occurrence of all possible events:

$$
v = \sum_{k=1}^{9} \sum_{i=1}^{n} v_i^k.
$$

$$(4.36)$$

At every time t, v represents the rate at which the next event is expected to occur. Therefore, the interarrival time between two subsequent events is an exponentially distributed random variable with

Box 4.14 *Continued*

mean v [229, 531]. The type of event that will occur is randomly selected among all possible events. Specifically, the probability of selecting event k in node i is equal to v_i^k/v. The state of the system is then updated according to the randomly selected event. The SSA is iterated until the expiration of the simulation horizon.

To estimate the probability of extinction of the epidemic, several realizations of the stochastic formulation of the model have been run (specifically, from October 20, 2010, to December 31, 2017). To project the trajectory of the epidemic in the future, rainfall scenarios had to be assumed. They were generated starting from the observed daily fields of precipitation estimates (15-year dataset, 1998–2012): each month (say, a year's May) of the rainfall time series used to force the model from January 2013 to December 2017 is obtained by randomly selecting (with replacement) among all the corresponding months (all the Mays) available. As a result, each sequence of generated rainfall is a standard bootstrapping of the observed data. This procedure, standard in hydrologic analysis, allows us to generate realistic space-time correlated rainfall fields [494].

Differently from a standard stochastic SIR model, the absence of infected individuals ($\mathbf{I} = 0$) is not an absorbing state of the dynamics of the system. In fact, the presence of bacteria that can survive outside the human host may allow for new infections even in the absence of other infected individuals. Thus Bertuzzo et al. [494] classified a trajectory as extinct when, from that time on, no new infections are observed in the simulated timespan. Summarizing, to estimate the probability of extinction, one needs to (1) sample a parameter set from the posterior distribution obtained in the calibration phase; (2) generate a rainfall scenario, as hydrologic drivers are known to very much affect reinfection patterns [427, 428]; and (3) simulate the epidemic using the stochastic formulation of the model. The previous points are iterated for $O(10^3)$ times, and the relative proportion of trajectories that go extinct is recorded, thus providing the probability of extinction.

Figure 4.26 shows the projection of the future course of the epidemic obtained using the stochastic formulation of the model forced by the generated rainfall scenarios. Data on incidence during 2013, which became available after the implementation of the model of Bertuzzo et al. [494], were also reported for comparison with the model forecast. Projected future patterns suggested that the

annual epidemic cycle was expected to show a lull phase during the dry winter months, followed by an increased incidence caused by spring rainfalls. The epidemic was expected to peak each year during autumn tropical rainfalls. This yearly cycle is progressively attenuated by the decreasing population exposure resulting from the bulk of the public health interventions. On average, the exposure rate is estimated to be reduced by 20, 50, and 55 percent at the end of 2010, 2011, and 2012, respectively. At the end of the simulated period, the exposure rate is, on average, approximatively 30 percent of its preepidemic value. The median incidence values simulated by the deterministic and stochastic formulations of the model during the calibration phase, are almost indistinguishable [494], thus supporting the initial hypothesis that during the calibration phase, the population of infected individuals was indeed large enough to neglect demographic stochasticity and to approximate state variables with continuous ones. Figure 4.26c shows that the probability that the epidemic would go extinct before the end of 2016 and 2017 was estimated to be 1 and 7 percent, respectively.

Before discussing the actual estimates, which are not exempt from uncertainties and limitations, it is worth highlighting the strength of the method by [494]. A combination of deterministic and stochastic approaches keeps the advantages of both while avoiding their drawbacks. Specifically, the deterministic framework allows for an efficient and reliable calibration of the model parameters, which are directly incorporated into the stochastic approach. The stochastic framework allows us to evaluate the probability of extinction of the epidemic exploring the variability induced by parameter uncertainty, demographic stochasticity, and rainfall fluctuations.

The results presented in Figure 4.25 show that the model is able to reproduce complex spatiotemporal patterns with a so far unseen level of detail, at least for the Haiti cholera epidemic. The sensitivity analysis (Figure 4.25c) shows that the model results in the calibration phase are robust with respect to parameter variations. Indeed, variations of the model output are

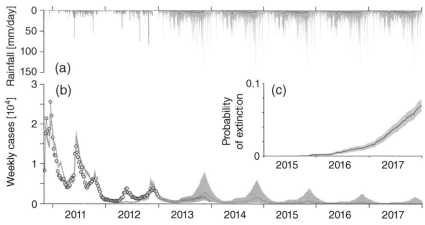

Figure 4.26 Projection of the future course of the epidemic obtained using the individual-based stochastic model. (a) Recorded rainfall patterns (gray bars) and median (green line) and 5th–95th percentile bounds (green shaded area) of the generated rainfall scenarios. (b) New weekly cholera cases as reported in the epidemiological records (gray and red circles) and simulated by the model. The green line and the green shaded area represent the median and the 5th–95th percentile bounds of the simulated trajectories, respectively. (c) Probability that the epidemic goes extinct before a certain time horizon (black line). The gray shaded area represents the 5th–95th percentile of the uncertainty estimated through a standard bootstrapping (random sampling with replacement) of the simulated trajectories. Figure after [494]

comparable to (or smaller than) the parameter variations. It is interesting to note the role of hydrologic transport, l. A positive variation of the rate at which bacteria are flushed away and lost into the ocean reduces disease incidence. The analysis also shows that the total incidence is not particularly sensitive to variations of the parameters controlling human mobility (m and D). This can be explained by the fact that human mobility plays a crucial role at the onset of an outbreak when the infection invades disease-free regions, whereas it may become less important looking at longer time horizons (27 months in this case), when the disease has already invaded the whole country and epidemic dynamics are mostly controlled by local factors.

The comparison with the incidence data of 2013 (Figure 4.26a), which were not used for calibration, provides the most compelling test for the predictive ability of the model. The pattern that actually occurred (a lull phase during winter followed by a slow increase during summer that finally peaks in autumn) was satisfactorily anticipated by our model forecast. The results presented in [494] strongly support the use of mathematical models for real-time prediction of the evolution of a cholera outbreak, yet areas where improvements are needed still remain.

For instance, often, models tend to underestimate late epidemic peaks revamping early outbreaks most often

due to heavy rainfall, with the associated washout of pathogens that accumulate in the environmental waters. In the Haitian case, one can notice, by analyzing the epidemiological records in detail, that the secondary peak was mostly localized in the capital Port-au-Prince. The comparative mismatch of the ability of the models to represent it could be caused by an erroneous estimation of rainfall intensity. However, a possible source of such mismatch could be local outbreaks caused by environmental or social factors not directly (or not entirely) linked to rainfall and therefore not accounted for in the deterministic model. Clearly, modifying the force of infection in both the deterministic and the stochastic models is easy in principle and hard to properly calibrate in view of the combined effects of the various sources of uncertainty in the processes and in the reference epidemiological data. Analysis of epidemiological reports at the much more detailed communal level [427] could be a possible avenue to reducing the structural errors of the models. Such data could shed light on specific transmission processes that are ironed out at the departmental level.

Despite the notable uncertainties, robust estimates indicated that the extinction of the epidemic in Haiti was a rather unlikely event by the year 2016 [494]. This result quantitatively confirmed field experts' feelings that cholera had not burned itself out by that time, nor had it a couple of years later [495] (see Figure 4.27 for an

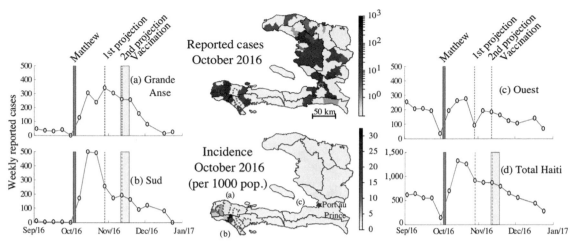

Figure 4.27 Reported cholera cases in Haiti after the passage of Hurricane Matthew (2016). Maps of the reported cholera cases and the associated incidence during October 2016 at the communal level with the detailed weekly dynamics for the departments most affected after Matthew – (a) Grande Anse, (b) Sud, and (c) Ouest – an (d) for the whole country. Figure after [495]

update on more recent data). The results described in Figure 4.27 indicate that the probability of the epidemic going extinct before the end of 2016 – computed two years ahead of time in the reference paper [494] – was of the order of 1 percent, as obviously confirmed by later observation. Such low probability of extinction was argued to highlight the need for more targeted and effective interventions to possibly stop cholera in Haiti [494]. Because this actually occurred [440], the same calibration could not be used for longer projections. However, the method is quite valuable for such types of investigation, as the low numbers of infected individuals approaching eradication prevent a continuous approximation suited to the quasi-determinism shown by large outbreaks that treat the state variables as continuous owing to the large numbers involved.

Spatially explicit mathematical modeling of epidemic cholera proves an increasingly useful tool to mainstream epidemiological practice, especially when coupled to assimilation strategies to consider the available epidemiological and environmental data. Quantifying the risk associated with each department, and possibly even communes, and evaluating alternative emergency management are now standard requirements. Significantly, recent work [495] suggests that spatially explicit models of the type predicated in this chapter are capable of assessing risk by ranking the implementation of alternative vaccination campaigns in terms of, say, the numbers of averted cholera cases.

As expected, the predicted impact was larger in the scenarios more favorable to cholera spread. Moreover, sequential assimilation of epidemiological data is fundamental to updating the model in near-real-time and tracking the epidemic dynamics. Outbreak surveillance systems are therefore key to optimizing cholera modeling and controlling interventions, and supporting high-quality surveillance and data sharing among partners is an important part of both short- and long-term strategies to improve the cholera prevention and control efforts [495]. Data sharing systems that allow near-real-time modeling work should be put in place to ensure the usefulness and timeliness of the modeling outputs as an additional element supporting the decision-making process [495].

Improving our understanding of the transmission dynamics during these periods would help in designing specific intervention strategies whose effects can be estimated by the proposed modeling framework. A real-time assessment not only of the future course of an outbreak but also of the effects of alternative intervention strategies is the next important development on which modern mathematical epidemiology should focus to become an essential tool for emergency management. It is hoped that the proposed approach will pave the way for future modeling-based recommendations to inform public health interventions in response to cholera outbreaks, helping to achieve the objective of reducing by 90 percent the number of deaths due to cholera by 2030 [532].

4.3.4 Mobile Phone Data, Tracking of Human Mobility, and the Spread of Infection

Big data and, in particular, mobile phone data are expected to revolutionize epidemiology, yet their full potential is still untapped. Here, we discuss, following [520], how an epidemiological model for epidemic cholera may account for the spatiotemporal patterns of human mobility derived by directly tracking properly anonymized mobile phone users. Such data allow us to investigate, with an unprecedented level of detail, patterns of human mobility that no traditional approach (like the gravity [105] or the radiation [533] models) would be capable of reproducing, for example, the effect that religious mass gatherings can have on the spread of waterborne diseases like cholera [520]. Identifying and understanding transmission hotspots generated by patterns of mobility of susceptibles or infected individuals open the way to implementing novel disease control strategies. In fact, the spatiotemporal evolution of human mobility and the related fluctuations of population density are known to be key drivers of the dynamics of infectious disease outbreaks. Understanding these dynamics, however, is usually limited by the lack of accurate data, especially in developing countries. Mobile phone call data provide a new, first-order source of information that allows tracking the evolution of mobility fluxes with high resolution in space and time [483] (although this paper used human fluxes as a proxy of infection patterns in what seems like a notable underexploitation of such data).

The rationale for tracking people's movements from mobile calls is as follows. Each time a phone emits or receives a call or text message, the antenna that the cell phone is logged into is registered by the service provider, along with the time of the event [534]. It is thus possible to track the movement of cell phone users as they advance from antenna to antenna. Suitably aggregated and properly anonymized to prevent privacy issues, a sample of these data can be used to estimate fluxes of people between areas in a region by assigning a set of antennas to each geographical area in the study domain. The resolution in time can be as high as the typical frequency of calls allows, whereas the spatial resolution is limited only by the typical distance between two antennas [534]. Using mobile phone records of a sufficiently large number of users, one can thus estimate human mobility fluxes with high accuracy, including spatiotemporal variability across a variety of scales [520] and without resorting to any particular model [483]. A number of recent studies focused on the use of mobile phone data to extract human mobility patterns in developing countries

at different scales in space and time [483, 535, 536]. Others compared the movement patterns extracted from mobile phone records to traditional data sources, such as censuses and surveys. Several studies dealt with the comparison with human mobility models (which we shall not review here). In the context of infectious disease spread in developing countries, this new source of information enables previously unseen kinds of analyses. Examples are the derivation of magnitude and destination of population fluxes following a sudden outbreak [537] and the quantification of the importance of human mobility and its seasonal variations on the spread of disease in terms of increased outbreak risk in and infectious pressure on connected areas [538–540].

With such premises, Finger et al. [520] have studied the cholera epidemic that spread throughout Senegal in 2005. A distinctive feature of this outbreak was its sudden flare. It started from the order of magnitude of hundreds of cases per week during the first three months of the year, localized in the region of Diourbel and surroundings, and abruptly jumped to thousands of cases at the end of March, rapidly spreading to 10 out of 11 regions of the country, with more than 27,000 reported cases [520]. Anecdotal evidence suggested that this first recorded peak of infections was related to a religious pilgrimage, the Grand Magal de Touba (GMdT), that took place in late March, when an estimated 3 million pilgrims traveled to Touba in the region of Diourbel. During later stages, the outbreak evolved, showing distinct dynamics in different regions of the country, rainfall and the associated floods being important drivers, especially in the capital city of Dakar. A spatially explicit, fully mechanistic model for the 2005 Senegal cholera outbreak, based on the formulation shown in Boxes 4.13 and 4.15, was then put together [520] where, in addition to human mobility, rainfall was considered as an important driver of disease transmission [25, 441]. No hydrologic connectivity yielded significant pathogen transport, so that flux was set to zero. The effect of overcrowding was considered by assuming an increase in exposure and contamination rates caused by an unusually high density of people and the related pressure on water and sanitation infrastructures (see Materials and Methods in [520]). Daily population fluxes between the 123 arrondissements of Senegal were estimated from a dataset of roughly 150,000 randomly selected mobile phone users tracked during the entire year of 2013, which is assumed as representative of year 2005 owing to the yearly frequency and the recurrent size of the mass gatherings [520]. The specific aim was testing the role played by human mobility and mass gatherings in the spread of a cholera epidemic.

Box 4.15 Human Mobility Matrices

The population is assumed to be in demographic equilibrium, with per capita birth and natural death rate μ. Equations of different nodes are coupled via a modified time-variant human mobility matrix, termed $M_{ij}(t)$ in [520], which is derived from the matrix Q_{ij} in Box 4.13, estimated directly from mobile phone data. To account for a possible underestimation of the number of people staying at their home nodes due to, for example, bias in mobile phone ownership, a calibration parameter c was introduced that relates the two matrices, as follows:

$$M_{ii}(t) = c\, Q_{ii}(t), \tag{4.37}$$

$$M_{ij}(t) = c_i'(t)\, Q_{ij}(t), \quad j \neq i, \tag{4.38}$$

$$c_i' = \frac{1 - c\, Q_{ii}(t)}{\sum_{k \neq i} Q_{ik}(t)}, \tag{4.39}$$

where $c_i'(t)$ ensures compatibility, that is, all rows add to 1.

Figure 4.28a shows the evolution of the estimated number of mobile people (i.e., people having left their home arrondissement on a given day) throughout the year 2013. Seasonal fluctuations, weekly patterns, and sudden peaks can be clearly identified. The latter correspond to mass gatherings, most notably the GMdT (which took place twice in 2013 [520]), and during which the number of people traveling outside their home arrondissement almost doubles with respect to an average day. Figure 4.28b shows the estimated fraction of people present in every arrondissement of Senegal during the GMdT. Major differences can be noted with respect to the yearly average (Figure 4.28c). People traveled to Touba, in fact, from all over the country, and the estimated number of people present during the GMdT in the arrondissement where the city is located was nearly 6 times its usual population.

Model results and estimated uncertainties of the best-performing model are shown in Figure 4.29 (total cases and the regions most severely hit). The values of the calibrated parameters are reported in [520]. The model accurately reproduces the important peak of cases in Diourbel coinciding with the GMdT (coefficient of determination between modeled and reported weekly cases $R^2 = 0.78$ in the region of Diourbel) as well as the spread of the disease throughout Senegal by pilgrims returning to their homes. The second peak, most probably related to the rainy season, is also well reproduced ($R^2 = 0.72$ in the region of Dakar). The overall value of R^2, computed

using all data points in all regions, is equal to 0.77 [520]. Figure 4.30 shows the spatial distribution of reported cases in the country during the GMdT and during two other key periods of the outbreak (the corresponding fit of the best-performing models is shown in [520]). A comparison of different models therein shows that the models, including both human mobility fluxes between arrondissements and the effect of overcrowding, outperform the other models. Including either of the two mechanisms individually was not sufficient to reproduce all features of the epidemic correctly, and a model adopting a calibrated gravity model performs poorly compared with models using mobile phone data to estimate human mobility. The inclusion of rainfall as a driver of the disease enabled the capture of the autumn peak in addition to the one related to the GMdT [520]. Both the correction of bias in mobile phone ownership and the calibration of the initial number of infected in Diourbel improved the model performance – with implications for general applications.

Potential effects of localized interventions in Touba during the GMdT, such as improving sanitation and access to clean drinking water, have been carried out. Under the assumptions of the model, these actions could have led to considerably lower numbers of new cases during the pilgrimage as well as all over the country during later stages of the outbreak [520]. For instance, a reduction of the rates of exposure and contamination by 10 percent (20 percent) in Touba during the GMdT could

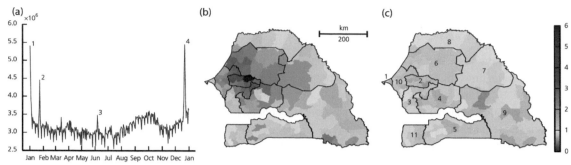

Figure 4.28 (a) Daily evolution of the total number of moving people (i.e., people leaving their home arrondissement) throughout 2013 estimated from mobile phone records. Numbered peaks correspond to the following mass gatherings: GMdT (1 and 4), Gamou de Tivaouane (2), and Magal de Kazu Rajab (3). (b, c) Number of people present in each arrondissement on December 22, 2013, (b) during the GMdT and (c) averaged over the year (c) divided by the number of people living there. Regions (according to the 2005 subdivision) are numbered as follows: Dakar (1), Diourbel (2), Fatick (3), Kaolack (4), Kolda (5), Louga (6), Matam (7), Saint-Louis (8), Tambacounda (9), Thiés (10), and Ziguinchor (11). Figure after [520]

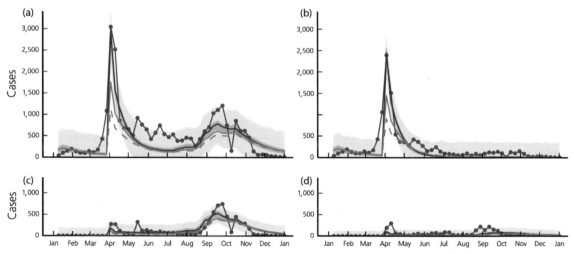

Figure 4.29 Reported (red line) and modeled number of new cases per week (a) for the entire country of Senegal, and for the regions of (b) Diourbel, (c) Dakar, and (d) Thiés during 2005. Blue lines correspond to runs of the model with the best posterior parameter set. Shaded bands correspond the 2.5–97.5 percentiles of the uncertainty related to parameter estimation (dark blue) and of the total uncertainty assuming Gaussian, homoscedastic error (light blue). Modeled cases under the assumption of a 10 percent (solid green line) and 20 percent (dashed green line) reduction in transmission in Touba during the GMdT are also shown. Figure after [520]

have led to a reduction of the total number of cases of 23 percent (38 percent) in Diourbel and 18 percent (34 percent) in the whole country [520].

Finger et al. [520] concluded that the case study of the 2005 Senegal cholera outbreak illustrated quite well the crucial role played by human mobility (and its spatiotemporal variability) in a cholera epidemic whose sudden flare and subsequent spread can be explained by the repercussions of a mass gathering that took place during the initial phase of the outbreak. No approach to quantify

human mobility other than mobile phone data analysis could have provided the required level of detail to capture such phenomena. In addition, the comparison of different models [520] showed that the actual epidemiological dynamic cannot be reproduced accurately without including proper time-varying mobility fluxes and the related effect of overcrowding.

Even if mobile phone data provide an excellent source of information about human mobility, several downsides still exist. One of them is the strong assumption made

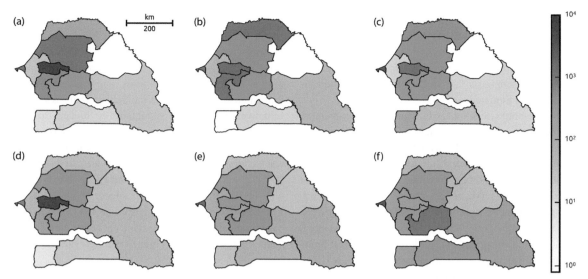

Figure 4.30 Spatial distribution of (a–c) reported and (d–f) modeled cases from March 28 to May 29, the first weeks after the GMdT (a, d), from June 30 to September 4 (b, e), and from September 5 to December 31 (c, f). Figure after [520]

when translating mobile phone records to human mobility patterns, especially considering that they are difficult to validate due to the lack of alternative data sources. Further studies comparing different methods and their underlying assumptions would be necessary to determine the sensitivity of the resulting mobility patterns (but see [541]). In addition, a potential source of inaccuracy in the analysis of mobile phone data is the possible presence of a bias in device ownership. A Kenya-based case study [538], in fact, has suggested that mobile phone owners are more likely to be wealthy, male, and well educated and that a bias exists between urban and rural populations. It would imply that urbanites with higher incomes tend to travel more often and farther, leading to overestimations of frequency and distance of trips [542]. This effect could be partially addressed by the introduction of a parameter accounting for some underrepresentation of people staying at their home node. The values taken by this parameter during calibration might indeed indicate the presence of a bias, but might also be due to the fact that long-distance human mobility has played a major role in the propagation of the outbreak only during the pilgrimage, whereas local factors, such as precipitation and flooding, might have been more important in later stages.

Additional sources of bias could arise from the fact that not all social classes are equally represented among the pilgrims [520] as well as from the uneven coverage of the mobile phone network between different areas of the country. The reconstruction of the 2005 mobility matrix

from that of 2013 in [520] was based on the implicit assumption that the general features of mobility patterns on relevant scales did not change significantly between the two years. Although several ways of reconstructing the 2005 mobility matrix have been compared in [520], their validity cannot be verified owing to the lack of alternative data sources. Among numerous factors that might have influenced mobility patterns is the cholera outbreak itself, which might have led to a behavioral change among individuals in 2005, in turn affecting the disease dynamics [520].

To wrap up, the study in [520] and related ones have demonstrated that mobile phone records allow for an accurate quantification of spatiotemporal fluctuations in human mobility, whether short term, seasonal, or during rare events, such as mass gatherings. The resulting mobility patterns undoubtedly allow for a deeper understanding of epidemiological dynamics. Their inclusion in epidemiological models is thus necessary. However, it is straightforward only for spatially explicit models, and thus may conditionally lead to higher accuracy with respect to other approaches. This constitutes an interesting avenue for further research (Chapter 5).

4.4 Endemic Schistosomiasis

Many important diseases of humans, particularly dangerous in tropical and subtropical regions, arise from infections by macroparasites (i.e., multicellular organisms that reproduce but do not multiply inside their host) [417]. The most common belong to the helminth and arthropod groups, and in particular include: flukes (the *trematodes*), tapeworms (the *cestodes*) and nematodes to mention a few. Schistosomiasis, on which we shall focus here, is caused by a trematode.

Vector-borne infections, which are due to the spread of pathogens by fleas, ticks and lice, may be water-based (in rather sophisticated ways, for example, *Onchocerciasis*, or river blindness, to name one) but are not treated here for the need to confine the perimeter of the possible detours. Macroparasites tend to exhibit much longer generation times than microparasites. They often possess complex life cycles involving two (or even more) obligatory host species [417]. Moreover, no direct multiplication usually occurs within the humans, their definitive host. At times, this may occur although at low rates. Because sexual reproduction often occurs within the human host, this process entails the production of transmission stages, such as eggs or larvae, which leave the host entering the environment through feces or urine to complete their maturation or development. Direct asexual reproduction occurs only in intermediate hosts for certain flukes [417]. What distinguishes mostly micro-from macro-parasite infections is the relative timescales on which the dynamics of parasite and host populations operate [417]. They are determined by comparison with the expected lifespans of the human host compared to those of the various developmental stages, however complex, of the parasites. Empirical observations help to that end. As we shall see in this section, the major differences in relevant timescales support important simplifications in the mathematical description on their transmission dynamics.

Here, we focus on how the river network structure allows us to render spatially explicit the temporal evolution of the transmission cycle of schistosomiasis (Figure 4.31 which displays information about the Senegal case), a parasitic disease emblematic of the interplay among spatially varying drivers and controls, and one where hydrologic controls in particular play a decisive role in the spreading of WB infections. Schistosomiasis, or *bilharzia*, is a chronic debilitating disease caused by parasitic worms of genus *Schistosoma* that affected an estimated 249 million people around the world in 2012. Estimates show that at least 220 million people required preventive treatment for schistosomiasis in 2017. A crushing 93 percent of these people live in Sub-Saharan Africa [419], where both the urinary and intestinal forms of the disease, caused by *S. haematobium*

and *S. mansoni*, respectively, are present. This figure has grown from 77 percent in 2006 [543]. For instance, both forms of schistosomiasis are reported in Burkina Faso since the early fifties, with measured prevalences prior to the implementation of Mass Drug Administration Campaigns (MDAs) within the Schistosomiasis Control Initiative (SCI) [544] systematically higher than 30 percent of the population [545, 546]. A north–south decreasing gradient was observed for urinary bilharzia and an opposite trend for the intestinal form of the disease [546]. The MDAs had an important effect on prevalence with immediate post-MDA prevalence levels ten times lower than the pretreatment baseline [547], but which have in many cases risen again in recent years, with some villages presenting pretreatment prevalence levels [548, 549].

Challenges to the successful control of the disease are manifold due to the complexity of the transmission cycle, which requires freshwater aquatic snails (*Bulinus* spp. or *Biomphalaria* spp. for *S. haematobium* and *S. mansoni*, respectively) as obligate intermediate hosts. The transmission cycle (Figure 4.31) consists of the excretion of parasite eggs from human to water bodies where they hatch into miracidia, the first larval stage, which infect the aquatic snail intermediate host. Asexual reproduction therein produces second-stage larvae called cercariae which infect humans through skin penetration. Once in the human host, they migrate in the system, mature into adult schistosomes and mate in the capillary surrounding the bladder or the intestine depending on the parasite genus leading to egg production and excretion. Environmental, climatic, ecological and socioeconomic factors drive the transmission cycle by conditioning both snail and human probability of infection [550].

Human mobility is known to play a major role in the persistence and the expansion of the disease within, and from, endemic areas in Africa [551–554] and Brazil [555]. Indeed people may become infected while performing their livelihood tasks away from home and importing the disease back in their home village. Furthermore infected migrants coming from endemic regions can introduce schistosomiasis in disease-free villages. This type of medium-to-long range contamination is indeed compatible with the successive focal transmission of the disease at the local level suggested by recent landscape genetics studies [556]. The socio-ecological and epidemiological specificities of the transmission of intestinal schistosomiasis in sub-Saharan Africa and Brazil may explain the different roles played by human mobility, a topic to be elucidated by future research. Given its importance for transmission in the African context, data regarding mobility patterns, possibly directly via cellular phone displacements, will be necessary to produce reliable predictive tools for disease elimination both in terms of reappearance and of prioritization of directed interventions.

4.4.1 Spatially Explicit Models of Schistosomiasis

The seminal model is the one proposed by Mcdonald [557] which includes the average burden of worms in humans, W, the prevalence of infection in snails, Y, and the densities (C and M) of cercariae and miracidia, respectively. Since then many refinements have been proposed.

Dynamical models of schistosomiasis [557, 558], and their spatial extension to connected environments [550], provide the opportunity to show the applicability of a general mathematical framework for analysing the disease invasion conditions and the resulting spatial patterns of human schistosomiasis which could inform control and elimination programs. Furthermore, social and hydrological connectivity for schistosomiasis have been discussed in the context of distributed human-snail contact sites [559, 560], and formally developed in agent-based [561] and connected metapopulation models, building on the seminal model of [557] for schistosomiasis. Gurarie and Seto [550] specifically highlighted the potential importance of both hydrological and social connectivity in small-scale environments. The exploration of invasion conditions could be of relevance in the perspective of sustainable elimination of schistosomiasis, and the inference of the spatial scale of the deployment of control measures in changing conditions of endemicity. The transition between morbidity-centred control programs and transmission-based elimination strategies would ideally require the deployment of surveillance-response mechanism entailing a higher degree of planning and decision-making than traditional mass drug administration campaigns [562], thus calling for new tools to support them. The design of improved sampling protocols for surveillance, and the investigation of transmission breakpoints have been recently highlighted as research priorities for helminth diseases modeling [563]. In line with such priorities, we propose a novel mathematical exploration of an established model [550] and an application of the geography of the disease to the context of Burkina Faso. Special emphasis is placed on linking disease spread and human mobility and water resource development as a proof of concept for the usefulness of the implementation of these kind of tools for macroparasitic waterborne diseases like schistosomiasis.

Human infections or zoonoses occurring through contacts with water contaminated with free-living macroparasite stages pose other challenges largely mediated by the abundance of intermediate hosts possibly subject to a complex ecology [564]. In the case of schistosomiasis, the disease transmission is controlled by contact with environmental freshwater infested with parasite larvae endowed with a complex life cycle and ecology [565–567]. Reinfection after treatment is a problem that plagues efforts to control parasites with complex transmission pathways. Low-cost, sustainable forms of intermediate host control coupled with drug distribution campaigns could reduce or locally eliminate the parasite [568, 569]. At regional scales, where the need for spatially explicit descriptions is particularly compelling, different communities are characterized by differential infection risks linked to their geographical and socioeconomic context. A recent global assessment of schistosomiasis control over the past century has suggested that targeting the snail intermediate host works best [570]. However, neither safe water supplies could completely prevent human contact with environmental freshwater, nor adequate sanitation could guarantee its generalized usage [569]. Moreover, agricultural, domestic, occupational and recreational tasks may foster contact with potentially infested water leading to nonnegligible risk factors even in contexts where water provisioning and sanitation are adequate [571]. This, coupled with differences in lifestyles, leads to complex determinants of transmission heterogeneity [572]. Social groups and ages, such as people whose routinary activites bring them in contact with water (say, fishermen and farmers, or consider cultural settings in which women are more prone to water contact than men) are also a matter of concern for macro-parasitic infections. Water-contact patterns may also vary over time as a response to seasonal changes in human activities, for example, related to agriculture and farming, resulting in temporal heterogeneity in schistosomiasis transmission. Attempts to incorporate socioeconomic classes and risk groups into waterborne disease models are quite a few both in applied [573] and theoretical studies [559]. Strong seasonal patterns in the transmission intensity and infection prevalence in both human and snail hosts still need a settlement within large-scale spatially explicit modeling frameworks.

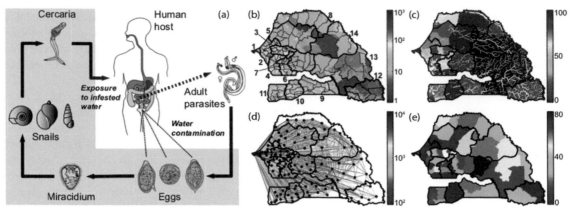

Figure 4.31 (a) A scheme of the schistosomiasis transmission cycle and (b–e) data for a spatially explicit model application to the whole of Senegal [540]. (a) Paired adult worms within the human host produce eggs (left to right: *S. mansoni, S. japonicum, S. haematobium*) that are shed through feces or urine and hatch into miracidia, which infect species-specific intermediate snail hosts (left to right: genus *Biomphalaria, Bulinus, Oncomelania*). Snails shed free-swimming cercariae that can penetrate the human skin. Once in the human host, they develop into reproductive schistosomes. (b) High-resolution population density map of Senegal [inhabitants/km². Regions are labeled as follows: Dakar (1), Thiés (2), Djourbel (3), Fatick (4), Louga (5), Kaolack (6), Kaffrine (7), Saint Louis (8), Kolda (9), Sédhiou (10), Zuguinchor (11), Kédougou (12), Tambacounda (13), Matan (14). (c) People living in rural settings [%] and rivers of Senegal indicating perennial/ephemeral regimes via the thickness of the white lines. (d) Human mobility fluxes in year 2013 [number of individuals], estimated from anonymized mobile phone records. Fluxes between any two arrondissments (say, i and j) is obtained through the human mobility matrix Q_{ij} (Section 4.4 and [540]; see also table 1 therein). Note that only fluxes ≥ 100 individuals are shown as links in the figure. (e) Prevalence of urogenital schistosomiasis [% of infected people]. Data, shown at the scale of health districts, cover the timespan 1996–2013 (see [540] for details on data sources). The drawings in panel (a) are from the Centers for Disease Control and Prevention (www.cdc.gov/parasites/schistosomiasis/biology.html. Figure after [540]

With this goal in mind, we study schistosomiasis spread and persistence in the context of Burkina Faso and Senegal with an emphasis on the roles of human mobility and water resources development.

4.4.2 A Spatially Explicit Model of Schistosomiasis in Burkina Faso

In Burkina Faso, the spatial distribution of the snail host is determined by the strong South-to-North climatic gradient of the transition from sudanian to sahelian regimes [574]. Furthermore, an extensive study of the ecology of both *Bulinus* spp. and *Biomphalaria pfeifferi* in Burkina Faso revealed the importance of both climatic zones and habitat type (including natural and man-made) on snail presence or absence [575]. Indeed *B. pfeifferi* was seldom found present North of the 12th parallel and absent North of the 14th [576], matching the observed historical range of intestinal schistosomiasis in the country [546]. Furthermore, the marked climatic seasonality in Burkina Faso determines the ephemerality of waterbodies and

the hydrological networks in which transmission occurs, thus strongly conditioning the epidemiological characteristics of the disease [545]. Not only do climatic and environmental conditions strongly determine the ecological ranges of the intermediate hosts of the disease, as aforesaid, but also underpin the patterns of human-water contacts during socioeconomic and domestic activities through which both exposure to infection and contamination of waterbodies occur [577–580]. In Burkina Faso agricultural activities (market gardening, rice culture) and fishing (in the vicinities of man-made reservoirs or rivers) are the main causes of human-water contacts for males, whereas domestic activities (laundry, dish washing, water collection) are the main factors of exposure to infection for women [580, 581]. Contamination and exposure for children are known to be mainly associated with recreational bathing in streams, temporary ponds and reservoirs, and participation in parental agricultural activities [577, 581]. Despite the seasonality of water availability, the seasonal variation of human-water contacts to our knowledge has not been addressed in Burkina Faso. Knowledge of the existence of the disease, its burden and transmission pathways has been reported to be poor in most of the settings that have been investigated [546, 548, 567, 577, 580].

As observed in many other schistosomiasis-affected countries, agricultural development and the construction of large-scale irrigation schemes have induced anthropogenic perturbations of the underlying natural matrix affecting schistosomiasis distribution [554, 582–586]. This fact is well illustrated by the construction of the Sourou valley dam at the northern border with Mali in the late 1980s, which resulted in the expansion of the ecological range of *Biomphalaria pfeifferi* along the Mouhoun river from the Bobo-Dioulasso region to the north [546, 584]. Arrival of intestinal bilharzia followed shortly, possibly brought in by migrants coming from the endemic South of the country to work in the rice paddies, yielding an increase in prevalence from virtually zero to more than 60 percent among school-aged children in villages located around the Sourou dam in about 10 years. With a five-fold increase in the number of small reservoirs in Burkina Faso in the past 60 years, and in the perspective

of further constructions of large dams [587], these observations highlight the need to explicitly address the impact of water resources development on schistosomiasis transmission [585, 588]. The interwoven complexity of economic development, water management, and biology requires appropriate prediction tools to attain an integrated control of schistosomiasis [589]: it must address the inherent contradiction between water resources development and livelihood preservation [590, 591] linked to the emergence and persistence of water-borne diseases [586]. Under this viewpoint large improvements can be made in terms of both model adequacy to the local context, in this case that of the country of Burkina Faso, and knowledge of the processes related to disease transmission, specifically human mobility and snail ecology.

The structure of the model employed is similar to the spatially explicit models described earlier in this chapter for cholera, the local model being of course replaced by the Macdonald equations for W, Y, C and M. It is based on the model proposed by Gurarie and Seto [550] and is described in Box 4.16.

Box 4.16 Schistosomiasis Transmission Model

The mechanistic process model proposed by [550] expanded the seminal spatially implicit model of schistosomiasis by [557]. The extension proposes a connected metapopulation network of n villages. The system of differential equations is expressed in terms of the mean worm burden in human populations, W_i, the prevalence of infection in the snail intermediate hosts, Y_i, and the densities of cercariae and miracidia, C_i and M_i, the two intermediate larval stages of the parasite. Connectivity is accounted for by human mobility and hydrologic transport of larvae (while neglecting snail mobility). The parameters of this model and the values used for analysis are detailed in [592]. With our notation, the model of [550] can be written as

$$\frac{dW_i}{dt} = a\left[(1-m_i)\theta_i C_i + m_i \sum_{j=1}^{n} Q_{ij}\theta_j C_j\right] - \gamma W_i,$$

$$\frac{dY_i}{dt} = bM_i(1-Y_i) - \nu Y_i,$$

$$\frac{dC_i}{dt} = \frac{\Pi_C}{V_i}N_i Y_i - \mu_C C_i - l_i^C C_i + \sum_{j=1}^{n} l_j^C P_{ji} S_{ji}^C \frac{V_j}{V_i} C_j,$$

$$\frac{dM_i}{dt} = \frac{\Pi_M}{V_i}\theta_i'\left[(1-m_i)H_i\frac{W_i}{2} + \sum_{j=1}^{n} m_j H_j \frac{W_j}{2}Q_{ji}\right] - \mu_M M_i$$

$$- l_i^M M_i + \sum_{j=1}^{n} l_j^M P_{ji} S_{ji}^M \frac{V_j}{V_i} M_j.$$

(4.40)

Box 4.16 *Continued*

Here at each node i, the rate of change in the mean worm burden is given by the difference between the intensity of cercarial infection expressed as a probability of infection upon contact a times the exposure to contaminated water θ_i and cercarial density, and worm mortality at given rate γ. Density of cercariae is dynamically determined by the per capita cercarial output from infected snails Π_C and the number of infected snails $N_i Y_i$ at time t, and limited by cercarial mortality μ_C. The rate of change in the prevalence of infected snails is given by the probability of infection of a susceptible snail $b(1 - Y_i)$ times miracidial density and limited by snail mortality ν. Finally, miracidial density is determined by the rate of miracidial output from humans Π_M, the contamination rate θ_i', and the number of pairs of adult schistosomes, given by half of the mean worm burden times the number of human hosts H_i divided by the volume of water V_i in which they are released (to obtain larval concentration in water). Human mobility is included by specific contact distribution fractions Q_{ij}. In the original version of [550], the connectivity matrix is set to be proportional to distance between nodes, that is, $Q_{ij} \propto e^{-\alpha d_{ij}}$, with α being a measure of zonal social connectivity. Hydrologic connectivity is implemented through a distance-dependent transport survival rate $S_{ij} = e^{-\beta d_{ij}}$ for each larval stage from upstream to downstream settlements with different volumes of contaminated water V_i. Model (4.40) differs slightly from the one proposed by [550] in the way it treats larval densities and human connectivity. Indeed, we explicitly introduce the volume of water V_i in which larvae are released to have the state variable expressed as a concentration per unit volume (rather than general unit habitat as expressed by [550]). However, in our analysis, we set all V_i to 1 so as to restrain the analysis on the connectivity parameters. Regarding human connectivity, we choose to keep traveler contacts and local contacts distinct by the mobility parameter m_I instead of distributing the whole local contact parameters θ_i as a function of population and distance, as proposed by [550] in the corresponding social contact matrix ω_{ij}. Indeed, in their approach, m_i is implicitly considered to be constant and proportional to $H_i / \sum_j H_j e^{-\alpha d_{ij}}$ for local contamination and exposure rates. By distinguishing between travelers and nontravelers, we assume that contamination and exposure rates depend only on the physical characteristics of the site and not on the traveler origin (captured by θ_i). This is reasonable when analyzing patterns on a national scale where the physical characteristics of the human settlements are heterogeneous (cities vs. towns vs. villages), and thus the contact rate at a visited location θ_j depends mainly on j. The main features of model Equation (4.40) are (1) well-mixed and stationary human and snail populations at each village i; (2) miracidial dispersal and cercarial uptake through human movement matrix Q_{ij} modulated by the fraction of moving people m_i; and (3) miracidial and cercarial dispersal through hydrologic connectivity matrices $P_{ji} S_{ji}^C$ and $P_{ji} S_{ji}^M$ modulated by the hydrologic transport rates $l_j^{C,M}$. By setting $\{m_i, l_j^{C,M}\} = 0 \; \forall i, j$, model (4.40) collapses to the local Macdonald model [557] expressed for each individual settlement. The local basic reproduction numbers are derived as $R_{0,i} = (ab\theta\theta' \Pi_C \Pi_M H_i N_i)/(2\gamma\nu\mu_C\mu_M V_i^2)$ (a complete derivation is given in [592]). In this simple case, parasite invasion of the disease-free system depends locally on the condition $R_{0,i} > 1$. The above-mentioned model [550] does not include detailed biological controls (worm mating probability, negative density feedback on within-host worm population) or immunological ones (acquired immunity of human hosts). Although this could prove inadequate for the study of a particular village close to endemic equilibrium, we deem these factors of secondary importance in the analysis of pathogen invasion conditions at the national level.

Box 4.16 *Continued*

The stability analysis of the model is carried out here for the reformulated version and extended to explore the spatial patterns of disease spread. Sensitivity analysis of model assumptions can be found in the original paper [550].

Model parameter description and values used in the analysis of intestinal schistosomiasis in Burkina Faso are reported in the original reference [592]. Note that in this case as well one needs to assign the mobility matrix Q_{ij} quantifying the fluxes from node j to node i. In the case described here, reliance has been made on the so-called radiation model [533] for its parsimony. Details are reported below. Key parameters are taken from the literature, specifically: the probability of successful miracidial infection in snails is set to $a = 10^{-5}$ [593]; the probability of successful cercarial infection in humans is $b = 10^{-5}$ [593]; the local exposure and contamination rates θ, θ' have been assigned a predefined range [550]; the cercarial emission rate/infected snail is $\Pi_C = 100$ [day^{-1}] [593]; the miracidial emission rate per worm pair is $\Pi_M = 300$ [day^{-1}] [593]; the per capita mortality rates of cercariae and miracidia are, respectively, $\mu_C = 3.04$ [day^{-1}], and $\mu_M = 0.91$ [day^{-1}] [558]; the per capita mortality rates of schistosome in host ($\gamma = 1/5$ [yr^{-1}]) and of snails $v = 1/0.1$ [yr^{-1}] are also taken from the range shown in [558].

Both urinary and intestinal forms of the disease are present with very different historical geographical coverages, governed by the ecological ranges of their respective species of intermediate host [546]. The intestinal form of the disease is used for the analysis, given the relevance of human movement in its geographical expansion in the country [545]. We define the context as the geographical extent of the country (Figure 4.32) and the distribution of human settlements that constitute the nodes of the connected network on which the disease is considered to spread. In addition to parameters taken from the literature, many others can be inferred directly from available data regarding the country, which are illustrated below.

Settlements and Population: H_i

The population of the settlements in Burkina Faso ranges from villages of less than 100 inhabitants to cities of more

than 100,000 people, the urban area of the capital Ouagadougou housing more than 700,000. A settlement was considered to be a city if its population H_i is bigger than a definition threshold of 10,000 inhabitants, and a village if H_i is smaller than 2,000 inhabitants. The model was applied to 10,592 settlements provided by a freely available database (courtesy of Humanitarian Response www.humanitarianresponse.inf). Population was obtained from assigning the sum of gridded Landscan estimates of population density to the Voronoi polygons resulting from the spatial configuration of settlements [592] (Figure 4.32).

Intermediate Host Distribution: N_i

The intermediate host of intestinal schistosomiasis in Burkina Faso is the aquatic gastropode *B. pfeifferi*. The ecological range of the species is determined by both climatic and hydrologic conditions, namely, the presence of agricultural infrastructure such as small and large reservoirs [584–586]. Presence and abundance of the host determine the viability of the transmission cycle of the disease. In the mathematical framework proposed [550, 592], the number of snails in a given node, N_i, is necessary for the analysis. Detailed snail counts at the settlement level in the country of Burkina Faso are ongoing in a number of villages, and are beyond the scopes of this book. For surveillance-response programs at the national level, a probability of presence is retained as a proxy of snail abundance. Probability of presence can be modeled for entire landscapes based on underlying environmental covariates, preserving information about heterogeneity across sites. The maximum entropy (MaxEnt) approach by [594] is one of such methods that has proven its worth in comparison to other predictive methods, and has already been used in the specific case of modeling the distribution of the intermediate hosts of the two genera of schistosomes across Africa [595]. Available data to implement MaxEnt modeling consisted of presence data obtained from the comprehensive literature review and field work of [575] in Burkina Faso, including a total of

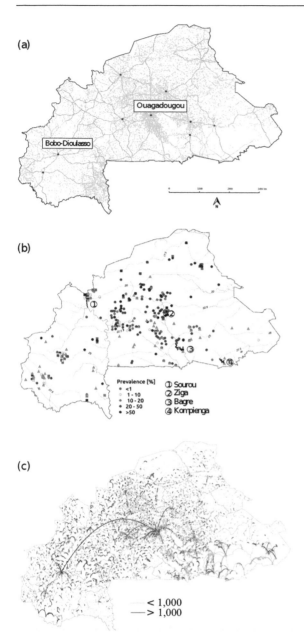

(a)

Ouagadougou

Bobo-Dioulasso

Figure 4.32 Model implementation for Burkina Faso. (a) Road network and settlement distribution. The principal transportation axes, the main cities (black points), including the capital, Ouagadougou, the regional hub, Bobo-Dioulasso, and smaller settlements (gray) are represented [592]. (b) Water resources and schistosomiasis. Only the four largest dams of the country are shown (numbered 1–4) along with the major rivers and mean monthly precipitation isolines (isohyets). Historical prevalence of intestinal(urinary) schistosomiasis is represented in color-coded rectangles (circles). Data were extracted from the Global Neglected Tropical Diseases Database (www .gntd.org) and consist of published parisotological studies in Burkina Faso from 1955 to 2007 (the average prevalence over survey years was taken in villages where multiple surveys were available). Green triangles represent presence data compiled by [575]. (c) Human mobility patterns. Fluxes were predicted using the radiation model and the population density of each node. Point size is proportional to the $\log_{10}$ of settlement population. Fluxes are divided into higher and lower than 1,000 travellers (evaluated taking all of people in the node travelling) and displayed in the main figure for settlements larger than 5,000 people [592]

64 unique sighting locations. For a detailed description of the sampling methodology we refer the reader to [567, 575, 592]. Modeling was performed in the freely available software provided by [594]. Selection of features and natural environmental covariates for prediction was done according to the method proposed by [595]. The anthropogenic covariate with most explanatory power is

the distance to dams or small reservoirs, $\mathscr{D}_{water}$, which was produced by calculating the Euclidean distance from each settlement to the closest water surface. Water surfaces were obtained at a 30m pixel resolution for all of Burkina Faso by Quadratic Discriminant Analysis of a mosaic of Landsat satellite images of 2014 (courtesy of the U.S. Geological Service) [592]. The resulting water surfaces were filtered and validated against the national spatial database of dams and small reservoirs [588]. Maps of the covariates used as inputs for MaxEnt are illustrated in the original reference [592]. The output of MaxEnt (probability of presence) was taken as being proportional to the site-specific parameter N_i during model implementation.

Human Mobility Model: Q_{ij}

No actual human mobility data, such as census or mobile phone records, is currently available for accurately estimating human mobility fluxes in Burkina Faso. Models of human mobility have grown popular and accurate with increasing access to big data on human behavior, and are valuable tools to overcome data scarcity despite the strong assumptions that need to be made regarding, for instance, travel means and accessibility [537, 542]. Here we implement the recently proposed radiation model which has proven to correctly reproduce mobility patterns at the national and regional scales. It was originally implemented in the context of the USA [533], and proved to hold well for intercity movement in West Africa [596]. Specifically, the radiation model expresses the probability Q_{ij} that a person traveling out of node i reaches node j as:

$$Q_{ij} = \frac{H_i H_j}{(H_i + s_{ij})(H_i + H_j + s_{ij})} \quad (4.41)$$

where $H_i[H_j]$ is the population size of the origin [destination] node, and s_{ij} is the total population living within a radius d_{ij} around the origin, excluding the origin and destination populations, d_{ij} is the distance between nodes i and j. The model has therefore the advantage of depending explicitly only on population density distribution in space to capture the structure of human mobility fluxes. The fraction of moving people, that is, people visiting another node during a short period of time, m_i, is therefore the only free parameter modulating the strength of mobility-related connectivity in the spatial spread of the disease. To our knowledge, no empirical estimate of m exists for Burkina Faso, although values of $m \to 1$ are unlikely. The vector of human population $\mathbf{H}$ and Euclidean distances were used to produce a $10,592 \times 10,592$ matrix Q_{ij} of human mobility connections. Three forms of human mobility can be observed in Figure 4.32: (1) strong local-level fluxes between low density populated areas, (2) large population centres interconnected by long-distance trips, and (3) medium-range fluxes in the

basin of attraction of the two major cities (Ouagadougou and Bobo-Dioulasso).

We note that other mobility models such as the gravity model imply different underlying assumptions, but with no validation data for Burkina Faso available these would suffer from the same weakness of verifiability as the radiation model. Needless to say, the availability of real mobility data would overcome the need for modeling as observed mobility fluxes could be used directly (with caveats) [520, 537].

Exposure and Contamination Rates: θ_i, θ_i'

Successful human infection by cercariae strongly depends on water contact frequency and duration, which in turn are determined by the socioeconomic activities of the populations [554, 580]. For this reason, urban areas are potentially less favorable schistosomiasis transmission sites given the lower human-to-water contact opportunities relative to the number of inhabitants. This needs to be encompassed into the single $\theta_i[\theta_i']$ parameters of exposure[contamination] rate for the whole population of a node i. Although there have been examples of recent increases in agriculture around small reservoirs and other water bodies in urban areas [597, 598], the proportion of such contact patterns is not deemed representative of the majority of urban settlements [554, 599]. In the absence of detailed and nationwide socioeconomic surveys assessing exposure and contamination rates in Burkina Faso, we have translated the reduced probability of having population-wide high contamination and exposure rates in urbanized settlements into an inverse-logistic formulation for the exposure/contamination rates (here assumed to coincide, i.e., $\theta = \theta'$, as in [550]) based on the size of each node. The hypothesis underlying this approach is therefore that, in Burkina Faso, densely populated and tertiarized urban centres should have low contamination/exposure rates with respect to small rural settlements where a number of socioeconomic activities entail human-water contacts [579, 580]. The resulting contact rate θ_i in node i is thus

$$\theta_i = \theta_{\text{MAX}} \left\{ 1 - \frac{1}{1 + a(\alpha)e^{-b(\alpha)S_{\log 10}(H_i)}} \right\}, \quad (4.42)$$

that is a sigmoid decreasing function of population size, where θ_{MAX} is a parameter indicating the maximum contact rate, while α controls the shape of the function. Specifically $a(\alpha)$ and $b(\alpha)$ are two positive coefficients that are selected so that cities have exposure equal to $\alpha\theta_{\text{MAX}}$, and imposing $\theta_i = 0.9 \cdot \theta_{\text{MAX}}$ for villages of exactly 2,000 inhabitants. In other words α can be seen as the percentage reduction in exposure/contamination rates due to urbanization for a settlement at the population threshold defining a city, here taken to be defined as a settlement of 10,000 inhabitants, with respect to the maximal rates observed in rural areas. $S_{\log 10}(H_i) = (\log_{10}(H_i) - \mu_{\log_{10}(\mathbf{H})})/(\sigma_{\log_{10}(\mathbf{H})})$ is the $\log_{10}$-transformed and standardized population density based on the mean, $\mu_{\log_{10}(\mathbf{H})}$, and the standard deviation, $\sigma_{\log_{10}(\mathbf{H})}$, of the log 10 transform of the population vector $\mathbf{H}$.

Hydrologic Connectivity

In Equation (4.40), hydrologic connectivity only concerns larval stages of the parasite. Survival times of both miracidia and cercariae are known to be on the order of hours [600], thus resulting in localized effects of this form of transport, as illustrated in [550] for a network of 15 connected villages. Since the context of the present analysis is all of Burkina Faso, we deemed it realistic to remove hydrologic transport of the larval stages of the parasite, since the only viable hydrologic transport routes at the national scale could be large rivers (Figure 4.32), which are known to be unsuitable environments for larvae that have limited lifespans. Nevertheless, all of the following analyses could be repeated including the hydrologic transport matrix P_{ij}. On the other hand a form of hydrologic connectivity that was not considered in [550] is the passive dispersion of the snail intermediate host in streams and rivers. Indeed it has been observed that water velocities greater than 33cm/s may dislodge *Bulinus* spp. individuals that may be transported downstream [575]. The only available data on this kind of phenomenon in the fairly flat topography of Burkina Faso is a study by Poda [575] which measured snail dispersion during the rainy season of the order of 1 snail/30min on the Nazinon river (Volta Rouge). Given that this result is based on only 4 samplings, and no information was collected on snail provenance, transport distance and variability, we considered that there was not enough evidence to justify modifying Equation (4.40) to include snail transport at this stage. Field data on the transport of the intermediate hosts of schistosomiasis in Burkina Faso would help elucidate its importance and relevance in full transmission models.

The analysis of the Burkina Faso space-explicit schistosomiasis model, whose parametrization we have just described, proceeds along the avenue already pursued for cholera. Endemic establishment of schistosomiasis can be studied by performing a stability analysis of the disease-free equilibrium (DFE, i.e., $W_i = Y_i = C_i = M_i = 0$). If the DFE is unstable, the disease can become endemic. Details are reported in Box 4.17, which shows that disease establishment is linked once again to the dominant eigenvalue g_0 of a suitable generalized reproduction matrix. This matrix depends on both human mobility and hydrological connections (the latters are neglected, however, in the reported case study).

Box 4.17 Parasite Invasion Conditions

Despite the complexity given by the spatially explicit formulation of model (4.40), a rigorous stability analysis can be used to determine spatially explicit conditions for pathogen invasion. As the system is positive and the disease-free equilibrium is characterized by null values of the state variables, the bifurcation can only occur via an exchange of stability. Specifically, the disease-free equilibrium switches from stable node to saddle through a transcritical bifurcation, at which the Jacobian has one zero eigenvalue [188, 437, 550, 601]. Parasite invasion is determined by the instability of the disease-free equilibrium (DFE) $\mathbf{X_0} = [\mathbf{0}_n, \mathbf{0}_n, \mathbf{0}_n, \mathbf{0}_n]^T$, with $\mathbf{0}_n$ a $1 \times n$ null vector, n, being the number of settlements. The stability properties of the DFE can be studied by analyzing the Jacobian linearized at the DFE, $\mathbf{J_0^*}$, which reads as follows:

$$\mathbf{J_0^*} = \begin{bmatrix} \mathscr{A} & \mathscr{B} \\ \mathscr{C} & \mathscr{D} \end{bmatrix}, \tag{4.43}$$

where

$$\mathscr{A} = \begin{bmatrix} -\gamma\mathbf{I} & 0 \\ 0 & -\nu\mathbf{I} \end{bmatrix} \qquad \mathscr{B} = \begin{bmatrix} a(\mathbf{I}-\mathbf{m}+\mathbf{mQ})\theta & 0 \\ 0 & b\mathbf{I} \end{bmatrix}$$

Box 4.17 *Continued*

$$\mathscr{C} = \begin{bmatrix} 0 & \Pi_C \mathbf{V}^{-1}\mathbf{N} \\ \frac{\Pi_M}{2}\mathbf{V}^{-1}\theta'(\mathbf{I} - \mathbf{m} + \mathbf{Q}^T\mathbf{m})\mathbf{H} & 0 \end{bmatrix}$$

$$\mathscr{D} = \begin{bmatrix} -\mu_C\mathbf{I} + \mathbf{T_C} & 0 \\ 0 & -\mu_M\mathbf{I} + \mathbf{T_M} \end{bmatrix}.$$

where $\mathbf{I}$ is the identity matrix; $\mathbf{m}$, θ, $\mathbf{V}$, $\mathbf{N}$, θ, and $\mathbf{H}$ are diagonal matrices whose nonzero elements are made up by the parameters m_i, θ_i, V_i, N_i, θ_i, and H_i, respectively; $\mathbf{Q} = [Q_{ij}]$ is the connectivity matrix for human mobility; $\mathbf{T_C} = (\mathbf{V}^{-1}\mathbf{P_C}^T\mathbf{V} - \mathbf{I})\mathbf{l_C}$ and $\mathbf{T_M} = (\mathbf{V}^{-1}\mathbf{P_M}^T\mathbf{V} - \mathbf{I})\mathbf{l_M}$, while $\mathbf{P_C} = [P_{ij}S_{ij}^C] = \mathbf{P} \circ \mathbf{S_C}$ (where $\circ$ is the Hadamard product); $\mathbf{P_M} = [P_{ij}S_{ij}^M] = \mathbf{P} \circ \mathbf{S_M}$ are the transport matrices accounting for hydrologic connectivity and larval survival during transport; and $\mathbf{l_C}$ and $\mathbf{l_M}$ are diagonal matrices whose nonzero elements are the local values of l_i^C and l_i^M, respectively. After some manipulations, the bifurcation condition $\det(\mathbf{J}_0^*) = 0$ can be reformulated as

$$\det\left(\mathbf{I} - (\mathbf{I} - \mathbf{m})^2\mathbf{R_0} + \mathbf{R_0^M}(\mathbf{m}, \mathbf{Q}) + \mathbf{T}(\mu_C, \mathbf{T_C}, \mu_M, \mathbf{T_M})\right) = 0. \tag{4.44}$$

We can define a generalized reproduction matrix (GRM) for our model of schistosomiasis transmission as the sum of three matrices. One depends on local dynamics only, the other two (nonlinearly) on spatial coupling mechanisms, reading

$$\mathbf{G_0} = (\mathbf{I} - \mathbf{m})^2\mathbf{R_0} + \mathbf{R_0^M}(\mathbf{m}, \mathbf{Q}) + \mathbf{T}(\mu_C, \mathbf{T_C}, \mu_M, \mathbf{T_M}). \tag{4.45}$$

Parasite invasion conditions can therefore be decomposed into spatially explicit local conditions $\mathbf{R_0}$ and connectivity-induced conditions $\mathbf{R_0^M}$ and $\mathbf{T}$ for mobility and hydrologic transport, respectively. $\mathbf{R_0}$ is a diagonal matrix whose nonzero elements are the local values R_{0i} of the basic reproduction number. $\mathbf{R_0^M}$ is a matrix depending on human mobility structure and magnitude, that is,

$$\mathbf{R_0^M}(\mathbf{m}, \mathbf{Q}) = \frac{ab\Pi_C\Pi_M}{2\gamma\nu\mu_C\mu_M}\mathbf{N}\mathbf{V}^{-1}\theta'\left[(\mathbf{I} - \mathbf{m})\mathbf{m}\mathbf{H}\mathbf{Q} + \mathbf{Q}^T\mathbf{H}\mathbf{m}(\mathbf{I} - \mathbf{m}) + \mathbf{Q}^T\mathbf{m}^2\mathbf{H}\mathbf{Q}\right]\theta\mathbf{V}^{-1}.$$

$\mathbf{T}$ is another matrix depending on larval death rates and transport through the hydrologic network:

$$\mathbf{T}(\mu_C, \mathbf{T_C}, \mu_M, \mathbf{T_M}) = \frac{1}{\mu_C}\mathbf{V}\mathbf{T_C}\mathbf{V}^{-1} + \frac{1}{\mu_M}\mathbf{N}\mathbf{T_M}\mathbf{N}^{-1} - \frac{1}{\mu_C\mu_M}\mathbf{N}\mathbf{T_M}\mathbf{V}\mathbf{N}^{-1}\mathbf{T_C}\mathbf{V}^{-1}.$$

Ensuing from Equations (4.44) and (4.45) the condition for parasite invasion can be stated in terms of the dominant eigenvalue g_0 of matrix $\mathbf{G_0}$, that is, $g_0 = \max_k(\lambda_k(\mathbf{G_0}))$, with disease spread occurring if $g_0 > 1$. Bearing the operational use of this condition in mind, it is important to note that control measures that successfully lead to a stable DFE (thus driving g_0 below 1), for instance, by providing improved sanitation or the use of molluscicides, do not imply immediate transmission interruption, and the country may instead present nonzero parasite loads in the population and in the environment for some time. In ecological terms, the time to reach pathogen extinction corresponds

Box 4.17 *Continued*

to the relaxation time of the system [88, 602], $t_{\mathscr{R}}$, which is determined by the characteristic timescale of pathogen spread dynamics. The relaxation time of schistosomiasis transmission is therefore set by the dominant eigenvalue g_0 as $t_{\mathscr{R}} = -1/|\log(g_0)|$ [447]. Based on this relation, it is possible to associate a given socioecological setting in the country (i.e., a given value of g_0) to an estimate of the time for transmission to effectively die out once conditions for pathogen extinction are met ($g_0 < 1$).

Furthermore bifurcation analysis for waterborne disease models proved to be able to capture spatial patterns of the spread of the pathogen [451]. It is possible to show that the dominant eigenvector $\mathbf{g_0}$ of the GRM can be used to study the geography of disease spread in the case of macroparasitic infection as well, at least close to the transcritical bifurcation. In fact, $\mathbf{g_0}$ corresponds to the cercarial component $\mathbf{C}$ of the vector of state variables for trajectories diverging from the (unstable) DFE (see Appendix 6.1.6). This corollary is of particular interest in the study of spatial patterns of disease spread. Indeed, close to an unstable DFE, the dominant eigenvector of matrix $\mathbf{J_0}^*$, $\lambda_{max}(\mathbf{J_0^*})$, pinpoints the directions in the state space along which the system trajectories will diverge from the equilibrium. In other words, computing $\lambda_{max}(\mathbf{J_0^*})$ enables the quantitative analysis of conditions for the parasite not only to invade a connected set of spatial locations but also to predict the direction of disease spread geographically in the time close to the onset. Of particular interest in the dominant eigenvector analysis is the mean worm burden compartment, $\mathbf{W}$, which, in the case of schistosomiasis, is an important measure of infection severity and the natural target of control interventions through antischistosomal treatments [603, 604]. At the bifurcation through which the DFE loses stability, spatial distribution of the mean worm burden can be computed in terms of $\mathbf{g_0}$ as $\mathbf{W} = (a/\gamma)(\mathbf{I} - \mathbf{m} + \mathbf{mQ})\theta\mathbf{g_0}$ [540].

The exercise reported here assumes that the structure of the human mobility network is accurately captured by the radiation model, thus here we first explore the model sensitivity to the intensity of mobility fluxes. Different values of the fraction of mobile population in the range $m \in [0,1]$ are considered. The relation between contamination and exposure rates and population density dependence also lacks ground-truthed data, and therefore it is explored through sigmoid contact functions (Eq. (4.42)) with $\alpha \in \{0.0001, 0.001, 0.01, 0.1\}$. Increasing α can be seen as increasing urban contamination and exposure rates. Not surprisingly, human mobility is found to strongly condition successful parasitic invasion of the system. Two opposite effects are illustrated in Figure 4.33 with a dilution effect for fractions of mobile people increasing from $m = 0$ to $m \approx 0.3$; and an increase in the risk of disease spread for $m > 0.3$. For all investigated levels of urban contamination and exposure rates the dominant eigenvalue g_0 of the GRM falls below 1 for intermediate ranges of the fraction of mobile people and low values of θ_{MAX}.

The parameter region in which the DFE is stable, preventing parasite invasion, is of similar size for all values of α except for high levels of urban contact rates ($\alpha = 0.1$) for which pathogen invasion is observed for a wider range in the parameter space [592]. This can be seen as the effect of coupling locations with high mobility (densely populated areas) with locations with high transmission potential (high contamination and exposure rates). For all subsequent analysis we therefore focused on the case that seemed the most realistic, namely, intermediate urban contamination and exposure rates ($\alpha = 0.01$).

In addition to the stability of the DFE, the dominant eigenvalue of the GRM contains information on the average time for transmission to fade out in the case of unfavorable conditions for pathogen spread (Figure 4.33). The average times to extinction based on relaxation time

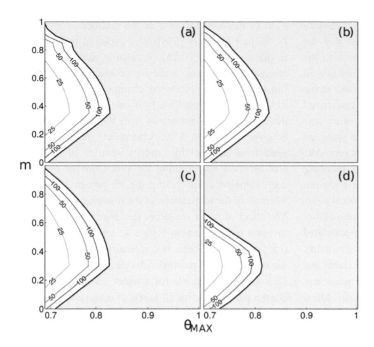

Figure 4.33 Pathogen invasion conditions and average time to extinction for the Burkina Faso settlement network. The values of the dominant eigenvalue g_0 of the generalized reproduction matrix are plotted for increasing values of α $(0.0001, 0.001, 0.01, 0.1)$ (respectively in panels (a–d)), against values of maximal contamination and exposure rates θ_{MAX} and the fraction of mobile people m. Yellow areas indicate parameter combinations permitting parasitic invasion. Contours to the left of the stability boundary give pathogen extinction time isolines $(5t_R)$ of 25, 50, and 100 years. Figure after [592]

isolines (constant $t_{\mathscr{R}}$) follow the shape of the stability boundary, with a logarithmic decrease of extinction time as a function of distance to the boundary $g_0 = 1$. The time to extinction is of the order of a century close to the stability line, and decreases sharply for small modifications of human mobility and contamination/exposure rates. Similarly to the stability conditions, the relaxation time is minimal at intermediate mobility levels for a given θ_{MAX}. Furthermore the shape of the time to extinction isolines suggests that, above the mobility threshold $m \approx 0.3$, reducing contamination/exposure rates would have a larger marginal reduction on the time to extinction than acting on mobility. If we contextualize these results in the framework of elimination programs it may be argued that an extinction time of 100 years may be considered an effectively stable transmission scenario from a public health management point of view. In this perspective $g_0 = 1$ could be inappropriate as an operational indicator of transmission characteristics, and the threshold value should be chosen based on an acceptable relaxation time from a public health standpoint. Moreover our remark on the effect of the two parameters on relaxation time should be reformulated in terms of costs for it to be projected into disease control strategy design, and although the latter is of great practical interest it is beyond the scope of this work.

The geographical pattern of disease spread is related (similarly to many of the cases illustrated before and according to theory for microparasitic disease [437]) to the dominant eigenvector associated with the dominant eigenvalue g_0. Box 4.17 provides the details. In the perspective of mathematical modeling, schistosomiasis differs from cholera-like diseases in that it could be considered as a slow–fast dynamical system [605], meaning having rate parameters spanning 3 orders of magnitude (see [592]). Indeed, by definition, the dominant eigenvector of the Jacobian of the linearised system approximates well the trajectory close to the DFE, but its predictive power of the endemic equilibrium remains to be verified. This is done by comparing the predicted spatial patterns of disease with the state of the system at the endemic equilibrium. Spatial patterns of disease spread are obtained from rescaling the dominant eigenvector $\mathbf{g_0}$ of the GRM to represent $\mathbf{W}$, the mean worm burden along model trajectories diverging from the DFE. The results [592] show that at the transcritical bifurcation the rescaled dominant eigenvector reproduces well the spatial patterns observed at the endemic equilibrium of the system.

We chose to study spatial patterns for three mobility regimes along the $g_0 = 1$ contour line by tuning the free parameters (m and θ_{MAX}) to place the system precisely at the transcritical bifurcation that determines the instability of the DFE (Figure 4.34). It is to be noted that the dominant eigenvector analysis holds only close to the bifurcation which represents conditions tending toward elimination, thus relevant to surveillance-response strategies, as opposed to areas of high endemicity characterized by $g_0 \gg 1$. Figure 4.35 illustrates the effect of connectivity on disease spread in comparison to local pathogen invasion condition, expressed by the local basic reproduction number $R_{0,i}$ (panel (b) in Figure 4.34, corresponding to point (1) in panel (a)). Local potential for pathogen invasion results from the interaction between local population density and snail presence. At low human mobility intensity the disease concentrates around the populated areas in the center of the country. By increasing mobility, the dilution effect causes the DFE to become stable and the pathogen cannot invade the country. For parameter combinations yielding pathogen invasion at intermediate human mobility ($m = 0.3$), the spatial patterns of disease spread shift toward the south-east with the appearance of a second hotspot in an area with favorable local conditions. The effect of mobility is most visible when the majority of people are mobile ($m = 0.65$). Indeed, the south-eastern hotspot takes over the central one, and the disease spreads to the whole south-eastern part of the country in areas where the local reproduction number is below 1, indicating unfavorable local conditions for transmission. The capacity to predict the spatial patterns of the disease close to the disease-free equilibrium could provide a tool for prioritizing surveillance sampling in those areas where schistosomiasis is predicted to pick up the most, thus where the reinvasion signal in the mean worm burden would be the strongest.

Relevant to the tenet of the book, water resources development directly impacts large-scale schistosomiasis transmission by altering the probability of presence of the intermediate snail host in the environment. This is explicitly accounted for in the approach through the mean distance to water $\mathscr{D}_{water}$, that is, a permanent water body whether natural or artificial, of a human settlement [592]. To investigate the side effects of water resources development on schistosomiasis in the country, the probability of presence was reevaluated using MaxEnt for alternative scenarios of water resources development by randomly removing existing reservoirs to consider different numbers of built dams expressed as a fraction ϵ of the existing ones and the stability of the DFE recomputed for suitable parameter combinations [592]. The evolution of water resources development in Burkina

Faso (a five-fold increase in the number of reservoirs in the past 60 years [588]) was explored by varying ϵ in the set $\{0.25, 0.5, 0.75\}$ of existing reservoirs, which represents increasing numbers of built reservoirs. An illustration of the predicted changes in the ecological range of the intermediate host due to water resources development, in comparison with the current scenario, is given in Figure 4.36. Uncertainty in the stability predictions induced by random removal of reservoirs was assessed by repeating the procedure 10 times for each value of ϵ and taking the 95 percent confidence intervals of the realizations of the resulting stability lines. The effect of water resources development on pathogen invasion is illustrated in Figure 4.36 for the case of $\epsilon = 0.5$; results for the other two scenarios are similar. For all the reduced water resources development scenarios, the DFE tends to be stable for a wider set of mobility and contact parameters, for all levels of urban contamination and exposure rates. Human mobility plays an important role in the stability of the DFE, that is, pathogen invasion conditions, in such alternative scenarios [592]. Indeed, the interplay between human mobility and water resources development is greatly conditioned by the level of urban contamination/exposure rates. For $\alpha < 0.1$ the effect of building dams is most detrimental for low levels of human mobility ($m < 0.25$), while the bifurcation lines tend to be close to the invasion condition of the current scenario for $m > 0.8$. On the other hand, for high levels of urban transmission of schistosomiasis ($\alpha = 0.1$), human mobility accentuates the negative impact of the increasing number of dams by augmenting the gap between the bifurcation curves of the alternative and current scenarios $m > 0.6$. Furthermore, uncertainty in the stability of the DFE increases with the fraction of mobile people for all αs. These observations can be seen as the contribution of less connected settlements, where contamination and exposure are high, to the overall (in)stability condition of the DFE. By increasing urban contamination and exposure rates, nodes with high connectivity also present high transmission, thus having a large impact on the stability of the system. When removing reservoirs that determine the number of snails in these key nodes, the stability region of the DFE expands. In other words, the system is more dependent on a small set of highly connected nodes, thus the construction of dams close to these points has a much larger effect on stability. The uncertainty associated to the location of the stability line is thus directly related to the random removal of the reservoirs influencing snail abundance in these key

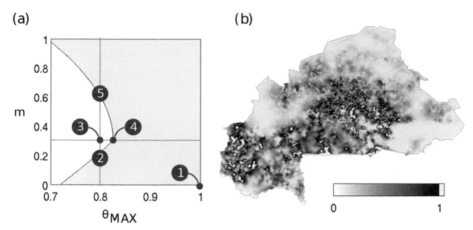

Figure 4.34 Pathogen invasion conditions and local transmission risk. (a) Stability diagram of the DFE in terms of contamination and exposure rates and the fraction of mobile people for intermediate urban contamination/exposure rates ($\alpha = 0.01$). Points in the white (yellow) area represent parameter combinations yielding a stable (unstable) DFE. Parasite invasion occurs when crossing the bifurcation curve (black line, $g_0 = 1$). (b) Local conditions of pathogen invasion risk in terms of R_0 for no human mobility and maximal contact rates (point 1 in panel (a)). Areas presenting $R_{0,i} > 1$ (in yellow) experience sustained pathogen invasion, and schistosomiasis is introduced in the node. Even for maximal contact rates, locally schistosomiasis-prone areas are restrained to the southwestern and central parts of the country. Figure after [592]

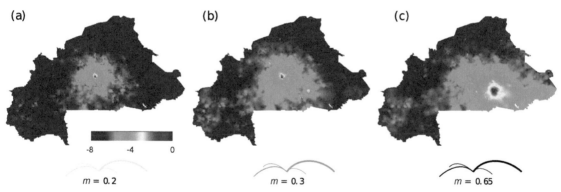

Figure 4.35 Human mobility and spatial patterns of the disease in the case of intermediate urban schistosomiasis. (a–c) Predicted $\log_{10}$-rescaled values of the mean worm burden compartment $\mathbf{W}$ for increasing levels of human mobility ($m = 0.2, 0.3, 0.65$) along the bifurcation line (points d–e–f in panel (a)) of Figure 4.34 and the $\theta_{MAX} = 0.8$ transect (panel (a)) of Figure 4.34. Figure after [592]

nodes. Human mobility therefore exacerbates the effect of water resources development on risk of pathogen invasion in the case of high urban contamination and exposure rates.

Although preliminary, the results taken from [592] illustrate the use of the proposed eigenvector analysis of the GRM to quantify the potential impacts of water resources development in a spatially explicit framework that includes connectivity mechanisms such as human

mobility. In particular, we concur with previous studies suggesting that strategies to mitigate negative effects on human health should become integral parts in the planning, implementation, and operation of future water resources development projects [586]. In particular, the interplay between the water resources development and connectivity induced by human mobility is deemed to be important in the risk of reemergence of the disease if elimination is attained.

(a)   (b)

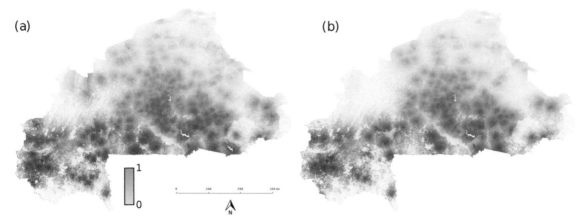

Figure 4.36 Impact of water resources on the ecological range of the intermediate host. (a) *B. pfeifferi* probability of presence for the current scenario of water resources development. (b) Probability of presence for an example of the alternative scenario with half the number of reservoirs ($\epsilon = 0.5$). The predicted species distribution of *B. pfeifferi* by the MaxEnt approach reveals the importance of distance to water surfaces along with the influence of the north-south precipitation/temperature gradient that prevents the snail from colonizing the northern parts of the territory. The example of one realization of the scenario with half the reservoirs illustrates the reduction of the ecological range of the intermediate host, a necessary but not sufficient condition for explaining the observed results. Figure after [592]

The study, though by no means exhausting, nevertheless has value for the tenet of this book. In fact, by implementing dynamical system analysis techniques for waterborne diseases, the roles of human mobility and water resources development in the spread of intestinal schistosomiasis in Burkina Faso were highlighted by proper mathematical techniques. Human mobility was shown to play a role not only in the degree of success of invasion of the network of human settlements by the pathogen but also in the spatial patterns of disease spread. For small fractions of mobile people, mobility induced a dilution effect of the parasite concentration in accessible waters, leading to an effective prevention of parasite invasions, in agreement with previous results derived at much smaller spatial scales. Even in settings unfavorable to pathogen invasion, human mobility had a large effect on average times to metapopulation extinction of the pathogen. Above a threshold value, mobility prompted a systematic exacerbation of the disease by shifting its hotspots from the most populated areas (even where local conditions warrant local reproduction numbers larger than one) to more transmission-prone areas with higher contamination and exposure rates and greater snail abundance. Moreover, predicting the spread of the disease based on water resources development was investigated by quantifying the modification of the stability conditions of the disease-free equilibrium resulting from

anthropogenic expansions of suitable habitats for *B. pfeifferi*, the snail acting as intermediate host in the life cycle of the parasite *S. mansoni*. By considering theoretical scenarios representative of preexisting levels of water resources development in the country, we showed that the marked increase in the number of dams prompted by water development projects favored pathogen invasion. For low urban schistosomiasis transmission, the greatest effects of water resources development were found to occur for low levels of human mobility, thus favoring localized transmission, which maximizes the effect of each additional dam built. Interestingly, intense human mobility was also shown to exacerbate the impact of the building of the dams in the case of high levels of urban contamination and exposure rates.

The exercise allows us to conclude that insight from the tools developed in this chapter could directly inform national and regional control measures in the perspective of elimination in terms of pathogen invasion conditions and initial spatial spread, thus contributing to two of the points brought to the fore as research priorities for helminth modeling [563]. Both prevention and intervention at regional scales, in fact, need to address the central role of human mobility in the perspective of the deployment of surveillance-response mechanisms and the localization of treatment measures. Controls of urban contamination and exposure rates, say, by reducing contamination rates by control programs like WASH (water, sanitation, and hygiene), could be measured against predicted disease burden reductions. Even at the current state of development, results of management alternatives obtained from mathematical models are potentially informative for epidemiological decisions in conditions of close-to-threshold transmission that will hopefully be in the near future the stage of elimination programs in sub-Saharan Africa and Burkina Faso. Indeed, the proposed theoretical framework is an opportunity for the design of control indicators based on timescales of pathogen extinction in addition to dynamical stability criteria.

4.4.3 An Integrated Study of Endemic Schistosomiasis in Senegal

We now report about another study carried out on large-scale patterns of schistosomiasis transmission [540]. The study focuses on the drivers of its geographical distribution in Senegal via a spatially explicit network model accounting for epidemiological dynamics driven by local socioeconomic and environmental conditions and human mobility. The model (Box 4.18) is parameterized by tapping several available geo-databases and, critically, a large dataset of mobile phone traces.

With respect to the Burkina Faso case study, there are some important developments: (1) human mobility is not modeled but rather is directly inferred from mobility phone call records; (2) human population is stratified according to different levels of parasite burden; (3) snail densities are considered and snails are classified as susceptible or infected; and (4) calibration is performed on available epidemiological data, and different model structures are compared via information criteria.

We shall show that the study reliably reproduces the observed spatial patterns of regional schistosomiasis prevalence throughout the country, provided that spatial heterogeneity and human mobility are suitably accounted for. Specifically, a fine-grained description of the socioeconomic and environmental heterogeneities involved in local disease transmission proved crucial to capturing the spatial variability of disease prevalence, while the inclusion of human mobility significantly improved the explanatory power of the model. Concerning human movement, the main finding, perhaps of general significance for the purposes of this book, is that moderate mobility may reduce disease prevalence, whereas either high or low mobility may result in major increases in prevalence of the infection. The effects of control strategies based on exposure and contamination reduction via improved access to safe water or educational campaigns have also been analyzed via these powerful tools – modeling matters, once tools are safely calibrated, for the scenarios they entail. The approach [540] can be seen as the first application of a truly integrative schistosomiasis transmission model at a whole-country scale. It is described in some detail in Box 4.18.

Box 4.18 An Integrative Mathematical Model for Senegal

Here we briefly describe the model and its data [540]. The reader is referred to the original contribution for full details on materials and methods.

The human population is subdivided into n communities (following, for example, administrative boundaries, health zones, or geographical divides). Within each community i, the resident human population (of size K_i) is considered to be "stratified" [606–609], that is, divided into different infection classes characterized by increasing parasite burden p (from $p = 0$ to some maximum abundance $p = P$). Except for this, the population is assumed to be well mixed within each community (no demographic/socioeconomic grouping). Let H_i^p be the abundance of individuals in community i who host exactly p parasites. Furthermore, let S_i and I_i be the densities of susceptible and infected snails in community i, and let C_i and M_i be the concentrations of cercariae and miracidia in the freshwater resources accessible to community i. Following the transmission cycle of schistosomiasis (Figure 4.31a), disease transmission can be described by the following set of $n(P + 5)$ differential equations:

Box 4.18 *Continued*

$$\dot{H}_i^0 = \mu_H(K_i - H_i^0) - \mathscr{F}_i H_i^0 + \gamma^1 H_i^1, \tag{4.46}$$

$$\dot{H}_i^p = \mathscr{F}_i H_i^{p-1} - (\mu_H + \alpha_H^p + \mathscr{F}_i + \gamma^p)H_i^p + \gamma^{p+1}H_i^{p+1} \qquad (0 < p < P), \tag{4.47}$$

$$\dot{H}_i^P = \mathscr{F}_i H_i^{P-1} - (\mu_H + \alpha_H^P + \gamma^P)H_i^P, \tag{4.48}$$

$$\dot{S}_i = \mu_S(N_i - S_i) - bM_i S_i, \tag{4.49}$$

$$\dot{I}_i = bM_i S_i - (\mu_S + \alpha_S)I_i, \tag{4.50}$$

$$\dot{C}_i = \pi_C I_i - \mu_C C_i, \tag{4.51}$$

$$\dot{M}_i = \mathscr{G}_i - \mu_M M_i. \tag{4.52}$$

The first $n(P + 1)$ equations of the model describe the dynamics of human hosts, the following $2n$ the dynamics of intermediate snail hosts, and the last $2n$ the dynamics of the larval stages of the parasite. Model variables and parameters are fully summarized in table 1 of [540], while a graphical representation of the transmission model is provided here for completeness (Figure 4.37).

As for the dynamics of human hosts, μ_H is the baseline per capita mortality rate, while $\mu_H K_i$ represents the total birthrate, here assumed to be constant (i.e., leading to a constant community size K_i in the absence of disease-induced mortality). Human hosts progress from one infection class to the following because of exposure to water infested with cercariae. Specifically, $\mathscr{F}_i = a\sum_{j=1}^n Q_{ij}\theta_j C_j$ is the force of infection for the inhabitants of community i: $\mathbf{Q} = [Q_{ij}]$ is a row-stochastic matrix (i.e., a matrix in which rows sum to one) that describes the probability that residents of community i travel to community j (possibly different from their home community as a result of human mobility); θ_j is the human exposure rate, that is, the rate at which human hosts either permanently or temporarily staying in community j are exposed to contaminated freshwater (exposure rate is assumed to be possibly community dependent, so as to account for the geographical heterogeneity of living conditions); and a is the probability that a cercaria successfully develops into a reproductive adult parasite following contact with a human host. The term γ^p represents the parasite resolution rate, that is, the transition rate from infection class p to infection class $p - 1$, because of the death of one parasite ($\gamma^p = p\mu_P$, with μ_P being the per capita schistosome mortality rate). Disease-related mortality in humans is accounted for by the term α_H^p, which describes increasing mortality for increasing parasite burden ($\alpha_H^p = p\alpha_H$, where α_H is the additional mortality rate possibly experienced by an infected host because of the presence of each adult parasite). As for the dynamics of snail hosts, μ_S is the baseline mortality rate, whereas $\mu_S N_i$ is the constant recruitment rate (local population size in the absence of the parasite is N_i). The parameter b represents the exposure rate of susceptible snails to miracidia in the freshwater environment. Exposure triggers a transition to the infected compartment (possible delays between exposure and onset of infectivity [610] are neglected here for the sake of model minimality). Infective snails suffer from an extra-mortality rate α_S. As for the dynamics of larval stages, cercariae are shed by infected snails at rate π_C and die at rate μ_C. Similarly, miracidia are shed by infected human hosts and die at rate μ_M; specifically, the total human contamination rate for community i is $\mathscr{G}_i = \pi_M \delta_i \sum_{j=1}^n Q_{ji}\mathscr{W}_j/2$, with π_M being the shedding rate of miracidia by infected humans, δ_i the possibly site-specific probability of contaminating accessible freshwater, and Q_{ji} the probability that inhabitants of community j come in contact with freshwater in community i. Shedding is

Box 4.18 *Continued*

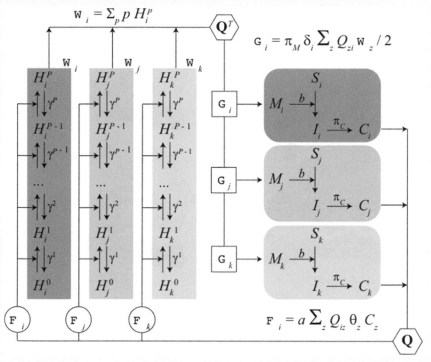

Figure 4.37 Schematic of the schistosomiasis transmission model. Rectangles represent the human components of three sample communities (say, i, j, and k, identified by different colors), stratified by infection class (H_i^p are hosts burdened with $0 \le p \le P$ parasites in community i). Rounded rectangles represent the freshwater components of the three communities, including (say, for community i) susceptible (S_i) and infected (I_i) snails, and the larval stages of the parasite (miracidia, M_i, and cercariae, C_i). Circles, squares, and hexagons indicate the force of infection for the different human communities, the human contribution to freshwater infestation, and mobility-related processes, respectively. See Figure 4.31a for a graphical depiction of the transmission cycle of the disease and table 1 in [540] for a complete description of the variables and parameters used in the model. Figure after [540]

assumed to be proportional to the total number $\mathscr{W}_j/2$ of adult parasite pairs undergoing sexual reproduction in the human hosts of community j, with $\mathscr{W}_j = \sum_{p=1}^{P} pH_j^p$. Note that this may represent an overestimate, especially at low parasite counts [609].

The model has been run [540] at the spatial scale of the arrondissements (third-level administrative units as of 2013). A high-resolution population density map (Figure 4.31b) has been used to obtain local values of K_i (Figure 4.38). To simplify the structure of the model, some equilibrium assumptions are made for the larval stages of the parasite. As a result, a synthetic exposure (β_i, accounting also for snail abundance) and contamination (χ_i) rates are introduced as composites of some of the parameters of the above system. These parameters are assumed to increase with the product (Figures 4.38b,c) between the fraction of people living in rural areas (ρ_i) and the availability of environmental freshwater (ω_i, measured as the total length of the rivers encompassed in each

Box 4.18 *Continued*

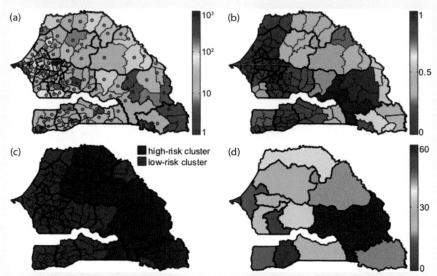

Figure 4.38 Data for model simulation and calibration. (a) Distribution of population abundance in third-level administrative units [thousands of inhabitants]. (b) Spatial distribution of the product between the rurality score ρ_i and the freshwater availability index ω_i (model setups with fine-grained environmental heterogeneity, M3 and M4). (c) Identification of high-risk versus low-risk transmission communities based on the clustering of $\rho_i\omega_i$ scores (model setups with coarse-grained environmental heterogeneity, M1 and M2). (d) Regional prevalence of urogenital schistosomiasis [% of infected people] according to the current estimates of the Senegalese Ministry of Health. Figure after [540]

spatial unit; Figure 4.31c), that is, $\beta_i = \beta_0(1 + \phi\rho_i\omega_i)$ and $\chi_i = \chi_0(1 + \xi\rho_i\omega_i)$. Human mobility is estimated from the anonymized movement routes of about 9 million Sonatel mobile phone users (corresponding to more than 60 percent of the Senegalese population) collected for one year, from January 1 to December 31, 2013. The entries of the mobility matrix **Q** are assumed to be proportional to the number of phone calls (call detail records, or CDR) made by users living in site i while being in site j (Figure 4.31d). The home site of each user is identified as the place where the most calls are made during night hours (7:00 PM to 7:00 AM).

From the model, it is straightforward to evaluate disease prevalence (namely, by assuming that a minimum number T of parasites per host is required for the infection to be clinically apparent), the APB (average parasite burden, a common measure of community-level infection intensity), and suitable indicators of parasite aggregation within each community, such as the dispersion index (defined as the ratio between the sample variance of the parasite distribution and the APB) and the aggregation parameter (obtained from fitting a negative binomial to the simulated parasite distribution). These quantities can be evaluated ex post, that is, as outputs of model simulations. Technical details are reported in [540].

The model is calibrated against regional estimates of urogenital schistosomiasis prevalence (Figure 4.38d) upscaled from the health district data available at the Senegalese Ministry of Health (Figure 4.31e). The prevalence of schistosomiasis in the country is periodically evaluated during national surveys conducted within the PNLB (Programme National de Lutte contre la Bilharzose). Model calibration is performed against the data that are currently in use at the Ministry of Health

Box 4.18 *Continued*

and that refer to surveys conducted through standard diagnostic techniques (urine testing via reagent strips, followed by filtration and microscopic examination of samples positive for haematuria) in schools selected from all of the 14 regions of Senegal between 1996 and 2013. Performing model calibration at the regional (rather than a finer) spatial scale is deemed to decrease the effects of the uncertainties possibly associated with census and/or epidemiological data. Details and references for model parameterization and calibration are reported in the supplementary information of the original reference [540], along with a description of some control strategies aimed at decreasing disease burden by preventing human exposure and contamination. As a closing note, it should be remarked that although used in this case to study schistosomiasis transmission in Senegal, this modeling framework can be applied with little change to other geographical regions, provided that suitable data for model calibration are available.

As argued above, spatial coupling mechanisms are important for the spread, persistence, and infection intensity of schistosomiasis [550, 592]. Parasites may in fact be carried in advective flows along canals – whether natural or artificial – and streams of any type as larvae, displaced between aquatic and riparian habitats inside snail hosts, or transported by human hosts as adult worms. While larval transport and snail movement may represent significant propagation pathways for the disease only over short spatial scales (e.g., in the order of hundreds of meters or long temporal windows because of habitat expansion following water resources development [586, 611]), human mobility can play a significant role in disease propagation within and from endemic areas [553, 555]. People can in fact be exposed to water infested with cercariae while visiting endemic regions and import the parasites back to their home communities; also, if infected, they can contribute to water contamination while traveling outside their home communities. Both mechanisms are expected to favor parasite dispersion and may even introduce schistosomes into villages that were previously disease-free [540].

In Senegal the urogenital form of the infection is widespread [540]. Schistosomiasis thus represents a major health problem, being the third disease after malaria and lymphatic filariasis in terms of years lived with disability. Because of the large spatial scale of interest, the exposure/contamination rates for the human host communities are assumed to be spatially heterogeneous to account for local differences in transmission risk [540]. For the same reason, human mobility is here retained as the most important mechanism for the spatial spread of the disease, as done

for the case of cholera induced by crowding called for by religious pilgrimages [520].

Human movement patterns were extracted [540] from a large dataset of anonymized CDRs (more than 15 billion records) made available by Sonatel, the largest Senegalese telecommunication provider (with a customer base of more than 9 million people). The model (Box 4.14) has been parameterized with georeferenced data on population abundance, socioeconomic conditions, and freshwater distribution and calibrated against the most up-to-date regional estimates of urogenital schistosomiasis prevalence currently available at the Senegalese Ministry of Health (Figures 4.31b–e and 4.38d after [540]). The scope of the work was to illustrate how the analysis of local heterogeneities in disease transmission could help guide resource allocation in the fight against the disease, with the overarching goal of showcasing how mathematical modeling can be used for societal development, namely, by assisting decision makers in the fight against a poverty-reinforcing infection like schistosomiasis [540].

Four model setups (M1–M4) were used [540] to investigate the role played by spatial heterogeneity and human mobility in schistosomiasis transmission. The different setups are characterized by either a coarse-grained (M1 and M2) or a fine-grained (M3 and M4) description of environmental heterogeneity and by either neglecting (M1 and M3) or taking into account (M2 and M4) human mobility. As for spatial heterogeneity, communities are grouped into two clusters according to transmission risk (either low or high) in M1 and M2, while they are endowed with site-specific exposure/contamination rates (depending on environmental and socioeconomic factors) in M3 and M4. As for human movement, the mobility matrix is set to be the identity matrix in M1 and M3, while its entries are estimated from CDRs in M2 and M4; note that in M1 and M3, the system describing schistosomiasis transmission reduces to a set of spatially disconnected local models. Full-fledged technical details on the evaluation of environmental heterogeneity and human mobility from georeferenced data are given in the original reference [540]. Of the four tested model setups [540], the one accounting for both a fine-grained description of spatially heterogeneous transmission risk and human mobility (M4) performs best in reproducing regional schistosomiasis prevalence values (Figure 4.39). The simulation results from this model are in good quantitative agreement with the available epidemiological data (Figure 4.39a, coefficient of determination data vs. model $R^2 = 0.76$). The average absolute data-model deviation is 6.0 percent, while the largest differences are found for the regions of Kaolack (6), Ziguinchor (11), and Kolda (9), where the model overestimates (Kaolack and Kolda) or underestimates (Ziguinchor) disease prevalence by more than 10 percent. Although calibrated with regional prevalence data, the reference model can project infection patterns throughout the country at the spatial scale of third-level administrative units (so-called arrondissements; Figure 4.39b) [540].

From an epidemiological standpoint, the model projects a country-wide schistosomiasis prevalence of about 21 percent, with a regional maximum in Tambacounda (59 percent of clinically infected people). Sensitivity analysis suggested that these figures are quite robust to (moderate) changes in model parameterization, with the baseline human exposure rate, the schistosome mortality rate, and the threshold for clinical infection in humans being responsible for the largest variations in model predictions. The model also projects an average parasite burden (APB) of approximately 7.2 parasites per person, with arrondissement-level values ranging between 2.3 and 12.7 with a bimodal frequency distribution, with marked peaks around 3 and 10 parasites per person [540]. Contrasting the reference model against a simulation with the same parameter values but no mobility proved useful to enucleate the impact of human mobility on the spatial patterns of schistosomiasis prevalence [540]. Regional disease prevalence is found to be higher in all regions but Dakar if human mobility is completely switched off. Instead, the impact of human mobility at the arrondissement level is relatively more diversified in space: the effects of the mobility switch-off are less pronounced in the westernmost part of the

country, where smaller increments (or even decrements in some arrondissements) of disease prevalence are predicted in the absence of mobility. More in general, the results of [540] suggest that it is possible to contrast the model with simulations in which the human mobility rate is artificially manipulated, that is, decreased or increased with respect to the country-wide estimate from CDR analysis (26 percent of mobile people in a one-year interval), requiring a suitable redistribution of mobility fluxes. According to model projections, country-scale APB shows an increasing trend with increasing mobility, while disease prevalence attains a well-defined minimum for intermediate levels of mobility – remarkably, close to the actual estimate of mobility obtained from CDRs [540]. Therefore, changes in mobility can alter within-host parasite distributions at the community level in a way that nontrivially influences disease prevalence. This finding can be explained by the fact that, under the current mobility scenario, the largest mobility fluxes are attracted by the most populated and urbanized regions (Dakar, Thiès, and Diourbel), where schistosomiasis transmission is quite low. Conversely, mobility may represent a risk factor for people living in prevalently urban areas, as found for the region of Dakar and several other areas in the western part of the country, where local infection prevalence is expected to be higher in the presence of mobility, clearly because of movement to/from rural areas where schistosomiasis thrives. Numerical simulations also show that, should propensity to moving increase with respect to what is currently inferred from CDRs, country-wide average schistosomiasis prevalence could become higher as well. Country-wide APB as a function of mobility shows a different trend, namely, a nearly monotonic increase for higher levels of mobility. This result seems to agree with other modeling studies that have reported a positive relationship between human mobility and transmission emergence, parasite burden, and disease spread [550]. Per se, however, APB may represent a relatively poor epidemiological indicator for schistosomiasis dynamics. By accounting for the stratification of the infection [607], instead, the framework described here [540] shows that human mobility may play a more complex role in the definition of the spatial patterns of schistosomiasis prevalence than previously thought, especially at small spatial scales.

The reference model was finally used to evaluate the effects of water, sanitation, and hygiene (WASH) or information, education, and communication (IEC)

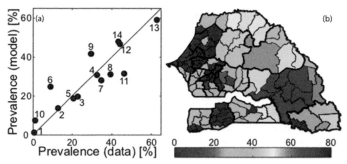

Figure 4.39 Reference model simulation and comparison with epidemiological evidence. (a) Quantitative agreement between simulated disease prevalence at the regional scale and the available data (labels as in Figure 4.31b) for the best-fit model accounting for fine-grained spatial heterogeneity in transmission risk and human mobility estimated from CDRs. (b) Projected schistosomiasis prevalence [% of people infected] at the scale of third-level administrative units. Calibrated parameter values are reported in table 1 of [540]. Figure after [540]

strategies aimed to reduce the burden of schistosomiasis in Senegal through prevention of human exposure and contamination. Concerning WASH [540], the model suggests that targeted interventions, prioritizing either high-risk communities (where schistosomiasis transmission is expected to be highest because of the synergistic effect of rural living conditions and abundance of freshwater environments) or high-prevalence communities, may be more effective than untargeted ones in reducing both the average and the maximum regional prevalence of infection. Specifically, risk-targeted actions are predicted to be the most effective option if the expected efficiency of the interventions is high; conversely, prevalence-targeted interventions may be more effective in reducing average disease prevalence for low expected efficiency. Different results are obtained with IEC campaigns. Untargeted actions may represent the most effective option to reduce average prevalence, namely, if the expected efficiency of the interventions is high and the plan involves at least $\approx$ 1 million people ($\approx$ 5 millions for maximum regional prevalence). In case of smaller-scale plans or low expected efficiency, targeted actions are predicted again to be more efficient; in this case, the effects of prioritizing high-risk versus high-prevalence communities depend on the planned extent of the intervention.

Current measures for fighting schistosomiasis are principally focused on chemotherapy, and Senegal has implemented a national control program (Programme National de Lutte contre la Bilharziose, PNLB) since 1999 [540]. The PNLB is still ongoing to date, but scarcity of large-scale data on treatment coverage and intervention effectiveness impairs a fair assessment

of the effects of any mass drug administration in country-wide models. Because chemotherapy does not confer permanent immunity, preventing infection by improving access to safe water and spreading awareness about disease transmission can represent a sustainable path toward reducing the burden of schistosomiasis. The analysis of control strategies based on water contact conforms with a meta-analysis of observational studies [612] that found that both safe water supplies and adequate sanitation are associated with significantly lower odds of schistosomiasis infection. However, safe water supplies may reduce contact with environmental water but cannot completely avert it; similarly, sanitation can prevent snail infection, but the availability of adequate sanitation does not guarantee its use. Modeling analyses of the type presented here strongly indicate that disease control efforts should be guided by a thorough understanding of the drivers that determine local heterogeneities in transmission risk, especially if resource availability is limited (a commonplace in the fight of neglected tropical diseases in developing countries) and/or a high efficiency of the interventions cannot be taken for granted. From a biological perspective, a detailed description of parasite–host interactions and in-host parasite biology [609], as well as of the ecology of the obligate intermediate snail host of schistosomes [567, 613], has yet to be integrated into current spatially explicit modeling frameworks. As the presence of snail hosts is a major determinant of transmission risk, accounting for the spatiotemporal variability of the environmental drivers (most notably, water temperature and rainfall [614]) that influence their distribution and abundance could greatly improve the explanatory power of the model. Particular attention should be devoted to studying the possible interplay between the seasonality of environmental signals and time-varying human mobility, which could introduce nontrivial effects on disease transmission [115]. Integrating the ecology of the intermediate host into the modeling framework described here is also deemed crucial to planning and optimizing unconventional intervention strategies based on biological snail control. All these observations stress the importance of a comprehensive approach to schistosomiasis management [562] that vitally includes the tools described in this chapter.

4.5 Proliferative Kidney Disease (PKD) in Salmonid Fish

One last example of application of the tools developed in this chapter that seems particularly fit to the focus on fluvial ecological corridors concerns the spread, possibly rapid and deadly, of a particular fish disease. Proliferative kidney disease (PKD) is a high-mortality pathology that critically affects freshwater salmonid populations. Infection is caused by the endoparasitic myxozoan *Tetracapsuloides bryosalmonae*, which cycles between freshwater bryozoans and salmonids exploiting the former as primary hosts. PKD is recognized as a major threat to both wild and farmed salmonid populations in Europe and North America. Mortality due to PKD in farmed fish ranges from 20 to 100 percent [410, 615–617]. The impacts of PKD on wild fish populations are generally poorly known, although PKD has been linked to long-term declines in Swiss brown trout populations [618]. PKD incidence and relevant fish mortality are strongly correlated with water temperature, therefore climate change is feared to extend the disease range to higher altitude and latitude regions with major consequences [619].

T. bryosalmonae has a complex life cycle that exploits freshwater bryozoans and salmonids as hosts (for a review, see [620]). Within bryozoans, the parasite can express either covert or overt infection stages. During covert infections, the parasite exists as nonvirulent single-cell stages. The transition to overt infection implies increase in virulence, with the formation of multicellular sacs from which thousands of *T. bryosalmonae* spores are released into water. Parasite transmission from bryozoans to fish can thus take place only during the overt infection stages. Peaks of overt infections have been observed

in late spring and autumn. Overt infection hampers bryozoan growth and elicits temporary castration, with a severe reduction in the production of statoblasts (asexually produced dormant propagules). Covertly infected bryozoans can produce infected statoblasts, thus allowing vertical transmission of *T. bryosalmonae*.

Parasite spores released into water infect fish through skin and gills. *T. bryosalmonae* subsequently enters the kidney of fish hosts, where it undergoes multiplication and differentiation, entailing a massive granulomatous nephritis with vascular necrosis. Spores developed in the lumen of kidney tubules are eventually excreted via urine and can infect bryozoans (not other fish). Although PKD seems to develop in all salmonids to a varying degree, life cycle completion may be highly species specific [621]. For example, in Europe, parasite spores infective to bryozoans develop in the brown trout (*Salmo trutta*) and the brook trout (*Salvelinus fontinalis*) but not in the rainbow trout (*Oncorhynchus mykiss*). Fish infected with PKD often die owing to secondary infections; however, PKD alone has been shown to cause mortality. Fish that do not die during the acute phase of the infection may become long-term carriers of the parasite. Such carriers are reportedly able to infect *Fredericella sultana*, one of the most common bryozoan hosts of *T. bryosalmonae*, for a period up to two years after exposure. PKD incidence and fish mortality are strongly correlated with water temperature, therefore climate change is feared to extend the disease range to higher altitude and latitude regions, with major consequences [619].

4.5.1 Hierarchy of Models

The first dynamical model of PKD epidemiology in local communities has recently been developed [621]. The local model accounts for local demographic and epidemiological dynamics of bryozoans and salmonids, explicitly incorporates the role of water temperature, and couples intra- and interseasonal dynamics. Figure 4.40 shows that during the warm season (left column), susceptible bryozoans (B_S) become covertly infected (B_C) after contact with spores (Z_F) released by infected fish. Infection in bryozoans cycles between covert and overt (B_O) states. B_S yield uninfected statoblasts S_S (i.e., asexually produced propagules), while B_C release both uninfected and infected (S_I) statoblasts. Infected bryozoans may clear the infection. Susceptible fish (F_S) are exposed to PKD after contact with spores (Z_B) released by B_O; after an incubation phase (F_E), fish can either become acutely

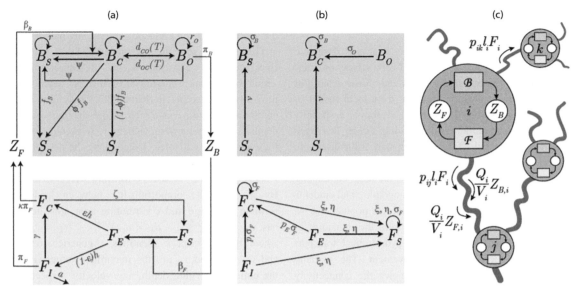

Figure 4.40 Schematic of the proliferative kidney disease model. State variables and parameters are briefly mentioned in the text, and fully detailed in tables S1 and S2 in [416]. (a) Local intra-annual dynamics. Natural fish mortality is independent of epidemiological status; therefore it is not displayed for the sake of readability. (b) Local interannual dynamics. (c) Spatially explicit framework, showing the river network as the substrate for ecological interactions and for the spread of the infection. Note that the main symbols are: B, for the bryozoan submodel; F, for the fish submodel. Several parameters depend on temperature. The main ones are the transition rates (denoted in the figure by d_{CO} and d_{OC}, respectively) for covert–overt and overt–covert infections, and the fish mortality rate. Figure adapted from [416, 621]

infected (F_I) or directly enter an asymptomatic carrier state (F_C). F_I can die owing to PKD or become long-term disease carriers. F_C may become susceptible again. Both F_I and F_C shed spores infective to bryozoans. At the beginning of a new warm season (central column), B_S comprise susceptible colonies that survived during winter and hatched uninfected statoblasts, S_S; similarly, B_C consist of survived colonies that were infected at the beginning of winter and of hatched infected statoblasts, S_I. F_S are composed by survived susceptible individuals and newborn fish from all classes. The abundance of F_C is determined by the number of individuals belonging to classes F_E, F_I, and F_C that survived through winter. Other classes are absent.

A stability analysis of this time-hybrid system, performed via Floquet theory [447, 450] (Appendix 6.1.4), suggests that, whenever the epidemiological parameters are set to realistic values, the introduction of *T. bryosalmonae* in a previously disease-free community will most likely trigger a PKD outbreak. A sensitivity analysis of the system shows that, when the disease becomes endemic, the impact of PKD on fish population

size is mostly controlled by the (temperature-dependent) rates of disease development in the fish host.

On the other hand, this first model was still a local one with no regard to the fundamental role played by spatial dynamics. Indeed, understanding how different local communities distributed in space interact with each other is crucial to possibly predicting the spread of a PKD epidemic and the effect of intervention strategies. From a theoretical viewpoint, as stated above, the condition that disease-free equilibria in all local communities be unstable is neither necessary nor sufficient for the occurrence of outbreaks when such communities are connected by different layers of connectivity affecting transmission [437]. While the effects of landscape and river network connectivity have been extensively evaluated for waterborne diseases affecting humans, the literature focusing on modeling processes of waterborne epizootics was scarce if not null in a spatial metacommunity framework. Yet the issue of pinpointing effective mitigation strategies for wildlife diseases such as PKD is becoming increasingly critical for environmental policy makers (see Section 5.6 for a noteworthy example of a recent

PKD outbreak that occurred in the Yellowstone River, Montana, USA).

For the above reason, Carraro et al. [416] set out to study the key influence of network effects on PKD mortality and prevalence patterns and on the celerity of disease propagation. Building on their local model of PKD transmission [621] mentioned above, a spatially explicit metacommunity framework has been developed to study the spatial effects in the spread of the disease in idealized stream networks [416]. These are described by a suitable connection matrix **D**. At the local community scale, the model accounts for demographic and epidemiological dynamics of bryozoan and fish populations. At the network scale, the model couples the dynamics of each local community through hydrological transport of parasite spores and fish movement. The spatially explicit model (Figure 4.40c) shows the connectivity of local models operating at the scale of a river reach of relatively homogeneous geomorphic features (i.e., a node) effectively carried out by the river network, the substrate for host and pathogen interactions (for mathematical details, see [416]). The model also explicitly accounts for water temperature variations that influence epidemiological parameters, for heterogeneity in habitat characteristics, and for hydrological conditions along a river network.

4.5.2 Spread of PKD in Idealized River Networks

The idealized stream network used in the simulations is obtained as an OCN (Box 1.3), which defines a tree – an object whose connectivity together with the directdness of the embedded graph warrants a unique path joining any two nodes – that spans the whole assigned landscape/area. We consider landscapes formed by square lattices of D_p^2 pixels whose side has length L_p. In the real world, however, a drainage path becomes a stream only when certain hydrological conditions are met. The simplest and most tested method assumes that pixels form a channel when their contributing area (a proxy of landscape-forming discharge) exceeds a certain threshold $A_{C,T}$ [416]. The network is then discretized into stretches, each of which is defined as a sequence $\mathscr{T}$ of channelized pixels starting from one pixel having either zero (river sources) or more than one channelized upstream pixels (confluences) and containing the downstream sequence of channelized pixels until another confluence or the outlet is reached. The obtained

network of stretches is an oriented graph suitable for the application of the PKD metacommunity model.

The second step in the generation of synthetic river networks consists in the definition of the geomorphological properties of each river stretch, using the relations illustrated in Section 2.2. Cross sections are approximated with rectangular shapes having width w, depth d, and average water velocity v. To account for how these geometric variables change along the network, we exploit the relationships $w \propto Q^{0.5}$, $d \propto Q^{0.4}$, and $v \propto Q^{0.1}$, where Q is the landscape-forming river discharge. By invoking the proportionality between landscape-forming discharge and contributing area at-a-site, one has $w_k = w_o A_k^{0.5}$, $d_k = d_o A_k^{0.4}$, $v_k = v_o A_k^{0.1}$, where the subscript k identifies a generic stretch k and w_o, d_o, and v_o are the maximum values (i.e., at the outlet) for width, depth, and velocity, respectively; $A_k = A_{C,i_k}/D_p^2$ is the normalized contributing area to stretch k; and subscript i_k identifies the last downstream pixel of stretch k (the last element of $\mathscr{T}_k$). The discharge at the outlet of the catchment reads, then, $Q_o = w_o d_o v_o$; the water volume in a generic stretch k is $V_k = w_k d_k L_{s,k}$, where the length $L_{s,k} = \sum_{i \in \mathscr{T}_k} L_i$ and $L_i = L_p$ if the flow direction along pixel i is parallel to a pixel side or $L_i = \sqrt{2}L_p$ if parallel to the pixel diagonal. The slope–area relationship allows one to evaluate the elevation of the network stretches. The elevation drop along a network pixel can indeed be expressed as $\Delta z_i = s_o A_i^{-0.5} L_i$, where s_o is the outlet slope. This relationship can be iteratively applied starting from the outlet pixel (where an arbitrary elevation z_o is imposed) toward all upstream paths, in order to reconstruct the elevation of each channelized pixel. The elevation of a stretch can be defined as the average elevation among all constituting pixels. Examples of OCNs obtained from changing only the location of the output pixel are shown in Figure 4.41. Scale parameters that define the metric of the river network were chosen to be representative of a prealpine catchment of around 1,000 km^2.

The seasonal cycle of stream water temperature typically follows that of air temperature, albeit being damped and possibly delayed. However, notable deviations can be observed in streams with large impoundments or lakes upstream, or when the thermal regime is dominated by ice/snow melting. For this exercise, we assume that water temperature at the outlet reach $T_o(\tau)$ follows a sinusoidal function with period equal to one year. To derive time series of water temperature for all network stretches, we further assume that water temperature mirrors the environmental

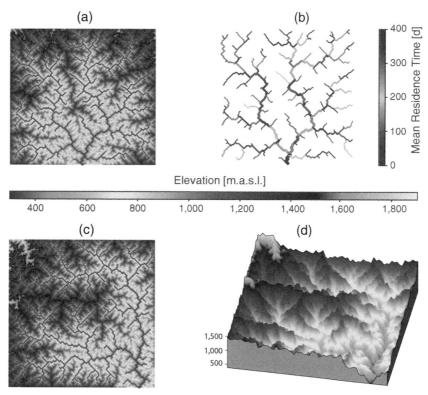

Figure 4.41 (a) Example of an optimal channel network. The elevation map has been obtained from extrapolating a deterministic slope–area law to unchanneled pixels as well; while this hypothesis is not generally valid in real landscapes, it has no implication for this work as discussed in Appendix 6.2 (this is tantamount to assuming that the characteristic unchanneled length, the inverse of the drainage density $[L^{-1}]$ [63], is of the order of the pixel size tiling the landscape). (b) Distribution of mean fish residence times for the OCN presented in panel (a) with average fish mobility $l_{avg} = 0.02$ $[d^{-1}]$. Mean residence times are computed by assuming reasonable parameters and spatially uniform fish density [416]. (c) A replica of OCN grown in the same domain with a different localization of the output pixel. (d) Tridimensional landscape generated by the OCN depicted in panel (c) [416].

lapse rate of air temperature. Water temperature in a generic stretch k is then $T_k(\tau) = T_o(\tau) + \Gamma_w \Delta z_k$, where Γ_w is the lapse rate and Δz_k the difference in elevation with respect to the outlet. Lapse rates for air temperature can range from about -9.8°C km^{-1} for dry air (dry adiabatic lapse rate, or DALR) to about -4.0°C km^{-1} for hot saturated air (saturated adiabatic lapse rate). We assume Γ_w equal to -6.5°C km^{-1}, a typical value that is used as global mean environmental lapse rate for air temperature [416]. Temperature influences demographic and epidemiological parameters of both bryozoans and fish. The main effects of a higher temperature are an increase in covert–overt transition rate and fish mortality (details in table S2 of [416]).

The fish mobility rate l_i can be thought of as the inverse of the population-average residence time within stretch i. In general, stretches may have different

geometric and physical characteristics (e.g., length, water volume and depth, velocity, fish carrying capacity) and thus l_i is expected to change across the river network; l_i is computed in such a way that the stationary state of the underlying diffusion process is a specific spatial distribution of fish abundances F_i. The underlying idea is that, in order to apply the model, first a distribution of fish abundance at carrying capacity is assigned according to the characteristics of each stretch, then a set of mobility rates l_i is derived so that fish movement leads, at steady state, to the desired abundance distribution. Once movement rules are determined (diffusion matrix **D**), the values of l_i such that a distribution of fish abundances F_i is an equilibrium state are obtained from solving the following linear system:

$$d_{ij} l_i F_i = d_{ji} l_j F_j \qquad \forall i \leq N_s, j \leq N_s, \qquad (4.53)$$

where N_S is the number of network nodes, that is, river stretches. Note that only $N_s - 1$ of the above equations are nontrivial identities: in fact, $d_{ij} \neq 0$ only if stretches i and j are directly connected, and every stretch has one downstream connection, with the exception of the outlet stretch. Fish are assumed to have equal probability of moving upstream or downstream. The system has ∞^1 solutions; indeed, if a set of l_i is a solution, also the same set multiplied by a scalar is a solution. It is thus possible to focus on a single solution by specifying the average mobility rate across the network (l_{avg}). Instead, spores are passively transported, hence their fluxes are proportional to the river flows.

The model has been applied to several replicas of OCN landscapes (see Box 1.3 and Appendix 6.2) (Figure 4.41). Results show how network connectivity and hydrological conditions critically control the spatial distribution of the prevalence of PKD and the celerity of invasion fronts in the upstream and downstream directions. Figure 4.42 shows how connectivity and fish mobility affect the distribution of disease prevalence (i.e., the fraction of infected fish) in a flat landscape, where temperature is chosen so that the total thermal energy of the river network equals that of the OCN landscape of Figure 4.41.

Two main types of patterns arise. Prevalence decreases with any mobility rate. However, prevalence is higher for generally higher mobility rates (Figures 4.42c,d). The sites with higher prevalence are those whose contributing area is higher, for any mobility rate (Figures 4.42b,c). In analogy, sites characterized by larger distance from the outlet (Figures 4.42a, d) are less exposed when fish mobility is small to negligible. We note that total contributing area is not only a proxy of landscape-forming discharge but also a proxy of the abundance of parasite spores that enter a given stretch – thus of the main driver of infections in the absence of fish movement. In this case, small headwater stretches in the proximity of the outlet are not invaded by the parasite (see magenta arrows in Figure 4.42b). Conversely, high mobility rates enhance the mixing process among local communities, with the result that the variability in prevalence among neighboring stretches is low and the average network prevalence increases. When mobility rates are tuned to more realistic values (i.e., by assuming the distribution of mean residence times of Figure 4.41b), the distribution of prevalence shows an intermediate behavior (blue lines in Figures 4.42c,d). Patterns of fish loss (Figure 4.42f) exhibit an analogous trend.

Elevation of river stretches is intimately related to the structure of the underlying river network (Appendix 6.2).

Indeed, the well-established and observed slope–drainage area relationship dictates that a network configuration is uniquely associated to a relative elevation distribution of river sites [74] (see [75] for the connection with OCNs in the general case). Thus network structure controls, as a byproduct, also the distribution of elevations, a proxy of mean water temperature and in turn of PKD prevalence [621]. Indeed, water temperature generally decreases with elevation and thus with the distance to the outlet. Temperature gradients tend thus to produce distributions of PKD prevalence akin to those discussed above driven by hydrological transport and fish mobility. The river network analyzed herein spans an elevation relief of about 1,000 m, which translates to about 6.5°C difference in mean water temperature. Larger networks with more pronounced elevation relief can possibly lead to more important effects of temperature gradients. In this case, the inclusion of an elevation gradient has a minor impact on the distribution of prevalence (Figure 4.42e); of course, this depends on the lapse rate chosen. As a consequence of diminished temperature, a decrease of prevalence in upstream stretches is observed, which is not compensated for by a corresponding increment in lower altitude stretches. This causes a small reduction in the average network prevalence (about −1.5 percent regardless of fish mobility). As for the distribution of fish loss (Figure 4.42f), the effects of nonuniform temperature are slightly accentuated. At the outlet, fish loss is generally higher (+4.3 percent when l_{avg} is null), but the fish loss does not change substantially with respect to the flat landscape case.

The invasion celerity of PKD in the downstream direction is mostly controlled by hydrological transport of spores, whereas fish mobility has only a marginal effect. For transmission parameters leading to high PKD prevalence (above 90 percent, a value sometimes observed in affected river systems [622, 623]), the disease can invade from tens to hundreds of kilometers of river within a single proliferation season, provided that all sites are equally suitable for fish and bryozoans. Upstream invasion of PKD from a region of the network close to the outlet can occur only via fish swimming against the flow direction. The corresponding celerity is much slower than the downstream one. With realistic values of fish mobility (e.g., $l_{avg} = 0.02$ d^{-1}), PKD can travel upstream only a few kilometers per season.

Also, a sensitivity analysis of disease propagation celerity with respect to the contamination rates (pathogen spores produced by both fish and bryozoans) and the fish mobility rate l_{avg} has been conducted. Downstream

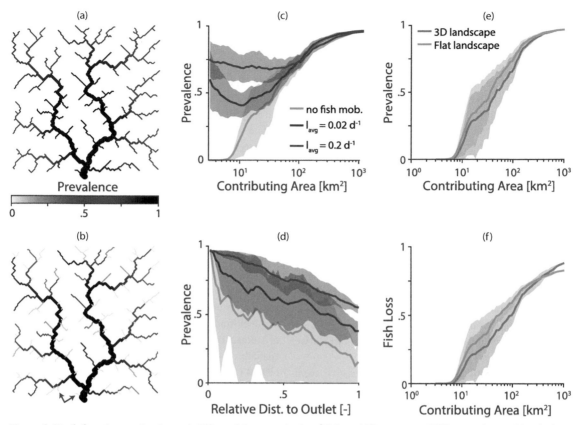

Figure 4.42 (left and central columns) Effect of the magnitude of fish mobility rates on PKD prevalence. Simulations are run for 50 years; the prevalence at the end of the fiftie-th season is shown. A flat landscape is assumed. (a) Prevalence map for a given OCN and fish mobility $l_{avg} = 0.2$ [d^{-1}] [416]. (b) Prevalence map in the absence of fish mobility. (c) Prevalence as a function of contributing area. For 10 different OCNs, prevalence at each stretch is evaluated. Solid lines represent mean trends; shaded areas identify 25th–75th percentiles of the distribution. (d) Prevalence as a function of relative distance to the outlet. (right column) Effect of elevation gradient on prevalence (e) and fish loss (f) when fish mobility is set to zero. Symbols are as in panel (c). Epidemiological parameters are set to their reference value [416]

propagation generally occurs after one to three seasons much faster than upstream propagation (which reaches a steady state after around 100 seasons), as a consequence of the bias in the hydrological transport of spores and of its fast dynamics. Expectedly, both the fish mobility and the contamination rates are positively correlated with propagation celerity in both directions, although the role of l_{avg} in downstream propagation is minor. When both contamination and mobility are small, rather obviously, PKD might not establish in the network. Note that, while the absence of PKD at the outlet stretch implies that the whole network is disease-free, this is not necessarily true with regard to the headwaters. Similar effects are produced by variations of fish or of bryozoan contamination rates.

Fish movement and hydrological transport within a river network can thus produce a heterogeneous distribution of PKD prevalence and fish loss even in the absence of spatial gradients of fish and bryozoan densities or of transmission rates [416]. The typical lifetime of PKD spores (around one day) allows them to travel along with the flow, and possibly infect fish, tens of kilometers downstream of the point where they are released. Stretches farther downstream thus collect spores from the whole (or a large portion of the) upstream area and are therefore more likely to exhibit higher PKD prevalence and fish loss. Conversely, headwaters and low-order streams are subject only to the spore load that is locally released and thus tend to be relatively less affected by PKD. Therefore, hydrological transport of spores tends to produce spatial patterns of disease prevalence correlated to the upstream drainage area. This dominant pattern can partially be affected by fish mobility. Indeed, a headwater connected directly to a high-order stream is subject to immigration of likely infected fish that foster local prevalence. Overall, fish mobility promotes the mixing between low- and high-order streams, resulting in a net increase of overall prevalence at the network scale.

4.5.3 Integrated Field, Laboratory, and Theoretical Studies of PKD Spread in a Swiss Prealpine River

The above results further our understanding of the drivers of fish distribution in riverine ecosystems and provide the basis for the development of intervention and management tools, which is one of the main tenets of this book. A real benchmark, however, came from a comprehensive field, laboratory, and theoretical study carried out on the Wigger catchment in Switzerland [624], which we describe here.

Given the complex life cycle of the causative agent and the significance of the ecological and environmental factors involved in its transmission [410], as we pointed out at the onset of this section, mathematical modeling is of paramount importance to understand the epidemiology of PKD in real river networks – to possibly assess the effectiveness of mitigation strategies. We move from the the epidemiological model for PKD described above [416, 621], expanded into a metacommunity framework fine-tuned to the specificities of the Wigger [624]. In particular, it accounts for heterogeneity in habitat characteristics and hydrothermal conditions along the river network. Here, the original model is further developed to incorporate age structure and spawning migration of the fish population. Young-of-the-year (YOY) fish are explicitly considered, because they might respond to infection differently from adults. The model is applied to the Swiss river Wigger (Figures 4.43a,b), where the brown trout population was comprehensively surveyed for three years (2014–2016) [624], together with *F. sultana* eDNA concentration (Figure 4.43c) and relevant environmental parameters such as water temperature. The river Wigger, located in the Swiss plateau, is a tributary of the river Aare and has a length of 48.11 km. It drains a watershed of 382.4 km^2, which has an elevation range between 396 and 1409 m a.s.l., at the Mount Napf (Figures 4.43a,b). The Wigger has been subject to endemic PKD for several years (see [622, 625]; a dataset of PKD

occurrence across Switzerland is freely available at the website of the Swiss Federal Geoportal [626]. The river network was extracted from a 25-m resolution digital terrain model using the Taudem routine in GIS software [215]. The geological characterization of the catchment (Figure 4.43e) was obtained from a vectorized geological map of Switzerland provided by the Swiss Federal Office of Topography (Swisstopo) [626]. Daily mean discharges measured by the Swiss Federal Office for the Environment (FOEN) in Zofingen (corresponding to site 3 of Figure 4.43a) were used to compute time series of discharge for all stretches, based on the assumption of proportionality between discharge and contributing area [63]. Data collection protocols are illustrated in Box 4.19.

Reliable predictions of the spread of infectious diseases and possible management strategies must be based on accurate assessments of the spatial distribution of the invertebrate host [627]. In the reference work [624], a novel framework was developed to estimate local *F. sultana* densities based on temporally repeated and spatially distributed quantitative environmental DNA (eDNA) point measurements. This framework was further developed in the more recent recent work [376] that was illustrated in Section 3.6.

The novel methodology proposed therein opened avenues for a generalized use of eDNA testing for predicting species occurrence and density in natural habitats and, especially, in aquatic ecosystems. Because the river acts as an integrator of spatially heterogeneous sources of genetic material [394], the eDNA concentration at a given river cross section results from a combination of dilution effects and decay processes [376, 402]. For the sake of easy readership, here we also summarize the main findings regarding bryozoan spatial distributions as inferred from eDNA measurements, already discussed in Section 3.6. Local species densities are correlated to site-specific covariates (Figure 4.44e). The so-obtained bryozoan density map is then used in the metacommunity PKD model.

Confirming the sensitivity of the eDNA based detection method, multiple manual field searches for bryozoan colonies on all of the 15 eDNA sampling sites were able to locate populations only in the three sites (#9, #13, #15) corresponding to the highest eDNA concentrations and the most frequent detections (see Figure 4.44d). Similarly, sites known to be bryozoan negative through exhaustive field searches (e.g., #5) were found consistently negative via eDNA. Local bryozoan density was positively correlated with the presence of moraines upstream (*GeoMrn*)

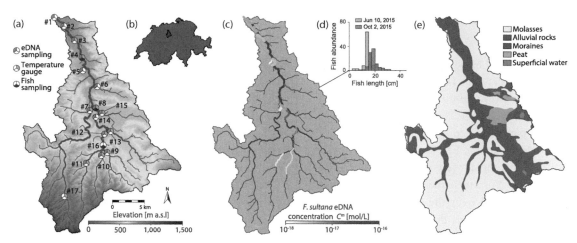

Figure 4.43 Overview of the study area and data. (a) Digital terrain model of the Wigger and extracted river network with locations of the sampling sites. (b) Position of the Wigger catchment in Switzerland. (c) Mean measured *F. sultana* eDNA concentration. Ungauged stretches are depicted in blue. (d) Results of fish sampling campaign on site #8 in 2015. (e) Geological characterization of the catchment. Figure after [624]

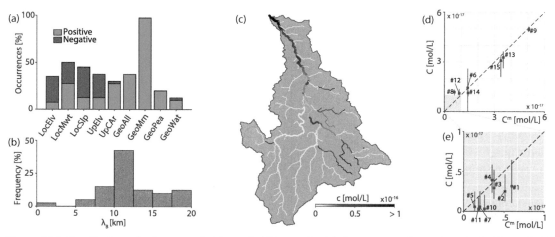

Figure 4.44 Results from the bryozoan habitat suitability study. (a) Occurrence of covariates in habitat suitability models. *Positive* and *negative* refer to the signs of the calibrated β coefficients. Covariates' abbreviations are *GeoAll*, percentage of alluvial rocks; *GeoMrn*, percentage of moraines; *GeoWat*, percentage of superficial water; *LocMwt*, local mean water temperature; *LocSlp*, local slope; *UpCAr*, contributing area; *UpElv*, upstream mean elevation. (b) Distribution of calibrated values of λ_B in habitat suitability models. (c) Map of local eDNA concentration obtained from averaging results from all habitat suitability models. (d) Modelled (C) versus observed (C^m) *F. sultana* eDNA concentration. Red lines identify 10th–90th percentile ranges of the distribution of all accepted models; squares represent values averaged over all accepted models. (e) Zoom from panel (d). Figure after [624]

(present with positive sign in 97.5 percent of the tested models (Figure 4.44a). Environmental covariates such as mean water temperature (*LocMwt*) and local slope (*LocSlp*) had a less clear effect on bryozoan density, as their correlation might be positive or negative depending on the particular model. Local (*LocElv*) and upstream mean elevation (*UpSlp*) are in most cases negatively associated with bryozoan density. In 42.5 percent of

the tested models, the calibrated values of the decay length λ_B lie between 10 and 12.5 km (Figure 4.44b). Overall, the capability of the models to reproduce the mean *F. sultana* eDNA concentrations measured in the 15 sampling sites (C^m) seems satisfactory (Figures 4.44d,e). The predicted spatial distribution of *F. sultana* eDNA local concentrations (averaged over the various models) is shown in Figure 4.44c.

Box 4.19 Data Collection Protocols for the Wigger Case Study

In the period May 2014–2015, stream water samples were collected in 15 different locations along the river network (Figure 4.43a). For each site, 21 500-mL samples were taken at approximately bi-weekly (or monthly during December, January, and February) intervals (except site #5, which was abandoned after 12 samples). The 500-mL samples were filtered onto 5-cm diameter GF/F filters and filter papers frozen at −80°C until extraction of eDNA (the collection method is described in [376]; also Hartikainen, Fontes, and Jokkela, personal communication).

Water samples were analysed via qPCR to detect and quantify *F. sultana* 18S rDNA concentration in water. For subsequent data analysis, the target DNA quantity estimated in each water sample was averaged over the 21 temporally distributed samples; the resulting mean concentration C^m (Figure 4.43c) was then used as input for the determination of bryozoan habitat suitability. Water temperature has been measured since July 2014 in 11 sites via HOBO TidbiT v2 data loggers. An additional temperature gauge (site #17) was added in September 2015. Two loggers, recording data at 15-min intervals, have been deployed per each site. Water temperature is known to affect bryozoan growth rate [621] and may also impact spore shedding [624].

All young of the year (YOY) (Box 4.20) trout sampled in this study were originated from natural reproduction, as there was no fish stocking in the Wigger in the period 2014–2016 (Aargau and Lucerne cantonal authorities, personal communication). Trout abundance estimation, collection of fish, and kidney sampling were performed according to the Swiss regulations, and the field setup was accepted by the relevant authority under the number LU05/14+. Fish were caught by electrofishing in early and late summer on sites #4, #8, and #16 (Figure 4.43a), over a distance of 100 m. An example of fish density assessment is shown in Figure 4.43d. During each sampling trip, 25 YOY brown trout were collected outside the stretch used for density assessment. When available, 5 adult brown trout adult (>1 year old) were sampled during the late summer field campaign [624].

Electrofishing was performed with an EFKO FEG8000 machine (8 kW power). Two runs of electrofishing per date and stretch were performed. For density assessment purpose, caught fish were kept in tanks supplied with oxygen, and fish were set back into the river after the second fishing run. Prior to release, fish length was measured. The trout were euthanized using 3-aminobenzoic acid ethyl ester (MS 222, Argent Chemical Laboratories), and length was recorded. The kidney was removed and preserved in RNAlater (Sigma Aldrich, Switzerland) for qPCR based assessment of *T. bryosalmonae* infection presence [624].

Field surveys assessed PKD prevalence in fish at three sites (#4, #8, and #16; Figure 4.43a) both in early (for YOY) and late summer (for both YOY and adults). In the downstream and intermediate sites (#4 and #8, respectively), the general pattern is that all fish of all age classes in all sampling dates are infected (Figure 4.45a), while in the upstream site #16, YOY prevalences in early summer were always lower than 100 percent. Notably, in the late 2015 summer sampling, no YOY were found at the usual location of the sampling site #16 (i.e., downstream of the junction of the three tributaries Änziwigger, Buechwigger, and Seewag), hence the sampling was shifted upstream of the confluence with Seewag (see circle in Figure 4.45e). None of the YOY sampled on that occasion tested positive for PKD [624].

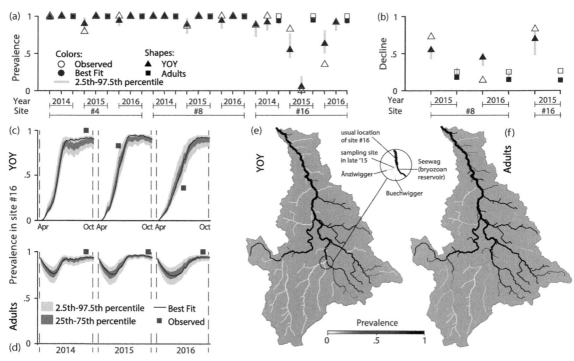

Figure 4.45 Results from the epidemiological metacommunity model. (top row) Results of calibration against (a) prevalence data; (b) seasonal fish decline, measured as the fractional decline of the estimated population size in late summer compared to early summer. In panel (a), the left point of each year group corresponds to YOY early summer sampling. The observed 0-prevalence value was actually measured in a stretch upstream of site 16 (see panel (e) and main text). (bottom left corner) Time evolution of modelled prevalence in site 16 for (c) YOY and (d) adult fish. The intra-seasonal model is run for 200 days starting on April 1. Note that the YOY prevalence sample in October 2015 is missing. (bottom right corner) Maps of best fit modeled PKD prevalence evaluated at the end of summer 2016 for (e) YOY and (f) adults. Panel (e) features a zoom on site 16 [624].

The epidemiological metacommunity model, developed as an evolution of the model discussed in the previous subsections, is illustrated in detail in Box 4.20. The reference set of parameters is fully detailed in the reference paper [376]. This model is capable of reproducing the observed patterns of PKD prevalence (Figure 4.45a) and, in particular, the late summer 100 percent prevalences observed for both age-classes at sites 4 and 8. The model forecasts prevalences close to 100 percent in large parts of the network in late summer for both adults (Figure 4.45f) and, to a lesser extent, YOY (Figure 4.45e). Prevalence in adults tends to be high during the whole season (Figure 4.45d), while the initially null prevalence level in YOY (Figure 4.45c) is due to the absence of vertical transmission of PKD in fish [624]. Remarkably, modeled PKD prevalence is lower in those parts of the network where there are no upstream stretches where predicted bryozoan abundance is high (Figure 4.44c). This result agrees with the

observed null prevalence upstream of the confluence with the Seewag: indeed, this tributary, unlike the Änziwigger and the Buechwigger, is characterized by high *F. sultana* density, according to the model of bryozoan suitability (see Figure 4.44c). The decrease of brown trout population size was estimated only for sites where two sampling campaigns in the same year were conducted. The decline in fish abundance is captured by the model (Figure 4.45b), despite a certain tendency toward underestimation. This could be attributed to undocumented fishing activity or other possible stress factors not included in the model at the current stage.

The results described above show that local abiotic and biotic conditions favoring parasite proliferation might vary in time owing to environmental change. In the case of parasites with complex life cycles, such conditions must remain conducive to the persistence of multiple susceptible host classes. For example, the correct species of snail and vertebrate hosts are required to coexist at appropriate points during the life cycle of *Schistosoma mansoni* parasites to sustain the transmission of schistosomiasis [543] (Section 4.4). Although *T. bryosalmonae* exhibits a similarly complex life cycle with no fish-to-fish transmission, a notable difference is that that long-term parasite persistence in the bryozoan populations is possible, even in the absence of the fish host [416, 628]. Parasite propagation along the budding growth of the bryozoan and incorporation into asexually produced resting stages create an effective parasite reservoir with frequent spill-over effects on, for example, stocked and highly susceptible fish. For this type of pathogen, any environmental change favoring establishment of the key reservoir host increases disease risk in all the other hosts, including those that may be economically relevant. It also greatly complicates eradication of the disease through management measures.

It is worthwhile also to remark that disease emergence may occur in various manners, through either range expansion of existing pathogens or appearance of new, more virulent agents in existing endemic ranges [629, 630]. Environmental change can also trigger the emergence of previously relatively avirulent, endemic parasites by altering the expression of virulence via, for example, temperature-linked effects on host immune function [631].

Box 4.20 Epidemiological Model of PKD

The metacommunity model (Figure 4.46), originally presented in [624], was modified to account for fish population age structure [624]. A brief description follows (*B* indicates bryozoan state variables, *Y* represents YOY, and *F* stands for adult fish). During the warm season (left column in Figure 4.46), susceptible bryozoans (B_S) become covertly infected (B_C) after exposure to spores (Z_F) released by infected fish. Infection in bryozoans cycles between covert and overt (B_O) states. B_S produces uninfected statoblasts S_S (i.e., asexually produced propagules), while B_C release both uninfected and infected (S_I) statoblasts. Infected bryozoans may clear the infection. Susceptible fish (Y_S, F_S) are exposed to infectious spores (Z_B) released by B_O. After an incubation phase (Y_E, F_E), fish can either develop acute PKD (Y_I, F_I) or directly enter an asymptomatic carrier state (Y_C, F_C). Y_I and F_I may die owing to PKD or else become long-term parasite carriers. Y_C and F_C may then become susceptible again. Y_I, F_I, Y_C, and F_C shed spores infective to bryozoans. At the beginning of a new warm season (right column), B_S comprises susceptible colonies that survived during winter and hatched S_S; similarly, B_C consists of survived colonies that were infected at the beginning of winter and hatched S_I. Y_S are constituted by newborn fish from all classes. F_S are composed by survived susceptible individuals. The abundance of F_C is determined by the number of individuals belonging to nonsusceptible classes that survived through winter. The abundance of the other classes is null.

The local epidemiological model is applied to each river stretch, considered as a node of an oriented graph spanning the whole river network. Connectivity between nodes entails passive hydrological transport of parasite spores (Z_B, Z_F) in the downstream direction and fish movement (modeled as a diffusive process) both upstream and downstream. Spatial heterogeneity in bryozoan and fish suitabilities is included: bryozoan biomass is expressed in dimensionless units, with local carrying capacities assumed proportional to the local bryozoan concentration estimated from eDNA data, c_i, and equal to unity for the stretch characterized by the highest c_i value (where c_i values are taken as in Figure 4.44c). As for both YOY and adults, fish density at equilibrium is proportional to the mean stretch depth [415], which corresponds to an appropriate parameterization of the effect of density dependence affecting YOY during winter. Mobility rates and the diffusion matrix are as in [416] such that, given any spatial distribution of fish at the beginning of the season, the system tends to reach a target equilibrium distribution (which assumes that fish density is proportional to the mean stretch depth) toward the end of the season, provided that the average fish mobility rate l_{avg} is large enough (see Eq. (4.53)). During winter, the number of newborn fish generated by the

Box 4.20 *Continued*

female adults living in each stretch is estimated according to a Ricker model [632]. As brown trout are subject to spawning migration to seek for suitable habitats and subsequent natal homing [633], newborn individuals generated by adults living in a specific stretch are assumed to hatch in suitable upstream stretches according to a gravity model [105]. For the sake of simplicity, the same set of epidemiological and mobility parameters was used for both YOY and adult fish.

The system of equations for the time-hybrid epidemiological model is adapted from [624]. A list of state variables is reported in Table 4.1; table S6 of [624] displays all model parameters and their reference values.

Intraseasonal Model

The set of ordinary differential equations describing the intraseason disease dynamics in stretch i reads

$$\frac{dB_{S,i}}{d\tau} = r_i \left[1 - \frac{\rho_i}{V_i}(B_{S,i} + B_{C,i} + B_{O,i}) \right] B_{S,i} - \beta_B \frac{Z_{F,i}}{V_i} B_{S,i}, \tag{4.54a}$$

$$\frac{dB_{C,i}}{d\tau} = r_i \left[1 - \frac{\rho_i}{V_i}(B_{S,i} + B_{C,i} + B_{O,i}) \right] B_{C,i} + \beta_B \frac{Z_{F,i}}{V_i} B_{S,i} - d_{CO,i} B_{C,i} + d_{OC,i} B_{O,i}, \tag{4.54b}$$

$$\frac{dB_{O,i}}{d\tau} = r_{O,i} \left[1 - \frac{\rho_i}{V_i}(B_{S,i} + B_{C,i} + B_{O,i}) \right] B_{O,i} + d_{CO,i} B_{C,i} - d_{OC,i} B_{O,i}, \tag{4.54c}$$

$$\frac{dS_{S,i}}{d\tau} = f_B B_{S,i} + \phi f_B B_{C,i}, \tag{4.54d}$$

$$\frac{dS_{I,i}}{d\tau} = (1 - \phi) f_B B_{C,i}, \tag{4.54e}$$

$$\frac{dY_{S,i}}{d\tau} = -\mu_F Y_{S,i} + \zeta Y_{C,i} - \beta_F \frac{Z_{B,i}}{V_i} Y_{S,i} + \sum_{k=1}^{N} d_{ki} l_k Y_{S,k} - \sum_{k=1}^{N} d_{ik} l_i Y_{S,i}, \tag{4.54f}$$

$$\frac{dY_{E,i}}{d\tau} = \beta_F \frac{Z_{B,i}}{V_i} Y_{S,i} - (\mu_F + h_i) Y_{E,i} + \sum_{k=1}^{N} d_{ki} l_k Y_{E,k} - \sum_{k=1}^{N} d_{ik} l_i Y_{E,i}, \tag{4.54g}$$

$$\frac{dY_{I,i}}{d\tau} = (1 - \epsilon) h_i Y_{E,i} - (\mu_F + a_i + \gamma) Y_{I,i} + \sum_{k=1}^{N} d_{ki} l_k Y_{I,k} - \sum_{k=1}^{N} d_{ij} l_i Y_{I,i}, \tag{4.54h}$$

$$\frac{dY_{C,i}}{d\tau} = \epsilon h_i Y_{E,i} + \gamma Y_{I,i} - (\mu_F + \zeta) Y_{C,i} + \sum_{k=1}^{N} d_{ki} l_k Y_{C,k} - \sum_{k=1}^{N} d_{ij} l_i Y_{C,i}, \tag{4.54i}$$

$$\frac{dF_{S,i}}{d\tau} = -\mu_F F_{S,i} + \zeta F_{C,i} - \beta_F \frac{Z_{B,i}}{V_i} F_{S,i} + \sum_{k=1}^{N} d_{ki} l_k F_{S,k} - \sum_{k=1}^{N} d_{ik} l_i F_{S,i}, \tag{4.54j}$$

$$\frac{dF_{E,i}}{d\tau} = \beta_F \frac{Z_{B,i}}{V_i} F_{S,i} - (\mu_F + h_i) F_{E,i} + \sum_{k=1}^{N} d_{ki} l_k F_{E,k} - \sum_{k=1}^{N} d_{ik} l_i F_{E,i}, \tag{4.54k}$$

Box 4.20 *Continued*

$$\frac{dF_{I,i}}{d\tau} = (1-\epsilon)h_i F_{E,i} - (\mu_F + a_i + \gamma)F_{I,i} + \sum_{k=1}^{N} d_{ki}l_k F_{I,k} - \sum_{k=1}^{N} d_{ij}l_i F_{I,i}, \tag{4.54l}$$

$$\frac{dF_{C,i}}{d\tau} = \epsilon h_i F_{E,i} + \gamma F_{I,i} - (\mu_F + \zeta)F_{C,i} + \sum_{k=1}^{N} d_{ki}l_k F_{C,k} - \sum_{k=1}^{N} d_{ij}l_i F_{C,i}, \tag{4.54m}$$

$$\frac{dZ_{B,i}}{d\tau} = \pi_B B_{O,i} - \mu_Z Z_{B,i} + \sum_{k=1}^{N} w_{ki}\frac{Q_k}{V_k}Z_{B,k} - \sum_{k=1}^{N} w_{ik}\frac{Q_i}{V_i}Z_{B,i}, \tag{4.54n}$$

$$\frac{dZ_{F,i}}{d\tau} = \pi_F(Y_{I,i} + F_{I,i}) + \kappa\pi_F(Y_{C,i} + F_{C,i}) - \mu_Z Z_{F,i} + \sum_{k=1}^{N} w_{ki}\frac{Q_k}{V_j}Z_{F,j} - \sum_{k=1}^{N} w_{ik}\frac{Q_i}{V_i}Z_{F,i}. \tag{4.54o}$$

For the sake of economy in the model formulation, we introduce two new state variables $Z_F^* = \beta_B Z_F$, $Z_B^* = \beta_F Z_B$, which we term equivalent spores. They represent the abundance of spores needed to infect a unit bryozoan biomass (a single susceptible fish) per unit time and unit water volume. Thanks to this assumption, the exposure rates β_B and β_F can be discarded and two synthetic contamination rates $\pi_B^* = \beta_F \pi_B$ and $\pi_F^* = \beta_B \pi_F$ are introduced, (see table S4 of [624]). Synthetic contamination rates π_B^* and π_F^* are then calibrated.

Interseasonal Model

The following difference equation system relates the state of the model variables at the end of a season (y) with that at the beginning of the following season $(y+1)$:

$$B_{S,i}(y+1) = \sigma_B B_S(y) + \nu S_S(y), \tag{4.55a}$$

$$B_{C,i}(y+1) = \sigma_B B_C(y) + \sigma_O B_O(y) + \nu S_I(y), \tag{4.55b}$$

$$B_{O,i}(y+1) = 0, \tag{4.55c}$$

$$S_{S,i}(y+1) = 0, \tag{4.55d}$$

$$S_{I,i}(y+1) = 0, \tag{4.55e}$$

$$Y_{S,i}(y+1) = \sum_{j\neq i} W_{ij}\eta\mathscr{F}_j(y)\exp\left(-\frac{\xi_j}{V_j}\mathscr{F}_j(y)\right), \tag{4.55f}$$

$$Y_{E,i}(y+1) = 0, \tag{4.55g}$$

$$Y_{I,i}(y+1) = 0, \tag{4.55h}$$

$$Y_{C,i}(y+1) = 0, \tag{4.55i}$$

$$F_{S,i}(y+1) = \sigma_F(Y_{S,i}(y) + F_{S,i}(y)), \tag{4.55j}$$

$$F_{E,i}(y+1) = 0, \tag{4.55k}$$

$$F_{I,i}(y+1) = 0, \tag{4.55l}$$

$$F_{C,i}(y+1) = \sigma_F\{p_E[Y_{E,i}(y) + F_{E,i}(y)] + p_I[Y_{I,i}(y) + F_{I,i}(y)] + Y_{C,i}(y) + F_{C,i}(y)\}, \tag{4.55m}$$

$$Z_{B,i}(y+1) = 0, \tag{4.55n}$$

$$Z_{F,i}(y+1) = 0, \tag{4.55o}$$

Box 4.20 *Continued*

where $\mathscr{F}_j(y) = Y_{S,j}(y) + F_{S,j}(y) + p_E \left[Y_{E,j}(y) + F_{E,j}(y)\right] + p_I \left[Y_{I,j}(y) + F_{I,j}(y)\right] + Y_{C,j}(y) + F_{C,j}(y)$, while W_{ji} is the fraction of newborns generated by adults living in i that hatch in j ($j \in \mathscr{U}_i$, where $\mathscr{U}_i$ contains all stretches upstream of i and i itself) and is calculated via a gravity model [105]

$$W_{ji} = \frac{W_j^A \, e^{-L_{ji}/\lambda_F}}{\sum_{j \in \mathscr{U}_i} W_j^A \, e^{-L_{ji}/\lambda_F}}, \tag{4.56}$$

where W_j^A is a dimensionless score for spawning suitability and λ_F the shape factor of the exponential kernel representing the deterrence factor. An exponential function was also used to express spawning suitability, $W_j^A = e^{-A_j/A_F}$, implying that eggs tend to be deposited in small, low-velocity stretches (as the upstream contributing area A_i is a proxy of stretch cross section and mean water velocity [63, 101]). The normalization parameters $A_F = 100$ km^2 and $\lambda_F = 2{,}000$ m are chosen such that, with the reference value $l_{avg} = 0.02$ d^{-1}, the distribution of YOY at the end of the season is sufficiently close to the equilibrium distribution. Note that the value of λ_F is congruent with observed values of typical migration distances covered by brown trout [624].

Fish Mobility

Movement rules for fish are assumed as in [416] (see also the previous part illustrating Eq. (4.53)).

The maximum flux of fish exiting from stretch i is $l_i F_i$ (see Eqs. (4.54f)–(4.54m)), where F_i is the local fish abundance regardless of the epidemiological class and age, while l_i is a mobility rate. The flux of fish from stretch i to stretch j reads $d_{ij} l_i F_i$, where **D** is a mobility matrix whose entries d_{ij} are positive when either w_{ij} or w_{ji} is equal to unity, and null otherwise. Furthermore, one has $\sum_{j=1}^{N_s} d_{ij} \leq 1$; d_{ij} expresses the relative probability that a fish exiting from stretch i chooses stretch j out of all stretches connected to i, either downstream or upstream. The strict inequality $\sum_{j=1}^{N_s} d_{ij} < 1$ holds when stretch i represents either a headwater or the outlet, where, assuming the river network as a closed system, only one direction for fish movement is allowed. The diffusion matrix henceforth used is derived thanks to the following assumptions: $\sum_{j=1}^{N_s} d_{ij} = 0.5$ if stretch i is a headwater or the outlet – for other stretches, fish have equal probability to move downstream or upstream; the probability to choose a given upstream stretch is proportional to its cross-sectional area.

The values of l_i such that a distribution of fish abundances F_i is an equilibrium state of the diffusion process are obtained from solving the following linear system:

$$d_{ij} l_i F_i = d_{ji} l_j F_j \qquad \forall i \leq N_s, j \leq N_s. \tag{4.57}$$

Only $N_s - 1$ of the above equations are not trivial identities: in fact, $d_{ij} \neq 0$ only if stretches i and j are directly connected, and every stretch has one downstream connection, with the exception of the outlet stretch. The system has ∞^1 solutions (i.e., if a set of l_i is a solution, then also $k l_i$, $k \in \mathscr{R}^+$ is a solution); a set of mobility rates for all stretches is thus defined by specifying its mean value l_{avg}.

Details on calibration and model simulations are in the original work [624], together with a vast discussion of the field data, including bryozoan habitat suitability and niche dimensions, prevalence data, brown trout density assessments, and details on water temperature measurements.

Table 4.1 List of variables for the epidemiological model reported in Box 4.20

Symbol	Variable
B_S	Biomass of susceptible bryozoans
B_C	Biomass of covertly infected bryozoans
B_O	Biomass of overtly infected bryozoans
S_S	Noninfected statoblast abundance
S_I	Infected statoblast abundance
Y_S	Abundance of susceptible YOY
Y_E	Abundance of exposed YOY
Y_I	Abundance of acutely infected YOY
Y_C	Abundance of carrier YOY
F_S	Abundance of susceptible adult fish
F_E	Abundance of exposed adult fish
F_I	Abundance of acutely infected adult fish
F_C	Abundance of carrier adult fish
Z_B	Abundance of spores released by bryozoans
Z_F	Abundance of spores released by fish

Note. All state variables are dimensionless.

The modeling approach [624] suggested a nexus between increased bryozoan density and PKD severity and occurrence, and thus the key predictive factor for PKD was the habitat suitability for the bryozoan *F. sultana*. A strong correlation was found [624] between the presence of upstream moraines and local abundance of *F. sultana*. This pattern, which was instrumental in generating the bryozoan density map for the Wigger, might imply that moraines create advantageous conditions for the proliferation of *F. sultana* by constituting the suitable substrate and/or by affecting geogenic solute concentrations, that is, the chemical properties of the stream environment. However, moraines are present in a limited area of the watershed with few sampling sites with notably high *F. sultana* eDNA concentration; therefore a spurious correlation cannot be excluded. Understanding the causality and mechanisms explaining why moraines are priming bryozoan presence requires further survey studies in other catchments and laboratory experiments. Knowledge on habitat requirements of bryozoans might be crucial for disease control purposes, as the strong link between PKD severity and bryozoan density suggests that PKD management might rely on control of bryozoan populations. Other possible strategies might count on the

production of resistant fish strains, also by eliminating fish stocking and thus allowing for selection of resistant fish strains in a natural way. In any case, suitability of abiotic habitat characteristics provides a basis for the presence of endemic diseases. In parasitic diseases like PKD, the key requirement is the suitability of the habitat for the key host species, here bryozoan *F. sultana*. For this type of pathogen, any environmental change favoring establishment of the key host increases the disease risk in the economically relevant host.

Regarding the epidemiological model (Box 4.20), it is noteworthy that, even without specifying different epidemiological parameters for YOY and adults, the model predicts a population size decline of YOY that is almost three times larger than that of adults. Therefore, the higher susceptibility of YOY to PKD revealed by the data can be explained by the not yet acquired immunity rather than by an intrinsic severity of PKD for young fish. The observation that decline in adults predicted by the model is lower than in observed values can be explained by an additional fishing mortality term not accounted for by the model (say, widespread anglers' impacts or other stress factors). Confidence intervals of fish decline are rather narrow because the main factor influencing adults' reduction is the natural mortality rate, which was kept constant in the calibration procedure. Other possible stress factors related to temperature increase were already taken into account by the calibration protocol (see [416, 624]), because the PKD-related mortality is expressed as a function of temperature. The seasonal decrease in the abundance of YOY observed at site 8 in 2016 was considerably low (see Figure 4.45b). It is remarkable that the model is actually capable of partially reproducing this trend by predicting a lower YOY decrease compared to that of 2015. This is likely due to the shorter time lapse between the two seasonal sampling campaigns and the cooler summer temperatures (see SI of [624] for details). Indeed, water temperature proves a key driver of PKD impacts on fish population abundance even though its influence on prevalence patterns is minor [416, 621, 624]. However, the model appears unable to reproduce the null prevalence observations upstream of site 16 in summer 2015, despite that the predicted values are relatively low (around 25 percent). A better fit is probably prevented due to the fact that the bryozoan density map of Figure 4.44c, which was obtained as the mean of all bryozoan suitability models, was kept constant in the calibration of the epidemiological model. An additional drawback to consider is that the fact that PKD can persist in bryozoan populations without the fish

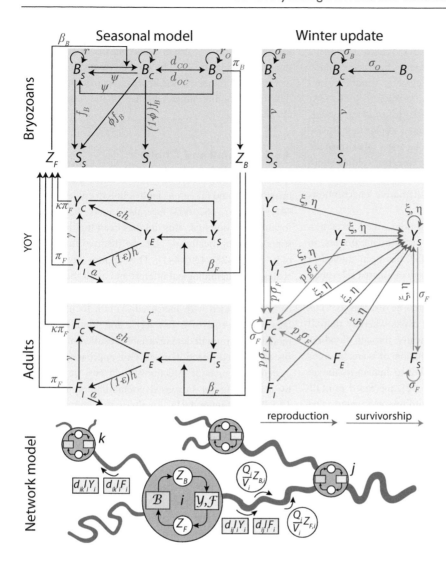

Figure 4.46 Schematic of the epidemiological model. (a) Intraseasonal local model. (b) Interseasonal local model. Parameters are indicated in gray. (c) Spatial model. $\mathscr{B}$, bryozoan submodel; $\mathscr{Y}\mathscr{F}$, fish submodel. All state variables and parameters are listed in [624]

is actually surprising. Most other multiple host parasites require all hosts available in the site. This puts special management constraints on the system, as one cannot clean a river from PKD by culling fish and restocking (as was done with *Gyrodactylus* in Norway [634]). Control of bryozoans, or rather managing the habitats so that they do not include key niche parameters, is debatable, as they are part of the ecosystem, and is perhaps feasible only in fish farms.

In conclusion, the comprehensive work described in this section (adapted from [624]) highlights the profound influence of an emerging aquatic disease on the abundance and seasonal demography of threatened and economically import fish stocks. Our integrated approaches resulted in a comprehensive spatial predictive framework of disease and identify key factors in driving

disease patterns in the wild – and nicely agree with the main aims of this book.

4.6 Of Hydrologic Drivers and Controls of WR Disease

Hydroclimatological forcings, the mobility of susceptible/infected individuals, and large-scale treatment are key drivers of waterborne disease transmission [428, 464, 550, 635]. The direct inclusion of these factors into spatially explicit mathematical models of epidemic cholera, to quote an example of paramount importance, has improved our understanding of complex disease patterns [30, 325, 428, 429, 435, 438, 440, 441, 467, 468, 494]. Spatially explicit mathematical models encapsulate remotely acquired, relevant descriptions of human settlements, waterways where pathogens and/or hosts disperse, movement between human communities, hydrologic and climate variables, and water resources development infrastructure [636–638]. Objective manipulation of such information yields characterizations of waterborne and water-related disease dynamics of unprecedented robustness that provide insight into the course of ongoing epidemics aiding short-term emergency management of health care resources and long-term assessments of alternative interventions [428, 440, 468].

As noted in the previous sections, the correlation between WR disease epidemics and climatic drivers, in particular seasonal tropical rainfall, has been studied in a variety of contexts owing to its documented relevance. Rainfall influences both cholera and schistosomiasis as well as other macroparasitic waterborne diseases [639]. A variety of potential mechanisms exists whereby rainfall may alter infection risk, specifically through flooding, leading to raw sewage contamination of water sources [640, 641] and enhanced exposure to human-to-human transmission due to crowding [642], increased rainfall-driven iron availability in environmental waters that enhances pathogen survival and expression of toxins [523, 643], and dry spells and decreased water levels leading to increased usage of unsafe water sources [473].

The general structure of spatially explicit models of waterborne disease that incorporates hydrological drivers is outlined in Figure 4.47. The dynamics of disease spread reflect the spatial heterogeneity of its drivers. To support this statement, in this section, reference is made to two paradigmatic cases of waterborne disease, one microparasitic (epidemic cholera) and the other macroparasitic (endemic schistosomiasis), taken as templates of very diverse infections that share the same matrix (human communities and waterborne pathogen) but cover a rather diverse spectrum of relevant epidemiological conditions. For instance, climate affects cholera transmission rates via correlation with environmental variables like rainfall and air/water temperature locally [477, 484, 523, 640, 641, 644–652], and at large scale [653–657]. At the other end of the spectrum, the links between climate and schistosomiasis are being actively investigated [595, 614, 658] because climate change might hinder progress toward control in endemic countries [659, 660]. Climate influences in any case both pathogen ecology and host exposure, and its prospective change is speculated to bear major impacts on the geography of waterborne and WR diseases in general [661].

The synthesis of the tools used to tackle spatial and temporal patterns of drivers and controls of WB disease reported in Figure 4.47 highlights rainfall's articulate driver roles – rather diverse for different diseases – whose description is the subject of the final section of this chapter.

4.6.1 Rainfall and Cholera

A clear correlation between rainfall and enhanced transmission is found in regions hit by cholera epidemics [440, 441, 473, 637, 662, 663] (see also Figure 4.2). Notably for the Haiti outbreak, this link has been highlighted and supported empirically [427], to be justified theoretically only later [428, 429, 441]. This applies at all spatial scales of epidemiological interest and at all the locations examined in Haiti. Intense rainfall events were significantly correlated with increased cholera incidence with lags of the order of a few days, and forcing dynamic models with rainfall data invariably resulted in improved fits of reported infection cases. One approach [428] suggested that bursts of infections could best be explained by accounting for increased contamination rates (flux $I \rightarrow W$; see Figure 4.21). An alternative approach [429] employed rainfall-dependent exposure rates ($S \rightarrow I$). The merits of both approaches, within spatially explicit deterministic or stochastic models, have recently been studied in detail [441].

The incorporation of rainfall effects, suitably generalized by using a nonlinear function that can increase or decrease the relative importance of the largest precipitation events, has been tested against daily epidemiological data collected during the 2015 cholera outbreak within the urban context of Juba, South Sudan [441]. That specific epidemic has been characterized by a particular intraseasonal double peak of the incidence in apparent relation with particularly strong rainfall events. Eight models were tested, which also included the possibility of human-to-human transmission. The structure of three of these models is described in Figure 4.48.

The results of the eight models that arise from this setting were compared on the basis of their ability to match the time series of daily reported cases during the 2015 cholera epidemic in Juba [441]. The degrees of freedom of the models vary from 7 to 12 for the deterministic model. Given the low number of daily reported cases and their ensuing variability, a stochastic equivalent of the deterministic ODE system was formulated as a continuous-time partially observed Markov process model, accounting for both demographic and

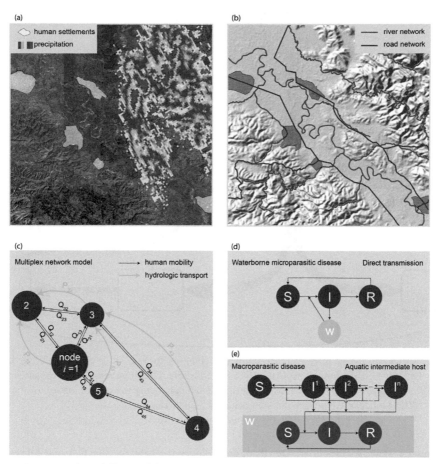

Figure 4.47 Development of spatially explicit models of waterborne disease. (a) Examples of information that can be remotely acquired and directly integrated into epidemiological models: locations of water bodies and human settlements along with proxies of population density (e.g., WorldPop, www.worldpop.org.uk); distributed precipitation estimates through satellites (e.g., Tropical Rainfall Measuring Mission, `http://trmm.gsfc.nasa.gov`) or radars (e.g., http://radar.weather.gov). (b) Ancillary spatially distributed information: river networks can be delineated by analyzing a digital elevation model of the area; road network databases (e.g., Openstreetmap, www.openstreetmap .org) can be used to estimate distance or travel time among human settlements. (c) Multiplex network model. The information described in panels (a) and (b) is synthesized into a network model where nodes represent human communities where epidemiological dynamics take place while different layers of edges represent hydrologic connectivity and human mobility. Q_{ij} represents the probability that an individual travels from node i to j, while P_{ij} is the probability that infective agents are transported from i to j through the hydrologic network. (d, e) Compartmental epidemiological models. The evolution of the number of susceptible S, infected I recovered R individuals, as well as the aquatic concentration of infective agents W, is computed for all the nodes of the network (or any generalization of classical SIR [susceptible, infected, recovered] models [424]). (d) An example of a model for a microparasitic disease with direct transmission (e.g., cholera) where new infections ($S \rightarrow I$) occur through contact with water contaminated by infected individuals and recovered individuals are temporarily immune. (e) An example of a model for macroparasitic disease models with an aquatic intermediate host (e.g., schistosomiasis). These models may differentiate between classes of infection ($I^1, I^2, \ldots, I^n$) and do not account for immunity. Parasite death within humans (due to either natural causes or medical treatment) progressively decreases the infectious state of the host. Infection in the intermediate host is typically treated as a microparasitic disease. Figure after [25]

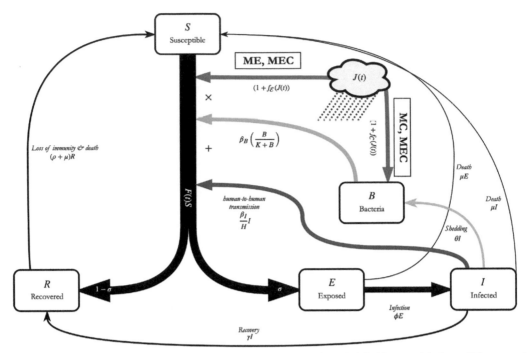

Figure 4.48 Flow diagram for alternative cholera models accounting for rainfall drivers, with three different variants indicated (namely, **ME** = SEIRB model using the formulation where rainfall increases the exposure to bacteria [429]; **MC** = SEIRB model using the approach accounting for rainfall enhancing the contamination of the water reservoir [428]; **MEC** = SEIRB model combining both approaches **MC** and **ME**). Figure after [441]

disease transmission stochasticities [494]. Details of the involved procedure for calibration, performed using a Markov chain Monte Carlo–based algorithm drawing samples from the posterior distribution of the parameters and inference on the stochastic model performed using a frequentist multiple iterated filtering algorithm, are described in the original paper [441]. Models were tested against the daily reported cases accounting for over- or under-reporting, assuming a Poisson distribution. They were compared using the Bayesian information criterion (BIC), Bayes factors, and the likelihood ratio test for the nested models [441].

Regarding the epidemiological model, the study cited here [441] introduced two innovations with respect to previous attempts to model cholera epidemics. First, the focus on daily incidence data, as opposed to weekly epidemiological reports commonly used in modeling works, justifies the introduction of a compartment of exposed individuals (Figure 4.48) to account for the incubation period of the disease and thus the lag between the possibly rainfall-driven infection process and the manifestation of the symptoms resulting in the time series of daily reported cases. Second, a nonlinear version of the rainfall

driver (namely, the function that relates contamination and/or exposure to rainfall intensity), in the form of a power law controlled by one parameter (the exponent), was introduced to tune the relative importance of rainfall intensities. Such formulation has the flexibility either to emphasize the impact of the largest rainfall events or to give equal weight to all nonzero rainfall intensities.

All model assumptions were compared for both deterministic and stochastic model types, to draw more general conclusions. The statistics and tests used to compare the model results [441] supported the significance of rainfall effects during the 2015 epidemic in Juba. The main results (Figure 4.49) show that for both model types, there exists a significant positive correlation of rainfall drivers, in particular because standard SEIRB models without rainfall driving prove unable to reproduce secondary peaks of reported cases that occurred in July during the recession period. All models considering rainfall instead show an increase in the number of cases in correspondence of the second epidemiological peak, evidently due to the large rainfall rates that occurred in the previous days (Figure 4.49). This marked difference in the simulated responses of models considering (or ignoring) rainfall leads to strong support for rainfall-driven models. Owing to the small variations among the likelihoods and BICs of rainfall-based models [441], however, it is not straightforward to draw conclusions on the best way to include the rainfall effect. Models with the minimum BIC were those considering the increase in exposure (model **ME**) for both the stochastic and the deterministic model types. Moreover, a fair comparison between the likelihoods of the two models' types (either deterministic or stochastic) shows that considering the stochasticity of the processes improves the model results, as foreseeable given the relatively low number of cases. This suggests that deterministic models should also include a stochastic term in the computation of the force of infection in analogous conditions.

The above results [441] have shown that rainfall-based models, both in their deterministic and stochastic formulations, generally outperform models that do not account for rainfall in real-life applications, confirming the conclusions drawn for the Haiti 2010 epidemic [428, 429]. In fact, classical SIRB models – whether local or spatial – are unable to reproduce the second epidemiological peak shown in the Juba data, thus suggesting that it was rainfall induced. Stronger support across model types for rainfall acting on increased exposure rather than on exacerbated water contamination was found for the Juba epidemic. These conclusions differ from those of the comparative study carried out for the Haiti epidemics [428], yet the relative role of sanitation in different contexts cannot be assessed in general. Thus we argue that the results shown above are context specific. Taken together, all studies on the subject uniformly stressed the importance of a systematic and comprehensive appraisal of transmission pathways, including hydrologic drivers, and other possible environmental forcings thereof, when embarking on modeling epidemic cholera anywhere in the world.

4.6.2 Rainfall and Schistosomiasis

The relationship between precipitation and schistosomiasis is far from obvious. It has been argued [614], in fact, that rainfall could not only boost disease transmission (especially in dry climates, where it is a key driver of habitat formation for the intermediate snail host) but also reduce it, for example, by increasing water flow (which in turn decreases habitat suitability for both the snails and the larval stages of the parasites). Rainfall can also affect human activities related to water contact, thus potentially altering exposure and transmission risk [664]. Also, the temporal fluctuations of rainfall

patterns may be particularly important in determining the seasonality of transmission [665]. Several studies in which the ecology of the intermediate snail hosts has been analyzed through field campaigns and geospatial models highlight the strength of the link between rainfall and schistosomiasis transmission [592, 595, 660, 664, 666]. A hydrology-driven assessment of intermediate host habitat suitability thus seems necessary [667]. Snail abundance, in fact, may depend on density feedbacks that, in turn, can be driven by hydrologic controls, especially in ephemeral hydrologic regimes like those typical of sub-Saharan Africa [567]. Therefore, a quantitative link exists between hydrological drivers and snail population dynamics, suggesting that statistical methods may provide reliable snail abundance projections [567] and that state-of-the-art mechanistic models of transmission [609] could be made dependent on habitat type (e.g., natural vs. man-made) and hydrological characteristics (e.g., ephemeral vs. permanent).

As noted in Section 4.4, water resources development (e.g., damming and irrigation) play a significant role in the increase of disease burden owing to habitat expansion of pathogens and/or hosts, especially for schistosomiasis [586, 639]. Habitats are shaped naturally by the hydrology–geomorphology connections [63], defining the suitability of certain intermediate host species key to the closure of the pathogens' life cycle and the survival of pathogens in natural environments [567, 609]. However, the natural habitats are increasingly altered by humans, thus making it necessary to include land-use change as one of the most important drivers not only of biodiversity loss but also of increased disease suceptibility. The development of agricultural practices is certainly one of the main causes of land-use change. Also, the larger and larger use of herbicides and fertilizers has been shown to alter the ecological food web and favor the proliferation of specific intermediate hosts (e.g., those of schistosomiasis [668]).

Where rainfall patterns, and hydrologic drivers in general, become totally discriminating is in defining the ecology of the host species. In fact, as seen in Section 4.4, certain freshwater snail species are intermediate hosts in the life cycle of parasites causing human schistosomiasis – recall that it is a neglected water-based disease, treatable but debilitating and poverty reinforcing, that affects about 150 million people yearly in sub-Saharan Africa alone. Local snail abundance, a key parameter determining the potential force of the infection due to water contact, is thus often the target of epidemiological control measures of schistosomiasis incidence. It seems

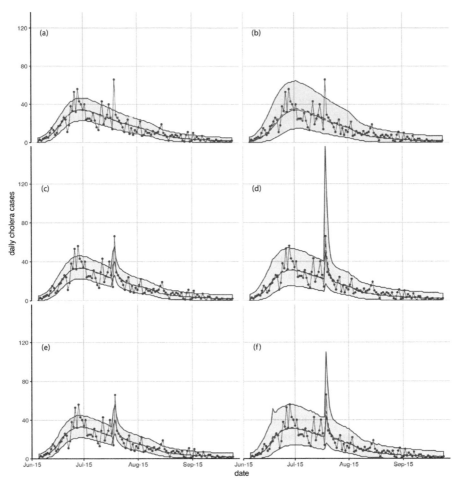

Figure 4.49 Simulations of the (a, b) **MC**, (c, d) **ME**, and (e, f) **MEC** models described in the caption of Figure 4.48. Simulations for the deterministic versions (a, c, e) are given by the mean (blue dashed line), median (blue full line), and 95 percent simulation envelopes (blue ribbon) of 100 simulations of the measurement model for each trajectory from 100 samples from the posteriors of model parameters against reported daily cholera cases (red line and dots). Simulations from the stochastic models (b, d, f) are given for 10,000 simulations of the stochastic process and measurement models. Figure after [441]

therefore appropriate here to review studies centered on the ecology of host snails, in particular, those guided by modeling frameworks that used field campaigns and theoretical models within natural/artificial water habitats across Burkina Faso's highly seasonal climatic zones [567]. This is in fact a significant case study where incidence of the disease is largest [669]. Snail abundance is shown to depend on hydrological controls and to obey density-dependent demographic evolution. Statistical methods based on model averaging yield reliable snail abundance projections.

We report now about field and theoretical studies on the ecology of the aquatic snails (*Bulinus* spp. and *Biomphalaria pfeifferi*) that, as seen in Section 4.4 and

Figure 4.31, serve as obligate intermediate hosts in the complex life cycle of the parasites causing human schistosomiasis. Snail abundance fosters disease transmission by emitting large quantities of cercariae once infected, and thus the dynamics of snail populations are critically important for schistosomiasis modeling and control. One faces, therefore, the need to single out hydrological drivers and density dependence (or lack of it) for ecological growth rates of local snail populations. This has been carried out in particular by contrasting novel ecological and environmental data (Figure 4.50) with various models of host demography [567]. Specifically, various natural and man-made habitats across Burkina Faso were targeted. Therein demographic models were

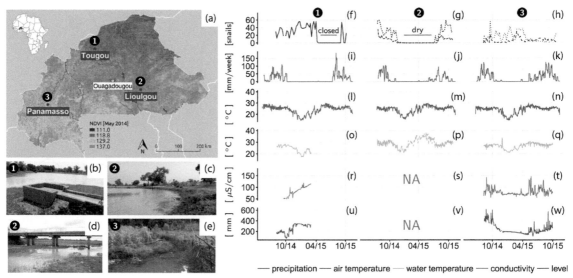

Figure 4.50 Ecological and environmental field data relating hydrologic drivers and the abundance of the snail host of the pathogen of schistosomiasis after [567]. (a) Situation map of the field sites (black dots) chosen along the south–north climatic gradient in Burkina Faso, here highlighted by values of the normalized difference vegetation index (see [567] for details) and the state capital (Ouagadougou). (b–e) Illustrations of the ephemeral or permanent habitats chosen for snail sampling in the experimental sites of Tougou, Lioulgou, and Panamasso. (f–w) Ecological and environmental data time series. Weekly snail relative abundance data are given per species and habitat: *Bulinus* spp. in a canal in Tougou (solid line) (f) and in a pond (short dashed line) and a stream (long dashed line) in Lioulgou (g) and *B. pfeifferi* (dashed line) and *Bulinus* spp. (dotted line) in a stream in Panamasso (h). Periods during which the habitat dried out are indicated for Lioulgou and Tougou (canal closed for operational reasons). Their respective environmental data are given in columnar format. Environmental data consist of (i–k) weekly cumulative precipitation and daily averages of (l–n) air temperature, (o–q) water temperature, (r–t) conductivity, and (u–w) water depth. Graphs containing the term "NA" indicate ephemeral habitats (S and V) (see also Section 5.3). Details on sampling and environmental monitoring methods are given in source references. Figure after [567]

ranked through formal model comparison via structural risk minimization [148]. This led to evaluations of the suitability of population models while clarifying the relevant covariates that explain empirical observations of snail abundance under the actual climatic forcings experienced by the various field sites.

Burkina Faso experiences pronounced seasonal climatic fluctuations along a south–north gradient (Figure 4.51). Seasonality, in turn, greatly affects the suitability of snail aquatic habitats [575]. Rainfall occurs mainly during the rainy season (July–October), with little or no rain falling during the rest of the year (Figure 4.50). Other details on climatic drivers are in [567]. Abundance data show that snail populations in different climatic zones and habitats present distinct seasonal patterns (Figures 4.50f–h). Of the three sites in which field study was carried out (Figure 4.50a), only Panamasso harbored the intermediate host of intestinal schistosomiasis (*B. pfeifferi*), whereas *Bulinus* spp. were found at all sites. Large populations were observed in irrigation canals in

Tougou during normal irrigation operation conditions (Figure 4.50), strongly suggestive of a causal reason for the purported nexus between water resources development and increased incidence of the disease [586]. In the ephemeral habitats of Lioulgou, snails were absent during most of the year, from December to July, and population peaked toward the end of the rainy season between September and October (Figure 4.50g). Opposite fluctuations were observed in the permanent stream of Panamasso for both *B. pfeifferi* and *Bulinus* spp., for which the maximum abundance was observed in the middle of the dry season between January and March. It is interesting to remark that in the 1990s, *Bulinus* spp. data were collected by Poda [546, 575, 577] in a small, man-made reservoir not far from Tougou as well as in a nearby temporary pond. While snails were quite abundant in the temporary pond only during the rainy season, the presence of snails in the reservoir extended to the dry season as well, similarly to the situation observed in Panamasso.

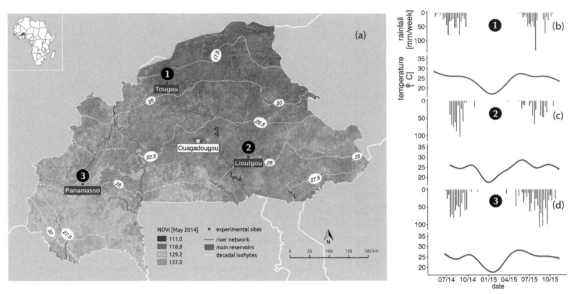

Figure 4.51 (a) Map of the south–north climatic gradient in Burkina Faso, highlighted by values of the Normalized Difference Vegetation Index (NDVI) and average decadal precipitation isolines. The field sites chosen for the present study are marked with red dots. (b–d) The right panels show measured cumulative weekly precipitation (blue bars) and average daily air temperatures (red lines) in the experimental sites of (b) Tougou, (c) Lioulgou, and (d) Panamasso. Note the different starting measurement dates in each experimental site. Figure after [567]

Population dynamics models were developed to interpret the field data [567]. Specifically, snail population dynamics are simulated by discrete-time demographic models inclusive of extrinsic environmental forcing. Specifically, if N_t denotes the abundance of the focus snail population in a given habitat at sampling time t, the simplest demographic model reads $N_{t+1} = \lambda N_t$, where λ corresponds to the finite intrinsic growth rate of the population between two sampling dates. Density feedback can be incorporated via a Ricker model in the form

$$N_{t+1} = \lambda N_t \exp(bN_t),$$

where b is a parameter that sets the feedback's strength [632]. The Ricker model thus reads $\log(N_{t+1}/N_t) = a + bN_t$, where log is the natural logarithm and $a = \log \lambda$. Decreased recruitment and/or increased death rates at high population densities obviously occur for $b < 0$. Exogenous forcing of environmental covariates, their lags, and lagged effects of abundance can be accounted for by [567]

$$\log\left(\frac{N_{t+1}}{N_t}\right) = a + \sum_{\tau_n=0}^{r_N} b_{\tau_N} N_{t-\tau_N} + \sum_{i=1}^{m}\sum_{\tau_X=0}^{r_X} c_{i,\tau_X} X^i_{t-\tau_X},$$

$$(4.58)$$

where b_{τ_N} and c_{i,τ_X} are the weights of the lagged population abundances $N_{t-\tau_N}$ and the environmental

covariate features $X^i_{t-\tau_X}$, taken at nonnegative integer time lags $\tau_N \le r_N$ and $\tau_X \le r_X$ (r_N, r_X are the maximum lags allowed for abundance and covariates). Lagged covariate features were taken in all possible combinations. Lagged abundance effects were only considered one at a time or all at once. Similarly, a second class of models that incorporate density feedback is the Gompertz scheme [567], where the feedback is taken as a function of the logarithm of population abundance $L_t = \log N_t$ rather than abundance itself [567]. The relevant demographic processes were also assumed to be subject to multiplicative log-normal white noise, resulting in additive Gaussian noise when using logarithms of snail abundance. A null random walk model was also considered (see [567] for details). Model results were then tested against all time series (a full account of the tested model structures is given in [567]).

According to a formal model selection procedure [567], density feedback proved significant across all species and habitats, although operating at different strengths, depending on habitat type, hydrological conditions, and specific snail species. A strong density feedback provided by the Ricker model was consistently selected only for *Bulinus* spp. in the ephemeral streams. Weaker density feedbacks were selected in the irrigation canals, in the temporary pond, and for the perennial streams investigated for both *Bulinus* spp. and *B. pfeifferi*. The strength of density feedbacks also varied across species and habitats.

Why does this matter for the tenet of this book? Because the study of the ecology of the intermediate hosts of schistosomes highlights the key role that hydrological drivers hold in parasite transmission dynamics. The findings above, and a few others whose review is beyond the scope of this book, have direct implications for the design of effective disease control programs. In fact, any improvement in our understanding of snail ecology – as implied by the prediction of their local population abundances in time – can directly refine spatially explicit disease transmission models by relaxing the key assumptions of stationary snail populations [540, 550, 592]. By providing empirical and theoretical evidence for discriminating ecological processes across climatic zones and habitat types, a truly ecohydrological framework accounts explicitly for the relevant environmental drivers, in particular, the kinds of climatic seasonality and habitat ephemerality typical of the Sahelian ecotone where the disease is rampant.

Studies encompassing multiple intermediate host species, habitats, and climatic zones are thus needed to generalize the above results to a framework that is context independent. Through these studies, the individual roles and the mutual interactions among drivers and controls could be sorted out, and snail population abundances and fluctuations could reasonably be assigned in space and time within a given fluvial habitat. In particular, the field evidence on how snail abundances fluctuate during the dry and rainy seasons, with opposite phases in the ephemeral habitats in the north and the perennial habitats in the south of Burkina Faso, is particularly suggestive.

Climatic conditions obviously determine the ephemerality of surface waters [670], resulting in the periodic disappearance of suitable habitats for schistosoma's intermediate hosts during the dry season (note that seasonality is survived through aestivation by *Bulinus* spp. but possibly not by *Biomphalaria* spp. in ephemeral water bodies [pond and stream in Lioulgou] [567]). Interestingly, the perennial stream in the Sudanian climatic region (Panamasso) presented opposite abundance fluctuations (for both snail species) with peaks (more noticeable for aestivating *Bulinus* spp.) during the dry season (March-April) and low population levels during the rainy season. A similar situation was found [546] in a man-made

reservoir and in the hydrography relevant to communities of farmers and fishermen in western Burkina Faso [578, 579]. This observation, if confirmed for every perennial stream suitable as a habitat for the snail hosts, would have consequences on predictive determinants of snail abundance. It denies validity to all-purpose stationary and homogeneous reference population sizes being used to date in large-scale disease pattern identification.

The phase shift between the different types of habitats (permanent vs. ephemeral and natural vs. man-made; see Section 5.3 for an account of the role of ephemerality) in the different seasonal climates of the country could have profound consequences for the logistics of national programs for schistosomiasis control due to the regional climatic conditions. In fact, the presence of hundreds of small reservoirs, which have been built in Burkina Faso to foster agricultural, social, and economic development, can extend considerably the period of the year in which intermediate hosts are present in the water points used for daily socioeconomic activities, with a peak of abundance in man-made habitats that occurs as the temporary ponds dry out.

Upscaling through the use of remotely sensed data is conceivable, in the perspective of providing predictions of the effectiveness of control interventions and insight on the impacts of climate change on schistosomiasis transmission (known to be operating at regional scales [671]). Taken together, the field and theoretical analyses described in this section provide a framework for forecasting the effects of rainfall regimes on snail abundance fluctuations, largely determined by habitat expansions and contractions. Dam construction and the related expansion of irrigation canals pursued to meet increasing demand and shrinking supply of water resources in sub-Saharan Africa raise therefore, great concerns about related increased incidences of schistosomiasis [586]. This poses, as noted in the preface, a significant challenge to environmental economic evaluations where sustainability includes the induced loss of natural capital. By linking quantitatively hydrological drivers to snail population dynamics through specific density feedback, the above results suggest that statistical methods provide reliable snail abundance projections and their fluctuations in space and time. Quantitative predictions on epidemiological effects of hydrologic regimes, water resources development, risk mapping, and the allocation of control measures thus appear within reach.

5 | Afterthoughts and Outlook

Upon wrapping up the material for this book, many afterthoughts came to mind, foremost ones related to what we could have addressed – for completeness, for due diligence, for predictions of what are bound to become hotspots of research – and didn't. Well aware of what a famous architect once said ("The safest way to sink a project is to broaden it"), we have decided to come to a close with the material we have dealt with so far – however incomplete and possibly inadequate it may be. The subject area is expanding, and plenty of knowledge is published in disparate outlets, so that the interested reader cannot easily gather it in any single place. Thus the time was ripe to assemble in one place the current knowledge of this field. Here, it seems appropriate to recap the conceptual thread that joins the material, evoked in the preface and, we hope, now evident to the reader who will have had the persistence to reach this point. This chapter also provides an outlook on what we believe will come next, either as applications, logical extensions, or altogether new research fields unfolding before our eyes – material that was great fun to put together. We truly hope that the reader will enjoy this discussion.

5.1 The Book's Design

As promised in the preface, we aimed at a research book that targeted graduate students and researchers in an emerging field that is being shaped across life and earth sciences. The technicalities, introduced at a fairly accessible level, at times have accelerated in, it is hoped, a logical flow. Several methodological boxes have been added to ease comprehension. Boxes contain heavy content at times – but boxes are boxes are boxes, and they can be skipped, unless the reader is directly interested in the specific topic.

Most challenging, we strived to merge problems and tools from ecology, hydrology, and geomorphology. Their practitioners currently constitute rather different groups, well separated academically and endowed with their field's rituals, tics, preferences, jargon, and sense of belonging. Our view is that this will have to change, as the challenges posed (academically, or to society and welfare at large) require all these fields' tools. We hope that this book will contribute somewhat to the creation of a common ground of widespread accessibility and acknowledgment – concepts and tools of one field immediately accessible to the others.

The interfaces of ecohydrology took some time to develop but are now clear in our view. River networks, and the landscapes they imply, are fundamental to defining the spatial substrate for ecological interactions, including the connectivity of many ecosystems and coupled human–natural systems. On the long march initiated more than 10 years ago, along which we have joined forces with an uncommon circulation of members of our research groups, we found inroads into seemingly unrelated, now acknowledged, common ground. We felt that no book was currently summarizing this fascinating interdisciplinary area. This book is therefore timely, given that freshwater is an increasingly limited resource and the effects of climate and land-use change on hydrology.

The theoretical tools – the authors' starting point in almost all their ventures – are drawn from disease and conservation ecology, epidemiology, biomathematics, environmental engineering, water resources, planetary health, and, perhaps, complex systems epitomized by a few relevant network problems we use (say, models of human mobility that mean so much to the spread of epidemics). Therefore, ultimately, we touch on sustainability science.

We see no logical or conceptual hiatus in the materials we have collated in this volume. Our lines of argument suggest that an integrated ecohydrological framework, focused on hydrologic controls on the biota, has recurrent features hinging on the spatially explicit description of the interactions imposed by a very particular, and quite deeply studied, substrate for ecological interactions: the dendritic ecological corridors put together by the ubiquitous river networks that carry so much weight for large-scale ecological functions. Such functioning proves relevant to a number of key ecological processes that control the spatial ecology of species and biodiversity in the river basin (Chapter 2), the population dynamics and biological invasions along waterways (Chapter 3), and the spread of waterborne disease (Chapter 4), very much via related (if not outright analogous) mechanisms, determinants, and descriptors. We thus went on to describe metapopulation persistence in fluvial ecosystems, metacommunity predictions of fish diversity patterns in large river basins, geomorphic controls imposed by the fluvial landscape on elevational gradients of species richness, zebra mussel invasions of an iconic river network, and major cholera epidemics and neglected range expansions of endemic schistosomiasis as a direct consequence of water resources management. Also, new frontiers opened up for epidemiological and anthropological study for which big data are now available, for example, studies on the spread of proliferative kidney disease in salmonid fish in the former case and for human mobility in the latter.

Our main tenet, built step by step in a crescendo that led to what we now consider a compelling argument, is that ecological processes occurring in fluvial landscapes are so constrained by hydrologic controls and by the morphology of the matrix for ecological interactions (notably, the directional dispersal embedded in fluvial and mobility networks, e.g., host–pathogen ones for disease transmission) that the spatial and temporal patterns in ecology are indelibly marked by them. Actually, in some cases, river networks are templates of such patterns. Decoding these signals requires spatial descriptions that have now spread over a remarkably broad range of applications, which we collected in this book, highlighting the coherent framework that produced them. In brief, our overarching theme is an investigation of how the physical structure of the environment affects biodiversity, species invasions, survival and extinction, and waterborne disease spread. Looking beyond the current perimeter of any given cultural landscape, we aimed at highlighting key research questions worth exploring. In this contamination process, we have exploited our strongly complementary skills, weaving together a work that we hope will be useful to novel investigations of biodiversity and conservation.

The present chapter deals with afterthoughts and an outlook. The former originate from the panic that typically strikes science book authors when coming to a close: Did we forget something critical? Have we proposed a balanced view? The answer to the second question is certainly negative, as we are – because of age, academic taste, or what have you – undoubtedly idiosynchratic in our choice of material. The outlook, a rather entertaining endeavor, is devoted to divining the directions that research in this area may take. Indeed, we look forward to a series of next-generation problems and approaches along the lines of attack and the tools described herein. Many an important topic, which we suggest here, appears ready to be attacked under the general framework presented in this book. The examples that the curious reader will find later in this chapter show features that at times may differ from the ones mentioned in previous chapters. Yet, seen at a distance, they can be related with appropriate modifications to the approach. We thus deem it appropriate to end the book with a selection of topics worth keeping an eye on, and with a brief description of some of the next-generation research questions we perceive to be of key importance in this milieu.

5.2 Outlook on Spatially Explicit Epidemiology of WR Disease

Several water-related (WR) diseases show common spreading dynamics and even similarities in waterborne ecological cycles of the host(s) or pathogens. Therefore we contend that the spatially explicit epidemiological framework initiated with epidemic cholera [325] can be extended to a number of other contexts, possibly of broad epidemiological significance. The sentiment about a common approachability stems also from the fact that we have used consistently the same framework for describing disease spread in ecological communities for three very different WR diseases: deadly epidemic cholera and a neglected debilitating disease like schistosomiasis in humans and the high-fatality spread of proliferative kidney disease in salmonid fish. Few impediments exist, therefore, to generalized uses for other WR diseases within the general context put forward in this book.

Interest in spatially explicit mathematical models of epidemics has grown considerably in the recent past [25, 485, 601, 610, 672, 673]. These models have the potential to account for heterogeneous disease drivers and thus to portray complex patterns of disease dynamics. Current scientific and technological advancements allow us to carry out widespread and relatively straightforward applications of general spatially explicit modeling techniques. Large-scale hydroclimatological and anthropogenic drivers (or their proxies) are key to detailed descriptions of WR disease dynamics and can often be inferred by satellite imagery and the tracking of human mobility via mobile phones. Many parasitic waterborne infections share drivers and transmission processes. However, their hosts and/or pathogens are usually characterized by vastly different disease life cycles, with major implications for the way we infer drivers and controls of the spread of infections. Lessons learned from epidemic cholera research, as suggested in Chapter 4, may be effective in guiding containment efforts also for WR parasitic and viral diseases.

Infections caused by viruses, bacteria, protozoans, flatworms, and roundworms are caused by ingestion of, or contact with, water contaminated by specific pathogens. These viral, bacterial, or parasitic infections share similar

hydroclimatological and socioeconomic drivers that control pathogen concentrations within the accessible water reservoir. Suffice here to recall once more that WR diseases still represent a major threat to human health, especially in low-income countries. Most of the burden of WR infections is attributable to an unsafe water supply, a lack of sanitation, and poor hygienic conditions, which may affect exposure and transmission rates either directly or indirectly. However, the dynamics of disease spread also largely reflect the spatial heterogeneity of disease drivers. In fact, several aspects of climate, such as rainfall and air and water temperatures, affect the transmission rates of WR diseases. Moreover, habitat suitability for pathogens and hosts in fluvial domains is controlled by the nexus of climate, hydrology, and geomorphology that shapes river cross sections (say, the at-a-station relationships between total contributing area and depth, width and velocity – Boxes 2.3 and 2.4) and the ephemerality of streamflows epitomized by the relative occurrence of dry-bed conditions [670]. This nexus controls the suitability of host species that are key to the completion of many parasites' life cycles and the survival of the pathogen in natural waters. Therefore, as we have claimed throughout this book, hydrology is an essential component of the spatial description of the transmission of many, if not all, WR diseases.

More on Macroparasites and Hydrology

Significant cases concern the prevalence of *Opisthorchis viverrini*, *Schistosoma mekongi*, and soil-transmitted helminths (STH), whose public health implications remain high in a number of countries especially in Southeast Asia, despite control efforts including mass drug administration and education and communication campaigns [674]. New approaches, quite possibly an outgrowth of the approaches described in this book, are required to advance helminth control. A little context is in order. Liver flukes (*O. viverrini*); blood flukes (*S. mekongi*); and STH, such as roundworm (*Ascaris lumbricoides*), whipworm (*Trichuris trichiura*), and two-hookworm species (*Ancylostoma duodenale*, *Necator americanus*) cause infections often endemic nationwide but most prevalent in floodplains crossed by major waterways like in the lowlands along the Mekong River, where fish are abundant and local inhabitants prefer to consume traditional dishes prepared with raw fish [674, 675]. Infections with these helminths negatively affect human health and well-being. For example, untreated or chronic infection with *O. viverrini* may lead to severe hepato-biliary morbidity, including cholangiocarcinoma, a fatal bile duct cancer. Chronic infection with *S. mekongi* may result in portal hypertension associated with periportal liver fibrosis. In certain sites, *O. viverrini* and *S. mekongi* are coendemic, further increasing the risk of hepatobiliary morbidity [674]. Finally, anemia and undernourishment are associated with long-lasting STH infections.

Helminths have complex life cycles, more than those presented (Section 4.3) for the various macroparasites that cause schistosomiasis. *O. viverrini*, for example, involves two aquatic intermediate hosts, namely, fresh-water snails (of the genus *Bithynia*) and freshwater fish (of the Cyprinidae family). Humans and other mammals are infected by eating raw or undercooked fish. The life cycle of *S. mekongi* involves humans and other mammals (such as dogs, pigs, and possibly rats) [674]. The *Neotricula aperta* snail serves as intermediate host. It lives in the crevices of submerged rocks in the Mekong River (inasmuch as freshwater bryozoans serve as pri-mary hosts for the pathogens of PKD; Section 4.5). Even in this case, similarly to the case of schistosomiasis, the cercariae emerge from the infected snails during daytime and lie under the water surface. Humans and animals are infected with this parasite via skin penetra-tion when they come into contact with infested waters, as in the case of schistosomiasis. Ecohealth research studies the prevalence and risk factors of the above macroparasitic infections in humans in the ecological environment where potential animal reservoirs and inter-mediate hosts, such as mollusks and fish, live in close connectivity.

A recent analysis of two population-based models of the transmission dynamics of the worm parasite *O. viver-rini* (whose life cycle includes humans, cats, and dogs as definitive hosts and snails and fish as intermediate hosts) includes two models [676]. The first model has only one definitive host (humans), while the second model has two additional hosts: the reservoir hosts, cats and dogs. Dis-tributions of the host-specific type-reproduction numbers show that humans are necessary to maintain transmission and can sustain transmission without additional reservoir hosts, suggesting that targeting humans should be suffi-cient to interrupt transmission of *O. viverrini*. Clearly, a spatially explicit version of the above models is in sight with a straightforward approach that follows those described in Chapter 4. A minor but significant modifica-tion would stem from a new kind of mobility affecting disease transmission: that of the raw fish, traveling as the catch of the day from fishermen's communities to

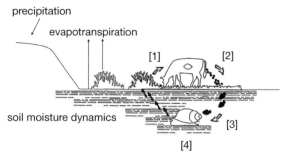

precipitation

evapotranspiration

soil moisture dynamics

Figure 5.1 A scheme of the liver fluke (*Fasciola hepat-ica*) life cycle, where an amphibious mud snail serves as the intermediate host. [1] Metacercariae survive at moderate temperatures; [2] eggs hatch at suitable temper-atures in thin films of moisture; [3] Myracidia float or swim in water, actively seeking a snail host; [4] snails live in muddy areas with no (or little) drainage. Figure after [677]

nearby markets, which might indeed propagate surviving pathogens.

Other kinds of hydroepidemiological models emerge, posing interesting novel challenges. For example, the majority of existing models for predicting disease risk in response to climate change are unsuitable for cap-turing impacts beyond historically observed variability and have limited ability to guide interventions. Beltrame et al. [677] integrated environmental and epidemiologi-cal processes into a new mechanistic model, taking the widespread parasitic disease of fasciolosis as an exam-ple (Figure 5.1). The model simulates environmental suitability for disease transmission by explicitly linking the parasite life cycle to key weather–water–environment conditions. Using epidemiological data, it was shown that the model can reproduce observed infection patterns in time and space for case studies in UK livestock farming [677].

It is interesting that an amphibious mud snail serves as intermediate host in the above example. This suggests another avenue for the study of hydrologic controls on infection disease cycles, that is, those involving the soil moisture dynamics equation, arguably the most funda-mental equation of hydrology [161]. This could very well be the case for STH.

Other cases of soil moisture–controlled disease have been studied, and the relevant hydrology may be significantly different. For example, the incidence of coccidioidomycosis (also called the valley fever, a sometimes lethal respiratory disease caused by a soil-borne fungus, *Coccidioides* spp., whose desiccation under dry spells causes release of airborne spores and their possible inhalation by humans and animals) fluctuates in relation to soil moisture levels from previous summers and falls [678]. Therefore, one may investigate crossing properties of certain soil moisture levels during antecedent seasons to produce scenarios of infections.

Vector-based diseases are of utmost epidemiological relevance and are in many cases water related. Mosquito-related disease (like malaria, dengue fever, chikungunya, and Zika) are water related because the life cycle of mosquitoes is inextricably linked to ephemeral or permanent water bodies. They have not been dealt with in this book, although, in some cases, hydrologic controls exist [679–681], even related to river networks in some cases. For example, possible inroads to the control of river blindness (onchocerciasis) may stem from ecohydrologic approaches of the kind studied in Chapter 4. Great strides have been made toward onchocerciasis elimination by mass drug administration [682]. It is no small matter. In fact, an estimated 25 million people are currently infected with onchocerciasis (a parasitic infection caused by the filarial nematode *Onchocerca volvulus* transmitted by the bites of *Simulium* blackflies [683]). The transmission dynamics of vector-borne diseases such as river blindness are underpinned by complex interactions between populations of vectors, hosts, and parasites. For onchocerciasis, experimental infections of the insect vector have been conducted to understand and quantify the processes determining vector competence, and indeed the life cycle of the vector suggests the need for very specific hydrologic environments [683]. Their modeling incorporates fundamental mechanistic processes and permits prediction of otherwise elusive epidemiological trends. Standard models of river blindness track the life histories of individual male and female adult blackflies, the vectors, and populations of microfilariae within individual human hosts. The related infection process is usually modeled deterministically, with seasonal variation in transmission defined by monthly biting rates (number of bites per person per month) where age- and sex-dependent heterogeneity in exposure to blackfly bites is considered, along with treatment compliance [684]. Clearly, habitat suitability for blackflies is tied to hydrologic conditions, and thus avenues for mapping the heterogeneity of exposure may be envisioned. As championed at length in this book, mathematical modeling provides a means to guide the design of interventions targeting these diseases, in terms of their likely effectiveness and cost-effectiveness in attaining control (and possibly elimination) goals [681]. The reader may also note that in this case, a spatially explicit epidemiology will find fertile ground for extensive applications.

More on Epidemic Cholera Modeling

As we have suggested in Section 4.3.4, mobile phone data are expected to revolutionize epidemiology [596]. In fact, with the increasing diffusion of mobile phones, which have become very widely used even in developing countries and are surprisingly less socially biased than commonly perceived [542], a new model-independent source of information about human mobility has emerged. As we have highlighted in Chapter 4, each time a phone emits or receives a call or text message, the antenna that the cell phone is logged into is registered by the service provider, along with the time of the event [534]. It is thus possible to track the movements of cell phone users as they advance from antenna to antenna. Suitably aggregated and properly anonymized to prevent privacy issues [685], a sample of these data can be used to estimate fluxes of people between areas in a region by assigning a set of antennas to each geographical area in the study domain (e.g., based on administrative boundaries) [520, 540]. The resolution in time is only limited by the typical frequency of calls, whereas the spatial resolution is limited only by the typical distance between two antennas – not necessarily small, yet appropriate for mobility fluxes that are relevant over larger scales [483] (but see [610]). Using mobile phone records of a sufficiently large number of users, one can thus estimate human mobility fluxes with high accuracy, including spatiotemporal variability across a variety of scales [536] and without resorting to any particular model like the gravity or the radiation models (Section 4.3.4).

A number of studies focused on the use of mobile phone data to extract human mobility patterns in developing countries at different scales in space and time, also for epidemiological purposes [535, 537–539, 542]. Others compare the movement patterns extracted from mobile phone records to traditional data sources such as censuses and surveys [686]. We have specifically focused in Section 4.3.4 on the unique role these tools acquire when considering the epidemiological impact of ephemeral mass gatherings, such as pilgrimages, sporting events, or music festivals, which can be critical in the spread of infectious diseases via various transmission routes [520]. In fact, when it comes to orofecally transmitted diseases, such as shigellosis or cholera, insufficient safe drinking water supply and sanitary infrastructure related to overcrowding are often the main causes of local disease outbreaks and subsequent spread by homecoming infected attendees [520]. To model the effect of mass gatherings, one needs to account for the spatiotemporal dynamics of human mobility and the associated short-term fluctuations of population distribution. Standard mobility models, gravity- or radiation-like, and static data sources, such as censuses or surveys, prove therefore largely unsuitable. Conversely, mobile phone records contain all the required information at the desired timescales and thus represent an excellent new (big) data source for epidemiological models.

The downside of a generalized use of phone-derived mobility models lies in the actual availability of data. In fact, providers tend to be shy in making them broadly available – purportedly for reasons of privacy but perhaps, we suspect, also for commercial reasons. Moreover, in a few cases, we have experienced firsthand an academically improper closure to mobility data sharing for strictly academic purposes, where supposedly learned institutions and their academic endpoints act as sinks rather than sources. At the same time, even reputed journals where the relevant raw mobility data (say, fluxes in space and time) should be publicly available *de facto* do not render them usable.

Identifying and understanding transmission hotspots opens the way to implementing novel disease-control strategies. One important lesson learned – largely by modeling studies (Chapter 4) – is that multiple networks affect the making of infection patterns of WR disease. Typically, as seen in the mathematics of spatially explicit models (see, e.g., Box 4.9), multiplex network approaches involve a hydrologic connectivity network (the WR connection) where pathogens can spread and contaminate local water reservoirs nonlocally; and the human mobility network through which critical pathogen loads are released in different nodes in time owing to mobility of susceptibles and asymptomatic infected individuals (recall that often the main source of pathogen loads during an outbreak is due to the attack rate – local new infection cases – caused by the displacement in areas of high exposure of symptomatic and asymptomatic infected individuals). Model results also show how concentrated efforts toward disease control in a transmission hotspot could have an important effect on the large-scale progression of an outbreak [520].

The results shown in Chapter 4 about the generalized reproduction numbers for spatially explicit models of WR disease epidemics – in particular the fact that local basic reproduction numbers $R_0 > 1$ (mediated from a spatially implicit model of disease transmission, i.e., lacking spatial connections of any kind) are neither necessary nor sufficient for the onset of disease outbreak [437] – are echoed by a recent study of data-driven contact networks [687]. Therein, in fact, it has been found that the classical concept of the basic reproduction number does not provide any conceptual understanding of the epidemic evolution in real epidemiological contexts. This departure, analogous to the one that we have derived exactly by studying the stability of disease-free conditions of WB infections (Section 4.2.4), is argued to be due not to behavioral changes or other exogenous epidemiological determinants. Rather, it was explained by the (clustered) contact structure of the population, that is, its spatial structure, revealed by the data analysis. This supports our picture of the geography of disease spread (Section 4.2.5). Therefore, evidence exists for a completely different route, in this case directly from incidence data [687], to estimate actual reproduction numbers that can be used operationally to characterize realistic epidemic dynamics. While this was supported by reference to studies on an influenza pandemic, we note that a similar study for WR disease spread is still lacking.

Where modeling efforts become essential is in identifying how best to deploy intervention measures like a finite batch of vaccines. One noteworthy case is the study of the potential impact of case-area targeted interventions in response to cholera outbreaks [688]. In fact, cholera prevention and control interventions targeted to neighbors of cholera cases (case-area targeted interventions [CATIs]), including improved water, sanitation, and hygiene; oral cholera vaccine (OCV); and prophylactic antibiotics, may be able to efficiently avert cholera cases and deaths while saving scarce resources during epidemics. It has been remarked that efforts to quickly implement target interventions to neighbors of reported cases have been made in recent outbreaks, but little empirical evidence has been provided on the effectiveness, efficiency, or ideal design of this approach. To provide practical guidance on how CATIs might be used by exploring key determinants of intervention impact, Finger et al. [688] have recently studied the mix of interventions (including ring size and timing) in simulated cholera epidemics fitted to data from an urban cholera epidemic in Africa. For this problem (Figure 5.2) a microsimulation model has been developed and calibrated contrasting the epidemic curve and the small-scale spatiotemporal clustering pattern of case households from a large 2011 cholera outbreak in N'Djamena, Chad (4,352 reported cases over 232 days) [688].

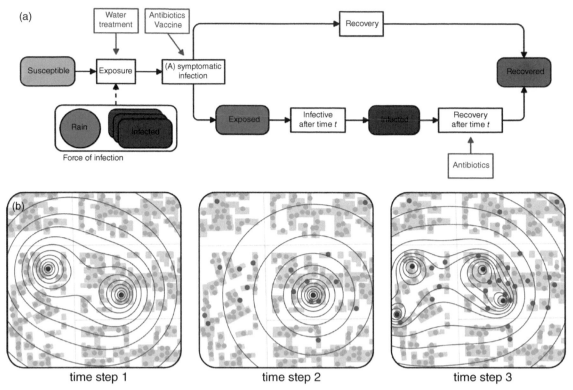

Figure 5.2 Schematic of the epidemiological model and evolution of the infectious state of inhabitants of a neighborhood. (a) Flow chart of the model representing the different epidemiological states a person can be in and the processes that lead to a change of state. The force of infection acting on a susceptible individual depends on the number of infected individuals and the distance to each of them as well as on rainfall during the last 10 days. Orange boxes represent pathways through which interventions (antibiotics, oral cholera vaccine, and point-of-use water treatment) influence the processes in the model. (b) Schematic of the evolution of the epidemiological state of the inhabitants of a neighborhood in N'Djamena during three time steps. The closer susceptible people (blue dots) live to an infected individual (red dots), the higher the force of infection (red contours) they face. Susceptible individuals can get symptomatically infected, which means that they get exposed (green dots) and go on to become infectious after their incubation period (red dots), thus contributing to the force of infection, or asymptomatically infected, in which case they are assumed to recover (purple dots). Infected individuals recover after a given duration. Between time steps 1 and 2, 1 infected person recovered, 4 susceptible individuals got exposed, and 14 susceptible individuals contracted an asymptomatic infection. At time step 3, the individuals infected at time step 1 have recovered, and all exposed individuals have become symptomatic. Figure after [688]

Mechanisms behind cholera persistence during lull periods still remain poorly understood, and specific research in this area combining field, laboratory, and modeling work is in need. Significant contributions in this area dealt with the 2010 Haiti cholera outbreak that was addressed in some detail in Chapter 4. For example, it has been observed that by mid-2014, cholera transmission seemed to persist only in the northern part of Haiti, whereas cholera appeared nearly extinct in the capital, Port-au-Prince, where it eventually exploded again in September 2014 [689]. Determining whether revamping of the outbreak was caused by local undetected cases, climatic drivers like tropical rainfall enhancing exposure and bacterial concentrations in the accessible environment, or else by reimportation of the disease from the north highlights the need for specific tools. Such tools include – besides model improvements already discussed at length – progress in surveillance by suitable cholera genotyping as a powerful means to fine-tune control strategies [689].

Most basic assumptions of the spread mechanisms have been directly transferred to stochastic, individual-based approaches, as we have seen in Box 4.14. In the case of the N'Djamena outbreak, all 993,500 inhabitants of the city were assigned a geographical location according to the population density estimated using remotely sensed built-up density as a proxy (in line with the epidemiological engineering outlined throughout Chapter 4). For instance, demographic processes, like births and deaths, were assumed to be negligible during the short time course of the outbreak, and each individual's state (e.g., susceptible, exposed, infectious, or recovered) is tracked during the outbreak (Figure 5.2a). Susceptible individuals are exposed to a spatially distributed force of infection originating from infectious individuals and decreasing with distance (Figure 5.2b). The force of infection is modulated by rainfall, which has been shown to be an important environmental driver of cholera epidemics in several settings (Section 4.6.1). Exposed individuals could become either symptomatically infected, after an incubation period with a mean duration of two days, or mildly/asymptomatically infected, in which case it is assumed that they do not significantly contribute to the force of infection [688]. Also, completely in line with the assumption of the large-scale deterministic models of Chapter 4, symptomatic infection is assumed to last for an average of five days before individuals recover (actually, shortening the duration of symptomatic infections by use of rice-based, rather than glucose-based, hydration may significantly affect infection patterns [519]). Overall, this approach requires four free parameters: the ratio of symptomatic to asymptomatic infections, a kernel independent transmission rate, a shape parameter of the power-law transmission kernel, and a coefficient governing the influence of rainfall [688]. Model calibration was performed by using a suitable Bayesian approach [688]. To evaluate the benefits (e.g., total averted symptomatic cases) and resource needs (e.g., number of people and number of clusters targeted through CATI) of different types of interventions, a total of 111 scenarios (and several sensitivity analyses) were carried out combining different intervention types, modes of allocation, and intervention starting times. The study explored the potential impact of CATIs in simulated epidemics. CATIs were implemented with realistic logistical delays after cases presented for care using different combinations of prophylactic antibiotics, OCV, and/or point-of-use water treatment (POUWT). Various options were considered: starting at different points during the epidemics and targeting rings of various radii around incident case households. The main findings therein [688] suggest that CATIs shorten the duration of epidemics and are more resource efficient

than mass campaigns. OCV was predicted to be the most effective single intervention, followed by POUWT and antibiotics. CATIs with OCV started early in an epidemic focusing on a 100-m radius around case households (Figure 5.2) were estimated to shorten epidemics by approximately 68 percent, with an 81 percent reduction in cases compared to uncontrolled epidemics [688]. Also, they led to a 44-fold reduction in the number of people needed to target to avert a single case of cholera, compared to mass campaigns in high-cholera-risk neighborhoods.

Why do these studies matter to future public health management? Because, even for studies based on a rigorous, data-driven approach, the relatively high uncertainty about the ways in which POUWT and antibiotic interventions reduce cholera risk, and the heterogeneity in outbreak dynamics from place to place, limits the precision and generalizability of quantitative estimates. Models can provide an efficient tool to complement intervention campaigns that prove particularly useful during the initial phase of an outbreak or to shorten the often protracted tails of cholera epidemics. Moreover, the fact that interventions targeted to neighbors of cholera cases can be an effective strategy to fight cholera epidemics implies that they may be particularly useful during the early phase of an outbreak, when the number of cases is still low. This in turn suggests that individual-based, spatially explicit stochastic models, different from the deterministic approaches pursued when epidemic figures become large and an effective determinism applies (Chapter 4), are in order. They also prove flexible in accommodating field evidence and performing calibration like what was carried out for the 2011 cholera outbreak in N'Djamena [688].

Finally, computational models of WB disease (e.g., epidemic cholera) transmission can provide objective insight into the course of an ongoing outbreak to aid decision-making on allocation of health care resources. However, models are typically designed, calibrated, and interpreted *post hoc*. At times, efforts were reported from teams from academia, field research and humanitarian organizations to model in near–real time cholera outbreaks (e.g., the revamping of the Haitian cholera outbreak after Hurricane Matthew in October 2016 [495]), to assess risk and to quantitatively estimate the efficacy of a (then) ongoing oral cholera vaccine (OCV) campaign. In fact, following the passage of Hurricane Matthew on cholera-struck Haiti in October 2016, a large OCV campaign targeting approximately 760,000 individuals was planned to minimize the risk of cholera transmission after the heavy hurricane rainfall. Model results were therefore translated into operational recommendations during the outbreak management [495]. The projections highlighted the departments that were at risk of a second epidemic wave, and supported the planned vaccination campaigns therein. Modeling projections provided estimates and prediction intervals of the actual number of averted cases due to OCV per each of the 140 Haitian communes. In perspective, this suggests the future importance of optimal control tools to complement modeling approaches of the type described in Chapter 4. Optimal control, in fact, is a systems engineering approach that in this context may be subsumed by the choice of the optimal deployment in space and time of intervention measures to control the spread of infections. A specific function to be minimized needs to be chosen (e.g., total number of averted cases). This seems particularly relevant in the case of the design of vaccination campaigns. In fact, optimal control constrained by a finite number of doses would aim at producing the best location and timing of their deployment. In the writers' view, this is one of the most significant future practical applications of the tools described in Chapter 4, whose predictive ability, as shown formally [435], is greatly enhanced by the use of spatially explicit data and results.

The experience gained by this modeling effort showed that state-of-the-art computational modeling and data-assimilation methods produced informative near real time projections of cholera incidence. This task also clearly showed that collaboration among modelers and field epidemiologists is indispensable to gaining fast access to field data and to translating model results into operational recommendations for emergency management during an outbreak. Future efforts should thus draw together multidisciplinary teams to ensure that model outputs are appropriately based, interpreted, and communicated.

The same concepts apply to the projected impact of geographic targeting of oral cholera vaccinations. One such study, based on evidence collected in sub-Saharan Africa, suggested that targeting substantially improves the cost-effectiveness and impact of oral vaccination campaigns [690].

5.3 Streamflow Ephemerality and Schistosomiasis (and Other WR Diseases) Control

The transmission of waterborne diseases hinges on the interactions between hydrology and ecology of hosts, vectors, and parasites, with the long-term absence of water constituting a strict lower bound. However, the link between spatiotemporal patterns of hydrological ephemerality and waterborne disease transmission is poorly understood and difficult to account for. The use of limited biophysical and hydroclimate information from otherwise data scarce regions is therefore needed to characterize, classify, and predict river network ephemerality in a spatially explicit framework. Large-scale ephemerality classification and prediction methodologies have been recently developed, specifically aiming at their epidemiological relevance [670]. They are based on monthly discharge data, water and energy availability, and remote-sensing measures of vegetation. These approaches must maintain a mechanistic link to catchment hydrologic processes: under what conditions can the intermediate hosts or the pathogen survive?

River network ephemerality is an especially important hydrological characteristic influencing waterborne and WR diseases, owing to its direct and indirect effects at different stages of the transmission cycle. The former include pathogen spread and survival, intermediate host ecology, human exposure to the pathogens, and the wealth of known contamination pathways of water supply sources. Indirectly, the changing connectivity of intermittent river networks (an important subject of hydrologic research; see, e.g., [691]) plays a role in the metapopulation dynamics of hosts, vectors, pathogens, and humans.

Examples of interest are related in particular to the ecology of intestinal and urogenital schistosomiasis caused by *S. mansoni* and *S. haematobium*, respectively (see Figure 5.3, after [670]). As seen in Chapter 4, the disease is contracted during human–water contact through skin penetration of motile free-living larvae, thus exposing local communities during domestic (washing, laundry), leisure (bathing), and livelihood-related activities (fishing, agriculture), with initial contamination occurring as schistosome eggs reach water bodies through feces or urine. The parasites' life cycle requires obligatory aquatic snail intermediate hosts, specifically from the genus *Biomphalaria* for the intestinal and *Bulinus* for the urogenital forms of the disease. Here is the key connection with hydrologic regimes: although both are capable of surviving periods of desiccation through aestivation, *Biomphalaria* snails are much less adapted than *Bulinus* to prolonged dry spells, which can severely increase snail mortality. It is known that the mean lifespan of *Biomphalaria pfeifferi* is about 40 days, according to laboratory estimates, whereas the species within the *Bulinus* genus have

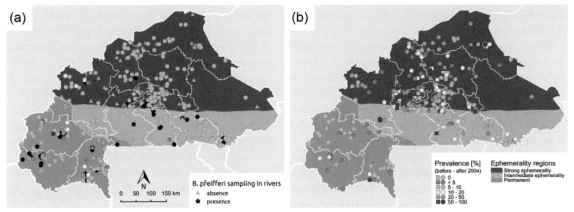

Figure 5.3 River network ephemerality and the geography of schistosomiasis. (a) Map of predicted ephemerality regions in Burkina Faso corresponding to common ephemerality classes found within such regions – along with the locations of historical malacological and parasitological data collections before (squares) and after (circles) the start of the national program of Mass Drug Administration (MDA) in the country (white lines indicate the 13 administrative regions). (b) Same as (a), except for urogenital schistosomiasis. The intermediate hosts of *Schistosoma haematobium* can be found throughout the country and therefore are not shown here. For an accurate interpretation of the references to color coding, the reader is referred to the original reference. Figure after [670]

evolved to survive beneath temporary ponds and streams that only have water for a couple of months per year. Therefore, hydrologic ephemerality critically determines local habitat suitability for the snail intermediate hosts, in particular for intestinal schistosomiasis, and also conditions human exposure/contamination by limiting the temporal window and the number of locations in which human–water contacts can occur.

Specifically, with reference to the context of Burkina Faso in sub-Saharan Africa, a relevant set of catchment covariates was extracted that includes the aridity index, annual runoff estimation using the Budyko framework, and hysteretical relations between precipitation and vegetation [670] (the Budyko framework is a widely used representation of the land–water balance that describes the mean annual partitioning of precipitation into streamflow and evaporation as a function of the ratio of the atmospheric water supply [precipitation] to water demand [potential evaporation]). Five ephemerality classes, from permanent to strongly ephemeral, were defined from the duration of zero-flow periods [670]. A strong south–north gradient in precipitation, temperature, and available water resources provides the key determinants. Therein, stream ephemerality was classified via (1) clustering the available hydrological data into different degrees of ephemerality, (2) building classification-relevant catchment characteristics, and (3) finally predicting ephemerality for any site of the whole river network (Figure 5.3). Using these classes, a gradient-boosted, tree-based

prediction yielded three distinct geographic regions of ephemerality directly connected with a strong epidemiological association – hydrologic ephemerality and spatial patterns of schistosomiasis proved much related (Figure 5.3, after [670]).

The observed ecological range of *Biomphalaria* snails, and thus of intestinal schistosomiasis, is generally found to coincide with the presence of quasi-permanent rivers and waterbodies in countries with strongly seasonal climates and ephemeral flow regimes within a suitable temperature range. Notably, the development of water reservoirs for irrigation is associated with higher schistosomiasis risk throughout sub-Saharan Africa [586]. The hydrologic regime (e.g. permanent vs. ephemeral) is undoubtedly a key determinant for the seasonal fluctuations of freshwater snail abundance, thus dam construction can have important consequences for the presence and abundance of these species and the parasites. On the other hand, construction of man-made reservoirs, often in response to water scarcity, can dramatically change local population densities and human–water contact patterns by fostering new economic activities such as "market gardening" in the dry season, fishing, or cattle herding. The new behaviors may prompt heterogeneous exposure and contamination leading to increases in the overall disease burden in the community.

A progressive loss of hydrologic ephemerality is a risk factor in the case of urogenital schistosomiasis. Factors that must be accounted for are chances of survival for intermediate hosts of *S. haematobium* even in extremely dry areas and the progressive concentration of human activities around long-lasting water contact points, such as man-made reservoirs, lakes, or large temporary ponds. Seasonal hydrologic connectivity may also play an important role in the dispersal of the snail intermediate hosts between different habitats, with potential implications for the focal control of snail populations and for the coevolution of snails and schistosomes.

While the importance of hydrologic ephemerality has already been identified separately by ecologists, sociologists, and epidemiologists, it has yet, to our knowledge, to be quantitatively implemented on a regular basis, as the above example shows. In many cases of epidemiological interest, this is due to the lack of an amenable approach for the spatial evaluation of river network ephemerality at regional scales. A better understanding of the hydrological underpinnings of large-scale ephemerality, including the likely extent of occurrence of dry channels, is therefore crucial to spatially explicit predictions of schistosomiasis (and other WR disease) transmission dynamics, with implications for disease control and elimination strategies [25]. In the case of the determinants of stream ephemerality, they frequently span multiple spatiotemporal scales across the climatic, vegetation, soil, and topographic features that characterize each river network. The importance of feedback between these variables for streamflow generation is notable, but our ability to predict the frequency and duration of zero-flow events (i.e., not simply the probability of discharge $Q = 0$ but also the probability of a given number of consecutive days with $Q = 0$) is rather limited in data-scarce regions. Thus defining the relative amount of dry-bed time (a true hydroperiod) poses important methodological challenges. Determining ephemerality has typically been a component of hydrologic regime classification studies that rely on the extraction of hydrological indices from streamflow data. This is also the case for the exact determination of the atom of probability at the origin, that is, the frequency of $Q = 0$ events, from flow duration curves [160]. In such approaches, the classification of ungauged rivers mostly relies on the prediction of relevant indices of hydrological regimes using remotely sensed catchment characteristics like contributing area and slope and climatic information like precipitation and temperature.

Downsides of refined methods, like index clustering [670], are related to their requirements of long time series of discharge data seldom existing in developing countries, which bear most of the WR disease burden. These requirements may hinder their applicability to data-scarce regions and to the needs of spatial context where waterborne diseases typically occur. These reasons could explain the limited number of studies concerning catchment classification in sub-Saharan Africa, despite its importance for water resources, ecological, and disease management – especially for studies on the effects of climate change.

The general nature of the underpinnings of the example shown in this section should be evident, as should be their relevance for predicting hydrologic controls on other WR diseases for which fluvial or reservoir waters provide a pathway for transmission of infections. On this we expect further important developments.

Highlighting hydrologic controls underpinning the habitat of hosts/pathogens of WR disease is a lively research field. For example, current strategies to interrupt schistosomiasis transmission emphasize the targeting of freshwater snails that are the necessary intermediate host of schistosome parasites (Chapter 4). To that end, the effects of prawn aquaculture on poverty alleviation and schistosomiasis control have proved significant [692]. Moreover, predictors of snail presence/abundance tend to be more stable in space and time than snail habitat suitability models [693] and nonetheless may still utilize remotely acquired and objectively manipulated environmental proxies. Remarkably, proxies (e.g., the area covered by floating, nonemergent vegetation relative to the size of the water contact area) proved distinctly more effective than hard-gained data on snail variables for predicting human infections. Unlike snail surveys, proxies are easily estimated from drone or satellite imagery [694].

Fieldwork underpinning smart interventions aimed at interrupting transmission cycles is rapidly picking up in the wake of the recent surge of interest in schistosomiasis as a model neglected disease. As an example of the emerging literature, evidence gathered in the Niger river valley highlighted a marked seasonality in both overall snail abundance and infection with *Schistosoma* spp. in *Bulinus truncatus*, the main intermediate host in the region [695]. Therein, monthly longitudinal surveys, representing rather intensive sampling efforts, have shown the kind of resolution that is needed to ascertain both temporal and spatial trends suited to informing effective interventions in regions endemic for the disease.

Biological invasions have been treated in detail for their dependence on hydrologic controls and the species-specific ecological dynamics (Section 3.2), whereas WR disease spread through waterways is the core business of Chapter 4. Incidentally, an outlook referring to biological invasions cannot ignore the role of hubs of connectivity and habitat distributions within the river network [696]. In fact, during the stages of an invasion, the interactions between the features of the invaded landscape and the internal dynamics of an introduced population have a crucial impact on establishment and spread. By combining simulations with the experimental introductions of *Trichogramma chilonis* (Hymenoptera) in artificial laboratory microcosms, it has been demonstrated that spread is hindered by clusters and accelerated by hubs, while at the same time being affected by small-population mechanisms such as Allee effects [696].

Relatedly, new perspectives on the use of tools and concepts described in Chapter 4 may concern the setup of integrative matrices for biomonitoring aquatic ecosystems. For example, *Cryptosporidium parvum, Toxoplasma gondii,* and *Giardia duodenalis* are pathogenic protozoa recognized as causal agents of waterborne disease outbreaks. To overcome limitations of protozoa detection in aquatic systems, the use of the zebra mussel (*Dreissena polymorpha*, the freshwater bivalve mollusk whose ecology has been described in detail in Chapter 3; see also [697]) has recently been proposed as a tool for biomonitoring protozoan contamination [698].

To achieve elimination of chronic disease like schistosomiasis, however, more sensitive diagnostic tools are needed to monitor progress of transmission interruption in the environment, especially in low-intensity infection areas. One promising avenue concerns the development of eDNA-based tools (Sections 3.6 and 5.6). For instance, a recent study [699] pinpointed the importance of efficiently detecting DNA traces of the parasite *S. mansoni* (Section 4.4) directly in the aquatic environment. Remarkably, the study has achieved a comparative analysis of the detectability of "true" eDNA at very low concentrations of cercariae and tested the field applicability of the method at known transmission sites by comparison of schistosome detection by conventional surveys (snail collection and cercariae shedding) with eDNA in water samples [699]. The substantial agreement between the methods when surveys provide a baseline and, critically, the detection of schistosome presence at sites where snail surveys failed suggest that eDNA will likely provide in the future a key tool for field epidemiology of WR disease.

5.4 Hydrologic Controls on Microbial Diversity and Beyond

Does river networks' structure affect the many by-products of metabolic activity and thus the global metabolism of a river basin? And if so, how? Carbon storage in soils, rivers, and biomass operated by various assemblages of bacterial communities under various stressors (functioning at the scale of river basins under heterogeneous, environmentally relevant conditions) is a topic of paramount importance that deserves thorough investigation.

It has been suggested that carbon storage in soils depends on cycles of oxic/anoxic conditions directly related to hydrologic drivers [700]. The makeup of microbial diversity and communities in the river basin is of great interest here, because it has direct implications for the type of metabolism these organisms (and thus the river basin as a whole) can ultimately carry out. Measuring how carbon storage in soils or sediments varies as a bacterial community evolves – subject to changing carbon sources or to varying redox regimes – is a question of great relevance and modernity that we believe will soon impact the hydrologic community. It is known, in fact, that the frequency of redox cycling imposed by hydrologic forcings (shifts in soil saturation, hydrodynamic shear affecting sediments) impacts the composition and activity of the affected microbial community [701].

Several mechanisms have been documented by which redox cycling may impact microbial carbon utilization. In a nutshell, (1) shifts in the microbial community in response to redox conditions favor the growth of specific organisms capable of utilizing the available carbon, (2) the availability of specific electron acceptors depending on the geochemical regime can influence the type of organic carbon that can be degraded, and (3) microbial evolution via point mutations (e.g., single nucleotide polymorphisms, SNPs) or large sequence variants (via horizontal gene transfer, HGT) results in changes to the genetic material that is retained under the relevant selection pressure. It is relevant for our scope that soil and sediments are now seen as active areas of microbial evolution and that geochemically dynamic environments favor evolutionary processes. Taking river networks as the object of large-scale testing of the hypothesis that the frequency of redox cycling impacts the rate of HGT and overall genome mutation, with particular focus on carbon utilization, seems to us of utmost importance. Moreover, it is completely related to the role of hydrologic controls because under highly dynamic redox conditions, fluctuations in organic carbon availability result in environmental conditions that require rapid adaptation. The faster the redox fluctuation is, the more versatile microorganisms need to be to retain competitiveness. We wonder whether measures of HGT and point mutations enabling key metabolic innovations that allow adaptations to natural fluvial environments are templates of the river basin structure at large scales.

Fluvial network organization imprints on microbial co-occurrence networks have represented a significant starting point for the general connections that we seek [163]. In fact, microbial communities modulate most biogeochemical processes on Earth. In streams and rivers, surface-attached and matrix-enclosed biofilms dominate microbial life. Despite the relevance of these biofilms for ecosystem processes (e.g., metabolism and nutrient cycling), it remains unclear how features inherent to stream and river networks affect the fundamental organization of biofilm communities in these ecosystems. In a first effort in this direction [163], analyses of biofilm diversity based on next-generation sequencing were combined with a hydrological model to show how fragmentation of microbial co-occurrence networks change alongstream. These types of analyses offer potential insight into the response of a microbial community's organization, also in the face of human pressures that increasingly change the hydrological regime and biodiversity dynamics in fluvial networks.

A legitimate question is whether the structure of the landscape inhabited by organisms – the fluvial domain – leaves an imprint on their ecological networks. Widder et al. [163] analyzed, based on pyrosequencing profiling of the biofilm communities in 114 streams, how features inherent to fluvial networks affect the co-occurrence networks that the microorganisms form in these biofilms. For a complete account, the reader is referred to the original paper [163]: in a nutshell, data comprised 955,691 sequences constituting 1,005 operational taxonomic units (OTUs) affiliated with 126 bacterial families, from which co-occurrence networks at OTU level and their fragmentation were computed. Figure 5.4 shows how the fragmentation of co-occurrence networks of biofilm communities (defined in the caption) may change with catchment size as a continuous and scalable parameter that varies with the hydrologic regime in the Ybbs catchment (Figures 2.7 and 2.8). Notable differences in fragmentation between small and large streams (Figure 5.4a,b) suggest that it is not driven by taxa richness.

What makes this and other studies relevant from the viewpoint of this book is the clear ecohydrological connection. In this case, fragmentation of co-occurrence networks was found to be significantly higher in biofilm communities downstream of confluences than in confluences upstream. This pattern, irrespective of stream size, suggests that hydrological disturbances and metacommunity dynamics are potential controls on the co-occurrence patterns of benthic biofilm communities in fluvial networks. The hydrological model of streamflow distributions (see Box 2.5) shows how the hydrological regime, a major control on benthic microbial life, changes from upstream to downstream as catchment size increases (Figure 5.4c,d). Co-occurrence network fragmentation along this hydrological gradient is somewhat counterintuitive, we believe. In fact, fragmentation is expected to parallel decreasing flow-induced disturbance downstream, whereas, for instance, the hydrological regime in small streams is characterized by notable temporal fluctuations and even by nonzero probability of dry-bed days ($\lambda/k < 1$) [163]. It should be remarked that changes in flowrates can physically alter benthic biofilms (e.g., by abrasion and erosion exerted by shear stresses), affecting their ecological functioning and community succession. Farther downstream, where contributing areas are larger, temporal flow variability is damped owing to reduced discharge fluctuations, a well-known empirical hydrologic fact, relieving physical stress on benthic biofilms. This overall scenario is quite in line with ecological field observations on food chain length, which in the river continuum scales with catchment size and flow variability in streams [113].

Mechanisms linked to hydrology and metacommunity dynamics abound in stream ecology, as supported by a notable body of theoretical and empirical evidence already reviewed in this book [6, 8, 14, 100, 702]. Many other facets of this important ecohydrological problem are arising, and it will be interesting to monitor developments in this area. For instance, although hydrological variation is indeed high in small-headwater streams, the constrained contributing area of these catchments limits the habitat and thus the size of the metacommunity, from which local biofilm communities assemble in these streams. Along with their relative isolation, headwater streams in fluvial networks play a fundamental ecological role, however constrained by dispersal limitation in small compared with large streams. The concerted effects of hydrological regime and dispersal limitation thus characterize the distinct composition of biofilm communities in streams – with implications, say, for larger beta diversities (Section 2.1) in headwaters [702]. The hierarchical structure of the river network pitches in via the nonlocal interactions posited by streamflow (in turn tied to total contributing area). For example, mid-size streams, progressively more affected by the larger size of the upstream metacommunity, exhibit reduced

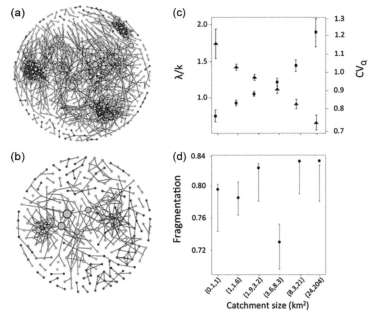

Figure 5.4 Microbial co-occurrence networks based on 454 pyrosequencings of benthic biofilm communities in the river Ybbs (Austria). Each node is an OTU whose size (diameter in the plot) represents its betweenness centrality [5] measuring the importance of the node in terms of connectivity and mean distance from any other node. If two nodes co-occur in the same site, a link (arc) is shown in the network. Co-occurrence networks from communities in (a) 50 upstream (<5 km^2; first- to third-order streams) and (b) 64 downstream (>5 km^2, second- to fifth-order streams) sites. Fragmentation of the co-occurrence networks from upstream communities is significantly lower than that from downstream communities [163]. Giant components are indicated in green for both co-occurrence networks; other colors denote smaller components. (c) Fragmentation patterns differ from patterns of the coefficient of variation of streamflows CV_Q (triangles) and the hydrologic parameter λ/k [160] (Section 2.1, Box 2.5) across the bins of size of nested catchments organized alongstream – the larger the catchment size, the further downstream the sampling site (as in [d]). (d) Fragmentation (95 percent confidence interval) is highest in small and large streams but significantly lower in mid-sized streams. Co-occurrence networks were computed from 10, 10, 14, 15, 14, and 14 communities for the six bins of catchment size (here shown in log scale), respectively. We used only sites upstream of confluences to avoid dependencies on (and effects of) fluvial confluences. Figure after [163]

streamflow fluctuations and high levels of biotic interactions and species sorting mediated by the local environment. Theory and experimental observations suggest enhanced biodiversity in more connected communities that occupy a central position in fluvial networks. The previously described work on the Ybbs river network [163] shows maximal values of alpha diversity in mid-sized streams, corroborating this commonsense notion. Farther downstream, at stable regimes of reduced fluctuations and long timescales for the variation of processes of interest, it is reasonable to assume that metacommunity dynamics rather than hydrology per se become the most important control, at least on the co-occurrence patterns of biofilms, because contributing area increases, and therefore the metacommunity from which microorganisms can immigrate increases as well.

The erratic flows that characterize the abundant and rapidly responding streams in the upstream catchments (i.e., at high values of λ/k; see Figure 5.4), hence the

microbial diversity they transport downstream, may shape the composition and dynamics of the downstream metacommunity. As was noted [163], asynchronous contributions of microbial diversity from these upstream catchments may affect dispersal and assembly dynamics of local communities downstream, possibly via neutral processes (Section 1.1.1).

The main importance of the results shown above lies in their context. In fact, they suggest the types of linkages that govern the organization of microbial communities in relation to flow dynamics across fluvial networks. It is therefore of great importance to promote an understanding of the consequences of the hydrologic controls on microbial interactions and of the persistence of biofilm communities, which are critical for stream ecosystem processes. All in all, this brief and partial outlook suggests that hydrology and metacommunity dynamics, both changing predictably across fluvial networks, affect stream ecology from food webs to the fragmentation of the microbial co-occurrence networks. Definitive results in this area will be of great importance because of generalized increasing anthropogenic pressures deteriorating stream ecosystem integrity and biodiversity.

5.5 Scaling of Carbon Sequestration and Fluvial Corridors: Global Issues

As we noted already, natural ecosystems often exhibit deep structural similarities emerging across scales of space, time, and organizational complexity [2]. Scale invariance of ecological patterns offers a powerful tool to make way for coherent, unified descriptions, especially in river networks whose geometry, topology, and geomorphology affect their role as ecologic substrates [63]. If the part and the whole of the substrate for interactions are statistically indistinguishable across many orders of magnitude, implications abound for the function of the embedded biota, as this book, we hope, has shown. Several areas where results are still missing do exist, however. One deals with scaling of carbon storage in inland waters, recently argued to play a central role in the global carbon cycle [152].

Metabolic principles of river basin organization have been, and will be, well sought (see, e.g., [703]). If we define the metabolism of a river basin as the set of processes through which the basin maintains its structure and responds to its environment, then several links with hydrology at large scales exist – meaning not simply the effects of a single organism but rather those of a large and diverse assemblage of organisms like vegetation at catchment scales. Green (or biotic) metabolism is measured by transpiration and blue (or abiotic) metabolism through runoff, with distinctions arising because of the sink in biofilms that proves quite significant [152]. In fact, currently, the "boundless carbon cycle" must be seen at a different angle, as the terrestrial biosphere is assumed to take up most of the carbon on land, but it is now clear that inland waters process large amounts of organic carbon in their networks [152]. Approximately 0.6 Pg C/yr are buried in inland water sediments, approximately equivalent to 20 percent of the carbon assumed to be uptaken by terrestrial biomass and soils. Still, these estimates do not include long-term net carbon burial in floodplains and other near-water landscapes that probably make for another significant flux [152]. It has been argued convincingly that the organic carbon from terrestrial ecosystems and its subsequent burial in inland waters represent a

redistribution of carbon sinks that must be taken into account in climate change mitigation strategies [152].

Linkages between flow regime, biota, and ecosystem processes have been recently summarized with reference to implications for river restoration [704]. Therein, as in Box 2.4, focus is on the degradation of fluvial ecosystems caused by alterations of the natural sequence of streamflows. Effective river restoration requires, in fact, the kind of mechanistic understanding of how flow regimes affect biota and ecosystem processes that we have addressed throughout all the chapters of this book. Space-time variations in flow regimes not only exert key controls on the composition, structure, and dynamics of communities (Chapter 2) but also critically influence ecosystem metabolism, nutrient uptake and transformation, and organic matter processing [704].

Other questions come to mind. Allometric scaling of metabolic rates with organismic size has been variously addressed in the biological literature and network theory under the banner of Kleiber's law [705]. Decades of research in field and theoretical ecology have unraveled a wide array of ecological scaling laws pertaining to the distribution of species, their abundances, and metabolic requirements [706]. These laws hold for most ecosystems on Earth and help predict how ecological communities assemble in the environment. Relatedly, physiological studies have shown that metabolic rates, B, scale with body mass, M, according to a universal power law, $B = cM^{\alpha}$ (where α is a scaling exponent and c a constant). This is Kleiber's law (KL) [705, 707, 708], which has been claimed to hold across more than 27 orders of magnitude of body mass – from molecules of the genetic code and metabolic machinery to whales and sequoias [709]. The metabolic power required to support life across that range spans more than 21 orders of magnitude and is fundamental for our understanding of ecology across scales. It is widely accepted that KL applies across species with $\alpha = 3/4$ in many groups of organisms, a predominance termed the central paradigm of comparative physiology [706]. Metabolic demand per unit mass thus decreases as body mass increases. For higher organisms, KL reflects both the ability of the organism's transport system to deliver metabolites to the tissues and the rate at which the tissues use them. The universality of the exponent 3/4, however, has been challenged by robust empirical evidence of a wealth of exponents α deviating from 3/4 and by arguments suggesting alterations of the power law relation. Implications of the structure of KL are claimed to include many fundamental processes, such as life history attributes (including development rates, mortality rate, age at maturity, lifespan, and population growth rate), population dynamics and interactions (including carrying capacity, strength of competition and predation, and patterns of species diversity), and ecosystem processes (including rates of biomass production and respiration). This led to a metabolic theory of ecology of broad aims [706] and to discussions of its limits and validity.

All this matters to metabolic principles of river network organization. In fact, the above results suggest by analogy a possible relation of overall metabolic rates to total cumulative area at any point of a river network [710]. In fact, in Euclidean geometry, a D-dimensional object, a 3D body or an almost 2D catchment, is characterized by a linear size L and a "body" size (i.e., mass M) that scale as $M \propto L^D$. To model the metabolic system of living organisms, it has been postulated that the fundamental processes of nutrient transfer at the microscopic level are independent of organism size. In a D-dimensional organism, the number of such transfer sites scales as L^D [710]. Each transfer site is fed with nutrients (e.g., through blood) by a central source through a network providing a route for the transport of the nutrients to the sites. The total amount of nutrients being delivered to the sites per unit time is the equivalent of the metabolic rate B and simply scales as the number of sites, or $B \propto L^D$. The total blood volume, proportional to the body size M, depends, in steady state of supply and demand of metabolites, on the structure of the transportation network. It is proportional to the sum of individual flowrates in the links or bonds that constitute the network. In a very broad class of efficient or simply directed networks, one has $M \propto L^{D+1}$ [710]. Thus the total blood volume increases faster than the metabolic rate B as the characteristic size scale of the organism increases. Thus larger organisms have a lower number of transfer sites (and hence B) per unit blood volume. Metabolic rates do not scale linearly with mass but rather as $B \propto M^{D/(D+1)}$, or $\alpha = D/(D+1)$ [710], a result suggesting that the fully 3D bodies should have $\alpha = 3/4$ and flat 2D organisms $\alpha = 2/3$.

For a river network, delivering water and sediments to maintain its function, one therefore argues that $D = 2$ and $\alpha \sim 2/3$. To test this, Maritan et al. [711] showed that total contributing area A_i at any site i plays the role of the basin metabolic rate B_i, whereas the analog of organismic mass at i, M_i, is defined by the quantity $M_i \propto \sum_{j \in \gamma(i)} A_j$, where $\gamma(i)$ indexes all paths connected to i, that is, the integral of all fluxes in steady state. The main result [711] is that indeed, through a vast

empirical testing, one retrieves $\alpha = 2/3$ with minimal scatter. Evidence also suggests that ensemble averaging of the allometric property (where individual realizations are different networks) leads to results in excellent accord with the limit scaling of efficient networks. Network-related allometric scaling in living organisms is thus regulated by metabolic supply–demand balance, where scaling features are robust to geometrical fluctuations of network properties.

This may develop into a research question deeply connected to the tenet of this book. Are there meaningful principles of equal metabolic rate per unit area throughout the basin structure? Such principles would have profound implications for large-scale geochemical cycles, and they seem likely because of the very spatial organization of river basin hydrologic dynamics. Do we have a definitive view on whether the basin structure itself leads to a power law for the probability distribution of metabolic rates, for example, transpiration from a randomly chosen subbasin? Empirical evidence has merely suggested that river basin metabolic activity is linked with the spatial organization that takes place around the drainage network and therefore with the mechanisms responsible for the fractal geometry of the network [703]. A new coevolutionary framework for biological, geomorphological, and hydrologic dynamics may very well be in sight.

Streams and rivers play a major role in the global carbon cycle because they collect, transform, and deliver terrestrial organic carbon to the oceans. The rate of dissolved organic carbon (DOC) removal depends on hydrological factors, primarily water depth and residence time. These factors change quite predictably within a river network, which assumes universal scaling of geomorphic features regardless of size, vegetation, exposed lithology, or climate [63]. Therefore, local DOC concentration and composition are the result of transformation and removal processes occurring in the whole upstream catchment. The nonlocal character of any local interactions, whether of physical, chemical, or biological nature, is subsumed by the master variable A_i, total contributing area at any point i in the river network.

Recently, theory of the form and scaling of river networks was combined with models [162, 712] and experiments [713] of DOC removal from streamwater to investigate how the structure of river networks and the related hydrological drivers control DOC dynamics. Therein, it was found that the physical process that shapes the topological and metric properties of river networks, imperfect minimization of total energy dissipation leading to dynamically accessible states (Appendix 6.2), leads to structures that are also more efficient in terms of total DOC removal per unit of streambed area. This unintended consequence of nature's operation echoes the increase of metapopulation capacity with evolving networks (Section 2.4).

The structure of river networks also induces a scaling of the DOC mass flux with the contributing area that does not depend on the particular network used for the simulation. This fact, which proves totally robust to spatial heterogeneity of model parameters, echoes another property of river networks as ecological corridors, the inevitability of the topology seen by the backbone for propagation of a biological invasion along a network that explained patterns of human migration (Sections 1.6). In this case, it was found that such scaling enables the derivation of removal patterns across a river network in terms of clearly identified biological, hydrological, and geomorphological factors [162]. In particular, the fraction of terrestrial DOC load removed by the river network scales with the catchment area A (and with the area of a region drained by multiple river networks). This result proves to be of particular importance (Figure 5.5).

It should be acknowledged that a direct comparison of scaling relationships with empirical data could be hindered by the fact that the framework in Figure 5.5 focuses on the terrestrial deliveries of DOC and ignores instream DOC sources [162]. The release of DOC from algae and microorganisms can greatly contribute to the overall DOC pool in streams during baseflow. Therefore, a sensible conclusion is that this autochthonous DOC is generally highly available to heterotrophic metabolism and is therefore thought to yield high uptake velocities. This labile DOC could easily facilitate the removal of terrestrial DOC, an interaction called priming [714]. Therefore, future modeling efforts should focus on combining the various carbon sources intrinsically diverging in bioavailability and hence in uptake velocities [162].

The scaling relationships investigated here may be useful for characterizing basal patterns of terrestrial DOC, the main source of organic carbon in numerous streams. Moreover, deviations from these patterns may indicate the relevance of autochthonous DOC. Other caveats and limitations of current approaches are discussed in the original paper [162].

Previous efforts to characterize the role of network structure and stream size in carbon and nutrient removal relied on the concepts of stream order and Horton's ratios [715]. As noted in [162], an advantage of the framework presented in this section is that it works with the full realistic set of metrics (e.g., individual reach length, width, contributing area) and the complete description of the structure of real (or synthetic) river networks rather than relying on average properties defined by stream order. Besides, the computation of mean stream length via Strahler's ordering is tricky and may lead to severely overestimated mean network lengths [67, 162, 712]. The availability and analysis of digital elevation models of the Earth's surface have rendered the use of concepts like stream order obsolete [63]. Indeed, contributing area proved to be the key variable describing the scaling of DOC removal patterns across river networks.

Scaling studies have furthered our understanding of carbon sinks in inland waters and opened new areas for ecohydrologic investigation. Low-hanging fruits are relaxations of simplistic assumptions, such as those of steady state hydrology. The major implication of those studies is that the impact of instream processes on carbon cycling can be clearly related to global scales – once freed of previous untenable simplifications. More research is also appearing on the contribution of mountain streams to global carbon fluxes. For instance, it has been recently found – by using insights from gas exchange in turbulent streams – that areal CO_2 evasion fluxes from mountain streams equal or exceed those reported from tropical and boreal streams, typically regarded as hotspots of aquatic carbon fluxes [716].

We may therefore confidently speak of metabolic regimes of flowing waters [717]. The production of organic carbon by aquatic photosynthesis, in fact, is a central property of stream ecosystems that has a defining influence on food webs and nutrient cycling rates. A sufficient operational understanding – for example, suited to the challenges of global predictions – of how these factors combine to determine primary productivity at the scale of whole river networks is still lacking. Recent simulations of a range of river networks have provided a first example of suitable approximation of river ecosystem productivity at broad scales by using concepts and tools central to the tenet of this book (e.g., the optimal channel network model; Section 6.2) [718]. Therein, hydrologically driven temporal variability, especially in reach-scale productivity regimes, is suggested to be central to the point that in some cases, small streams and certain time periods disproportionately influence overall river network production [718]. Emergent general patterns of primary production at river-network scales, and, in particular, defined envelopes of possible productivity regimes, are thus in sight. Stay tuned.

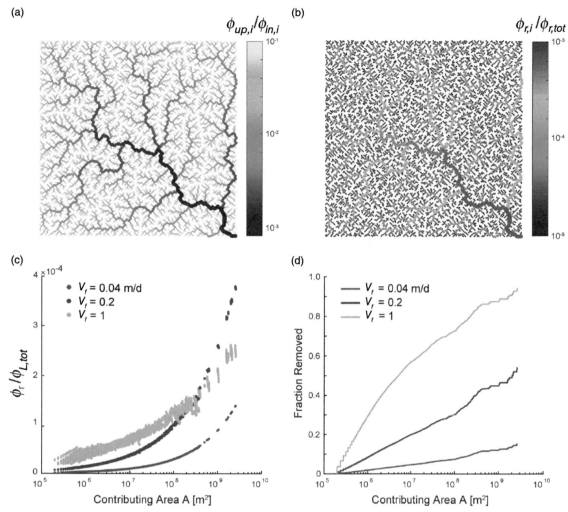

Figure 5.5 Distribution of DOC removal along optimal channel networks (OCNs). (a, b) The OCNs used for the simulations. Color codes represent (a) the fraction of mass input to each reach that is removed by that same reach and (b) the contribution of each reach to the total network-scale removal. (c) Removal flux of each reach, $\phi_{r,i}$, normalized by the total load, termed $\phi_{L,tot}$, as a function of the master variable for stream ecology, total contributing area A at any reach. (d) Fraction of the total load that is removed by streams with contributing area smaller than A. Top panels show results for an uptake velocity $V_f = 0.2$ [m/d]. (The latter is defined as the ratio between the removal of a given element per unit of streambed surface area [M/L^2 T] and its concentration in the water column [M/L^3]. The model thus assumes that most of the processing occurs at water–sediment (benthic biota) or water–air (i.e., photodegradation) interfaces.) For every reach i of the river network, the downstream mass flux $\phi_{r,i}$ [M/T] can be expressed as a function of input flux, $\phi_{in,i}$, and removal via $\phi_{r,i} = \phi_{in,i}\exp[-V_f\tau_i/d_i]$, where d_i is the depth of the water column and $\tau_i = w_i d_i L_i/Q_i$ is the residence time of water within reach i. V_f is a biological measure independent of hydrological conditions because it is based on per unit area removal and is well suited for comparing biological activity in streams of different sizes [162]. Thus, even at constant V_f, the removal rate decreases downstream as water depth increases. Moreover, average water velocity typically increases (mildly) downstream (Box 3.27), thus reducing residence time τ_i for equal reach lengths L_i. Therefore, hydrological and geomorphological drivers can exert a critical control on elemental removal at the river network scale. Other parameters are reported in the reference article [162]

5.6 eDNA, Species Dispersal and River Networks – What's Up?

As discussed in Chapter 3 (Section 3.6), environmental DNA (eDNA), present as loose fragments, shed cells, or microscopic organisms, can be extracted with relative ease from their matrix, whether water or soil (see [719] for a review of methods for collection, extraction, and detection and [387] for the application of this powerful technique to parasitology). Its identification may be used to track the presence of target species or even the composition of entire communities, and its extended use provides a rich new source of temporal data for research questions that have so far been overlooked [720]. Approaches using eDNA for qualitative species detection have already proved their worth for species management and biodiversity conservation by substantially improving field evidence in a replicable and consistent manner (Section 3.6), besides facilitating the detection of rare, invasive, or parasitic species.

Specifically, in Chapter 3, we have claimed that eDNA measurements in river waters provide a snapshot record of the species present upstream but also that the interpretation of its signals is a rather complex issue blurred by many uncertainties [376]. The key filtering process that allows us to underpin spatial patterns of its sources is the fact that, once released to the environment, eDNA undergoes selective decay. Nucleic acids incur progressive damage during hydrological advection, retention, and resuspension, resulting in alterations that affect eDNA detection in whatever environmental sample – which in turn involves a role for hydrologic controls. Most importantly, eDNA has polydisperse properties due to its origin in diverse organic sources (e.g., spores, cells, tissues, feces), which complicates the evaluation of decay rates and the analysis of the possible source – in a truly remarkable inverse problem examined in Section 3.6. As we discussed in that section, the eDNA sampled at any point within a dendritic network of sources is the outcome of diffuse eDNA release from points upstream, modified by decay processes during transport that are governed by network connectivity, in which each path to the observation point may be described by different hydromorphological conditions. As a result, while it is relatively straightforward to link a positive test with the presence of the target species at some (unknown) distance upstream, quantification of species densities and the location of populations is currently impossible because, besides a number of potentially confounding factors affecting eDNA shedding (e.g., animal behav-

ior, movement, physiology, and size [376]), it requires consideration of the effects of the dynamics of eDNA transport along river branches and the deconvolution of the hierarchical aggregation of the various network branches. In part, the latter gap has been filled [376].

To conclude our outlook, it thus seems worth examining a number of simplifying assumptions in the reference model [376] with a view to possible forthcoming extensions and generalizations.

One open issue is the fact that the framework in Carraro et al. [376] postulates that eDNA production within a river stretch is homogeneous in space and constant in time. It is important to note that this assumption is not as restrictive as it might at first seem, whether from a spatial or a temporal perspective – but also that it can be relaxed. In fact, in a spatial perspective, the size of a given river stretch can be tuned to the species and the spatial setting at stake. For instance, finer discretizations of the river network domain could be easily employed if needed, either relying on remotely sensed resolution of heterogeneous features of relevance or simply by using finer digital terrain maps (Figure 1.3). This would entail much longer computational time for any transport model to run, but would not add any further unknowns to the equations presented here (related to the use of a species distribution model) and thus would not deter us from chasing such descriptions.

From a temporal perspective, it is well known that eDNA production is highly sensitive to time and life phases of the shedding organisms: for instance, spawning activity in fish engenders the release of high concentrations of mitochondrial eDNA within milt [405]. In the example shown in Section 3.6, both *F. sultana* and *T. bryosalmonae* typically proliferate during the warm season. Nonetheless, the time scales of eDNA transport and decay (as pointed out by the distribution of travel times through the whole network; see Box 3.28 [376]) are generally much shorter than those characteristic of the biology of the organisms. This provides an opportunity to infer temporal changes in eDNA production across a catchment, provided that multiple measurements are taken at each measuring site in an appropriate time window, for example, one day in the Wigger catchment, where the fieldwork had been carried out for the example of Section 3.6. In the form presented there, we refrained from using a time-varying approach in particular owing to the lack of available replicates for the same site and the same day. Proper model-guided design of experiments would allow for interpretation of time-dependent measured eDNA signals. However, solving the transport equation (inclusive of advection, hydrodynamic dispersion, and decay) from a pulsating source poses no major theoretical problems, especially in view of the moment-generating techniques examined in Box 3.24. This will likely be addressed in the future.

Furthermore, the possibility of inferring upstream species distributions rests on the hypothesis that eDNA production is proportional to local species' biomass. While a positive correlation between eDNA production and biomass has been widely reported (e.g., in carp [395, 404]), the assumption of proportionality between the two variables (say, regardless of the species' fitness or habitat suitability) might be somewhat questionable. For example, Klymus et al. [404] found that the type and the quantity of food consumed can foster a 10-fold increase in the amount of eDNA shed by carp. Thereby, to conveniently apply the framework as presented in Section 3.6, one must ensure that, during the period when eDNA sampling is performed, all target organisms across the surveyed watershed experience reasonably similar conditions, so that the assumption of proportionality between eDNA production and biomass might prove plausible. Clearly, more sophisticated assumptions may be put forth as soon as field measurements provide the necessary resolution.

A basic introduction to how a fluvial landscape impedes, or creates, resistance to the dispersal of organisms has been provided in Chapter 2 in the context of metacommunity or metapopulation analyses, where a key role was found for the network of connectivities set by the nature of the substrate for ecological interactions. This obviously implies gene flow, and in a broader sense, one may consider how, in general, spatially structured ecological networks represent spatial landscape–genetic relationships (see [721] for a recent review). Looking forward, the ability to model landscape–genetic data will likely be the center of much interest in the future, in particular through the lens of modeling spatially explicit processes that may affect genetic structure, thus potentially providing insight into the evolutionary processes that generated ecological networks as well as valuable information about the optimal characteristics of conservation corridors [721, 722].

A significant conceptual change in the assessment of biodiversity scenarios, for example, exploring the viability of a complex landscape for given species under changing drivers (Chapter 2), concerns the predictive rewiring of spatial networks of interactions prompted by human-mediated dispersal [723]. Research on this topic is rapidly accelerating, and its results call for extensions seen from the angle explored for many topics

dealt with in this book – for example, the effects of directional dispersal whose imprinting is unavoidable (Chapter 1). The main tenet [723] is that humans fundamentally affect dispersal by altering landscapes and vectors or directly transporting individuals, thereby modifying the selective advantages available to ecological competition. As a result, human-mediated dispersal may modify substantially the deterrence distance allowed to potential colonizer species and possibly change dispersal paths altogether (Section 2.2.8). This, as we have seen elsewhere in this book by addressing, for instance, the impact on disease spread by human mobility (Chapter 4), has major consequences. For example, scenarios of altered dispersal kernels explicitly linked to human factors may have ecological and evolutionary consequences of fundamental importance [723].

Exploring patterns in diversity, spatial distribution, and temporal dynamics of metacommunities in continuous stream habitats is challenging. River networks have been termed the habitat template [724] because the scaling structure of stream networks and their embedded geomorphological attributes [63, 101] determine their features. Combination of detailed field studies with modeling of dispersal proves necessary for improved understanding of metacommunity dynamics in stream networks and of the impact of anthropic alterations (like damming) [725]. Because most metacommunity-level processes are likely to happen at the stream network level, further research on the effects of network structure will be needed in this area as well. Overall, separation of the effect of dispersal processes from local-scale community dynamics has shown its worth in yielding a mechanistic understanding of the assembly of fish communities in stream networks, which may also enhance the effectiveness of restoration efforts [724, 725].

Because we have shown throughout this book (in particular, in Section 2.5) that biodiversity patterns are governed by landscape structure, one wonders about the evolution of dispersal strategies of residing organisms. In fact, the landscape often changes, and dispersal strategies must evolve with it. We expect much forthcoming work in this direction, in particular, based on metacommunity models that allow for dispersal evolution. As one recent example, they were implemented in river networks with different structures, mimicking the geomorphological dynamics of fluvial landscape [726]. Therein, it has been shown that for a given dispersal kernel, a more compact network structure, where local communities are closer to one another, results in distinctive biodiversity patterns (characteristic of a more well-mixed environment). When dispersal evolution is present, however, organisms adopt more local dispersal strategies in a more compact network, counteracting the effects of the better-mixed environment. The combined effects lead to biodiversity patterns different from the case where dispersal evolution is absent [726]. Such findings have underscored the importance of considering the interplay between the evolution of dispersal, landscape, and biodiversity patterns when studying and managing biodiversity in changing landscapes. The development of evolutionarily stable strategies on river networks [727], seen as both ecological and evolutionary substrates, is therefore a research subject where we expect a surge of activity in the near future.

A necessary ingredient to all of the above is a solid assessment of metapopulation stability in branching river networks based on the recurrent scaling structures, shown by the river network regardless of size, climate, vegetation, and exposed lithology [63]. Remarkably, recent theoretical studies showed that fluvial structures are indeed strong stabilizers as a consequence of pure probabilistic processes [728], adding yet another layer to the results of Chapter 1. Echoing past debates on the relative interrelations between complexity and stability of ecological networks [729], these results have recently been supported from a different angle, concluding that spatial variation in branch size promotes metapopulation persistence in dendritic river networks [730]. Both branch-size variation and species traits interact to determine species persistence, thus demonstrating the ecological significance of their interplay [730]. On this issue as well we expect much forthcoming work.

Many specific assumptions will be needed to suit the biology of different source organisms functional to specific studies of river networks as ecological corridors via eDNA sampling. For example, we have noted only in passing in Section 3.6 that in the case of sessile species colonizing fluvial habitats, the source habitat can be considered the riverbed area of the river stretch at hand. This is not always the case. The source of eDNA shed from terrestrial species that colonize unchanneled areas draining into an arbitrary node is a function of the directly contributing area [63]. Contributing areas to each river reach may be vital to determining the size of the fluvial habitat [8] and hence of key features of life history, population, or ecosystem attributes. Variations on the basic theme may be many, and they are left to future developments.

Also, in Section 3.6, damage rates of eDNA fragments were assumed as constant for a given species. However,

several studies (see [398] for a review) showed that decay rates are highly sensitive to DNA characteristics (e.g., conformation, containment by membranes and length; see also [731]), abiotic (temperature, light, oxygen, pH, sediments, salinity) as well as biotic (microbial community composition, extracellular enzymes) factors. Small differences in environmental conditions might prompt high variations of decay rates. Note that, in this regard, the Bayesian approach adopted by [376] is beneficial, as parameters (in the case discussed in Section 3.6, the parameter τ quantifying the travel time within a given fluvial path) are treated as random variables. Thereby, in this approach, the variability of τ is subsumed by its posterior distribution. However, we need to acknowledge that an issue might arise when strong local differences in abiotic or biotic factors across the watershed are present, so that the different paths from sources to sampling site would be characterized by different decay rates. Hence, we suggest limiting the application of the framework to catchments where spatial variability in environmental conditions can reasonably be assumed to remain relatively small with respect to decay regimes. An alternative approach would consist in explicitly embodying such spatial variability of τ directly in the model. For instance, this could be achieved by specifying path-varying decay rates τ_{ij} that could be parameterized as a function of environmental factors (by following an approach akin to that of [376] for eDNA production).

In the reference approach [376], the hydrological model is based on some rather crude assumptions, in particular, the proportionality between water discharge and contributing area (which allows us to compute $\overline{Q_j}$ for all j given a discharge time series at a network node; Section 3.6) and the hypothesis of large ($W \gg h$, where W, h are, respectively, mean channel width and depth), nearly rectangular river cross sections in uniform flow conditions under spatially constant Manning's roughness coefficient. These hypotheses are used to compute average path velocities $\overline{v_{ij}}$. The former assumption generally holds for landscape-forming (or bankfull) discharges roughly corresponding to approximately two-year return period floods (Box 2.4 and [63]). The same relation may be applied for mean annual discharges but can be thought of as reasonable also for finer-scale (e.g., daily) averages in catchments whose drainage area does not exceed the order of magnitude of 10^2 km^2. The latter assumption might fail at correctly reproducing the velocity fields of headwater catchments, where river cross sections cannot be approximated as large and rectangular and riverbed composition is generally coarser.

The New York Times

Tiny Invader, Deadly to Fish, Shuts Down a River in Montana

By JIM ROBBINS   AUG. 23, 2016

Figure 5.6 A clipping from the *New York Times*, August 26, 2016

The proportionality of cumulative drainage area and relevant discharges hinges on the assumption of spatially (nearly) uniform runoff-forming precipitation, untenable as an assumption as the catchment grows large compared to the correlation scale of significant precipitation events. However, at large spatial scales, it may very well be that a large part of the eDNA signal decays below detection limits and thus at-a-station measurements become blind to a portion of the catchment. In any case, the message that should come clearly from these notes is that an approach fitting all field conditions does not exist, and tuning hydrological, geomorphological, and biological information on the scales at hand is of paramount importance. In any case, if available, more details on the hydrological characterization of the case study catchment can be implemented to extend the original formulation [376]. Such a refinement is likely to be immaterial in improving this framework's ability to infer species distributions, considering all the (much larger) uncertainties in eDNA shedding and decay rates, as discussed above. However, sensible experimental design in such a case should foresee suitable nested measurements guided by the model.

In any case, among the preoccupations that we have for the ecohydrological community, we do not harbor that of seeing it deprived of new challenges, and we safely conclude that the eDNA studies will contribute to keeping it busy for quite some time. As a real challenge, we would like to come to a close by mentioning the threats posed by climate change to entire populations, for instance, those of iconic fish in mountain streams. A notable increase in the incidence of proliferative kidney disease in northern Europe has been recently documented with potentially deadly consequences [732], such as the possible disappearance of the brown trout from Alpine rivers and streams owing to a small increase in average air temperature [733]. Figure 5.6 illustrates our final point well: a recent outbreak in the Yellowstone River (Montana, United States) fostered an abnormal kill of mountain whitefish (*Prosopium williamsoni*) to the point that local wildlife officials temporarily shut down a 300-km-long stretch of the river to all recreational activities in a bid to impede the parasite's spread. We strongly believe that the tools we have illustrated in this book will serve to limit, if not to avoid, such disasters.

6 | Appendices

This chapter contains three appendices that we decided to move to a unique location rather than scattering them throughout the text. Indeed, they are needed in a number of places, whether for singling out the Fisher–Kolmogorov speed of a traveling wave resulting from an equation of diffusion subject to logistic population growth along a linear channel (Chapter 1) or along a fractal (Chapter 3) or when assessing the exchange of stability between a disease-free equilibrium condition and a positive endemic state (Chapter 4). This also allows us a somewhat more extended treatment than suitable within a differently oriented text. The second appendix deals with a relatively long detour on the origins, rationale, and broad applicability of the optimal channel network (OCN) concept that we have used so extensively in this book. This was originally confined to a box in Chapter 1, which – along with the evolving angles through which one would look at the exploitation of OCN replicas – had grown far longer than a box would allow. We have thus decided to leave a terse account (Box 1.2) referring to the relatively long discussion here, which includes some proof of its statistical validity. Finally, a third appendix section deals with examples of numerical solutions to a system of coupled ODEs representing a SIRB model (Chapter 4) that is central to much of our disease spread modeling efforts. It has been famously said that engineering is computation – and alas! we are engineers. Thus, it seemed useful to show how simple codes may tackle complex problems. In any case, the direct computation of how state variables of coupled nonlinear ODEs evolve can be reduced to a somewhat common numerical problem, which we exemplify here. We hope that the reader will find this methodological detour useful – we certainly have had fun in putting it together.

6.1 Stability of Dynamical Systems and Bifurcation Analysis

In this appendix we will provide the fundamentals of stability theory and the basic notions of nonlinear analysis with reference to both time-continuous and time-discrete dynamical models. An extensive textbook dealing with nonlinear analysis of dynamical systems is the one by Kuznetsov [188]. A deeper review of the topics that will be treated in this appendix can be found in [734], while numerical methods are reviewed in [735].

6.1.1 Stability of Linear Systems

We will consider linear systems of the kind

$$\frac{dz}{dt} = \mathbf{A}(t)z(t) + \mathbf{B}(t)u(t)$$

in continuous time t and

$$z(t+1) = \mathbf{A}(t)z(t) + \mathbf{B}(t)u(t)$$

in discrete time. The state vector $z(t)$ has dimension n, $\mathbf{A}$ is a matrix of dimension $n \times n$, $u(t)$ is the input vector of dimension m (representing external forcings), and $\mathbf{B}$ is a matrix of dimension $n \times m$. The solutions to the above equations are as follows:

$$z(t) = \mathbf{\Phi}\left(t, t_0\right) z\left(t_0\right) + \int_{t_0}^{t} \mathbf{\Phi}(t, \tau)\mathbf{B}(\tau)u(\tau)d\tau \quad (6.1)$$

$$z(t) = \mathbf{\Phi}\left(t, t_0\right) z\left(t_0\right) + \sum_{\tau=t_0}^{t-1} \mathbf{\Phi}(t, \tau+1)\mathbf{B}(\tau)u(\tau), \quad (6.2)$$

where $\mathbf{\Phi}\left(t, t_0\right)$ is termed the state transition matrix and has the following properties (from now on, $\mathbf{I}$ will indicate the identity matrix):

$$\mathbf{\Phi}\left(t, t\right) = \mathbf{I}$$

$$\frac{d}{dt}\mathbf{\Phi}\left(t, \tau\right) = \mathbf{A}(t)\mathbf{\Phi}\left(t, \tau\right) \quad \text{or} \quad \mathbf{\Phi}\left(t+1, \tau\right) = \mathbf{A}(t)\mathbf{\Phi}\left(t, \tau\right).$$

Equations (6.1) and (6.2) show that the solution is the sum of two terms: the first (natural response) depends only on the initial conditions $z(t_0)$; the second (forced or driven response) depends only on the input $u(t)$. Because the problem of stability is to study whether a certain motion of the system (for a given initial condition $z(t_0)$ of the state and a given forcing function $u(t)$) can asymptotically absorb a perturbation of its initial condition, it is clear that we can discard the forced response and focus only on the natural response. In fact, the forced response is not influenced by the perturbation. Therefore,

the stability properties of linear systems (whatever the motion we are considering) depend only on the matrix $\mathbf{A}(t)$. We can thus consider the autonomous systems

$$\frac{dz}{dt} = \mathbf{A}(t)z(t)$$

and

$$z(t+1) = \mathbf{A}(t)z(t);$$

$z = 0$ is always an equilibrium (fixed point) of the equations. The study of stability actually amounts to understanding whether perturbations of $z = 0$ will be absorbed in the long run.

6.1.2 Linear Time-Invariant Systems

We first consider the important case of linear time-invariant systems, namely, $\mathbf{A}(t) = \mathbf{A}$ constant in time. In this case, the state transition matrices are

$$\mathbf{\Phi}(t, t_0) = \exp[\mathbf{A}(t - t_0)]$$

for the continuous-time model and

$$\mathbf{\Phi}(t, t_0) = \mathbf{A}^{t-t_0}$$

for the discrete time one. The stability properties are completely determined by the eigenvalues of $\mathbf{A}$, namely, the solutions $\lambda_1, \lambda_2, \ldots \lambda_n$ of the characteristic equation $\det(\lambda\mathbf{I} - \mathbf{A}) = 0$, where $\det(\mathbf{M})$ is the determinant of a matrix $\mathbf{M}$. We recall that the determinant and the trace of $\mathbf{A}$ are linked to the eigenvalues via the relationships $\det(\mathbf{A}) = \lambda_1 \times \lambda_2 \times \cdots \times \lambda_n$ and $\text{Tr}(\mathbf{A}) = \lambda_1 + \lambda_2 + \cdots + \lambda_n$. Also, if one eigenvalue is not real, there exists another eigenvalue that is its complex conjugate. If no eigenvalue of $\mathbf{A}$ is equal to 0, then the matrix is nonsingular, and the equation $\mathbf{A}z = 0$ has the unique solution $z = 0$, which is the only fixed point of the system. Otherwise, there exists an infinity of equilibria.

To each eigenvalue we associate an eigenvector that satisfies the equation $\mathbf{A}z_i = \lambda_i z_i$, $i = 1, 2, \ldots, n$. Eigenvectors have the important property that if the initial condition belongs to a certain eigenvector, the ensuing motion of the system remains on that eigenvector and the rate of convergence or divergence is given by the corresponding eigenvalue, because the motion is governed by the simple equations $dz/dt = \lambda_i z$ and $z(t+1) = \lambda_i z(t)$. If the initial condition is not on an eigenvector, the ensuing motion is a linear combination of the motions that take place on the eigenvectors. It is thus easy to derive the conditions for the asymptotic stability of the system (namely, all the perturbations of $z = 0$ will give rise to motions that will tend to 0 for $t \to \infty$). In fact, it suffices that all the motions taking place on the eigenvectors converge to 0. This occurs if and only if (1) in continuous-time systems, all the real parts of the eigenvalues $Re(\lambda_i)$ are negative and (2) in discrete-time systems, the moduli of all the eigenvalues $|\lambda_i|$ are smaller than 1. When the eigenvalues are complex numbers, the corresponding eigenvectors are also complex; however, each pair of conjugate eigenvalues/eigenvectors identifies a real plane on which the corresponding motions will spiral toward the origin. It is also easy to show that the system is exponentially unstable (that is, at least one perturbation of $z = 0$ results in a motion whose distance from the origin increases exponentially) when (1) in continuous-time systems, the real part of at least one eigenvalue (or of a pair of conjugate eigenvalues) is positive and (2) in discrete-time systems, the modulus of at least one eigenvalue (or of a pair of conjugate eigenvalues) is larger than 1. If we assume that no eigenvalue has a real part equal to zero in time-continuous systems (or modulus equal to 1 in time-discrete systems), then the subspaces containing the asymptotically stable eigenvectors and the unstable eigenvectors are called the stable subspace and the unstable subspace, respectively. Their sum is the entire n-dimensional space.

One important concept is that of dominant eigenvalue. In continuous time the dominant eigenvalue (or pair of eigenvalues in the case of complex conjugate eigenvalues) is the one with the largest real part. In discrete time systems the dominant eigenvalue (or pair of eigenvalues in the case of complex eigenvalues) is the one with the largest modulus. As t goes to infinity, the motion of a linear time-invariant system tends to flatten along the eigenvector (or plane, in the case of paired complex eigenvectors) corresponding to the dominant eigenvalue.

To further our understanding of the above concepts, let us consider the simple case of time-continuous two-dimensional linear systems, that is,

$$\frac{dz_1}{dt} = a_{11}z_1 + a_{12}z_2 \qquad \frac{dz_2}{dt} = a_{21}z_1 + a_{22}z_2, \qquad (6.3)$$

which can be written in matrix form as $dz/dt = \mathbf{A}z$, where

$$\mathbf{A} = \begin{bmatrix} a_{11} & a_{12} \\ a_{21} & a_{22} \end{bmatrix}.$$

The characteristic equation is

$$\det(\lambda \mathbf{I} - \mathbf{A}) = 0 \quad \to \quad \lambda^2 - \mathrm{Tr}(\mathbf{A})\lambda + \det(\mathbf{A}) = 0. \quad (6.4)$$

Since $\det(\mathbf{A}) = \lambda_1\lambda_2$ and $\mathrm{Tr}(\mathbf{A}) = \lambda_1 + \lambda_2$, the conditions for asymptotic stability are $\det(\mathbf{A}) > 0, \mathrm{Tr}(\mathbf{A}) < 0$. In this case, if $\Delta = \det(A)^2 - 4\mathrm{Tr}(A)$ is negative, the eigenvalues

are complex conjugates and the approach to the origin is a spiral (stable focus); if Δ is positive, the eigenvalues are both real and negative and the approach to the origin tends to the dominant eigenvector (stable node). If $\det(\mathbf{A}) < 0$, one eigenvalue is real and negative, the other real and positive. In this case the origin is unstable and is called a saddle; motions are converging along the eigenvector corresponding to the negative eigenvalue and diverging along the other eigenvector, which corresponds to the positive dominant eigenvalue. Finally, if $\det(\mathbf{A}) > 0, \mathrm{Tr}(\mathbf{A}) > 0$, then we have an unstable node or an unstable focus, depending on the sign of Δ. The possible shapes of the equilibria for linear systems can be grouped parametrically in relation to values of $\det(\mathbf{A})$, $\mathrm{Tr}(\mathbf{A})$, and Δ. They are shown in Figure 6.1; note that real eigenvalues imply real eigenvectors and that the dominant eigenvector (corresponding to the maximum eigenvalue) retains the physical meaning of identifying the direction of approach to the equilibrium (attraction, repulsion). Of course, the cases $\det(\mathbf{A}) = 0, \mathrm{Tr}(\mathbf{A}) = 0$, $\Delta = 0$ give rise to degenerate situations that are also displayed in the figure.

We can also exploit the two-dimensional case to introduce the concept of bifurcation (which plays a fundamental role in nonlinear analysis) in a simple way. Suppose that the matrix $\mathbf{A}$ depends on a parameter p and we want to study how the dynamical behavior of the system varies when the value of p changes. First, assume that the parameter p varies between the values p_1 and p_2 and that the dependence of A on p is such that the trace and the determinant of $\mathbf{A}$ follow the pathway shown in Figure 6.1. At p_1 the system converges toward a stable node. Little variations of p do not cause a qualitative change of the dynamic behavior of the system: it is still characterized by a stable node. In other words, the trajectory portraits are topologically equivalent (i.e., they can be obtained from one another through a simple deformation of the trajectories). Larger variations of p cause the change of a stable node into a stable focus when the pathway crosses the curve $\Delta = 0$; this corresponds to the eigenvalues turning from real into complex. However, this does not represent a change of the qualitative behavior of the system: trajectories can still be obtained via a continuous deformation, and the system is in any case asymptotically stable. Instead, when the parameter p assumes the value p_H, the real part of the two complex eigenvalues becomes zero, and for larger values of p, the system becomes unstable and characterized by an unstable focus: the trajectory portrait is not topologically equivalent to those corresponding to $p < p_H$. We say that at p_H the

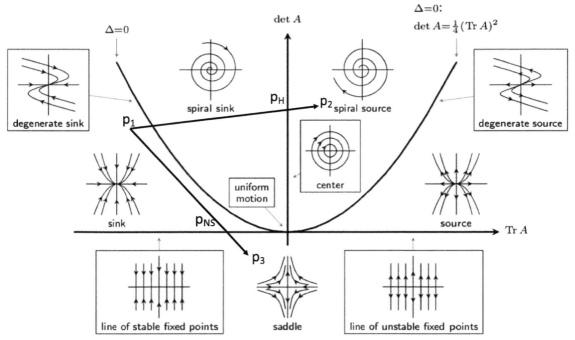

Figure 6.1 Sketch of possible trajectories (phase portrait) of two-dimensional linear systems. The classification is based on the values of the trace and determinant of A. Sink and source are other terms for stable node and unstable node, spiral sink/source for stable/unstable focus. When $det(A) = 0$, there is an infinity of equilibria. The directions of the eigenvectors are the two coordinate axes. Also reported are possible pathways of bifurcation. Variable p_{NS} indicates a saddle node bifurcation, p_H a Hopf bifurcation.

system undergoes a bifurcation – more precisely, a Hopf bifurcation.

A second possibility is that the parameter p varies between the values p_1 and p_3 and that the dependence of A on p is such that the trace and the determinant of A follow the pathway shown in Figure 6.1. In this case, the system is characterized by a stable node until p assumes the value p_{NS}, at which $det(A) = 0$, and thus one of the real eigenvalues becomes zero. For values of p larger than p_{NS}, one of the eigenvalues turns from negative into positive. The system is thus characterized by a saddle. Therefore, the bifurcation taking place at p_{NS} is called a saddle node bifurcation.

As for time-discrete systems, things are a bit more complex. First consider the one-dimensional case

$$z(t + 1) = a\, z(t)$$

and make the simple remark that obviously a coincides with the unique eigenvalue. Contrary to time-continuous systems, there can be oscillations even in the one-dimensional case, because it suffices that a be negative. The values of z alternate between being positive and being negative. Oscillations are damped if $|a|$ is smaller than 1; otherwise, their amplitude increases with time. The period (when $a = 1$) or pseudo-period (when $a \neq 1$) is 2.

Second, consider the two-dimensional case $z(t + 1) = A z(t)$, A being a 2×2 matrix. If both eigenvalues are real, oscillations are again linked to the fact that one or both eigenvalues are negative. Oscillations are damped if both eigenvalues have modulus smaller than 1. If the eigenvalues are complex conjugate, the system displays oscillations, but they are not of the kind that we find when eigenvalues are negative: these latter always have a period or pseudo-period equal to 2. Let the conjugate pair be given by $\chi \exp(\pm j\omega)$, where χ is the modulus and j the imaginary unit. Solutions are of the kind $\chi^k [a_1 \cos(\omega t) + a_2 \sin(\omega t)]$, with a_1 and a_2 depending on initial conditions. Therefore, only if $\omega = (2\pi)/n$ is the solution n-periodic or n-pseudo-periodic. Consider the case $\chi = 1$: if $\omega \neq (2\pi)/n$, the oscillations will never come back to the initial conditions but will forever stay on a closed line (actually an ellipse) around the origin.

6.1.3 Positive Systems and Perron–Frobenius Theory

A special case that deserves attention in the context of this book is that of positive systems, namely, those systems in which, for a given nonnegative initial condition $z(t_0) \geq 0$, the ensuing motion $z(t)$ remains nonnegative forever. Population numbers, rainfall, body size, and biomass are just a few examples of the many environmental variables that can never take on negative values. A time-discrete linear system is obviously positive if and only if the matrix $\mathbf{A}$ is nonnegative, that is, all its elements are nonnegative. Instead, a time-continuous linear system is positive if and only if the matrix $\mathbf{A}$ is Metzler, that is, if all its off-diagonal entries are nonnegative. In this way, the time derivative of a state variable that is very close to being zero, while the other state variables are positive, is always nonnegative.

Positive systems have important properties regarding their eigenvalues and eigenvectors. First of all, we make the very reasonable assumption that the system is irreducible, namely, that it does not consist of unconnected subsystems, or equivalently that the graph associated with matrix $\mathbf{A}$ is strongly connected. Then, the Perron–Frobenius theorem asserts that the dominant eigenvalue of a nonnegative irreducible matrix $\mathbf{A}$ is real, positive, and simple (i.e., a simple root of the characteristic equation). Moreover, the components of the dominant eigenvector are all positive. Therefore, a nonnegative irreducible time-discrete system is stable if the Perron–Frobenius eigenvalue is smaller than 1, and unstable otherwise. Also, all the motions tend to the Perron–Frobenius eigenvector. As for a time-continuous system, one can exploit the properties of Metzler matrices, which are similar to those of Perron–Frobenius matrices for time-discrete systems. In fact, it turns out that the dominant eigenvalue (the one with the largest real part) is real and simple. If it is negative, the system is stable, if it is positive, the system is unstable. The corresponding dominant eigenvector is positive, and all the motions tend to the dominant eigenvector converging to or diverging from the origin.

6.1.4 Time-Varying Periodic Systems

The stability of general time-varying linear systems is outside the scope of this book. We will consider only the special but important case of periodic systems, namely, those characterized by $\mathbf{A}(t + T) = \mathbf{A}(t)$ for any time t and a given period T. The analysis is based on the idea that,

equivalently, we can study the stability of the so-called (linear) Poincaré map $z(t) \mapsto z(t + T)$.

For time-discrete systems, the analysis is quite simple, because it suffices to consider the matrix $\mathbf{A}_T = \mathbf{A}(T - 1) \mathbf{A}(T - 2) \ldots \mathbf{A}(0)$ and determine whether the dominant eigenvalue's modulus is smaller or larger than 1. In fact, let $t = k + nT$, where k varies between 0 and $T - 1$ and n is any positive integer. Then $z(t + T) = z(k + (n + 1)T) = \mathbf{A}_T^k z(k + nT) = \mathbf{A}_T^k z(t)$, where $\mathbf{A}_T^k$ is a matrix that is the kth circular permutation of $\mathbf{A}(T-1)\mathbf{A}(T-2)\ldots\mathbf{A}(0)$. For instance, $\mathbf{A}_T^0 = \mathbf{A}_T$, $\mathbf{A}_T^1 = \mathbf{A}(0)\ldots\mathbf{A}(2)\mathbf{A}(1)$. Because the products $\mathbf{CD}$ and $\mathbf{DC}$ of any two square matrices $\mathbf{C}$ and $\mathbf{D}$ have the same eigenvalues, it turns out that the eigenvalues of all the A_T^k are the same, irrespective of k. Thus, it is sufficient to consider the stability of the Poincaré map $z(0) \mapsto z(T)$ and examine whether the dominant eigenvalue of $\mathbf{A}_T$ is smaller or larger than 1.

The stability analysis of time-continuous periodic systems is a bit more complex and related to the so-called Floquet theory. In this case, $z(t + T) = \mathbf{\Phi}(t + T, t)\, z(t) = \mathbf{\Psi}(t)z(t)$. The state-transition matrix $\mathbf{\Psi}(t)$ is called the monodromy matrix at time t. It is possible to prove that the eigenvalues of $\mathbf{\Psi}(t)$ are the same for any t [449], so it is sufficient to consider the Poincaré map $z(0) \mapsto z(T)$ and the eigenvalues, for instance, of $\mathbf{\Psi}(T)$. To calculate $\mathbf{\Psi}(T)$, one can remember the properties of state-transition matrices. Thus, $\mathbf{\Psi}(T)$ can be obtained as the solution at time T of the matrix equation

$$\frac{d}{dt}\mathbf{\Psi}(t) = \mathbf{A}(t)\mathbf{\Psi}(t), \qquad \mathbf{\Psi}(0) = \mathbf{I}.$$

In practice, one must integrate over one period T (usually with a suitable numerical method) the differential equation $dz/dt = \mathbf{A}(t)z(t)$ n times with n different initial conditions. Specifically, in each of the initial conditions, one of the n components of the state $z(0)$ is set to 1, and all the other components are set to 0. The stability is thus related to the dominant eigenvalue of the monodromy matrix $\mathbf{\Psi}(T)$. If its modulus is smaller than 1, the periodic system is asymptotically stable; if it is larger than 1, the system is exponentially unstable.

As for oscillations around the origin, the remarks that we made for time-discrete invariant systems apply. It suffices to apply those considerations to $\mathbf{A}_T$ or to the monodromy matrix. Negative eigenvalues correspond to oscillations of period or pseudo-period equal to 2 T, pairs of conjugate eigenvalues to oscillations that in general are not periodic.

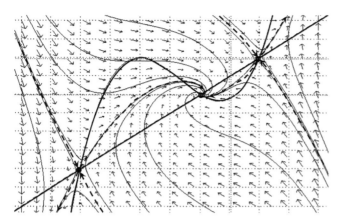

Figure 6.2 Trajectories of a two-dimensional nonlinear time-continuous system with three equilibria (indicated by dots): two saddles and a stable focus. Solid lines indicate the isoclines $\dot{z}_1 = 0$, $\dot{z}_2 = 0$, which are crossed horizontally or vertically by trajectories. Little arrows indicate the direction and intensity of the vectors dz/dt, which are tangent to the trajectories. Dashed curves are the stable and unstable manifolds of the two saddles.

6.1.5 Stability of Nonlinear Systems

We will consider autonomous nonlinear systems described by differential or difference equations of the kind

$$\frac{dz}{dt} = f(z(t)) \tag{6.5}$$

$$z(t+1) = f(z(t)), \tag{6.6}$$

where f is a continuous vector function mapping the n-dimensional state vector $z(t)$ into another n-dimensional vector that might be the time derivative of z or the state after one time unit. In what follows, we shall address, in sequence, the stability of equilibria of nonlinear systems (Section 6.1.6), of their cycles (Section 6.1.7), and of bifurcations (Section 6.1.8).

6.1.6 Stability of Equilibria of Nonlinear Systems

The equilibria (or steady states or fixed points) of systems (6.5) and (6.6) are provided by the equations $f(z) = 0$ or $z = f(z)$. Differently from linear systems that always have the origin as equilibrium and possibly an infinity of equilibria in some cases, nonlinear systems can have a finite multiplicity of equilibria characterized by different stability: one equilibrium might be a stable node, another one an unstable focus, and so on. Figure 6.2 shows an example with three steady states.

To study the stability of equilibria, one can use the method of linearization, which consists in approximating the dynamics of a nonlinear system near each equilibrium with a linear system. To this end, consider $\mathbf{J}$, the Jacobian of the system

$$\mathbf{J} = \begin{bmatrix} a_{11} & a_{12} & \cdots & a_{1n} \\ a_{21} & a_{22} & \cdots & a_{2n} \\ \cdot & \cdot & \cdot & \cdot \\ a_{n1} & a_{n2} & \cdots & a_{nn} \end{bmatrix},$$

where

$$a_{ij} = \frac{\partial f_i}{\partial z_j}.$$

Let $\bar{z}$ be an equilibrium of system (6.5) or (6.6). Compute the Jacobian at the equilibrium by evaluating the functions $\partial f_i / \partial z_j$ at $\bar{z}$, thus obtaining a constant matrix $\mathbf{J}_{\bar{z}}$. Then the stability of $\bar{z}$ can be studied by analyzing the stability of the corresponding linear system. Introduce the perturbation $\Delta z(t) = z(t) - \bar{z}$. The linearized system is

$$\frac{d}{dt}\Delta z = \mathbf{J}_{\bar{z}}\Delta z$$

in continuous time and

$$\Delta z(t+1) = \mathbf{J}_{\bar{z}}\Delta z(t)$$

in discrete time. The stability of each equilibrium is related to the eigenvalues of the corresponding Jacobian $\mathbf{J}_{\bar{z}}$. In time-continuous nonlinear systems, the steady state $\bar{z}$ is asymptotically stable if the real parts of all the eigenvalues are negative and unstable if at least one eigenvalue has a positive real part. In time-discrete systems, the steady state $\bar{z}$ is asymptotically stable if the moduli of all the eigenvalues are smaller than 1 and unstable if at least one eigenvalue has modulus larger than 1. It should be remarked that when one or more of the eigenvalues has a null real part (or a modulus equal to 1), the stability of the equilibrium cannot be ascertained by linearization: the equilibrium might be stable or unstable because stability depends on the terms of order greater than 1 in the Taylor expansion of f around $\bar{z}$. In this case, we say that the equilibrium is nonhyperbolic [120].

If the system is time continuous and the equilibrium is hyperbolic, we can associate to it one stable manifold and one unstable manifold. Time-continuous systems are reversible (while time-discrete ones may not be reversible because two different states at time t can be mapped into the same state at time $t + 1$), which means that it is possible to integrate the equations backward in time. Suppose then that the corresponding Jacobian has k eigenvalues with negative real part and $n - k$ eigenvalues with positive real part. Then there exists a k-dimensional manifold tangent to the stable subspace of $\mathbf{J}_{\bar{z}}$ such that any motion starting from a point of the manifold remains in the manifold and tends to $\bar{z}$ for $t \to \infty$. In the same way, there exists an n-k-dimensional manifold tangent to the unstable subspace of $\mathbf{J}_{\bar{z}}$ such that any motion arriving at time 0 to a point of the manifold was in the manifold for $t \leq 0$ and tends to $\bar{z}$ for $t \to -\infty$. The manifolds are called the stable and unstable manifolds of the equilibrium.

Consider the simple case of time-continuous two-dimensional nonlinear systems. Eigenvalues of the Jacobian determine the nature of the equilibria. If $\text{Re}(\lambda_1, \lambda_2) \neq 0$, the fixed point is hyperbolic (stable, unstable, or a saddle). The asymptotic stability condition is $\text{Re}(\lambda_1, \lambda_2) < 0$, implying $tr\,(\mathbf{J}_{\bar{z}}) < 0$ and $det\,(\mathbf{J}_{\bar{z}}) > 0$, whereas the exponential instability condition requires that at least one eigenvalue have positive real part. Figure 6.3 shows an example with a stable node and a saddle. The eigenvectors of the linearized system are tangent to stable/unstable manifolds.

If all the equilibria of a nonlinear system are hyperbolic, the global picture of the system's orbits (the so-called phase plane in two-dimensional systems) can be obtained by "sticking" together the orbits of the linearized systems. Figure 6.2 is actually an example of this kind. Note that the dashed lines indicate the stable and unstable manifolds of the two saddles. In particular, note that they in part coincide with trajectories that join the saddles to the stable focus. These joining trajectories are called heteroclinic and can play an important role in the analysis of nonlinear systems. Moreover, note that the stable manifolds of the two saddles delimit the basin of attraction of the stable equilibrium, namely, the set of all the initial conditions from which the trajectory will asymptotically converge to the stable focus.

6.1.7 Stability of Cycles in Nonlinear Systems

In addition to equilibria, other particular orbits deserve being considered in nonlinear systems. In particular, nonlinear time-continuous systems can display isolated periodic orbits (limit cycles). This is not possible in linear systems, where the only case with periodic orbits is that of a center (see Figure 6.1), which is not hyperbolic and is surrounded by an infinite number of cycles. Cycles in nonlinear systems can be stable or unstable (Figure 6.4). There is a difference with equilibria that is important: significant perturbations of initial conditions must be transversal to the cycle. In fact, if the perturbed initial condition lies on the cycle itself, then the motion will necessarily remain on the cycle and the perturbation will neither increase nor decrease. Therefore, stability corresponds to all transversal perturbations being absorbed for $t \to \infty$. If there exist transversal perturbations that are not absorbed, the cycle is unstable. In this latter case, if the system dimension is at least 3, the cycle might be saddle-like, which means that there are transversal perturbations lying in the stable manifold of the cycle that will asymptotically decrease to zero, while any other transversal perturbation will diverge and tend to flatten along the unstable manifold of the cycle (Figure 6.5). The sum of the dimensions of the two manifolds is obviously $n - 1$.

The stability of cycles can be ascertained again by suitable linearization. Let $\bar{z}(t)$ be the periodic motion of period T. Compute the Jacobian along the periodic orbit and obtain a periodic matrix $\mathbf{A}(t)$. Introduce the perturbation $\Delta z(t) = z(t) - \bar{z}(t)$. Then the corresponding linearized system is the periodic linear system

saddle   **stable node**

Figure 6.3 Stable and unstable manifolds in a two-dimensional system. (right) The case of a stable node; the stable subspace of the linearized system is the whole plane, which contains the two stable eigenvectors and obviously coincides with the stable manifold of the nonlinear system. (left) The case of a saddle; the stable subspace of the linearized system is the converging eigenvector, while the unstable subspace is the diverging eigenvector; the black solid curves are the stable and unstable manifolds of the nonlinear system, which are tangent to the respective subspaces.

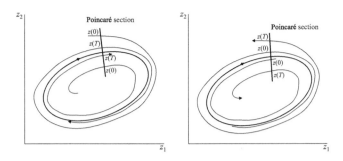

Figure 6.4 Stable and unstable limit cycles with Poincaré sections.

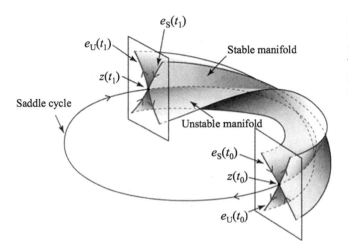

Figure 6.5 Stable and unstable manifolds of a saddle cycle; $e_S(t_0)$ and $e_U(t_0)$ are sections of the stable and unstable manifolds at time t_0. They are tangent to the eigenvalues of the Poincaré map $z(t_0) \mapsto z(t_0 + T)$. Same for time t_1.

$$\frac{d\Delta z}{dt} = \mathbf{A}(t)\Delta z(t),$$

whose stability analysis provides the key to understanding the stability of the cycle. In fact, it suffices to consider the monodromy matrix $\mathbf{\Psi}(T)$ of $\mathbf{A}(t)$ and inspect its eigenvalues. One eigenvalue is always equal to +1 because it corresponds to perturbations tangent to the cycle that neither increase or decrease. If all the other eigenvalues have modulus smaller than 1, the cycle of the nonlinear system is asymptotically stable. If at least one eigenvalue of the monodromy matrix has a modulus larger than 1, the cycle is unstable. The cycle is saddle-like if some eigenvalues of the monodromy matrix are larger than 1 and some are smaller than 1. The cycle is hyperbolic if the only eigenvalue of $\mathbf{\Psi}(T)$ with unitary modulus is the one corresponding to tangent perturbations. In practice, one studies the stability of an equilibrium in the Poincaré map corresponding to the limit cycle.

It is worthwhile to note that exactly the same approach can be used in nonlinear systems that are periodically forced, namely, of the kind

$$\dot{z} = f(z,t) \text{ with } f(z,t+T) = f(z,t).$$

Obviously, these systems in general can display periodic orbits $\tilde{z}(t)$ of the same period T of the exogenous forcing. The stability of these cycles can be analyzed again by computing the Jacobian along the periodic orbit, thus obtaining a periodic matrix for the linearized system, from which one can compute the monodromy matrix and examine its eigenvalues. It must be remarked, however, that the response of a nonlinear system to periodic forcing is not necessarily a cycle of the same period. Nonlinearities may for instance elicit a subharmonic response (namely, with a period that is a multiple of T) or even irregular chaotic oscillations. One very well known example is that of the Duffing oscillator, a damped harmonic oscillator in which the spring stiffness is not linear.

Consider now a nonlinear system that is characterized by both hyperbolic equilibria and hyperbolic cycles. Once again, the global picture of the system's orbits (phase plane in two-dimensional systems) can be obtained by "sticking" together the orbits of the linearized systems, be they time invariant or periodic. Figure 6.6 shows an example with two attractors (equilibrium A and limit cycle D), one saddle (equilibrium B), and one repellor (equilibrium C).

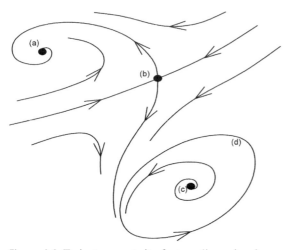

Figure 6.6 Trajectory portrait of a two-dimensional non-linear system exhibiting (a) one stable focus, (b) one saddle, (c) one unstable focus, and (d) one stable limit cycle.

6.1.8 Bifurcations in Nonlinear Systems

We will now very briefly investigate the changes in the dynamical behaviour of a nonlinear system that depends on parameters such as

$$\frac{dz}{dt} = f(z, p) \tag{6.7}$$

$$z(t+1) = f(z(t), p), \tag{6.8}$$

where p is a vector of parameters. For simplicity we will consider only the case of one parameter. In this case, the appropriate theory is that of the so-called codimension 1 bifurcations [188]. The goal of nonlinear analysis is that of determining the values of p at which there is a qualitative change in the system dynamics. We say that the qualitative behavior is the same at p_1 and at p_2 if the orbit portraits at p_1 and p_2 are topologically equivalent (i.e., they can be obtained from one another through a simple deformation of the orbits). The values of p at which the behavior changes are called bifurcations. Figure 6.7 shows an example: for $p < p_{NS}$ the qualitative behavior is the same with all the phase portraits being topologically equivalent. However, at $p = p_{NS}$, there is a qualitative change, and the phase portraits for $p > p_{NS}$ are topologically equivalent between themselves, not with the phase portraits corresponding to $p < p_{NS}$.

We have already met that concept in linear systems (see Figure 6.1) where the bifurcations were very simple, for instance, a stable node becoming a saddle. In

nonlinear systems, bifurcations can be much more complicated, because, for instance, a stable limit cycle might become unstable or disappear. Actually, the situation is even more complicated because nonlinear systems may display attractors (or repellors or saddles) that are not equilibria or cycles. More precisely, there can exist quasi-periodic attractors or chaotic attractors. A quasi-periodic attractor is, for instance, a torus that contains orbits that are the sum of two periodic oscillations with periods whose ratio is irrational; orbits that have initial conditions outside the torus but close to it are attracted to the torus, and vice versa if the torus is a repellor. A chaotic attractor consists of irregular oscillations that display a continuous Fourier spectrum, which makes the oscillating motions of the system quite similar to realizations of a stochastic process: as they are in any case generated by a deterministic equation, the term deterministic chaos is used. Very often, chaotic attractors have a noninteger, fractal dimension. Orbits that have initial conditions outside the fractal set but close to it are attracted to the fractal set, and vice versa if there is a chaotic repellor. We will not delve into these complications; we refer the reader to [188] for a thorough account and to [734] for a good introduction.

We restrict our analysis to systems that display only equilibria or cycles and describe only those simple bifurcations that pertain to the material of this book. In particular, all the bifurcations we illustrate are local bifurcations, that is, involving the degeneracy of some eigenvalue of the Jacobian or of the monodromy matrix associated with equilibria or cycles.

Bifurcations of Equilibria

Bifurcations of equilibria can also be seen as collisions of equilibria. In fact, suppose that one stable and one unstable equilibrium collide for a certain value of the parameter (e.g., p_{NS} in Figure 6.7). The stable equilibrium is characterized by all the eigenvalues having negative real part or modulus smaller than 1; the unstable equilibrium is characterized by having at least one real eigenvalue (or a pair of complex conjugate eigenvalues) having positive real part or modulus larger than 1. When the equilibria collide, the new equilibrium, which results from the fusion of a stable with an unstable equilibrium, must be degenerate by continuity because the eigenvalues are continuous functions of p. In time-continuous systems, this corresponds to the dominant real eigenvalue of the two equilibria becoming zero when they collide or the dominant pair of complex conjugate eigenvalues becoming purely imaginary. In time-discrete systems, this corresponds to the dominant real eigenvalue

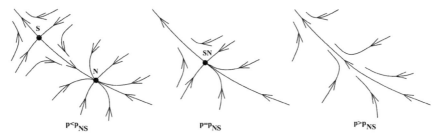

Figure 6.7 Example of bifurcation in a nonlinear system (saddle node bifurcation). For values of the parameter $p <$ p_{NS} the systems exhibits one stable node and one saddle, with some trajectories converging to the stable node and some trajectories first approaching the saddle and then diverging to infinity or to another attractor not shown in the figure. For $p > p_{NS}$, the two equilibria do not exist anymore, and all the trajectories diverge; for $p = p_{NS}$, there is a degeneracy: the equilibrium SN is not hyperbolic.

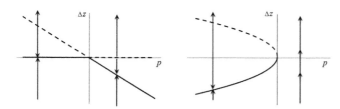

Figure 6.8 Bifurcation diagrams for the normal forms of (left) transcritical and (right) saddle node.

becoming $+1$ or -1 or to the dominant pair of complex conjugate eigenvalues hitting the unit circle.

Let us start with time-continuous systems. The first possibility is that the dominant real eigenvalue becomes zero, which corresponds to the determinant of the Jacobian becoming zero, $det\,(\mathbf{J}_{\bar{z}}) = 0$, where $\bar{z}$ is one or the other of the two equilibria. It is possible to make a change of coordinates in such a way as to consider only the dynamics in one dimension (that corresponding to the dominant eigenvector). Therefore, all the dynamics can be studied by considering equivalent one-dimensional equations, which are called normal forms. Also, it is possible to rescale the parameter p so that the bifurcation occurs at $p = 0$. If we make the hypothesis that the second-order terms in Taylor's expansion of $f(z,0)$ around the equilibrium $\bar{z}$ never vanish, then we obtain the normal forms of the two simplest bifurcations

$$\Delta\dot{z} = p\Delta z + \Delta z^2 \text{ (transcritical bifurcation)}$$

$$\Delta\dot{z} = p + \Delta z^2 \text{ (saddle node bifurcation)},$$

where by suitable further rescaling of p the coefficient of the second-order term has been set to $+1$. In the first normal form (transcritical bifurcation or exchange of stability), for values of p larger or smaller than the bifurcation threshold $p = 0$, there exist two equilibria: $\Delta z = 0$ and $\Delta z = -p$. At $p = 0$ they exchange stability (see Figure 6.8): for $p < 0$, $\Delta z = 0$ is stable, while

for $p > 0$, it is unstable. In the second normal form (saddle node or tangent or fold bifurcation), there are no equilibria for positive p, while negative p corresponds to two equilibria of which one is unstable and the other is stable. The unstable equilibrium is called a saddle because in general, the system is thought of as being multidimensional, and obviously all the eigenvalues other than the dominant one must have negative real parts. For increasing p the saddle and the node become closer and closer until they collide and disappear. Note that there is no way of distinguishing the transcritical from the saddle node by inspecting the Jacobian at $z = \bar{z}$: both bifurcations corresponding to $det\,(\mathbf{\Psi}_{\bar{z}}) = 0$. However, one can solve the equation $f(z,p) = 0$ for values of p slightly larger and slightly smaller than the critical value at which the determinant vanishes: if there are always two real solutions (coinciding at the bifurcation value of p), then the bifurcation is an exchange of stability.

The second possibility is that the dominant pair of conjugate eigenvalues hits the imaginary axis. We have already met this kind of bifurcation in two-dimensional linear systems (see Figure 6.1). It is called Hopf bifurcation. Two remarks are necessary. First, while in two-dimensional nonlinear systems, this bifurcation is still spotted by looking at the trace of the Jacobian and finding whether it vanishes, this is no longer true for systems of dimension larger than 2 (both linear and nonlinear): one must trace the eigenvalues and see if the dominant

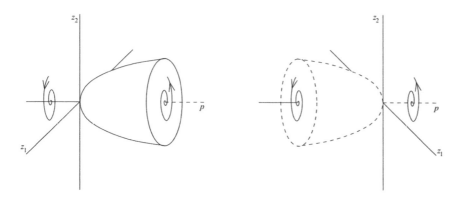

Figure 6.9 Hopf bifurcation diagram: (left) supercritical and (right) subcritical Hopf.

pair crosses the imaginary axis. Second, unstable spiraling equilibria (foci) in linear systems always imply perturbations that diverge to infinity. In nonlinear systems, it may not be so: perturbations may imply motions that converge to a stable cycle.

Even in the Hopf case, we can resort to normal forms. In fact, one can make a change of coordinates and consider only the dynamics in a two-dimensional manifold, the one corresponding to the dominant eigenvectors. The manifold can then be rectified to a plane of coordinates $\Delta z_1 = z_1 - \bar{z}_1, \Delta z_2 = z_2 - \bar{z}_2$. It is then convenient to resort to polar coordinates: the distance from the equilibrium $\rho(t)$ and the angle $\theta(t)$. The normal form turns out to be given by

$$\dot{\rho} = p\rho + c\rho^3$$
$$\dot{\theta} = \omega,$$

where the parameter p has been rescaled so that the Hopf bifurcation occurs at $p = 0$. The eigenvalues of the normal form are $p \pm j\omega$. The second equation shows that the solution (if the equilibrium is perturbed) in any case spirals around the equilibrium at constant angular speed ω, which is the imaginary part of the eigenvalues at the bifurcation. The first equation provides the stability of the equilibrium: if $p < 0$, the perturbed distance decreases with time and the equilibrium is stable; if, instead, $p > 0$, the equilibrium is unstable. However, there are two possible cases, as shown by Figure 6.9: in the supercritical case, the unstable equilibrium corresponding to positive p is surrounded by a stable limit cycle; in the subcritical case, the stable equilibrium is surrounded by an unstable cycle. Super- or subcriticality is linked to the sign of the cubic term c. If it is positive, we have the subcritical case, because for positive p, the derivative of ρ is always positive. Instead, if c is negative, $\rho = \sqrt{-p/c}$ is a stable steady state of the first equation, which clearly corresponds to a stable limit cycle. Note that, obviously, all this holds for

small variations of p around the bifurcation value. The sign of c cannot be determined by analyzing the Jacobian $\mathbf{J}_{\bar{z}}$ and, in general, is not so easy to evaluate. Of course, one can always change the parameter a little bit so that the dominant eigenvalues have a small positive real part and simulate to see whether perturbations of the equilibrium will go to a limit cycle, but if the solution does not converge, one can always suspect that the simulation has not been sufficiently long.

As for time-discrete systems, we will not consider in detail the case in which a pair of conjugate complex eigenvalues hits the unit circle. The corresponding bifurcation is called Neimark–Sacker. We have already seen in linear systems that the corresponding motion may be not periodic, although it can lie on a closed curve in the state space. It suffices to consider the two cases when the dominant eigenvalue at bifurcation is $+1$ or -1.

When the dominant eigenvalue is $+1$, the resulting bifurcations are quite similar to those of time-continuous systems. In particular, if we make the assumption that nonlinearity is dominated by second-order terms, we obtain again the transcritical (exchange of stability) bifurcation and the tangent bifurcation. If we rescale p so that $p = 0$ is the bifurcation value and change coordinates so as to consider the direction of the dominant eigenvector, the normal forms are

$$z(t+1) = (1+p)z(t) - z^2(t)$$
$$z(t+1) = p + z(t) - z^2(t)$$

for the transcritical and the tangent bifurcation, respectively. They are shown in Figure 6.10.

When the dominant eigenvalue is -1, the resulting bifurcation is called flip or period-doubling bifurcation. We had already seen that phenomenon in linear systems: when the eigenvalue is -1, there exist permanent oscillations of period 2. As a paradigm of this kind of bifurcation, we can still use the normal form of the

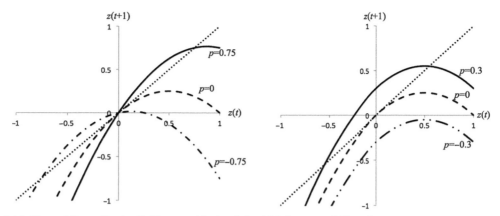

Figure 6.10 Normal forms for the (left) transcritical and the (right) tangent bifurcations.

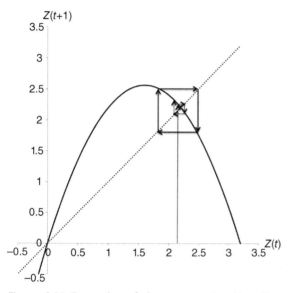

Figure 6.11 Dynamics of the system $z(t + 1) = (1 + p)z(t) - z^2(t)$ for $p = 2.2$. The cobweb shows that the equilibrium $z = 2.2$ is unstable, and dynamics converge toward the displayed period-2 cycle.

transcritical bifurcation $z(t + 1) = (1 + p)z(t) - z^2(t)$ and consider values of the parameter p that are much larger than 0, that is, the threshold value for the transcritical. In fact, the equilibrium $z = p$, which is stable for positive and small ps, becomes unstable for $p > 2$, because the corresponding eigenvalue of the linearized systems is $1 - p$. At $p = 2$ the eigenvalue is exactly -1. The behavior of the system for p slightly larger than 2 is shown in Figure 6.11. The equilibrium $z = p$ is unstable, as shown by the cobweb, while the oscillation of period 2 (the cycle displayed in the figure) is asymptotically stable. As an equilibrium in discrete time can be considered an

oscillation of period 1, this bifurcation is also termed period doubling.

Bifurcations of Cycles

The study of the bifurcations of cycles in time-continuous systems is greatly facilitated by the illustration of the previous results. In fact, the stability of cycles is basically related to the stability of equilibria in the corresponding Poincaré map. Therefore, to search for bifurcations, one must examine the eigenvalues of the monodromy matrix. It should be remembered that one of the eigenvalues is always equal to +1 because it is related to the perturbations tangent to the cycle. Bifurcations are then linked to the other $n - 1$ eigenvalues and in particular to the dominant eigenvalue or dominant pair of conjugate complex eigenvalues. We do not consider all the possible bifurcations but focus on the commonest.

If we remember that bifurcations can be seen as collisions, it is clear that collisions of cycles can be analyzed as collisions of equilibria in the time-discrete Poincaré map. Therefore, transcritical and tangent bifurcations that we studied above for time-discrete systems also apply to limit cycles. The corresponding transversal eigenvalue of the monodromy matrix is +1 at the bifurcation. The tangent bifurcation is more common and is described in Figure 6.12. For $p < p_T$ the system displays two cycles C_S and C_U, respectively stable and unstable. C_U surrounds a stable focus F. Thus the system has two attractors, the stable focus and the cycle C_S. The unstable cycle C_U is the boundary of the basin of attraction of F. At $p = p_T$ the two cycles collide, and correspondingly, the Poincaré map has a tangent bifurcation. For $p > p_T$ the cycles disappear, leaving the stable focus as the unique attractor.

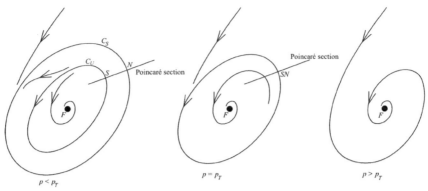

Figure 6.12 Tangent bifurcation of cycles. C_S and C_U collide at $p = p_T$ and then disappear. F is a stable focus, S the saddle of the Poincaré map, N the stable node.

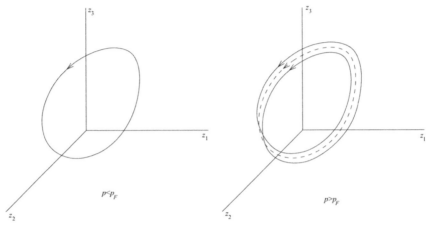

Figure 6.13 Period-doubling bifurcation of cycles. At $p = p_F$ the limit cycle of period T, which was stable for $p < p_F$, becomes unstable (dash) and a new cycle of period $2T$ arises.

When the transversal dominant eigenvalue of the Poincaré map is -1, the corresponding bifurcation of cycles is the period-doubling bifurcation. In fact, the Poincaré map undergoes a flip bifurcation of a stable equilibrium, and a cycle of period 2 arises. This kind of bifurcation is illustrated in Figure 6.13. For $p < p_F$, there exists a stable limit cycle of period T. For $p > p_F$, there exist two cycles: one is unstable and of period T, whereas the other is stable and of period $2T$.

6.2 Optimal Channel Networks and Geomorphological Statistical Mechanics

Moving from the exact result that drainage network configurations minimizing total energy dissipation are stationary solutions of the general equation describing landscape evolution [257], it was argued [63–66] that,

at least in the fluvial landscape, nature works through imperfect searches for dynamically accessible optimal configurations and that purely random or deterministic constructs are unsuitable to properly describing natural network forms. In fact, optimal networks are loopless configurations spanning (i.e., connecting) all sites only under precise physical requirements that arise under the constraints imposed by continuity. In the case of rivers, every spanning tree proves a local minimum of total energy dissipation. This is stated in a theorem form applicable to generic networks, suggesting that other branching structures occurring in nature (e.g., scale-free and looping) may possibly arise through optimality to different selective pressures. Thus, one recurrent self-organized mechanism for the dynamic origin of fractal forms is the robust striving for imperfect optimality that we see embedded in many natural patterns, chiefly and foremost hydrologic ones.

6.2.1 Formulation of the Mathematical Problem

Here, we review the model of river networks known as optimal channel networks (OCNs) [65, 66]. The OCN model was originally based on the *ansatz* that configurations occurring in nature minimize a functional describing total energy dissipation and on the derivation of an explicit form for such a functional. The latter uses, locally, landscape-forming flowrates Q_i and the drop in potential energy to define energy dissipation $Q_i \Delta z_i$, that is, approximated by $Q_i \sim A_i$ (where A_i is total contributing area at i) and by the drop in elevation $\Delta z_i \sim A_i^{\gamma-1}$ [63]. Spanning, loopless network configurations characterized by minimum energy dissipation are obtained by selecting the configuration, say, s, that minimizes the functional:

$$H_\gamma(s) = \sum_{i=1}^{N} Q_i \Delta z_i \propto \sum_{i=1}^{N} A_i^\gamma, \qquad (6.9)$$

where i spans the lattice, say, of N sites. Given that $A_i = \sum_j W_{j,i} A_j + 1$, where $W_{j,i} = W_{j,i}(s)$ is the generic element of a connectivity matrix (equal to 1 if $j \to i$ through a drainage direction, and 0 otherwise), the configuration s determines uniquely on a spanning tree the values of A_i. In this case, $W_{j,i}$ is the element of an actual adjacency matrix (where a site is connected directly only to its nearest neighbors by a drainage direction), spanning the connectivity of every node j to i. It proves crucial [63–66, 257] that a concave functional as defined by Equation (6.9) is warranted (i.e., $\gamma < 1$), and this stems directly from the physics of the problem subsumed by the slope–area relation $\Delta z_i \sim A_i^{\gamma-1}$ (i.e., slopes decrease as catchment area, and thus landscape-forming discharges, increases) [63]. Empirically, this result is also verified by a large body of field observations in fluvial domains (see, e.g., [70, 71] in the context of channel initiation).

The global minimum (i.e., the ground state) of the functional in Equation (6.9) is exactly characterized by known mean-field exponents [736], and one might expect to approach the mean-field behavior by reaching a stable local minimum upon careful annealing of the system. This is in fact the case. The proof of the above is not trivial: any stationary solution of the landscape evolution equation must locally satisfy the relationship $|\vec{\nabla} z_i| \propto A_i^{-1/2}$ between flux/area and the topographic gradient at any point i, and gradients are approximated by Δz_i, the largest drop in elevation at i. One can thus uniquely associate any landscapes with an oriented (spanning and loopless) graph on a lattice and reconstruct the field of cumulative areas $\{A_i\}$ corresponding to a given oriented

spanning graph. Note also that we wish to emphasize the dependence on the configuration $s = \{A_1, A_2, \ldots, A_N\}$ that the system assumes on the features of the oriented spanning graph associated with the landscape topography z through its gradients ∇z that uniquely defines total contributing areas $A_i, i = 1, N$ in an N-site lattice. Totally surprising is that empirically observed scaling features are only reproduced by myopic searches of dynamically accessible configurations that cannot get rid of the inheritance of extant constraints. The scaling structure (topological and metric) of configurations characterized by suboptimal energy dissipation matches extremely well those observed in nature and differs significantly from the ground states [64, 67, 69]. What are the implications of a worse energetic performance and yet a better representation of natural networks performed by cold ($T = 0$) OCNs? An explanation has been provided moving from the role of constraints in the evolution of a river landscape under nonlocal erosion (that is, using cumulated area A at every site). In fact, channel networks cannot disregard initial conditions, or geological quenched randomness, because they have an imprinting on the evolution of the final structure. The optimization that nature seems to perform in organizing of the parts and the whole of the river basin cannot be farsighted, that is, capable of evolving in a manner that completely disregards initial conditions by allowing for major migration of divides in the search for a more stable configuration evolving through transient, very unfavorable conditions [63, 64]. The evidence clearly indicates that the type of optimization that nature performs must be myopic, that is, willing to accept changes only if their impact is favorable immediately rather than in some ill-defined long run. Such results posit that the natural process of network evolution is unlikely to allow for dramatic changes of the initial configuration, which, in the short run, will decrease the fitness of the configuration but which will provide paths for later improvement leading to an overall better state.

Optimal arrangements of network structures and branching patterns thus result from the direct minimization of the functional in Equation (6.9). As stated above, OCNs for a given domain are a connected path spanning all its sites that minimizes $H_\gamma(s)$ without postulating predefined features, for example, the number of sources or the link lengths. The basic problem is screening all spanning trees in a given domain. This is a task where, interestingly, the number of degrees of freedom prevents complete enumeration as long as the size of the network becomes appreciable – the computational problem is termed NP-complete (or computationally unfeasible for the number of operations needed, which grows exponentially with the system size), akin to the celebrated traveling salesman problem [63]. In such cases, random search for testing configurations is unavoidable – and rewarding, as we detail later.

6.2.2 Comparative Geomorphological Studies

One key problem is the assessment of the robustness of OCN configurations selected by any minimum-seeking procedure. This has been studied [63] with respect to the strategy for minimum search, the role of initial conditions, the robustness of the functional dependence on γ, the role of lattice anisotropies, and the effects of "quenched" randomness [66]. The basic optimization strategies are similar to algorithms developed in the context of nonnumerable (NP-complete) problems where the exponential growth of possible configurations prevents complete enumeration. Iterated random searches work best in that context [63]. The basic algorithm proceeds as follows. An initial network configuration, s, is chosen as a spanning tree on the grid to drain an overall area made up by N sites. This defines an orientation and a connection for each pixel stating to which of the eight neighboring pixels its area is draining, neighbors being assumed at unit distance from their centroid. This in turn needed both preliminary and a posteriori speculations on whether a triangular lattice – with six neighboring nodes – or an anisotropic scheme in which diagonal connections were weighted by a $\sqrt{2}$ factor would be a better model of local interactions [63]. A scalar state variable, $A_i(t)$, denotes the total area at a point i at stage t of the optimization process,

$$A_i(t) = \sum_j W_{ij}(t)\, A_j(t) + 1,$$

where W_{ij} is the (now dynamic) functional operator that has the connectivity matrix as its static counterpart by

$$W_{ij}(t) = \begin{cases} 1, & \text{if } i,j \text{ are connected} \\ 0, & \text{otherwise} \end{cases}$$

(that is, $W_{ij}(t)$ implies that $j \to i$ is a drainage direction). Note that j spans the eight neighboring pixels of the arbitrary ith site. The unit mass added refers to the area representative of the actual site as a proxy of the distributed injection term. From the initial configuration (stage $t = 0$), the basic strategy consists of drawing a site at random and perturbing the system by assigning a change δW_{ij}, that is, by modifying at random its connection to the former

receiving pixel. Hence $W_{ij}(t + 1) = W_{ij}(t) + \delta W_{ij}$. This corresponds to perturbing the configuration s ($s \to s'$). Adjusting to such a local modification, all aggregated areas A_i are modified in the downstream region until the original and the modified paths reconvene. The change is accepted if the modified value of $H_\gamma(s')$ is lowered by the random change ($H_\gamma(s') < H_\gamma(s)$) and no loops are formed. Loops are excluded on a rigorous basis, as it was shown exactly that they lead to energetically unfavorable configurations (for the functional in equation with $\gamma < 1$, every tree is a local minimum of total energy expenditure [69]). As the new configuration is adopted as a base configuration, the process is iterated. Otherwise, the change is discarded (if $H_\gamma(s') \geq H_\gamma(s)$), and the t-stage configuration s is perturbed again. The procedure leads to a configuration for which no improvement on total energy expenditure appears after a fixed (and large) number of iterations, that is, an OCN. The whole process may or may not be reset and restarted from the same initial configuration. This is done several times to allow the random process a fair chance of capturing nonlocal minima – should they be of interest. The configuration attaining the lowest energy dissipation among the trials described before is chosen as the OCN. Instructive visual schemes of the progress of the basic selection algorithm are illustrated elsewhere [63, 72].

This basic procedure, at times termed the Lin (or the greedy) approach because of the similarities with the N-city traveling salesman algorithm [63], respects the rules of a fair search for approximate solutions but is apt to yield trapping in local minimum energy. Variants of the basic algorithm, implemented to test the importance of the strategy for minimum search, include the Metropolis algorithm [737]. This implies multiple simultaneous perturbations and simulated annealing schemes engineered to avoid trapping of the configuration into local minima. This is done by accepting perturbations of the current configuration ($s \to s'$) even if they yield $H_\gamma(s') \geq H_\gamma(s)$ with a probability depending on a state parameter T. In practice, the probability of acceptance of the perturbation is given by the Metropolis rule, that is, it is 1 if the resulting change corresponds to $H_\gamma(s') < H_\gamma(s)$) or, if $\Delta H = H_\gamma(s') - H_\gamma(s) \geq 0$ and $e^{-\Delta H/T} > R$, and 0 otherwise (R is a random number $\in (0,1)$). To carry out proper annealing, one makes changes in the parameter T from relatively high values at the start to low values toward the end of the analysis. Clearly, for high values of T, the likelihood of accepting unfavorable changes is high, whereas for $T \to 0$ the rule is equal to the basic algorithm. A "cooling" schedule for decreasing values of T as the procedure evolves is thus required [63].

Optimal arrangements of network structures and branching patterns thus result from the direct minimization of the functional in Equation (6.9). The basic operational problem to obtain OCNs for a given domain is to find the connected path s draining it that minimizes $H_\gamma(s)$ without postulating predefined features, for example, the number of sources or the link lengths. Random perturbations of an initial structure imply disconnecting and reorienting a single link at a time [63]. They lead to new configurations that are accepted, details aside, if they lower total energy expenditure – that is, the functional in Equation (6.9) – iterated until many perturbations are unable to prompt change by finding better configurations. Loops possibly generated by the random configuration search in the fitness landscape were excluded at first without a rigorous basis. Only later was it shown exactly that they lead to energetically unfavorable configurations [69, 74, 257] for realistic values of γ derived from slope–area empirical evidence. Boundary conditions are required for the evolving optimal trees, as outlet(s) must be imposed (single or multiple outlets along drainage lines) as well as no-flux or periodic boundary conditions [63, 72].

Interesting issues emerge on the statics, dynamics, and complexity of OCNs [64]. Exact results (reviewed in [64]) exist on the existence of many dynamically accessible stable states, the practical impossibility of pointing out a priori the most stable feasible state among all metastable states without an evolutionary account of the history of the current configuration of the system, and the hierarchical structure and the universality class of dynamically accessible states. Although the above are features that river networks share with other natural complex systems, the extent of observations and comparative analyses, the exact relation to the general evolution equations, and the broad range of scales involved suggest their interest as a general model system of how certain natural systems work [738]. Thus, one recurrent self-organized mechanism for the dynamic origin of fractal forms is the robust striving for imperfect optimality that we see embedded in many natural patterns, especially hydrologic ones.

The OCN model has been thoroughly explored, as it has produced various interesting results – and wonderful replicas of dendritic forms (see, e.g., Figures 1.2 and 6.14) whose statistical structure is indistinguishable from empirical ones regardless of vegetation cover, exposed lithology, geology, and climate.

Figure 6.14 Example of a multiple-outlet OCN, here mimicking a $2,631$ km^2 drainage area, used to compute the scaling of dissolved organic carbon removal in river networks (redrawn from [162]). OCNs have consistently been used to obtain statistically identical and yet rather different replicas of fluvial networks for studies in ecology, hydrology, and geomorphology.

For example, ensemble averages of the exceedance probability of total contributing area $P(\geq a)$ (the relative count of the number of pixels whose area is larger than or equal to a) obtained by a greedy algorithm (i.e., accepting only changes that decrease total energy dissipation), in a Monte Carlo setting where the single outlet imposed as boundary condition changes for every realization, show that they obey power laws $P(A \geq a)$ akin to those observed in the field via DEMs (see below). Note that here and in the following, A is the random variable "total contributing area at any site within the catchment." Suffice here to mention that A is sampled by $A_i, i = 1, \ldots, N$ for a lattice of N sites. One notes that $\sum_{i=1}^{N} A_i \neq N$, to distinguish the random variable A from other quantities, for instance, the area drained by each boundary site in a multiple-outlet OCN where we allow each site to be an outlet. For such a case, for instance, in a square lattice of L sites, the random variable is sampled by the total contributing area at each boundary site j consisting of $4L$ entities, and one has $\sum_{j=1}^{4L} A_j = N$, reflecting a rather different probability distribution.

Recursive statistical properties of local minima of the functional in Equation (6.9) prove robust with respect to initial and boundary conditions to the insertion of quenched randomness or of variously correlated heterogeneities like those induced by spatial effects in the injection field. One case deals with altering the linkage between total contributing area and landscape-forming discharges via a heterogeneous injection field r_i that redefines "area" (i.e., landscape-forming flowrate) as

$$A_i = \sum_j W_{ij} A_j + r_i$$

with the previous symbols' definitions (see [64]). A truly remarkable result is that local minima exhibit critical behavior characterized by scaling exponents indistinguishable from those observed in nature and different from those of the ground state [63, 67, 736]. Local and global minima of OCNs for multiple-outlet settings show that results obtained by chance-dominated constructs, such as dendrites generated by Eden growth (a self-avoiding random walk), lead to suboptimal structures analog to the ones attainable by enforcing only pointwise optimality. Eden growth is a benchmark because of its chance-dominated selection (no necessity is implied by the self-avoiding random-walk dynamics, and only treelike structures are selected by construction). Such structures were initially thought of as capturing the essentials of natural selection [101]. That turned out to be an artifact of nondistinctive comparative tests of tree structures [67] echoed by the "statistical inevitability" of Horton's laws frequently referred to in the hydrologic literature [68]. Indeed, if topological measures alone (say, Horton numbers of bifurcation and length imposed on Strahler's ordering, or Tokunaga matrices) are used to sort out the fine comparative properties of trees, one can be misled into finding spurious similarities with natural forms [67]. Exercises comparing constructs that have indistinguishable topological features even though they differ with the naked eye are revealing [69]. Rather, distinctive statistics of tree forms must rely on linked scaling exponents of aggregated areas, lengths, and elongation [67].

Exact proofs are available, such as in the noteworthy case of Peano's basin [121, 126], where topological measures match those of real basins and OCNs but fail to satisfy the requirements for aggregation and elongation. More subtle – but equally clear – is the failure of random walk – type models [101] or topologically random networks [63] to comply with stringent statistical comparisons [67]. Note that the latter models were especially influential in suggesting that chance alone was behind the recurrence of natural patterns, because of the equal likelihood of any network configurations

implied by the topologically random model. Instead, their purported similarity with natural patterns is now seen as an artifact of lenient comparative tools because statistical properties and "laws" derived in that context are inevitable for spanning trees. This applies to certain properties of directed networks as well [710]. Analytical results complete a view of dynamically accessible optimal states [63, 67, 103, 736].

Exact properties for the global minimum of the functional $H_\gamma(s)$ in Equation (6.9) have been obtained. Here, we mention for brevity only a few results pertaining to the total contributing area distribution $P(A \geq a)$ and the upstream length L distribution from any point i (described below for the standard case $\gamma = 1/2$). For the limit case $H_1(s)$, we denote with x_i the along-stream length of the pathway connecting the ith site to the outlet. It is straightforward to show that $\sum_i A_i \propto \sum_i x_i$ [63, 103, 710]. Thus the minimization of energy dissipation for $\gamma = 1$ corresponds to the minimization of the weighted path connecting every site to the outlet, that is, the mean distance to the outlet, and the global minimum is the most direct network. The configurations yielding a minimum of $H_1(s)$ are realized on a large subclass of the set of spanning trees – all the directed ones where every link has positive projection along the diagonal oriented toward the outlet. The $\gamma = 1$ case gives a minimum energy scaling $E \sim L^3$ (where L is the characteristic linear size of the lattice). This follows from the observation that any directed network corresponds to the Scheidegger model of river networks [63, 67] where all directed trees are equally probable by construction. Such a model, exactly solved, can be mapped into a model of mass aggregation with injection, later shown to map exactly the time activity of the Abelian sandpile model of self-organized criticality [738]. The corresponding scaling exponents are $\beta = 1/3, \psi = 1/2$, and $h = 2/3$. All directed trees are equally probable, having the same mean distance to the outlet, and each stream behaves like a single random walk. The $\gamma = 0$ case, instead, implies the minimization of $H_0(s)$, the total weighted length of the spanning tree. Every configuration has the same energy, because every spanning tree has the same number of links ($L^2 - 1$ for a $L \times L$ square lattice). The minimum energy E gives $E \sim L^2$ for each network, in analogy to the problem of random spanning trees, and the related scaling exponents for a square lattice are [736] $\beta = 3/8, \psi = 3/5$, and $h = 5/8$.

Necessity is thus at work in the selection of natural networks because all spanning trees are not equally likely, as shown by local minima of Equation (6.9) obtained by the greedy algorithm. The fine features of the resulting trees match perfectly those found in nature [63, 67]. These results, obtained by accepting configurations perturbed by disconnecting and reorienting a single link at a time only if the new configuration lowers total energy dissipation, entail a myopic search capable of exploring only essentially similar configurations. Once the Metropolis procedure [737] with a slow schedule of decreasing temperatures is unleashed, however, a massive annealing strategy implies that OCNs approach the ground state characterized by mean-field scaling exponents. Even with the naked eye, the differences in the aggregation structure and in the regularity of the selected landforms are striking [69].

6.2.3 Feasible Optimality

Feasible optimality (i.e., dynamically accessible, that is, coping with constraints imposed by initial and boundary conditions and various forms of heterogeneity) found in the river landscape echoes various results of statistical physics. In fact, the progressive departure of the exponent β from the typical observational value of 0.43 ± 0.02 [63] along with refinements of the minimum search procedure suggests a remarkable property of the fitness landscape defined by the functional (6.9). In fact, the scaling exponents of rapidly annealed minima on relatively unconstrained OCNs (say, subject to periodic boundary conditions) are consistently found in the range $\beta = 0.48 \pm 0.01$, whereas the limit value 0.50 ± 0.005, corresponding to the ground state [736], is consistently obtained only for the least constrained arrangements [63], that is, subject to multiple outlets and periodic boundary conditions and accessible through a schedule of slowly decreasing temperatures of the annealing scheme – effectively preventing any legacy of the initial configurations. Such results forcefully suggest that nature works by settling into imperfect stable states into which the evolutionary dynamics get at least temporarily trapped. Each constraint affects selectively the feasible optimal state, that is, to a different degree, depending on its constraining severity, matching the observation of consistent scaling exponents with minor but detectable variations in describing the morphology of different fluvial basins [63]. Different fractal signatures embedded in linked scaling exponents thus suggest that evolution is adaptive to the climatic, lithologic, vegetational, and geologic environments. The worse energetic performance and yet the better representation of natural networks by suboptimal OCNs imply a defining role for geologic constraints in the evolution of a channel network. In fact,

channel networks cannot change widely regardless of their initial conditions, because such conditions leave long-lived geomorphic signatures [262]. Therefore, OCNs clarify the possible extent of the legacy issue. This is a long-standing question in geomorphology: whether the topography you see once you look across a landscape is in balance with current climate-driven processes or else contains relict signatures of past climates in regions where such processes have only changed intensity rather than character [262]. Finally, OCNs are spanning loopless configurations only under precise physical requirements that arise under the constraints imposed by river dynamics – every spanning tree is exactly a local minimum of total energy dissipation.

How did we assess the deep and consistent similarities of fluvial forms with the results of OCN simulations? Many empirical geomorphological features can be analyzed in the greatest detail with the widespread availability of digital terrain maps (DTMs) (see Figure 1.3). Channeled portions of the landscape are extracted from elevation fields through the exceedance of geomorphological thresholds of slope-dependent areas (Figure 1.2). A benchmark statistic concerns the master variable for river network geomorphology and stream ecology. In fact, total contributing area A_i at any point i of a lattice is related to the gradient of the elevation field (the topographic slope) at that point:

$$| \vec{\nabla} z_i | \propto A_i^{\gamma-1},$$

where observed values of γ lie inevitably around $1/2$, as typically observed in runoff-generating areas [63]. Slope-area relations indeed provide a powerful synthesis of the local physics. The probability distributions of relevant random variables (such as total contributing area A at any site or the upstream distance L from any site of the network to the farthest source) are characterized by finite-size scaling [67, 103]:

$$P(A \geq a) = a^{-\beta} \, F\left(\frac{a}{A_{\max}}\right), \qquad \textbf{(6.10)}$$

where $A_{\max}$ is the area at the catchment closure that obviously places an upper limit on the distribution (Figure 6.15). For the random variable L defined by the upstream length at any site,

$$P(L \geq \ell) = \ell^{-\psi} \, G\left(\frac{a}{A_{\max}^h}\right)$$

where β, ψ are suitable scaling coefficients and h is Hack's coefficient ($h \approx 0.57$) [63], $F(x), G(x)$ are suitable cutoff functions (Figure 6.15). The empirically

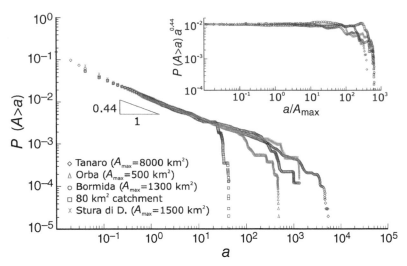

Figure 6.15 Statistical evidence of finite size scaling of total contributing catchment area in the fluvial landscape. Empirical estimates of $P(A \geq a)$ are derived by computing all values of A at all sites via the connectivity matrix extracted from the topography. The probability $P(A \geq a)$ is estimated by counting the relative proportion of sites anywhere in the catchment whose total contributing area exceeds a, here expressed in square kilometers. Empirical plot for five nested subcatchments of different maximum area A_{max} indicated in the legend. (inset) Collapse plot of the probability distribution once properly rescaled. The nested catchments shown here belong to the Tanaro river basin (Italy). Figure after [64]

defined scaling exponents, from vast comparative analyses [63], lie in a narrow (and related [67]) range $\beta = 0.42 \div 0.45$ and $\psi = 0.70 \div 0.80$, respectively.

Figure 6.15 shows one significant example of empirical verification of Equation (6.10) [64]. One notes the following:

- The *ansatz* in Equation (6.10) postulates that if the embedding domain $A_{max} \to \infty$, exact scale invariance is achieved; that is, $P(A \geq a) = a^{-\beta} F(a/A_{max}) \to Ca^{-\beta}$ (where C is a constant) if $F(x \to 0) = C$ and $F(\infty) = 0$.

- To avoid binning issues, the cumulative probability of exceedance $P(A \geq a)$ is plotted instead of its density for five nested subcatchments of a relatively large basin ($\sim 3,000$ km^2) [5]. In fact, the scaling exponents are related; that is, if $P(\geq x) = Cx^{-\beta}$, then its pdf is $p(x) = -dP(\geq x)/dx = C\beta x^{-(1+\beta)}$. The estimation of β follows by maximum likelihood from the largest sample of values of $P(\geq x_i)$ belonging to the power law regime, $x_i, i = 1, N$, with $x \in [x_{min}, \infty]$ (in practice, x_i ranks by relative frequency from $1/N$ for A_{max} to 1 for the pixel unit Δx^2, always equaled or exceeded). In the DTM case, one has that the minimum size x_{min} is the one for which $P(\geq x_{min}) = 1$, that is, $x_{min} = \Delta x^2$, where Δx is the pixel size used to extract the network.

The result (excluding areas clearly belonging to the decaying tail of the distribution for which $a \sim A_{max}$) is [5]

$$\beta = \frac{N}{\sum_{i=1}^{N} \log(x_i/\Delta x^2)},$$

which provides, in the test case, the value of $\beta = 0.44$ shown in Figure 6.15. Issues of determining the error of the estimate are discussed elsewhere, and β is here estimated at 0.44 ± 0.01 [64].

- The data shown in Figure 6.15 show clearly a truncated power law character, emphasized by the remarkable collapse of the rescaled plot (inset) defining the finite-size scaling effect induced by the cutoff $F(x)$ imposed on the power law (for $P(\geq a) = a^{-\beta} F(a/A_{max})$, consistency requires that $F(x) \to 0$ for $x \to \infty$ and $F(x) \to$ const for $x \to 0$).

- The factor imposing a cutoff in an otherwise scale-invariant power law behavior is the maximum area at each catchment closure, which limits selectively the sample size at areas next to the maximum. The empirical cutoff function is estimated from the plots (inset of Figure 6.15)

$$P(A \geq a) \ a^{0.44} \qquad \text{versus} \qquad \frac{a}{A_{max}^{\zeta}},$$

noting that the various curves (corresponding to the different subcatchments of maximum areas A_{max}) would collapse onto a single curve if the functional dependence were truly $F(a/A_{max}^{\zeta})$. To that end, one tweaks the value of ζ to minimize a measure of the differences obtained by the collapse of the various curves [739]. The result in this case is $\zeta = 1.00 \pm 0.002$ [64]. Thus Equation (6.10) is recovered [64].

- $F(x)$ is shown to behave properly, either by visual check of the extent of the collapse onto a single curve regardless of widely varying values of A_{max} or by the verification of the consistency argument ($F(0) = const$ and $F(1) \sim 0$, let alone $F(\infty) = 0$).

In analogy, the finite-size scaling *ansatz* provides a compelling observational proof of self-similarity [63] and a strong version of Hack's law relating the largest upstream stream length L_{max} to its total cumulative area A_{max}: $L_{max} \propto A_{max}^{h}$, with $h \approx 0.57$. Hack's coefficient h is a measure of the elongation of catchment shapes and an unmistakable signature of the fractal geometry of the river basin [121]. The finite-size scaling argument requires the linkage of aggregation and elongation, yielding [103]

$$\beta = 1 - h,$$

a framework in which the accepted standard independently observed values $\beta = 0.43$ and $h = 0.57$ obviously fit perfectly. Scaling in the river basin has been documented in many other geomorphological indicators, including exact limit properties [63]. Such measures have been used for comparative analyses with patterns derived from evolution and selection [64].

Optimal arrangements of network structures and branching patterns result from the direct minimization of the functional in Equation (6.9). As pointed out before, OCNs are a connected path s that minimizes $H_{\gamma}(s)$ without postulating predefined features, for example, the number of sources or the link lengths [63]. As we have seen, random perturbations of an initial structure imply disconnecting and reorienting a single link at a time. Accepted perturbations lead to new configurations that are accepted, details aside, if they lower total energy expenditure–iterated until many perturbations are unable to prompt change by finding better configurations. Loops possibly generated by the random configuration search in the fitness landscape were excluded at first without a rigorous basis. Only later was it shown exactly that they lead to energetically unfavorable configurations [74, 257]. Boundary conditions are required for the

evolving optimal trees, as outlet(s) (single or multiple outlets along drainage lines) as well as no-flux or periodic boundary conditions must be imposed (Figure 6.14).

Figure 6.16 shows the progressive departure of the exponent β from the typical observational value of 0.43 ± 0.02 along with refinements of the minimum search procedure [64, 74]. The scaling exponents of rapidly annealed minima on highly constrained OCNs are consistently found in the range $\beta = 0.43 \pm 0.01$ [67], whereas the limit value 0.50 ± 0.005, corresponding to the ground state [736], is consistently obtained only for the least constrained arrangements (Figure 6.16b), that is, subject to multiple outlets and periodic boundary conditions and through a schedule of very slowly decreasing temperatures in the annealing scheme to avoid any legacy of the initial configuration. Each constraint affects the feasible optimal state to a different degree, depending on its severity, matching the observation of consistent scaling exponents with minor but detectable variations in describing the morphology of different fluvial basins.

Different fractal signatures embedded in linked scaling exponents thus suggest that evolution is adaptive to the climatic, lithologic, vegetational, and geologic environments [64]. The worse energetic performance and yet the better representation of natural networks by suboptimal OCNs imply a defining role for geologic constraints in the evolution of a channel network. In fact, channel networks cannot change widely regardless of their initial conditions, because these conditions leave long-lived geomorphic signatures [262].

A question, arguably of general validity for open dissipative systems with many degrees of freedom, is whether the optimization that nature performs in the organization of the parts and the whole could really be farsighted, that is, capable of evolving in a manner that completely disregards initial conditions. This would allow for major changes of structural features in the search for more stable configurations, even though it would necessarily involve evolution through many unfavorable states. The experiments shown in Figure 6.16 suggest that the type of optimization that nature performs, at least in the fluvial landscape, is myopic. The proof that river networks are not free to explore extended regions of their fitness landscapes suggests that nature might not search for global minima in general when striving for optimality but rather trades to settle in for dynamically accessible local minima whose features are still scaling, although differently from the universal features of inaccessible ground states – and matching superbly those of real landforms.

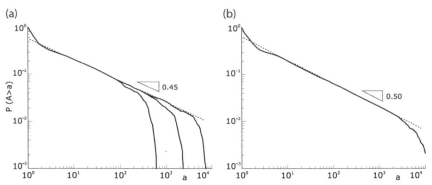

Figure 6.16 Ensemble mean of the statistics of aggregated area in 100 realizations of OCNs generated with different boundary conditions. (a) Plot of $P(A \geq a)$ obtained through ensemble averaging of the largest networks developing within 32×32, 64×64, and 128×128 lattices with multiple-outlet domains with lateral periodic boundary conditions (no-flux condition at the top boundary). Here, local minima are found through a greedy search procedure. Note that the imposed periodicity relaxes constraints on the aggregation of the accessible minimum energy configurations. (b) As above for the largest lattice. Here, the minimum search procedure is a carefully annealed Metropolis algorithm [737]. Note that the ensemble mean exponent is $\beta = 0.5 \pm 0.005$, matching the characteristics of the ground state. Figure after [64]

6.2.4 Examples of OCNs and Their Landscapes

Examples of local and global minima of OCNs for a multiple-outlet setting are reported in Figure 6.17. Figure 6.17a shows the results obtained by an algorithm called Eden growth generated by a self-avoiding random walk [63]. These types of algorithms generate structures that have been very influential in geomorphology because they led to the early suggestion that natural forms in the fluvial landscapes were chance dominated. Eden growth is one of them. It is known [63] to lead to suboptimal structures analogous to the ones attainable by enforcing only pointwise optimality. Eden growth is a benchmark because of its chance-dominated selection (no necessity is implied by the self-avoiding random walk dynamics, and only treelike structures are selected by construction). Such structures were initially thought as capturing the essentials of natural selection. That turned out to be an artifact of nondistinctive tests on tree structure [67] echoed by the purported statistical inevitability of Horton's laws [68]. Indeed, if topological measures alone (say, Horton numbers of bifurcation and length imposed on Strahler's ordering, or Tokunaga matrices) are used to sort out the fine comparative properties of trees, one can be misled into finding spurious similarities with natural forms. The exercise of comparing Figures 6.17a,b is revealing, as the two constructs have indistinguishable topological features, although they differ even with the naked eye [63, 64, 67].

Distinctive statistics of tree forms must rely on linked scaling exponents of aggregated areas, lengths, and elongation [67]. Exact proofs are available, such as in the noteworthy case of Peano's basin [66, 126, 127], where topological measures match those of real basins and OCNs but fail to satisfy the requirements for aggregation and elongation. More subtle – but equally clear – is the failure of random walk-type models or topologically random networks to comply with stringent statistical comparisons [63]. Note that the latter models were especially influential in suggesting that chance alone was behind the recurrence of natural patterns, because of the equal likelihood of any network configurations implied by the topologically random model. Instead, their purported similarity with natural patterns is now seen as an artifact of lenient comparative tools because statistical properties and "laws" derived in that context are inevitable for spanning trees. This applies to certain properties of directed networks as well [710].

Necessity is at work in the selection of natural networks because all spanning trees are not equally likely. Figure 6.17b shows a local minimum of the functional in Equation (6.9) (with $\gamma = 1/2$) obtained by a greedy algorithm. The fine features of the trees match perfectly those found in nature. These results, obtained by accepting configurations perturbed by disconnecting and reorienting a single link at a time only if the new configuration lowers total energy dissipation, entail a myopic search capable of exploring only close configurations. Figure 6.17c is instead obtained through the Metropolis

(a)   (b)   (c)

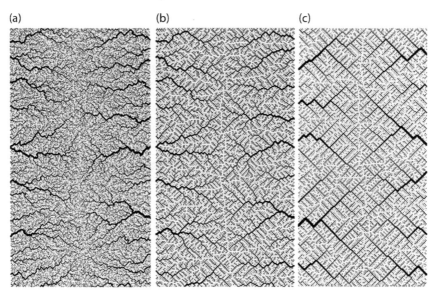

Figure 6.17 Different degrees of optimality in multiple-outlet networks obtained in the same rectangular domain imply substantial modifications only detected by truly distinctive statistics. (a) Optimality enforced only locally and treelike constraints settle the system into unrealistic shapes whose topological structure is still indistinguishable from the one found in real rivers and in dynamically accessible states of total energy minimization: Eden growth patterns of self-avoiding random walks filling the domain. (b) An imperfect OCN (i.e., $T = 0$ in a Metropolis scheme) leading to a local minimum of total energy dissipation. Note that OCNs bear long-lived signatures of the initial condition owing to the myopic search procedure but actually reproduce perfectly, besides topology, also the aggregation and elongation structures observed in real river landscapes [63]. (c) Ground-state OCNs obtained through simulated annealing using a very slow schedule of decreasing temperatures T. The reaching of the ground state is confirmed by the matching of the exact mean-field exponents with those calculated for (c). Note also the unrealistic regularity of the drainage patterns. Figure after [64, 67]

procedure [737], where a massive annealing strategy has been implemented. This implies that many energetically unfavorable changes have been accepted owing to a schedule of slowly decreased temperatures reaching a ground state characterized by mean-field scaling exponents [736], here matched perfectly. Even with the naked eye, however, the differences in the aggregation structure and in the regularity of the selected landforms are striking. Note that all three networks have identical topological features indistinguishable from those found in nature, proving even visually their inability to be distinctive from network structures [67].

OCNs entail the useful concept of replicas of ecological substrates. In fact, the random search procedure needed to access the rich structure of the fitness landscape of trees, each being a local minimum of the functional Equation (6.9), produces for any search a different outcome, yet characterized by identical statistics. This means that 100 OCNs may be seen as 100 independent realizations of landscape evolution within the same domain.

To construct the landscape associated to an OCN, we follow the procedure described in [73] and the theoretical insight developed later to justify it [72, 74, 75]. Specifically, the algorithm starts from the outlets (sites placed at the lowest elevation) and travels upstream, assigning at each step an elevation gain $\Delta z_i \sim A_i^{\gamma-1}$, as predicted by the slope–area relation, which is assumed to apply deterministically. Obviously there would be no problem in prescribing a noisy slope–area relation following proper geomorphological insight [63]. Figure 6.18 shows two OCNs and their landscapes.

It should be remarked that every OCN is a so-called natural river tree, in the sense that there exists a height function z_i such that the flow directions are always directed along steepest descent [74]. This result formalized earlier empirical–computational results, such as the fact that a Peano network is not an OCN [66], a result originally claimed by analyzing its function $P(A \geq a)$ and using Peano's tree as an initial condition to evolve OCNs. In that case, brute force construction of a landscape proves impossible. Recently, a mathematical study of

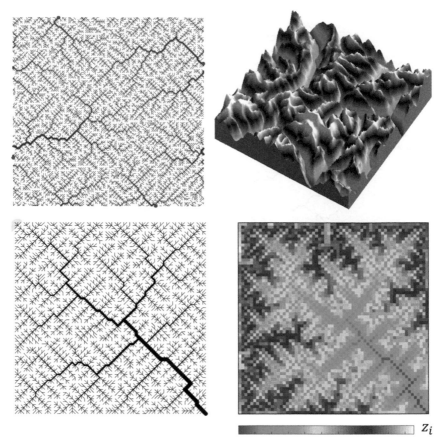

Figure 6.18 Example of construction of a landscapes associated to an OCN. (top left) Multiple-outlet OCN onto which a landscape is constructed. The OCN provides directly the total cumulative area A_i at any site i of the lattice. By starting from a site j (e.g., an outlet with $z_j = 0$), landscape elevation of the single node i connected to j is obtained by the recursive rule $z_i = z_j + \Delta z_i$, where $\Delta z_i = cA_i^{-1/2}$, where the constant c simply rescales the relief. (top right) A 3D rendition of the landscape associated with the OCN on the left and a specific choice of the constant c. (bottom left) A single-outlet OCN. (bottom right) A color-code 3D landscape associated with it. Figure after [75, 229]

natural river trees in an arbitrary graph defined precisely the problem in terms of forbidden substructures, called k-path obstacles, and of OCNs on d-dimensional lattices, improving earlier results by determining the minimum energy up to a constant factor for every $d > 2$ [75]. These results considerably extended our capabilities in environmental statistical mechanics and ecology, because proper independent and statistically indistinguishable replicas can be generated of 3D substrates for ecological interactions generated by the fluvial basin.

6.2.5 Exact Results

A number of other exact results are summarized in [64]. An interesting question is how networks resulting from

general erosional dynamics are related to the landforms arising from minimization of total energy dissipation [74]. Specifically, any landscape reconstructed from an optimal configuration using the slope-area relation – a constituting equation for the OCN concept that is deeply rooted in field evidence [63, 70, 71] – has been shown to be a stationary solution of the general landscape evolution equation in the small gradient approximation [74]. Superficially, this might seem obvious, because the relation between topographic gradients and cumulative area is locally verified by construction. One must notice, however, that the slope-area relation alone does not imply stationarity because the water flow determining erosion may not (and in general will not) be in the direction of the steepest descent in the reconstructed landscape. Thus,

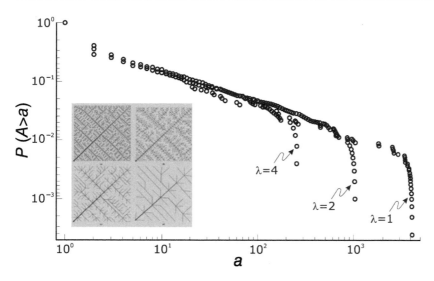

Figure 6.19 Effects of coarse graining, at different levels λ, of the probability of exceedence of total cumulative area A for the OCNs shown in the inset. The invariance of the probability distribution under coarse graining, a signature of the scaling aggregation structure, is remarkable. (inset) Renormalized OCNs with $\lambda = 1$ (upper left), $\lambda = 2$ (upper right), $\lambda = 4$ (lower left), and $\lambda = 8$ (lower right). Figure after [63, 64]

OCNs consist of the configurations s, which are local minima of Equation (6.9) in the sense specified below: two configurations, s and s', are close if one can move from one to the other just by changing the direction of a single link. A configuration s is said to be a local minimum of the functional in Equation (6.9) if each of the close configurations s' corresponds to greater energy expended. Note that not all changes are allowed in the sense that the new graph again needs to be loopless. Thus, a local minimum is a stable configuration under a single link flip dynamics, that is, a dynamic in which only one link can be flipped at a given time only when the move does not create loops and decreases the functional Equation (6.9). Any elevation field obtained by enforcing the slope-area relation to a configuration minimizing at least locally the functional Equation (6.9) is a stationary solution of the landscape evolution equation yielding the slope-area rule at steady state. This, in turn, implies that the landscape reconstructed from an optimal drainage network with the slope-area rule is consistent with the fact that the flow must follow steepest descent. The proof has been found in [74] and reviewed and put in context a number of times [64].

One interesting issue, thoroughly discussed elsewhere [63, 64] but worth mentioning here, is the invariance of OCN landscapes under coarse graining. Because optimal energy expenditure is the foundation of the OCN concept, its variation under a change in the scale of observation of the landscape is of considerable importance. Here, it suffices to coarse grain the site of a lattice of size L in squares of side L/λ, where the initial OCN is described at a resolution of pixels of side length L ($\lambda = 1$).

The conserved property is elevation: the 3D structure of the OCN is assigned everywhere through the slope-area relationship $\Delta z_i = A_i^{-1/2}$, where Δz_i approximates the topographic gradient at i. The original sites are then grouped in squares of side L/λ, with $\lambda \geq 2$ (Figure 6.19), such that, from an initial number of pixels N, one coarse grains the description of the terrain to a total number of pixels N/λ^2. The elevation of each of the new sites (a larger pixel of side λL) is computed as the average elevation of the λ^2 constituent pixels of side length L. From this coarse-grained 3D landscape, a new drainage network is drawn following as flow directions the lines of steepest descent, as in the original construct.

The above transformation preserves the mean elevation of the basin. Examples of OCNs progressively coarse grained according to the previous rules are shown in the inset of Figure 6.19. The key result highlighted therein is that the probability distribution of aggregated area is effectively invariant under coarse graining; that is, $P(A \geq a)^{(\lambda)} \approx P(A \geq a)$. Under such a premise, one obtains that total energy expenditure E of OCNs (which is proportional to the mean elevation of the OCN landscape) scales under coarse graining as $E^{(\lambda)} \propto \lambda^{-\delta} E$ with $\delta = 2(1 - \beta) \sim 1.2 > 0$ [63, 64]. These results paved the way for a subtle demonstration that every tree is a local minimum of energy expenditure [74] and especially the fact that minimization of free energy (where entropy is properly defined as the logarithm of the number of trees with the same energy) leads to the same configurations obtained by minimizing Equation (6.9) in the thermodynamic limit $L \to \infty$. The latter is practically reached very soon (say, $L \geq 30$ pixels) [740].

6.3 Computational Tools for Waterborne Disease Spread

As the reader will have noticed, the general scheme for most of the problems described in this book may be reduced to the following problem:

$$\frac{d\mathbf{x}}{dt} = \mathbf{F}(t, \mathbf{x}), \qquad \mathbf{x}(t_0) = \mathbf{x}_0, \tag{6.11}$$

where t is the independent time variable, $\mathbf{x}$ is a vector of independent state variables (whose determination is the goal of the computations), and $\mathbf{F}(t, \mathbf{x})$ defines the properties of the coupled ODEs. The mathematical problem is specified when the vector of functions on the right-hand side of Equation (6.11), $\mathbf{F}(t, \mathbf{x})$, is set and the initial conditions at t_0 are given.

MATLAB's standard solver for ordinary differential equations (ODEs) is the function ode45. It uses a fourth-order Runge–Kutta scheme with five base points and variable step size. This entails a stable and accurate numerical scheme more reliable for nonlinear problems than a first-order Euler time-marching scheme using only information at time t to project solutions at time $t + \Delta t$. The first step toward a solution is obviously the proper definition of the ODEs governing the system's behavior given a proper set of initial conditions. Whether parameters and forcings need be read and/or calibrated is another issue, which we do not cover here.

6.3.1 A Solver for a Spatially Implicit SIRB Model

Suffice for our example to note that we shall concentrate on a simple – yet structurally as complex as any other problem dealt with in this book – system of four coupled ODEs representing a standard spatially implicit SIRB model (Chapter 4). Note that it is often beneficial to produce a visual representation of the trajectories of the state variables, as done in what follows. The code has been kindly provided by Damiano Pasetto.

For simplicity, the code describes a relatively standard case where we consider a fraction σ of asymptomatic infections in a population H (Section 4.3) and reduce this to a case that corresponds to each of the nodal equations in Box 4.13, where human mobility ($m = Q_{ij} = 0$) and pathogen transport through waterways ($l = P_{ij} = 0$) are switched off. Thus one may safely remove the nodal index (e.g., $S_i(t) = S(t)$) and write the following set of coupled ODEs:

$$\frac{dS}{dt} = \mu(H - S(t)) - \mathscr{F}(t)S(t) + \rho R(t), \tag{6.12}$$

$$\frac{dI}{dt} = \sigma \mathscr{F}(t)S(t) - (\gamma + \mu + \alpha)I(t), \tag{6.13}$$

$$\frac{dR}{dt} = (1 - \sigma)\mathscr{F}(t)S_i + \gamma I(t) - (\rho + \mu)R(t), \tag{6.14}$$

$$\frac{dB}{dt} = -\mu_B B(t) + \theta[1 + \lambda J(t)]I(t), \tag{6.15}$$

where the usual parameters' meanings are employed as in Box 4.13, and in addition, the force of the infection is defined by $\mathscr{F}(t) = \beta_0 B(t)/(K + B(t))$. $J(t)$ is a sequence of rainfall pulses; H is the size of the human population, assumed to be constant over time. Note that in the code below, the parameter values are simply assigned. Note also that the computation of the total number of cumulated cases ($C(t)$) is succinctly defined by the related equation $dC/dt = \sigma \mathscr{F}S(t)$. The initial conditions (reminiscent of the Haitian epidemics; Section 4.3) are $I(0) = 2100$, $R(0 = I(0)(1 - \sigma)/\sigma$, $S(0) = H - I(0) - R(0)$, and $B(0) = \theta I(0)$.

In the following, the reader will find the MATLAB script that performs the solution of the problem described in Equations (6.12)–(6.15). The code(s) may be downloaded from the ECHO Lab website (www.echo.epfl/downloads) in a format ready to be executed. The scripts of the codes are shown in Figures 6.20, 6.21, and 6.22. The outputs of the code are the plots shown in Figures 6.24 and 6.23.

6.3.2 Computing Conditions for Transient Epidemics in Spatially Explicit Systems

As an example of spatially explicit problems described in Chapter 4, here we provide the implementation of a spatial SIRB/SIRW model. The reference material is in [453]. The code has been kindly provided by Lorenzo Mari.

The mathematical model describes local epidemiological, demographic, and ecological processes; pathogen transport along water systems; and the effects of short-term human mobility on disease propagation. Network nodes represent n human communities of assigned population size, arranged in a given spatial setting and connected by hydrological pathways and human mobility [453]. Let $S_i(t)$ and $I_i(t)$ be the local abundances of susceptible and infected individuals in each node i of a network at time t, and let $B_i(t)$ be the concentration of pathogens in the local water reservoirs to which human

```
 1  %%%%%%%%%%%%%%%%%%%%%%%%%%%%%%%%%%%%%%%%%%%%%%%%%%%%%%%%%%%%%%%%%%%%%%%%
 2  %% SIRB model
 3  %
 4  function dydt=SIRB(t,y,p,J)
 5  %
 6  %%%%%%%%%%%%%%%%%%%%%%%%%%%%%%%%%%%%%%%%%%%%%%%%%%%%%%%%%%%%%%%%%%%%%%%%
 7  % Function describing the 1-Dimensional SIRB model
 8  %
 9  % Inputs:
10  %    t                   time
11  %    y                   vector containing values of S,I,R,B,C at time t
12  %    p                   Structure containing the required parameters
13  %    J                   Structure containing the dates and values of
14  %                        precipitation
15  % Outputs:
16  %    dydt                derivatives at time t.
17  %%%%%%%%%%%%%%%%%%%%%%%%%%%%%%%%%%%%%%%%%%%%%%%%%%%%%%%%%%%%%%%%%%%%%%%%
18
19  % state variables
20  S=y(1);
21  I=y(2);
22  R=y(3);
23  B=y(4);
24  CumI=y(5);
25
26  % find rainfall at time t
27  indexJ=find(J.dates>=t,1,'first');   % index of precipitation vector
28  Jv=J.values(indexJ);                 % value of precipitation
29
30  % compute force of infection at time t
31  FI = p.beta0*B/(p.K+B);
32
33  % model equations
34  dSdt =            -FI*S +  p.mu*(p.H-S)   + p.rho*R;
35  dIdt =      p.sigma*FI*S - (p.mu + p.alpha + p.gamma)*I;
36  dRdt =(1-p.sigma)*FI*S - (p.mu + p.rho)*R + p.gamma *I;
37  dBdt = p.theta*(1+p.lambda*Jv)*I - p.muB*B;
38  dCumIdt=  p.sigma*FI*S;
39
40  % output
41  dydt=[dSdt;dIdt;dRdt;dBdt;dCumIdt];
42  end
```

Figure 6.20 MATLAB function describing the SIRB equations. This function can be used in any MATLAB solver for ODEs (e.g., ode45).

communities have access. Epidemiological dynamics and pathogen transport over the hydrological and human mobility networks can be described by the following set of $3n$ ordinary differential equations [453]:

$$\frac{dS_i}{dt} = \mu(N_i - S_i) - \left[(1 - m_i^S)\beta_i \frac{B_i}{K + B_i}\right.$$

$$\left. + m_i^S \sum_{j=1}^{n} Q_{ij}\beta_j \frac{B_j}{K + B_j}\right] S_i, \qquad (6.16)$$

$$\frac{dI_i}{dt} = \left[(1 - m_i^S)\beta_i \frac{B_i}{K + B_i} + m_i^S \sum_{j=1}^{n} Q_{ij}\beta_j \frac{B_j}{K + B_j}\right] S_i$$

$$- (\gamma + \mu + \delta)I_i, \qquad (6.17)$$

$$\frac{dB_i}{dt} = -(v_i + l_i)B_i + \frac{1}{W_i}\sum_{j=1}^{n} l_j P_{ji} W_j B_j$$

$$+ \frac{p_i}{W_i}\left[(1 - m_i^I)I_i(t) + \sum_{j=1}^{n} m_j^I Q_{ij}I_j\right]. \qquad (6.18)$$

The spatial model is described as follows [453]. For the human host population, the dynamics of the susceptible compartment in each community (the first equation) are described as a balance between population demography and infections due to dose-dependent exposure to the pathogen. The host population, if uninfected, is assumed to be at demographic equilibrium N_i, with μ being the human mortality rate. The parameter β_i represents the

```
1    close all
2    clear
3    clc
4    %%%%%%%%%%%%%%%%%%%%%%%%%%%%%%%%%%%%%%%%%%%%%%%%%%%%%%%%%%%%%%%%%%%%%%%%%%%%%%%
5    %
6    % Example of solution of SIR model with MATLAB
7    %
8    %%%%%%%%%%%%%%%%%%%%%%%%%%%%%%%%%%%%%%%%%%%%%%%%%%%%%%%%%%%%%%%%%%%%%%%%%%%%%%%
9    %% define model parameters
10   p.H     = 10e6;              % population
11   p.mu    = 1/(65*365);        % natural mortality rate      [1/d]
12   p.rho   = 1/(7*365);         % rate of loss of immunity    [1/d]
13   p.gamma = 1/5;               % rate of loss of infection   [1/d]
14   p.alpha = 0.01/5;            % cholera mortality rate       [1/d]
15   p.theta = 1e6;               % rate of contamination / cumulative exposure   [1/d]
16   p.K     = 1e12;              % scaling factor for dose-dependent bacterial threshold for
         infection
17   p.lambda= 0.1;               % precipitation washout factor [1/ [J] ]
18   p.beta0 = 0.01;              % maximum exposure coefficient
19   p.sigma = 0.3;               % percentage of symptomatic infectious individuals
20   p.muB   = 1/(8);             % bacterial mortality rate [1/d]
21
22   %% times for simulation
23   t0 = datenum(2010,10,01);
24   tf = datenum(2016,12,01);
25   tspan=t0:1:tf;
26
27   %% define initial conditions for S0, I0, R0, B0,C0 ;
28   % (steady state IC considered here)
29   I0 = 2100; % initial cases in haiti in 2010
30   R0 = I0*(1-p.sigma)/p.sigma;
31   S0 = p.H-I0-R0;
32   B0 = p.theta*I0;
33   CumI0 = I0;
34   y0=[S0; I0; R0; B0; CumI0];
35
36   %% precipitation
37   % usually precipitation is read in input
38   % here it is generated using a seasonal trend and random perturbations
39   J.dates=tspan;
40   J.values=(5+10*sin((tspan-datenum(year(tspan),1,1)+90)/(240*2*pi)))...
41       .*(2+2*randn(1,length(tspan)));
42   J.values(J.values<0)=0;
43
44   %% resolve using ODE45 on function SIRB
45   options= odeset('NonNegative',ones(1,length(y0)));
46   [t,y]=ode45(@(t,y) SIRB(t,y,p,J), tspan, y0,options);
47
48   S=y(:,1);
49   I=y(:,2);
50   R=y(:,3);
51   B=y(:,4);
52   CumI=y(:,5);
53   cases=diff(CumI); % daily cases
54   % plot the results
```

Figure 6.21 Main MATLAB script for the solution of the SIRB model (Figure 6.20) using the built-in function ode45. The code to plot the results is shown in Figure 6.22.

site-specific rate of exposure to contaminated waters, and the fraction $B_i/(K + B_i)$ is the dose–response function describing the probability of becoming infected due to exposure to a concentration B_i of pathogens (with K being the standard half-saturation constant; Box 4.13). Exposure to contaminated water for susceptible individuals of community i can occur either in their home community (with probability $1 - m_i^S$, with m_i^S being

```
1   %% plot results - S, I, R and B
2   fig1=figure(1);
3   set(gcf,'color','white')
4
5   h(1)=plot(t,S,'-b');
6   ax(1) = gca;
7   set(ax,'box','off','color','none')
8   set(ax(1),'YaxisLocation','left')
9   hold on
10  h(2)=plot(t,I,'--r');
11  h(3)=plot(t,R,'-g');
12  datetick('x','mmm-yy','keeplimits','keepticks')
13  xlim([t0,tf]);
14
15  ax(2)=axes('Position',get(ax(1),'Position'));
16  h(4)=plot(t,B,'-m','Parent',ax(2));
17  box off
18  set(ax(2),'YaxisLocation','right', ...
19      'Color','none', 'XTickLabel',[],'Xlim',[t0,tf]);
20  linkaxes([ax(1),ax(2)],'x');
21  set(ax(1),'YScale','log','YLim',[1 p.H])
22
23  legend(h,'S','I','R','B')
24
25  yl(1)=get(ax(1),'YLabel');
26  yl(2)=get(ax(2),'YLabel');
27  set(yl(1),'String','S I R')
28  set(yl(2),'String','B')
29
30  %% plot cases and precipitation
31  cases=diff(CumI);
32
33  fig2=figure(2);
34  set(gcf,'color','white')
35
36  subplot(2,1,1)
37  plot(J.dates,J.values,'-b');
38  set(gca,'ydir','reverse')
39  xlim([t0,tf]);
40  datetick('x','mmm-yy','keeplimits','keepticks')
41  ylabel('daily precipitation (mm)')
42
43  subplot(2,1,2)
44  plot(t(2:end),cases,'-k');
45  xlim([t0,tf]);
46  datetick('x','mmm-yy','keeplimits','keepticks')
47  ylabel('daily cases')
48  xlabel('dates')
```

Figure 6.22 MATLAB code designed to solve the coupled ODEs of a standard SIRB model and to plot the results obtained. The source file may be downloaded at www.echo.epfl/downloads (courtesy of Damiano Pasetto)

the overall probability of exposure outside the home site i) or elsewhere (with probability $m_i^S Q_{ij}$, where Q_{ij} represents the probability that water contacts taking place outside the home site i occur in site $j \neq i$, and $Q_{ii} = 0$). Other routes of infection, such as fast human-to-human transmission, which have been proposed for WB diseases like cholera, are here neglected for simplicity but could be dealt with within the same modeling framework.

The evolution of the infected compartment (second equation) quantifies the balance between newly infected individuals and losses due to recovery or natural/pathogen-induced mortality, with δ and γ being the rates of disease-induced mortality and recovery from infection, respectively. Note that the recovered compartment is not modeled explicitly, so that individuals who recover from the acute phase of disease are simply removed from the population, as they are conferred lifelong immunity to reinfection. It has been argued [453] that the loss of acquired immunity is unlikely to influence transient, short-term epidemic dynamics, which are the main focus

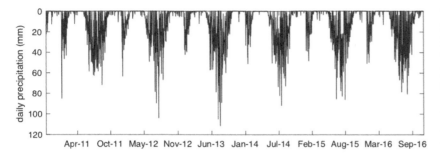

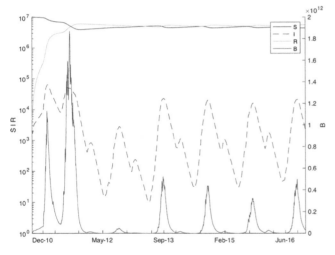

Figure 6.23 Output figure 1 of the code in Figure 6.22. (top) Generated realistic patterns of driving rainfall consisting of a seasonal trend and a random perturbation (code lines 39–42). (bottom) Computed total number of daily cases. Figure courtesy of Damiano Pasetto

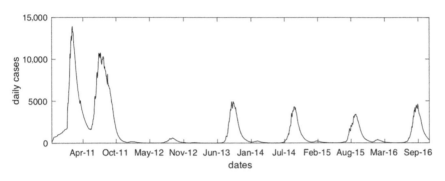

Figure 6.24 Output figure 2 of the code in Figure 6.22. Color coding enphasizes the variations in the pools of susceptibles ($S(t)$), infected ($I(t)$), and recovered ($R(t)$) individuals. Also shown is the evolution of the bacterial concentration ($B(t)$) in the water reservoir, driven by rainfall washout of newly shed bacteria from the actual infectives. Parameters are listed in the code (lines 10–20). Figure courtesy of Damiano Pasetto

of the present exercise. Although simplistic, the choice of neglecting the dynamics of recovered individuals is reasonable for the problem exemplified here.

The pathogen population is determined by the dynamics of the local concentrations of pathogens in the aquatic environment (third equation), given by a balance between water contamination, pathogen mortality, and hydrological transport. Pathogens are released in water (e.g., excreted) by infected individuals from either the local community, with probability $(1 - m_i^I)$, with m_i^I being the overall probability of contamination outside

the home site i, or elsewhere, with probability $m_j^I Q_{ji}$ at a site-specific rate p_i and immediately diluted in a well-mixed local water reservoir of size W_i (see, e.g., Box 4.5). Free-living pathogens are assumed to die at rate v_i. They can also move between any two neighboring nodes of the hydrological network (say, from i to j) at rate l_i and with probability P_{ij}. Some of the parameters of the sample model (namely, those related to human demography, μ, and the physiological response to the disease, δ, γ, K) are assumed to be constant over the spatial scales considered in this example. All the other parameters are allowed

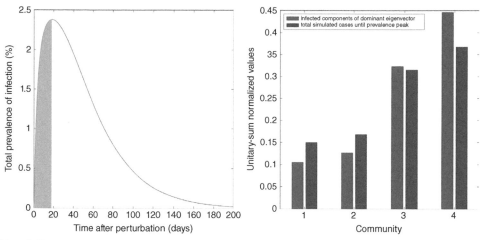

Figure 6.25 (left) Evolution of the prevalence of infection after a perturbation at time $t = 0$ of the disease-free equilibrium (DFE) of the mathematical problem stated in Equations (6.16)–(6.18). Parameters are assigned in the enclosed code. (right) A node-by-node comparison of the numerical solution for the total number of infections up to the prevalence peak (shaded area in [left]), and the infected component of the dominant eigenvector [453]. Figure courtesy of Lorenzo Mari

to be possibly site dependent. One notes that setting $m_i^S = m_i^I = 0$ and $l_i = 0$ for all i would produce a set of n disconnected models, as seen in the example in Section 6.3.1, that is, accounting only for local disease transmission processes.

As shown in Box 4.11, while conditions for endemicity are determined via stability analysis (Appendix 6.1), epidemicity thresholds may be defined via a generalized reactivity analysis, a method that allows us to study the nature and the possible damping of short-term instability properties of ecological systems subject to a pulse perturbation [454]. Understanding the drivers of waterborne and water-related disease dynamics over timescales that may be relevant to epidemic and/or endemic transmission has been argued to represent a challenge of most waterborne epidemiology applications [452]. Suffice here to recall that the generalized reactivity analysis requires the definition of a suitable linear output transformation $\mathbf{y} = \mathbf{C}\mathbf{x}$ for the linearized (or linear) system $d\mathbf{x}/dt = \mathbf{A}\mathbf{x}$, where $\mathbf{C}$ is a real, full-rank $m \times n$ ($m \leq n$) matrix defining a set of independent linear combinations of the system's state variables. A stable equilibrium point is defined as g-reactive if there exist small perturbations that can lead to a transient growth in the Euclidean norm of the output. Mari et al. [463] showed that the classification of a stable equilibrium point as g-reactive is based on the sign of the dominant eigenvalue of a matrix that can be easily obtained from the output and state matrices. Specifically, a stable steady state is g-reactive if

$$\lambda_{\max}\left[H(\mathbf{C}^T\mathbf{C}\mathbf{A})\right] > 0,$$

where $\lambda_{\max}(\mathbf{B})$ is, as usual, the maximum eigenvalue of the arbitrary matrix $\mathbf{B}$ and $H(\mathbf{C}^T\mathbf{C}\mathbf{A})$ is the Hermitian part of the matrix $\mathbf{C}^T\mathbf{C}\mathbf{A}$. The maximum eigenvalue is real because the matrix $H(\mathbf{C}^T\mathbf{C}\mathbf{A})$ is real and symmetric. If this condition is met, then some perturbations $\mathbf{x}_0$ exist that are temporarily amplified in the system output. These perturbations belong to the so-called g-reactivity basin (Box 4.11). Trajectories that originate in a neighborhood of the stable yet g-reactive steady state will generate a transient epidemic wave, while perturbations that lie outside the basin will monotonically decay in the system output without producing an outbreak. Therefore, initial conditions may indeed play a crucial role in the development of short-term epidemics. In this framework, all models of epidemic spread, whether spatially implicit or explicit, may be reanalyzed [452, 463], as shown by the attached codes. In the following, we provide MATLAB scripts for the evaluation of the stability and g-reactivity problems of the DFE of the system in Equations (6.16)–(6.18), and we perform numerical simulations of the model.

The first script computes the stability of the system and the leading eigenvalues for a simple network with four nodes. Figure 6.25 shows the graphical output of the script. The same script can be used to study more complex spatial settings. As an example, an OCN with ~ 400 nodes is used in Figure 6.26.

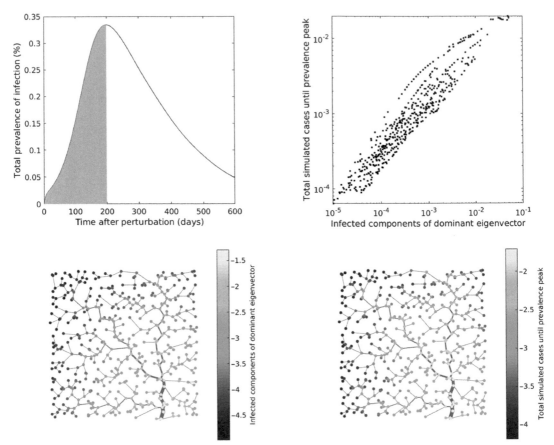

Figure 6.26 The same codes may be used to project the leading eigenvector on complex networks, in this case, OCNs (Section 6.2). (top left) Total prevalence, summed up for all nodes, where the shaded area represents the total number of cases up to the peak of the prevalence, to be computed also by direct integration of the model Equations (6.16)–(6.18) (bottom right). (top right) Scatterplot showing the close correlation of total simulated cases versus the values of the infection components of the dominant eigenvector. (bottom left) Spatially explicit plot of the color-coded infected components of the dominant eigenvector. Figure courtesy of Lorenzo Mari

```
1   % A script to evaluate the asymptotic stability properties of the
2   % disease-free equilibrium (DFE) of the spatially explicit
3   % Susceptible-Infected-Bacteria (SIB) model described in Chapter 4 based
4   % on its generalized reproduction matrix (G0)
5
6   clearvars
7   close all
8   clc
9
10  %%%%%%%%%%%%%%%%%%%%%%%%%%%%%%%%%%%%%%%%%%%%%%%%%%%%%%%%%%%%%%%%%%%%%%%%%%%%%%%
11  % baseline parameter values
12  % note: p is a data structure containing all model parameters
13  %%%%%%%%%%%%%%%%%%%%%%%%%%%%%%%%%%%%%%%%%%%%%%%%%%%%%%%%%%%%%%%%%%%%%%%%%%%%%%%
```

```
14
15  p.N = 1;                              % population  size  without  disease
16  p.mu = 1 / 65 / 365;                  % human  mortality  rate (1/day)
17  p.beta = 1;                           % exposure  rate (1/day)
18  p.delta = 4e-4;                       % disease-induced  mortality (1/day)
19  p.gamma = 1 / 5;                      % recovery  rate (1/day)
20  p.theta = 4e-2;                       % rescaled  contamination  rate (1/day)
21  p.W = 1;                              % water  reservoir  size
22  p.nu = 1 / 30;                        % pathogen  mortality  rate (1/day)
23  p.l = 1 / 3;                          % pathogen  transport  rate (1/day)
24  p.mS = 0.2;                           % mobility  of  susceptible  human  hosts
25  p.mI = 0.05;                          % mobility  of  infected  human  hosts
26
27  %%%%%%%%%%%%%%%%%%%%%%%%%%%%%%%%%%%%%%%%%%%%%%%%%%%%%%%%%%%%%%%%%%%%%%%%%%%%%
28  % hydrological  connectivity  and  human  mobility
29  % note: P and Q must be defined as (sub-)stochastic  matrices  of  size  n x n
30  %%%%%%%%%%%%%%%%%%%%%%%%%%%%%%%%%%%%%%%%%%%%%%%%%%%%%%%%%%%%%%%%%%%%%%%%%%%%%
31
32  p.n = 4;                              % number  of  communities
33  p.P = [0, 0, 1, 0;                    % hydrological  connectivity  matrix
34      0, 0, 1, 0;
35      0.05, 0.05, 0, 0.9;
36      0, 0, 0.1, 0];
37  p.Q = [0, 0.5, 0.4, 0.1;              % human  mobility  matrix
38      0.3, 0, 0.5, 0.2;
39      1 / 3, 1 / 3, 0, 1 / 3;
40      0.2, 0.2, 0.6, 0];
41
42  %%%%%%%%%%%%%%%%%%%%%%%%%%%%%%%%%%%%%%%%%%%%%%%%%%%%%%%%%%%%%%%%%%%%%%%%%%%%%
43  % note: here, all parameters are assumed to be spatially  homogeneous, yet
44  % this  script  can  be  used  also  if  some/all  parameters  are  spatially
45  % distributed; in that case, the community-specific  values  of  each
46  % spatially  heterogeneous  parameter  must  be  specified  as  the  nonzero
47  % elements  of  a  diagonal  matrix
48  %%%%%%%%%%%%%%%%%%%%%%%%%%%%%%%%%%%%%%%%%%%%%%%%%%%%%%%%%%%%%%%%%%%%%%%%%%%%%
49
50  U = diag(ones(p.n, 1));               % identity  matrix  of  size  n
51  Z = zeros(p.n);                       % zero  matrix  of  size  n
52  p.N = p.N * U; p.mu = p.mu * U;             % -|
53  p.beta = p.beta * U; p.delta = p.delta * U; %  |
54  p.gamma = p.gamma * U; p.theta = p.theta * U; %  | scalar  matrices
55  p.W = p.W * U; p.nu = p.nu * U; p.l = p.l * U; %  | of  size  n
56  p.mS = p.mS * U; p.mI = p.mI * U;           %  |
57  p.phi = p.mu + p.delta + p.gamma;           % -|
58
59  %%%%%%%%%%%%%%%%%%%%%%%%%%%%%%%%%%%%%%%%%%%%%%%%%%%%%%%%%%%%%%%%%%%%%%%%%%%%%
60  % asymptotic  stability  of  DFE
61  %%%%%%%%%%%%%%%%%%%%%%%%%%%%%%%%%%%%%%%%%%%%%%%%%%%%%%%%%%%%%%%%%%%%%%%%%%%%%
```

```matlab
62
63  % build G0 matrix at DFE
64  T0 = p.nu \ (p.W \ p.P' * p.W - U) * p.l;
65  R0 = (p.phi * p.nu) \ p.theta / p.W * (U - p.mI) * (U - p.mS) * p.N * p.beta;
66  R0_S = (p.phi * p.nu) \ p.theta / p.W * (U - p.mI) * p.mS * p.N * p.Q * p.beta
        ;
67  R0_I = (p.phi * p.nu) \ p.theta / p.W * p.Q' * p.mI * (U - p.mS) * p.N * p.
        beta;
68  R0_SI = (p.phi * p.nu) \ p.theta / p.W * p.Q' * p.mI * p.mS * p.N * p.Q * p.
        beta;
69  G0 = T0 + R0 + R0_S + R0_I + R0_SI;
70
71  % evaluate dominant eigenvalue of G0
72  [V, D] = eig(G0);                        % eigenvectors and eigenvalues of G0
73  [lambda_max_G0, maxind] = ...           % dominant eigenvalue of G0
74                     max(real(diag(D)));
75  display(lambda_max_G0);                  % DFE unstable if lambda_max_G0 > 1
76  clear D
77
78  % human-readable output
79  if lambda_max_G0 > 1
80      msg = 'The DFE is asymptotically unstable (lambda_max_G0 > 1)';
81  elseif lambda_max_G0 < 1
82      msg = 'The DFE is asymptotically stable (lambda_max_G0 < 1)';
83  else
84      msg = 'Asymptotic stability of the DFE cannot be decided (lambda_max_G0 =
            1)';
85  end
86  disp(msg)
87  clear msg
88
89  %%%%%%%%%%%%%%%%%%%%%%%%%%%%%%%%%%%%%%%%%%%%%%%%%%%%%%%%%%%%%%%%%%%%%%%%%%%%%%%
90  % analysis of the dominant eigenvector
91  %%%%%%%%%%%%%%%%%%%%%%%%%%%%%%%%%%%%%%%%%%%%%%%%%%%%%%%%%%%%%%%%%%%%%%%%%%%%%%%
92
93  g0 = abs(V(:, maxind));                  % dominant eigenvector of G0
94  B_eig = p.W \ g0;                        % bacterial components of g0
95  I_eig = p.phi \ ...                      % infected components of g0
96      ((U - p.mS) * p.N + p.mS * p.N * p.Q) * p.beta * B_eig;
97  clear V maxind g0
98
99  %%%%%%%%%%%%%%%%%%%%%%%%%%%%%%%%%%%%%%%%%%%%%%%%%%%%%%%%%%%%%%%%%%%%%%%%%%%%%%%
100 % model simulation
101 % note: here, a random bacterial perturbation to the DFE is used as initial
102 % condition for simulating the model; different choices can be specified
103 % with a different definition of x0
104 %%%%%%%%%%%%%%%%%%%%%%%%%%%%%%%%%%%%%%%%%%%%%%%%%%%%%%%%%%%%%%%%%%%%%%%%%%%%%%%
```

```
105
106  % timespan and initial condition
107  tspan = 0 : 200;                          % simulation timespan (days)
108  rng('default')                            % for reproducibility
109  x0 = [diag(p.N); zeros(p.n, 1);           % initial condition (4n x 1 vector)
110      rand(p.n, 1) / 1e2; zeros(p.n, 1)];
111  options = odeset('NonNegative', 1 : 4 * p.n);      % for positivity
112
113  % numerical integration
114  [t, x] = ode45(@(t, x) spatialSIB_ODEs(t, x, p), tspan, x0, options);
115  S_t_i = x(:, 1 : p.n);                     % susceptibles
116  I_t_i = x(:, p.n + 1 : 2 * p.n);           % infected hosts
117  B_t_i = x(:, 2 * p.n + 1 : 3 * p.n);       % bacteria
118  C_t_i = x(:, 3 * p.n + 1 : end);           % cumulated cases
119  clear x options
120
121  % comparing eigenvector with simulation
122  [~, ind_Imax] = max(sum(I_t_i, 2));        % prevalence peak
123  [r, v] = corrcoef(I_eig, C_t_i(ind_Imax, :));
124  r = r(1, 2); v = v(1, 2);                  % Pearson r
125  disp(['Outbreak geography: eigenvector vs simulation r = ', num2str(r), ...
126      ' (p-value = ', num2str(v), ')'])
127
128  % plotting
129  subplot(121); hold on
130  prev_t = sum(I_t_i,2) / sum(diag(p.N)) * 100;
131  plot(t, prev_t, 'k')
132  area(t(1 : ind_Imax), prev_t(1 : ind_Imax), ...
133      'FaceColor', [3/4 3/4 3/4], 'EdgeColor', 'n')
134  xlabel('Time after perturbation (days)')
135  ylabel('Total prevalence of infection (%)')
136  axis square; box on
137
138  subplot(122)
139  bar([I_eig / sum(I_eig), C_t_i(ind_Imax, :)' / sum(C_t_i(ind_Imax, :))])
140  xlabel('Community')
141  ylabel('Unitary-sum normalized values')
142  legend('infected components of dominant eigenvector', ...
143      'total simulated cases until prevalence peak', 'Location', 'northwest')
144  axis square; box on
145  clear ind_Imax
```

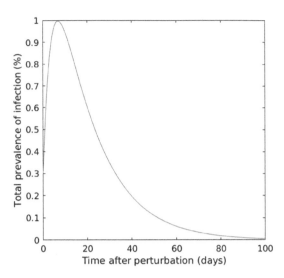

Figure 6.27 Evolution of the prevalence of infections after a perturbation at time $t = 0$ of the disease-free equilibrium (DFE) of the mathematical problem stated in Equation (6.16)–(6.18). Parameters are assigned in the enclosed code. Figure courtesy of Lorenzo Mari

The second script that we provide evaluates the generalized reactivity properties of the DFE of the model for the simple four-node examples already discussed.

Figure 6.27 shows the graphical output of the script, namely, the temporal dynamics of the infections in the metacommunity.

```
1   % A script to evaluate the short-term (in)stability properties of the
2   % disease-free equilibrium (DFE) of the spatially explicit
3   % Susceptible-Infected-Bacteria (SIB) model described in Chapter 4 based
4   % on generalized reactivity analysis
5
6   clearvars
7   close all
8   clc
9
10  %%%%%%%%%%%%%%%%%%%%%%%%%%%%%%%%%%%%%%%%%%%%%%%%%%%%%%%%%%%%%%%%%%%%%%%%%%%
11  % baseline parameter values
12  % note: p is a data structure containing all model parameters
13  %%%%%%%%%%%%%%%%%%%%%%%%%%%%%%%%%%%%%%%%%%%%%%%%%%%%%%%%%%%%%%%%%%%%%%%%%%%
14
15  p.N = 1;                          % population size without disease
16  p.mu = 1 / 65 / 365;              % human mortality rate (1/day)
17  p.beta = 1;                       % exposure rate (1/day)
18  p.delta = 4e-4;                   % disease-induced mortality (1/day)
19  p.gamma = 1 / 5;                  % recovery rate (1/day)
20  p.theta = 2e-2;                   % rescaled contamination rate (1/day)
```

```
21  p.W = 1;                              % water reservoir size
22  p.nu = 1 / 30;                        % pathogen mortality rate (1/day)
23  p.l = 1 / 3;                          % pathogen transport rate (1/day)
24  p.mS = 0.2;                           % mobility of susceptible human hosts
25  p.mI = 0.05;                          % mobility of infected human hosts
26
27  %%%%%%%%%%%%%%%%%%%%%%%%%%%%%%%%%%%%%%%%%%%%%%%%%%%%%%%%%%%%%%%%%%%%%%%%%%%%%%%
28  % hydrological connectivity and human mobility
29  % note: P and Q must be defined as (sub-)stochastic matrices of size n x n
30  %%%%%%%%%%%%%%%%%%%%%%%%%%%%%%%%%%%%%%%%%%%%%%%%%%%%%%%%%%%%%%%%%%%%%%%%%%%%%%%
31
32  p.n = 4;                              % number of communities
33  p.P = [0, 0, 1, 0;                    % hydrological connectivity matrix
34        0, 0, 1, 0;
35        0.05, 0.05, 0, 0.9;
36        0, 0, 0.1, 0];
37  p.Q = [0, 0.5, 0.4, 0.1;             % human mobility matrix
38        0.3, 0, 0.5, 0.2;
39        1 / 3, 1 / 3, 0, 1 / 3;
40        0.2, 0.2, 0.6, 0];
41
42  %%%%%%%%%%%%%%%%%%%%%%%%%%%%%%%%%%%%%%%%%%%%%%%%%%%%%%%%%%%%%%%%%%%%%%%%%%%%%%%
43  % output transformation
44  %%%%%%%%%%%%%%%%%%%%%%%%%%%%%%%%%%%%%%%%%%%%%%%%%%%%%%%%%%%%%%%%%%%%%%%%%%%%%%%
45
46  p.cI = 1;                            % weight of infected hosts
47  p.cB = 1;                            % weight of pathogen concentration
48
49  %%%%%%%%%%%%%%%%%%%%%%%%%%%%%%%%%%%%%%%%%%%%%%%%%%%%%%%%%%%%%%%%%%%%%%%%%%%%%%%
50  % note: here, all parameters are assumed to be spatially homogeneous, yet
51  % this script can be used also if some/all parameters are spatially
52  % distributed; in that case, the community-specific values of each
53  % spatially heterogeneous parameter must be specified as the nonzero
54  % elements of a diagonal matrix
55  %%%%%%%%%%%%%%%%%%%%%%%%%%%%%%%%%%%%%%%%%%%%%%%%%%%%%%%%%%%%%%%%%%%%%%%%%%%%%%%
56
57  U = diag(ones(p.n, 1));              % identity matrix of size n
58  Z = zeros(p.n);                      % zero matrix of size n
59  p.N = p.N * U; p.mu = p.mu * U;                      % -|
60  p.beta = p.beta * U; p.delta = p.delta * U;          %  |
61  p.gamma = p.gamma * U; p.theta = p.theta * U;        %  | scalar matrices
62  p.W = p.W * U; p.nu = p.nu * U; p.l = p.l * U;       %  | of size n
63  p.mS = p.mS * U; p.mI = p.mI * U;                    %  |
64  p.phi = p.mu + p.delta + p.gamma;                    % -|
65  C = [Z, p.cI * U, Z; Z, Z, p.cB * U]; % output matrix
66  clear p.cI p.cB
```

```
67
68  %%%%%%%%%%%%%%%%%%%%%%%%%%%%%%%%%%%%%%%%%%%%%%%%%%%%%%%%%%%%%%%%%%%%%%%%%%
69  % asymptotic stability of DFE
70  %%%%%%%%%%%%%%%%%%%%%%%%%%%%%%%%%%%%%%%%%%%%%%%%%%%%%%%%%%%%%%%%%%%%%%%%%%
71
72  % build Jacobian matrix at DFE
73  J0_13 = - ((U - p.mS) * p.N + p.mS * p.N * p.Q) * p.beta;
74  J0 = [- p.mu, Z, J0_13;
75        Z, - p.phi, - J0_13;
76        Z, p.theta / p.W * (U - p.mI + p.Q' * p.mI), ...
77        - p.nu - (U - p.W \ p.P' * p.W) * p.l];
78  clear J0_13
79
80  % evaluate dominant eigenvalue of J0
81  lambda_max_J0 = max(real(eig(J0)));      % dominant eigenvalue of J0
82  display(lambda_max_J0);                   % DFE unstable if lambda_max_J0 > 0
83
84  %%%%%%%%%%%%%%%%%%%%%%%%%%%%%%%%%%%%%%%%%%%%%%%%%%%%%%%%%%%%%%%%%%%%%%%%%%
85  % generalized reactivity
86  %%%%%%%%%%%%%%%%%%%%%%%%%%%%%%%%%%%%%%%%%%%%%%%%%%%%%%%%%%%%%%%%%%%%%%%%%%
87
88  % build Hermitian matrix at DFE
89  H0_temp = C * J0 * pinv(C);
90  H0 = (H0_temp + H0_temp') / 2;
91  clear H0_temp
92
93  % evaluate g-reactivity
94  [V, D] = eig(H0);                         % eigenvectors and eigenvalues of H0
95  [lambda_max_H0, maxind] = ...             % dominant eigenvalue of H0
96                       max(diag(D));
97  display(lambda_max_H0);                    % DFE g-reactive if lambda_max_H0 > 0
98
99  % human-readable output
100 if lambda_max_J0 > 0
101     msg = ['The DFE is asymptotically unstable (lambda_max_J0 > 0), ' ...
102            'thus also g-reactive (lambda_max_H0 > 0)'];
103 elseif lambda_max_J0 < 0
104     if lambda_max_H0 > 0
105         msg = ['The DFE is asymptotically stable (lambda_max_J0 < 0), ' ...
106                'yet g-reactive (lambda_max_H0 > 0)'];
107     else
108         msg = ['The DFE is asymptotically stable (lambda_max_J0 < 0), ' ...
109                'and non-g-reactive (lambda_max_H0 <= 0)'];
110     end
111 else
```

```
112      if  lambda_max_H0 > 0
113          msg = ['The DFE is g-reactive (lambda_max_H0 > 0), ' ...
114                  'but its asymptotic stability cannot be decided (lambda_max_J0
                         = 1)'];
115      else
116          msg = ['The DFE is non-g-reactive (lambda_max_H0 <= 0), ' ...
117                  'but its asymptotic stability cannot be decided (lambda_max_J0
                         = 1)'];
118      end
119  end
120  disp(msg)
121  clear msg
122
123  % optimal perturbation at time 0
124  opt_pert_t0 = abs(V(:, maxind));
125  clear V D maxind
126
127  %%%%%%%%%%%%%%%%%%%%%%%%%%%%%%%%%%%%%%%%%%%%%%%%%%%%%%%%%%%%%%%%%%%%%%%%%%%%%%
128  % model simulation
129  % note: here, the spatial signature of the optimal perturbation at time 0
130  % with an assigned total prevalence of disease in the human population is
131  % used as initial condition for simulating the model; different choices can
132  % be specified with a different definition of x0
133  %%%%%%%%%%%%%%%%%%%%%%%%%%%%%%%%%%%%%%%%%%%%%%%%%%%%%%%%%%%%%%%%%%%%%%%%%%%%%%
134
135  % timespan and initial condition
136  tspan = 0 : 100;                        % simulation timespan (days)
137  prev0 = 0.3;                            % initial total prevalence (%)
138  Ntot = sum(diag(p.N));                  % total population size
139  x0 = [diag(p.N);                        % initial condition (3 n x 1 vector)
140        opt_pert_t0 / sum(opt_pert_t0(1 : p.n)) * prev0 / 100 * Ntot;
141        zeros(p.n, 1)];
142  options = odeset('NonNegative', 1 : 4 * p.n);      % for positivity
143
144  % numerical integration
145  [t, x] = ode45(@(t, x) spatialSIB_ODEs(t, x, p), tspan, x0, options);
146  S_t_i = x(:, 1 : p.n);                  % susceptibles
147  I_t_i = x(:, p.n + 1 : 2 * p.n);        % infected hosts
148  B_t_i = x(:, 2 * p.n + 1 : 3 * p.n);    % bacteria
149  C_t_i = x(:, 3 * p.n + 1 : end);        % cumulated cases
150  clear x options
151
152  % plotting
153  plot(t, sum(I_t_i,2) / Ntot * 100)
154  xlabel('Time after perturbation (days)')
155  ylabel('Total prevalence of infection (%)')
156  axis square; box on
```

A third MATLAB file is needed to carry out the numerical integration of the system of Equations (6.16)–(6.18). This MATLAB function is used by the two scripts listed above and can be used for systems characterized by an arbitrary level of spatial complexity.

```matlab
function dx = spatialSIB_ODEs(~, x, p)
% function to evaluate the spatially explicit ordinary differential
% equations (ODEs) of the Susceptible-Infected-Bacteria (SIB) model
% described in Chapter 4

    % state variables (n x 1 vectors)
    S = x(1 : p.n);                 % susceptible population
    I = x(p.n + 1 : 2 * p.n);       % infected population
    B = x(2 * p.n + 1 : 3 * p.n);   % pathogen concentration

    % evaluating some useful quantities
    U = diag(ones(p.n, 1));
    recruitment = diag(p.mu * p.N);
    FoI = diag((U - p.mS + p.mS * p.Q) * p.beta * (B ./ (1 + B)));
    hydrotransport = (U - p.W \ p.P' * p.W) * p.l;
    contamination = p.theta / p.W * (U - p.mI + p.Q' * p.mI);

    % SIB model equations
    dSdt = recruitment - (p.mu + FoI) * S;
    dIdt = FoI * S - p.phi * I;
    dBdt = contamination * I - (p.nu + hydrotransport) * B;
    dCdt = FoI * S;                 % cumulated cases

    % assembling the function output
    dx = [dSdt; dIdt; dBdt; dCdt];
end
```

References

[1] Dasgupta, P. The nature of economic development and the economic development of nature. *Economic & Political Weekly 47* (2013), 38–51.

[2] Levin, S. The problem of pattern and scale in ecology. *Ecology 76* (1992), 1943–1963.

[3] Southwood, T., May, R., and Sugihara, G. Observations on related ecological exponents. *Proceedings of the National Academy of Sciences of the USA 103*, 1 (2006), 6931–6933.

[4] Banavar, J., Damuth, J., Maritan, A., and Rinaldo, A. Scaling in ecosystems and the linkage of macroecological laws. *Physical Review Letters 98*, 1 (2007), 068104.

[5] Newman, M. Power laws, Pareto distributions and Zipf's law. *Contemporary Physics 46*, 5 (2005), 323–351.

[6] Fagan, W. Connectivity, fragmentation, and extinction risk in dendritic metapopulations. *Ecology 83*, 12 (2002), 3243–3249.

[7] Bertuzzo, E., Maritan, A., Gatto, M., Rodriguez-Iturbe, I., and Rinaldo, A. River networks and ecological corridors: reactive transport on fractals, migration fronts, hydrochory. *Water Resources Research 43* (2007), W04419.

[8] Muneepeerakul, R., Bertuzzo, E., Lynch, H., Fagan, W., Rinaldo, A., and Rodriguez-Iturbe, I. Neutral metacommunity models predict fish diversity patterns in the Mississippi–Missouri basin. *Nature 453*, 7192 (2008), 220–224.

[9] Bertuzzo, E., Suweis, S., Mari, L., Maritan, A., Rodriguez-Iturbe, I., and Rinaldo, A. Spatial effects for species persistence and implications for biodiversity. *Proceedings of the National Academy of Sciences of the USA 108*, 11 (2011), 4346–4351.

[10] Chisholm, C., Lindo, Z., and Gonzalez, A. Metacommunity diversity depends on network connectivity arrangement in heterogeneous landscapes. *Ecography 34* (2011), 415–424.

[11] Ricklefs, R. Community diversity – relative roles of local and regional processes. *Science 235*, 4785 (1987), 167–171.

[12] Holyoak, M., and Lawler, S. The contribution of laboratory experiments on protists to understanding population and metapopulation dynamics. In *Advances in Ecological Research 37: Population Dynamics and Laboratory Ecology*. vol. 37. Academic Press, 2005, pp. 245–271.

[13] Altermatt, F., Bieger, A., Carrara, F., Rinaldo, A., and Holyoak, M. Effects of connectivity and recurrent local disturbances on community

structure and population density in experimental metacommunities. *PLoS ONE 6*, 4 (2011), e19525.

[14] Carrara, F., Altermatt, F., Rodriguez-Iturbe, I., and Rinaldo, A. Dendritic connectivity controls biodiversity patterns in experimental metacommunities. *Proceedings of the National Academy of Sciences of the USA 109* (2012), 5761–5766.

[15] Giometto, A., et al. Generalized receptor law governs phototaxis in the phytoplankton *Euglena gracilis. Proceedings of the National Academy of Sciences of the USA 112* (2015), 7045–7050.

[16] Giometto, A., Altermatt, F., and Rinaldo, A. Demographic stochasticity and resource autocorrelation control biological invasions in heterogeneous landscapes. *Oikos 126* (2017), 1554–1563.

[17] Morrissey, M., and de Kerckhove, D. The maintenance of genetic variation due to asymmetric gene flow in dendritic metapopulations. *American Naturalist 174* (2009), 875–889.

[18] Clarke, A., Mac Nally, R., Bond, N., and Lake, P. Macroinvertebrate diversity in headwater streams: a review. *Freshwater Biology 53* (2008), 1707–1721.

[19] Brown, B., and Swan, C. Dendritic network structure constrains metacommunity properties in riverine ecosystems. *Journal of Animal Ecology 79* (2010), 571–580.

[20] Finn, D., Bonada, N., Murria, C., and Hughes, J. Small but mighty: headwaters are vital to stream network biodiversity at two levels of organization. *Journal of the North American Benthological Society 30* (2011), 963–980.

[21] Rodriguez-Iturbe, I., Muneepeerakul, R., Bertuzzo, E., Levin, S., and Rinaldo, A. River networks as ecological corridors: a complex systems perspective for integrating hydrologic, geomorphologic, and ecologic dynamics. *Water Resources Research 45* (2009), W01413.

[22] Ammermann, A., and Cavalli-Sforza, L. *The Neolithic Transition and the Genetics of Population in Europe*. Princeton University Press.

[23] Battin, T., Kaplan, L., Newbold, J., and Hansen, C. Contributions of microbial biofilms to ecosystem processes in stream mesocosms. *Nature 426* (2003), 439–442.

[24] Battin, T., et al. Biophysical controls on organic carbon fluxes in fluvial networks. *Nature Geosciences 1* (2008), 95–100.

[25] Rinaldo, A., Bertuzzo, E., Blokesch, M., Mari, L., and Gatto, M. Modeling key drivers of cholera transmission dynamics provides new perspectives on parasitology. *Trends in Parasitology 33*, 8 (2017), 587–599.

[26] Raymond, P., and Bauer, J. Riverine export of aged terrestrial organic matter to the north atlantic ocean. *Nature 409* (2001), 497–501.

[27] Alexander, R., Jones, P., Boyer, E., and Smith, R. Effect of stream channel size on the delivery of nitrogen to the Gulf of Mexico. *Nature 403* (2000), 758–761.

[28] D'Odorico, P., Laio, F., Porporato, A., Ridolfi, L., Rinaldo, A., and Rodriguez-Iturbe, I. Ecohydrology of terrestrial ecosystems. *BioScience 11* (2010), 898–907.

[29] Mari, L., Bertuzzo, E., Casagrandi, R., Gatto, M., Levin, S. A., Rodriguez-Iturbe, I., and Rinaldo, A. Hydrologic controls and anthropogenic drivers of the zebra mussel invasion of the Mississippi–Missouri river system. *Water Resources Research 47* (2011), W03523.

[30] Bertuzzo, E., et al. Prediction of the spatial evolution and effects of control measures for the unfolding Haiti cholera outbreak. *Geophysical Research Letters 38* (2011), L06403.

[31] Pielou, E. *Mathematical Ecology*. Wiley, 1977.

[32] Vellend, M. Conceptual synthesis in community ecology. *Quarterly Review of Biology 85*, 2 (2010), 183–206.

[33] Caswell, H. Community structure: a neutral model analysis. *Ecological Monographs 46*, 3 (1976), 327–354.

[34] Hubbell, S. *The Unified Theory of Biodiversity and Biogeography*. Princeton University Press, 2001.

[35] Economo, E., and Keitt, T. Species diversity in neutral metacommunities: a network approach. *Ecology Letters 11*, 1 (2008), 52–62.

[36] Nee, S. The neutral theory of biodiversity: do the numbers add up? *Functional Ecology 19*, 1 (2005), 173–176.

[37] McGill, B., Maurer, B., and Weiser, M. Empirical evaluation of neutral theory. *Ecology 87*, 6 (2006), 1411–1423.

[38] Purves, D., and Turnbull, L. Different but equal: the implausible assumption at the heart of neutral

theory. *Journal of Animal Ecology 79*, 6 (2010), 1215–1225.

[39] Chave, J., Muller-Landau, H., and Levin, S. Comparing classical community models: theoretical consequences for patterns of diversity. *American Naturalist 159*, 1 (2002), 1–23.

[40] Muneepeerakul, R., Weitz, J., Levin, S., Rinaldo, A., and Rodriguez-Iturbe, I. A neutral metapopulation model of biodiversity in river networks. *Journal of Theoretical Biology 245* (2007), 351–363.

[41] Kimura, M. Evolutionary rate at the molecular level. *Nature 217* (1968), 624–626.

[42] Kimura, M. *The Neutral Theory of Molecular Evolution*. Cambridge University Press, 1983.

[43] Hubbell, S. Tree dispersion, abundance and diversity in a tropical dry forest. *Science 203* (1979), 1299–1309.

[44] Kendall, D. On some modes of population growth leading to R. A. Fisher's logarithmic series distribution. *Biometrika 35*, 1/2 (1948), 6–15.

[45] Karlin, S., and McGregor, J. The number of mutant forms maintained in a population. *Proceedings of the 5th Berkeley Symposium of Mathematical Statistics and Probability 4* (1967), 415–438.

[46] Watterson, G. The sampling theory of selectively neutral alleles. *Advances in Applied Probability 6*, 3 (1974), 463–488.

[47] Magurran, A. *Ecological Diversity and Its Measurement*. Princeton University Press, 1988.

[48] Fisher, R., Corbet, A., and Williams, C. The relation between the number of species and the number of individuals in a random sample of an animal population. *Journal of Animal Ecology 12* (1943), 42–58.

[49] Volkov, I., Banavar, J., Hubbell, S., and Maritan, A. Neutral theory and relative species abundance in ecology. *Nature 424*, 6952 (2003), 1035–1037.

[50] Lande, R. Risks of population extinction from demographic and environmental stochasticity and random catastrophes. *American Naturalist 142* (1993), 911–927.

[51] Purves, D., and Pacala, S. Ecological drift in niche-structured communities: Neutral pattern does not imply neutral process. In *Biotic Interactions in the Tropics*, D. Burslem, M. Pinard, and S. Hartley, eds. Cambridge University Press, 2005, pp. 107–138.

[52] Condit, R. Beta-diversity in tropical forest trees. *Science 295* (2002), 666–669.

[53] Holt, R. Emergent neutrality. *Trends in Ecology and Evolution 21* (2006), 451–457.

[54] McGill, B. A test of the unified neutral theory of biodiversity. *Nature 422* (2003), 881–885.

[55] Chave, J. Neutral theory and community ecology. *Ecology Letters 7* (2004), 241–253.

[56] Dornelas, M., Connolly, S., and Hughes, T. Coral reef diversity refutes the neutral theory of biodiversity. *Nature 440* (2006), 80–82.

[57] Etienne, R., and Olff, H. Confronting different models of community structure to species-abundance data: a Bayesian model comparison. *Ecology Letters 8* (2005), 493–504.

[58] Volkov, I., Banavar, J., Hubbell, S., and Maritan, A. Patterns of relative species abundance in ecology. *Nature 450* (2007), 45–49.

[59] Guisan, A., Thuiller, W., and Zimmermann, N. *Habitat Suitability and Distribution Models*. Cambridge University Press, 2018.

[60] Diamond, J. The present, past and future of human-caused extinctions. *Philosophical Transactions of the Royal Society B 325*, 1228 (1989), 469–477.

[61] Burness, G., Diamond, J., and Flannery, T. Dinosaurs, dragons, and dwarfs: the evolution of maximal body size. *Proceedings of the National Academy of Sciences of the USA 98* (2001), 14518–14523.

[62] Grant, P., and Grant, B. The secondary contact phase of allopatric speciation in Darwin finches. *Proceedings of the National Academy of Sciences of the USA 106* (2009), 20141–20148.

[63] Rodriguez-Iturbe, I., and Rinaldo, A. *Fractal River Basins: Chance and Self-Organization*. Cambridge University Press, 2001.

[64] Rinaldo, A., Rigon, R., Banavar, J., Maritan, A., and Rodriguez-Iturbe, I. Evolution and selection of river networks: statics, dynamics, and complexity. *Proceedings of the National Academy of Sciences of the USA 111*, 7 (2014), 2417–2424.

[65] Rodriguez-Iturbe, I., Rinaldo, A., Rigon, R., Bras, R., Ijjasz-Vasquez, E., and Marani, A. Fractal structures as least energy patterns – the case of river networks. *Geophysical Research Letters 19*, 9 (1992), 889–892.

[66] Rinaldo, A., Rodriguez-Iturbe, I., Rigon, R., Bras, R., Vasquez, E., and Marani, A. Minimum energy

and fractal structures of drainage networks. *Water Resources Research 28*, 9 (1992), 2183–2195.

[67] Rinaldo, A., Rigon, R., and Rodriguez-Iturbe, I. Channel networks. *Annual Review of Earth and Planetary Sciences 26* (1999), 289–306.

[68] Kirchner, J. Statistical inevitability of Horton's laws and the apparent randomness of stream channel networks. *Geology 21* (1993), 591–594.

[69] Rinaldo, A., Banavar, J., and Maritan, A. Trees, networks and hydrology. *Water Resources Research 42* (2006), 88–93.

[70] Montgomery, D., and Dietrich, W. Where do channels begin? *Nature 336* (1988), 232–234.

[71] Montgomery, D., and Dietrich, W. Channel initiation and the problem of landscape scale. *Science 255* (1992), 826–830.

[72] Briggs, L., and Krishnamoorthy, M. Exploring network scaling through variations on optimal channel networks. *Proceedings of the National Academy of Sciences of the USA 110* (2013), 19295–19300.

[73] Rigon, R., Rinaldo, A., and Rodriguez-Iturbe, I. On landscape self-organization. *Journal of Geophysical Research 99* (1994), 11971–11993.

[74] Banavar, J., Colaiori, F., Flammini, A., Maritan, A., and Rinaldo, A. Scaling, optimality, and landscape evolution. *Journal of Statistical Physics 104* (2001), 1–48.

[75] Balister, P., et al. River landscapes and optimal channel networks. *Proceedings of the National Academy of Sciences of the USA 115* (2018), 6548–6553.

[76] Whittaker, R. H. Evolution and measurement of species diversity. *Taxon 21*, 2 (1972), 213–251.

[77] Anderson, M., et al. Navigating the multiple meanings of β-diversity: a roadmap for the practicing ecologist. *Ecology Letters 14*, 1 (2011), 19–28.

[78] Grant, E., Lowe, W., and Fagan, W. Living in the branches: population dynamics and ecological processes in dendritic networks. *Ecology Letters 10* (2007), 165–175.

[79] Luczkovich, J., Borgatti, S., Johnson, J., and Everett, M. Defining and measuring trophic role similarity in food webs using regular equivalence. *Journal of Theoretical Biology 220* (2003), 303–321.

[80] MacArthur, R., and Wilson, E. *The Theory of Island Biogeography*. Princeton University Press, 1967.

[81] Levins, R. Some demographic and genetic consequences of environmental heterogeneity for biological control. *Bulletin of the Entomogical Society of America 15* (1969), 237–240.

[82] Jablonski, D. Extinction and the spatial dynamics of biodiversity. *Proceedings of the National Academy of Sciences of the USA 105* (2008), 11528–11535.

[83] Chandrasekhar, S. Stochastic problems in physics and astronomy. *Reviews of Modern Physics 15*, 1 (1943), 1–89.

[84] Durrett, R., and Levin, S. Spatial models for species-area curves. *Journal of Theoretical Biology 179*, 2 (1996), 119–127.

[85] Pigolotti, S., Flammini, A., Marsili, M., and Maritan, A. Species lifetime distribution for simple models of ecologies. *Proceedings of the National Academy of Sciences of the USA 102* (2005), 147–151.

[86] Brown, J. *Macroecology*. University of Chicago Press, 1995.

[87] Kerr, B., Riley, M., Feldman, M., and Bohannan, B. Local dispersal promotes biodiversity in a real-life game of rock-paper-scissors. *Nature 418*, 6894 (2002), 171–174.

[88] Tilman, D., May, R., Lehman, C., and Nowak, M. Habitat destruction and the extinction debt. *Nature 371*, 6492 (1994), 65–66.

[89] Suweis, S., Bertuzzo, E., Mari, L., Maritan, A., and Rinaldo, A. On species persistence-time distributions. *Journal of Theoretical Biology 303* (2012), 15–24.

[90] Keitt, T., and Stanley, H. Dynamics of North American breeding bird populations. *Nature 393*, 6682 (1998), 257–260.

[91] Department of the Interior, Geological Survey Patuxent Wildlife Research Center. *North American breeding bird survey*. www.pwrc.usgs.gov/bbs (2008).

[92] Adler, P., Tyburczy, W., and Lauenroth, W. Long-term mapped quadrats from Kansas prairie: demographic information for herbaceous plants. *Ecology 88*, 10 (2007), 2673–2673.

[93] Rodriguez-Iturbe, I., Cox, D., and Isham, V. Some models for rainfall based on stochastic

point-processes. *Proceedings of the Royal Society A 410*, 1839 (1987), 269–288.

[94] de Aguiar, M., Baranger, M., Baptestini, E. M., Kaufman, L., and Bar-Yam, Y. Global patterns of speciation and diversity. *Nature 460*, 7253 (2009), 384–387.

[95] Moyle, P., and Chech, J. *An Introduction to Ichthyology*, 5th ed. Benjamin Cummings, 2003.

[96] Benda, L., et al. The network dynamics hypothesis: how channel networks structure riverine habitats. *Bioscience 54* (2004), 384–U98.

[97] Harvey, E., Gounand, I., Fronhofer, E., and Altermatt, F. Population turnover reverses classic island biogeography predictions in river-like landscapes. *BioRXiv* (2018), 1–28.

[98] Srivastava, D., et al. Are natural microcosms useful model systems for ecology? *Trends in Ecology and Evolution 19* (2004), 379–384.

[99] Altermatt, F., Schreiber, S., and Holyoak, M. Interactive effects of disturbance and dispersal directionality on species richness and composition in metacommunities. *Ecology 92* (2011), 859–870.

[100] Carrara, F., Rinaldo, A., Giometto, A., and Altermatt, F. Complex interaction of dendritic connectivity and hierarchical patch size on biodiversity in river-like landscapes. *American Naturalist 183*, *1* (2014), 13–25.

[101] Leopold, L., Wolman, M., and Miller, J. *Fluvial Processes in Geomorphology*. Freeman, 1964.

[102] Zaoli, S., Giometto, A., Maritan, A., and Rinaldo, A. Covariations in ecological scaling laws fostered by community dynamics. *Proceedings of the National Academy of Sciences of the USA 114* (2017), 10672–10677.

[103] Maritan, A., Rinaldo, A., Rodriguez-Iturbe, I., Rigon, R., and Giacometti, A. Scaling in river networks. *Physical Review E 53* (1996), 1501–1513.

[104] Cardinale, B. Biodiversity improves water quality through niche partitioning. *Nature 472* (2011), 86–91.

[105] Erlander, S., and Stewart, N. F. *The Gravity Model in Transportation Analysis – Theory and Extensions*. VSP Books, 1990.

[106] Altermatt, F., Pajunen, V., and Ebert, D. Climate change affects colonization dynamics in a metacommunity of three daphnia species. *Global Change Biology 14* (2008), 1209–1220.

[107] De Bie, T., De Meester, L., Brendonck, L., and Martens, K. E. Body size and dispersal mode as key traits determining metacommunity structure of aquatic organisms. *Ecology 15* (2012), 740–747.

[108] Wilson, J., Dormontt, E., Prentis, P., Lowe, A., and Richardson, D. Something in the way you move: dispersal pathways affect invasion success. *Trends in Ecology and Evolution 24* (2009), 136–144.

[109] Gonzalez, A., Lawton, J., Gilbert, F., Blackburn, T., and Evans-Freke, I. Metapopulation dynamics, abundance, and distribution in a microecosystem. *Science 281* (1998), 2045–2047.

[110] Gonzalez, A., Rayfield, B., and Lindo, Z. The disentangled bank: how loss of habitat fragments and disassembles ecological networks. *American Journal of Botany 98* (2011), 503–516.

[111] Matthiessen, B., and Hillebrand, H. Dispersal frequency affects local biomass production by controlling local diversity. *Ecology Letters 9* (2006), 652–662.

[112] Mouquet, N., and Loreau, M. Community patterns in source/sink metacommunities. *American Naturalist 162* (2003), 544–557.

[113] Vannote, R. L., Minshall, G., Cummins, K., Sedell, J., and Cushing, C. River continuum concept. *Canadian Journal of Fisheries and Aquatic Sciences 37* (1998), 130–137.

[114] Haddad, N. M., Holyoak, M., Mata, T., Davies, K, F., Melbourne, A., and Preston, K. Species traits predict the effects of disturbance and productivity on diversity. *Ecology Letters 11* (2008), 348–356.

[115] Mari, L., Casagrandi, R., Bertuzzo, E., Rinaldo, A., and Gatto, M. Metapopulation persistence and species spread in river networks. *Ecology Letters 17*, *4* (2014), 426–434.

[116] Cadotte, M., Mai, D., Jantz, S., Collins, M., Keele, M., and Drake, J. On testing the competition-colonization trade-off in a multispecies assemblage. *American Naturalist 168* (2006), 704–709.

[117] Livingston, G., et al. Competition–colonization dynamics in experimental bacterial metacommunities. *Nature Communications 3* (2012), doi:10.1038/ncomms2239.

[118] Campos, D., Fort, J., and Mendez, V. Transport on fractal river networks: application to migration

fronts. *Theoretical Population Biology 69* (2006), 88–93.

[119] Kolmogorov, A., Petrovsky, I., and Piscounov, N. Étude de l'équation de la diffusion avec croissance de la quantité de matiére et son application á un probléme biologique. *Moscow University Mathematics Bulletin 1* (1937), 1–25.

[120] Murray, J. *Mathematical Biology I: An Introduction.* Springer, 2004.

[121] Mandelbrot, B. *The Fractal Geometry of Nature.* Henry Holt, 1983.

[122] Fort, J., and Méndez, V. Time-delayed theory of the neolithic transition in Europe. *Physical Review Letters 82* (1999), 867–870.

[123] Flanders, S. *Atlas of American Migration.* Facts of Life, 1988.

[124] Faragher, J. *Women and Men on the Overland Trail.* Yale University Press, 1979.

[125] Fisher, R. The wave of advance of advantageous genes. *Annals of Eugenics 7* (1937), 355–369.

[126] Marani, A., Rigon, R., and Rinaldo, A. A note on fractal channel networks. *Water Resources Research 27* (1991), 3041–3049.

[127] Colaiori, F., Flammini, A., Banavar, J., and Maritan, A. Analytical and numerical study of optimal channel networks. *Physical Review E 55* (1997), 1298–1310.

[128] Peano, G. Sur une courbe qui remplit toute une aire plane. *Math. Ann. 36* (1890), 157–160.

[129] Méndez, V., Campos, D., and Fedotov, S. Front propagation in reaction-dispersal models with finite jump speed. *Physical Review E 70* (2004), 036121.

[130] Méndez, V., Campos, D., and Fedotov, S. Analysis of front in reaction-dispersal processes. *Physical Review E 70* (2004), 066129.

[131] Campos, D., and Mendez, V. Reaction-diffusion wavefronts on comblike structures. *Physical Review E 71* (2005), 31–39.

[132] Hughes, B. *Random Walks and Random Environments.* Random Walks 1. Oxford University Press, 1995.

[133] Mathan, O., and Havlin, S. Mean first-passage time on loopless aggregates. *Physical Review A 40* (1990), 6573–6579.

[134] Van den Broeck, C. Waiting time for random walks on regular and fractal lattices. *Physical Review Letters 62* (1989), 1421–1424.

[135] Flammini, A., and Colaiori, F. Exact analysis of the Peano basin. *Journal of Physics A: Mathematical and General 29* (1996), 6701–6708.

[136] Lotka, A. *Elements of Mathematical Biology.* Dover, 1956.

[137] Van den Broeck, C. A new sample of males linked from the public use microdata sample of the 1850 US federal census of population to the 1860 US federal census. *Historical Methods 29* (1996), 41–156.

[138] Ackland, G., Signizer, M., Stratford, K., and Cohen, M. Cultural hitchhiking on the wave of advance of beneficial technologies. *Proceedings of the National Academy of Sciences of the USA 104* (2007), 8714–8719.

[139] Fang, Y., and Jawitz, J. The evolution of human population distance to water in the USA from 1790 to 2010. *Nature Communications 10* (2019), 1–8.

[140] NatureServe. *Distribution of native US fishes by watershed.* Tech. rep., USGS, 2004.

[141] Seaber, P., Kapinos, F., and Knapp, G. *Distribution of native US fishes by watershed.* Tech. rep., USGS, 2004.

[142] Guegan, J., Lek, S., and Oberdorff, T. Energy availability and habitat heterogeneity predict global riverine fish diversity. *Nature 39* (1998), 382–384.

[143] Angermeier, P., and Winston, M. Local vs. regional influences on local diversity in stream fish communities of virginia. *Ecology 79* (1998), 911–927.

[144] Oberdorff, T., Guegan, J., and Hugueny, B. Global scale patterns of fish species richness in rivers. *Ecography 18* (1995), 345–352.

[145] Levin, S., Muller-Landau, H., Nathan, R., and Chave, J. The ecology and evolution of seed dispersal: a theoretical perspective. *Annual Review of Ecology and Systematics 34* (2003), 575–604.

[146] Gebert, W., Graczyk, D., and Krug, W. *Average Annual Runoff in the United States.* Tech. rep., USGS, 1987.

[147] Akaike, H. A new look at the statistical model identification. *IEEE Trans. Automat. Control 19* (1974), 716–723.

[148] Corani, G., and Gatto, M. Structural risk minimization: a robust method for density-dependence detection and model selection. *Ecography 30*, 2 (2007), 400–416.

[149] Casagrandi, R., and Gatto, M. A persistence criterion for metapopulations. *Theoretical Population Biology 61* (2002), 115–125.

[150] Campbell Grant, E., Nichols, J., Lowe, W., and Fagan, W. Use of multiple dispersal pathways facilitates amphibian persistence in stream networks. *Proceedings of the National Academy of Sciences of the USA 107* (2010), 6936–6940.

[151] Hack, J. Studies of longitudinal profiles in Virginia and Maryland. *US Geological Survey Professional Paper* 294-B (1957), 1–21.

[152] Battin, T., et al. The boundless carbon cycle. *Nature Geosciences 2* (2009), 598–600.

[153] Poff, N., et al. The natural flow regime. *Bioscience 47* (1997), 760–784.

[154] Ceola, S., Bertuzzo, E., Dinger, G., Battin, T., Montanari, A., and Rinaldo, A. Hydrologic variability affects invertebrate grazing on phototrophic biofilms in stream microcosms. *PLoS ONE 8* (2014), e60629.

[155] Ceola, S., Bertuzzo, E., Dinger, G., Battin, T., Montanari, A., and Rinaldo, A. Hydrologic controls on basin-scale distribution of benthic invertebrates. *Water Resources Research 50* (2014), W015112.

[156] Poff, N., Olden, J., Merritt, D., and Pepin, D. Homogenization of regional river dynamics by dams and global biodiversity implications. *Proceedings of the National Academy of Sciences of the USA 104* (2007), 5732–5737.

[157] Botter, G., Basso, S., Porporato, A., Rodriguez-Iturbe, I., and Rinaldo, A. Natural streamflow regime alterations: damming of the Piave river basin (Italy). *Water Resources Research 46* (2010), W06522.

[158] Kupferberg, S., et al. Effects of flow regimes altered by dams on survival, population declines, and range-wide losses of California river-breeding frogs. *Conservation Biology 26* (2012), 513–524.

[159] Botter, G., Basso, S., Rodriguez-Iturbe, I., and Rinaldo, A. Resilience of river flow regimes. *Proceedings of the National Academy of Sciences of the USA 110* (2013), 12925–12930.

[160] Botter, G., Porporato, A., Rodriguez-Iturbe, I., and Rinaldo, A. Basin-scale soil moisture dynamics and the probabilistic characterization of carrier hydrologic flows: slow, leaching-prone components of the hydrologic response. *Water Resources Research 43* (2007), W06404.

[161] Rodriguez-Iturbe, I., Cox, D., and Isham, V. Probabilistic modelling of water balance at a point: the role of climate, soil and vegetation. *Proceedings of the Royal Society A 455* (1999), 3789–3805.

[162] Bertuzzo, E., Helton, A., Hall, R., and Battin, T. Scaling of dissolved organic carbon removal in river networks. *Advances in Water Resources 110* (2018), 136–146.

[163] Widder, S., et al. Fluvial network organization imprints on microbial co-occurrence networks. *Proceedings of the National Academy of Sciences of the USA 111* (2014), 12799–12804.

[164] Campbell Grant, E. Structural complexity, movement bias, and metapopulation extinction risk in dendritic ecological networks. *Journal of the North American Benthological Society 30* (2011), 252–258.

[165] Speirs, D., and Gurney, W. Population persistence in rivers and estuaries. *Ecology 82* (2001), 1219–1237.

[166] Lutscher, F., Nisbet, R., and Pachepsky, E. Population persistence in the face of advection. *Theoretical Ecology 3* (2010), 271–284.

[167] Müller, K. *Investigations on the organic drift in North Swedish streams*. Tech. rep., Institute of Freshwater Research, Drottningholm. 133–148 pages.

[168] Müller, K. The colonization cycle of freshwater insects. *Oecologia 53* (1982), 202–207.

[169] Waters, T. The drift of stream insects. *Annual Review of Entomology 17* (1972), 253–272.

[170] Reynolds, C., Carling, P., and Beven, K. Flow in river channels: new insights into hydraulic retention. *Archiv für Hydrobiologie 121* (1991), 171–179.

[171] Lancaster, J., and Hildrew, A. Characterising instream flow refugia. *Canadian Journal of Fisheries and Aquatic Science 50* (1993), 1663–1675.

[172] Lancaster, J., and Hildrew, A. Flow refugia and the microdistribution of lotic macroinvertebrates. *Journal of the North American Benthological Society 12* (1993), 385–393.

[173] Fischer, H., List, N., Koh, R., Imberger, J., and Brooks, N. *Mixing in Inland and Coastal Waters*. Academic Press, 1979.

[174] Rinaldo, A., Marani, A., and Rigon, R. Geomorphological dispersion. *Water Resources Research 27* (1991), 513–525.

[175] Peterson, E., et al. Modelling dendritic ecological networks in space: an integrated network perspective. *Ecology Letters 16* (2013), 707–719.

[176] Pachepsky, E., Lutscher, F., Nisbet, R., and Lewis, M. Persistence, spread and the drift paradox. *Theoretical Population Biology 67* (2005), 61–73.

[177] Lutscher, F., Pachepsky, E., and Lewis, M. The effect of dispersal patterns on stream populations. *SIAM Journal on Applied Mathematics 65* (2005), 1305–1327.

[178] Blasco-Costa, I., Waters, J. M., and Poulin, R. Swimming against the current: genetic structure, host mobility and the drift paradox in trematode parasites. *Molecular Ecology 21* (2012), 207–271.

[179] Goldberg, E., Lynch, H., Neubert, M., and Fagan, W. Effects of branching spatial structure and life history on the asymptotic growth rate of a population. *Theoretical Ecology 3* (2010), 137–152.

[180] Ramirez, J. Population persistence under advection-diffusion in river networks. *Mathematical Biology 65* (2012), 919–942.

[181] Collier, K., and Smith, B. Dispersal of adult caddisflies (Trichoptera) into forests alongside three New Zealand streams. *Hydrobiologia 361* (1998), 53–65.

[182] Didham, R., Blakely, T., Ewers, R., Hitchings, T., Ward, J., and Winterbourn, M. Horizontal and vertical structuring in the dispersal of adult aquatic insects in a fragmented landscape. *Fundamental and Applied Limnology 180* (2012), 27–40.

[183] Hanski, I., and Ovaskainen, O. The metapopulation capacity of a fragmented landscape. *Nature 404* (2000), 755–758.

[184] Casagrandi, R., and Gatto, M. A mesoscale approach to extinction risk in fragmented habitats. *Nature 400* (1999), 560–562.

[185] Farina, L., and Rinaldi, S. *Positive Linear Systems: Theory and Applications*. Wiley Interscience.

[186] Silvester, J. Determinants of block matrices. *The Mathematical Gazette 84* (2000), 460–467.

[187] Gantmacher, F. *Theory of Matrices*. AMS Chelsea, 1959.

[188] Kuznetsov, Y. A. *Elements of Applied Bifurcation Theory* (3rd ed.). Springer, 2004.

[189] Casagrandi, R., and Gatto, M. The intermediate dispersal principle in spatially explicit metapopulations. *Journal of Theoretical Biology 239* (2006), 22–32.

[190] Hanski, I., and Ovaskainen, O. Extinction debt at extinction threshold. *Conservation Biology 16*, 3 (2002), 666–673.

[191] Organ, J. Studies of the local distribution, life history, and population dynamics of the salamander genus *Desmognathus* in Virginia. *Ecological Monographs 31* (1961), 189–220.

[192] Benson, L., and Pearson, R. Drift and upstream movement in Yaccabine creek, an Australian tropical stream. *Hydrobiologia 153* (1987), 225–239.

[193] Mackay, R. Colonization by lotic macroinvertebrates: a review of processes and patterns. *Canadian Journal of Fisheries and Aquatic Sciences 49* (1992), 617–628.

[194] Williams, D., and Williams, N. The upstream/downstream movement paradox of lotic invertebrates: quantitative evidence from Welsh mountain stream. *Freshwater Biology 30* (1993), 199–218.

[195] Jackson, J., McElravvy, E., and Resh, V. Long-term movements of self-marked caddisfly larvae (Trichoptera: Sericostomatidae) in a California coastal mountain stream. *Freshwater Biology 42* (1999), 525–536.

[196] Sode, A., and Wiberg-Larsen, P. Dispersal of adult Trichoptera at a Danish forest brook. *Freshwater Biology 30* (1993), 439–446.

[197] Kovats, Z., Ciborowski, J., and Corkum, L. Inland dispersal of adult aquatic insects. *Freshwater Biology 36* (1996), 265–276.

[198] Caudill, C. Measuring dispersal in a metapopulation using stable isotope enrichment: high rates of sex-biased dispersal between patches in a mayfly metapopulation. *Oikos 101* (2003), 624–630.

[199] Petersen, I., Winterbottom, J., Orton, S., Friberg, N., Speirs, A. H. D., and Gurney, W. Emergence and lateral dispersal of adult Plecoptera and Trichoptera from Broadstone Stream, UK. *Freshwater Biology 42* (1999), 401–416.

[200] Kopp, M., Jenschke, J., and Gabriel, W. Exact compensation of stream drift as an evolutionarily stable strategy. *Oikos 92* (2001), 522–530.

[201] Briers, R., Cariss, H., and Gee, J. Dispersal of adult stoneflies (Plecoptera) from upland streams draining catchments with contrasting land-use. *Archiv für Hydrobiologie 155* (2002), 627–644.

[202] Macneale, K., Peckarsky, B., and Likens, G. Stable iosopes identify dispersal patterns of stonefly populations living along stream corridors. *Freshwater Biology 50* (2005), 1117–1130.

[203] Sweeney, B., Funk, D., and Vannote, R. Population genetic structure of two mayflies (*Ephemerella subvaria, Eurylophella verisimilis*) in the Delaware River drainage basin. *Journal of the North American Benthological Society 5* (1986), 253–262.

[204] Sweeney, B., Funk, D., and Vannote, R. Genetic variation in stream mayfly (Insecta: Ephemeroptera) populations in eastern North America. *Annals of the Entomological Society of America 80* (1987), 600–612.

[205] Jackson, J., and Resh, V. Variation in genetic structure among populations of the caddisfly *Helicopsyche borealis* from three streams in northern California, USA. *Freshwater Biology 27* (1992), 29–42.

[206] Schmidt, J., Hughes, J., and Bunn, S. Gene flow among conspecific populations of *Baetis* (Ephemeroptera): adult flight and larval drift. *Journal of the North American Benthological Society 14* (1995), 147–157.

[207] Bunn, S., and Hughes, J. Dispersal and recruitment in streams: evidence from genetic studies. *Journal of the North American Benthological Society 16* (1997), 338–346.

[208] Gibbs, H., Gibbs, K., Siebenmann, M., and Collins, L. Genetic differentiation among populations of the rare mayfly *Siphlonisca aerodromia* Needham. *Journal of the North American Benthological Society 17* (1998), 461–474.

[209] Miller, M., Blinn, D., and Keim, P. Correlation between observed dispersal capabilities and patterns of genetic differentiation in populations of four aquatic insect species from the Arizona White Mountains. *Freshwater Biology 47* (2002), 1660–1673.

[210] Chaput-Bardy, A., Lemaire, C., Picard, D., and Secondi, J. In-stream and overland dispersal across a river network influences gene flow in a freshwater insect, *Calopteryx splendens*. *Molecular Ecology 17* (2008), 3496–3505.

[211] Labonne, J., Ravigne, V., Parisi, B., and Gaucherel, C. Linking dendritic network structures to population demogenetics: the downside of connectivity. *Oikos 17* (2008), 1479–1490.

[212] Chaput-Bardy, A., Fleurant, C., Lemaire, C., and Secondi, J. Modelling the effect of in-stream and overland dispersal on gene flow in river networks. *Ecological Modelling 220* (2009), 3589–3598.

[213] Burnham, K., and Anderson, D. *Model Selection and Multimodel Inference: A Practical Information-Theoretic Approach*. Springer, 2002.

[214] Band, L. Topographic partition of watersheds with digital elevation models. *Water Resources Research 22* (1986), 15–24.

[215] Tarboton, D. G. A new method for the determination of flow directions and upslope areas in grid digital elevation models. *Water Resources Research 33*, 2 (1997), 309–319.

[216] Johnson, A., Hatfield, C., and Milne, B. Simulated diffusion dynamics in river networks. *Ecological Modelling 83* (1995), 311–325.

[217] Hanski, I. *Metapopulation Ecology*. Oxford University Press, 1999.

[218] Marquet, P., Nones, R. Q., Abades, S., Labra, F., Tognelli, M., Arim, M., and Rivadeneira, M. Scaling and power-laws in ecological systems. *Journal of Experimental Biology 208* (2005), 1749–1769.

[219] Allan, J., and Castillo, M. *Stream Ecology*. Springer, 2007.

[220] Kuussaari, M., et al.. Extinction debt: a challenge for biodiversity conservation. *Trends in Ecology and Evolution 24* (2009), 564–571.

[221] Hylander, K., and Ehrlén, J. The mechanisms causing extinction debts. *Trends in Ecology and Evolution 28* (2013), 341–346.

[222] Lutscher, F., Lewis, M., and McCauley, E. Effects of heterogeneity on spread and persistence in rivers. *Bulletin of Mathematical Biology 68* (2006), 2129–2160.

[223] Lutscher, F., and Seo, G. The effect of temporal variability on persistence conditions in rivers. *Journal of Theoretical Biology 283* (2011), 53–59.

[224] Klausmeier, C. Floquet theory: a useful tool for understanding nonequilibrium dynamics. *Theoretical Ecology 1* (2008), 153–161.

[225] Ferrière, R., and Gatto, M. Lyapunov exponents and the mathematics of invasion in oscillatory or chaotic populations. *Theoretical Population Biology 48* (1995), 126–171.

[226] White, J., Botsford, L., Hastings, A., and Largier, J. Population persistence in marine reserve networks: incorporating spatial heterogeneities in larval dispersal. *Marine Ecology Progress Series 398* (2010), 49–67.

[227] Aiken, C., and Navarrete, S. Environmental fluctuations and asymmetrical dispersal: generalized stability theory for studying metapopulation persistence and marine protected areas. *Marine Ecology Progress Series 428* (2011), 77–88.

[228] Naeem, S., Duffy, J., and Zavaleta, E. The functions of biological diversity in an age of extinction. *Science 336* (2012), 1401–1406.

[229] Bertuzzo, E., Carrara, F., Mari, L., Altermatt, F., Rodriguez-Iturbe, I., and Rinaldo, A. Geomorphic controls on elevational gradients of species richness. *Proceedings of the National Academy of Sciences of the USA 113* (2016), 1737–1742.

[230] Colwell, R., Rahbek, C., and Gotelli, N. The mid-domain effect and species richness patterns: what have we learned so far? *American Naturalist 163* (2004), E1–E23.

[231] Gaston, K. Global patterns in biodiversity. *Nature 405* (2000), 220–227.

[232] Körner, C. Why are there global gradients in species richness? Mountains might hold the answer. *Trends in Ecology and Evolution 15* (2000), 513–514.

[233] Körner, C. The use of altitude in ecological research. *Trends in Ecology and Evolution 22*, 11 (2007), 569–574.

[234] Lomolino, M. Elevation gradients of species-density: historical and prospective views. *Global Ecology and Biogeography 10*, 1 (2001), 3–13.

[235] McCain, C. M., and Grytnes, J.-A. Elevational gradients in species richness. *Encylcopedia of Life Sciences 15* (2010), 1–10.

[236] Nogues-Bravo, D., Araujo, M., Romdal, T., and Rahbek, C. Scale effects and human impact on the elevational species richness gradients. *Nature 453* (2008), 216–219.

[237] Rahbek, C. The role of spatial scale and the perception of large-scale species-richness patterns. *Ecology Letters 8* (2005), 224–239.

[238] Kraft, N., et al. Disentangling the drivers of beta diversity along latitudinal and elevational gradients. *Science 333* (2011), 1755–1758.

[239] Rahbek, C. The elevational gradient of species richness – a uniform pattern. *Ecography 18* (1995), 200–205.

[240] Sanders, N. Elevational gradients in ant species richness: area, geometry, and Rapoport's rule. *Ecography 25* (2002), 25–32.

[241] Rosenzweig, M. *Species Diversity in Space and Time*. Cambridge University Press, 1995.

[242] Romdal, T., and Grytnes, J. An indirect area effect on elevational species richness patterns. *Ecography 30* (2007), 440–448.

[243] Hutchinson, G. Population studies, animal ecology and demography: concluding remarks. *Cold Spring Harbor Symposia on Quantitative Biology 22* (1957), 415–427.

[244] Marani, M., Da Lio, C., and D'Alpaos, A. Vegetation engineers marsh morphology through multiple competing stable states. *Proceedings of the National Academy of Sciences of the USA 110* (2013), 3259–3263.

[245] Nieto-Lugilde, D., et al. Tree cover at fine and coarse spatial grains interacts with shade tolerance to shape plant species distributions across the alps. *Ecography 38* (2015), 578–589.

[246] Kearney, M. Habitat, environment and niche. *Oikos 115* (2006), 3119–3131.

[247] Chase, J., and Leibold, M. *Ecological Niches*. University of Chicago Press, 2003

[248] Diamond, J. Ecological consequences of island colonisation by south-west Pacific birds. I. Types of niche shift. *Proceedings of the National Academy of Sciences of the USA 67* (1970), 529–536.

[249] Tilman, D. Niche tradeoffs, neutrality, and community structure: A stochastic theory of resource competition, invasion, and community assembly. *Proceedings of the National Academy of Sciences of the USA 101*, 30 (2004), 10854–10861.

[250] Rosindell, J., Hubbell, S., and Etienne, R. The unified neutral theory of biodiversity and biogeography at age ten. *Trends in Ecology and Evolution 26* (2011), 340–348.

[251] Hanski, I. Metapopulation dynamics. *Nature 396* (1998), 41–49.

[252] Rybicki, J., and Hanski, I. Species–area relationships and extinctions caused by habitat loss and fragmentation. *Ecology Letters 16* (2013), 27–38.

[253] Bertuzzo, E., Rodriguez-Iturbe, I., and Rinaldo, A. Metapopulation capacity of evolving fluvial landscapes. *Water Resources Research 51* (2015), 2696–2706.

[254] Ovaskainen, O., and Hanski, I. Spatially structured metapopulation models: global and local assessment of metapopulation capacity. *Theoretical Population Biology 60* (2001), 281–302.

[255] Ovaskainen, O., and Hanski, I. Transient dynamics in metapopulation response to perturbation. *Theoretical Population Biology 61* (2002), 285–295.

[256] Newman, M. *Networks: An Introduction.* Oxford University Press, 2010.

[257] Banavar, J., Colaiori, F., Flammini, A., Maritan, A., and Rinaldo, A. Topology of the fittest transportation network. *Physical Review Letters 84* (2000), 4745–4748.

[258] Fraser, D., Lippe, C., and Bernatchez, L. Consequences of unequal population size, asymmetric gene flow and sex-biased dispersal on population structure in brook charr (*Salvelinus fontinalis*). *Molecular Ecology 13* (2004), 67–80.

[259] Haddad, N. Corridor and distance effects on interpatch movements: a landscape experiment with butterflies. *Ecological Applications 9* (1999), 612–622.

[260] Markwith, S., and Scanlon, M. Multiscale analysis of *Hymenocallis coronaria* (Amaryllidaceae) genetic diversity, genetic structure, and gene movement under the influence of unidirectional stream flow. *American Journal of Botany 94* (2007), 151–160.

[261] Pianka, E. On *r* and *k* selection. *American Naturalist 104* (1970), 592–597.

[262] Rinaldo, A., Dietrich, W., Rigon, R., Vogel, G., and Rodriguez-Iturbe, I. Geomorphological signatures of varying climate. *Nature 374* (1995), 632–635.

[263] Giezendanner, J., Bertuzzo, E., Pasetto, D., Guisan, A., and Rinaldo, A. A minimalist model of extinction and range dynamics of virtual mountain species driven by warming temperatures. *PLoS ONE 4* (2019), e0213775.

[264] Elsen, P. R., and Tingley, M. W. Global mountain topography and the fate of montane species under climate change. *Nature Climate Change 5*, August (2015), 5–10.

[265] Parmesan, C., and Yohe, G. A globally coherent fingerprint of climate change impacts across natural systems. *Nature 421* (2003), 37–42.

[266] Parmesan, C. Ecological and evolutionary responses to recent climate change. *Annual Review of Ecology, Evolution, and Systematics 37* (2012), 637–669.

[267] Chen, I., Hill, J., Ohlemueller, R., Roy, D., and Thomas, C. Rapid range shift of species associated with high levels of climate warming. *Science 20* (2011), 1024–1026.

[268] Field, C., et al., eds. *IPCC Climate Change 2014: Impacts, Adaptation, and Vulnerability.* Cambridge University Press, 2014.

[269] Hanski, I. *Messages from Islands: A Global Biodiversity Tour.* University of Chicago Press, 2016.

[270] Rumpf, S., et al. Range dynamics of mountain plants decrease with elevation. *Proceedings of the National Academy of Sciences of the USA 115* (2018), 1–6.

[271] McCain, C. M. Area and mammalian elevational diversity. *Ecology 88*, 1 (2007), 76–86.

[272] McCain, C., and Colwell, R. Assessing montane biodiversity from discordant shifts in temperature and precipitation in a changing climate. *Ecology Letters 14* (2007), 1236–1245.

[273] Dullinger, S., et al. Extinction debt of high-mountain plants under twenty-first-century climate change. *Nature Climate Change 2*, 8 (2012), 619–622.

[274] Guisan, A., and Theurillat, J.-P. J.-P. Assessing alpine plant vulnerability to climate change: a modeling perspective. *Integrated Assessment 1*, 1 (2001), 307–320.

[275] Theurillat, J.-P., and Guisan, A. Potential impact of climate change on vegetation in the European

Alps: a review. *Climatic Change 50* (2001), 77–109.

[276] Marquet, P. A., et al. On theory in ecology. *Bioscience 64*, 8 (2014), 701–710.

[277] Lenoir, J., Gegout, J., Marquet, P., de Ruffray, P., and Brisse, H. A significant upwardshift in plant species optimum elevation during the 20th century. *Science 320*, 5884 (2008), 1768–1771.

[278] Engler, R., et al. Predicting future distributions of mountain plants under climate change: does dispersal capacity matter? *Ecography 32* (2009), 34–45.

[279] Barry, R., and Chorley, R. *Atmosphere, Weather and Climate*. Routledge, 2009.

[280] Ovaskainen, O. Metapopulation dynamics in highly fragmented landscapes. In *Ecology, Genetics and Evolution of Metapopulations*. Elsevier, 2004, pp. 73–103.

[281] Tischendorf, L., Bender, D. J., and Fahrig, L. Evaluation of patch isolation metrics in mosaic landscapes for specialist vs. generalist dispersers. *Landscape Ecology 18* (2003), 41–50.

[282] Elton, C. *The Ecology of Invasions by Animals and Plants*. Methuen, 1958.

[283] Shigesada, N., and Kawasaki, K. *Biological Invasions: Theory and Practice*. Oxford University Press, 1997.

[284] Clobert, J., Danchin, E., Dhondt, A., and Nichols, J. *Dispersal*. Oxford University Press, 2001.

[285] Okubo, A., and Levin, S. *Diffusion and Ecological Problems: Modern Perspectives*. Springer, 2002.

[286] Méndez, V., Fedotov, S., and Horsthemke, W. *Reaction-Transport Systems*. Springer, 2010.

[287] Schick, R., and Lindley, S. Directed connectivity among fish populations in a riverine network. *Journal of Applied Ecology 44* (2007), 1116–1126.

[288] Melbourne, B., and Hastings, A. Highly variable spread rates in replicated biological invasions: fundamental limits to predictability. *Science 325*, 5947 (2009), 1536–1539.

[289] Giometto, A., Rinaldo, A., Carrara, F., and Altermatt, F. Emerging predictable features of replicated biological invasion fronts. *Proceedings of the National Academy of Sciences of the USA 111*, 1 (2014), 297–301.

[290] Taylor, G. Diffusion by continuous movements. *Proceedings of the London Mathematical Society A* 20 (1921), 196–211.

[291] Britton, N. *Reaction-Diffusion Equations and Their Applications to Biology*. Academic Press, 1986.

[292] Newmark, W. Species–area relationship and its determinants for mammals in western North American national parks. *Biological Journal of the Linnean Society 28* (1986), 83–98.

[293] Teschl, G. *Ordinary Differential Equations and Dynamical Systems*. American Mathematical Society, 2012.

[294] Skellam, J. Random dispersal in theoretical populations. *Biometrika 38* (1951), 196–218.

[295] Erickson, J. M. The displacement of native ant species by the introduced argentine ant *Iridomyrmex humilis* Mayr. *Psyche 78* (1971), 257–266.

[296] Grosholz, E. Contrasting rates of spread for introduced species in terrestrial and marine systems. *Ecology 77*, 6 (1996), 1680–1686.

[297] Liggett, T. M. *Interacting Particle Systems*. Springer.

[298] Liggett, T. M. Stochastic models for large interacting systems and related correlation inequalities. *Proceedings of the National Academy of Sciences of the USA 107*, 38 (2010), 16413–16419.

[299] Durrett, R., and Levin, S. Stochastic spatial models: a user's guide to ecological applications. *Philosophical Transactions of the Royal Society of London. Series B: Biological Sciences 343*, 1305 (1994), 329–350.

[300] Schinazi, R. *Classical and Spatial Stochastic Processes with Applications to Biology*. Birkhäuser, 2014.

[301] Neuhauser, C. Mathematical challenges in spatial ecology. *Notices of the American Mathematical Society 48*, 11 (2001), 1304–1314.

[302] Shi, Z. *Branching Random Walks*. Springer International, 2015.

[303] Durrett, R., and Neuhauser, C. Particle systems and reaction-diffusion equations. *Annals of Probability 22* (1994), 289–333.

[304] Ellner, S., Sasaki, A., Haraguchi, Y., and Matsuda, H. Speed of invasion in lattice population models:

pair-edge approximation. *Journal of Mathematical Biology 36* (1998), 469—484.

[305] Egan, D. *The Death and Life of the Great Lakes.* W. W. Norton, 2018.

[306] Stokstad, E. Feared quagga mussel turns up in western United States. *Science 315* (2007), 453–454.

[307] Stoeckel, J., Schneider, D., Soeken, L., Blodgett, K., and Sparks, R. Larval dynamics of a riverine metapopulation: implications for zebra mussel recruitment, dispersal, and control in a large-river system. *Journal of the North American Benthological Society 16* (1997), 586–601.

[308] Casagrandi, R., Mari, L., and Gatto, M. Modelling the local dynamics of the zebra mussel (*Dreissena polymorpha*). *Freshwater Biology 52* (2007), 1223–1238.

[309] Carlton, J. Dispersal mechanisms of the zebra mussel (*Dreissena polymorpha*). Pages 677–697 In *Zebra Mussels: Biology, Impact, and Control*, T. F. Nalepa and D. W. Schloesser, eds. CRC Press, 1992, pp. 677–697.

[310] Allen, Y. C., and Ramcharan, C. W. *Dreissena* distribution in commercial waterways of the US: using failed invasions to identify limiting factors. *Canadian Journal of Fisheries and Aquatic Sciences 58* (2001), 898–907.

[311] Chase, M. E., and Bailey, R. The ecology of the zebra mussel (*Dreissena polymorpha*) in the lower Great Lakes of North America: I. Population dynamics and growth. *Journal of Great Lakes Research 293* (1999), 657–660.

[312] Stoeckel, J., Padilla, D., Schneider, D., and Rehmann, C. Laboratory culture of *Dreissena polymorpha* larvae: spawning success, adult fecundity, and larval mortality patterns. *Canadian Journal of Zoology 82* (2004), 1436–1443.

[313] Stoeckel, J., Rehmann, C., Schneider, D., and Padilla, D. Retention and supply of zebra mussel larvae in a large river system: importance of an upstream lake. *Freshwater Biology 49* (2004), 919–930.

[314] Mari, L., Casagrandi, R., Pisani, M., Pucci, E., and Gatto, M. When will the zebra mussel reach Florence? A model for the spread of *Dreissena polymorpha* in the Arno water system (Italy). *Ecohydrology 2* (2009), 428–439.

[315] Mackie, G. L., and Schloesser, D. W. Comparative biology of zebra mussels in Europe and North America: an overview. *American Zoologist 36* (1996), 244–258.

[316] Lewis, M. A., and Kareiva, P. Allee dynamics and the spread of invading organisms. *Theoretical Population Biology 43* (1993), 141–158.

[317] Kot, M., Lewis, M. A., and van den Driessche, P. Dispersal data and the spread of invading organisms. *Ecology 77* (1996), 2027–2042.

[318] Leung, B., Drake, J. M., and Lodge, D. M. Predicting invasions: propagule pressure and the gravity of Allee effects. *Ecology 85* (2004), 1651–1660.

[319] Potapov, A. B., and Lewis, M. A. Allee effect and control of lake system invasion. *Bulletin of Mathematical Biology 70* (2008), 1371–1397.

[320] MacIsaac, H. J. Potential abiotic and biotic impacts of zebra mussels on the inland waters of North America. *American Zoologist 36* (1996), 287–299.

[321] Strayer, D. L., and Malcom, H. M. Long-term demography of a zebra mussel (*Dreissena polymorpha*) population. *Freshwater Biology 51* (2006), 117–130.

[322] Mantecca, P., Vailati, G., Garibaldi, L., and Bacchetta, R. Depth effects on zebra mussel reproduction. *Malacologia 45* (2003), 109–120.

[323] Schneider, D. W., Stoeckel, J. A., Rehmann, C. R., Douglas Blodgett, K., Sparks, R. E., and Padilla, D. K. A developmental bottleneck in dispersing larvae: implications for spatial population dynamics. *Ecology Letters 6* (2003), 352–360.

[324] Pachepsky, E., Nisbet, R. M., and Murdoch, W. W. Between discrete and continuous: consumer–resource dynamics with synchronized reproduction. *Ecology 89* (2008), 280–288.

[325] Bertuzzo, E., Azaele, S., Maritan, A., Gatto, M., Rodriguez-Iturbe, I., and Rinaldo, A. On the space-time evolution of a cholera epidemic. *Water Resources Research 44* (2008), W01424.

[326] MacIsaac, H. J., Sprules, W. G., and Leach, J. H. Ingestion of small-bodied zooplankton by zebra mussels (*Dreissena polymorpha*): can cannibalism on larvae influence population dynamics? *Canadian Journal of Fisheries and Aquatic Sciences 48* (1991), 2051–2060.

[327] Keevin, T. M., Yarbrough, R. E., and Miller, A. C. Long-distance dispersal of zebra mussels *Dreissena polymorpha* attached to hulls of

commercial vessels. *Journal of Freshwater Ecology 7* (1992), 437–437.

[328] Schneider, D. W., Ellis, C. D., and Cummings, K. S. A transportation model assessment of the risk to native mussel communities from zebra mussel spread. *Conservation Biology 12* (1998), 788–800.

[329] Buchan, L. A. J., and Padilla, D. K. Estimating the probability of long-distance overland dispersal of invading aquatic species. *Ecological Applications 9* (1999), 254–263.

[330] Bossenbroek, J. M., Kraft, C. E., and Nekola, J. C. Prediction of long-distance dispersal using gravity models: zebra mussel invasion of inland lakes. *Ecological Applications 11* (2001), 1778–1788.

[331] Bossenbroek, J. M., Johnson, L. E., Peters, B., and Lodge, D. M. Forecasting the expansion of zebra mussels in the United States. *Conservation Biology 21* (2007), 800–810.

[332] Johnson, L. E., and Carlton, J. T. Post-establishment spread in large-scale invasions: dispersal mechanisms of the zebra mussel *Dreissena polymorpha*. *Ecology 77* (1996), 1686–1690.

[333] Sprung, M. The other life: an account of present knowledge of the larval phase of *dreissena polymorpha*. In *Zebra Mussels: Biology, Impacts and Control*, T. F. Nalepa and D. W. Schloesser, eds. Lewis, 1993, pp. 39–53.

[334] Hastings, A., et al. The spatial spread of invasions: new developments in theory. *Ecology Letters 8*, 1 (2005), 91–101.

[335] Neubert, M., Kot, M., and Lewis, M. Invasion speed in fluctuating environments. *Proceedings of the Royal Society Series B 267* (2000), 1603–1610.

[336] Andow, D., Kareiva, P., Levin, S., and Okubo, A. Spread of invading organisms. *Landscape Ecology 4*, 2/3 (1990), 177–188.

[337] Volpert, V., and Petrovskii, S. Reaction-diffusion waves in biology. *Physics of Life Reviews 6*, 4 (2009), 267–310.

[338] Lubina, J., and Levin, S. The spread of a reinvading species: range expansion in the California sea otter. *American Naturalist 131*, 4 (1988), 526–543.

[339] Ellner, S., and Schreiber, S. Temporally variable dispersal and demography can accelerate the spread of invading species. *Theoretical Population Biology 82* (2012), 283–298.

[340] Xin, J. Front propagation in heterogeneous media. *SIAM Rev. 42* (2000), 161–230.

[341] Cantrell, R., Cosner, C., and Lou, Y. Evolutionary stability of ideal free dispersal strategies in patchy environments. *Journal of Mathematical Biology 65* (2011), 943–965.

[342] Berestycki, H., Nadin, G., Perthame, B., and Ryzhik, L. The non-local fisher-kpp equation: travelling wavesand steady states. *Nonlinearity 22* (2009), 2813.

[343] Coulan, A., and Roquejoffre, J. Transition between linear and exponential propagation in fisher-kpp type reaction-diffusion equations. *Communications in Partial Differential Equations 37* (2012), 2029–2049.

[344] Benichou, O., Calvez, V., Meunier, N., and Voituriez, R. Front acceleration by dynamic selection in fisher population waves. *Physical Review Letters 86* (2012), 041908.

[345] Alfaro, M., Coville, J., and Raoul, G. Travelling waves in a nonlocal equation as a model for a population structured by a space variable and a phenotypic trait. *Communications in Partial Differential Equations 38* (2013), 2126–2154.

[346] Berestycki, N., Mouhot, L., and Raoul, G. Existence of self-accelerating fronts for a non-local reation-diffusion equation. arXiv:1512.00903v2 math.AP (2018).

[347] Bilton, D., Freeland, J., and Okamura, B. Dispersal in freshwater invertebrates. *Annual Review of Ecology and Systematics 32* (2001), 88–93.

[348] Holmes, E. Are diffusion models too simple? A comparison with telegraph models of invasion. *American Naturalist 142* (1993), 779–795.

[349] With, K., and Christ, T. Critical thresholds in species responses to landscape structure. *Ecology 76* (1995), 2446–2459.

[350] With, K. The landscape ecology of invasive spread. *Conservation Biology 16* (2002), 1192–1203.

[351] Dewhirst, S., and Lutscher, F. Dispersal in heterogeneous habitats: spatial scales and approximate rates of spread. *Ecology 90* (2009), 1338–1345.

[352] Bergelson, J., et al. Rates of weed spread in spatially heterogeneous environments. *Ecology 74* (1994), 999–1011.

[353] Bailey, D., Otten, W., and Gilligan, C. Saprotrophic invasion by the soil-borne fungal plant pathogen *Rhizoctonia solani* and percolation thresholds. *New Phytologist 146* (2000), 535–544.

[354] Williams, J., et al. The influence of evolution on population spread through patchy landscapes. *American Naturalist 188* (2016), 15–26.

[355] Fronhofer, E., et al. Information use shapes the dynamics of range expansions into environmental gradients. *Global Ecology and Biogeography 26* (2017), 400–411.

[356] Keller, E., and Segel, L. Initiation of slime mold aggregation viewed as an instability. *Journal of Theoretical Biology 26* (1970), 399–415.

[357] Keller, E., and Segel, L. Model for chemotaxis. *Journal of Theoretical Biology 30* (1971), 225–234.

[358] Tindall, M., Maini, P., Porter, S., and Armitage, J. Overview of mathematical approaches used to model bacterial chemotaxis II: Bacterial populations. *Bulletin of Mathematical Biology 70* (2008), 1570–1607.

[359] Giometto, A., Nelson, D., and Murray, A. Physical interactions reduce the power of natural selection in growing yeast colonies. *Proceedings of the National Academy of Sciences of the USA 115* (2018), 11448–11453.

[360] Dornic, I., et al. Integration of Langevin equations with multiplicative noise and the viability of field theories for absorbing phase transitions. *Physical Review Letters 94* (2005), 100601.

[361] Bonachela, J., et al. Patchiness and demographic noise in three ecological examples. *Journal of Statistical Physics 148* (2012), 723–739.

[362] Hallatscheck, O., and Korolev, K. Fisher waves in the strong noise limit. *Physical Review Letters 103* (2009), 108103.

[363] Van Dyck, H., and Baguette, M. Dispersal behavior in fragmented landscapes: routine or special movements. *Basic and Applied Ecology 6* (2005), 535–545.

[364] Borger, L., et al. Are there general mechanisms of animal home range behaviour? A review and prospects for future research. *Ecology Letters 11* (2008), 637–650.

[365] Méndez, V., et al. Speed of reaction-diffusion fronts in spatially heterogeneous media. *Physical Review E 68* (2003), 041105.

[366] Urban, M., et al. A toad more traveled: the hetereogeneous invasion dynamics of cane toads in Australia. *American Naturalist 171* (2008), 134–148.

[367] Mack, R., Simberloff, D., Lonsdale, W., Evans, H., Clout, M., and Bazzaz, F. Biotic invasions: causes, epidemiology, global consequences, and control. *Ecological Applications 3* (2000), 689–710.

[368] Fjellheim, A., Raddum, G., and Barlaup, B. Dispersal, growth and mortality of brown trout (*Salmo trutta* L.) stocked in a regulated West Norwegian river. *Regulated Rivers: Research & Management 10* (1995), 137–145.

[369] Kahler, T. H., Roni, P., and Quinn, T. Summer movement and growth of juvenile anadromous salmonids in small western Washington streams. *Canadian Journal of Fisheries and Aquatic Sciences 58* (2001), 1947–1956.

[370] Mortensen, E. The population dynamics of young trout (*Salmo trutta* L.) in a Danish brook. *Journal of Fish Biology 10* (1977), 23–33.

[371] Knouft, J. H., and Spotila, J. Assessment of movements of resident stream brown trout (*Salmo trutta* L.) among contiguous sections of stream. *Ecology of Freshwater Fish 11* (2002), 85–92.

[372] Jonsson, N. Influence of water flow, water temperature and light on fish migration in rivers. *Nordic Journal of Freshwater Research 66* (1991), 2–35.

[373] Jonsson, B., and Jonsson, N. *Ecology of Atlantic Salmon and Brown Trout: Habitat as a Template for Life Histories*. Springer, 2011.

[374] Barquin, J., et al. Assessing the conservation status of alder-ash alluvial forest and Atlantic salmon in the Natura 2000 river network of Cantabria, northern Spain. *River Conservation and Management 66* (2012), 193–210.

[375] Rinaldo, A., Vogel, G., Rigon, R., and Rodriguez-Iturbe, I. Can one gauge the shape of a basin? *Water Resources Research 31* (1995), 1119–1127.

[376] Carraro, L., Bertuzzo, E., Hartikainen, H., Jokkela, J., and Rinaldo, A. Estimating species distribution and abundance in river networks using environmental DNA. *Proceedings of the*

National Academy of Sciences of the USA 115 (2018), 11724–11729.

[377] Rodriguez-Iturbe, I., and Valdes, J. The geomorphologic structure of hydrologic response. *Water Resources Research 15* (1979), 1409–1420.

[378] Gupta, V., Waymire, E., and Wang, C. A representation of an IUH from geomorphology. *Water Resources Research 16* (1980), 885–862.

[379] Rinaldo, A., Botter, G., Bertuzzo, E., Uccelli, A., Settin, T., and Marani, M. Transport at basin scale. 1. Theoretical framework. *Hydrology and Earth System Science 10* (2006), 19–29.

[380] Shreve, R. Stream lengths and basin areas in topologically random networks. *Journal of Geology 77* (1969), 397–414.

[381] Kirkby, M. Tests of the random model and its application to basin hydrology. *Earth Surface Processes and Landforms 1* (1976), 197–212.

[382] Rinaldo, A., and Marani, A. Basin scale model of solute transport. *Water Resources Research 23*, 11 (1987), 2107–2118.

[383] Bak, P., and Chen, K. Self-organized criticality. *Scientific American 46* (1991), 52–61.

[384] Taberlet, P., Coissac, E., Hajibabaei, M., and Rieseberg, L. H. Environmental DNA. *Molecular Ecology 21*, 8 (2012), 1789–1793.

[385] Thomsen, P. F., and Willerslev, E. Environmental DNA – an emerging tool in conservation for monitoring past and present biodiversity. *Biological Conservation 183* (2015), 4–18.

[386] Pace, N. R. A molecular view of microbial diversity and the biosphere. *Science 276*, 5313 (1997), 734–740.

[387] Bass, D., Stentiford, G. D., Littlewood, D. T. J., and Hartikainen, H. Diverse applications of environmental DNA methods in parasitology. *Trends in Parasitology 31*, 10 (2015), 499–513.

[388] Bohmann, K., et al. Environmental DNA for wildlife biology and biodiversity monitoring. *Trends in Ecology and Evolution 29*, 6 (2014), 358–367.

[389] Kelly, R. P., et al. Harnessing DNA to improve environmental management. *Science 344*, 6191 (2014), 1455–1456.

[390] Yoccoz, N. G. The future of environmental DNA in ecology. *Molecular Ecology 21*, 8 (2012), 2031–2038.

[391] Ficetola, G. F., Miaud, C., Pompanon, F., and Taberlet, P. Species detection using environmental DNA from water samples. *Biology Letters 4*, 4 (2008), 423–425.

[392] Jerde, C. L., Mahon, A. R., Chadderton, W. L., and Lodge, D. M. "Sight-unseen" detection of rare aquatic species using environmental DNA. *Conservation Letters 4*, 2 (2011), 150–157.

[393] Dejean, T., Valentini, A., Miquel, C., Taberlet, P., Bellemain, E., and Miaud, C. Improved detection of an alien invasive species through environmental DNA barcoding: the example of the American bullfrog *Lithobates catesbeianus*. *Journal of Applied Ecology 49*, 4 (2012), 953–959.

[394] Mächler, E., Deiner, K., Steinmann, P., and Altermatt, F. Utility of environmental DNA for monitoring rare and indicator macroinvertebrate species. *Freshwater Science 33*, 4 (2014), 1174–1183.

[395] Takahara, T., Minamoto, T., Yamanaka, H., Doi, H., and Kawabata, Z. Estimation of fish biomass using environmental DNA. *PLoS ONE 7*, 4 (2012), e35868.

[396] Huver, J. R., Koprivnikar, J., Johnson, P. T. J., and Whyard, S. Development and application of an eDNA method to detect and quantify a pathogenic parasite in aquatic ecosystems. *Ecological Applications 25*, 4 (2015), 991–1002.

[397] Olds, B., et al. Estimating species richness using environmental DNA. *Ecology and Evolution 6*, 12 (2016), 4214–4226.

[398] Barnes, M., Turner, C., Jerde, C., Renshaw, M., Chadderton, W., and Lodge, D. Environmental conditions influence eDNA persistence in aquatic systems. *Environmental Science and Technology 48*, 3 (2014), 1819–1827.

[399] Lance, R. F., et al. Experimental observations on the decay of environmental DNA from bighead and silver carps. *Management of Biological Invasions 8*, 3 (2017), 343–359.

[400] Jerde, C. L., et al. Influence of stream bottom substrate on retention and transport of vertebrate environmental DNA. *Environmental Science and Technology 50*, 16 (2016), 8770–8779.

[401] Shogren, A. J., et al. Controls on eDNA movement in streams: transport, retention, and resuspension. *Scientific Reports 7* (2017), 5065.

[402] Deiner, K., and Altermatt, F. Transport distance of invertebrate environmental DNA in a natural river. *PLoS ONE 9*, 2 (2014), e88786.

[403] Wilcox, T. M., McKelvey, K. S., Young, M. K., Lowe, W. H., and Schwartz, M. K. Environmental DNA particle size distribution from brook trout (*Salvelinus fontinalis*). *Conservation Genetics Resources 7*, 3 (2015), 639–641.

[404] Klymus, K., Richter, C., Chapman, D., and Paukert, C. Quantification of eDNA shedding rates from invasive bighead carp *Hypophthalmichthys nobilis* and silver carp *Hypophthalmichthys molitrix*. *Biological Conservation 183*, 1 (2015), 77–84.

[405] Bylemans, J., Furlan, E., Hardy, C., McGuffie, P., Lintermans, M., and Gleeson, D. An environmental DNA-based method for monitoring spawning activity: a case study, using the endangered macquarie perch (*Macquaria australasica*). *Methods in Ecology and Evolution 8*, 5 (2017), 646–655.

[406] Sansom, B. J., and Sassoubre, L. M. Environmental DNA (eDNA) shedding and decay rates to model freshwater mussel eDNA transport in a river. *Environmental Science and Technology 51*, 24 (2017), 14244–14253.

[407] Pfister, L., et al. The rivers are alive: on the potential for diatoms as a tracer of water source and hydrological connectivity. *Hydrological Processes 23*, 19 (2009), 2841–2845.

[408] Pilgrim, D. Isochrones of travel time and distribution of flood storage from a tracer study on a small watershed. *Water Resources Research 13*, 3 (1977), 587–595.

[409] Burkhardt-Holm, P., et al. Where have all the fish gone? *Environmental Science and Technology 39*, 21 (2005), 441A–447A.

[410] Okamura, B., Hartikainen, H., Schmidt-Posthaus, H., and Wahli, T. Life cycle complexity, environmental change and the emerging status of salmonid proliferative kidney disease. *Freshwater Biology 56*, 4 (2011), 735–753.

[411] James, F. *Statistical Methods in Experimental Physics*. World Scientific, 2006.

[412] Hastie, T., Tibshirani, R., and Friedman, J. *The Elements of Statistical Learning*. Springer, 2001.

[413] Roberts, G., and Rosenthal, J. Examples of adaptive MCMC. *Journal of Computational and Graphical Statistics 18*, 2 (2009), 349–367.

[414] Tops, S., and Okamura, B. Infection of bryozoans by *Tetracapsuloides bryosalmonae* at sites endemic for salmonid proliferative kidney disease.

Diseases of Aquatic Organisms 57, 3 (2003), 221–226.

[415] Heggenes, J., Bagliniare, J. L., and Cunjak, R. A. Spatial niche variability for young Atlantic salmon (*Salmo salar*) and brown trout (*S. trutta*) in heterogeneous streams. *Ecology of Freshwater Fish 8*, 1 (1999), 1–21.

[416] Carraro, L., Mari, L., Gatto, M., Rinaldo, A., and Bertuzzo, E. Spread of proliferative kidney disease in fish along stream networks: a spatial metacommunity framework. *Freshwater Biology 63* (2018), 114–127.

[417] Anderson, M., and May, R. *Infectious Diseases of Humans: Dynamics and Control*. Oxford University Press, 2008.

[418] Morens, D., Folkers, G., and Fauci, A. S. The challenge of emerging and re-emerging infectious diseases. *Nature 430* (2004), 242–249.

[419] World Health Organization. *Fact sheet no. 115: schistosomiasis*. Tech. rep., 2014.

[420] World Health Organization. *Preventing diarrhoea through better water, sanitation and hygiene*. Tech. rep., 2014.

[421] World Health Organization. *Global Health Observatory*. Tech. rep., 2014.

[422] Jones, K. E., et al. Global trends in emerging infectious diseases. *Nature 451* (2008), 990–994.

[423] Daszak, P., Cunningham, A. A., and Hyatt, A. D. Emerging infectious diseases of wildlife – threats to biodiversity and human health. *Science 287* (2000), 443–449.

[424] Heesterbeek, J., and Roberts, M. Mathematical models for microparasites of wildlife. In *Ecology of Infectious Diseases in Natural Populations*, B. Grenfell and A. P. Dobson, eds., Cambridge University Press, pp. 90–122.

[425] Pacini, F. Osservazioni microscopiche e deduzioni patologiche sul cholera asiatico. *Gazzetta Medica Italiana Federativa Toscana 4* (1854), 397–401, 405–412.

[426] Rebaudet, S., Gazin, P., Barrais, R., Moore, S., Rossignol, E., Barthelemy, N., Gaudart, J., Boncy, J., Magloire, R., and Piarroux, R. The dry season in Haiti: a window of opportunity to eliminate cholera. *PLoS Currents Outbreaks 1* (2013).

[427] Gaudart, J., et al. Spatio-temporal dynamics of cholera during the first year of the epidemic in Haiti. *PLoS Neglected Tropical Diseases 7* (2013), e2145.

[428] Rinaldo, A., et al. Reassessment of the 2010–2011 Haiti cholera outbreak and rainfall-driven multiseason projections. *Proceedings of the National Academy of Sciences of the USA 109* (2012), 6602–6607.

[429] Eisenberg, M. C., Kujbida, G., Tuite, A. R., Fisman, D. N., and Tien, J. H. Examining rainfall and cholera dynamics in Haiti using statistical and dynamic modeling approaches. *Epidemics 5* (2013), 197–207.

[430] Anderson, R. M., and May, R. M. Regulation and stability of host–parasite population interactions: I. Regulatory processes. *Journal of Animal Ecology 47* (1978), 219–247.

[431] Hudson, P., Dobson, A., and Newborn, D. Regulation and stability of a free-living host-parasite system: *Trichostrongylus tenuis* in red grouse. I. Monitoring and parasite reduction experiments. *Journal of Animal Ecology 61* (1992), 477–486.

[432] Dobson, A., and Hudson, P. J. Regulation and stability of a free-living host-parasite system, *Trichostrongylus tenuis* in red grouse. II: Population models. *Journal of Animal Ecology 61* (1992), 487–498.

[433] Capasso, V., and Paveri-Fontana, S. A mathematical model for the 1973 cholera epidemic in the European Mediterranean region. *Revue d'Epidemiologie et de Sante Publique 27* (1979), 121–132.

[434] Codeço, C. Endemic and epidemic dynamics of cholera: the role of the aquatic reservoir. *GBMC Infectious Diseases 1* (2001), 1–12.

[435] Mari, L., Bertuzzo, E., Finger, F., Casagrandi, R., Gatto, M., and Rinaldo, A. On the predictive ability of mechanistic models for the Haitian cholera epidemic. *Journal of the Royal Society Interface 12* (2015), 20140840.

[436] Mukandavire, Z., Liao, S., Wang, J., Gaff, H., Smith, D. L., and Morris, J. G. Estimating the reproductive numbers for the 2008–2009 cholera outbreaks in Zimbabwe. *Proceedings of the National Academy of Sciences of the USA 108*, 21 (2011), 8767–8772.

[437] Gatto, M., et al. Generalized reproduction numbers and the prediction of patterns in waterborne disease. *Proceedings of the National Academy of Sciences of the USA 48* (2012), 19703–19708.

[438] Mari, L., Bertuzzo, E., Righetto, L., Casagrandi, R., Gatto, M., Rodriguez-Iturbe, I., and Rinaldo, A. Modelling cholera epidemics: the role of waterways, human mobility and sanitation. *Journal of the Royal Society Interface 9* (2012), 376–388.

[439] Watts, D., and Strogatz, S. Collective dynamics of small-world networks. *Nature 393* (1998), 440–442.

[440] Pasetto, D., Finger, F., Rinaldo, A., and Bertuzzo, E. Real-time projections of cholera outbreaks through data assimilation and rainfall forecasting. *Advances in Water Resources 37* (2017), e1006127.

[441] Lemaitre, J., Pasetto, D., Perez-Saez, J., Sciarra, C., Wamala, J., and Rinaldo, A. Rainfall as a driver of epidemic cholera: comparative model assessments of the effect of intra-seasonal precipitation events. *Acta Tropica XX* (2019), 235–243.

[442] Diekmann, O., Heesterbeek, J. A. P., and Roberts, M. G. The construction of next-generation matrices for compartmental epidemic models. *Journal of the Royal Society Interface 7* (2010), 873–885.

[443] Roberts, M. G., and Heesterbeek, J. A new method for estimating the effort required to control an infectious disease. *Proceedings of the Royal Society B 270* (2003), 1359–1364.

[444] Lopez, L. F., Coutinho, F. A. B., Burattini, M. N., and Massad, E. Threshold conditions for infection persistence in complex host–vectors interactions. *CR Biologies 325* (2002), 1073–1084.

[445] Diekmann, O., Heesterbeek, J., and Metz, J. On the definition and the computation of the basic reproduction ratio R_0 in models for infectious diseases in heterogeneous populations. *Journal of Mathematical Biology 28*, 4 (1990), 365–382.

[446] Bertuzzo, E., Casagrandi, R., Gatto, M., Rodriguez-Iturbe, I., and Rinaldo, A. On spatially explicit models of cholera epidemics. *Journal of the Royal Society Interface 7*, 43 (2010), 321–333.

[447] Mari, L., Casagrandi, R., Bertuzzo, E., Rinaldo, A., and Gatto, M. Floquet theory for seasonal environmental forcing of spatially explicit waterborne epidemics. *Theoretical Ecology 7*, 4 (2014), 351–365.

[448] Neubert, M., and Caswell, H. Alternatives to resilience for measuring the responses of

ecological systems to perturbations. *Ecology 78* (1997), 653–665.

[449] Bittanti, S., and Colaneri, P. *Periodic Systems, Filtering and Control.* Springer, 2008.

[450] Bacaër, N. Approximation of the basic reproduction number R_0 for vector-borne diseases with a periodic vector population. *Bulletin of Mathematical Biology 69* (2007), 1067–1091.

[451] Gatto, M., et al. Spatially explicit conditions for waterborne pathogen invasion. *American Naturalist 182* (2013), 328–346.

[452] Mari, L., Casagrandi, R., Rinaldo, A., and Gatto, M. Epidemicity thresholds for water-borne and water-related diseases. *Journal of Theoretical Biology 447* (2018), 126–138.

[453] Mari, L., Casagrandi, R., Bertuzzo, E., Rinaldo, A., and Gatto, M. Conditions for transient epidemics of waterborne disease in spatially explicit systems. *Royal Society Open Science 6* (2019), 181517.

[454] Caswell, H., and Neubert, M. Reactivity and transient dynamics of discrete-time ecological systems. *Journal of Difference Equations and Applications 2* (2005), 295–310.

[455] Heffernan, J., Smith, R., and Wahl, L. Perspectives on the basic reproductive ratio. *Journal of the Royal Society Interface 2* (2005), 281–293.

[456] Hastings, A. Timescales, dynamics, and ecological understanding. *Ecology 91* (2010), 3471–3480.

[457] Neubert, M., Klanjscek, T., and Caswell, H. Reactivity and transient dynamics of predator-prey and food web models. *Ecological Modelling 139* (2004), 29–38.

[458] Neubert, M., Caswell, H., and Murray, J. Transient dynamics and pattern formation: reactivity is necessary for Turing instabilities. *Mathematical Biosciences 175*, 8 (2002), 1–11.

[459] Hastings, A. Transients: the key to long-term ecological understanding? *Trends in Ecology and Evolution 19* (2004), 39–45.

[460] Hosack, G., Rossignol, P., and van den Driessche, P. The control of vector-borne disease epidemics. *Journal of Theoretical Biology 255* (2008), 16–25.

[461] Tang, S., and Allesina, S. Reactivity and stability of large ecosystems. *Frontiers in Ecology and Evolution 2* (2014), 21–35.

[462] Suweis, S., Grilli, J., Banavar, J., Allesina, S., and Maritan, A. Effect of localization on the stability of mutualistic ecological networks. *Nature Communications 6* (2015), 10179.

[463] Mari, L., Casagrandi, R., Rinaldo, A., and Gatto, M. A generalized definition of reactivity for ecological systems and the problem of transient species dynamics. *Methods in Ecology and Evolution 8*, 11 (2017), 1574–1584.

[464] Tien, J., Shuai, Z. S., Eisenberg, M. C., and van den Driessche, P. Disease invasion on community networks with environmental pathogen movement. *Journal of Mathematical Biology 70* (2015), 1065–1092.

[465] Frerichs, R. R., Keim, P. S., Barrais, R., and Piarroux, R. Nepalese origin of cholera epidemic in Haiti. *Clinical Microbiology and Infection 18*, 6 (2012), 158–163.

[466] Pan American Health Organization. *Haiti cholera outbreak data.* Tech. rep., 2011.

[467] Tuite, A., Tien, J., Eisenberg, M., Earns, D. J. D., Ma, J., and Fisman, D. N. Cholera epidemic in Haiti, 2010: using a transmission model to explain spatial spread of disease and identify optimal control interventions. *Annals of Internal Medicine 154* (2011), 593–601.

[468] Chao, D. L., Halloran, M. E., and Longini, I. M. Vaccination strategies for epidemic cholera in Haiti with implications for the developing world. *Proceedings of the National Academy of Sciences of the USA 108* (2011), 7081–7085.

[469] Andrews, J. R., and Basu, S. Transmission dynamics and control of cholera in Haiti: an epidemic model. *Lancet 377* (2011), 1248–1252.

[470] Clark, J. A new method for estimating the effort required to control an infectious disease. *Science 293* (2001), 657–660.

[471] Azman, A., and Lessler, J. Reactive vaccination in the presence of disease hotspots. *Proceedings of the Royal Society Series B 282*, 1 (2014), 1–13.

[472] Scobie, H. M., et al. Safe water, sanitation, hygiene, and a cholera vaccine. *Lancet 387*, 1 (2016), 28–29.

[473] Rebaudet, S., Sudre, B., Faucher, B., and Piarroux, R. Environmental determinants of cholera outbreaks in inland Africa: a systematic review of main transmission foci and propagation routes. *Journal of Infectious Diseases 208* (2013), S46–54.

[474] Centers for Disease Control and Prevention. *Defeating cholera: clinical presentation and management for Haiti cholera outbreak.* Tech. rep., 2010.

[475] Dunkle, S., et al. Epidemic cholera in a crowded urban environment, Port-au-Prince, Haiti. *Emerging Infectious Diseases 17* (2011), 2143–2146.

[476] Chin, C., et al. The origin of the Haitian cholera outbreak strain. *New England Journal of Medicine 364* (2011), 33–42.

[477] Colwell, R. R. Global climate and infectious disease: the cholera paradigm. *Science 274* (1996), 2025–2031.

[478] King, A., Ionides, E., Pascual, M., and Bouma, M. Inapparent infections and cholera dynamics. *Nature 454* (2008), 877–880.

[479] Weil, A., et al. Frequency of reexposure to *Vibrio cholerae* O1 evaluated by subsequent vibriocidal titer rise after an episode of severe cholera in a highly endemic area in Bangladesh. *American Journal of Tropical Medicine and Hygiene 87* (2012), 921–926.

[480] Piarroux, R., et al. Understanding the cholera epidemic, Haiti. *Emerging Infectious Diseases 17* (2011), 1161–1168.

[481] Eubank, S., et al. Modelling disease outbreaks in realistic urban social networks. *Nature 180–184* (2004), 429.

[482] Riley, S. Large-scale spatial-transmission models of infectious disease. *Science 316* (2007), 1298–1301.

[483] Bengtsson, L., et al. Using mobile phone data to predict the spatial spread of cholera. *Scientific Reports 5* (2015), 8923.

[484] Koelle, K., Rodó, X., Pascual, M., Yunus, M., and Mostafa, G. Refractory periods and climate forcing in cholera dynamics. *Nature 436* (2005), 696–700.

[485] Colizza, V., Barrat, A., Barthèlemy, M., and Vespignani, A. The role of the airline transportation network in the prediction and predictability of global epidemics. *Proceedings of the National Academy of Sciences of the USA 103* (2006), 2015–2020.

[486] Longini, I. M., Jr. A mathematical model for predicting the geographic spread of new infectious agents. *Mathematical Biosciences 90* (1988), 367–383.

[487] Eggo, R. M., Cauchemez, S., and Ferguson, N. M. Spatial dynamics of the 1918 influenza pandemic in England, Wales and the United States. *Journal of the Royal Society Interface 55* (2010), 233–243.

[488] Thomas, R. Reproduction rates in multiregion modeling systems for HIV/AIDS. *Journal of Regional Science 39* (1999), 359–385.

[489] Ferrari, M. J., et al. The dynamics of measles in sub-Saharan Africa. *Nature 451* (2008), 679–684.

[490] Merrell, D., et al. Host-induced epidemic spread of the cholera bacterium. *Nature 417* (2002), 642–645.

[491] Alam, A., et al. Hyperinfecttivity of human-passaged *V. cholerae* can be modeled by growth in the infant mouse. *Infection and Immunity 73* (2005), 6674–6679.

[492] Hartley, D., Morris, J., and Smith, D. Hyperinfectivity: a critical element of *Vibrio cholerae* to cause epidemics? *PLoS Medicine 3* (2006), 63–69.

[493] Pascual, M., Koelle, K., and Dobson, A. P. Hyperinfectivity in cholera: a new mechanism for an old epidemiological model? *PLoS Medicine 3* (2006), 931–938.

[494] Bertuzzo, E., Finger, F., Mari, L., Gatto, M., and Rinaldo, A. On the probability of extinction of the Haiti cholera epidemic. *Stochastic Environmental Research and Risk Assessment 30* (2016), 2043–2055.

[495] Pasetto, D., et al. Near real-time forecasting for cholera decision making in Haiti after Hurricane Matthew. *PLoS Computational Biology 14* (2018), e1006127.

[496] Vrugt, J., and Robinson, B. Improved evolutionary optimization from genetically adaptive multimethod search. *Proceedings of the National Academy of Sciences of the USA 104* (2007), 708–711.

[497] ter Braak, C., and Vrugt, J. Differential evolution Markov chain with snooker updater and fewer chains. *Statistics and Computing 18* (2008), 435–446.

[498] Pascual, M., Bouma, M., and Dobson, A. Predicting endemic cholera: the role of climate variability and disease dynamics. *Climate Research 36* (2008), 131–140.

[499] Ali, M., et al. Natural cholera infection-derived immunity in an endemic setting. *Journal of*

Infectious Diseases 204 (2011), 912–918.

[500] Cash, R., et al. Response of man to infection with *V. cholerae* 1. Clinical, serologic, and bacteriological response to a known inoculum. *Journal of Infectious Diseases 129* (1974), 45–52.

[501] Longini, I. M., Jr., Yunus, M., Zaman, K., Siddique, A. K., Sack, R. B., and Nizam, A. Epidemic and endemic cholera trends over a 33-year period in Bangladesh. *Journal of Infectious Diseases 186* (2002), 246–251.

[502] Chauduri, A., and De, S. Cholera and blood groups. *Lancet 2* (1977), 404–405.

[503] Harris, J., et al. Blood group, immunity, and risk of infection with *Vibrio cholerae* in an area of endemicity. *Infection and Immunity 73* (2005), 7422–7427.

[504] Sack, D., Sack, R., Nair, G., and Siddique, A. K. Cholera. *Lancet 377* (2004), 223–233.

[505] Ivers, L., et al. Five complementary interventions to slow cholera: Haiti. *Lancet 376* (2010), 2048–2051.

[506] Farmer, P. Meeting cholera's challenge to Haiti and the world: a joint statement on cholera prevention and care. *PLoS Neglected Tropical Diseases 5* (2011), e1145.

[507] Chaignat, C., et al. Cholera in disasters: do vaccines prompt new hopes? *Expert Review of Vaccines 7* (2008), 431–435.

[508] Cyranoski, D. Cholera vaccine plan splits experts. *Nature 469* (2011), 273–274.

[509] World Health Organization. *Fact sheet n. 107 August 2011*. Tech. rep., Regional Office for the Eastern Mediterranean, 2014.

[510] Nelson, E., Harris, J., Morris, J., Calderwood, S., and Camilli, A. Cholera transmission: the host, pathogen and bacteriophage dynamic. *Nature Reviews Microbiology 7* (2009), 693–702.

[511] Harris, J., et al. Immunologic responses to *V. cholerae* in patients co-infected with intestinal parasites in Bangladesh. *PLoS Neglected Tropical Diseases 3* (2009), e403.

[512] Mwansa, J., et al. Multiply antibiotic-resistant *Vibrio cholerae* O1 biotype El Tor strains emerge during cholera outbreaks in zambia. *Epidemiology & Infection 135* (2007), 847–853.

[513] Kitaoka, M., Miyata, S. T., Unterweger, D., and Pukatzki, S. Antibiotic resistance mechanisms of *Vibrio cholerae* micro-epidemiology of urinary schistosomiasis in Zanzibar: local risk factors associated with distribution of infections among schoolchildren and relevance for control. *Journal of Medical Microbiology 60* (2011), 397–407.

[514] Beaber, J., Hochhut, B., and Waldor, M. SOS response promotes horizontal dissemination of antibiotic resistance genes. *Nature 427* (2004), 72–74.

[515] Guerin, E., et al. The SOS response controls integron recombination. *Science 324* (2009), 1034–1035.

[516] Meibom, K., Blokesch, M., Dolganov, N., Wu, C., and Schoolnik, G. Chitin induces natural competence in *V. cholerae*. *Science 310* (2005), 1824–1827.

[517] Suckow, G., Seitz, P., and Blokesch, M. Quorum sensing contributes to natural transformation of *Vibrio cholerae* in a species-specific manner. *Journal of Bacteriology 193* (2011), 4914–4924.

[518] Borgeaud, S., Metzger, L., Scrignani, T., and Blokesch, M. The type VI secretion system of *V. cholerae* fosters horizontal gene transfer. *Science 347* (2015), 63–67.

[519] Kühn, J., et al. Glucose- but not rice-based oral rehydration therapy enhances the production of virulence determinants in the human pathogen *V. cholerae*. *PLoS Neglected Tropical Diseases 8* (2014), e3347.

[520] Finger, F., et al. Mobile phone data highlights the role of mass gatherings in the spreading of cholera outbreaks. *Proceedings of the National Academy of Sciences of the USA 113* (2016), 6421–6426.

[521] Diekmann, O., and Heesterbeek, J. A. P. *Mathematical Epidemiology of Infectious Diseases*. Wiley, 2000

[522] Barzilay, E., et al. Cholera surveillance during the Haiti epidemic: the first 2 years. *New England Journal of Medicine 368* (2013), 599–609.

[523] Lipp, E., Huq, A., and Colwell, R. Effects of global climate on infectious disease: the cholera model. *Clinical Microbiology Reviews 15* (2002), 757–762.

[524] Islam, M., et al Role of cyanobacteria in the persistence of *Vibrio cholerae* O139 in saline microcosms. *Canadian Journal of Microbiology 50* (2004), 127–131.

[525] Hill, V., et al. Toxigenic *Vibrio cholerae* O1 in water and seafood, Haiti. *Emerging Infectious Diseases 17* (2011), 2147–2150.

[526] Azaele, S., Maritan, A., Bertuzzo, E., Rodriguez-Iturbe, I., and Rinaldo, A. Stochastic dynamics of cholera epidemics. *Physical Review E 81*, 5 (2010), 051901.

[527] Angulo, J., Yu, H., Langousis, A., Madrid, A., and Christakos, G. Modeling of space-time infectious disease spread under conditions of uncertainty. *International Journal of Geographical Information Science 26* (2012), 1751–1772.

[528] Angulo, J., et al. Spatiotemporal infectious disease modeling: a BME-SIR approach. *PLoS ONE 8* (2013), e72168.

[529] Vrugt, J., et al. Accelerating Markov chain Monte Carlo simulation by differential evolution with self-adaptive randomized subspace sampling. *International Journal of Nonlinear Sciences and Numerical Simulation 10* (2009), 271–288.

[530] World Health Organization. *Fact sheet no. 115: prevention and control of cholera outbreaks: WHO policy and recommendations*. Tech. rep., 2010.

[531] Gillespie, D. Exact stochastic simulation of coupled chemical reactions. *Journal of Physical Chemistry 81* (1977), 2340–2361.

[532] Zaidi, A. Make plans to eliminate cholera outbreaks. *Nature 550* (2017), 28980651.

[533] Simini, F., González, M. C., Maritan, A., and Barabási, A. L. A universal model for mobility and migration patterns. *Nature 484* (2012), 96–100.

[534] Candia, J., et al. Uncovering individual and collective human dynamics from mobile phone records. *Journal of Physics A: Mathematical and Theoretical 41* (2008), 224015.

[535] Lu, X., Bengtsson, L., and Holme, P. Predictability of population displacement after the 2010 Haiti earthquake. *Proceedings of the National Academy of Sciences of the USA 109* (2012), 11576–11581.

[536] Perkins, T., et al. Theory and data for simulating fine-scale human movement in an urban environment. *Journal of the Royal Society Interface 11* (2014), 20140642.

[537] Lu, X., Wetter, E., Bharti, N., Tatem, A. J., and Bengtsson, L. Approaching the limit of predictability in human mobility. *Scientific Reports 3* (2013), 2923.

[538] Wesolowski, A., Eagle, N., Tatem, A. J., Smith, D. L., Noor, A. M., Snow, R. W., and Buckee, C. Quantifying the impact of human mobility on malaria. *Science 6104* (2012), 267–270.

[539] Wesolowski, A., et al. Quantifying seasonal population fluxes driving rubella transmission dynamics using mobile phone data. *Proceedings of the National Academy of Sciences of the USA 112* (2015), 11114–11119.

[540] Mari, L., et al. Big-data-driven modeling unveils country-wide drivers of endemic schistosomiasis. *Scientific Reports 7*, 1 (2017), 489.

[541] Panigutti, C., Tizzoni, M., Bajardi, P., Smoreda, Z., and Colizza, V. Assessing the use of mobile phone data to describe recurrent mobility patterns in spatial epidemic models. *Royal Society Open Science 4* (2017), 160950.

[542] Wesolowski, A., Eagle, N., Noor, A. M., Snow, R. W., and Buckee, C. O. The impact of biases in mobile phone ownership on estimates of human mobility. *Journal of the Royal Society Interface 10*, 81 (2013), 20120986.

[543] Gryseels, B., Polman, K., Clerinx, J., and Kestens, L. Human schistosomiasis. *Lancet 368*, 9541 (2006), 1106–1118.

[544] Garba, A., Touré, S., Dembelé, R., Bosque-Oliva, E., and Fenwick, A. Implementation of national schistosomiasis control programmes in West Africa. *Trends in Parasitology 22*, 7 (2006), 322–6.

[545] Poda, J.-N., et al. Profil parasitologique de la schistosomose urinaire du complexe hydroagricole du Sourou au Burkina Faso . *Société de pathologie exotique 94*, 1 (2001), 21–24.

[546] Poda, J., Traoré, A., and Sondo, B. L'endémie bilharzienne au Burkina Faso. *Société de Pathologie Exotique 97*, 1 (2004), 47–52.

[547] Koukounari, A., et al. *Schistosoma haematobium* infection and morbidity before and after large-scale administration of praziquantel in Burkina Faso. *Journal of Infectious Diseases 196*, 5 (2007), 659–669.

[548] Anonymous. *Rapport du suivi-évaluation de vingt deux (22) sites sentinelles pour le contrôle de la schistosomiase et les vers intestinaux*. Tech. rep., Programme National de Lutte Contre les

Maladies Tropicales Négligées, Ministère de la Santé du Burkina Faso, Ouagadougou (BF), 2013. Unpublished data.

[549] Bagayan, M., et al. Evolution recente de la schistosomiase au Burkina Faso: cas de 11 regions sanitaires. Unpublished data.

[550] Gurarie, D., and Seto, E. Y. W. Connectivity sustains disease transmission in environments with low potential for endemicity: modelling schistosomiasis with hydrologic and social connectivities. *Journal of the Royal Society Interface 6* (2009), 495–508.

[551] Bella, H., de C. Marshall, T., Omer, A., and Vaughan, J. Migrant workers and schistosomiasis in the Gezira, Sudan. *Transactions of the Royal Society of Tropical Medicine and Hygiene 74*, 1 (1980), 36–39.

[552] Appleton, C., Ngxongo, S., Braack, L., and Le Sueur, D. *Schistosoma mansoni* in migrants entering South Africa from Moçambique a threat to public health in north-eastern KwaZulu-Natal? *South African Medical Journal 86*, 4 (1996), 350–353.

[553] Cetron, M. S., et al. Schistosomiasis in Lake Malawi. *Lancet 348*, 9037 (1996), 1274–1278.

[554] Bruun, B., and Aagaard-Hansen, J. *The social context of schistosomiasis and its control: an introduction and annotated bibliography.* Tech. rep., World Health Organization.

[555] Kloos, H., Correa-Oliveira, R., dos Reis, D. C., Rodrigues, E. W., Monteiro, L. A. S., and Gazzinelli, A. The role of population movement in the epidemiology and control of schistosomiasis in Brazil: a preliminary typology of population movement. *Memórias do Instituto Oswaldo Cruz 105*, 4 (2010), 578–586.

[556] Criscione, C. D., et al. Landscape genetics reveals focal transmission of a human macroparasite. *PLoS Neglected Tropical Diseases 4*, 4 (2010), e665.

[557] Macdonald, G. The dynamics of helminth infections, with special reference to schistosomes. *Transactions of the Royal Society of Tropical Medicine and Hygiene 59* (1965), 489–506.

[558] May, R. M. and Anderson, R. Population biology of infectious diseases: Part II. *Nature 280* (1979), 455–461.

[559] Barbour, A. Macdonald's model and the transmission of bilharzia. *Transactions of the*

Royal Society of Tropical Medicine and Hygiene 72, 1 (1978), 1–6.

[560] Woolhouse, M. On the application of mathematical models of schistosome transmission dynamics. I. Natural transmission. *Acta Tropica 49*, 4 (1991), 241–270.

[561] Hu, H., Gong, P., and Xu, B. Spatially explicit agent-based modelling for schistosomiasis transmission: human-environment interaction simulation and control strategy assessment. *Epidemics 2*, 2 (2010), 49–65.

[562] Rollinson, D., et al. Time to set the agenda for schistosomiasis elimination. *Acta Tropica 128*, 2 (2013), 423–440.

[563] Basáñez, et al. A research agenda for helminth diseases of humans: modelling for control and elimination. *PLoS Neglected Tropical Diseases 6*, 4 (2012), e1548.

[564] Colley, D., Bustinduy, A., Secor, W., and King, C. Human schistosomiasis. *Lancet 383* (2014), 2253–2264.

[565] Sokolow, S., Lafferty, K., and Kuris, A. Regulation of laboratory populations of snails *Biomphalaria* and *Bulinus* spp. by river prawns, *Macrobrachium* spp. (Decapoda, Palaemonidae): Implications for control of schistosomiasis. *Acta Tropica 132* (2014), 64–74.

[566] Swartz, S., De Leo, G., Wood, C., and Sokolow, S. Infection with schistosome parasites in snails leads to increased predation by prawns: implications for human schistosomiasis control. *Journal of Experimental Biology 218* (2015), 3962–3967.

[567] Perez-Saez, J., Mande, T., Ceperley, N., Bertuzzo, E., Mari, L., Gatto, M., and Rinaldo, A. Hydrology and density feedbacks control the ecology of the intermediate hosts of schistosomiasis across habitats in seasonal climates. *Proceedings of the National Academy of Sciences of the USA 113* (2016), 6427–6432.

[568] Sokolow, S., et al. Reduced transmission of human schistosomiasis after restoration of a native river prawn that preys on the snail intermediate host. *Proceedings of the National Academy of Sciences of the USA 112* (2015), 9650–9655.

[569] Grimes, J., Croll, D., Harrison, W., Utzinger, J., Freeman, M., and Templeton, M. The roles of water, sanitation and hygiene in reducing

schistosomiasis: a review. *Parasites & Vectors 8* (2015), 156.

[570] Sokolow, S., et al. Global assessment of schistosomiasis control over the past century shows targeting the snail intermediate host works best. *PLoS Neglected Tropical Diseases 10* (2016), e0004794.

[571] Spear, R. Internal versus external determinants of *Schistosoma japonicum* transmission in irrigated agricultural villages. *Journal of the Royal Society Interface 9* (2012), 272–282.

[572] Mari, L., Ciddio, M., Casagrandi, R., Perez-Saez, J., Bertuzzo, E., Rinaldo, A., Sokolow, S., De Leo, G., and Gatto, M. Heterogeneity in schistosomiasis transmission dynamics. *Journal of Theoretical Biology 432* (2017), 87–99.

[573] Remais, J. V., Zhong, B., Carlton, E., and Spear, R. Model approaches for estimating the influence of time varying socio-environmental factors on macroparasite transmission in two endemic regions. *Epidemics 1* (2009), 213–220.

[574] Sivakumar, M., and Faustin, G. *Agroclimatology of West Africa: Burkina Faso.* Information Bulletin 23. International Crops Research Institute for the Semi-Arid Tropics, 1987.

[575] Poda, J.-N. *Distribution spatiale des hôtes intermediaires des schistosomes au Burkina Faso: Facteurs influençant la dynamique des populations de* Bulinus truncatus rohlfsi *Classin, 1886 et de* Bulinus senegalensis *Muller, 1781.* Ph.D. thesis,

[576] Poda, J.-N., Sellin, B., Sawadogo, L., and Sanogo, S. Distribution spatiale des mollusques hôtes intermédiaires potentiels des schistosomes et de leurs biotopes au Burkina Faso. *OCCGE INFO 101* (1994), 12–19.

[577] Poda, J.-N., et al. Les schistosomoses au complexe hydroagricole du Sourou au Burkina Faso: situation et modèle de transmission. *OCCGE INFO* (2006), www.sifee.org/static/uploaded/Files/ressources/actes-des-colloques/bamako/session-2/A_Poda_etal_comm.pdf.

[578] Zongo, D., Kabré, B., Poda, J.-N., and Dianou, D. Schistosomiasis among farmers and fisherman in the west part of Burkina Faso (west africa). *Journal of Biological Sciences 8*, 2 (2008), 482–485.

[579] Zongo, D., Kabre, B. G., Dayeri, D., Savadogo, B., and Poda, J. N. Comparative study of

schistosomiasis transmission (urinary and intestinal forms) at ten sites in Burkina Faso (in sub-Saharan Africa). *Médecine et Santé Tropicales 22*, 3 (2012), 323–9.

[580] Kpoda, N. W., Sorgho, H., Poda, J.-N., Ouédraogo, J. B., and Kabré, G. B. Endémie bilharzienne à *Schistosoma mansoni* à la vallée du Kou : caractérisation du système de transmission et impact socioéconomique. *Comptes Rendus Biologies 336*, 5 (2013), 284–288.

[581] Cecchi, P. *Les petits barrages au Burkina Faso : un vecteur du changement social et de mutations des réalités rurales.* Tech. rep., Small Reservoirs Project.

[582] Kloos, H. Water resources development and schistosomiasis ecology in the Awash Valley, Ethiopia. *Social Science & Medicine 20*, 6 (1985), 609–625.

[583] Hunter, J. M. Inherited burden of disease: agricultural dams and the persistence of bloody urine (*Schistosomiasis hematobium*) in the Upper East Region of Ghana, 1959–1997. *Social Science & Medicine 56*, 2 (2003), 219–234.

[584] Poda, J.-N., Sondo, B., and Parent, G. Influence des hydro-aménagements sur la distribution des bilharzioses et de leurs hôtes intermédiaires au Burkina Faso. *Cahiers d'Études et de Recherches Francophones / Santé 13*, 1 (2003), 49–53.

[585] Boelee, E., and Madsen, H. *Irrigation and schistosomiasis in Africa: ecological aspects.* Tech. rep., International Water Management Institute, Colombo, Sri Lanka, 2006. (IWMI Research Report 99).

[586] Steinmann, P., Keiser, J., Bos, R., Tanner, M., and Utzinger, J. Schistosomiasis and water resources development: systematic review, meta-analysis, and estimates of people at risk. *Lancet Infectious Diseases 6*, 7 (2006), 411–425.

[587] Barbier, B., Yacouba, H., Maïga, A. H., Mahé, G., and Paturel, J.-E. Le retour des grands investissements hydrauliques en Afrique de l'Ouest: les perspectives et les enjeux. *Géocarrefour 1-2* (2009), 31–41.

[588] Cecchi, P., Meunier-Nikiema, A., Moiroux, N., Sanou, B., and Bougaire, F. *Why an atlas of lakes and reservoirs in Burkina Faso?* Tech. rep. III, Small Reservoirs Project. 1–20 pages.

[589] Utzinger, J., N'Goran, E. K., Caffrey, C. R., and Keiser, J. From innovation to application:

social–ecological context, diagnostics, drugs and integrated control of schistosomiasis. *Acta Tropica 120* (2011), S121–S137.

[590] Südmeier-Rieux, K., Masundire, H., Rizvi, A., and Rietbergen, R. *Ecosystems, Livelihoods and Disasters: An Integrated Approach to Disaster Risk Management.* IUCN, 2006.

[591] Abdelhak, S., Sulaiman, J., and Mohd, S. The missing link in understanding and assessing vulnerability to poverty: a conceptual framework. *Trends in Applied Sciences Research 7*, 4 (2012), 256–272.

[592] Perez-Saez, et al. A theoretical analysis of the geography of schistosomiasis in Burkina Faso highlights the roles of human mobility and water resources development in disease transmission. *PLoS Neglected Tropical Diseases 9* (2015), e0004127.

[593] Feng, Z., Eppert, A., Milner, F. A., and Minchella, D. J. Estimation of parameters governing the transmission dynamics of schistosomes. *Applied Mathematics Letters 17*, 10 (2004), 1105–1112.

[594] Phillips, S. J., Anderson, R. P., and Schapire, R. E. Maximum entropy modeling of species geographic distributions. *Ecological Modelling 190*, 3 (2006), 231–259.

[595] Stensgaard, A., et al. Large-scale determinants of intestinal schistosomiasis and intermediate host snail distribution across Africa: does climate matter? *Acta Tropica 128* (2013), 378–390.

[596] Palchykov, V., Mitrović, M., Jo, H.-H., Saramäki, J., and Pan, R. K. Inferring human mobility using communication patterns. *Scientific Reports 4* (2014), 6174.

[597] Compaoré, G., and Kaboré, I. Gestion urbaine et environnement: l'exemple de Ouagadougou (Burkina Faso). *Villes du Sud et environnement* 3 (1997), 80–99.

[598] Kêdowidé, C. M. G., Sedogo, M. P., and Cissé, G. Dynamique spatio temporelle de l'agriculture urbaine à Ouagadougou: Cas du Maraîchage comme une activité montante de stratégie de survie. *VertigO 10*, 2 (2010).

[599] Ernould, J. C., Kaman, A., Labbo, R., Couret, D., and Chippaux, J. P. Recent urban growth and urinary schistosomiasis in Niamey, Niger. *Tropical Medicine and International Health 5*, 6 (2000), 431–437.

[600] Anderson, R. M., Mercer, J. G., Wilson, R. A., and Carter, N. P. Transmission of *Schistosoma mansoni* from man to snail: experimental studies of miracidial survival and infectivity in relation to larval age, water temperature, host size and host age. *Parasitology 85*, 2 (1982), 339–360.

[601] Rohani, P., Earn, D., and Grenfell, B. Opposite patterns of synchrony in sympatric disease metapopulations. *Science 286* (1999), 968–971.

[602] Diamond, J. M. Biogeographic kinetics: estimation of relaxation times for avifaunas of Southwest Pacific islands. *Proceedings of the National Academy of Sciences of the USA 69*, 11 (1972), 3199–3203.

[603] Fenwick, A., Rollinson, D., and Southgate, V. Implementation of human schistosomiasis control: challenges and prospects. *Advances in Parasitology 61* (2006), 567–622.

[604] Utzinger, J., and de Savigny, D. Control of neglected tropical diseases: integrated chemotherapy and beyond. *PLoS Medicine 3*, 5 (2006), e112.

[605] Auger, P., Charles, S., Viala, M., and Poggiale, J.-C. Aggregation and emergence in ecological modelling: integration of ecological levels. *Ecological Modelling 127*, 1 (2000), 11–20.

[606] Gurarie, D., and King, C. Heterogeneous model of schistosomiasis transmission and long-term control: the combined influence of spatial variation and age-dependent factors on optimal allocation of drug therapy. *Parasitology 130* (2005), 49–65.

[607] Gurarie, D., King, C. H., and Wang, X. A. A new approach to modelling schistosomiasis transmission based on stratified worm burden. *Parasitology 137* (2010), 1951–1965.

[608] Gurarie, D., and King, C. H. Population biology of Schistosoma mating, aggregation, and transmission breakpoints: more reliable model analysis for the end-game in communities at risk. *PLoS ONE 9* (2014), e115875.

[609] Gurarie, D., King, C. H., Yoon, N., and Li, E. Refined stratified-worm-burden models that incorporate specific biological features of human and snail hosts provide better estimates of *Schistosoma* diagnosis, transmission, and control. *Parasites & Vectors 9* (2016), 428–431.

[610] Ciddio, M., Mari, L., Sokolow, S., De Leo, G., Casagrandi, R., and Gatto, M. The spatial spread

of schistosomiasis: a multidimensional network model applied to Saint-Louis region, Senegal. *Advances in Water Resources* 108 (2017), 406–415.

[611] Clennon, J. A., King, C. H., Muchiri, E. M., and Kitron, U. Hydrological modelling of snail dispersal patterns in Msambweni, Kenya and potential resurgence of *Schistosoma haematobium* transmission. *Parasitology 134*, Pt 5 (2007), 683–93.

[612] Grimes, J. E. T., Croll, D., Harrison, W. E., Utzinger, J., Freeman, M. C., and Templeton, M. R. The relationship between water, sanitation and schistosomiasis: a systematic review and meta-analysis. *PLoS Neglected Tropical Diseases 8*, 12 (2014), e3296.

[613] Gurarie, D., King, C. H., Yoon, N., Alsallaq, R., and Wang, X. Seasonal dynamics of snail populations in coastal Kenya: model calibration and snail control. *Advances in Water Resources 108* (2017), 397–405.

[614] McCreesh, N., and Booth, M. Challenges in predicting the effects of climate change on *Schistosoma mansoni* and *Schistosoma haematobium* transmission potential. *Trends in Parasitology 29* (2013), 548–555.

[615] Ferguson, H., and Ball, H. Epidemiological aspects of proliferative kidney disease amongst rainbow trout *Salmo gairdneri* Richardson in Northern Ireland. *Journal of Fish Diseases 2*, 3 (1979), 219–225.

[616] Clifton-Hadley, R. S., Richards, R. H., and Bucke, D. Proliferative kidney disease (PKD) in rainbow trout *Salmo gairdneri*: further observations on the effects of water temperature. *Aquaculture 55*, 3 (1986), 165–171.

[617] Feist, S. W., and Longshaw, M. Phylum Myxozoa. In *Fish Diseases and Disorders* vol. 1, P. T. K. Woo ed. CABI Publishing, Wallingford, UK, 230–296.

[618] Borsuk, M. E., Reichert, P., Peter, A., Schager, E., and Burkhardt-Holm, P. Assessing the decline of brown trout (*Salmo trutta*) in Swiss rivers using a Bayesian probability network. *Ecological Modelling 192*, 1–2 (2006), 224–244.

[619] Hari, R., Livingstone, D. M., Siber, R., Burkhardt-Holm, P., and Güttinger, H. Consequences of climatic change for water temperature and brown trout populations in alpine rivers and streams. *Global Change Biology 12*, 1 (2006), 10–26.

[620] Hartikainen, H., and Okamura, B. Ecology and evolution of malacosporean-bryozoan interactions. In *Myxozoan Evolution, Ecology and Development*, B. Okamura, A. Gruhl, and J.L. Bartholomew, eds. Springer, 2015, pp. 201–216.

[621] Carraro, L., et al. An epidemiological model for proliferative kidney disease in salmonid populations. *Parasites & Vectors 9* (2016), 487.

[622] Wahli, T., et al. Proliferative kidney disease in Switzerland: current state of knowledge. *Journal of Fish Diseases 25*, 8 (2002), 491–500.

[623] Wahli, T., Bernet, D., Steiner, P. A., and Schmidt-Posthaus, H. Geographic distribution of *Tetracapsuloides bryosalmonae* infected fish in Swiss rivers: an update. *Aquatic Sciences 69*, 1 (2007), 3–10.

[624] Carraro, L., et al. An integrated field, laboratory and theoretical study of PKD spread in a Swiss prealpine river. *Proceedings of the National Academy of Sciences of the USA 114* (2017), 11992–11997.

[625] Burkhardt-Holm, P., Peter, A., and Segner, H. Decline of fish catch in Switzerland: Project Fishnet – a balance between analysis and synthesis. *Aquatic Sciences 64*, 1 (2002), 36–54.

[626] Federal Office of Topography Swisstopo. Geological map of Switzerland 1:500000. www .geocat.ch/geonetwork/srv/eng/md.viewer#/full_ view/ca917a71-dcc9-44b6-8804-823c694be516, 2005.

[627] Alexander, J. D., Bartholomew, J. L., Wright, K. A., Som, N. A., and Hetrick, N. J. Integrating models to predict distribution of the invertebrate host of myxosporean parasites. *Freshwater Science 35*, 4 (2016), 1263–1275.

[628] Fontes, I., Hartikainen, H., Taylor, N., and Okamura, B. Conditional persistence and tolerance characterize endoparasite–colonial host interactions. *Parasitology 144*, 8 (2017), 1052–1063.

[629] Keesing, F., et al. Impacts of biodiversity on the emergence and transmission of infectious diseases. *Nature 468*, 7324 (2010), 647–652.

[630] Penczykowski, R. M., Hall, S. R., Civitello, D. J., and Duffy, M. A. Habitat structure and ecological drivers of disease. *Limnology and Oceanography 59*, 2 (2014), 340–348.

[631] Engering, A., Hogerwerf, L., and Slingenbergh, J. Pathogen–host–environment interplay and disease emergence. *Emerging Microbes & Infections 2*, 2 (2013), e5.

[632] Ricker, W. E. Stock and recruitment. *Journal of the Fish Resources Board of Canada 11*, 5 (1954), 559–623.

[633] Frank, B. M., Gimenez, O., and Baret, P. V. Assessing brown trout (*Salmo trutta*) spawning movements with multistate capture-recapture models: a case study in a fully controlled Belgian brook. *Canadian Journal of Fisheries and Aquatic Sciences 69*, 6 (2012), 1091–1104.

[634] Johnsen, B., and Jenser, A. The Gyrodactylus story in Norway. *Aquaculture 98*, 1–3 (1991), 289–302.

[635] Tien, J. H., and Earn, D. J. D. Multiple transmission pathways and disease dynamics in a waterborne pathogen model. *Bulletin of Mathematical Biology 72* (2010), 1506–1533.

[636] Jutla, A. S., Akanda, A. S., Griffiths, J. K., Colwell, R. R., and Islam, S. Warming oceans, phytoplankton and zooplankton blooms, and river discharge: implications for cholera outbreaks. *American Journal of Tropical Medicine and Hygiene 85* (2011), 303–308.

[637] Jutla, A. S., et al. Environmental factors influencing epidemic cholera. *American Journal of Tropical Medicine and Hygiene 89* (2013), 597–607.

[638] Finger, F., et al. Cholera in the Lake Kivu region (DRC): integrating remote sensing and spatially explicit epidemiological modeling. *Water Resources Research 50* (2014), 5624–5637.

[639] Bergquist, R., Yang, G. J., Knopp, S., Utzinger, J., and Tanner, M. Surveillance and response: tools and approaches for the elimination stage of neglected tropical diseases. *Acta Tropica 141* (2015), 229–234.

[640] Ruiz-Moreno, D., Pascual, M., Bouma, M., Dobson, A. P., and Cash, B. Cholera seasonality in Madras (1901–1940): dual role for rainfall in endemic and epidemic regions. *EcoHealth 4* (2007), 52–62.

[641] Hashizume, M., et al. The effect of rainfall on the incidence of cholera in Bangladesh. *Epidemiology 19* (2008), 103–110.

[642] Boelee, E., et al. Options for water storage and rainwater harvesting to improve health and resilience against climate change in Africa. *Regional Environmental Change 13* (2013), 509–519.

[643] Faruque, S. M., et al. Self-limiting nature of seasonal cholera epidemics: role of host-mediated amplification of phage. *Proceedings of the National Academy of Sciences of the USA 102* (2005), 6119–6124.

[644] Pascual, M., Rodó, X., Ellner, S. P., Colwell, R. R., and Bouma, M. J. Cholera dynamics and El Niño Southern Oscillation. *Science 289* (2000), 1766–1769.

[645] Pascual, M., Bouma, M., and Dobson, A. Cholera and climate: revisiting the quantitative evidence. *Microbes and Infection 4* (2002), 237–245.

[646] de Magny, et al. Cholera outbreak in Senegal in 2005: was climate a factor? *PLoS ONE 7* (2012), e44577.

[647] Emch, M., Feldacker, C., Islam, M. S., and Ali, M. Seasonality of cholera from 1974 to 2005: a review of global patterns. *International Journal of Health Geographics 7* (2008), 31.

[648] Jutla, A. S., Akanda, A. S., and Islam, S. A framework for predicting endemic cholera using satellite derived environmental determinants. *Environmental Modelling & Software 47* (2013), 148–158.

[649] Reiner, R. C., Jr., King, A. A., Emch, M., Yunus, M., Faruque, A. S., and Pascual, M. Highly localized sensitivity to climate forcing drives endemic cholera in a megacity. *Proceedings of the National Academy of Sciences of the USA 109* (2012), 2033–2036.

[650] Rodo, X., et al. Climate change and infectious diseases: can we meet the needs for better prediction? *Climatic Change 118* (2013), 625–640.

[651] Ramírez, I. J., and Grady, S. C. El Niño, climate, and cholera associations in Piura, Peru, 1991–2001: a wavelet analysis. *EcoHealth 13* (2016), 83–99.

[652] Vezzulli, L., et al. Climate influence on *Vibrio* and associated human diseases during the past half-century in the coastal North Atlantic. *Proceedings of the National Academy of Sciences of the USA 113* (2016), 5062–5071.

[653] Baker-Austin, C., Trinanes, J. A., Taylor, N. G. H., Hartnell, R., Siitonen, A., and Martinez-Urtaza, J. Emerging *Vibrio* risk at high latitudes in response to ocean warming. *Nature Climate Change 3* (2013), 73–77.

[654] Vezzulli, L., Colwell, R., and Pruzzo, C. Ocean warming and spread of pathogenic vibrios in the aquatic environment. *Microbial Ecology 65* (2013), 817–825.

[655] Cash, B. A., Rodó, X., Emch, M., Yunus, M., Faruque, A. S. G., and Pascual, M. Cholera and shigellosis: different epidemiology but similar responses to climate variability. *PLoS ONE 9* (2014), e107223.

[656] Escobar, L. E., et al. A global map of suitability for coastal *Vibrio cholerae* under current and future climate conditions. *Acta Tropica 149* (2015), 202–211.

[657] Vezzulli, L., Pezzati, E., Brettar, I., Höfle, M., and Pruzzo, C. Effects of global warming on *Vibrio* ecology. *Microbiology Spectrum 3* (2015), 0004–2014.

[658] Pedersen, U. B., et al. Modelling spatial distribution of snails transmitting parasitic worms with importance to human and animal health and analysis of distributional changes in relation to climate. *Geospatial Health 8* (2014), 335–343.

[659] Wang, W., Dai, J., and Liang, Y. Apropos: factors impacting on progress towards elimination of transmission of *Schistosomiasis japonica* in China. *Parasites & Vectors 7* (2014), 408.

[660] Stensgaard, A., Booth, M., Nikulin, G., and McCreesh, N. Combining process-based and correlative models improves predictions of climate change effects on *Schistosoma mansoni* transmission in eastern Africa. *Geospatial Health 11* (2016), 94–101.

[661] Zhou, X., et al. Potential impact of climate change on schistosomiasis transmission in China. *American Journal of Tropical Medicine and Hygiene 78* (2008), 188–194.

[662] de Magny, et al. Environmental signatures associated with cholera epidemics. *Proceedings of the National Academy of Sciences of the USA 105* (2008), 17676–17681.

[663] Rebaudet, S., Sudre, B., Faucher, B., and Piarroux, R. Cholera in coastal Africa: a systematic review of its heterogeneous environmental determinants. *Journal of Infectious Diseases 208* (2013), S98–S106.

[664] Lai, Y., et al. Spatial distribution of schistosomiasis and treatment needs in sub-Saharan Africa: a systematic review and geostatistical analysis. *Lancet Infectious Diseases 15* (2015), 927–940.

[665] McCreesh, N., Nikulin, G., and Booth, M. Predicting the effects of climate change on *Schistosoma mansoni* transmission in eastern Africa. *Parasites & Vectors 8* (2015), 4.

[666] Hu, Y., et al. Spatial pattern of schistosomiasis in Xingzi, Jiangxi Province, China: the effects of environmental factors. *Parasites & Vectors 6* (2013), 214.

[667] Wu, X. H., Zhang, S. Q., Xu, X. J., and Huang, Y. X. Effect of floods on the transmission of schistosomiasis in the Yangtze River valley, People's Republic of China. *Parasitology International 57* (2008), 271–276.

[668] Rohr, J., et al. Agrochemicals increase trematode infections in a declining amphibian species. *Nature 455* (2008), 1235–1239.

[669] Ouedraogo, H., et al. Schistosomiasis in school-age children in Burkina Faso after a decade of preventive chemotherapy. *Bulletin of the World Health Organization 94*, 1 (2016), 37–45.

[670] Perez-Saez, J., Mande, T., Ceperley, N., and Rinaldo, A. Classification and prediction of river network ephemerality and its relevance for waterborne disease epidemiology. *Advances in Water Resources 110* (2017), 263–278.

[671] Simoonga, C., et al. Remote sensing, geographical information system and spatial analysis for schistosomiasis epidemiology and ecology in Africa. *Parasitology 136* (2009), 1683–1693.

[672] Bajardi, P., Poletto, C., Ramasco, J., Tizzoni, M., Colizza, V., and Vespignani, A. Human mobility networks, travel restrictions, and the global spread of 2009 H1N1 pandemic. *PLoS ONE 6*, 1 (2011), 1–8.

[673] Merler, S., Ajelli, N., Fumanelli, L., Aleta, A., Moreno, Y., and Vespignani, A. Spatiotemporal spread of the 2014 outbreak of ebola virus disease in Liberia and the effectiveness of non-pharmaceutical interventions: a computational modelling analysis. *Lancet Infectious Diseases 15* (2015), 204–211.

[674] Vonghachack, Y., Odermatt, P., Taisayyavong, K., Phounsavath, S., Akkhavong, K., and Sayasone, S. Transmission of *Opisthorchis viverrini*, *Schistosoma mekongi* and soil-transmitted helminths on the Mekong islands, Southern Lao PDR. *Infectious Diseases of Poverty 131* (2017), 1–15.

[675] Forrer, A., et al. Spatial distribution of, and risk factors for *Opisthorchis viverrini* infection in Southern Lao PDR. *PLoS Neglected Tropical Diseases 6* (2012), e1481.

[676] Buerli, C., Harbrecht, H., Odermatt, P., Sayasone, S., and Chitnis, N. Mathematical analysis of the transmission dynamics of the liver fluke, *Opisthorchis viverrini. Journal of Theoretical Biology 439* (2018), 181–194.

[677] Beltrame, L., et al. A mechanistic hydro-epidemiological model of liver fluke risk. *Journal of the Royal Society Interface 15* (2018), 20180072.

[678] Coopersmith, E., Bell, J., Benedict, K., Shriber, J., McCotter, O., and Cosh, M. Relating coccidioidomycosis (valley fever) incidence to soil moisture conditions. *GeoHealth 1* (2017), 51–63.

[679] Bomblies, A., Duchemin, J., and Eltahir, E. Hydrology of malaria: model development and application to a Sahelian village. *Water Resources Research 44* (2008), W12445.

[680] Yamana, T., and Eltahir, E. Incorporating the effects of humidity in a mechanistic model of *Anopheles gambiae* mosquito population dynamics in the Sahel region of Africa. *Parasites & Vectors 6* (2013), 235.

[681] Whittaker, C., et al. Loa loa: more than meets the eye? *Trends in Parasitology 34* (2019), 254–262.

[682] Verver, S., et al. How can Onchocerciasis elimination in Africa be accelerated? Modeling the impact of increased Ivermectin treatment frequency and complementary vector control. *Clinical Infectious Diseases 66* (2018), S267–S274.

[683] Colebunders, R., et al. From river blindness control to elimination: bridge over troubled water. *Infectious Diseases of Poverty 21* (2018), 1–15.

[684] Basáñez, M., et al. River blindness: mathematical models for control and elimination. *Advances in Parasitology 94* (2016), 247–341.

[685] de Montjoye, Y., Hidalgo, C., Verleysen, M., and Blondel, V. Unique in the crowd: the privacy bounds of human mobility. *Scientific Reports 3* (2013), 1376–1380.

[686] Wesolowski, A., et al. Quantifying travel behavior for infectious disease research: a comparison of data from surveys and mobile phones. *Scientific Reports 4* (2014), 5678.

[687] Liu, Q., Ajelli, M., Aleta, A., Merler, S., Moreno, Y., and Vespignani, A. Measurability of the epidemic reproduction number in data-driven contact networks. *Proceedings of the National Academy of Sciences of the USA 115*, 50 (2018), 12680–12685.

[688] Finger, F., et al. The potential impact of case-area targeted interventions in response to cholera outbreaks: a modeling study. *PLoS Medicine 15* (2018), e1002509.

[689] Rebaudet, S., et al. Epidemiological and molecular forensics of cholera recurrence in Haiti. *Scientific Reports 9* (2019), 1164–1175.

[690] Lee, E., Azman, A., Kaminsky, J., and Lessler, J. The projected impact of geographic targeting of oral cholera vaccination in sub-Saharan Africa: a modeling study. *PLoS Medicine 16*, 12 (2019), e1003003.

[691] Garbin, S., Alessi Celegon, E., Fanton, S., and Botter, G. Hydrological controls on river network connectivity. *Royal Society Open Science 6* (2019), 181428.

[692] Hoover, C., et al. Modelled effect of prawn aquaculture on poverty alleviation and schistosomiasis control. *Nature Sustainability 2* (2019), 611–620.

[693] Perez-Saez, J., Mande, T., and Rinaldo, A. Space and time predictions of schistosomiasis snail host population dynamics across hydrologic regimes in Burkina Faso. *Geospatial Health 14* (2019), 306–313.

[694] Wood, C., et al. Precision mapping of snail habitat provides a powerful indicator of human schistosomiasis transmission. *Proceedings of the National Academy of Sciences of the USA 116*, 46 (2019), 23182–23191.

[695] Rabone, M., et al. Freshwater snails of biomedical importance in the Niger River Valley: evidence of temporal and spatial patterns in abundance, distribution and infection with *Schistosoma spp. Parasites & Vectors 12* (2019), 498–518.

[696] Morel-Journel, T., Rais Assa, C., Meilleret, L., and Vercken, E. It's all about connections: hubs and invasion in habitat networks. *Ecology Letters 22* (2019), 313–321.

[697] Streyer, D., Fisher, D., Hamilton, S., Malcom, H., Pace, M., and Solomon, C. Long-term variability and density dependence in Hudson River *Dreissena* populations. *Freshwater Biology 65*, 3 (2019), 474–489.

[698] Geba, E., et al. Use of the bivalve *Dreissena polymorpha* as a biomonitoring tool to reflect the protozoan load in freshwater bodies. *Water Research 170* (2020), 115297.

[699] Sengupta, M., et al. Environmental DNA for improved detection and environmental surveillance of schistosomiasis. *Proceedings of the National Academy of Sciences of the USA 116*, 18 (2019), 8931–8940.

[700] Magnabosco, C., et al. The biomass and biodiversity of the continental subsurface. *Nature Geosciences 11* (2018), 707–711.

[701] Simkus, D., et al. Variations in microbial carbon sources and cycling in the deep continental subsurface. *Geochimica et Cosmochimica Acta 173* (2016), 264–283.

[702] Besemer, K., Singer, G., Quince, C., Bertuzzo, E., Sloan, W., and Battin, T. Headwaters are critical reservoirs of microbial diversity for fluvial networks. *Proceedings of the Royal Society B 280*, 1771 (2013), 20131760.

[703] Rodriguez-Iturbe, I., Caylor, K., and Rinaldo, A. Metabolic principles of river basin organization. *Proceedings of the National Academy of Sciences of the USA 108* (2011), 11751–11755.

[704] Palmer, M., and Ruhi, A. Linkages between flow regime, biota, and ecosystem processes: implications for river restoration. *Science 365* (2019), eaaw2087.

[705] Kleiber, M. Body size and metabolic rate. *Physiology Review 27* (1947), 511–541.

[706] Brown, J., Gilloly, J., Allen, A., Savage, V., and West, G. Toward a metabolic theory of ecology. *Ecology 85* (2004), 1171–1789.

[707] Calder, W. *Size, Function, and Life History.* Harvard University Press, 1984.

[708] McMahon, P., and Bonner, J. *On Size and Life.* Scientific American, 1983.

[709] West, G., and Brown, J. Life's universal scaling laws. *Physics Today 57* (2004), 36–42.

[710] Banavar, J., Maritan, A., and Rinaldo, A. Size and form in efficient transportation networks. *Nature 399* (1999), 130–132.

[711] Maritan, A., Rigon, R., Banavar, J., and Rinaldo, A. Network allometry. *Geophysical Research Letters 29* (2002), 1–4.

[712] Helton, A., Hall, R., and Bertuzzo, E. How network structure affects nitrogen removal by streams. *Freshwater Biology 63* (2019), 128–140.

[713] Wollheim, W., Stewart, R., Aiken, G., Butler, K., Morse, N., and Salisbury, J. Removal of terrestrial DOC in aquatic ecosystems of a temperate river network. *Geophysical Research Letters 42* (2015), 6671–6679.

[714] Hotchkiss, E., and Hall, R. Whole-stream ^{13}C tracer addition reveals distinct fates of newly fixed carbon. *Ecology 92* (2015), 403–416.

[715] Raymond, P., Saiers, J., and Sobczak, W. Hydrological and biogeochemical controls on watershed dissolved organic matter transport: pulse-shunt concept. *Ecology 97* (2016), 5–16.

[716] Horgby, A., et al. Unexpected large evasion fluxes of carbon dioxide from turbulent streams draining the world's mountains. *Nature Communications 10* (2019), 4888.

[717] Bernhardt, E., et al. The metabolic regimes of flowing waters. *Limnology and Oceanography 63* (2018), S99–S118.

[718] Koenig, L. E., Helton, A., Savoy, P., Bertuzzo, E., Heffernan, J., Hall, R., and Bernhardt, E. Emergent productivity regimes of river networks. *Limnology and Oceanography Letters 4* (2019), 173–181.

[719] Tsuji, S., Takahara, T., Doi, H., Shibata, N., and Yamanaka, H. The detection of aquatic macroorganisms using environmental DNA analysis – a review of methods for collection, extraction, and detection. *Environmental DNA 1*, 2 (2019), 99–108.

[720] Bálint, M., et al. Environmental DNA time series in ecology. *Trends in Ecology & Evolution 33* (2018), 945–957.

[721] Peterson, E., Hanks, E., Hooten, M., Ver Hoef, J., and Fortin, J. Spatially structured statistical network models for landscape genetics. *Ecological Monographs 89*, 2 (2019), e01355.

[722] Pilger, T., Gido, K., Propst, D., Whitney, J., and Turner, T. River network architecture, genetic effective size and distributional patterns predict

differences in genetic structure across species in a dryland stream fish community. *Molecular Ecology 26* (2017), 2687–2697.

[723] Bullock, J., et al. Human-mediated dispersal and the rewiring of spatial networks. *Trends in Ecology & Evolution 33*, 12 (2018), 958–970.

[724] Erös, T. Scaling fish metacommunities in stream networks: synthesis and future research avenues. *Community Ecology 18*, 1 (2017), 72–86.

[725] Gonzáles-Ferreira, A., Bertuzzo, E., Barquín, J., Carraro, L., Alonso, C., and Rinaldo, A. Effects of altered river network connectivity on the distribution of *Salmo trutta*: insights from a metapopulation model. *Freshwater Biology 64* (2019), 1877–1895.

[726] Muneepeerakul, R., Bertuzzo, E., Rinaldo, A., and Rodriguez-Iturbe, I. Evolving biodiversity patterns in changing river networks. *Journal of Theoretical Biology 462* (2019), 418–424.

[727] Maynard Smith, J., and Price, G. The logic of animal conflict. *Nature 246* (1973), 15–18.

[728] Terui, A., Ishiyama, N., Urabe, U., Ono, S., Finlay, J., and Nakamura, F. Metapopulation stability in branching river networks. *Proceedings of the National Academy of Sciences of the USA 115* (2018), 5963–5969.

[729] May, R. *Stability and Complexity of Model Ecosystems*. Princeton University Press, 1972.

[730] Ma, C., Shen, Y., Bearup, D., Fagan, W., and Liao, J. Spatial variation in branch size promotes metapopulation persistence in dendritic river networks. *Freshwater Biology 65*, 3 (2020), 426–434.

[731] Bylemans, J., Furlan, E. M., Gleeson, D. M., Hardy, C. M., and Duncan, R. P. Does size matter? An experimental evaluation of the relative abundance and decay rates of aquatic environmental DNA. *Environmental Science and Technology 52*, 11 (2018), 6408–6416.

[732] Bruneaux, M., et al. Parasite infection and decreased thermal tolerance: impact of proliferative kidney disease on a wild salmonid fish in the context of climate change. *Functional Ecology 31* (2017), 216–226.

[733] Debes, P., Gross, R., and Vasemägi, A. Quantitative genetic variation in, and environmental effects on pathogen resistance and temperature-dependent disease severity in a wild trout. *American Naturalist 190* (2017), 244–265.

[734] Dercole, F., and Rinaldi, S. *Dynamical Systems and Their Bifurcations*. IEEE-Wiley Press, 2011.

[735] Meijer, H. G. E., Dercole, F., and Oldeman, B. E. *Numerical Bifurcation Analysis*. Springer, 2009.

[736] Maritan, A., Colaiori, F., Flammini, A., Cieplank, M., and Banavar, J. Universality classes of optimal channel networks. *Science 272* (1996), 984–986.

[737] Metropolis, N., Rosenbluth, A. W., Rosenbluth, M. N., Teller, A. H., and Teller, E. Equations of state calculations by fast computing machines. *Journal of Chemical Physics 21* (1996), 1087–1092.

[738] Bak, P. *How Nature Works: The Science of Self-Organized Criticality*. Springer, 1996.

[739] Bhattacharjee, S., and Seno, F. A measure of data collapse for scaling. *Journal of Physics A Math Gen 34*, 33 (2001), 6375–6380.

[740] Rinaldo, A., et al. Thermodynamics of fractal networks. *Physical Review Letters 76* (1996), 3364–3367.

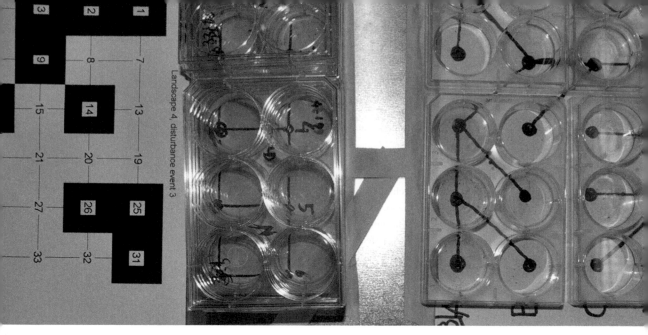

Index